Computational Music Science

Series editors

Guerino Mazzola
Moreno Andreatta

More information about this series at http://www.springer.com/series/8349

Guerino Mazzola

The Topos of Music
I: Theory

Geometric Logic, Classification, Harmony, Counterpoint, Motives, Rhythm

Second Edition

Springer

Guerino Mazzola
School of Music
University of Minnesota
Minneapolis, MN, USA

This is the first of four volumes in the set "The Topos of Music" in the series Computational Music Science. Volume II is ISBN 978-3-319-64443-1, Volume III is ISBN 978-3-319-64479-0, Volume IV is ISBN 978-3-319-64494-3. The first edition of The Topos of Music appeared in one volume in the Birkhäuser program ISBN 978-3-7643-5731-3.

ISSN 1868-0305 ISSN 1868-0313 (electronic)
Computational Music Science
ISBN 978-3-030-09717-2 ISBN 978-3-319-64364-9 (eBook)
https://doi.org/10.1007/978-3-319-64364-9

Mathematics Subject Classification (2010): 14K99, 18A99, 57P99, 18F20, 20C99

Printed on acid-free paper

This Springer imprint is published by the registered company Springer International Publishing AG part of Springer Nature.
The registered company address is: Gewerbestrasse 11, 6330 Cham, Switzerland

Preface to the Second Edition

Comprendre, c'est
attraper le geste
et pouvoir continuer
Jean Cavaillès [181, p. 186]

A major reason for a second edition of *The Topos of Music*—besides the simple fact that the first edition is now sold out—goes back to spring 2002, when I was completing its first edition, published in fall 2002. I was asked to give a talk in the MaMuX seminar of the IRCAM in Paris, to explain how I applied the mathematics of *The Topos of Music* to my free jazz improvisations.

While preparing my talk I realized that despite the presence of mathematical music theory the decisive generator of my instant compositions was the gestural deployment of formulas, the "action painting" of musical thoughts, not the abstract formulas in their static facticity. First and foremost this was a shocking insight in view of the forthcoming publication of the formulaic setup in *The Topos of Music.*

Fortunately, I knew from Hermann Hesse that "every end is a beginning"[1], which meant in my case that the end of a scientific development as traced in the book's first edition initiated the next step: a music theory of gestures. It goes without saying that this new phase would not destroy the previous research, but incorporate it as the stratum of facticity in an extended ontology of embodiment, where facts are the output of processes and their gestural generators.

In the sequel, I discovered that I was far from being the first scholar and artist to discover the crucial role of gestures in music. For instance, free jazz pianist Cecil Tayor, music philosopher Theodor Wiesengrund Adorno, or lateral thinker Paul Valéry had clearly stressed the dancing essence of art, an insight that I had embodied in my own pianist's art, but never understood on an intellectual level.

Of course, I could not be satisfied by the very existence of gesture philosophy or gestural practice, just as I could not accept traditional music and performance theory when I started my enterprise of mathematical music theory in 1978. The gesturally colored thoughts and actions needed a rigorous conceptualization in the same vein as my efforts before the first edition of *The Topos of Music.*

In 2002, I was in the privileged position to work in the multimedia division of Peter Stucki at the Institut für Informatik of the University of Zurich. I had excellent PhD students, and we could, with one of them, Stefan Müller, realize a first experimental software for the gestural representation of a pianist's hand, a work presented at the ICMC conference in 2003 [772].

This experimental preliminary work was then taken as a point of departure for a mathematical theory of musical gestures. I presented this theory in a course in spring 2005 at École normale supérieure in Paris, a course that was later in 2007 taken as the material basis of my French book *La vérité de beau dans la musique* [718]. The first publication of a formally more evolved mathematical theory of musical gestures was

[1] Actually, he says that "Jeder Anfang ist ein Ende," but the reverse is immediate.

written with co-author Moreno Andreatta in 2007 [720]. This date could be called the birthday of a valid mathematical theory of musical gestures.

Until the publication of this second edition of *The Topos of Music*, several important conceptual extensions of the mathematical theory of musical gestures, models of musical gestural processes, as well as a number of theorems have been published. The decade from date of birth to the presence proved that the mathematical theory of musical gestures is an important added value to the theory described in the first edition of *The Topos of Music*.

We can however not state that this theory of gestures is in a complete state, quite the opposite is true: The coming years will reveal important news of theoretical as well as practical nature. So why did we make the decision to publish the present state of the art? The first argument is that the present state is rich enough to define concrete new directions, be it in music theory, such as harmony or counterpoint, be it in performance theory, or be it in the understanding of embodiment in the making of music. The second argument is that the present material, roughly 500 pages of new material, is ample enough to present a book's stature. And the third and very important argument is that we would like to communicate the state of the art in the spirit of Cavaillès: *Understanding is catching the gesture and being able to continue.* The co-authors of the gesture theory part, René Guitart, Jocelyn Ho, Alex Lubet, Maria Mannone, Matt Rahaim, and Florian Thalmann, are a wonderful confirmation of this philosophy. So let us continue!

Here is a summary of the new material and its authorship. Whenever I don't mention the author, it is my own contribution, all others are mentioned explicitly.

Until Part XIV, nearly everything is as in the book's first edition, refer to the preface of that edition (also included in this edition) for detailed summaries. The only new content—besides errata corrections—is Chapter 45 in Part XI, which is a shortened version of a paper [110] on a statistical analysis of Chopin's Prélude op. 28, No. 4, written with Jan Beran, Robert Goswitz, and Patrizio Mazzola.

Gesture theory starts with Part XV: Gesture Philosophy for Music. Chapter 57 gives an overview of philosophical aspects of gestures, including works by Jean-Claude Schmitt, Vilém Flusser, Michel Guérin, Adam Kendon, David McNeill, Juhani Pallasmaa, André Chastel, Émile Benveniste, Marie-Dominique Popelard, and Anthony Wall. In Chapter 58, we discuss the presemiotic approch to gestures in the French perspective of Maurice Merleau-Ponty, Gilles Deleuze, Jean Cavaillès, Charles Alunni, and Gilles Châtelet. Paul Valéry is also referenced in Section 59.4.

Chapter 59 deals with gestural aspects in cognitive science. After a review of Embodied AI and anthropology, Alex Lubet in Section 59.5 introduces us to gestural disability studies, focusing on two famous disabled pianists: Horace Parlan and Oscar Peterson (in his last years). Then in Section 59.6 Lubet reflects on perception of musical gesture as being inherently synaesthetic.

Chapter 60 concludes this part with a review of musical models of gesturality as proposed by Wolfgang Graeser, Theodor W. Adorno, Neil P. McAngus Todd, David Lewin, Robert Hatten, Marcelo Wanderley, Claude Cadoz, and Marc Leman.

Part XVI introduces the mathematics of gestures. Chapter 61 presents the mathematical concept of a gesture in a topological space and states the Diamond Conjecture, which deals with a hypothetical big space that unites algebraic and topological categories. Chapter 62 extends the theory from gestures in topological spaces to gestures in topological categories and introduces functorial gestures, i.e., functors on topological categories with values in the category of gestures, similar to functorial compositions in the previous theory.

Chapter 63 presents a generalized singular homology, where cubes are replaced by general hypergestures. Hypergesture homology applies to a gestural model of counterpoint and to a gestural refinement of performance stemma theory.

Chapter 64 presents—similar to Chapter 63—Stokes' Theorem for hypergestures. This theorem applies to problems in gestural modulation theory.

Chapters 65 and 66 discuss categories of local and global compositions, processes/networks, and gestures, together with their functorial relationships. This triple typology composition/process/gesture corresponds to the ontological dimension of embodiment with its three coordinates facts/processes/gestures.

In Sections 67.1–67.7 of Chapter 67, René Guitart develops a fascinating and demanding mathematical model of mathematical creativity, where thought is viewed as an algebra of gestures.

In Section 61.14, we, Maria Mannone and Guerino Mazzola, present a group-theoretical model of Georg Wilhelm Friedrich Hegel's initial discourse in his *Wissenschaft der Logik*, a model that applies to the Yoneda concept of creativity [726, Chapter 19.2]. We illustrate the method with an experimental composition by Mannone. This discussion extends over Sections 67.8–67.15 and completes Chapter 67.

Part XVII deals with concept architectures and software for musical gesture theory. Chapter 68 explains the denotator formalism for gestures over topological categories. Chapter 69 is a summary of the Java-based RUBATO® Composer software [739], written to sketch the framework of Chapters 70–74, where Florian Thalmann presents his gesture-oriented software component, the BigBang rubette. This discourse again follows the coordinates of the dimension of embodiment: facts, processes, gestures, which in this situation specify to: visualization and sonification of denotators (Chapter 71), BigBang's operation graph (Chapter 72), and gestural interaction and gesturalization (Chapter 73). In the final Chapter 74 of this part, Thalmann discusses musical examples.

Part XVIII is entitled *The Multiverse Perspective* because it opens up the relationship of gesture theory with string theory in theoretical physics. After a critical review of Hermann Hesse's *Glasperlenspiel* with regard to its gestural deficiencies, we, Mazzola and Mannone, develop the Euler-Lagrange formalism of world-sheets for musical gestures. This theory extends to functorial global gestures over global topological categories.

Part XIX is dedicated to applications of gesture theory to a number of musical themes.

Chapter 79 deals with singular gesture homology being applied to counterpoint.

Chapter 80 introduces a gestural restatement of modulation theory, applying in particular Stokes' Theorem for hypergestures.

Chapter 81 applies gesture theory to a gestural performance stemma theory.

Chapter 82 is written by Jocelyn Ho as a creative presentation of composition and analysis as embodied gestures in an inter-corporeal world. She presents two compositions, Toru Takemitsu's *Rain Tree Sketch II* and her own composition *Sheng* for piano, smartphones, and fixed playback.

Chapter 83 is Mannone's analysis and classification of a conductor's movements from the viewpoint of gesture theory.

Chapter 84 is a review of gestural aspects that were developed in *Flow, Gesture, and Spaces in Free Jazz* [721].

Chapter 85 is written by Matt Rahaim and presents the gestural approach to understanding Hindustani music in its vocal gesturality.

Chapter 86 is a first approach, written by Mannone, to a future theory of vocal gestures. The short addendum was written by Mazzola.

The Appendix has been enriched by additional complements on mathematics (Chapter J) plus complements on physics (Chapter K).

The Leitfaden III has been added to the original Leitfaden I & II for the gestural chapters.

The ToM_CD has been updated, containing now the present book's pdf ToposOfMusic.pdf. However the original CD is no longer added to the book, instead the ToM_CD can be downloaded from

`www.encyclospace.org/special/ToM_CD.zip`.

Concerning the division of the now very large book into parts, this is the split:

- Volume I: *Theory*, Prefaces and Table of Contents, Parts I to VII
- Volume II: *Performance*, Parts VIII to XIV
- Volume III: *Gestures*, Parts XV to XIX
- Volume IV: *Roots*, Appendices

My sincere acknowledgments go to my co-authors and to Springer's Ronan Nugent and Frank Holzwarth as well as to Birkhäuser's Thomas Hempfling.

Minneapolis, October 2017 Guerino Mazzola

Preface

Man kann
einen jeden Begriff,
einen jeden Titel,
darunter viele Erkenntnisse gehören,
einen logischen Ort nennen.
Immanuel Kant [519, p. B 324]

This book's title subject, *The Topos of Music*, has been chosen to communicate a double message: First, the Greek word "topos" ($\tau\acute{o}\pi o\varsigma$ = location, site) alludes to the logical and transcendental location of the concept of music in the sense of Aristotle's [40, 1154] and Kant's [519, p. B 324] topic. This view deals with the question of *where music is situated* ***as*** *a concept*—and hence with the underlying ontological problem: *What is the type of being and existence of music?* The second message is a more technical understanding insofar as the system of musical signs can be associated with the mathematical theory of *topoi*, which realizes a powerful synthesis of geometric and logical theories. It laid the foundation of a thorough geometrization of logic and has been successful in central issues of algebraic geometry (Grothendieck, Deligne), independence proofs and intuitionistic logic (Cohen, Lawvere, Kripke).

But this second message is intimately entwined with the first since the present concept framework of the musical sign system is technically based on topos theory, so the topos of music receives its topos-theoretic foundation. In this perspective, the double message of the book's title in fact condenses to a unified intention: to unite philosophical insight with mathematical explicitness.

According to Birkhäuser's initial plan in 1996, this book was first conceived as an English translation of my former book *Geometrie der Töne* [682], since the German original had suffered from its restricted access to the international public. However, the scientific progress since 1989, when it was written, has been considerable in theory and technology. We have known new subjects, such as the denotator concept framework, performance theory, and new software platforms for composition, analysis, and performance, such as RUBATO® or OpenMusic. Modeling concepts via the denotator approach in fact results from an intense collaboration of mathematicians and computer scientists in the object-oriented programming paradigm and supported by several international research grants.

Also, the scientific acceptance of mathematical music theory has grown since its beginnings in the late 1970s. As the first acceptance of mathematical music theory was testified to by von Karajan's legendary Ostersymposium "Musik und Mathematik" in 1984 in Salzburg [383], so is the significantly improved present status of acceptance testified to by the Fourth Diderot Forum on Mathematics and Music [711] in Paris, Vienna, and Lisbon 1999, which was organized by the European Mathematical Society. The corresponding extension of collaborative efforts in particular entail the inclusion of works by other research groups in this book, such as the "American Set Theory", the Swedish school of performance research at Stockholm's KTH, or the research on computer-aided composition at the IRCAM in Paris.

Therefore, as a result of these revised conditions, *The Topos of Music* appears as a vastly extended English update of the original work. The extension is visibly traced in the following parts which are new with respect to [682]: Part II exposes the theory of denotators and forms, part V introduces the topological theories of rhythms and motives, part VIII introduces the structure theory of performance, part IX deals with the expressive semantics of performance in the language of performance operators and stemmata (genealogical trees of successively refined performance), part X is devoted to the description of the RUBATO® software platform for representation, analysis, composition, and performance, part XI presents a statistical analysis of musical analysis, part XII concludes the subject of performance with an inverse performance theory, in fact a first formalization of the problem of music criticism.

This does however not mean that the other parts are just translations of the German text. Considerable progress has been made in most fields, except the last part XIV which reproduces the status quo in [682]. In particular, the local and global theories have been thoroughly functorialized and thereby introduce an ontological depth and variability of concepts, techniques, and results, which by far transcend the semiotically naive geometric approach in [682]. The present theory is as different from the traditional geometric conceptualization as is Grothendieck's topos theoretic algebraic geometry from classical algebraic geometry in the spirit of Segre, van der Waerden, or Zariski.

Beyond this topos-theoretic generalization, the denotator language also introduces a fairly exceptional technique of circular concept constructions. This more precisely is rooted in Finsler's pioneering work in foundations of set theory [322], a thread which has been rediscovered in modern theoretical computer sciences [5]. The present state of denotator theory rightly could be termed a Galois theory of concepts in the sense that circular definitions of concepts play the role of conceptual equations (corresponding to algebraic equations in algebraic Galois theory), the solutions of which are concepts instead of algebraic numbers.

Accordingly, the mathematical apparatus has been vastly extended, not only in the field of topos theory and its intuitionistic logic, but also with regard to general and algebraic topology, ordinary and partial differential equations, Pólya theory, statistics, multiaffine algebra and functorial algebraic geometry. It is mandatory that these technicalities had to be placed in a more elaborate semiotic perspective. However, this book does not cover the full range of music semiotics, for which the reader is referred to [703]. Of course, such an extension on the technical level has consequences for the readability of the theory. In view of the present volume of over 1300 pages, we could however not even make the attempt to approach a non-technical presentation. This subject is left to subsequent efforts. The critical reader may put the question whether music is really that complex. The answer is yes, and the reason is straightforward: We cannot pretend that Bach, Haydn, Mozart, or Beethoven, just to name some of the most prominent composers, are outstanding geniuses and have elaborated masterworks of eternal value, without trying to understand such singular creations with adequate tools, and this means: of adequate depth and power. After all, understanding God's 'composition', the material universe, cannot be approached without the most sophisticated tools as they have been elaborated in physics, chemistry, and molecular biology.

So who is recommended to read this book? A first category of readers is evidently the working scientist in the fields of mathematical music theory, the soft- and hardware engineer in music informatics, but also the mathematician who is interested in new applications from the above fields of pure mathematics. A second category are those theoretical mathematicians or computer scientists interested in the Galois theory of concepts; they may discover interesting unsolved problems. A third category of potential readers are all those who really want to get an idea of what music is about, of how one may conceptualize and turn into language the "ineffable" in music for the common language. Those who insist on the dogma that precision and beauty contradict each other, and that mathematics only produces tautologies and therefore must fail when aiming at substantial knowledge, should not read such a book.

Despite the technical character of *The Topos of Music*, there are at least four different approaches to its reading. To begin with, one may read it as a philosophical text, concentrating on the qualitative passages, surfing over technical portions and leaving those paragraphs to others. One may also take the book as a dictionary for computational musicology, including its concept framework and the lists of musical objects and processes (such as modulation degrees, contrapuntal steps) in the appendices. Observe however, that not all existing important lists have been included. For example, the list of all-interval series and the list of self-addressed chords are omitted, the reader may find these lists in other publications. Thirdly, the working scientist will have to read the full-fledged technicalities. And last, but not least, one may take the book as a source for ideas of how to go on with the whole subject of music. The GPL (General Public License[2]) software sources in the appended CD-ROM may support further development.

The prerequisites to a more in-depth reading of this book are these. Generally speaking, a good acquaintance with formal reasoning as mathematics (including formal logic) preconizes, is a conditio sine qua non. As to musicology and music theory, the familiarity with elementary concepts, like chords, motives, rhythm, and also musical notation, as well as a real interest in understanding music and not simply (ab)using it, are recommended. For the more computer-oriented passages, familiarity with the paradigm of object-oriented programming is profitable. We have not included the appendix on mathematical basics because it should help the reader get familiar with mathematics, but as an orientation in fields where the specialized mathematician possibly needs a specification of concepts and notation. The appendix was also included to expose the spectrum of mathematics which is needed to tackle the formal problems of computational musicology. It is by no means an overkill of mathematization: We have even omitted some non-trivial fields, such as statistics or Lambda calculus, for which we have to apologize.

There are different supporting instances to facilitate orientation in this book. To begin with, the table of contents and an extensive subject and name index may help find one's key-words. Further, following the list of contents, a leitfaden (on page xlvii) is included for a generic navigation. Each chapter and section is headed by a summary that offers a first orientation about specific contents. Finally, the book is also available as a file `ToposOfMusic.pdf` with bookmarks and active cross-references in the appended CD-ROM (see page xlix for its contents). This version is also attractive because the figures' colors are visible only in this version.

In order to obtain a consistent first reading, we recommend chapters 1 to 5, and then appendix A: Common Parameter Spaces (appendix B is not mandatory here, though it gives a good and not so technical overview of auditory physiology). After that, the reader may go on with chapter 6 on denotators and then follow the outline of the leitfaden (see page xlvii).

This book could not have been realized without the engaged support of nineteen collaborators and contributors. Above all, my PhD students Stefan Göller and Stefan Müller at the MultiMedia Laboratory of the Department of Information Technology at the University of Zurich have collaborated in the production of this book on the levels of the LaTeX installation, the final production of hundreds of figures, and the contributions sections 20.2 through 20.5 (Göller) and sections 47.3 through 47.3.6.2 (Müller). My special gratitude goes to their truly collaborative spirit.

Contributions to this book have been delivered by (in alphabetic order): By Carlos Agon, and Gérard Assayag (both IRCAM) with their precious Lambda-calculus-oriented presentation of the object-oriented programming principles in the composition software OpenMusic described in chapter 52, Moreno Andreatta (IRCAM) with an elucidating discourse on the American Set Theory in section 11.5.2 and section 16.3, Jan Beran (Universität Konstanz) with his contribution to the compositional strategies in his original composition [103] in section 11.5.1.1, as well as with his inspiring work on statistics as reported in chapters 43 and 44, Chantal Buteau (Universität and ETH Zürich) with her detailed review of chapter 22, Roberto Ferretti (ETH Zürich) with his progressive contributions to the algebraic geometry of inverse performance theory in sections 39.8 and 47.2, Anja Fleischer (Technische Universität Berlin) with her short but critical preliminaries in chapter 23, Harald Fripertinger (Universität Graz) with his 'killer' formulas concerning enumeration of finite local and global compositions in sections 11.4, 16.2.2 and appendix C.3.6, Jörg Garbers (Technische Universität Berlin) with his portation of the RUBATO® application to Mac OS X, as documented in the screenshots in chapters 40, 41, Werner Hemmert (Infineon) with a very up-to-date presentation of room acoustics in section A.1.1.1 and auditory physiology in appendix B.1 (we would have loved to include more of his knowledge), Michael Leyton (DIMACS, Rutgers University) with a formidable cover figure entitled "Dark Theory", a beautiful subtitle to this book, as well as with innumerable discussions around time and its reduction to symmetries as presented in chapter 48, Emilio Lluis Puebla (UNAM, Mexico City)

[2] A legal matter file is contained in the book's CD-ROM, see page xlix.

with his unique and engaged promotion and dissipation of mathematical music theory on the American continent, especially also in the preparation and critical review of this book, Mariana Montiel Hernandez (UNAM, Mexico City) with her critical review of the theory of circular forms and denotators in section 6.5 and appendix G.2.2.1, Thomas Noll (Technische Universität Berlin) with his substantial contributions to the functorial theory of compositions, and for his revolutionary rebuilding of Riemann's harmony and its relations to counterpoint, Joachim Stange-Elbe (Universität Osnabrück) with a very clear and innovative description of his outstanding RUBATO® performance of Bach's contrapunctus III in the *Art of Fugue* in sections 42.2 through 42.4.3, Hans Straub with his adventurous extensions of classical cadence theory in section 26.2.2 and his classification of four-element motives in appendix O.4, and, last but not least, Oliver Zahorka (Out Media Design), my former collaborator and chief programmer of the NeXT RUBATO® application, which has contributed so much to the success of the Zürich school of performance theory. To all of them, I owe my deepest gratitude and recognition for their sweat and tears.

My sincere acknowledgments go to Alexander Grothendieck, whose encouraging letters and, no doubt, awe inspiring revolution in mathematical thinking has given me so much in isolated phases of this enterprise. My acknowledgments also go to my engaged mentor Peter Stucki, director of the MultiMedia Laboratory of the Department of Information Technology at the University of Zurich; without his support, this book would have seen its birthday years later, if ever. My thanks also go to my brother Silvio, who once again (he did it already for my first book [670]) supported the final review efforts by an ideal environment in his villa in Vulpera. My thanks also go to the unbureaucratic management of the book's production by Birkhäuser's lector Thomas Hempfling and the very patient copy editor Edwin Beschler. All these beautiful supports would have failed without my wife Christina's infinite understanding and vital environment—if this book is a trace of humanity, it is also, and strongly, hers.

Vulpera, June 2002 Guerino Mazzola

Volume I Contents

Part I Introduction and Orientation

Part II Navigation on Concept Spaces

Part III Local Theory

Part IV Global Theory

Part VII Counterpoint

Book Set Contents

Part I Introduction and Orientation

Part II Navigation on Concept Spaces

Part III Local Theory

Part IV Global Theory

Part XVII Concept Architectures and Software for Gesture Theory

Part XIX Gestures in Music and Performance Theory, and in Ethnomusicology

Part XXIV References and Index

Leitfaden

Leitfaden I & II

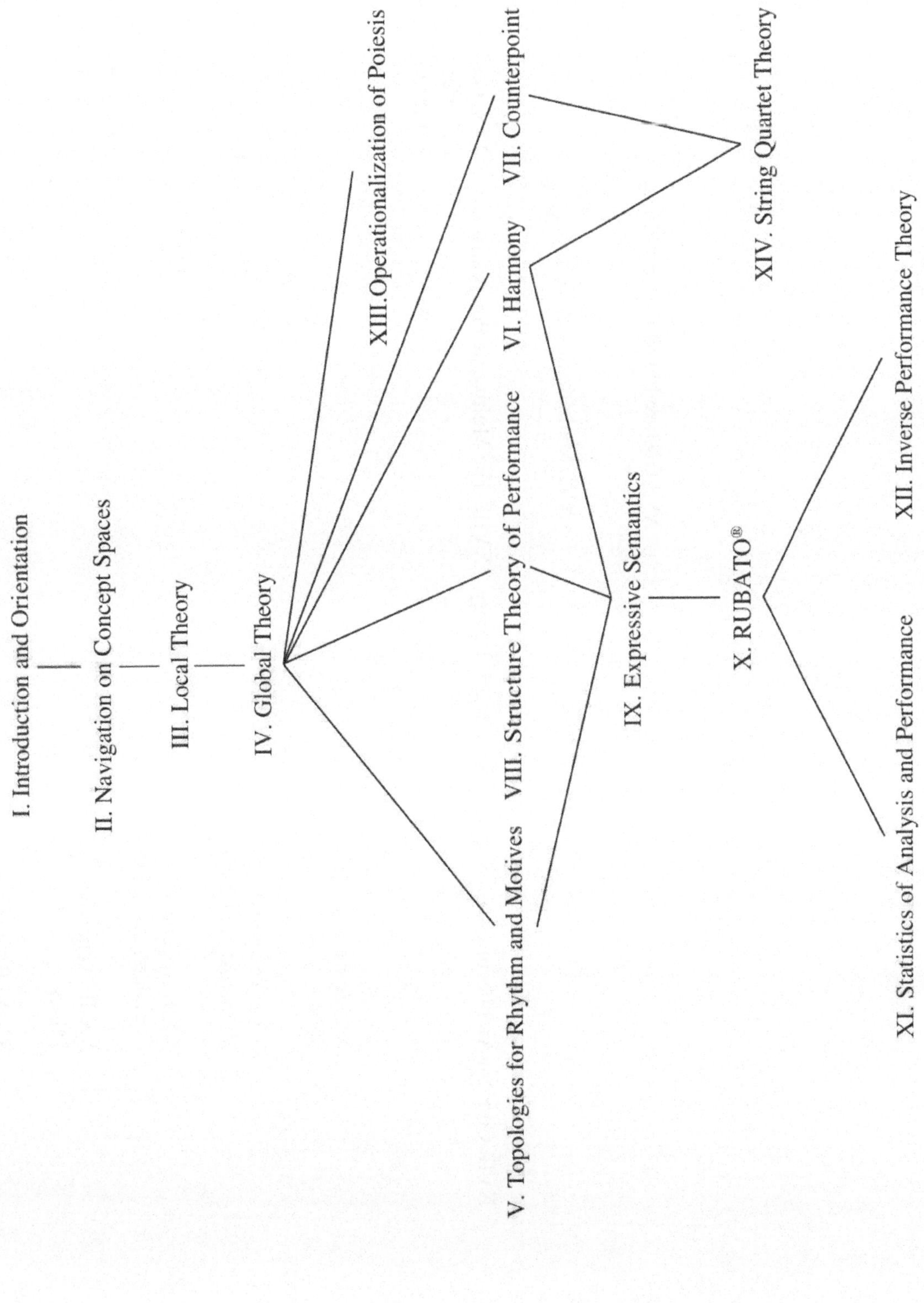

Leitfaden III

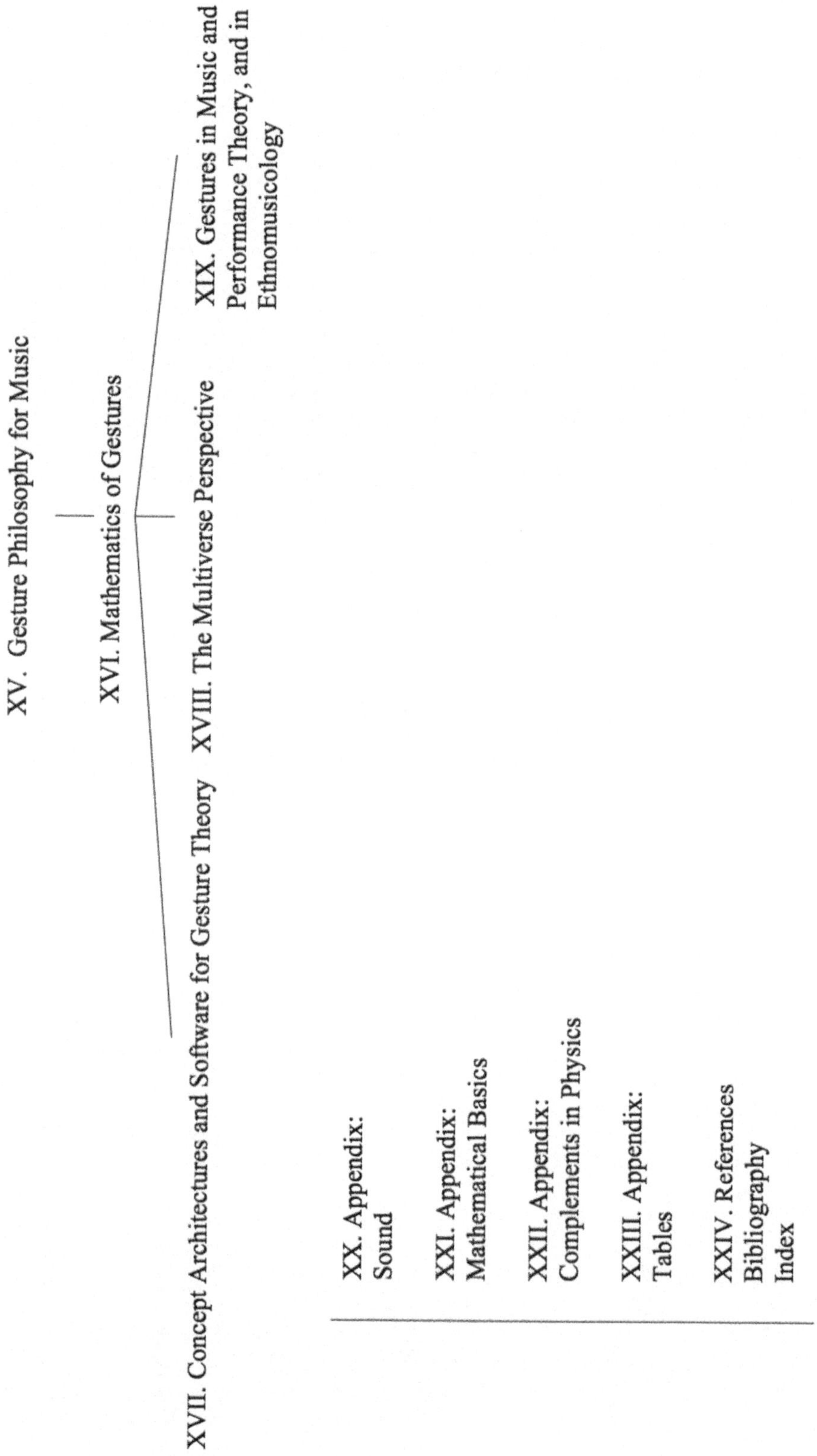

ToM_CD

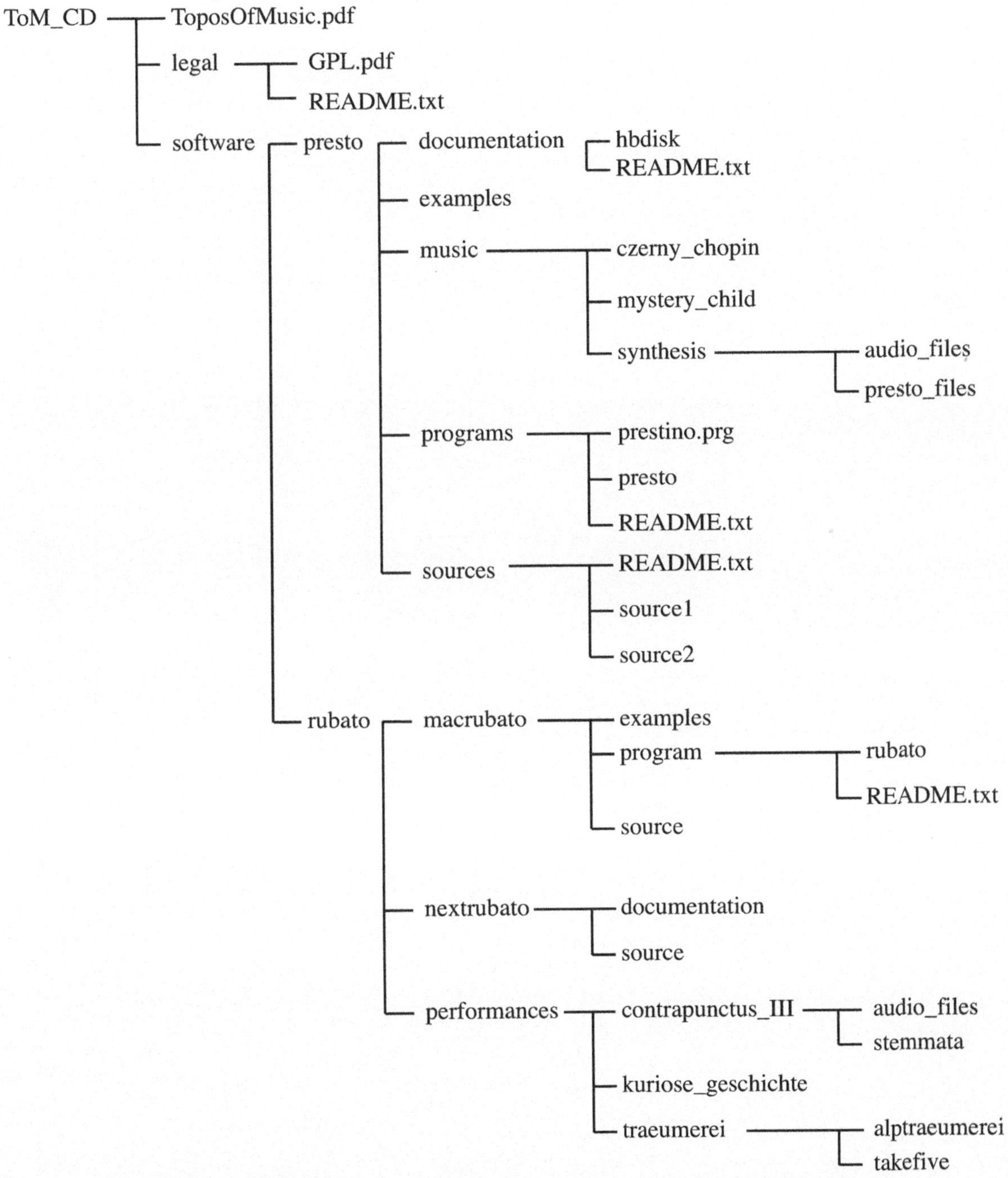
ToM_CD
ToposOfMusic.pdf
legal
GPL.pdf
README.txt
software
presto
documentation
hbdisk
README.txt
examples
music
czerny_chopin
mystery_child
synthesis
audio_files
presto_files
programs
prestino.prg
presto
README.txt
sources
README.txt
source1
source2
rubato
macrubato
examples
program
rubato
README.txt
source
nextrubato
documentation
source
performances
contrapunctus_III
audio_files
stemmata
kuriose_geschichte
traeumerei
alptraeumerei
takefive

Part I

Introduction and Orientation

Check for updates

1

What Is Music About?

Musik ist das ganze Leben.
Rudolf Wille [1133]

Summary. This chapter describes the overall extension of music-related activities in the space-time of human existence. The basic scope of the systematization as described in this book is declared. Our selection of corresponding fundamental scientific domains is not meant as a qualitative judgment over other scientific or artistic domains; it merely names the scientific pillars of the musical realm.

– Σ –

This book deals with the topos, the very concept of music. However, music appears in a wide range of human realms. It is a universal phenomenon of symbolic and physical, formal and emotional, individual and social, systematic and historic presence. Therefore, developing a concept of music should not get off with a definition *ex machina*, but offer propaedeutic orientation tools in order to make the reader *understand why* certain conceptual mechanisms or definitions are built. The need for such a support reveals a strong distinction from chemistry, physics or other natural sciences. The point is not that these sciences are dispensed from the fundamental question of what they are about. Rather the characteristic difference to musicology—and other humanities—is that natural sciences offer a fast access to effective activities, to the "working scientist" paradigm: Realization of one's scientific status by doing science.

"Doing" musicology is much less easy. The historical overhead in musicology is an index of the effort requested to get off the ground, to navigate within a safe concept framework. Later in this book we shall discuss some of the reasons for this deficiency, but for the time being, we just notice that musicology still lacks a stable concept framework, see [703] for a detailed discussion of the "unscientific status of musicology". It is one of the main goals of this work to develop such a framework. In fact—as a result of a lengthy concept development—an explicit concept framework for musicology will only be achieved in section 67.6.1.4. We shall, however, not deal with the full reality of music, as it appears in psychological, physiological, social, religious and political contexts. For a semiotically motivated systematization of music which comprises these aspects, refer to [703].

In this vein we want to start the preliminary orientation by presenting the *fundamental activities* related to music, and then their foundation upon *established scientific research fields*.

1.1 Fundamental Activities

Summary. The musical realm distributes among four types of artistic or scientific action and reflection: production, reception, documentation, and communication. These activities testify to a universally ramified presence of music in culture.

– Σ –

G. Mazzola, *The Topos of Music I: Theory*, Computational Music Science,
https://doi.org/10.1007/978-3-319-64364-9_1

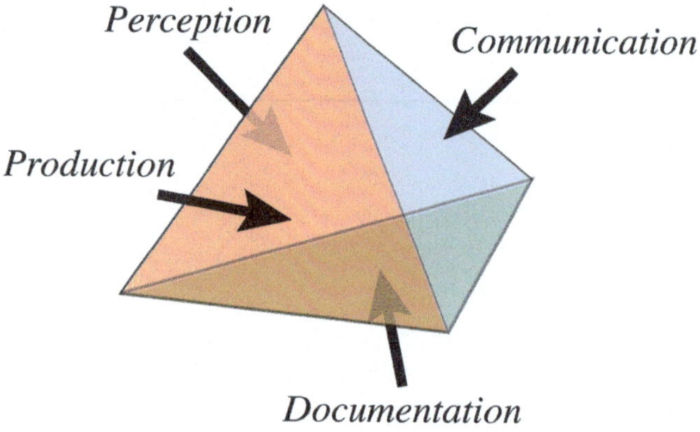

Fig. 1.1. The four fundamental activities in music are visualized as sides of a tetrahedron.

The following classification of music-related activities probably applies to many cultural fields. We recognize that developing a consciousness of their presence *in music* sheds a particular light on this subject. This is due to the fact that music traditionally relies on a strong connection between artistic facticity and intellectual reflection. In most sciences, and even arts, such an intertwinement is an exotic aspect, but musicology has to deal with it and this makes the case a very special one. The four activities are as follows: production, reception, documentation, and communication. For the sake of coherent representation, and because of the discussion in section 1.2, we represent these activities as sides of a tetrahedron, see figure 1.1.

These concepts are understood as follows:

→ *Production* refers to whatever is recognized as being on the side of the 'making' of music, and this on any level or reality. For example, we can make a sound on an instrument, or write or let a computer software 'write' a composition in a mental reality, or make an expressive performance. But we cannot 'make' an analysis in this sense, since this kind of activity receives a given musical body.

→ *Reception* refers to whatever is recognized as being on the side of 'taking' something from music on any level or reality. For example, we may hear a sound or let a machine decompose it into its partials, or make an analysis of a musical score—by hand or supported by a computer. But we may not perform a composition in the sense of a creative restatement. Whatever is on the side of a production of some new musical entity is understood as being on the productive side.

→ *Documentation* deals with whatever is related to bringing musical facts into more or less permanent sign systems. Musical notation, data base systems for musical objects, tradition of musical texts, the question of how and on which media ethnomusical performances should be stored, all this is subject to music documentation.

→ *Communication* is related to whatever happens when the three preceding activities are put into relation with each other. Production leads to documents, documents are retrieved and interpreted by instrumentalists or musicologists, live production of musical sounds by improvising musicians is perceived by other musicians of the band, then answered, and so on. Communication deals with the omnidirectional processes between production, documentation, and reception.

We do not contend that these four fundamental activities represent the only possible classification scheme, but they are broad enough to evidence the immense variety of perspectives when dealing with music. And we insist on the fact that all these activities are of comparable relevance. Let us make this clear. Since music is an art where time plays an omnipresent role, the making of music, its physical production in performance cannot be separated from the product as a preliminary task; it is substantial. The same holds for reception. On the other hand, documentation is not a necessary evil. It evokes the fundamental question of identity of an artistic work: Can we say that Beethoven's Fifth is that determined text? Or is performance intrinsically part of the Fifth's identity? Trying to understand music under permanent abstraction from any one of these aspects cannot succeed. And, above all, they are *activities* and not passive contemplation, a basic fact that we shall discuss in depth in chapter 5.

It is further remarkable that the synopsis of these equally important music activities draws a picture of perfect encyclopedic universality. In the "Encyclopédie" of Denis Diderot and Jean le Rond d'Alembert, we have three basic conceptual coordinates [238, 48] which go back to the global geography of science as designed by Francis Bacon in [58]: Reason/Philosophy, Documentation/History, and Imagination/Arts. If we compare these entities to the music activities, they correspond as follows: *Imagination/Arts* is the counterpart of *production*; the arts express a creative activity of making a work from one's imagination. *Reason/Philosophy* is the counterpart of *reception*; it deals with what in German is better conceived as "Vernunft", which means to perceive, receive, understand, observe something which is presented to your intellect. *Documentation/History* is verbally the same as in our list. What we have not inserted so far is the activity of *communication*. Now, the *genuine* encyclopedic concern is communication of the global orbit of human knowledge. Without this driving motor, reason, imagination, and documentation are worthless abstracta.

Summarizing, we can state that the main concerns in the classical encyclopedia of the Age of Enlightenment are congruent to the four basic music activities. This is not only a comfortable *raison d'être* for music as a cultural phenomenon, it also has deep consequences for the construction of a viable concept framework. Such a body must comprise a variety of ontological[1] perspectives and cannot refrain from contents in favor of formal virtuosity. The system of predicates to be developed in chapter 18 will account for this specification.

1.2 Fundamental Scientific Domains

Summary. The fundamental scientific domains relating to the activities described in the preceding section include semiotics, physics, mathematics, and psychology. This is explicated and justified—in particular with regard to the only apparent elimination of the historical perspective which dominates traditional musicology. The relation between the two lists of activities and domains is discussed.

– Σ –

To begin with, we have to make clear what "fundamental scientific domains" should signify. We may view the collection of sciences as being a hierarchy of disciplines with its particular relations being defined by a substantial dependency. For instance, chemistry is dominated by physics, but not vice versa. And every natural science is dominated by mathematics. Many of the hierarchical relations are quite problematic, blurred or otherwise unclear by the very definition of basics of particular disciplines; we have to accept such general uncertainties and will proceed in full consciousness of the provisional state of the art of epistemological classification. Now, if one has to locate a special research subject, such as music, within the hierarchy of disciplines, one is looking for a minimal set, a kind of *'basis'* of disciplines which are necessary to cover the subject. In this sense, we argue that four sciences: semiotics, physics, mathematics, and psychology, are such a 'basis'.

Let us explain this selection. We have to show that the four disciplines are sufficient to cover the field of music and that no one of these sciences is superfluous in that it either is not involved at all, or that it is only involved via a proper sub-discipline. At this point, we cannot circumvent a provisional description of what music is about, though this is only a first approximation in order to delimitate the scientific overall extension.

Definition 1 (First provisional version) *Music is a system of signs composed of complex forms which may be represented by physical sounds, and which in this way mediate between mental and psychic contents.*

To deal with such a system, semiotics are naturally evoked. To describe these forms, mathematics is the adequate language. To represent these forms on the physical level, physics are indispensable. And

[1] Ontology is the philosophical discipline dealing with the concept of "being".

to understand the psychic contents, psychology is the predestined science. Is this selection a "basic" one as claimed above? There are several critical points which we should clarify. First, the historic dimension of music seems to be neglected. However, semiotics has a strong competence in diachronic study of a sign system. Music history is prominently the *history of a system of significant signs* (much like diachronic linguistics), and here, diachronic semiotics is the adequate research field, whereas general history would be too generic in view of the given subject. Musical forms are not only mediated via physical representation. They have a strong properly mental ("symbolic") existence, and this evokes mathematics as a directly involved research aspect. Physics has a prominent role in music, and this not only with respect to sound production (classical acoustics and physics of musical instruments), but also in computer science which is concerned with data representation and processing in music—be it on the generic level of software or on the level of advanced digital sound synthesis and analysis. Hence, we do 'absorb' computer science and sound engineering within physics, together with the mathematical formalisms in theoretical computer science. Finally, social aspects of music are viewed as objects of social psychology on one, and socio-semiotics on the other hand. Also pedagogical problems (music education) are subsumed under the label of psychology since it is the human psyche which is educated.

It is an extremely important point to recognize the *relative autonomy* of these four research fields in their descriptive force. This means that we do not, and never will hope or hypothesize reductionism whatsoever. There is no interest to view psychological reality as a surface of neurophysiological depth structure. In this sense, our selection is not pragmatic but ontological. We come back to this theme in chapter 2.

To complete the picture, we should try to fix the relative positions of the four basic sciences with respect to the four fundamental activities of production, reception, documentation, and communication. It is evident that activity is more or less related to some of the four scientific disciplines. If we 'label' each vertex of the tetrahedron shown in figure 1.1 by a basic science, it turns out that every triangular activity side is 'spanned' by three of four vertices. If we pay attention to the obvious semantics of this configuration, we have to place the labels in such a way that each activity is delimited by the three most relevant sciences for its execution. We propose the labeling drawn in figure 1.2 and leave it to the reader to judge our selection.

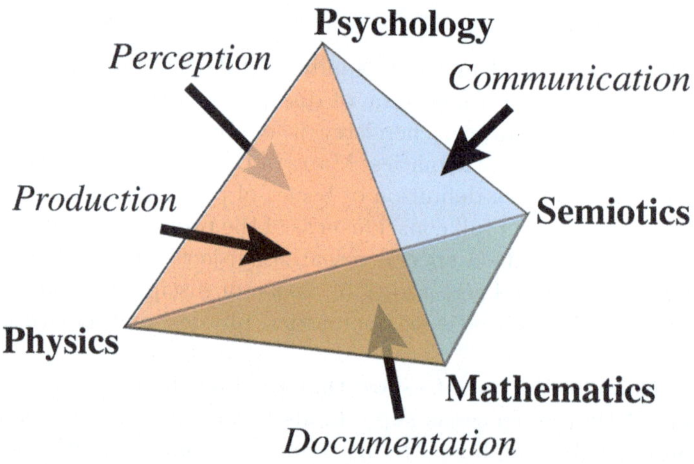

Fig. 1.2. The four fundamental activities, resp. their counterparts of basic sciences in music, are visualized as sides, resp. vertices, of a tetrahedron.

Let us simply give one example: Documentation is concerned with signs, their formal representation, and the physical storage techniques. The psychological aspect is less substantial, whence documentation is 'spanned' by the respective vertices. We do, however, not overdress this elegant configuration as a "magic" tetrahedron, the visualization merely serves as a concise and relatively well positioned synopsis of the overall situation, where the question "What is music about?" can be initiated. And it is also a very practical *orientation scheme for designing future music research policy and strategy.*

2

Topography

Die Linien des Lebens sind verschieden
Wie Wege sind, und wie der Berge Gränzen.
Was wir hier sind, kann dort ein Gott ergänzen
Mit Harmonien und ewigem Lohn und Frieden.
Friedrich Hölderlin: Die Linien des Lebens... (1843)

Summary. This chapter deals with an ontological orientation in the subject of music. It was already contended in chapter 1 that music is communication, has meaning and mediates on the physical level between its mental and psychological levels. Such an orientation is topographic in nature since it offers a number of ontological "dimensions" and "coordinates" to profile musicological discourse and helps avoiding misplaced or blurred arguments. The topography involves three mutually independent dimensions: communication, reality, and semiosis. The local nature of this orientation scheme is discussed.

– Σ –

Whoever is dealing with formal, logical, mathematical or computational methods in music does not necessarily have to be an expert in music philosophy. Nonetheless, these methods require a sufficiently refined orientation within the complex ontology of music in order to avoid erroneous interpretations of results obtained by use of these exact methods. It is well known that the precision of mathematical results, together with a poor knowledge about the delicate ontology of music may provoke a dogmatism for which mathematics is unjustly made responsible.

On the other hand, the *extensive* concept framework which we want to set up cannot even be approached without a firm pointer to musical ontology. In the following "topography of music" we shall sketch an ontological model without any dogmatic or unmodifiable character, though constantly available as a powerful coordinate system to locate the problems within the organism of music.

In view of the generic (and first provisional) delimitation of the concept of music (Definition 1), it is proposed to set up from the very beginning a "three-dimensional" ontology, i.e. an ontological system which tells rather *where* the concept of music subsists than *what* its being is like. Let us shortly deviate on this philosophical line to justify the somewhat particular approach. To be clear, we are not concerned with music as such (a Kantian "Ding an sich") but with music that we localize and conceive *within* our knowledge system. Following Immanuel Kant [519, p. B 324], a concept is a local place, a topos in the knowledge space. Ontology then questions the place of the concept, the space-coordination of the music topos.

We do, in other words, stress three ontological coordinates as an intrinsic unity pointing at the concept of music. These coordinates are

- *REALITY*
- *COMMUNICATION*
- *SEMIOSIS*[1]

[1] Semiosis refers to the fact that music deals with a system of signs.

G. Mazzola, *The Topos of Music I: Theory*, Computational Music Science,
https://doi.org/10.1007/978-3-319-64364-9_2

We shall hereafter (Sections 2.1, 2.2, 2.3) introduce them in detail. To understand music as a whole, you have to specify *simultaneously* its levels of reality, its semiotic character, and its communicative extension. Being a fact of music means having these three perspectives or ontological coordinates. Omitting any one of these determinants is an abstraction (though not an aberration) from the full ontology.

It might be argued that this topographical reduction of ontology simply delegates the difficult question to the ontological coordinates: instead of asking for one ontological specification, we have to examine three partial specifications. The argument is correct, but it is also true that these partials are simpler, more elementary. Dealing with semiosis *without* being concerned with its level of reality (and vice versa) is a huge advantage. Giving the ontology a topographical "turnaround" simplifies the problem. Whether the overall ontological question is settled by this procedure is not answered here. Let us instead begin with a closer inspection of the three proposed coordinates—we shall come back to this subject after that in Section 2.5.

2.1 Layers of Reality

Summary. The dimension of musical reality involves *physical*, *psychological*, and *mental* layers. It is not question of reducing one of these realities to the others: Either of them has an autonomous existence which can at most be transformed into others, but not eliminated. The great majority of reality-specific phenomena cannot even be translated into external layers without substantial deficiencies.

– Σ –

Music takes place on a wide range of realities. They may be grouped into physical, mental and psychological levels. Differentiation of realities is crucial for avoiding widespread misunderstandings about the nature of musical facts.

A representative example of this problem is Fourier's theorem, roughly stating that every periodic function is a unique sum of sinoidal components. Its a priori status is a mental one, a theorem of pure mathematics. In musical acoustics it is often claimed that—according to Fourier's theorem—a sound "is" composed of "pure" sinoidal partials. However, there is no physical law to support this claim. Without a specific link to physics, Fourier's statement is just one of an infinity of mathematically equivalent orthonormal decompositions based on "pure" functions of completely general character, see [268, ch. VI]. To give the claim a physical status, it would be necessary to refer to a concrete dynamical system, such as the cochlea of the inner ear (see Appendix, Section B.1), which is physically sensitive to the first seven partials in Fourier's sense.

Methodologically, there is no reason nor is it ontologically possible to reduce one reality to others. The problem is rather to describe the *transformation rules* from the manifestation of a phenomenon in one reality to its correspondences within the others. To be clear, a neurophysiological transformation ("explanation") of a psychological phenomenon does not, however, conserve the psychological ontology of the phenomenon. The specific phenomenon within the psychological topos *corresponds* to another phenomenon within the physiological topos. But ontologically, the phenomena do not collapse.

We now give an overview of the three fundamental topoi of reality and their specific characters.

2.1.1 Physical Reality

Music is essentially manifested as an acoustical phenomenon, made by means of special instruments and perceived by humans. Nonetheless, its acoustical characteristics are less—if at all—condensed within a unique physical sound quality than in the physical input-output systems for sound processing. The use of physical sounds is a function of very special devices for synthesis and analysis. This is reflected in the fact that to this date, there is no generally accepted classification method for musical sounds, see Section 12.3 for a thorough discussion. This is not due to missing synthesis or analysis methods and techniques. Besides classical analog sound synthesis methods as they are realized on musical instruments, there are various digital sound synthesis methods, see appendix A.1.2.

The problem is rather that classification of musical sounds is arbitrary *without* reference to their semantic potential, see Section 12.3.3. Physical reality of music is only relevant as an interface between 'expressive' and 'impressive' dynamical systems. All physical sounds are equally natural, be they produced by a live violin performance, a computer driven synthesizer via loudspeakers, or by the tape patchwork of musique concrète.

On the other hand, the central receptive system for music is the human auditory system: Peripherical and inner ear, auditory nerve, its path through multiple relays stations of the brain stem, the neo- and archicortical centers for auditory processing and memory, such as Heschl's gyrus and the hippocampal formation, see appendix B. This extremely complex physiological system is far from being understood. Even though some insights into the dynamics of the cochlear subsystem do exist, it is not known which analysis of the musical sounds takes place on the higher cortical levels. It is not even clear how the elementary pitch property of an ordinary tone is recognized [162]. This means that on the cognitive level, human sound analysis is not yet understood. Therefore, reference to particular sound representation models is good for synthesis options, and for speculative models of cognitive science [593], but *not* as a firm reference to human sound processing.

2.1.2 Mental Reality

Just like mathematical, logical and poetical constructions, musical creations germinate as autonomous mental entities. It is a common misunderstanding that musical notation is an awkward form to designate physical sound entities. Being a trace of intrinsically human activity, the phenomenological surface of music is linked to mental schemes which we call scores: oral or written text frames of extra-physical specification. Scores are mental guidelines to an ensemble of musical objects. They reflect the fact that music is composed as well as analyzed on a purely mental level. Obviously, scores do point at physical realization, but only as a *projection* of a mental stratum into physical reality. A fact of harmony or counterpoint is an abstractum much the same as an ideal triangle in geometry. In this sense, playing a chord on a piano corresponds to drawing a triangle on a sheet of paper.

2.1.3 Psychological Reality

Besides its physical manifestation and its mental framework, music fundamentally expresses emotional states of its creators and emotionally affects its listeners. This was already known to Pythagoreans [1080] and defined as a central issue of music by René Descartes [262], see [255]. Such an emotional reality of music is neither subordinate nor abusive, on the contrary: to music lovers the emotional response is a dominant aspect, individually, and socially. It is not by chance that song titles such as "It don't mean a thing if it ain't got that swing" de facto reduce musical meaning to an emotional category, the feeling of swing. Like other realities of music, the psychological dimension cannot and needn't be reduced to others, it is a manifestation of an autonomous, irreducible ontology.

2.2 Molino's Communication Stream

Summary. Communication is the second dimension of our topography. Following Jean Molino, it splits into a three-part stream starting from the *poietic* instance of the *creator* of a work of art, traversing the mediating *neutral datum* of the virtual *work*, and ending on the receptive side of the *listener's esthesic* perception.

– Σ –

According to Jean Molino [758], we describe the tripartite communicative character of music, see Figure 2.1. Molino's scheme is an abstraction to the essentials; they may be complex and interlocked in concrete situations, see Section 2.5. Its structure is comparable to Oskar Bätschmann's "grosses abstrakt-reales Bezugssystem der Auslegung" which he proposed as a hermeneutical framework for the analytical discourse in fine arts [90]. Like Molino's scheme, Bätschmann's construction issued from the insight that a dispassionate discourse on arts, be it "parler peinture" or "parler musique", cannot succeed without considering their communicative structure.

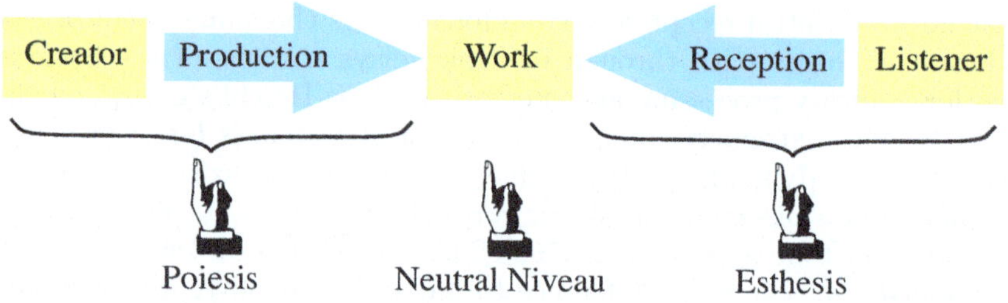

Fig. 2.1. The three-part scheme of music communication following Molino.

Molino's scheme partitions the communication process into three instances: creator, work, and listener. To these, three types of analytical discourse correspond: poiesis, neutral level, and esthesis. These concepts are technical terms which have to be explicated in the following. In particular, Molino does not prejudice such a thing as causality between creator and listener by means of the work, or receptive passivity of the listener towards work and creator.[2]

2.2.1 Creator and Poietic Level

To begin with, a *creator* instance subsumes all factors which are essential and sufficient for the production of a musical work. This level describes the "sender" instance of the "message", classically realized by the composer. The corresponding analytical discourse of the work's *poiesis* has been introduced by Etienne Gilson [366]. According to the Greek etymology ($\pi o \iota \varepsilon \iota \nu$ = to make), "poiesis" relates to the one who makes the work of art. Poiesis is concerned with the individual condition of the creator as a result of its history of development as well as with the role of the broader socio-cultural frame of the work. In specific cultural contexts or in a more refined discourse about art production, the creator instance may as well be the musician or the performing artist. In jazz, for example, the improvisational aspect is a genuine making of the music, and the performance of a classical piece of music is a creational act.

Poiesis is a sharp analytical instrument. Already the very definition of a sound by means of its defining parameters, such as pitch, color, etc. bears a strong poietical flavor—quite contrary to the naive and widespread belief. For example pitch (logarithm of frequency) is no invariant of the sound, it usually results from the sound construction as a product of a periodic vibration of a determined frequency and of an envelope (see appendix A.1.2.1). Neither frequency of the periodic vibration nor the envelope are uniquely determined by their product. And, worse, not even the frequency of an ideal periodic vibration is uniquely determined from the vibration as such. In fact, Fourier's theorem only states uniqueness of its coefficients *if* the fundamental period is given. But the latter is not automatically inscripted in a vibration. Such parameters are part of poiesis: the way of making the sound, not the "neutral" sound object. This shows that poiesis is not restricted to psychological aspects of "how to fabricate a composition", it can relate to completely "objective" physical or mathematical facts.

[2] We should like to relate these concepts to the activities of production, reception, documentation, and communication, as described in Section 1.1. Activities are instances of what here is denoted by "communication". What we call "poietic level" corresponds to "production activity", "reception" is on the side of "esthesic level", "documentation" relates to "neutral level", and the activity of communication corresponds to the transition processes between all the three levels in Molino's system. We have nonetheless maintained these concept frameworks since Molino's technical terminology does not reflect the aspect of overall activity in music.

2.2.2 Work and Neutral Level

The *work* is the turning point of musical communication. It subsumes whatever determines directly the production of sound events within a specific context. Its identification depends upon the contract of creator (sender) and listener (receiver) on the common object of consideration. Musicology has difficulties with the work concept since its determinants seem to be of fuzzy nature. This relates to the dramatic difference between music and fine arts. As a matter of fact, a painting is determined down to its humidity and temperature—at least from the moment it has gained 'eternal' appreciation and allocation in a professional museum. In music, the definition of a work is left with ample variants. Decisions about including sound color or tuning prescriptions in Bach's "Wohltemperiertes Klavier" do in any case redefine a new and different work. And above that, such a decision transcends the strict definition and establishes a *valuation* of the work in the light of historical parameters of Bach's creativeness.

In this context, a *score* may be viewed as being the *organizational scheme* of a work. In fact, physical realization of music is always based upon mental schemes which we call scores: oral or written text frames of extra-physical specification. Apart from classical western traditions where this fact is evident, the score concept is also adequate for describing traditions which are more towards ethnomusicological antipodes. First example: In the music of Noh theater [542], there are different score instances, e.g. for vocal utai music denoted in melodic units (fushi) to the right of texts, or for the hayashi notation systems for flutes and drums. Second example: The improvisational culture of jazz—which in its making only marginally relates to traditional western scores—is based on the concept of the interior score ("partition intérieure" [979]). This means that, even for free jazz improvisers, there is an interior reference system of lexical character, together with a selection code which guides performance. The fundamental fact behind the basic role of the score concept for music is that human organization in a complex time-space of acoustical and gestural nature cannot be executed without an interpersonal spiritual orientation common to the responsible participants.

Following Molino, the analytical discourse which—independently of the selection of the tools—is strictly oriented towards the given work is termed *neutral level.* In particular, the historical justification for a work's delimitation is *not* part of the neutral level. Molino's concept has only attracted controversies since it—faultily—seems to imply the preference of a particular analytical tool/method to the exclusion of other possible approaches. But such a thing as the "unique ideal interpretation" is precisely the famous *unicorn of hermeneutics*—a superfluous illusion. A momentous analytical perspective is just one among many possible accounts of the work, and neutral analysis is the variety of *all* these perspectives.

Such a work may be the divine creation of a medieval tone system, a work of computer music which is explicated as far as its acoustical details, or the beat of waves on the open waters of the sea, no matter: Neutral level means propaedeutic analytical precision work preceding any valuation whatsoever. Evidently, exact methods will play an eminent role in realizing such a task.

2.2.3 Listener and Esthesic Level

The *listener* is the instance which perceives and interprets a given work. It is subject to the same type of determinants as the creator. Coordinates of a listener will vary from case to case and coincide only punctually with each other or with the creator's coordinates. The characteristic difference between listener(s) and creator is the time arrow towards the common work. Whereas the creator produces a work, the listener perceives and interprets an already existing work. From the moment of a work's existence, the creator's existence is fixed whereas the number of the listeners' existences grows as a function of those who are engaged in the work.

According to a proposal of Paul Valéry, [1072] the analytical discourse on the listener is termed *esthesis*, following the Greek term $\alpha\iota\sigma\theta\eta\sigma\iota\sigma$ (perception), in order to avoid confusion with esthetics, the study of beauty. Esthesis deals with perceptive valuation of a work by the listener. This valuation of the work's attributes is realized as a function of the listener's individually variable position. This interpretative valuation from a determined perspective is not less active than the creator's activity in the creation of the work. And this not only regarding the artistic performance, but quite generally. Like poiesis, esthesis is integral part of the analytical elements constituting the existence of a work.

The distinction between poietical and esthesic levels is fundamental for a sound conceptualization in musicology. Only in very rare cases can we expect the conservation of results from a poietical discourse within the esthesic point of view, and vice versa. Nonetheless, we observe again and again the faulty trial to implant historically and systematically clearcut *poietical* concepts, such as "pitch", into an esthesic discourse. The famous paradigmatic justification of such a malfunction of analysis is known under the dictum that "the ear is the highest musical instance". On one hand, such a dogmatic phantom as "the ear" does not exist, is incompatible with the multiplicity of listening cultures. On the other, the ear fails as a metonymy of cortical music processing: there is a huge transformation process between the auditory cortex and the ear's cochlear system.

Poietics cannot be taken as the ideal access to a music work. There is no thing such as a causal relation between creator and work. The latter is a result which may be generated by very different processes. Within the communication process, the creator above all differs from the listener by his/her position in time. This fact becomes manifest when we imagine the work's production process as a fictional unfolding in a *retrograde* movie. This visualizes a well-known insight from fine arts theories that the artist/creator is the first observer. In this way, retrograde poiesis is embedded in esthesis. And vice versa, the esthesic discourse may be restated as a manifold of variants of imaginary retrograde poieses.

Example 1 The problem of symmetries in music is a good illustration of the communication-sensitive aspect. A classical conflict concerning the role of retrograde in music arises from the observation that this construction "cannot be heard and thus is a problematic feature". Communication coordinates make this discussion more transparent: The retrograde construction as a poietic technique is a common compositional tool. It fits into the toolkit of contrapuntal constructs for organizing the compositional corpus. On the other hand, the esthesic perspective of the retrograde is concerned with the question whether and how clearly such a construct can be decoded by the listener. This latter question is a completely different topic and cannot be identified with the former. More precisely, the role of the retrograde as an organizational instance is not a function of its perceptibility as an isolated structure. The psychological question of whether a retrograde can be perceived is rather this: "Can a retrograde be distinguished from random?" Finally, retrograde structures may be recognized as objective facts within the neutral level of the score without being either constructed be the composer or consciously perceived by the listener. Summarizing, the communicative coordinates help localizing and thereby making more precise the musicological discourse.

2.2.3.1 The Problem of Identity

From the preceding communication-theoretic considerations, the *identity of a work* is triply stratified. To begin with, we envisage a work that is given before any analysis is performed. We call this data its *abstract* identity. This first identity is then enriched to yield the *neutral identity* as a result of the multiplicity of neutral, valuation-free analyses. This identity is the basis for a variety of esthesic—in particular: retrograde poietic—valuations. The esthesic identity is built from all these esthesic valuations and their mutual relations.

Accordingly, a work is only identified when the *infinite* process of esthesic valuation, built upon abstract and neutral identification, is completed. This yields at least one justification for an incessant performance practice in classical music, but see Section 32.2 for more details on performance and identification.

2.3 Semiosis

Summary. Independently on what is communicated, and on which level of reality this takes place, music intrinsically involves complex signification processes. The generic setup of semiotics understands a sign as being a tripartite object. It consists of a *significant* expression, inducing the *signification* act of translation, and thereby producing the expression's content, the *significate*. We give a short review of dichotomies of structuralist semiotics. Musical semiosis reveals a complex concatenation of meta- and connotation-layers in the sense of Louis Hjelmslev. Rather than being absent, musical meaning is distributed over a sequence of semiotic subsystems.

– Σ –

By use of a highly developed textuality of musical notation as well as by the very intention of musical expression, music is structured as a complex system of signs. This is not only a marginal aspect: Music is one of the most developed non-linguistic systems of signs. We shall first present the elementary sign character of musical objects (Sections 2.3.1, 2.3.2, and 2.3.3), and then (2.3.4) review the overall semiotic perspective of music.

2.3.1 Expressions

Already the earliest medieval music notation is motivated by the very nature of the graphical neumes: etymologically as well as substantially they are gestural hints pointing at movements in pitch and rhythm. This coincides with the latin etymology of sign: signare = to point at, give a hint. Neumes are *aliquid pro aliquo*, they express something in the sense of semiotics. Besides and beyond musical notation, music is often viewed as an expression of emotions, spiritual contents or gestural units. In any case, music has a phenomenological surface that is organized in a spatio-temporal syntax. Albeit more complex than linguistic syntax, the musical syntax shares some of its characteristics, see Section 2.3.4.

2.3.2 Content

According to the famous dictum of Eduard Hanslick [438], we learn that: "Der Inhalt der Musik sind tönend bewegte Formen." This evidences that the notated complex of musical graphems is not the content but points at some kind of sounding content: they *mean* something. Hanslick's characterization is a minimal semantic setup but *some kind of content* can be identified. The remarkable aspect of Hanslick's approach is that it associates musical content with mathematical content. As a matter of fact, a triangle is a mathematical object that essentially reduces to form. But mathematicians do associate it with a content, usually with a platonic entity addressed by abstract symbols or by drawings with precise quantifications.

2.3.3 The Process of Signification

Signification is the most important instance for the realization of a sign. It is responsible for the transformation from the expressive surface to the hidden meaning. The fact that performance is such a central issue of music gives a strong proof of the qualified presence of signification in music. Without the laborious effort of turning significant expressions into the reality of their meaning, music is not what it is meant to be. For musical signs, this semiotic process bears a highly differentiated structure which is sensitive to Ferdinand de Saussure's dichotomies [86, 933], we review these items in the following Section 2.3.4.

2.3.4 A Short Overview of Music Semiotics

Semiotics of music is a complex subject which cannot be dealt with in this context. For a full account, we refer to [703]. This does, however, not dispense us from a short overview of this subject since there is no hope to understand music without a minimum of semiotical prerequisites. We shall describe music semiotics from the perspective of structuralist semiology as it has been sketched by Roland Barthes [86] as a generalization of the linguistic theory of Ferdinand de Saussure [933]. We pursue this project according to André Martinet's *principle of relevance* [655]. It states that selection of materials and methods has to be justified by relevant criteria in the sense of points of view towards the actual object—music in our case.

Let us recall that Barthes' position does not mean that music is interpreted as being a type of language. On the contrary, we shall repeatedly stress significant differences between the music system and linguistics.

To say that a system is semiotic means that it is articulate according to the fundamental stratification of signs into significant, signification, and significate layers (Saussure's signifiant/signification/signifié). This stratification articulates the process of production of meaning (the signification), which points from the expressive surface of the mediator (the significant), to the underlying meaning (the significate). Slightly deviating from Saussure's original approach, the existentiality of the significate is that of a concept, i.e. it is part of a "lexical body", of a system of sayable things ($\lambda\varepsilon\kappa\tau o\nu$ in the stoic tradition, in contrast to the "real" things or to the psychological imagination, the latter being Saussure's preferred significates).

Like every existing semiotic system, music is embedded in history. If we investigate the system's state at a determined moment, we deal with *synchronic* analysis. Of course, the concept of a moment really means a short duration wherein the system remains fixed. If, instead, the system's objects are observed in their temporal development, we deal with *diachronic* analysis. For example, the etymology of the concept of consonance is a diachronic subject, whereas looking at semantics of the augmented triad in Franz Liszt's "Faust" symphony points at a synchronic analysis. Evidently, sychronicity and diachronicity are not strictly separable since—just as with physics—the space-time continuum of music is indecomposable.

The special class of *shifter* or *deictic* signs is very important in music and is characterized as follows: Their significate transcends the lexical reality and penetrates the unsayable existence of the system's user. Whereas, for example, the concept of a sonata form is lexical, the emotional significate of music can neither be "lexicalized" by decree, nor verbalized whatsoever—contrary to numerous flopped attempts. In linguistics, shifters (such as "I", "here", "now") are exceptional signs. In music deixis, such as emotional signification, is the standard situation. But this does not mean that the lexical part of the system is marginal. On the contrary, a rich deixis cannot succeed without a rich lexical support. It is evidently a difficult business to "hit" the unsayable by use of an arsenal of lexical approximation pointers.

As a function of the type of signification process, Saussure distinguishes between *arbitrary* and *motivated* signification. Signification is arbitrary if it is defined by pure convention. For example, the signification "highten a pitch by a semitone!" of the sharp # is pure convention, whereas the signification of a tremolated kettledrum beat towards the significate "thunder" is motivated as onomatopoeia. Notice that the arbitrariness of musical notation is a major source of the elitarian music culture; fortunately, a huge number of semiotically well-motivated music software has started to break this artificial obstacle to the world of music. A standard means for diminishing arbitrariness is the use of system immanent logic or of instances exterior to the system which are at disposal as given evidences. The latter is the case if our system lives in the context of other systems (e.g. software environments) which may help to operationalize it. A famous example of system immanent logic is Roman Jakobson's *poetical function*. It enables production of significates by means of "projection of the paradigmatic axis into the syntagmatic axis" (see below and Sections 11.6.1, 49.2.1).

As a whole, the system has sign character but it is composed from single signs. If two signs are adjacent in space or time they are called contiguous. Contiguity is "in praesentia". It is a fact of concrete nature, no further abstraction is needed. A chain or network[3] of signs in successive contiguity is called syntagm, in linguistics also: syntagmatix axis. A progression of degrees or a contrapuntal interval sequence are syntagmata in time; a texture of four parts (soprano, alto, tenor, bass) or a discantus which is set against a cantus firmus voice, these are syntagmata in pitch space. As a consequence, syntagmatic articulation of music looks quite different from linguistics. Whereas linguistic syntagm is one-dimensional (in time), the music syntagm extends simultaneously in several dimensions.

If two signs are similar then they are called in *apposition.* (Martinet uses the term "opposition" in linguistics but in music this binary term is too narrow.) Similarity may regard any of the three layers—significant, signification, or significate. The collection of all signs which are in apposition to a given one define the *paradigm* of that sign. Apposition is a relation "in absentia". It requires a high degree of abstraction, we come back to this important difference to the syntagmatic contiguity in Section 50.4. For example, tones which are perceived with loudness equal to that of a reference tone define the paradigm of loudness of this tone. Two such tones do not necessarily yield the same loudness perception (see appendix B.2)!

According to Louis Hjelmslev [471], a given semiotic system can appear as a sign-theoretic instance (significant, signification, or significate) of another such system. If it appears as a significant layer, we speak about *connotation.* The significant system is the termed *denotative level*, where as the superior system is termed *connotative level.* If, conversely, a system occupies the significate layer of another system, the superior

[3] This extension transcends the usual definition, we allow a multiplicity of simultaneous concatenations, not only one chain such as the case in ordinary language.

system is termed *metalevel* or *metasystem*, whereas the inferior system represents the *objectlevel* or *objectsystem*. Generalizing Hjelmslev's approach, one should also envisage the case where a system appears as the *signification layer* of its supersystem. In general, semiotic systems are complex interlockings of connotative, metasystemic, and signification layers. For example, the double articulation in language: grapheme ⇒ acoustical significate ⇒ conceptual significate can also be observed in music. Here, in Hanslick's terminology, we have the double articulation Note ⇒ Ton ⇒ tönend bewegte Form. On the other hand, grammar of language and harmony condivide their metasystemic character [156, 802]. For an overview of the entire music system articulation in the sense of Hjelmslev, see [703].

The last of the famous Saussurean dichotomies—it is also decisive for distinguishing music from language—concerns *language* against *speech* (Saussure: langage/parole, Barthes: structure/procès). This means the contrast between general reglementation and concrete syntagms and paradigms as far as dialects and revolutionary poetical invention. In order to avoid confusion with the linguistic context, we shall adopt the dichotomy of (music) *norm* /(music) *process* (corresponding to language/speech). This dichotomy shows a strong difference to the linguistic dichotomy in so far as such a thing as a music norm is neither a synchronic nor a diachronic constant. And emotional significates of music by definition forbid any normative fixation, they are music processes, "sound speech" ("Klangrede") as they used to call it. On the other hand, the musical denotation level in classical score notation is extremely normative, and as such a source for elite formation whose control is often confused with musicality. The analphabet differs from the score illiterate in so far as production of musical meaning has very little to do with score notation whereas competence in linguistic articulation is intimately related to lexical competence.

The Hjelmslev articulation of music [703] shows that musical meaning is not concentrated on a semiotic spot, it is on the contrary distributed among several system layers of denotation and connotation. In other words, musical meaning is the result of a complex development of signs, it grows from simple categories of signification to more and more all-embracing and even transcendental categories as they appear in highly spiritual musical expression.

2.4 The Cube of Local Topography

Summary. Putting together the three dimensions of musical topography, the topographic cube is defined. While each of the described dimensions splits into three "coordinate" values, the cube involves $3^3 = 27$ distinguished topographic locations. Their role in guiding a sophisticated discourse on music is discussed and exemplified. However, this cube's orientation is of local nature, it does not provide a global point of view.

– Σ –

After presenting the ontological dimensions of reality, communication, and semiosis, we may view these contributions as coordinate axes of a more complete musical ontology. This means that an ontological localization can be interpreted as a point in a three-dimensional cube spanned by the axes of reality, communication, and semiosis, each of them being articulated in three "values":

- reality: physical–psychological–mental.
- communication: creator–work–listener
- semiosis: significant–signification–significate

This suggests a geometric representation of the ontological variety as a "topographic cube" of musical ontology, see Figure 2.2. To describe the ontological position of a musical object, we have to specify its three coordinates in reality, communication, and semiosis. Each coordinate may take one of three values as listed above. This produces a set of $3^3 = 27$ possible topographic locations. Of course, it is not necessary that a general object be localized at a single location, it may as well occupy any subset of the cube. So, the 27 points are just the elementary positions from which more complex ontological situations are composed.

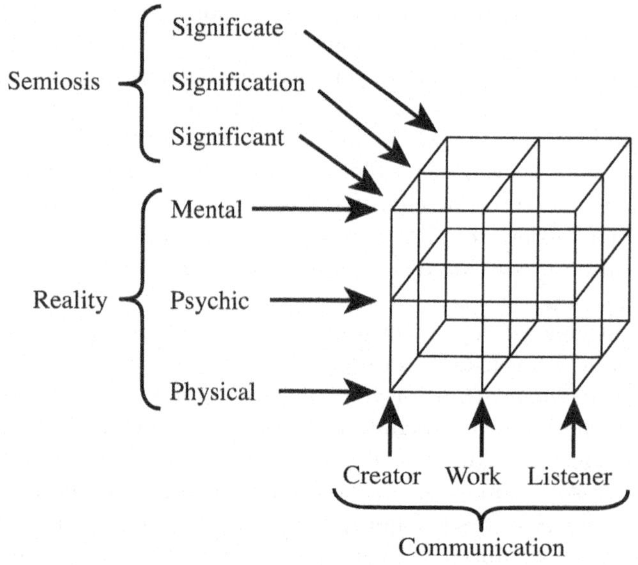

Fig. 2.2. The cube of musical ontology.

To illustrate this topography, we present a classical text of music critique. It is the review of a concert by Franz Liszt written by Ludwig Rellstab in Berlin's «Vossischen Zeitung», cited following Knapp [539, p.86]:

Unter dieser Erweckung der vorteilhaftesten Eindrücke setzte er sich an das Instrument. Jetzt wird ein neuer Geist in ihm lebendig. Er lebt die Musikstücke in sich, die er vorträgt. Während er mit erstaunenswürdigster Gewalt der Mechanik eigentlich alles leistet, was bisher von irgend jemand einzelnem bezwungen worden ist, und außerdem noch ein ganzes Füllhorn neuer Erfindungen, völlig ungekannter Effekte und mechanischer Kombinationen vor uns ausschüttet, so daß die aufs höchste gespannte Erwartung und Forderung sich weit überflügelt sieht: bleibt doch der eigentümliche Geist, den er diesen Formen einhaucht, das bei weitem anzie-hendere, anregendere und fesselndere Element. Diese geistige Bedeutsamkeit seines Kunstwerkes prägt sich aber auf das lebendigste in seiner Persönlichkeit aus. Die Affekte seines Spiels werden zu Affekten seiner leidenschaftlich aufgetürmten Seele und finden in seiner Physiognomie und Haltung den treuesten Spiegel. Seine künstlerische Leistung wird zugleich eine Tatsche des Inneren, sie bleibt nicht getrennt von ihm, sondern wirkt in dem mächtigen Bündnis mit dem Geist, der sie erzeugt.

Rellstab addresses all three levels of reality. Words such as "Instrument", "Gewalt der Mechanik", "mechanische Kombination", "Haltung", and "Physiognomie" refer to physical reality.—Expres-sions such as "Affekt seines Spiels", "Affekte der leidenschaftlich aufgetürmten Seele", "anregend", "fesselnd", and "lebt die Musik in sich" refer to psychological reality.—The mental level is addressed in expressions such as "neuer Geist wird lebendig", "Geist wird den Formen eingehaucht", "geistige Bedeutsamkeit", "Bündnis mit dem Geist", and "Tatsache des Inneren".

The communicative dimension specifies in its poietical coordinate with expressions such as "trägt vor", "*haucht* Geist *ein*", Geist "erzeugt" künstlerische Leistung", "Erfindung (Füll-horn)".—The neutral level is addressed in passages such as "Musikstücke vortragen", "Geist den Formen einhauchen", and "Kunstwerk". The esthesic coordinate is addressed by allusions like "wecket Eindrücke", "vor uns ausschüttet", "Erwartung (der Hörer)", "Forderung", "anregend", and "fesselnd".

The semiotic system is instantiated when Rellstab talks about the significant coordinate in expressions such as "Musikstück", "Kunstwerk", "Form", "Persönlichkeit als Gefäß für die Bedeutung", "Physiognomie", and "Haltung". (Observe that "Haltung", for instance, simultaneously addresses physical reality and the semiotic significant.)—Signification is traced in expressions such as "lebt die Musikstücke", "geistige

Bedeutsamkeit", "einhauchen", and "in dem mächtigen Bündnis mit dem Geist".—Finally, significates are addressed in "Leidenschaft", "Geist", "Affekt", and "Überflügelung der Erwartung und Forderung".

2.5 Topographical Navigation

Summary. The topographic cube offers a local and recursive orientation. Hence the corresponding discursive navigation receives a ramified path structure: A priori, each topographic location may open or participate in another topographic variety. The local and recursive nature of such a ramified navigation is described and exemplified.

– Σ –

Observe that the topographic specification of a fact of music is a local resp. recursive one in the following sense which one may visualize as a *conceptual zoom-in effect*:

- Parts of a sign may be entire signs of their own right. For example, the significate of a sign in a metasystem—by definition—is an entire sign of the objectsystem.
- Also, in the communication chain, the performing artist is a creator for the auditory, but he/she includes an entire communication process, starting from the composer, and being communicated through the score.
- Third, regarding levels of reality, an acoustic sound is essentially a physical entity, but its description refers to mental instances, such as real numbers for parameter values.

In other words, the topographic cube yields a local conceptual orientation, and by regression, the topographical description of a fact of music may open a complex tree of ramifications, each knot being loaded by a localization within the cube. Supposedly, there is no consistent ontology without such a self-referential regression. In particular, it is not necessary to introduce such a thing as a "topographic metacube" for the description of metamusical facts (e.g. harmony syntactics) since a metalanguage precisely *means* recursiveness on the level of the significate.

3

Musical Ontology

Musik ist ohne Begriffe.
Hans Heinrich Eggebrecht [232, p.192]

Summary. This chapter introduces the difficult subject of musical ontology. Such a discussion is substantial for a reconciliation of traditional musicology and innovative perspectives in cognitive and computational musicology.

– Σ –

It is a common argument of traditional musicologists against cognitive and computational methods that the full extent of musical being cannot be grasped by these methods. It is claimed that the very depth and transcendence of music are beyond any analytical and quantitative reasonment.

We do not condivide such a credo for the following reasons. Above all, cognitive and computational methods are neither restricted to quantification, nor to simple analysis. Rather modern mathematics and logic is an issue of conceptual explicitness; simple numerical quantification is a very special subject of old-style mathematics that has been overridden by structure theories since the development of set theory, modern algebra, topology, and geometric logic in the first half of the twentieth century. Further, semiotics of general sign systems has proven that formal explicitness on the levels of significants and signification processes does not prevent significates from important *added value* of depth and transcendence.

In other words, depth and transcendence are not missed by computational and cognitive approaches. Rather we can observe an increasing distribution of aspects of depth and transcendence among the conceptual topography. This provokes a crucial question of ontology: Can the problem of whatness be tackled by a differentiated discussion of the problem of whereness? At this point, we should make more precise the ontological problem.

In fact, it is not a question of specifying the music's whatness as a Kantian "Ding an sich", but of how we *conceive* music: "What is the concept of music?" But a concept is—again in Kant's words [519]—a topos, a logical place. That is, asking for the concept of music involves a topos in a particular 'concept space'. Recall [849] that platonic ideas live in an ontological fundus, the hyperouranios topos, which is an explicit topographic site: Relative position and hierarchical location of its instances serve as basements for the whatness of ideas.

It is therefore a legitimate procedure to open the ontological discourse by a view of the conceptual topography of music.

3.1 Where Is Music?

Summary. The difficult localization of musical existence is due to three factors: (1) the interweaved usage of topographically distinct locations in the classical ontological discourse; (2) the tendency of classical musicologists to override topographical complexity by all-embracing breviloquent approaches in order to

G. Mazzola, *The Topos of Music I: Theory*, Computational Music Science,
https://doi.org/10.1007/978-3-319-64364-9_3

uphold the discourse; (3) the genuinely related activities of thinking and playing—music is one of the best operationalized fields of human cognition. We discuss and exemplify these factors.

– Σ –

A closer look at musicological terminology reveals that many concepts are topographically blurred since their breviloquent function inhibits differentiation. For example, the concept pairing of consonance and dissonance is of complex reality (see Appendix B.3). It relates to psychic categories, to mental compositional principles, and to physical reality of hearing and cortical sound processing. Questioning this concept pairing *without* topographical differentiation (of reality coordinates in our case) yields a confusion of meanings and creates bunches of pseudo-problems (see loc. cit.). Historically, one can understand such a difficulty as a capitulation in view of exorbitant complexity, but systematically, it is necessary to distribute a concept over its topographical specializations. This procedure at first sight destroys comfortable unity and simplicity, but science is not about comfort, it is about truth. One may then, after establishing a differentiated and *adequate* terminology, return to unity by conceptual constructions which include complexity rather than repress it.

A most dramatic breviloquent repression strategy within traditional musicology was proposed by Hans Heinrich Eggebrecht on various occasions [291, 292]. Centered around the fundamental question, where to localize music ontologically, he concludes that music resides in the "I", a kind of super-breviloquent subject icon. This conclusion is deduced from a discussion of where music could be found elsewhere: in the work, in the composer, etc. Eggebrecht rightly observes that music is not exclusively in the work, nor is it exclusively in the composer, etc. Whence he draws the erroneous conclusion that only a kind of abstract ego may comprise the totality of music. It cannot be discussed in detail, how radical a declaration of scientific bankruptcy such a consequence does represent. But it is evident that the foundation of musical ontology in a whatever abstract subject *opens any issue of subjectivity* and prevents scientific discourse whenever it suits corresponding needs. A musicological discourse which is founded in the subject can go on forever, but it is for nothing.

A third difficulty for locating musical ontology is a fundamental, but difficult relation between thinking and making music. From the very beginning of European music theory in the Pythagorean tradition, playing an instrument (the monochord in those early days) was a method to gain evidence of the transcendent truth of music. It was contended that a metaphysical tetractys symbol was the core of musical ontology, and by playing the monochord, we could access that transcendent being [672, 1080]. This approach was not just a comfortable way to understand music, it was the unique solution to gain understanding. In our days, this is still a strong belief among musicologists, as the following proposition by Helga de la Motte [256, p.232] shows: "Musikalisches Denken ist grundsätzlich Denken in Musik." Not only is it admitted that 'thinking music by examples' is useful, no, it is fundamentally so! This is, of course, in the line of Eggebrecht's belief that music is without concepts. Thinking music is thinking *in* music, beyond words, *ineffable*, to rephrase it with the words of Diana Raffman [875].

On the one hand, this problematic stems from a poor localization on the semiotic axis. It is not recognized that signs of non-verbal type—mathematical formalisms, for example—could effectively resolve simple *verbal* ineffability. Sticking to common language, the dominant part of traditional musicological language, is the wrong strategy. No non-trivial insight in modern physics could be obtained via common language—not to speak of modern mathematics. We shall see in part VIII on performance that performance nuances, one of Raffman's advanced arguments for ineffability of music, can be perfectly formulated in the mathematical language of vector fields. In other words, thinking music by its direct invocation is first of all the proof of a defective language. The claim that something is *definitely ineffable* presupposes knowledge of any possible language, an absurdity.

And it is wrong to believe that music is a special issue of science in that it deals with objects that point at non-mental strata of reality. Psychology, for example, deals with emotions, physics deals with elementary particles. All these objects share aspects that transcend human conceptualization. But we may very well conceive them in a system of knowledge, model their behavior with remarkable success for our cognition. Music is not less and not more accessible than physics. But we have to establish a sophisticated system of signs in order to grasp the significates, common language is not the tool for a musical concept space.

On the other hand, music is substantially tied to its communicative dimension: Unlike physics, music was always pronouncedly *distributed* over poiesis, neutral niveau, and esthesis. However, this aspect is completely disjoint from the question of ineffability. It simply stresses the need for a communicative differentiation when dealing with music. Forgetting about the making of music, boiling it down to an object independently of its making and perceiving is a basic error, just as it is an error to make physical experiments without quantum mechanical uncertainty and complementarity. A famous defective argument which stems from the confusion of communicative coordinates is the judgment of contrapuntal symmetries in music. For instance, the retrograde movement has a completely different existentiality as a compositional (poietic) tool, than when discussed as a (esthesic) fact of perception, we have already discussed this issue in example 1 of Section 2.2.3. Summarizing, we can state that most of the musical objects cannot be treated properly outside their topographically specified ontological modality.

3.2 Depth and Complexity

Summary. The ramified discourse based on orientation via local topographic cubes is described as a tool for generating controlled complexity in understanding music. The fundamental pointer character of this tree structure is compared to pretended depth in traditional approaches. The latter is identified as a rhetoric chain of external reference pointers.

– Σ –

Both, traditional musicology and mathematics, use the word "depth", however in significantly different ways. To begin with, mathematicians speak of a "deep theorem" or a "deep theory" to indicate that it marks a result of a complex development, and that this result gives a compact, pregnant insight into the addressed complexity.

For example, Fermat's last theorem[1] is a deep result since it involves an incredible complexity of mathematical structures, but it is a very compact, in fact elementary, statement. In this sense depth involves complexity, but rather in an *encapsulated* form; its surface shows a simple appearance.

Semiotically speaking, mathematical depth deals with simple significants that are built from complex, viz ramified and recursive signification to connotation and/or metasystem levels.

What, then, is the difference to depth in musicology? Strictly speaking, it is *pretended* encapsulation without the license to access contents. Let us make this clear in the following example. Eggebrecht, an authority of traditionalist German musicology, has stated that "music is without concepts" (see the catchword heading this chapter [232, p.192]). This is a prototypical deep statement since the reader is supposed to know that Eggebrecht is a scientific authority who has written a large number of books and papers on music. In fact, the citation is one of seven final theses in a book entitled "Was ist Musik?" which Eggebrecht and the other German pope of traditionalist musicology, Carl Dahlhaus, have written to discuss the 'ultimate and deepest' questions about music. At first sight, Eggebrecht's statement provokes a contradiction: How can somebody write hundreds of pages about music and simultaneously claim that conceptualization must fail? The solution is *depth*: We have to recognize that Eggebrecht has gone an incredibly long and audacious way in search of the 'real thing', but alas! has to report that—ultimately—the real thing is beyond conceptual comprehension. We, his humble pupils and admirers have to accept that music is an unreachable secret and He, The Master, does rightly encapsulate a deep insight. This capsule can—however—not be opened, the encapsulation remains locked, and this is mandatory for two reasons.

First, by the pronounced authority, He, The Master has experienced the horrific failure of concepts and we, the pupils, should just accept and admire this report from hell. Second, the capsule's key has been lost, no trace down to the very proof of the *pretended* conceptual failure is at hand. The text [232, p.192] does not give any hint, it simply states the verdict. Such a type of "depth" is completely standard in musicology. You give an encapsulated statement and prevent any access to the (presumably) hidden complexity. Whereas in mathematics, the access to complexity is possible and its realization eventually yields insight, the musicological encapsulation sticks to a pointer to insight, however, the pointer points into the

[1] This theorem is one of the most famous and hardest results in the history of mathematics. It was proven by Andrew Wiles in 1994 and states that any non-trivial integer solution of the equation $x^n + y^n = z^n$ implies $n = 1$ or $n = 2$.

void. You cannot obtain a follow-up, information flow breaks down, and verification of the claim is not offered on the base of explicit complexity. Rather is the dark path to the hidden (pretended) rationale ornamented by metaphors. In our case, Eggebrecht terminates by stigmatizing music with mysterious weirdness, thus turning depth into mystery—and science into fabulation. As a basic rhetoric device, this kind of void pointer has been used in one of the most ambitious handbooks of traditional musicology [234]. The discourse of important parts of that handbook relies on void pointers to never ending chains of implicit (!) external references. Fortunately, musicology is not doomed to void pointer depth. If applied with all its ramifications,

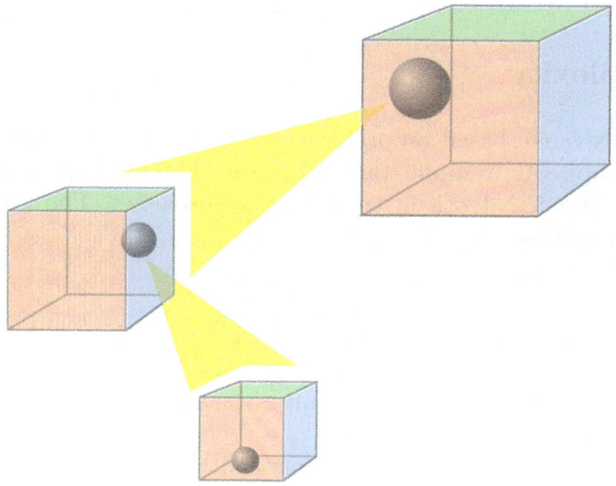

Fig. 3.1. The topographic cube of musical ontology has local character. A concept of music (represented by balls in this figure) may open different localizations with successively refined concept analysis.

the topographic cube (see Section 2.4 and Figure 3.1) offers an effective tool of complexity which can help hammer out depth in the mathematical sense, i.e. without void encapsulation. In Section 2.4, we have already introduced the local character of ontological topography. It means that the decision to attribute to a given situation specific topographic coordinates is no obstruction to other coordinates in a successive step of refined analysis, see Figure 3.1. But it does not mean that this recursion from one local perspective to the next tends to more elementary and eventually "atomic" ontology. Ontological atomism is neither desired nor even possible as a foundation of knowledge: Being has no elements, only perspectives. The cube of topography is a local coordinate system for such perspectives, it offers orientation, but no roots.

Example 2 On the string of communication, the performing artist is on the esthesic coordinate with respect to the composer. He or she receives the work and analyzes it as a mental entity. This complex identity of the performer is then encapsulated and taken as a new communication unity of the concrete performance. In this role, the performer stays on the poietic side and produces a work of physical reality, namely the sounding performance. This latter is an expression of the artist's reflection and will now transmit the message to the listeners of a concert hall, say. In this description, the listener is still encapsulated as an esthesic instance, but we should open it with respect to what is going on when understanding a performance of a work of art. In this perspective—the inner dialog of the listener with his or her consciousness—the acoustical performance is again transformed into a mental structure and interpreted along the personal lines of musical insight. The result is a new type of work of mental and/or psychic reality which the listener will communicate internally or externally in a further processing of the musical experience. Etc.–

We have just sketched some en- and decapsulation processes on the communicative dimension, but the reader can easily extend this flip-flop of perspectives on any other topographic dimension. It becomes clear now that opening ontological perspectives is by no means an approach to ontological roots, but it *is* a clarification and gives ample orientation which void pointers definitely *do* refuse.

It is one of the main tasks of part II to develop a profound and detailed navigation formalism through music, a formalism which provides encapsulation and reliable pointers to hidden complexity at the same time. But a word of caution is imperative: Our task cannot be attained without a considerable amount of technical machinery—nihil ex nihilo.

4

Models and Experiments in Musicology

Plato's pessimistic picture of empirical observation caused him to deny the validity of physical models and was largely responsible for the eclipse of empiricism for 2,000 years.
Peter Wegner [1109]

Summary. This chapter introduces the paradigm of experimental humanities. Such a perspective is basic to all computational, resp. effective, methods in musicology. We discuss the parallel to the epochal Galilean change from speculative to experimental natural sciences.

– Σ –

In the final chapter of this first introductory part, it is necessary to review the overall epistemology of musicology since the preceding ontological topography has questioned traditional standpoints in a measure that does not allow of uncritical takeover of epistemological fundamentals. Above all, it is the requirement of free access to (encapsuled) complexity that creates a boundless analytical attitude: It is no longer possible to 'cultivate' private, inaccessible regions of knowledge.

This is not only due to the fact that music belongs to humans and no longer appears as a revelation of divinities of whatever flavor. It is also due to the massively improved arsenal of information and communication technology where the knowledge space becomes a concretely accessible site, and any lack of precise or explicit information is immediately blamed.

This situation enforces a new, fundamentally interactive understanding of human knowledge. The paradigm of contemplative science which was essentially promoted by religious constraints [706, 945] has to be abandoned, knowledge is constantly updated and extended. The global knowledge space has become an ocean which is in permanent metamorphosis, on which we navigate and experiment in a spirit of dynamical space-time [705].

We shall make these points more explicit in the following sections. It is useful for the understanding of new paradigms in musicology, however, to have a closer look at the development of modern experimental science which prepared the Galilean revolution. In this preliminary discussion, we refer to Isabelle Stengers' article [971, p.395]. The case of Galileo Galilei is important for musicology since it is intimately related to a common problem: definition of musical tempo and physical velocity. Galileo was concerned with the definition of instantaneous velocity. Recall that, for historical reasons, he could not yet refer to calculus and define velocity as derivative ds/dt of space $s(t)$ as a function of time t. He had to develop the very *concept of instantaneous velocity* since the traditional concept of uniformly accelerated motion, as it was understood by medieval scientists such as Nicholas Oresme, could only view it as a succession of portions of uniform motion, changing discontinuously their constant velocity after a short duration. Galileo's concern was not just another speculative concept but resulted from intense observation of the movement of physical bodies on

G. Mazzola, *The Topos of Music I: Theory*, Computational Music Science,
https://doi.org/10.1007/978-3-319-64364-9_4

an inclined plane. His proposals (dated around 1604) rather aimed at measuring (by impact) local velocity in a determined moment of time.

Fig. 4.1. Galileo's method concentrated on observation and experimental interaction with nature. This detail of a fresco by Giuseppe Bezzuoli (Museo di Storia della Scienza, Florence) shows the famous experiment of a ball running down an inclined plane (right) and the traditional consultation of Aristotelian works (left). Reproduced with permission of: Saulo Bambi - Museo di Storia Naturale Università di Firenze, Sez. di Zoologia "La Specola"- Italia.

In other words, Galileo's approach was essentially built upon observation and measurement, it did not rely on abstract speculative reflections. The turning point here is precisely the passage from speculative encapsulation to explicit accessibility by 'doing' science. Measuring the body's mechanical impact invokes a momentous property of that body. Though Galileo could not make this property explicit on mathematical terms, he succeeded in discovering a physical one-to-one correspondence to instantaneous velocity. The interesting point in this procedure is a fundamentally *operational* method: thinking by doing. In fact, if one views Galileo's two-hundred-page notices written around 1604, it becomes clear that he was constantly dialoguing with his experiments, tables, number lists, and diagrams—and not with Aristotle's authoritative writings—in order to construct a homogeneous concept space of physical movement, see Figure 4.1.

We have insisted on this episode since the musicological analog of velocity, musical tempo, shows an astonishingly parallel concept history—though nearly five hundred years delayed from physics.

In traditional European musicology [243, p. 317], we encounter a concept of tempo which corresponds to Oresme's setting: Tempo is the quotient of musical duration (measured in quarters) and physical duration (measured in minutes), this yields the classical Mälzel metronome m.m. in quarters per minute (see Section 33.1 for a detailed discussion of tempo). It is essential that this concept does *not* include instantaneous tempo in the sense of a derivative, traditional musicologists deny the musical relevance of such a concept, see Figure 4.2.

This refutation stems from the completely speculative handling of tempo in traditional musicology. Many of the tempo-related concepts, such as fermatas, ritardandi, and accelerandi, are encapsulated locked objects, and theorists would not know how to give a workable description of such tempo phenomena. The opening and explication of these void pointers are left to musicians and programmers of music software. Instead of learning from Galileo's experimental observations theorists insist on a discrete tempo fantasm a la Oresme—despite the rich language of modern mathematics.

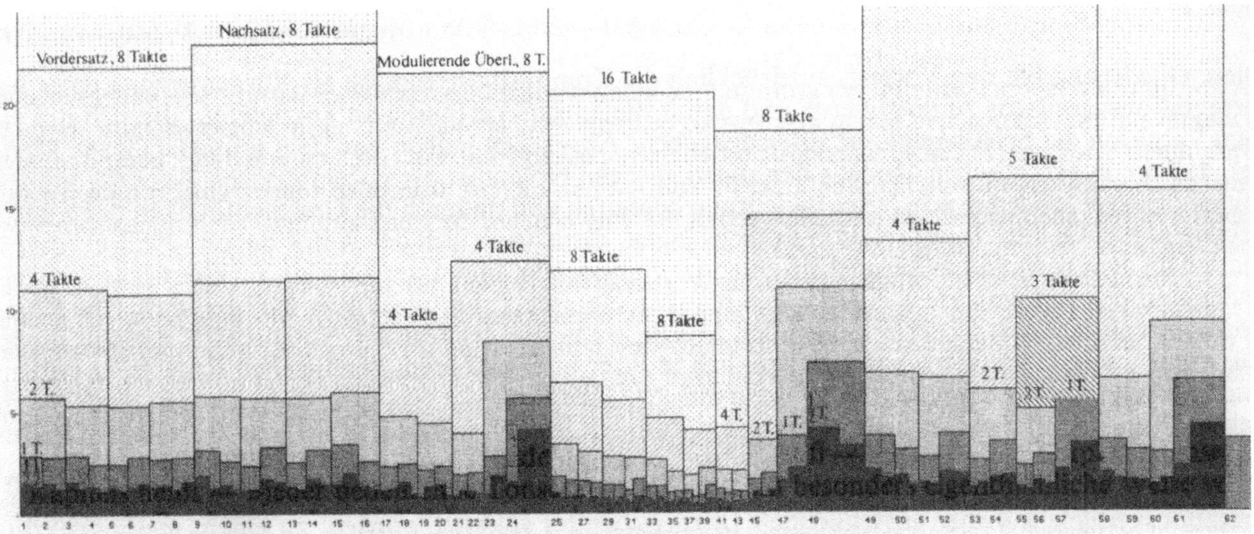

Fig. 4.2. The figure shows a typical musicological 'digitized' version of the tempo concept, as used by Hermann Gottschewski [243], reproduced with the author's permission. The rectangular regions define regions of constant tempo (visualized here by reciprocal duration values). Since no instantaneous tempo concept is given, the author obtains different averaged tempo regions, according to the respective time window.

This parallels physics to musicology in a basic issue. Precisely as Galileo, modern *working* musicians and engineers are at the cutting edge of explicitness and leave irrelevant speculations to feuilletonism where they are well-placed. We learn from this double experimental revolution that it is not banausic artisanship that enforces opening of concept capsules, but the mere desire to redeem conceptual pointers, and to drive out at last void rhetorics.

Galileo's revolution was the answer to pretended depth of rhetoric discourse. Just as medieval medicine was traditionally split into speculative academics (without license to practice!) and artisanal surgery (practised by charlatans), physics was split into mechanical engineering and speculative philosophy. Scientists such as physicist Galileo and the French surgean Ambroise Paré had to 'dissect' the concepts and—hitherto tabooed—human bodies to explicate complexity [916]. We shall now take a closer look at the epistemological implications of this revolution.

4.1 Interior and Exterior Nature

Summary. Reasons for the categorical distinction between interior and exterior nature. Can human thought be an experimental field?

– Σ –

Experimental sciences are classically related to empirical exterior nature. According to a traditional opinion condivided by Immanuel Kant [519], experiments are understood as interrogations of a passive witness. Nature is passive and will answer to the hearing in an objective way, i.e. independently of the interrogator's subject. These two characteristics, passivity and objective response, have since been recognized as erroneous idealizations. Physical experiments are substantially interactive processes, this was evidenced by uncertainty and complementarity principles of quantum mechanics.

But physical nature is also responsive on a more conceptual level. Each response to an experiment may alter the theoretical position and the concepts which drive the experiments. This is exactly Galileo's method, there are no platonic ideas which we have to retrace from empirical reality, the concepts are in incessant metamorphosis and mutate as a result of experimental cognition. As with genetics, only the fittest concepts can survive the experimental struggle.

In what respect is then inner nature so fundamentally different to exterior nature? It is said that inner human nature is subjective and that there is, consequently, no such a thing as objective facts. But the argument of subjectivity is misleading. It is definitely not true that we have a more indepth control of inner nature because of subjectivity. The inner human nature is a vast field of instances which we do not any better access than exterior objects. Subjectivity is only a minor part of the inner nature. Let us give two examples:

In *mathematics*, it is admitted that the Peano axioms for number theory[1] are trivial creations of the human mind. But the impression vanishes immediately when abording such a simple question as the famous and still unsolved Goldbach conjecture[2]: Can every even natural number be written as the sum of two prime numbers? This makes evident that the complexity of our mental creations is not more under control than physical objects. We believe that the Peano axioms are under control. But in their immediate conceptual vicinity, there are statements which do escape our mental power.

In *musical composition*, we typically deal with the creation of a score. But it is an illusion to believe that the richness of such a creation can be controlled by the composer. His or her contribution is but a *germ* of an incredible complexity that implies combinatorial processes, interpretation strategies, semiotic stratification and so on. In other words, a composition is quite the same as a piece of exterior nature, the part of subjectivity vanishes in front of an exorbitant structural variety.

So, what is the difference between exterior and interior nature? It relies on the more apparant, though not fundamental, autonomy of exterior 'natural' phenomena. But a second view reveals an inner nature—especially on the level of man-made universes such as mathematics and music—which is not less complex and uncontrollable than exterior nature. *As a piece of uncontrolled nature, Bach's "Art of the Fugue" is not fundamentally different from a supernova out in interstellar space.*

4.2 What Is a Musicological Experiment?

Summary. The Pythagorean tradition and the general concept of a music instrument. Candidates for experimental layers. Testing the congruence of a scientific model with corresponding material, such as a composition, performance, or analysis.

– Σ –

We know from Section 3.1 that Pythagorean tradition offers a type of "thinking by doing". Can we view this paradigm as an archaic version of Galileo's experimental approach? We could if it provided us with a conceptual laboratory where concepts could be fabricated while making music. But this fails, playing the monochord was nothing more and nothing less than a passive verification of already fixed ideas, in fact the metaphysical tetractys. It was a kind of doing on the ground of prefabricated thoughts.

Nonetheless, the insistence on the aspect of activity in thinking music has survived all speculative assaults from tetractys to Keplerian harmony. The shape in which we still encounter this method has of course changed. No longer can it be found under the auspices of divinities, it has been saved by composers, the experimental scientists of music. This is somewhat exaggerated since composers often condivide the status of medieval charlatans. But in their best performance, such as with Pierre Boulez, Giorgy Ligeti, or Iannis Xenakis, they really appear as prophets of a future type of experimental musicologists.

What happens is that these forerunners are left alone by a speculative void-pointed musicology. Already did Arnold Schönberg quit the theoretical background in view of the speculative overhead that could not cope

[1] (1) The number 0 (zero) is a natural number; (2) every natural number a has a unique successor a^+ among the natural numbers; (3) if $a^+ = b^+$, then $a = b$; (4) 0 is not a successor; (5) if a property ϕ for natural numbers is such that 1) $\phi(0)$ and 2) $\phi(a)$ implies $\phi(a^+)$, for every a, then the property holds for all natural numbers.

[2] Conjecture established by German mathematician Christian Goldbach, 1690-1764.

with requirements of effective conceptualization in harmony[3]. The so-called “emancipation of dissonance” is among others the emancipation of experimental musicology from speculative rhetorics. It is not by chance that Boulez demanded to blow up all opera houses.

So what is a musicological experiment? It must clearly include both, thinking and playing, reflection and action. But in the form of mutual inspection, i.e. it should neither be a passive sonification of given speculations nor should it reduce to speculation without musical realization. So the decisive criterion is a bidirectional dialog between a flexible conceptualization and an intelligent sonic realization. Such a dialog can only succeed if the transformation channels between thinking and playing are explicit, operational and highly efficient.

These requirements are necessary, but not sufficient to conceive such a thing as a musicological experiment. We are still missing the adaptation of the crucial concept of experience to the context of humanities. It is by no means clear that such a category can subsist within inner nature. This will be dealt with in the next section.

4.3 Questions—Experiments of the Mind

Summary. Experience in the humanities. Separating speculative questions from musicological experiments. Necessary and sufficient conditions for questions to be accessible on the level of the experimental setup. Improvement of scientific conceptualization by constraints from experimental options.

– Σ –

The naive view of experience in natural sciences evokes a bunch of facts in the space-time of exterior nature. But this is not what an experimental setup is dealing with; already in its very data processing are the ‘naked facts’ codified in numbers, figures, and all kinds of symbols. What is accessible to scientific reflection is already a body of information in a well-structured concept space. Measurement of physical events includes a transformation into *data streams within a mental knowledge space*. Experience then boils down to a complex navigation process along this data stream, and under use of statistical orientation tools and theoretical hypotheses to be verified or falsified.

When put together with the foregoing approach to the experimental setup, this yields a concept of a dialog between a set of data that are embedded in a concept space, and a set of analytical tools living in the same concept or knowledge space. In this perspective, the question underlying the experiment appears as a search mission towards a statement which fits the theoretical orientation defined by hypotheses and evaluation tools.

In other words, *the experiment turns out to be a search mission while navigating on a concept space*, the navigation being driven by two ‘dialog partners’: the data stream and the hypotheses and evaluation tools. But such a generic restatement of the concept of an experiment has *erased any specific reference to exterior nature*. We just deal with navigation on a data stream issued from a specific research object and a bunch of analytical orientation devices. This entire process is located on a (mental) knowledge space.

Under this perspective, an experiment may naturally occur in the context of humanities. Navigation may thus be restated as interactive interpretation of a given human work, be it music, painting, literature or whatever can be subsumed under the general structure of a verbal or non-verbal semiotic ‘text’. Navigation on such a text crystallizes in form of a search for structural coherence between the variety of the text’s syntax and semantics on one side, and corresponding hypotheses and (text) theories as a background of orientation on the other. Observe that the concept of empirical experience has been absorbed by the concept of navigation on a stream in a concept or knowledge space. The parameter of external physical time is no longer relevant, navigation unfolds on an ‘abstract’ space, and time is replaced by an abstract ‘curve parameter’ of the navigation trajectory.

[3] This does not contradict his book on harmony [948] which is much less a warmed-up catechism than a critical review of failures of harmony. However, Schönberg’s treatise on harmony did rather stress than solve those problems.

We should stress that the experimental dialog is strongly interactive in that there is no priority of theories over 'data streams'. A theory may completely mutate under the influence of new directions in artistic creativity. And conversely, the inner nature is no "nature morte", it is in constant renewal.

This being the case, it is important to review the role of speculative void pointer concepts under the experimental paradigm. In the situation of an explicit dialog between given data and theoretical instances, void pointer concepts are immediately ruled out. For example, an encapsulated *locked* concept of dominant, tonic and subdominant in function harmony cannot resolve any task of harmonic analysis for a concrete score since it asks for harmonic function values of *any possible chord* that may theoretically appear in the given score. In cases where the function value(s) cannot be determined, the data stream cannot be navigated and the experiment breaks down. Galilei could not work with the medieval velocity chimaeras since they evidently provided no navigation through the attentive observation of balls running down the inclined plane. *Experiments are excellent exterminators of empty pointers: Their dialog immediately lays bare deficiencies hidden in locked concept pointers.*

The experiment, if viewed as a question of the mind, not as an autistic reflection on locked concepts, rather as a dialoguing navigation, can help us understand what happens when we build concepts and theories.

4.4 New Scientific Paradigms and Collaboratories

Summary. Mutual influence of communicative networks and conceptual precision. Doing musicological experiments on a distributed laboratory. Redefining scientific competence from the communicative point of view.

– Σ –

From the preceding discussion it is straightforward that musicology (and humanities in general) in its experimental profile cannot subsist in the private ambience of classical humanities. The experimental navigation dialog is demanding for all partners, and the very nature of dialog enforces data exchange on powerful channels. Socially restated, such a type of science has to be realized in a strongly collaborative style. On the level of institutions, this demands the structure of collaboratories[4], as introduced by the US Department of Energy. Just as physicists could not survive in isolated research units, musicologists have to initiate intense communication within research in order to succeed. We insist on communication *while doing research*, not merely in noble and non-binding conference small talk. Collaboration is meant as a working style, not as a title for informal politeness.

Moreover, the communicative network of collaborative science enforces a radical conceptual precision. This is not only about the formal constraints when dealing with computer programs, it is really about communicating as a working method. Learning by *doing* science can only succeed on the basis of unlimited access to encapsulated complexity. Evidently, scientific competence must then be rebuilt on the fundament of conceptual communicativeness while private regions of knowledge will lose their historically grown honorability.

We should keep in mind that the entire navigation and orientation metaphor in the preceding discussion of the knowledge or concept space is not merely general knowledge science slang, but more concretely an adequate expression of the topographical (in fact: topological) setting which was developed in chapter 2. Navigation happens to move along (local) coordinates of topographic cubes, and we shall learn in part II that *the general metaphor of a concept space can be realized in form of a veritable geometric space.*

[4] The term "collaboratory" was coined by Bill Wulf in 1993 and is merged from "collaboration" and "laboratory", see [706] for further information.

Part II

Navigation on Concept Spaces

5

Navigation

Verlassen wir das strenge Labyrinth der zementierten Begriffe,
und ergehen wir uns ungezwungen in improvisierten Architekturen,
und dies selbst dann, wenn wir reden.
Pierre Boulez [138, II, p.78]

Summary. Well-conceived information produces knowledge. But conceiving means building concepts, and this amounts to having well-structured access modes. This enforces the development of concept-oriented access modalities to given information, including associated concept spaces. Music is an excellent field to exemplify such knowledge spaces. It has become virtually impossible to navigate through music information without developing powerful concept formats.

– Σ –

We initiate this part by a reflection on conceptual navigation since the concept spaces which will be defined in the subsequent chapters are strongly motivated by universal orientation paradigms while surfing on encyclopedias of inhomogeneous music-related information and knowledge.

More precisely, observe that knowledge involves two components: information and its mental organization. Whereas boiling down knowledge to mere information would mean getting drowned in a sea of amorphous substance, abstraction from information towards pure mental organization would mean getting stuck on Georg Wilhelm Friedrich Hegel's germs of logic, i.e. the zero state of philosophy as exposed in the first pages of [452]. Both reductions cannot cope with what knowledge deals with, viz building *concepts of something.* Building concepts then amounts to conceiving this data, organizing the access to information and doing this by use of well-defined access modes. Knowledge has an object, and it reaches that object via its concept. Conceiving means being able to navigate in a conceptual coordinate system to attend that something. In other words, pure information is substantial, but it is uncontrollable without a conceptual form, a coordinate system where we may place and retrieve substance.

This is precisely why so-called digital information has nothing to do with knowledge whatsoever, and why therefore the digital paradigm is completely irrelevant to the yoga of electronic age: Digital substance, reduced to the $BIT = \{0, 1\}$ of two substance values, 0 and 1, or OFF and ON, the name does not matter, is just a minimal substance set, but it cannot *per se* be responsible for any kind of knowledge. In fact, a digital record is worthless without being organized in a concept form, i.e. as encoded knowledge. 'Understanding', i.e., decoding information is needed in order to control and manage its knowledge potential. The digital age is *not* centered around "Bits and Bytes" but around their accessibility and handling, in short: *data navigation.*

From the above we learn that a powerful concept system for any field of knowledge must provide us with a thorough navigation method that works on an extensive concept space. And from the preceding discussion regarding depth and complexity (Section 3.2) we deduce that navigation must have access to any encapsulated concept, and permission to navigate on any possible "concept path".

It is essential for a successful concept navigation methodology in music to realize a highly interactive and non-authoritative, autonomous orientation and decision strategy. Rather than ending in obscure concept

G. Mazzola, *The Topos of Music I: Theory*, Computational Music Science,
https://doi.org/10.1007/978-3-319-64364-9_5

mist conceptual navigation should lay bare semantical lacunae, and this means developing a language for the open-ended navigation if ever this is the state of the art. In no case is it any longer acceptable to be navigated by dogmatic magisterial guides. This is what the Boulez citation heading this chapter suggests: To create a conceptual environment that allows quasi-improvisational and free-will driven discourse on music. Boulez must have felt that "cemented concept labyrinths" no longer cope with the requirements of advanced musicology, be it on the creative, or be it on the reflexive side.

Now, the term "improvisational" in Boulez' statement is somewhat misleading. It suggests an antagonist to the score-dominated way of making music. Like improvisation it should be interactive—this is to be retained. But it is much more and much more important than a way of making music. The key word is " cemented". Refraining from cemented concept architectures means that the concept architecture has become dynamic, ever-changing and soft. Concrete is a rigid material, hard and insensitive to whatever. Boulez' vision is that of a flexible concept environment, rather than improvisational it should be termed dynamic, i.e. interactive and adapting to changing demands.

The problem is how to realize such a dynamic concept architecture without loosing rigor and reliability. We shall propose such an architecture in chapter 6. For the time being, let us concentrate on the generic navigation paradigm as suggested by Boulez and stressed in the ongoing transformation from energy to information society. In particular we should reconsider navigation as an interactive interpenetration of knowledge agents and data bases. This latter idea turns out to be one of the deepest concerns of music in the making of our own world.

5.1 Music in the EncycloSpace

Summary. We discuss the navigation problem on music data in relation to the general concept of an encyclopedic knowledge space, the *EncycloSpace*. The latter is introduced as an upgrading of the classical concept of an encyclopedia, as conceived by the French encyclopedists Denis Diderot and Jean le Rond d'Alembert. The upgrading is characterized by three changes: (1) the static cosmos is replaced by a dynamically developing data organism in space-time; (2) the passive "speculum mundi", i.e. the purely receptive view, is replaced by an interactive instrumental relation to the data organism; (3) the textually oriented alphabetic ordering is generalized to a universal navigational orientation. Consequently, encyclopedic navigation splits into a receptive and a productive variant.

– Σ –

In [706], a general definition of an encyclopedic knowledge space, called EncycloSpace, was given. Let us shortly recall that definition and then explain the concept's genealogy and ingredients:

Definition 2 EncycloSpace *is the topological corpus of global human knowledge which evolves dynamically in a virtual space-time, is coupled to human knowledge production in an interactive and ontological way, and allows of unrestricted navigation according to universal orientation within a hypermedially represented concept space.*

According to Sylvain Auroux' analysis of Denis Diderot's Encyclopédie [48], the encyclopedic principles of the Age of Enlightenment can be viewed as a combination of the *completeness* of a dictionary, the *unity* of philosophical synopsis, and the *discoursivity* of a mathematically inspired ordered representation (the alphabetic ordering of words), see figure 5.1.

In a critical review of these encyclopedic characteristics, the Encyclopédie is recognized as being a faithful realization of the medieval idea of a "speculum mundi": "L'encyclopédie assume par sa fixité relative le rôle d'un miroir du monde naturel, tel qu'il peu s'offrir aux sujets connaissants." This insight specifies the attribute of completeness, symbolized by the orbit, as a fixed cosmos. This is a traditional static world view, similar to the photography of a stationary system. However, such a snapshot cannot claim to represent a world where knowledge is rapidly and incessantly growing. We have to accept that the purely spatial coordinates of a virtual cosmos (in Diderot's words: "connaissances éparses sur la surface de la terre") must

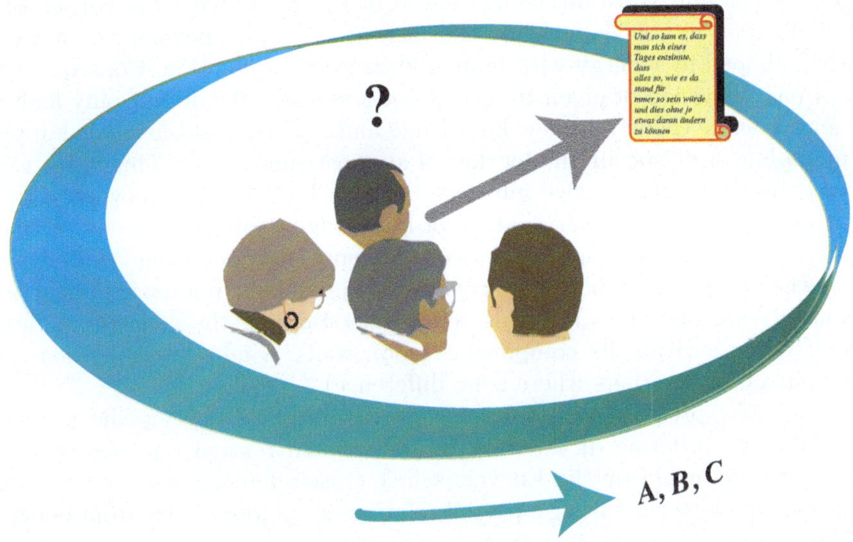

Fig. 5.1. The principles of classical encyclopedia: completeness (complete circumference of the circle), unity (the circle's perfect shape), and discoursivity (alphabetical discourse along the circle's line).

be completed by the time coordinate. We have to deal with a dynamic universe of knowledge embedded in virtual space-time, and subjected to laws of coupled synchronic and diachronic nature (see Section 2.3.4 for these semiotic concepts). The development of knowledge is not an accumulation of essentially immutable and isolated time-slices. See figure 5.2 for the following discussion.

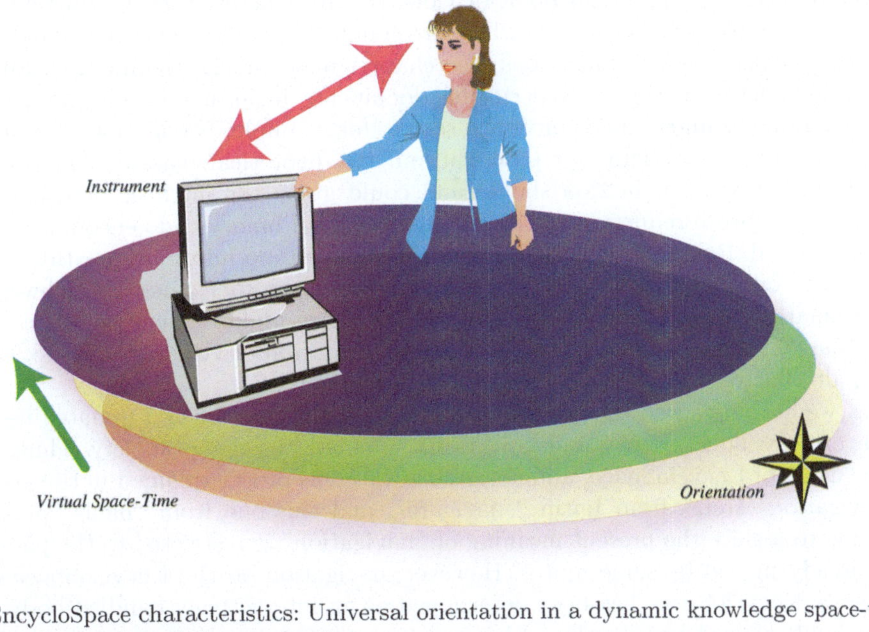

Fig. 5.2. The EncycloSpace characteristics: Universal orientation in a dynamic knowledge space-time which is interactively coupled to human knowledge production.

Further, the change of this dynamic system evidently does not happen by an autonomous activity: It is a result of a constant and substantial interaction of humanity with the corpus of knowledge. Much like political and civilization dynamics or brute continental shift, the encyclopedic body is an open system which is incessantly reshaped in synergy with human knowledge production. Consequently, the metaphor of a "speculum", i.e. a passive vision of given things, is not adequate. We do not only look at prefixed things, we do also define and redefine them, add new knowledge and relativize old-fashioned approaches. One could then replace the metaphor of a "speculum" by that of an "instrumentum". This latter is perfectly adequate to the computer as a bidirectionally active interface between humans and knowledge bases. This aspect is substantial since it also questions the very nature of knowledge. Concepts are no eternal entities who live out in a platonic sky (hyperouranios topos). They rather represent operational units with an ontology that is rightly defined by the very accessibility of conceptual components. (In a more radical perspective, it could be argued that concepts are operators.) In this vein, understanding the concept of "house", for example, boils down to the way you activate its components (roof, walls, windows, etc.) when asked to prove your understanding (to yourself or to others, there is no difference).

Finally, the order of representation of knowledge within the Encyclopédie is twofold: On the very surface, it is the alphabetic ordering dictated by the traditional textual representation. More in-depth, it is the geographic orientation given by the Encyclopédie's cross-reference system [48, p.321 f]. Whereas the former is strictly linear, the latter is rather a partial relation of "pointers", far from being an ordering in the technical sense. This is also due to the fact that the concepts are not cast in a formal language and hence, beyond the alphabetic ordering no intrinsic organization is visible on the representational level. This point is not just a marginal note concerning formal aspects of knowledge. Rather do we envisage the in-depth question of the structure of the space where concepts live. We know from Plato's allegory of the cave [849], from Aristotle's "Topic" [40] and from Immanuel Kant's comment on Aristotle's "Topic" in "Kritik der reinen Vernunft" [519], that the spatial metaphor for the ontology of concepts is crucial for any effective discussion of the system of concepts. In this sense, ontology of concepts is essentially topology: a study of the topoi where concepts subsist. In this theoretical sense, alphabetic navigation is not sufficient, conceptual navigation must provide tools and paradigms which transcend the textual alphabetism and include generic principles of answering to primordial discursive questions: Where do I come from? Where am I? In what direction do I proceed?

From a more operative and practical point of view, the alphabetic ordering inherent in the linguistic text paradigm must be completed by orderings which are genuinely related to non-textual data formats, such as geometric spaces, or spaces of set collections (see chapter 6 for precise definitions). Such more general data types emerge in a natural way in hypermedia documents including visual and acoustic instances, such as pictures, diagrams, sounds, music or dance scores, for example. Navigation orderings have to cope with hypermedia orientation paradigms for practical and for theoretical reasons: There is no reason why the completely arbitrary textual reduction should and could grasp the intrinsic ordering of concepts and thoughts. To be clear, the brute reduction to the definitely textual binary code encompasses virtually every possible information record. But it is a different business to understand information—this is definitely not in the reach of the binary code. "Worüber man nicht sprechen kann, darüber muss man schweigen." This killer sentence which terminates Ludwig Wittgenstein's "tractatus logico-philosophicus" [1138] collapses in the age of hypermedia discourse. It should be replaced by a recommendation of visual/geometric, acoustic/musical, or haptic/gestural alternatives to textual dead ends.

Summarizing, we have reviewed the concept of Diderot's Encyclopédie and updated its three attributes to meet the needs of a dynamic, active, and universally oriented knowledge society. Thus, we may baptize this update by the name of EncycloSpace, and determine the concept as explained in the above Definition 2.

Though "navigation" stems from Latin " navigare", and this one from "navis" plus "agere", which means giving motion to a ship, the present meaning of "navigation" is restricted to the passive steering of a vehicle which is already moved by some motor. However, navigation on the EncycloSpace must regain the original activity, since EncycloSpace has been determined as an interactively handled body of knowledge, a body which is not built, altered and developed by mysterious forces but by our genuine *agitation* upon the object. Therefore, the original etymology of "navigation" is restituted: We shall distinguish between *receptive* and *productive* navigation. Receptive navigation means moving around within the body of knowledge without changing it. In contrast, productive navigation does change the body of knowledge. We will discuss these two modalities in the following sections.

5.2 Receptive Navigation

Summary. Receptive navigation leaves the EncycloSpace unaltered. It deals with universal orderings to find and represent knowledge for optimized understanding. Since musical data are of very inhomogeneous and general types, universal navigation concepts are essential in innovative musicology.

– Σ –

Receptive navigation is the common situation in classical encyclopedias. One is looking for a bunch of information in a more or less well-determined, and *immutable* field of knowledge. The main requirement in this situation is an *optimal orientation environment.* In case of a simple name search, the alphabetic ordering is adequate: We just leaf through the linear alphabetic ordering of words. But this does not cover less trivial search problems. For instance, if we are looking for all chords in the score of Schumann's "Träumerei", say, the alphabetic orientation crashes. Surfing through the set of chords in a determined score is a new situation, it requires ordering principles for chords as special types of note sets. The EncycloSpace features should comprise a representation of all chords built from the 463 notes of the "Träumerei". They should then be able to order these chords according to *universal* ordering principles since we cannot (1) redefine from scratch orderings among EncycloSpace objects and (2) build individual orientation frames for each particular configuration. Given such a universal ordering, every chord could be retrieved according to corresponding intuitive orientation paradigms.

We recognize that inspecting such an ordered chord list is only the last link of an extended chain of encylopedic objects which finally leads to the required chords. We would typically start on the level of composer names, then—once "Robert Schumann" has been found—descend to piano works, then to "Kinderszenen", and finally to "Träumerei". At this point, we will have to surf through different types of musical score signs. Once arrived on the level of notes, representing ordered lists of chords requires non-alphabetic ordering principles. In fact, piano notes are points in a space of four dimensions, pitch, onset, duration, and loudness. Such a space requires different approaches to ordering—and chords do so *a fortiori.* Universal orderings among EncycloSpace objects is a central feature for receptive navigation because it yields orientation and thusly prepares the field for recruitment of added knowledge which is to be created from productive navigation.

Whatever are the searched for objects, it turns out to be crucial for valid receptive navigation to be provided with universal ordering principles: *orientation must be generic, and generically representable on the visual interface.*—This will be an important feature of the denotator system to be introduced in chapter 6.

5.3 Productive Navigation

Summary. Productive navigation interacts with the EncycloSpace such that the latter is enriched by a surplus value of added knowledge. The EncycloSpace is a repertoire as well as a laboratory. Productive navigation is a central issue in extending musicological and musical knowledge.

– Σ –

In the above example of an ordered list of chords, we supposed the existence of such a list on the EncycloSpace. However, it is not probable that all such lists of chords drawn from scores of the piano music repertoire are incorporated in the EncycloSpace. More probably most of what is required by an interested user will not be allocated. This can be the case on very different levels of knowledge. In the simplest level, it can be a numerical information we want to obtain, e.g.: "How many quarter notes are contained in Thelonious Monk's original tune "Blue Monk"?" This information usually is not part of the knowledge base, but it can easily be calculated. Searching for the answer is a trivial extension of knowledge, but it differs from receptive navigation: Navigation which is defined by the question and terminates upon its answer produces a (tiny) extension of the EncycloSpace.

On a less trivial level, the calculation of complex data, such as the number of (1) chords in the "Träumerei" which (2) contain more than three notes, which (3) are dominant in F major, and which (4) contain a diminished third, can require too much calculation time to allow forgetting about the result. In this case, we try to save the result and access it as soon as new navigation trips ask for this information.

On a still more invasive level, it may turn out that we have built new concepts, such as a special type of musical motives which meet non-standard conditions. Suppose that such a concept is named "Non-Standard-Motif". We do not want to be excluded from navigation because this concept is not yet part of the EncycloSpace's vocabulary! In other words, Boulez' requirement of an "improvisational" discourse/navigation must include free vocabulary extension and then the operation upon the extended vocabulary.

Summarizing, we see that productive navigation responds to the demand for dynamic navigation, interacting with the body of knowledge, and incessantly producing new data to be added to the given EncycloSpace. Evidently, such an extended functionality adds to the original repertory character of the EncycloSpace the character of a laboratory where questions are transformed into mental experiments as introduced in chapter 4. Besides the technical problem of how to realize such a functionality without losing control, we are also confronted with the level of accessibility of extended EncycloSpace contents by third parties. But this is not a question of fundamental ordering, rather technology should be concerned with the social aspect of navigation-induced knowledge extension.

Boulez' idea of an improvisational concept navigation provokes fundamental questions of communication which are well known from music when improvisation started to penetrate the music market and musicology as a science. In fact, most jazz productions essentially rely on sound tracks which override pure score or songbook data. The introduction of sound files as a major trace of and reference to the musical creativity is responsible for the very existence of the present jazz culture. Though technologically low level, the mere possibility to access ad infinitum an improvised piece of music initiated a discourse about music which was never possible before. And it enforced the relativization of classical scores as an identification of the musical work.

On the level of conceptual navigation, we experience the same effect: Classical 'scores of knowledge', i.e. hard-coded knowledge data bases in books and encyclopedias, are being confronted with dynamic records of ever moving knowledge streams. 'Improvisation' within knowledge can now be traced and integrated in the global stream of knowledge production. No need to stress the fundamental transformation of knowledge sociology under such perspectives.

We would like to insist on this background throughout the present book since it exerts an incredible impact on how we recognize important from irrelevant aspects of formal conceptualization in future musicology.

6

Denotators

C'est peut-être dans la façon de se
représenter l'esprit qu'on pourrait déchiffrer l'esprit.
Paul Valéry [1073, III]

Summary. This chapter introduces the universal data format of denotators to describe musical objects. Denotators generalize the structures of local compositions in mathematical music theory as well as the data model used in the RUBATO® software. They do, however, *not* include deeper semantic layers which are related to physical, psychic, social or religious meaning. Their semiotic structure represents an elaborate denotative baseline and resides in a purely mental level usually attributed to mathematical objects.

– Σ –

This chapter introduces the formal framework of the entire book. We believe that the preceding chapters have given a huge motivation to elaborate a formally rigorous concept framework. Therefore, we shall no longer comment on the general purpose of this issue. Let us instead describe the genealogy of the denotator concept. After the publication of the German precursor [682] of this volume, and in connection with the development of the composition software *presto*® [680], it was believed that the concept framework of mathematical music theory was sufficiently generic to handle formal issues in computational systematic musicology. This framework was built upon so-called *local and global compositions*. However, the theoretical and program design groundwork for the analysis and performance platform RUBATO® taught the developers that these concepts had several severe drawbacks:

1. They were centered around a special type of mathematical structures (essentially finite point sets in particular parameter spaces) which could not comprise less theoretical objects, such as to include attributes with character strings for names, or other non-numerical data.
2. Also were they 'hard-coded' in the sense that introduction of new concept types was not possible. In particular it was not possible to define concepts by recursion to already given concepts.
3. The mathematical flavor was not compatible with programming requirements for data bases.
4. The naming policy did not allow clear distinction between objects and names (in a strictly semiotic sense).
5. The so-called functorial point of view was not taken seriously enough. Intuitively, this means that we had to stick to one fixed 'ontological' point of view.
6. The objects were not well distinguished from the spaces they live in. This hindered recombination of object selections—which now is possible, and meets the requirements of productive navigation.

The denotator concept framework will eliminate these defects. But it will also require a rather complex concept scheme. We shall develop the details in several steps, starting with a more down-to-earth version of the denotator framework (Section 6.1), and then extending to the universally valid generic setup.

G. Mazzola, *The Topos of Music I: Theory*, Computational Music Science,
https://doi.org/10.1007/978-3-319-64364-9_6

Denotators are conceived according to the general criteria of EncycloSpace navigation, so observe that we do not follow mathematical tradition but navigation methodology, including 'mathematical navigation' as a special case.

Also we should take care of *abstraction from any deeper semantic loading with denotators.* The reader should postpone semantic issues to Chapter 18, the present discussion is uniquely concerned with the universal denotation formalism. Notwithstanding their formal power, denotators are radically pre-semantic, at least regarding to 'real' meaning of physical, psychic, social or religious character.

6.1 Universal Concept Formats

Summary. Discussion of three principles of universal data format typology: unity, completeness, and discursivity. These are realized by a typology which is recursive, of extensive ramification, and open to unrestricted modular data recombination. The denotator concept is modeled following the Aristotelian pairing of substance and form or of the geometric paradigm of point and space, respectively. Each of these components: substance and form are arranged according to the structure of signs. We discuss the fundamental pointer nature of points as basic concepts of axiomatized semiotics.

– Σ –

From the discussion of encyclopedic characteristics: unity, completeness, and discoursivity, we deduce three denotator attributes which cope with these characteristics.

encyclopedic characteristic	$\Longrightarrow$	*denotator attribute*
unity	$\Longrightarrow$	**recursive** typology
completeness	$\Longrightarrow$	extensive **ramification** modes
discoursivity	$\Longrightarrow$	**order** and **recombination**

The idea is that unity in creating concepts is achieved by means of recursive constructions. One builds new denotators from old ones by systematic reuse. Completeness must be guaranteed by extensive ramification modalities when stepping down along recursive paths. Each ramification mode ensures a particular way of stepping downwards in recursivity. Finally, discoursivity is met by means of free construction and recombination of denotators and by universal orderings among these objects. Unlike with usual data base management systems [1093] there is no fixed ensemble of possible denotator constructs. Figure 6.1 illustrates the denotator 'flow chart'.

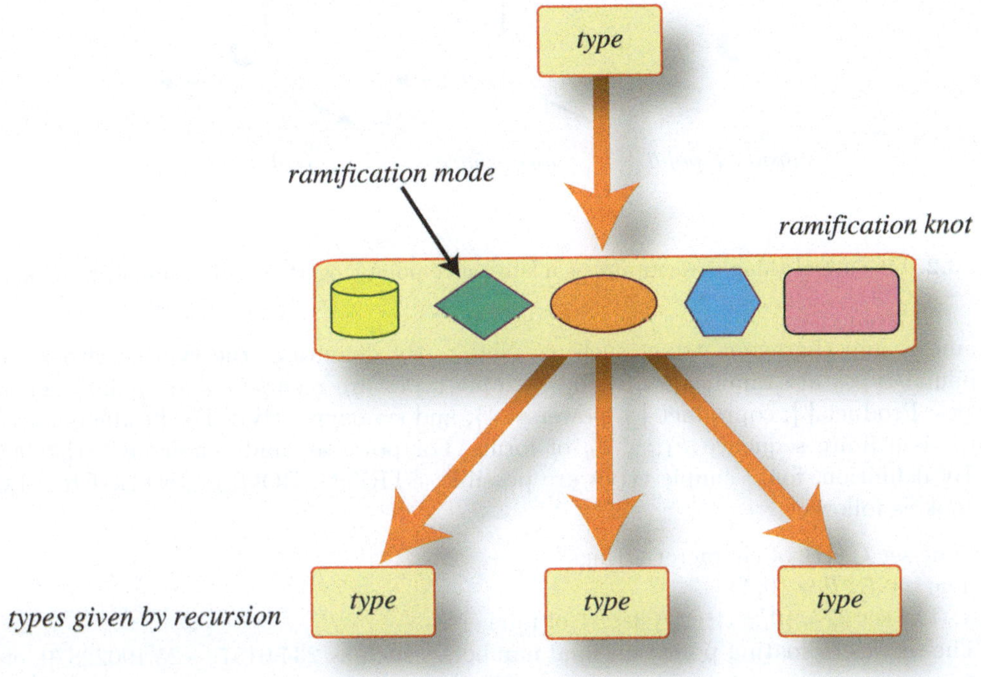

Fig. 6.1. Denotators are designed to cope with universal encyclopedic characteristics, such as unity, completeness, and discoursivity. This flow chart visualizes the corresponding denotator attributes. A type refers to other types in a recursive way through different ramification modes as indicated in the ramification knot. Construction of new and recombination of given denotators is non-restrictive. The system of denotators can be ordered in a universal way.

6.1.1 First Naive Approach to Denotators

Summary. A first approach to denotators is given without requiring further mathematical background.

$- \Sigma -$

The universality of our claim implies that we cannot just set up a generic space and take the objects as being "points" in such a space. Rather is it necessary that each point carries its space with it like a snail carries its shell. Referring to Aristotle's description of real things as a combination of substance and form, we view a denotator as being a substance-point within its form-space, see Figure 6.2.

The naive definition[1] of a denotator is split into a recursive definition of the form-space, and then, based upon this concept, the substance-points are defined. The recursive structure of a form

is a triple, consisting of Form-Name, Type, and Coordinator, denoted in this way:

$$Form\text{-}Name \rightarrow Type(Coordinator). \tag{6.1}$$

[1] In the mathematically rigorous framework of form semiotics as described in Appendix G.5.3, the naive approach is related to the topos **Ens** of sets.

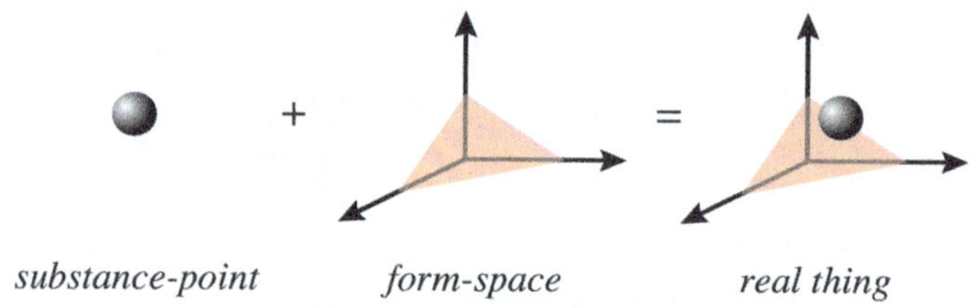

Fig. 6.2. Like a real thing, a denotator is a substance-point, together with a proper form-space.

The Form-Name is any character string, such as 'Note'[2]. By definition, the type is either compound or simple. Compound types are ramification modes, simple types are recursive start points. There are four[3] compound types: Product $\prod$, coproduct $\coprod$, powerset $\{\}$, and synonymy **Syn**. For products and coproducts, the coordinator is a finite sequence F_1,..., F_n of forms. For powerset and synonymy[4], the coordinator is one form F. By definition, four[5] simple types are possible: STRING, BOOLE, INTEGER, FLOAT. Their coordinators look as follows:

- STRING: The set **CHR** of character strings[6].
- BOOLE: The set $BIT = \{0, 1\}$.
- INTEGER: The set $\mathbb{Z} = \{0, \pm 1, \pm 2, \pm 3, \ldots\}$ of integers.
- FLOAT: The set $\mathbb{R}$ of (floating point) decimal numbers[7], such as 234.0157, -25.19022879, etc.

Since coordinators are uniquely determined for simple forms, we shall omit the coordinator bracket in these cases, i.e. INTEGER($\mathbb{Z}$) simplifies to INTEGER, etc.

The general naming policy for forms is that two different forms should bear different names. This enables us to identify a form with its name (the name is a "key" in the terminology of database theory [1093]). For example, if we have the product form

$$\text{'Piano-Note'} \to \prod(\text{Onset}, \text{Pitch}, \text{Loudness}, \text{Duration}), \tag{6.2}$$

with coordinators Onset, Pitch, Loudness, and Duration, then we want to give it the shorthand name Piano-Note. In this sense, we often adopt the bracketless denotation

$$\text{Piano-Note} \to \prod(\text{Onset}, \text{Pitch}, \text{Loudness}, \text{Duration}) \tag{6.3}$$

if no confusion is likely[8].

With this naming convention in mind we introduce the following simple forms as a first set of ad hoc examples for this 'naive' introduction:

- Onset→FLOAT,

[2] The special '...'quotation marks reduce the word to a character string, i.e. the sequence of characters without any further meaning.
[3] These are just the ones which can be easily understood within this naive introduction. Later, we shall generalize the types, but no fundamentally different ideas will be added.
[4] In the rigorous framework, synonymy is superfluous and may be mimicked by a product with one single factor.
[5] These simple domains will be vastly extended, the present selection is made with regard to practical and programming-oriented usage.
[6] We may take the standard set of ASCII characters, but from an informal point of view, just think of usual symbols, such as letters a, b,..., figures 0, 1, 2,..., and other characters, such as brackets, etc.
[7] We do not consider infinite decimal digits, such as $\pi = 3.1415926\ldots$, in this naive setup.
[8] In the general approach to denotator theory, names are also denotators, and the present distinction is obsolete, see Appendix G.5.3.

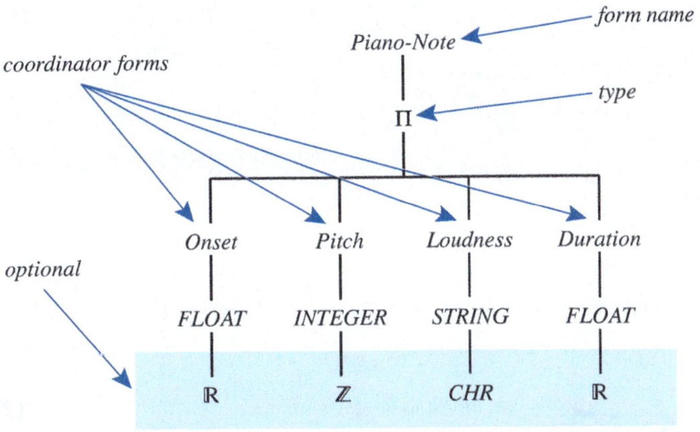

Fig. 6.3. The form Piano-Note in graphical representation, including the recursive explication of its coordinator forms Onset, Pitch, Loudness, and Duration.

- Pitch→INTEGER,
- Loudness→STRING,
- Duration→FLOAT.

A rigorous expression for such a form would read 'Onset'→ SIMPLE($\mathbb{R}$), for example, and the form itself would be called Onset.

If we deal with complex compound forms and wish to represent them in a more graphical way, the symbolism can be shown in a vertical arrangement, as illustrated for the Piano-Note form in Figure 6.3. Observe that presently, products and coproducts or powersets and synonymies, respectively, are not further distinguished. The former couple has a sequence of forms, the latter one a single form as coordinator. However, the specific difference will emerge when we introduce the substance-points on the respective form-spaces.

In our naive setting, a denotator is a triple, consisting of the denotator name Denotator-Name, the form Form, and the coordinates Coordinates; it is denoted by

$$Denotator\text{-}Name \rightsquigarrow Form(Coordinates). \tag{6.4}$$

Again, *Denotator-Name* is a character string, *Form* is a form, and *Coordinates* is an object which is determined according to the form's type. If the type is simple then, by definition, Coordinates is an element of the coordinator of the form. For example, if Type = INTEGER then Coordinates is an integer number, i.e. an element of $\mathbb{Z}$. For example, a denotator for pitch middle C on a keyboard could look like 'C'⇝Pitch(60), see Figure 6.4 for concrete examples of simple denotators. Again, if there is no risk of confusion, we shall identify a denotator by its denotator name, writing C or C⇝Pitch(60) instead of 'C'⇝Pitch(60), for example. For denotators we cannot however stick to the naming policy of forms since in general, denotator names are of secondary importance. This is so because denotators usually occur in large numbers (such as notes in a score), and we are forced to distinguish them by their coordinates rather than by names. Very often it will even be reasonable to reduce the denotator name to the empty string ''[9].

[9] Be careful in distinguishing the empty string '' from the string ' ' consisting of one empty space, sometimes also denoted by ␣!

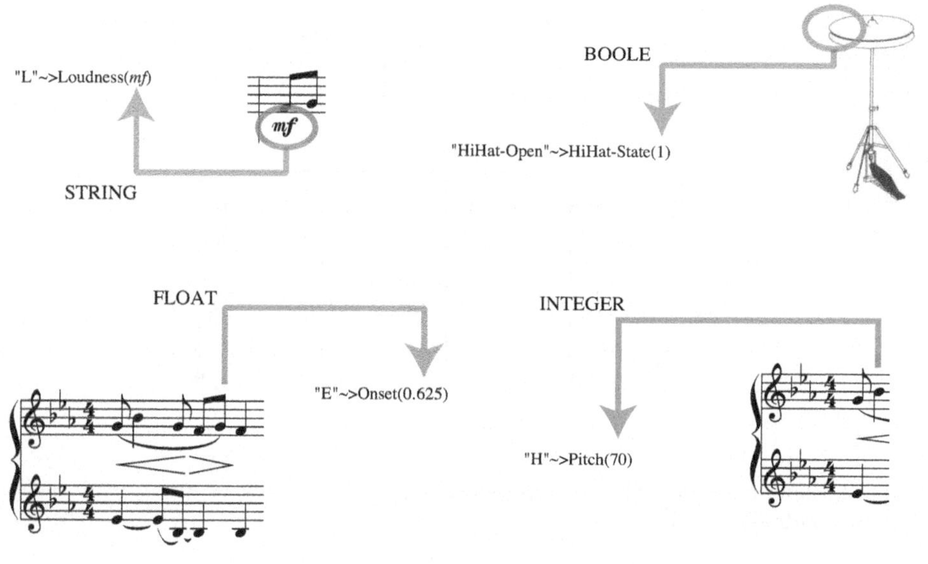

Fig. 6.4. Examples of simple denotators.

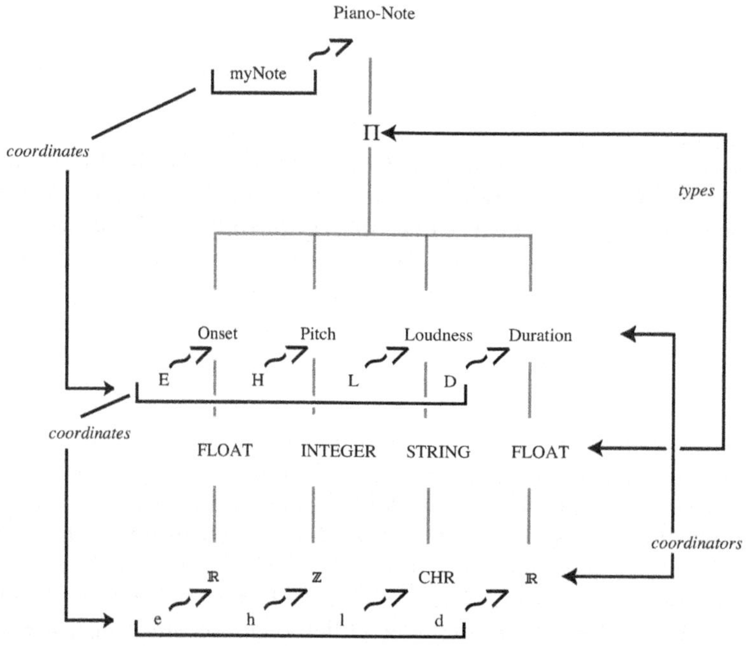

Fig. 6.5. A product denotator named 'myNote' with its coordinates in a recursively complete graphic representation.

Let Type be compound[10]. For *product*, and if[11] Coordinator $= F_1, \dots F_n$, we define Coordinates as being any sequence $f_1, \dots f_n$ of denotators with forms $F_1, \dots F_n$, respectively, i.e. $f_1 = f_1$-Name $\rightsquigarrow F_1(f_1$-

[10] In this naive setup, coordinates are also denotators. This is however different from the formal setup, but there, we need conceptual circularity on the level of forms and denotator namings. We have preferred this naive approach to ease understanding in a first approach.

[11] In this book, we make a logical, though not very common usage of the ellipsis symbol "...": it means that one has started with a sequence of symbol combinations which follows an evident law, such as $1, 2, \dots n$, or $a_1 + a_2 + \dots a_n$.

Coordinates), ..., $f_n = f_n$-Name $\rightsquigarrow F_n(f_n$-Coordinates). For example, if we are given a denotator named 'myNote'[12], and with form Piano-Note, the coordinates are four denotators E, H, L, D with forms Onset, Pitch, Loudness, Duration and Coordinates e, h, l, d in the coordinators $\mathbb{R}, \mathbb{Z}, \mathbf{CHR}, \mathbb{R}$ respectively:

$$\begin{aligned} &\text{myNote} \rightsquigarrow \text{Piano-Note}(E, H, L, D), \\ &E \rightsquigarrow \text{Onset}(e), H \rightsquigarrow \text{Pitch}(h), L \rightsquigarrow \text{Loudness}(l), \\ &D \rightsquigarrow \text{Duration}(d). \end{aligned} \tag{6.5}$$

Figure 6.5 shows the graphic representation of the Piano-Note form (see Figure 6.3) with the above denotator myNote inserted.

If Type is *coproduct*, and if Coordinator $= F_1, \ldots F_n$, we define Coordinates as being any one denotator f_i of form F_i, for an index $i = 1, ..., n$, i.e.

$$f_i = f_i\text{-}Name \rightsquigarrow F_i(f_i\text{-}Coordinates). \tag{6.6}$$

In the above example with $F_1, F_2, F_3, F_4 =$ Onset, Pitch, Loudness, Duration, let

$$PianoSelector \to \coprod(Onset, Pitch, Loudness, Duration)$$

be the corresponding coproduct form. Then, with this form, a denotator named 'mySelection', given by 'mySelection'$\rightsquigarrow$PianoSelector(f), is defined by the selection of either $f \rightsquigarrow$Onset(ϕ) or $f \rightsquigarrow$Pitch(ϕ) or $f \rightsquigarrow$Loudness(ϕ) or $f \rightsquigarrow$Duration(ϕ), ϕ being an element of $\mathbb{R}, \mathbb{Z}, \mathbf{CHR}, \mathbb{R}$, respectively.

If Type is *powerset*, and if Coordinator $= F$, we define Coordinates as being a finite set S of denotators, all having one and the same form F. For example, if we consider the form Chord$\to\{\}$(Pitch), then a denotator myChord$\rightsquigarrow$Chord(S) is defined by a set S of pitches, i.e. a chord in the classical terminology. For a finite set $S = \{f_1, ..., f_k\}$, we usually omit the curled set brackets in this notation and simply write $\{\}(f_1, ..., f_k)$ instead of $\{\}(\{f_1, ..., f_k\})$. For example, a second degree chord[13] $\mathrm{II_C}$ in C major consisting of pitch denotators D$\rightsquigarrow$Pitch(62), F$\rightsquigarrow$Pitch(65), and A$\rightsquigarrow$Pitch(69) would be denoted by $\mathrm{II_C}\rightsquigarrow\{\}$(D,F,A).

If Type is *synonymy*, and if Coordinator $= F$, we define Coordinates as being a denotator f with form F. Hence, if the denotator's form is New$\to$**Syn**(F), then a denotator with denotator name 'myNewName' is myNewName$\rightsquigarrow$New(f). Essentially, this means renaming f by 'myNewName' and doing this within the form which is essentially F, but with a new name 'New'. Besides a renaming of the space and the point, nothing really happens—unfortunately, this is a very common situation in the humanities. Notice, however, that synonymy is not symmetric, i.e. the new denotator is recursively later than the old one. This is essential for understanding the construction history of concepts.

If a denotator of form *AForm* and coordinates *myCoordinates* is given the empty name '', we shall denote it by

$$AForm(myCoordinates) \tag{6.7}$$

instead of ''$\rightsquigarrow AForm(myCoordinates)$. But we shall not introduce analogous notation for empty-named forms since form names are essential as an identification instance.

The evidence is built upon the starting unit, such as "1," or "a_1+" in our examples, and then the following unit, such as "2," or "a_2+", and then inducing the following units to be denoted, such as "3," or "a_3+", "4," or "a_4+", etc., until the sequence is terminated by the last symbol, such as "n" or "a_n" in our examples. The ellipsis means that the building law is repeated, and as such, it is a meta-sign referring to the inductive offset. Therefore the more common notation $1, 2, \ldots, n$, or $a_1 + a_2 + \ldots + a_n$ is not correct. In the limit, for $n = 3$, it would imply a notation such as $1, 2, , 3$ or $a_1 + a_2 + +a_3$, which is nonsense. Moreover, in complicated indexing situation, the common notation would be overloaded.

[12] If possible, we name denotators with small initials, but exceptions may occur.

[13] We shall, of course, give rigorous definitions of all musicological terms later in this book.

6.1.2 Interpretations and Comments

Summary. Interpretations of the above concepts and comments on non-trivial implications are given.

– Σ –

Pointer Character. To begin with, let us stress the pointer character of the denotator concept. The underlying form concept is self-referential: In general, a form is defined by another (bunch of) form(s), only the reference type is explicit. Recursion *is supposed* to stop at a simple form, and there, we are left with a selection of substance from a common repertoire, viz mathematics and computer programming data types (numbers, digits, strings). Similarly with denotators. We contend that this scheme of recursive pointers is fundamental in human concept construction. Understanding such a concept means navigating on its recursive ramification tree.

This is the way we usually execute our understanding of a concept, e.g. "house": We say that a house consists of a roof, a set of windows and doors, walls, etc. We point at components of the concept. If necessary, we penetrate such components, and prosecute their ramification trees, etc. But we do *not* point automatically at any possible knot of the entire concept tree! If the concept "house" appears in a discourse, we only delve into a minimal level of explicitness. In our setup, this means that we just take the name, or, if empty, the form of a denotator if this specification suffices. Only in case some request is made for more details do we unveil deeper concept levels and follow the path down to coordinators and coordinates.

Such a pointer scheme realizes a very economic handling of concepts. You only "unpack" what is really needed, the rest remains either referred by name or even completely hidden. This approach is similar to the paradigm of object-oriented programming, more precisely, to encapsulation of object data. They have an identity[14], but the contents are not unveiled until really needed.

Circularity. Denotators have the outstanding property that they are open to circular definitions, and this is a fundamental extension of concept building rules with respect to the apparently founded recursivity of denotator structures. Let us give a common example to make the point explicit. We want to define the concept of a book with chapters, subchapters, subsubchapters etc. Evidently, a book can have an arbitrary number of chapters, and the depth of sectioning and subsectioning a book is not limited. But we can say that a book always consists of title, text, and a set of chapters.

In turn, each chapter again consists of the same data type, except that we have to enumerate the chapters. We thereof deduce a natural ordering of appearance of the particular chapters. Let us therefore complete the above list by a number:

1. number
2. title
3. text
4. a set of chapters

The problem with this definition is that we have to declare the "set of chapters". In the context of denotators, this problem is solved in the following way: The basic form is termed 'book', and it has this structure:

$$\begin{aligned} 'book' &\to \prod(No., title, text, chapters), \\ 'No.' &\to INTEGER, \\ 'title' &\to STRING, \\ 'text' &\to STRING, \\ 'chapters' &\to \{\}(book). \end{aligned} \tag{6.8}$$

We have made self-reference to the form *book* within the powerset factor *chapters* of the four-fold product. However, this circularity of the form *book* is not necessarily inherited by substance-points within the form-space. Since a *book*-formed denotator

[14] Usually a pointer to some memory address.

$$myBook \rightsquigarrow book(..., myChapters)$$

has any set of chapters, we may give it the empty set $\varnothing$ of coordinates:

$$myChapters \rightsquigarrow chapters().$$

In this case, no further specification is needed, and the denotator is completely determined. This may happen anywhere in the recursive regression of such a denotator and therefore, any finite book tree can be captured in the sense we expect from practice.

Nonetheless, circularity of forms and denotators may occur in a much less inoffensive way which takes the very basis of concept construction into task. We come back to this issue in Section 6.7.

Meeting Encyclopedism. Let us recapitulate the encyclopedic characteristics and their realization on the denotator level. Recursive typology meets *unity* since we are given a unified construction principle of concepts; this is the yoga of pointers. It is, of course, not possible to demonstrate in a formal sense that our ramification modalities are *complete*. But in view of the rigorous formal setup to be exposed in Section 6.2, we can say that the most general mathematical construction principles of products, coproducts[15] and powersets, together with the epistemological factotum of synonymy, gather everything which is known to be relevant for formation of mathematical and database structures[16]. And *discoursivity* is guaranteed by two measures: First, complete freedom of form and denotator construction guarantees recombination. Even operations of *semantic completion* are feasible in the following sense. Suppose, for example, that an ethnological pitch-related form has not yet been made fully explicit and is merely 'sketched' in a simple, STRING-typed form $UnknownPitchType \to$ STRING within its recursion tree. The denotators of form *UnknownPitchType* have only a string of characters, no further contents are available. It may happen in a more in-depth research development that the pitch structure can be specified beyond strings. Then the form *UnknownPitchType* may be replaced by

$$KnownPitchType \to \prod(UnknownPitchType, \text{Pitch}), \tag{6.9}$$

or even by form

$$KnownPitchType \to \mathbf{Syn}(\text{Pitch}), \tag{6.10}$$

where Pitch is the integer-valued form discussed in formula (6.2). The first form relates every denotator pitch string to its numerical value, whereas the second form is a restatement of pitch data via a synonym of the already given Pitch form. Whatever procedure is selected, the old form *UnknownPitchType* can be replaced by the new *KnownPitchType* form in all forms which make use of it—just as in genetic engineering. And *UnknownPitchType*-formed denotators can keep their names, we only need to replace them by synonymous denotators which point at numerical pitch values.

For example, it may happen that a pitch is just named 'F♯', but no special value in terms of frequency is specified. This is a common situation in music theory. Then, in a more explicit situation—of instrumentation, say—this 'symbolic' data should be realized in a specific tuning. This is precisely the need to replace or enrich the string 'F♯' by more involved information, e.g. *KnownPitchType*('F♯', Pitch(66)) as related to keyboard middle C Pitch(60).

The second measure to meet discoursivity is a universally defined ordering principle on denotators. This provides us with an 'omnipresent' orientation for the discourse on denotators. Observe that this ordering system does not restrict to lexical alphabetic ordering but extends to more natural geometric constructions. We shall introduce (still naively, see Section 6.8 for the rigorous version) this topic in the following section.

[15] Products will then generalize to so-called limits, coproducts generalize to colimits.

[16] It is astonishing that database theorists have not yet learned to make systematic use of mathematical category theory, although the latter is being used in theoretical computer science.

6.1.3 Ordering Denotators and 'Concept Leafing'

Summary. We give a naive approach to the universally defined linear orderings among denotators and illustrate the results for musically meaningful denotators.

– Σ –

The question of what d'Alembert called the "encyclopedic ordering" is fundamental to any encyclopedic enterprise. We contend that the 'EncycloSpace of denotators' should be provided with a linear ordering $<$ (see Section C.2 for its definition). Here are three points why such a requirement is indeed fundamental:

- One should be able to have a universal orientation, something that could be termed 'leafing' on the denotator EncycloSpace.
- Orderings on denotators should be a germ for their representation as 'points in coordinator spaces'.
- A denotator search algorithm has to be defined in a universal way, and preceding visualization, since search engines must a priori be able to invoke this algorithm.

It was an interesting conflict within Diderot's Encyclopédie that the alphabetic ordering was so low level with respect to the order of ideas. This is above all due to the informal presentation of ideas—no database concepts were available in 1750.

We introduce the linear ordering on denotators and their forms by recursion, i.e. by inheritance from already defined orderings on the coordinates and their coordinators. To simplify the discussion, we shall stick to non-circular forms. To begin with, if we look at *simple coordinators*, we are provided with well-known linear orderings: The coordinator **CHR** of STRING has the *lexicographic ordering* realized in every dictionary (see example 94 in Appendix C.2 for this concept). The coordinator BIT of BOOLE is ordered by $0 < 1$, integers and decimal numbers are given their standard orderings.

To define a linear ordering among denotators D with

$$D\text{-}Name \rightsquigarrow D\text{-}Form(D\text{-}Coordinates), \tag{6.11}$$

we proceed as follows: Suppose that we have defined linear order relation $<^*$ for couples of denotators regardless of their names: $Form1(Coordinates1) <^* Form2(Coordinates2)$. Then we can easily extend the ordering lexicographically with priority on $<^*$:

$$'D1\text{-}Name' \rightsquigarrow Form1(Coordinates1) < 'D2\text{-}Name' \rightsquigarrow Form2(Coordinates2)$$

if and only if either

$$Form1(Coordinates1) <^* Form2(Coordinates2)$$

or

$$Form1(Coordinates1) = Form2(Coordinates2),$$

and in the latter case,

$$'D1\text{-}Name' < 'D2\text{-}Name'$$

in the lexicographic ordering on **CHR**. Therefore, we can concentrate on couples of denotators regardless of their names. The same holds for forms: If we have settled forms without name, the lexical naming ordering just refines the nameless ordering. So we may forget about names and concentrate on the structure $Type(F_1, ...F_n)$ of a form. Again, we proceed lexicographically with priority to the type. First, let us order the types as follows:

$$\text{BOOLE} < \text{INTEGER} < \text{STRING} < \text{FLOAT} < \textbf{Syn} < \prod < \coprod < \{\} \tag{6.12}$$

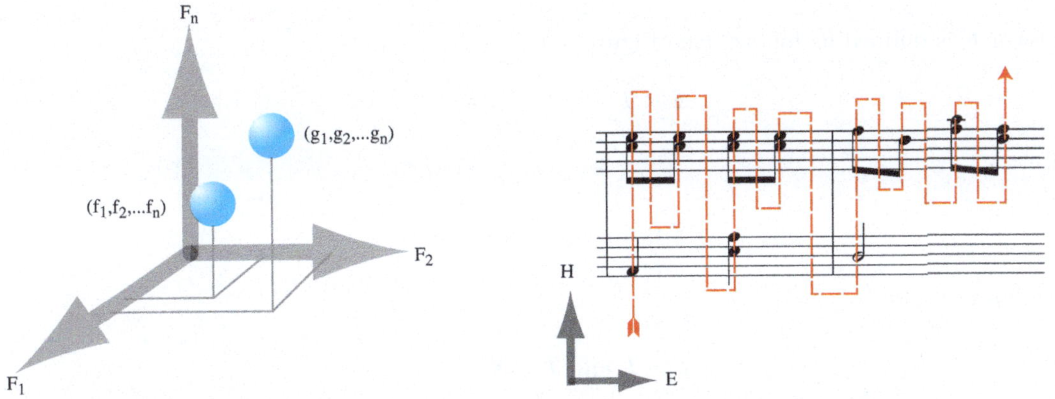

Fig. 6.6. Linear ordering on product type. To the right, we have an example of piano notes in their linear ordering indicated by the dotted arrow.

Then, if the type is fixed, we set[17]

$$Type(F_1, ...F_n) < Type(G_1, ...G_m) \text{ iff } F_1, ...F_n < G_1, ...G_m \tag{6.13}$$

for the lexicographic ordering on the words on the alphabet of forms. This settles the order relation between forms, and we are left with the order relation $Form(Coordinates1) <^* Form(Coordinates2)$. The simple forms are already settled by the introductory remark on the standard orderings on BIT, **CHR**, $\mathbb{Z}$, and $\mathbb{R}$.

We now turn to compound denotators and make the general recursion hypothesis: On coordinates, a linear ordering is defined. For **Syn**, we just inherit the ordering on the coordinator. For $\prod$, we take the lexicographic ordering: $\prod(f_1, \ldots f_n) < \prod(g_1, \ldots g_n)$ iff we have $f_i < g_i$ for the first index i where the coordinates differ, see Figure 6.6. For the coproduct $\coprod$, we set $\coprod(f_i) < \coprod(g_j)$ iff either $i < j$ or $i = j$, and then $f_i < g_i$. Intuitively, this is the situation we know from a library, see Figure 6.7.

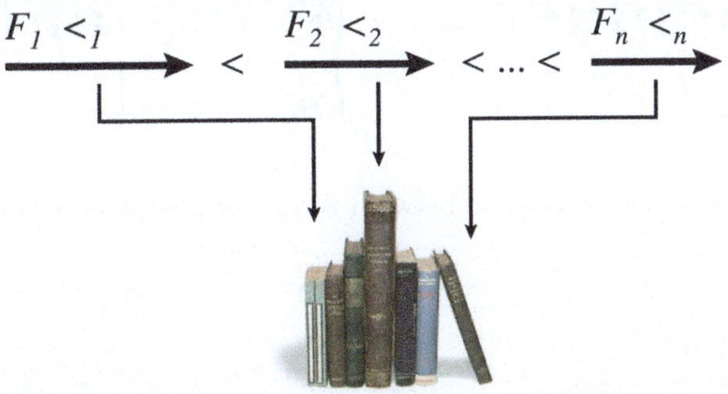

Fig. 6.7. Coproduct ordering is similar to the ordering within a library. The books F_i are numbered from 1 to n and arranged in this order on the shelf. Within each book, we have a given linear ordering of entities.

Finally, the powerset ordering[18] between different sets

[17] "iff" is mathematical shorthand for "if and only if".
[18] See Appendix C, Proposition 77, for a proof that the poweset ordering is linear.

$$S = \{\}(f_1, \dots f_n) \text{ and } T = \{\}(g_1, \dots g_m)$$

of given form F is defined as follows (see Figure 6.8):

$$S < T$$

iff either

$$S \text{ is a proper subset of } T$$

or both difference sets

$$S - T \text{ and } T - S$$

are non-empty, and

$$Max(S - T) < Max(T - S),$$

the maxima being taken on the linear ordering of the form F.

Intuitively, we may interpret the linear ordering among denotators as being a tool for conceptual leafing, the usual running through key-words being extended to typically leafing through geometrically shaped coordinator spaces with many coordinate axes (like vector spaces). Further, the user can "dive" into deeper coordinator layers, though never getting lost within hierarchical ramifications since linear ordering is a much stronger ordering principle than hierarchies.

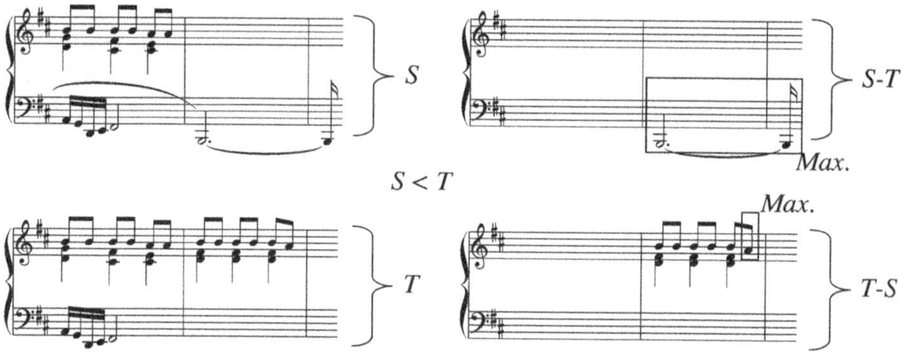

Fig. 6.8. The linear order relation between two sets S and T of notes; we have S < T.

6.2 Forms

Summary. Forms are introduced as the universal format of spaces. They are defined in a recursive way, carry a *name*, a recursive ramification *type*, a diagram of recursive references, the *coordinator*, and the so-called *space functor identifier* of the form. Simple types define the beginning of recursive ramification. The ramification type is either simple, synonym, limit, colimit, or powerset. The coordinator diagram is defined and discussed in detail.

– Σ –

The rest of this chapter is devoted to the generic and formal setup for forms and denotators. We shall, however, not change the surface of our terminology. Denotators and forms are the naive concepts for those readers who have no necessity to view the generic setup. The latter is intimately related to the so-called functorial point of view (see Appendix G.1.2). We shall nonetheless adopt the standard attitude of modern algebraic geometry and talk about "points" even though the space where the points are positioned is not a fixed one but may vary as a function of an entire parameter system of 'addresses'. For mathematical reasons which will become clear soon, the naive perspective will be called the *zero (address) perspective* if a clear-cut distinction from the functorial perspective is necessary.

6.2.1 Variable Addresses

Summary. The functorial approach is introduced as a variable address question.

– Σ –

Let us start with an elementary example. In Section 6.1.1, we have introduced the form

$$\text{Pitch} \rightarrow \text{INTEGER}$$

of simple type INTEGER = INTEGER($\mathbb{Z}$) and corresponding denotators which look like

$$c\text{-Pitch} \rightsquigarrow \text{Pitch}(c), c \in \mathbb{Z}.$$

The reason, why we say that such a denotator has "zero address" is that the set $0_\mathbb{Z}@\mathbb{Z}$ of affine homomorphisms from the zero module $0_\mathbb{Z}$ to the integers $\mathbb{Z}$ (Appendix E) reduces to the translations in $\mathbb{Z}$ and therefore is in bijection[19] with $\mathbb{Z}$, each $z \in \mathbb{Z}$ giving rise to the translation $e^z : 0 \mapsto z$. Whence the expression "zero address" to specify the set of denotators c-pitch with coordinates in $c \in \mathbb{Z}$. What is the interest in generalizing this zero address which presently looks like an algebraic overhead to catch the elements in modules?

First, there are deep reasons relying on universal constructions of new denotators from given ones: Without the functorial perspective some of these constructions would not be possible; we shall discuss this issue in Section 8.3.3. Second, there are important approaches in mathematical music theory which already make use of variable addresses, for example in harmony, see Chapters 24 through 26. But let us give a simple example as an ad hoc justification.

The above Pitch form describes integer-valued pitch denotators sitting on the zero address. This would be sufficient to grasp the chromatic pitches if we took the integers for labels of semitone steps, on a well-tempered[20] keyboard, say. But this is a somewhat reduced view of pitch in music. In score notation, we not only have the effective key number but additional information about alterations stemming from tonal frames. For example, if the denotator 66-pitch were attached to the 6th semitone above middle C, this could be denoted by F♯ or G♭, depending upon the tonal context. In order to express this refinement, an adequate non-zero address provides us the necessary data. In fact, if we take $\mathbb{Z}$ instead of $0_\mathbb{Z}$, the affine homomorphisms $f = e^a \cdot b : \mathbb{Z} \rightarrow \mathbb{Z}$ in $\mathbb{Z}@\mathbb{Z}$ are in one-to-one correspondence with couples $(a, b) \in \mathbb{Z}^2$ where $a = f(0), b = f(1) - f(0)$. In other words, a $\mathbb{Z}$-addressed object $f \in \mathbb{Z}@\mathbb{Z}$ gives rise to two data: the "base pitch" $a = f(0)$ and the "shifted pitch" $f(1) = b + f(0)$ staying b semitones above the base pitch, see Figure 6.9. Intuitively speaking, the $\mathbb{Z}$-addressed objects are the affine images of the unit arrow $0 \Rightarrow 1$ in $\mathbb{Z}$. And these denote precisely what we were looking for: A base pitch (= arrow tail) and a shifted pitch (= arrow head). In musical terms, our example 66-pitch could be refined to the pitch alteration F♯ by the arrow object $f = e^{65} \cdot 1$ or to the pitch alteration G♭ with arrow object $f = e^{67} \cdot -1$. This means that we have refined the zero-address object

$$0 \quad \mapsto \quad 66$$

[19] This is true in complete generality: For any module M, we have a canonical bijection $0_\mathbb{Z}@M \xrightarrow{\sim} M$ (Appendix E.3).

[20] Recall that all musicological terms will be introduced with care later. Here we only use examples for motivating the denotator concept.

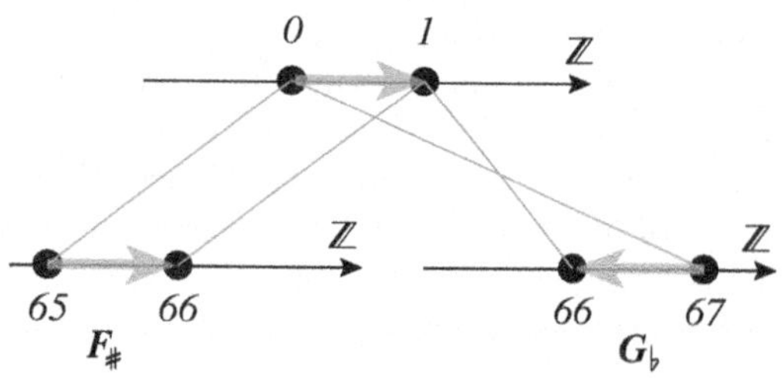

Fig. 6.9. Two different $\mathbb{Z}$-addressed points define the musical notations F♯ and G♭, respectively, as "arrow" objects in $\mathbb{Z}$.

to the $\mathbb{Z}$-addressed "arrow" object

$$0 \Rightarrow 1 \quad \mapsto \quad 65 \Rightarrow 66$$

for F♯ or

$$0 \Rightarrow 1 \quad \mapsto \quad 67 \Rightarrow 66$$

for G♭, respectively[21]. Moreover, zero-addressed and $\mathbb{Z}$-addressed points are canonically related: On one hand, the unique affine homomorphism $zero : \mathbb{Z} \to 0_{\mathbb{Z}}$ transforms every zero-addressed point $z : 0_{\mathbb{Z}} \to \mathbb{Z}$ into the composed $\mathbb{Z}$-addressed point $z \circ zero : \mathbb{Z} \to \mathbb{Z}$. If z stands for the denotator c-pitch, $z \circ zero$ is the arrow object which sends the arrow $0 \Rightarrow 1$ to the zero-length arrow $c \Rightarrow c$. On the other hand, if we look at the zero-addressed points $tail := e^0 \cdot 0$ and $head := e^1 \cdot 0$, a $\mathbb{Z}$-addressed point $f = e^a \cdot b$ yields f's tail point $f \circ tail$ and f's head point $f \circ head$, respectively.

Summarizing, adding variable addresses do enrich the expressivity of denotators, the relations between different addresses add canonical connections to the objects which are introduced on different address levels.

6.2.2 Formal Definition

Summary. The definition of a form is given, including name, type, coordinator, and the space functor identifier.

– Σ –

Recall (Appendix G) that $\mathbf{Mod}^@$ denotes the category of contravariant functors $Fu : \mathbf{Mod} \to \mathbf{Ens}$ from the category **Mod** of modules to the category **Ens** of sets. Recall also that for functors Fu in $\mathbf{Mod}^@$, a module M which is an argument of Fu is called an *address of* Fu, and a morphism $f : M \to N$ of modules is called an *address change*. Recall the notation $M@Fu$ for the value of Fu at address M. Recall finally the subobject classifier Ω in the topos $\mathbf{Mod}^@$133.

Definition 3 *A form*[22] *F is a quatruple $F = (NF, TF, CF, IF)$ where*

(i) NF is a string of ASCII characters; it is called the name *of F and denoted by $N(F)$.*

[21] Again, for any module M, the $\mathbb{Z}$-addressed points of M, i.e. the elements of $\mathbb{Z}@M$, are in bijection with the "arrows" $m \Rightarrow n$ in M, see Appendix E.

[22] A completely general definition of a form which implies names which are also denotators, is given in Appendix G.5.3. This generalization is necessary for more flexible "global" name spaces. Here, we stick to a naming convention which restricts to *ASCII* names.

(*ii*) *TF is one of the symbols*[23]
1. **Simple**,
2. **Syn**,[24]
3. **Power**,
4. **Limit**,
5. **Colimit**;
it is called the type *of F and denoted by* $T(F)$.
(*iii*) *CF is one of the following objects according to the previous symbols:*
A. For **Simple**, *CF is a module M,*
B. for **Syn**, *and* **Power**, *CF is a form,*
C. for **Limit** *and* **Colimit**, *CF is a diagram* $\mathbf{D}$ *of forms;*
it is called the coordinator *of F and denoted by* $C(F)$. *The diagram* $\mathbf{D}$ *is a diagram of functors* $Fun(F_i)$, *as defined in (iv), for a family* $(F_i)_i$ *of forms.*
(*iv*) *IF is a monomorphism of functors* $IF : Fu \rightarrowtail X$ *in* $\mathbf{Mod}^@$, *with this data:*
1. For **Simple**, $X = @M$,
2. for **Syn**, $X = Fun(CF)$,
3. for **Power**, $X = \Omega^{Fun(CF)}$,
4. for **Limit**, $X = lim(\mathbf{D})$,
5. for **Colimit**, $X = colim(\mathbf{D})$;
it is called the identifier *of F and denoted by* $I(F)$, *whereas its domain Fu is called the* space (functor) *of F and denoted by* $Fun(F)$. *The codomain of the identifier is called the* frame space *of the form.*

To denote a form F, we inherit the notation of the naive setup and add the identifier below the arrow:

$$Name \underset{Identifier}{\longrightarrow} Type(Coordinator). \tag{6.14}$$

Naming. Evidently, the naming formalism is identical with the naive one. We shall also adopt the policy that different forms should bear different names.

Typology. The typology is somewhat more complicated: For the naive approach, simple types were restrained to a selection of four sample coordinators ($\mathbf{CHR}, BIT, \mathbb{Z}, and\ \mathbb{R}$) whereas here, simple type means selecting any coordinator module M. We shall make this more explicit in Section 6.4. Philosophically, this means that the basic substance spaces are now taken from the category of modules and not merely from four types STRING, BOOLE, INTEGER, and FLOAT. It is plausible that the vast domain of modules[25] is large enough to cover once for all the needs of musicology, mathematics, and computer science for grounding space substance. As to the other types, synonymy looks quite the same as with the naive approach: We just have to go back one step in recursion. The real typological difference resides in the last three types, **Power**, **Limit**, and **Colimit** which generalize the naive types powerset, product, and coproduct. We are going to describe all the types and their characteristics in Section 6.2.3.

Coordinator. The coordinator looks quite familiar for simple type, instead of one of the four simple coordinators in the naive setting, we are given a module. For synonymy and power types, we have the same data as in the naive setting. For limits and colimits, we are given a diagram of forms. This generalizes the naive setup where we just had a finite sequence of forms in the coordinator, whereas a diagram would contain a number of "arrows" between members of such a sequence. See Appendix G for the concept of a diagram in a category. In our context, a form diagram has its vertexes in the set of form names and evaluates to the corresponding form functors, but see Appendix G.5.3 for a rigorous formalism.

---

[23] They turn out to symbolize five types of operators, but in the definition, we just need the symbols, i.e., the character strings.

[24] In the generic definition of a form, the synonym form type is superfluous. We however maintain this type for semiotic reasons: synonymy is a proper type of reference which is meant to be different from any other reference mode.

[25] Observe that modules can have any commutative or non-commutative coefficient rings!

Identifier. There is, however, a remarkable difference in that we add the new instance of an "identifier" to the naive setup. Why? In the naive setup, we had a unique choice of simple spaces: the four coordinators ($\mathbf{CHR}, BIT, \mathbb{Z}, \mathbb{R}$). In general, this cannot be guaranteed, many equivalent modules or higher constructs (functors) may intervene, so we have to integrate them without losing their structural identification. The latter is taken care of by the identifier monomorphism whereas the functor $Fun(F)$ of the form, i.e. the identifier's domain designates the concrete structure. For example, we may encounter several one-dimensional real vector spaces V, but essentially, they are all isomorphic to $\mathbb{R}$. Such an isomorphism $I(F) : V \xrightarrow{\sim} \mathbb{R}$ is denoted by the identifier[26].

A basic comment on the structure of Definition 3 is necessary. The concept of a "form" refers to four ingredients: Name, Type, Coordinator, and Identifier. Whereas the first two specifications are inoffensive, the last two share *circularity*: They refer to the form concept to be defined, at least in the cases (iii), B. and C. as well as (iv), 2. to 5. In mathematical definitions, circularity is admitted only if it is embedded in a recursive definition structure, i.e. if one is given an ordinal number associated with every instance of the concept in question that appears within the definition. For example, the product of natural numbers $a * b$ is defined according to the size of b: For $b = 0$, we set $a * 0 = 0$, and $a * (b + 1)$ is defined to be $(a * b) + a$. But in set theory, a set Z defined by $Z := \{Z\}$ is a problematic object, since no ordinal "size" number is given. In "normal" set theory, such non-regular, circular sets are excluded by the axiom of regularity [561]. We come back to this issue in Section 6.5. For a formal definition of logically consistent domains of forms, see Appendix G.5.3.

6.2.3 Discussion of the Form Typology

Summary. We discuss the characteristics of the form types and compare them to the naive variants.

– Σ –

Simple. A form F of simple type consists of a name $N(F)$, a coordinator module M, a functor $Fun(F)$ in $\mathbf{Mod}^{@}$, and an identifier monomorphism[27] $I(F) : Fun(F) \rightarrowtail @M$. Since in the naive setup, we did not have to change addresses, the module M (one of four sample modules) was all we needed to describe simple type forms.

Syn. This form type adds a new name $N(F)$ to an already known coordinator form $C(F)$, and a functor monomorphism $Fun(F) \rightarrowtail Fun(C(F))$.

Power. To relate the naive powerset type to **Power**, let Fu be a functor of $\mathbf{Mod}^{@}$ and consider the functor 2^{Fu} of $\mathbf{Mod}^{@}$, defined by $M@2^{Fu} := 2^{M@Fu}$, the powerset of the Fu-value at address M. For an address change $f : N \to M$, we use the set map $f@Fu : M@Fu \to N@Fu$ to define $f@2^{Fu} : M@2^{Fu} \to N@2^{Fu} : S \mapsto f@Fu(S) =: S \cdot f$.

Recall (Appendix G.3, examples 129 and 127) that Ω^{Fu} evaluates to $M@\Omega^{Fu} \xrightarrow{\sim} Sub(@M \times Fu)$ at address M. To connect the naive powerset type and the general one, we define a natural transformation

$$\hat{?} : 2^{Fu} \to \Omega^{Fu} \tag{6.15}$$

[26] Historically, the identifier problem arose when transcribing the software-oriented PrediBase data base management system [1147] to an abstract formalism. In fact, object-oriented programming languages ask for default values of instance variables in order to initialize objects. This initialization process means that you have to write down a first concrete value to be instantiated without further activities from the user's part. If you cannot stick to a unique domain of values, for decimal numbers (floats), say, the default value will tell what kind of representative of the domain you have selected. If you write down 0.0 as a default value, you make precise which kind of zero representation you have taken. What you are really doing is pointing at a particular representative of "the" $0_{\mathbb{R}} \in \mathbb{R}$, i.e. you sort of point at the particular 0.0: $0_{\mathbb{R}} \mapsto 0.0$, and this is what our monomorphism takes care of: It identifies a representative with "the abstract object". Indeed, quite often the identifier is not only mono but even iso. However, isomorphisms would create serious existence problems for circular forms, see Appendix G.2.2.1 for this subject.

[27] If the identifier is iso, this means that Fu is representable, and that it is represented by $I(F)$. In general, Fu is isomorphic to a sieve (Appendix G.3.1) in M.

by the following subfunctor $\hat{S} \subset @M \times Fu$ for $S \in M@2^{Fu}$. For a given address N, the subfunctor $\hat{S}$ is defined by

$$N@\hat{S} := \{(u \in N@M, v \in N@Fu) | v \in S \cdot u\}. \tag{6.16}$$

It is easily seen that the transformation (6.15) is also natural in Fu, i.e. if $h : Fu \to Gu$ is a natural transformation, the following canonical diagram is commutative:

$$\begin{array}{ccc} 2^{Fu} & \xrightarrow{\hat{?}} & \Omega^{Fu} \\ {\scriptstyle 2^h}\downarrow & & \downarrow{\scriptstyle \Omega^h} \\ 2^{Gu} & \xrightarrow{\hat{?}} & \Omega^{Gu} \end{array} \tag{6.17}$$

We shall see in Section 6.3.1, item "Coordinates", that this transformation connects naive powerset denotators to those of **Power** form.

Limit. A "diagram of forms" is just a diagram in the category $\mathbf{Mod}^{@}$, via the functor $Fun(F)$ of a form F (see Appendix G.1.2, Definition 165 for diagrams in categories). We can also speak of the *category of forms*, the morphisms being the morphisms of the forms' functors, so the terminology "diagram of forms" is also formally correct. The relation to the naive setting is that a product $\prod F_i$ of forms is a limit for a discrete diagram (no arrows) whose vertexes are the indices i of the product's family $(F_i)_{i=1,\dots n}$.

Colimit. Equally, a coproduct $\coprod F_i$ of forms is a colimit for a discrete diagram whose vertexes are the indices i of the product's family $(F_i)_{i=1,\dots n}$.

We would like to stress that all these constructions presuppose that the reference to the coordinator's forms and their functors is possible, i.e. that these objects all do exist[28]. We come back to this problem in Section 6.5.

6.3 Denotators

Summary. The formal concept of a denotator as being a form plus a substance point is introduced. Implicitly, the point concept is also recursively defined by its reference to the recursive structure of the underlying form space that was introduced in the preceding section.

– Σ –

6.3.1 Formal Definition of a Denotator

Summary. Based on the definition of a form (Section 6.2.2), we define denotators.

– Σ –

Definition 4 *Let M be an address. A M-*addressed denotator *is a triple $D = (ND, FD, CD)$ where*

(i) ND is a string of ASCII characters; it is called the name *of D and denoted by $N(D)$.*
(ii) FD is a form; it is called the form *of D and denoted by $F(D)$.*
(iii) CD is an element of $M@Fun(F(D))$; it is called the coordinates[29] *of D and denoted by $CT(D)$.*

According to the naive setup, a denotator is denoted by

$$Name : Address \rightsquigarrow Form(Coordinates) \tag{6.18}$$

[28] Following Paul Finsler [322], an object's existence is equivalent to the possibility to think the objects without provoking any contradiction to classical principles of logic: identity, contradiction, and excluded third.

[29] This is a plural in singular mode.

So, in the full notation, a denotator D with form F is symbolized as follows:

$$N(D) : Address \rightsquigarrow N(F) \underset{Identifier}{\longrightarrow} Type(Coordinator)(Coordinates), \tag{6.19}$$

but this clumsy writing is only used if absolutely unavoidable. Again, as already described for the naive setup, if the denotator name is empty (which happens often), we shall write

$$myAddress \rightsquigarrow AForm(myCoordinates) \tag{6.20}$$

for a denotator of form $AForm$ and coordinates $myCoordinates$, or even

$$AForm(myCoordinates) \tag{6.21}$$

if $myAddress$ is also clear (in fact, the zero address in the naive context).

Naming. The naming policy is identical with the naive situation. Recall that we can have the empty denotator name $''$.

Address. The address module M is an important generalization, however, we shall not overstress it since in the everyday language we speak of denotators independently of their address, as if there were just one ambient space instead of an entire space functor.

Form. The form of a denotator englobes the whole recursion information as well as the form's functor. It is the latter which contains the coordinates.

Coordinates. The coordinates are one "point"[30] or element of the form's functor *at* the given address M. Let us look at the shape of coordinates as a function of the particular form:

1. For **Simple**, the coordinator is a module N, and the coordinates $CT(D)$ are identified (via the identifier!) with an element of $M@N$, i.e. a morphism $CT(D) : M \to N$ of modules. In the case of the zero address $M = 0_{\mathbb{Z}}$, such a morphism clearly identifies with a 'real' element of N (Appendix E.3).
2. For **Syn**, the coordinates $CT(D)$ identify with an element of $M@Fun(C(F(D)))$. So the coordinates are described by recursion to the coordinator $C(F(D))$.
3. For **Power**, the coordinates $CT(D)$ identify with an element of $M@\Omega^{Fun(C(F))} \xrightarrow{\sim} Sub(@M \times Fun(C(F)))$, i.e. a subfunctor of $@M \times Fun(C(F))$. But we know from the discussion of form typology in Section 6.2.3 that each set $S \in M@2^{Fun(C(F))}$ of M-addressed points in $Fun(C(F))$ gives rise to a subfunctor $\hat{S}$ of $@M \times Fun(C(F))$. So the zero-addressed coordinates in the naive sense give rise to coordinates in the general setup. We also see that an M-addressed naive coordinate set $S \subset M@Fun(C(F))$ is just a set of "points" at a general but *fixed* address M. Passing to the functorial setup means switching to an entire system of sets $?@\hat{S}$, where $\hat{S}$ is a 'functor of subsets' in $M@\Omega^{Fun(C(F))}$.
4. For **Limit**, suppose that the diagram $\mathbf{D}$ of forms has vertex forms F_i and vertex functors $Fu_i := Fun(F_i)$ at vertexes i, as well as natural transformations $m(f) : Fu_{tail(f)} \to Fu_{head(f)}$ for arrows f in $\mathbf{D}$. Recall (Appendix G.2.1, Theorem 68) that at a given address M, we have a natural isomorphism

$$M@lim(\mathbf{D}) \xrightarrow{\sim}$$
$$\{x \in \prod_i M@Fu_i \mid m(f)(x_{tail(f)}) = x_{head(f)}, \text{ all arrows } f \text{of } \mathbf{D}\}$$

so that the coordinates of **Limit** denotators are special tuples in the product of all $M@Fu_i$. In other words, the limit denotators are canonically related to the product denotators from the naive setup. Therefore the general setup does resemble the naive one as being focused on a general, but fixed address.

[30] In algebraic geometry, an element of $M@F$ for a functor $F \in \mathbf{Mod}^@$ is called an "M-valued point of F".

5. For **Colimit**, let us inherit the notation of the preceding situation. Recall (Appendix G.2.1, Theorem 68) that for a given address M, we have the equivalence relation $\sim$ on the coproduct $\coprod_i M@Fu_i$, generated by the relation $x \sim y$, $x \in M@Fu_i$ and $y \in M@Fu_j$, iff there is an arrow f in $\mathbf{D}$ with $i = tail(f)$, $j = head(f)$, and $m(f)(x) = y$. We then have a natural isomorphism

$$M@colim(\mathbf{D}) \xrightarrow{\sim} \coprod_i M@Fu_i/\sim$$

and recognize that the naive coproduct is just the basic space of the **Colimit** denotator space before dividing through the equivalence relation $\sim$, and again, selecting a general, but fixed address M. So—up to general addressing—the coordinates of a **Colimit** denotator are those of the coproduct *modulo* an equivalence relation defined by the diagram's arrows.

If $f : F \to G$ is a morphism of forms, and if $D : X{\rightsquigarrow}F(C)$ is a denotator, we canonically have the *f-image of D* $f(D) : X{\rightsquigarrow}G(X@f(C))$. Superficially the denotator definition is not recursive, but we recognize from the above discussion of the coordinates that a denotator's coordinates are just as recursive as the involved forms. In the naive setup this was less hidden because of the missing identifier (the identifier was 'set to identity'). Again, a denotator bears a fundamental pointer character: It can only be understood by pointing down to the coordinates which are distributed among the various recursive forms.

6.4 Anchoring Forms in Modules

Summary. This section deals with recursive foundation of forms in simple spaces. It presents a justification for choosing modules and affine transformations as basic space types. First examples are presented and discussed in the light of musical and musicological requirements.

– Σ –

Since modules are anchor structures for denotators, we should comment on this decision: Why is the module structure a good choice? There are mathematical and musicological, practical and theoretical reasons. We want to expose them generically and by use of common examples. We have transferred the purely mathematical description of a module to Appendix E, we have also compiled a number of common parameter spaces for musical objects in Appendix A. Here, we refer to these chapters and concentrate on the justification discourse.

To begin with, let us shortly digress on the relation between mathematical structure and musical meaning. One could require that a determinate structural instance (such as additive closure of modules) has to play a musical or musicological role in all situations where the structure is present. However, such a requirement would be too restrictive since mathematics should make available to music what may possibly happen in music, and not what happens in any case. In other words, mathematics must furnish a minimal conceptual and theoretical region where the possible applications can be located.

We shall encounter cases where mathematical structure overhead seems to appear. We then obviously have to reflect on possible musical or musicological interpretations of this overhead. This is a delicate question since it may lead to contrary answers depending on its topographic context. In particular with regard to poietics, more concretely: in the process of generating a determinate composition, a mathematical structure may serve as a vehicle of the sound material's organization that otherwise has never been used, e.g. aleatorics with John Cage [552], Fibonacci numbers[31] with Karlheinz Stockhausen [458], or fractals with Mesias Maiguashca [461]. It cannot be decided a priori which mathematical structures carry musical meaning. In this sense, a module structure is but an adequate formal frame for an empirical material which *presently* guarantees a broad field of applications—in the future it may even be enlarged. It is however not true that reflecting

[31] Named after Italian mathematician Leonardo Fibonacci (1180-1250 (?)), the Fibonacci sequence $(x_i)_{i=1,2,3,...}$ is defined by $x_1 = x_2 = 1$ and recursively by $x_{i+2} = x_i + x_{i+1}$. The limit number $\lim_{i\to\infty} x_i/x_{i+1} \approx 0.618036$ is the famous *golden section.*

musical and musicological thinking via mathematical structuring may lead to "foreign infiltration in music" as it was argued in [784]. Rather does it catalyze the fundamental dialectic process of how music sees itself.

Before looking at concrete examples, let us view the general point of taking modules. In order to handle denotators in music, we need not only consider "points in spaces" but also algebraic operations which relate points to each other and allow creation of new points out of given ones. In music, sound objects are not understood as being isolated points in amorphous spaces but parts of an entire relational system, and this insight is the reason for introducing modules. They are the adequate structure to grasp this formal mechanism of building new from old points. We should also keep in mind in the following that it is not necessary (and sometimes not possible) to look at all addresses for simple denotators, but we will pick up some of them to illustrate their role.

6.4.1 First Examples and Comments on Modules in Music

Summary. Musicological aspects of the algebraic structure of a module and of affine transformations among modules are discussed. We presuppose the knowledge of module structure and of the examples from the appendices A and E. We also illustrate the use of modules for more general denotators which may intervene for not strictly musical purposes, such as text elements, for example.

– Σ –

1. *BIT*. This set is also known as $BIT = \mathbb{Z}_2 = \mathbb{Z}/2\mathbb{Z}$, the integers modulo 2, see Appendix D, example 105. It has an additive structure which turns the elements into logical units used in music instrument editing. In fact, look at the simple 'self-addressed' denotators $D(a,b) : BIT \rightsquigarrow Switch(e^a \cdot b)$ of form $Switch \underset{Id}{\longrightarrow}$ **Simple**(BIT), having coordinates $e^a \cdot b \in BIT@BIT$. Such a denotator $D(a,1)$ is the a-translation on BIT and represents its additive structure. If we view the elements s of BIT as being the states of an ON-OFF switch on a musical instrument, the denotators $D(a,b)$ can be viewed as operators on the states of the switch: We have[32]
$$D(a,1)(s) = \begin{cases} s & \text{if } a = 0, \\ 1 + s = \neg s & \text{if } a = 1. \end{cases} \tag{6.22}$$
So $a = 0$ leaves the state s of the button whereas $a = 1$ changes it. Observe that we also have
$$D(a,0)(s) = a, \tag{6.23}$$
and this means that the operator $D(a,0)$ is a *reset* to the state a of the switch. More generally, the theory of switching circuits and automata [487] can be executed on denotators which are built upon BIT as a simple 'module basis'.
2. **CHR**. In the naive setup, the set of character strings $'x_1x_2 \ldots x_n'$ from a character alphabet (the set $ASCII$ of ASCII characters in our case) is not a module, but it can be embedded in a module according to the very comfortable and common mathematical structure of a monoid algebra (see Appendix D.1.1). Its construction presupposes an alphabet A (generalizing $A = ASCII$) and a commutative ring R of 'coefficients'. We first build the free monoid $\langle A \rangle$ of words $'x_1x_2 \ldots x_n'$ over the alphabet A. The monoid algebra $R\langle A \rangle$ consists of all formal sums $\sum_{w \in \langle A \rangle} r_w w$ with only finitely many coefficients $r_w \neq 0$. In particular, the words $w = 1w$ are elements of the monoid algebra, and we have $\mathbf{CHR} \subset \mathbb{Z}\langle ASCII \rangle$. The structure of $\mathbb{Z}\langle ASCII \rangle$ as a $\mathbb{Z}$-module is more than the simple names of denotators and forms—for example—require. But the use of formal sums is straightforward: We can take the names of all forms which appear in a determined description of the denotators (notes, bar-lines, pauses, time and key signature, composer's name, etc.) of a score S and attribute to these names N the multiplicity $a_N \in \mathbb{Z}$ of their appearance in denotators of S. This defines a multiplicity-sensitive *list of forms* in S, i.e.

[32] The symbol $\neg s$ means the negation of state s, see also Appendix G.5.

$$FormList(S) := \sum_{N=\text{form name in S}} a_N N. \tag{6.24}$$

We obtain the effective calculation of the score's form list by successive addition: If S splits into two disjoint parts S_1 and S_2, we have $FormList(S) = FormList(S_1) + FormList(S_2)$. Further, if $FormList(S)$ and $FormList(T)$ are two such lists, the difference $FormList(S) - FormList(T)$ in general produces negative coefficients which measure the relative amount of form names in S, when compared to T. From this we learn that account of words from a determined alphabet can make use of the corresponding monoid algebra over some appropriate ring to perform its calculations.

3. The real numbers $\mathbb{R}$. Recall (Appendix E.2.1, example 107) that $\mathbb{R}$ is a module[33] over the subfield of rational numbers $\mathbb{Q}$ which we denote by $\mathbb{R}_{[\mathbb{Q}]}$. We want to view this module as being the coordinator of a simple pitch form

$$MathPitch \underset{Id}{\longrightarrow} \mathbf{Simple}(\mathbb{R}_{[\mathbb{Q}]}), \tag{6.25}$$

as discussed in Appendix A.2.3. In accordance with that discussion, we interpret denotators $MathPitch(h)$ as having the mathematical pitch coordinate

$$h = h(f) = u \cdot \ln(f) + v$$

associated with frequency f and a couple of normalization coefficients u, v. In this view, the ratio

$$h(f) - h(g) = u \cdot \ln(f/g)$$

of two frequencies f and g becomes a difference of MathPitch coordinates. This transformation reflects the auditory experience that the frequency ratio f/g is perceived as a "pitch distance":

Fact 1 *The additive structure of $\mathbb{R}_{[\mathbb{Q}]}$ is the basis of thinking in pitch distances.*

Further, scalar multiplication with a rational number a/b splits into a product of three special cases: multiplication by

- -1,
- a natural number $p \geqslant 0$,
- $1/q$, q a positive natural number.

What is the musical meaning of these three operations?

- Multiplication of a pitch difference $h(f) - h(g)$ by -1 yields $-1 \cdot (h(f) - h(g)) = h(g) - h(f)$, we look from the second pitch to the first instead of looking from the first to the second, i.e. the roles of the two pitches are *exchanged*.
- For $p \geqslant 0$, we have $p \cdot (h(f) - h(g)) = p$-fold *juxtaposition* of $h(f) - h(g)$; the special case $p = 0$ means that we switch to the zero distance.
- $1/q \cdot (h(f) - h(g))$ means that we seek a distance which—after a q-fold juxtaposition—yields the original distance $h(f) - h(g)$.

Fact 2 *Scalar multiplication in the pitch module splits into role exchange, p-fold juxtaposition, and q-fold division of pitch distances.*

Remark 1 The signification of a mathematical operation often becomes evident after a 'split' into special cases from which the general case results via appropriate conjunction. This is based on the principle that musical meaning is conserved after such 'recombination'. This is genuine musical thinking! The example of the contrapuntal operation of "retrograde inversion" rightly demonstrates that the operation is only 'understood' via its decomposition into "retrograde" and "inversion", and *not* as such. We come back to this very important subject in Section 8.3.

[33] In fact an infinite-dimensional vector space.

4. The Euler module $\mathbb{Q}^3$ over the field $\mathbb{Q}$ of rationals. This module is introduced in Appendix A.2.3 and represents one of the most classical mathematical structures in mathematical music theory. We also view this module as being the coordinator of the simple form

$$EulerModule \underset{Id}{\longrightarrow} \mathbf{Simple}(\mathbb{Q}^3). \tag{6.26}$$

Recall that we have a $\mathbb{Q}$-linear embedding

$$\langle pv| : \mathbb{Q}^3 \to \mathbb{R}_{[\mathbb{Q}]} : x \mapsto \langle pv|x\rangle \tag{6.27}$$

of the Euler module in the pitch module, where $pv = (\ln_b(2), \ln_b(3), \ln_b(5))$ is the "prime vector" relative to a logarithm base b, and where the scalar product $\langle pv|x\rangle$ is the above pitch $h(f)$ associated with frequency f. Therefore we have a corresponding morphism of forms

$$E2M : EulerModule \to MathPitch \tag{6.28}$$

induced by the above linear embedding and its canonical extension to the forms' functors

$$@\langle pv| : @\mathbb{Q}^3 \to @\mathbb{R}_{[\mathbb{Q}]}. \tag{6.29}$$

As in the preceding example, the additive structure becomes musically significant via differences $x - y$ of Euler points x and y (Euler points are just elements of the Euler module). In fact, we have

$$\langle pv|x\rangle - \langle pv|y\rangle = \langle pv|x - y\rangle, \tag{6.30}$$

and the difference of pitch corresponds to the difference of Euler points. With the same transformation, we can carry over the musical signification of scalar multiplication from pitch to Euler points.
But the Euler module has more in its structure, and this stems from music history. If $o = (1, 0, 0)$, $q = (0, 1, 0)$, and $t = (0, 0, 1)$ are the canonical base vectors, every Euler point $d = (r, s, u)$ can be written uniquely as a linear combination

$$d = r \cdot o + s \cdot q + u \cdot t. \tag{6.31}$$

On the other hand, the fifth and major third (Appendix A.2.3) are associated with the Euler points $q - o$ and $t - 2 \cdot o$. Hence, every difference $x - y$ can be written as

$$d = (r + s + 2u) \cdot o + s \cdot (q - o) + u \cdot (t - 2 \cdot o), \tag{6.32}$$

and hence be generated from the octave, fifth, and major third by role exchange, juxtaposition and division, according to fact 2. Here, exchange of roles means that distances are juxtaposed 'downward' instead of 'upward'. Division $1/12 \cdot o$ defines the distance for semitones in the 12-tempered system, whereas $1/2 \cdot (t - 2 \cdot o)$ means whole tone distance between tonic and third in mediante tuning (see Appendix A.2.3). If we work within just tuning, any distance $d = x - y$ can be produced by successive juxtaposition of octave, fifth, and major third, upward or downward.
In the tradition of just tuning, the following Euler point differences have special names [1089]:

$$\begin{aligned} &\text{fifth comma } (= \text{Pythagorean comma})\\ &Kq = 12 \cdot (q - o) - 7 \cdot o\\ &\quad = 12 \cdot q - 19 \cdot o \end{aligned} \tag{6.33}$$

$$\begin{aligned} &\text{third comma } (= \text{syntonic comma})\\ &Kt = (t - 2 \cdot o) + 2 \cdot o - 4 \cdot (q - o)\\ &\quad = t - 4 \cdot q + 4 \cdot o \end{aligned} \tag{6.34}$$

Historically, these commata arose from the 12-division of the octave in semitone steps. In this context, an octave equals 12 semitone steps, a fifth equals 7 semitone steps, and a major third equals 4 semitone steps. Therefore there was interest in the difference between least common multiple of the numbers of octave steps and fifth steps (= fifth comma) as well as of third and fifth steps module octave (=third comma). With these identifications we have:

$$12 \text{ fifths} = 7 \text{ octaves (semitone steps)} \quad (6.35)$$
$$4 \text{ fifths} = 2 \text{ octaves } + 1 \text{ third (semitone steps)} \quad (6.36)$$

One finds the following approximate values:

$$\text{fifth comma } \approx 23.46 \text{ Cent} \quad (6.37)$$
$$\text{third comma } \approx -21.51 \text{ Cent} \quad (6.38)$$

Mathematically, these commata figures are not interesting since we know from Appendix A.2.3 that arbitrary small pitch differences can be produced by integer linear combinations of the three vectors o, q, and t. Commata will only become interesting in harmony of just tuning (see Section 24.1).

5. Direct sums. Often, new forms are built from simple forms by products, i.e. limits without arrows. These are no longer simple forms, but they are very similar to simple forms. Let us make this precise for the prototypical case of the product of two forms $F \underset{Id}{\longrightarrow} \mathbf{Simple}(M)$ and $G \underset{Id}{\longrightarrow} \mathbf{Simple}(N)$ with coordinator R-module M and S-module N. We can then build two forms

$$F \times G \underset{@(M \oplus N) \tilde{\to} @M \times @N}{\longrightarrow} \mathbf{Limit}(M, N) \quad (6.39)$$

$$F \oplus G \underset{Id_{@(M \oplus N)}}{\longrightarrow} \mathbf{Simple}(M \oplus N) \quad (6.40)$$

where the first identifier is the canonical isomorphism related to the universal property of direct sums of modules. Clearly, these forms are isomorphic under the identity of $@(M \oplus N)$. Under this canonical isomorphism, we may identify these forms and also identify[34] $F_1 \times \ldots F_n$ with $F_1 \oplus \ldots F_n$ if no confusion is likely.
With this in mind, we may reinterpret the additive structure induced by direct sums of modules: If a product $F \times G$ of simple forms F and G is built as in (6.39), the algebraic a priori isolation of its factors is overridden by a common module where the factors can be compared on an algebraic level. For example, if n, m are two positive natural numbers, we can take

$$OnMod_m \underset{Id}{\longrightarrow} \mathbf{Simple}(\mathbb{Z}_m) \quad (6.41)$$

$$PiMod_n \underset{Id}{\longrightarrow} \mathbf{Simple}(\mathbb{Z}_n) \quad (6.42)$$

with $\mathbb{Z}$-modules $\mathbb{Z}_n$ and $\mathbb{Z}_m$ as coordinators. Consider the simplified form

$$OnPiMod_{m,n} = OnMod_m \times PiMod_n. \quad (6.43)$$

Recall from Appendix A.2.3 that $PiMod_n$ can be viewed as the form whose coordinator $\mathbb{Z}_n$ denotes the pitch classes in n-tempered tuning, whereas $OnMod_m$ can be viewed as the form whose coordinator $\mathbb{Z}_m$ denotes the m-cyclic metrical onsets. Putting these two a priori autonomous aspects of a musical note together, we obtain a common representation of onset and pitch within the module $\mathbb{Z}_m \oplus \mathbb{Z}_n$. The coordinator of the form $OnPiMod_{m,n}$ is positioned on a torus and denotes a type of motivic points, i.e. its denotators $OnPiMod_{m,n}(x, y)$ are points of what musicology considers in motif analysis (see Section 7.2.3). Now, the algebraic connection of onset and pitch data in this symbolization becomes evident from the self-addressed denotators

$$S(f) : \mathbb{Z}_m \oplus \mathbb{Z}_n \rightsquigarrow OnPiMod_{m,n}(f), \quad (6.44)$$

[34] We adopt the saying that "a compound form F *simplifies* to a simple form G" each time we build a simple form G which is isomorphic to a compound form F.

which evaluate to affine endomorphisms f of $\mathbb{Z}_m \oplus \mathbb{Z}_n$. These denotators $S(f)$ put into affine relation both, onset and pitch data, and thus initiate a *comparative theory of motivic points and consequently of motives* as we shall see in Chapter 11, Section 11.3.8, and more in detail in Section 22.4. The paradigm of comparison will be furnished by the denotators $S(f)$.

Concluding this section, we should review the role of mathematics and semantics in the form concept. The mathematical anchoring of forms in modules is a clear creed to mathematics, but it is also a relativization of mathematical structures in that they need to be enriched: Mathematics is only the 'significate' of the form sign, and the type, identifier and name are equally important to build the entire sign. The conceptual genealogy can be anchored in mathematics, but it is substantially more than that. For instance, the naming of forms is completely different from naming mathematical modules. If we have two module namings $M := \mathbb{Z}^4$, and $N := \mathbb{Z}^4$, then it can be stated that "M and N are identical modules". But for two forms $F \underset{Id}{\longrightarrow} \mathbf{Simple}(\mathbb{Z}^4)$ and $G \underset{Id}{\longrightarrow} \mathbf{Simple}(\mathbb{Z}^4)$, we cannot say that "$F$ and G are identical forms". However, we could say that, with the above notations, $F \underset{Id}{\longrightarrow} \mathbf{Simple}(M)$ and $F \underset{Id}{\longrightarrow} \mathbf{Simple}(N)$ are identical forms: The naming difference on the mathematical level is irrelevant.

The names are essential, and this stems from semiotics, not from mathematics. We shall see in Section 6.5 that there are fundamental differences between forms and straight mathematics. They transcend simple naming conventions in that form names are the only access to forms in circular definitions. Their very existence can be verified only via their name's function as a sign's significant.

6.5 Regular and Circular Forms

Summary. We discuss the existence problem of a form if in its definition recursion is non-circular, i.e. terminates on the level of simple forms, or if such a recursion is not possible.

– Σ –

As already announced, the existence question of a form is not trivial. We have to distinguish two cases: Regular and circular forms. Here is the definition of a regular form:

Definition 5 *Let ω be an ordinal number*[35].

(*i*) *A form is* regular of level $\omega = 0$ *iff it is simple.*
(*ii*) *Suppose that regular forms of all levels $\mu < \omega$ have been defined. Then a form is* regular of level ω *iff all its coordinator forms are regular of levels $\mu < \omega$, and if ω is the successor of all the coordinators' levels.*
(*iii*) *A form is* regular *iff there is an ω such that the form is regular of level ω.*
(*iv*) *A form is called* circular *iff it is not regular.*

So the recursive definition of a regular form is built upon its level, and we may define regular forms without any further complication. This can however not work for circular forms. Let us look at some catastrophes which may or may not occur in these cases:

Circular Synonymy. The simplest example of circular synonymy is a form of this shape:

$$'SynCirc' \underset{f:Fu \rightarrowtail Fu}{\longrightarrow} \mathbf{Syn}(SynCirc), \tag{6.45}$$

where a form $SynCirc$ is synonymous to itself and therefore the identifier is a monomorphism f of the form's functor Fu. Such a form is identified by its name and identifier, i.e. it reduces to a triple $('SynCirc', Fu, f)$ of a character string, an object in $\mathbf{Mod}^{@}$, and a mono endomorphism of this object.

[35] See [561], but in most cases, we may just think of ω being a natural number.

Circular Limits. The prototype of a circular form of limit type is this:

$$'LimCirc' \underset{f:Fu \rightarrowtail Fu \times Fun(G)}{\longrightarrow} \mathbf{Limit}(LimCirc, G), \tag{6.46}$$

where the circular form *LimCirc* appears in a product with another (supposedly inoffensive) form G. If we define the form's identifier by

$$Fu := Fun(G)^{\mathbb{N}}, \text{ and} \tag{6.47}$$

$$f : Fu \xrightarrow{\sim} Fu \times Fun(G), \tag{6.48}$$

where, for an address M,

$$\begin{aligned} M@f : M@Fu &\to M@Fu \times M@Fun(G) \\ (g_i)_{0 \leqslant i} &\mapsto ((g_{i+1})_{0 \leqslant i}, g_0). \end{aligned} \tag{6.49}$$

From this example we learn that there are infinitely many[36] realizations—on the level of f or of Fu—of the above circular limit form.

Circular Colimits. In complete analogy to the preceding limit case, we may look at this form[37]:

$$'ColimCirc' \underset{f:Fu \xrightarrow{\sim} Fu \sqcup Fun(G)}{\longrightarrow} \mathbf{Colimit}(ColimCirc, G), \tag{6.50}$$

with the disjoint union $\sqcup$ of functors in its identifier codomain. This form can be realized by

$$Fu := \coprod_{\mathbb{N}} Fun(G), \text{ and} \tag{6.51}$$

$$f : Fu \xrightarrow{\sim} Fu \sqcup Fun(G), \tag{6.52}$$

which is induced by the cofactor isomorphisms

$$f|_i = \begin{cases} Id : Fun(G)_i \xrightarrow{\sim} Fun(G)_{i-1} & \text{for } i > 0, \\ Id : Fun(G)_i \xrightarrow{\sim} Fun(G) & \text{for } i = 0, \end{cases} \tag{6.53}$$

$Fun(G)_i$ being the cofactor of index i in the coproduct. Again, clearly, there are infinitely many realizations of the above circular colimit form.

Circular Power. This one can be inspected on a prototype of shape

$$'PowerCirc' \underset{f:Fu \rightarrowtail \Omega^{Fu}}{\longrightarrow} \mathbf{Power}(PowerCirc). \tag{6.54}$$

But we have the canonical monomorphism

$$f = sg = \hat{?} \cdot \{\} : Fu \rightarrowtail 2^{Fu} \rightarrowtail \Omega^{Fu} \tag{6.55}$$

which is composed from the canonical singleton morphism $\{\}$ (Appendix G.2.2.1, Lemma 102) and $\hat{?}$ (see Equation (6.15)) and serves as identifier for any functor Fu. But observe that this situation forbids that the identifier becomes an isomorphism. In fact, such an isomorphisms f cannot exist since the calculus of cardinalities [561] categorically excludes set injections of type $2^A \rightarrowtail A$ for any set A.

Let us look at the example 6.8 from the naive setup. To restate that example in the formal context, we consider

$$'book' \underset{f}{\longrightarrow} \mathbf{Limit}(No., title, text, chapters) \tag{6.56}$$

[36] There are even infinitely many mutually non-isomorphic realizations.

[37] If a diagram has no arrows and is finite, we may denote the factors as for products in the naive approach.

with identifier

$$f : Fu \rightarrowtail Fun(No.) \times Fun(title) \times Fun(text) \times Fun(chapters) \tag{6.57}$$

and the factor forms

$$'No.' \underset{Fun(No.)\rightarrowtail @\mathbb{Z}}{\longrightarrow} \mathbf{Simple}(\mathbb{Z}), \tag{6.58}$$

$$'title' \underset{Fun(title)\rightarrowtail @\mathbb{Z}\langle ASCII\rangle}{\longrightarrow} \mathbf{Simple}(\mathbb{Z}\langle ASCII\rangle), \tag{6.59}$$

$$'text' \underset{Fun(text)\rightarrowtail @\mathbb{Z}\langle ASCII\rangle}{\longrightarrow} \mathbf{Simple}(\mathbb{Z}\langle ASCII\rangle), \tag{6.60}$$

$$'chapters' \underset{Fin(Fu)\rightarrowtail \Omega^{Fu}}{\longrightarrow} \mathbf{Power}(book), \tag{6.61}$$

where $Fin(Fu)$ is the subfunctor of 2^{Fu} of all finite subsets of $N@Fu$ at address N, and which canonically injects to Ω^{Fu}. Setting

$$H := Fun(No.) \times Fun(title) \times Fun(text), \tag{6.62}$$

we want to define a functor Fu such that f is an isomorphism

$$f : Fu \xrightarrow{\sim} H \times Fin(Fu), \tag{6.63}$$

this would guarantee that any text data (codified in H) and any finite set of chapters (codified in $Fin(Fu)$) can be reached by a book-form denotator. The point is that with the naive setup, there was just the zero address. In our situation, we have to find a solution which is functorial, and this is guaranteed by the second isomorphism in Proposition 115 in Appendix G. But notice that we do not have uniqueness of such a solution!

However, the above limitation technique to prevent the power sets to appear in full is not always sufficient to avoid non-standard situations with circular forms. Consider the example defined by these data:

$$F \underset{f}{\longrightarrow} \mathbf{Limit}(\mathbf{D}), \tag{6.64}$$

$$\mathbf{D} = \bullet \underset{g}{\rightarrow} P, \text{ where} \tag{6.65}$$

$$P \underset{Id}{\longrightarrow} \mathbf{Power}(F) \text{ and} \tag{6.66}$$

$$N@g : \bullet \mapsto N@\widehat{Fun}(F) \in Sub(@N \times Fun(F)) \tag{6.67}$$

for the final functor $\bullet$ and an address N. We deduce the functorial set equation

$$N@f : N@Fun(F) \xrightarrow{\sim} \{N@Fun(F)\} \tag{6.68}$$

which is fulfilled by any circular set $C = \{C\}$ and the constant functor $N@Fun(F) := C$.

6.6 Regular Denotators

Summary. This section is devoted to a first example set of denotators, including most of the common object types encountered in music scores and in music-theoretical contexts. They are structurally characterized by non-circular recursivity: These denotators do not make recurrent reuse of reference denotators—forms or coordinates—within their coordinators.

– Σ –

Definition 6 *A* regular *denotator is a denotator of a regular form.*

It cannot be the scope of this chapter to present an exhaustive list of musicologically relevant regular denotators. This would even not be the philosophy of the denotator formalism. We have to insist that denotators can and should be introduced again and again according to the specific need. There is no reason whatsoever to believe that a terminology should be limited in its concrete exemplifications: Music is a dynamic field of human culture and cannot be understood upon any historically fixed conceptual flashes. After all, denotators have been introduced to cope with all the options of dynamic conceptualization. Nonetheless, we should present a fairly representative list of denotators for a reasonable arsenal of musicological concepts. As theory evolves, we shall add more and more examples to meet the specific problematic. The lesson from this chapter is above all to learn the technique of building concepts via denotators.

Throughout the following explications the reader should bear in mind the prerequisites about common spaces in musicology as discussed in Appendix A.

RegDen-1. To begin with, we need a common numeric representation of elementary *mental* tone parameters as they appear in common European music *score* notation. These are three *basis* parameters onset, pitch, and loudness and three corresponding *pianola* parameters duration, glissando, and crescendo. A common space functor for each of them is the real numbers $\mathbb{R}$ (viewed as a real vector space). We then have this arsenal of simple forms:

$$Onset \underset{Id}{\longrightarrow} \mathbf{Simple}(\mathbb{R}), \tag{6.69}$$

$$Pitch \underset{Id}{\longrightarrow} \mathbf{Simple}(\mathbb{R}), \tag{6.70}$$

$$Loudness \underset{Id}{\longrightarrow} \mathbf{Simple}(\mathbb{R}), \tag{6.71}$$

$$Duration \underset{Id}{\longrightarrow} \mathbf{Simple}(\mathbb{R}), \tag{6.72}$$

$$Glissando \underset{Id}{\longrightarrow} \mathbf{Simple}(\mathbb{R}), \tag{6.73}$$

$$Crescendo \underset{Id}{\longrightarrow} \mathbf{Simple}(\mathbb{R}). \tag{6.74}$$

We inherit the notation from naive denotators and write $Onset(ON)$, etc. for an onset with coordinate ON if the address is clear. Without further comment, the address is always zero. Also we shall often use the names E, H, L, D, G, C for denotators of forms *Onset, Pitch, Loudness, Duration, Glissando, Crescendo* and also write $E = 3.57$ or even $Onset = 3.57$ for an *Onset* denotator of coordinate $3.57 \in 0_{\mathbb{Z}}@\mathbb{R}$, for example. But pay attention in distinguishing this sloppy notation from an equation between forms!
If we want to restrict the space functors of such simple forms, we specify the subspace[38] S of $@\mathbb{R}$ and then write

$$Onset|_S \underset{S \rightarrowtail @\mathbb{R}}{\longrightarrow} \mathbf{Syn}(Onset), \text{ etc.} \tag{6.75}$$

For example, if we restrict pitch to integer values, we simply write $Pitch|_{\mathbb{Z}} \longrightarrow \mathbf{Syn}(Pitch)$ since the other details are clear. In this notation, the middle C pitch of the MIDI code is $Pitch|_{\mathbb{Z}} = 60$ or $H = 60$ if the form $Pitch|_{\mathbb{Z}}$ is clear. Observe that an inclusion $T \rightarrowtail S$ of subspaces of $@\mathbb{R}$ implies inclusions of respective forms

$$Onset|_T \rightarrowtail Onset|_S \rightarrowtail Onset, \text{ etc.} \tag{6.76}$$

From Section 6.3.1 we know that a form morphism also transplants the respective denotators. In our situation, if we are given a denotator $Onset|_{\mathbb{Z}} = 123$, for example, then we may view it as a rational denotator and simply write $Onset|_{\mathbb{Q}} = 123/1 = 123$ since no confusion is likely.
In particular, the canonical (di)morphism $\mathbb{R}_{[\mathbb{Q}]} \rightarrowtail \mathbb{R}$ defines an isomorphism

[38] We often omit the "functor" specification and more intuitively speak about "spaces" when we mean "space functors".

$$MathPitch \xrightarrow{\sim} Pitch|_{@\mathbb{R}_{[\mathbb{Q}]}} \tag{6.77}$$

with the mathematical pitch space defined in (6.25).
With the identifications discussed in Section 6.4.1, item 5 in mind, we shall simplify any direct product of the above forms to the associated simple form, e.g.

$$Onset \times Pitch \times Loudness \times Duration \xrightarrow{\sim}$$
$$Onset \oplus Pitch \oplus Loudness \oplus Duration, \tag{6.78}$$

the latter being a form with coordinator $\mathbb{R}^4$.

RegDen-2. *Symbolic pitch* is often used, e.g. in the notation "$A_\sharp$". A quantitative pitch may be associated as a function of specific tuning paradigms. Thus we need a "neutral" symbolic notation which we realize by the obvious form

$$PitchSymb \underset{Id}{\longrightarrow} \mathbf{Simple}(\mathbb{Z}\langle ASCII\rangle). \tag{6.79}$$

RegDen-3. For performance purposes, musical *tone parameters in the physical domain* which correspond to the above mental parameters are required. We denote them by the following self-explanatory forms:

$$PhysOnset \underset{Id}{\longrightarrow} \mathbf{Simple}(\mathbb{R}), \tag{6.80}$$
$$PhysPitch \underset{Id}{\longrightarrow} \mathbf{Simple}(\mathbb{R}), \tag{6.81}$$
$$PhysLoudness \underset{Id}{\longrightarrow} \mathbf{Simple}(\mathbb{R}), \tag{6.82}$$
$$PhysDuration \underset{Id}{\longrightarrow} \mathbf{Simple}(\mathbb{R}), \tag{6.83}$$
$$PhysGlissando \underset{Id}{\longrightarrow} \mathbf{Simple}(\mathbb{R}), \tag{6.84}$$
$$PhysCrescendo \underset{Id}{\longrightarrow} \mathbf{Simple}(\mathbb{R}). \tag{6.85}$$

Corresponding restrictions as shown in the mental situation above are obvious, though not so commonly used.

RegDen-4. For *symbolic absolute* or *relative dynamics* such as "*mf*" or "*crescendo*", we have to take a dynamic symbol (word), onset, and duration parameters. First, define the form for dynamic symbols:

$$DynSymb \underset{Id}{\longrightarrow} \mathbf{Simple}(\mathbb{Z}\langle ASCII\rangle) \tag{6.86}$$

This gives these product forms:

$$AbsDyn \longrightarrow \mathbf{Limit}(DynSymb, Onset) \tag{6.87}$$
$$RelDyn \longrightarrow \mathbf{Limit}(DynSymb, Onset, Duration) \tag{6.88}$$

for absolute and relative dynamics. We shall see in Section 18.4 that the quantitative meaning of absolute and relative symbolic dynamics is quite non-trivial.

RegDen-5. *Bar-lines* are onsets with no other properties, the only specificity is the name. So we have

$$BarLine \underset{Id}{\longrightarrow} \mathbf{Syn}(Onset), \tag{6.89}$$

and the reader will recognize that we could have set

$$BarLine^* \underset{Id}{\longrightarrow} \mathbf{Simple}(\mathbb{R}), \tag{6.90}$$

and get an isomorphic form. However, synonymy guarantees the common ground of onset which yields the bar-line concept. This type of sophistication is specific to the formalism of forms and allows musicologically adequate constructs; we therefore reject the unspecific form BarLine*.

RegDen-6. A *pause* is determined by its onset and duration. Thus, we may set

$$Pause \underset{Id}{\longrightarrow} \mathbf{Syn}(Onset \oplus Duration). \tag{6.91}$$

RegDen-7. The *time signature* x/y is defined by numerator x and denominator y and has a determined onset origin. It is, however, not a mathematical fraction, since the denominator relates to metrical meaning. We thus may set the product form

$$TimeSig \underset{Id}{\longrightarrow} \mathbf{Limit}(Duration|_{\mathbb{Z}}, Duration|_{\mathbb{Z}}, Onset), \tag{6.92}$$

where a denotator $TimeSig(x, y, e)$ means that the time signature x/y has been positioned at origin e, usually identified with the first bar-line onset.

RegDen-8. To denote *instrument names*, we may construct a generic naming form

$$InstruName \underset{Id}{\longrightarrow} \mathbf{Simple}(\mathbb{Z}\langle ASCII\rangle), \tag{6.93}$$

and we are in state of defining an orchestral instrumentation

$$myOrchestra : 0_{\mathbb{Z}} \rightsquigarrow InstruName(\sum_i^n a_i Instr_i) \tag{6.94}$$

or shorter $myOrchestra = \sum_i^n a_i Instr_i$, with a_i copies of instrument named $Instr_i$. For example, a string quartet has the orchestral instrumentation denotator

$$myStringQuartet = 2'Violin' + 1'Viola' + 1'Violoncello'. \tag{6.95}$$

If we do not want to specify one orchestral configuration, but leave it to the user to be more specific, a generic form of the following type

$$OrchSet \underset{Fin(\mathbb{Z}\langle ASCII\rangle) \rightarrowtail \Omega^{@\mathbb{Z}\langle ASCII\rangle}}{\longrightarrow} \mathbf{Power}(InstruName) \tag{6.96}$$

is adequate. A zero-addressed denotator then is a finite set

$$myOrchSet = \{O_1, O_2, \ldots O_k\} \tag{6.97}$$

consisting of k orchestra configurations.

RegDen-9. Let us give two "small" examples for *non-zero-addressed denotators.* Later (e.g. in Section 8.3.4), we shall devote in depth discussions to this subject. The first small example relates to pitch modulo octave, i.e. to the form $PiMod_{12}$ introduced in (6.42). Whereas zero-addressed denotators

$$PC_{12}(r) : 0_{\mathbb{Z}} \rightsquigarrow PiMod_{12}(r) \tag{6.98}$$

are evidently viewed as 12-pitch classes, more generally addressed denotators deserve special interpretation. They were extensively studied by Thomas Noll [802] in the frame of generalized function theory, see Chapter 25 for a detailed discussion of this theory. This study concentrated on $\mathbb{Z}_{12}$- resp. "self"-addressed denotators

$$[a, b] : \mathbb{Z}_{12} \rightsquigarrow PiMod_{12}(e^a \cdot b) \tag{6.99}$$

for $a, b \in \mathbb{Z}_{12}$, i.e. $e^a \cdot b \in \mathbb{Z}_{12}@\mathbb{Z}_{12}$. Self-addressed pitch classes generalize the 12-pitch classes by the unique projection $! : \mathbb{Z}_{12} \to 0_{\mathbb{Z}}$ of $\mathbb{Z}$-modules. We have an injection

$$!@\mathbb{Z}_{12} : 0_{\mathbb{Z}}@\mathbb{Z}_{12} \rightarrowtail \mathbb{Z}_{12}@\mathbb{Z}_{12} \text{ which induces} \tag{6.100}$$

$$PC_{12}(r) \mapsto [r, 0]. \tag{6.101}$$

In other words, pitch classes are interpreted as constant self-addressed pitch classes. This change-of-address technique is the germ of a far-reaching generalization of harmony, see Chapter 24. We shall see there that self-addressed pitch denotators appear also in the context of Euler modules (6.26), i.e. as denotators of shape

$$[a, b] : \mathbb{Q}^3 \rightsquigarrow EulerModule(e^a \cdot b) \tag{6.102}$$

with $e^a \cdot b \in \mathbb{Q}^3 @ \mathbb{Q}^3$.

The second small example relates to the onset space *Onset* and its rational restriction $Onset|_{\mathbb{Q}}$ as considered in the theories of David Lewin [605] and Dan Tudor Vuza [1094]. In [605], Lewin considers *time spans.* These are pairs $(a, x) \in \mathbb{R} \times \mathbb{R}_+$ in the sense of time objects. The first coordinate is a zero-addressed onset in *Onset*, the second is a kind of onset difference, but viewed as a multiplicative quantity of relative increase. The structural meaning of this object is best understood via the transformation rules set up by Lewin. Given two time spans $(a, x), (b, y)$, they are compared by the "interval" which connects them, i.e. the function

$$int((a, x), (b, y)) = ((b - a)/x, y/x).$$

In other words, the first time span (a, x) is transformed into the second one (b, y) by an interval $int = (i, p) = ((b - a)/x, y/x)$, and the law is

$$(b, y) = (a, x)(i, p) = (a + xi, xp).$$

This is the known law of composition of affine transformations, i.e. we should rewrite it under the form

$$e^b \cdot y = e^a \cdot x \cdot e^i \cdot p = e^{a+xi} \cdot xp.$$

In other words, Lewin's time spans $t = (a, x)$ are just self-addressed onset denotators $t : \mathbb{R} \rightsquigarrow Onset(t)$. Later, when we know morphisms between denotators (8.2), the above interval action will be recognized as an address change, whereas Lewin's remark that the interval $int((a, x), (b, y))$ is invariant under time dilatation is a special case of the fact that for any affine automorphism $f = e^u \cdot v : \mathbb{R} \to \mathbb{R}$ of the ambient space $\mathbb{R}$ of the *Onset* form, the interval does not change if applied to the transformed objects, more precisely:

$$int((a, x), (b, y)) = int((u, v)(a, x), (u, v)(b, y)).$$

But this is clear since

$$e^b \cdot y = e^a \cdot x \cdot e^i \cdot p$$

implies

$$f \cdot e^b \cdot y = f \cdot e^a \cdot x \cdot e^i \cdot p,$$

the required equation with transformed objects and the same interval $int = e^i \cdot p$. Anticipating the notation of morphisms between denotators in (8.2), the above equation rewrites as $1/int(fe^b \cdot y) = fe^a \cdot x$, for all automorphisms f.

RegDen-10. To describe a *Fourier* decomposition of a periodic sound of frequency f, we may consider the simple form

$$Fourier \underset{Id}{\longrightarrow} \mathbf{Simple}(\mathbb{R} \oplus \mathbb{R}^{2\times\mathbb{N}}). \tag{6.103}$$

We then have zero-addressed denotators

$$myFourier = (f, (A_i, Ph_i)_{0 \leqslant i}) \tag{6.104}$$

which represent the periodic sound function[39]

[39] With the necessary convergence conditions [268]. Often, one also normalizes a periodic sound function to a maximum value $Max(sd(myFourier)) = 1$.

$$sd(myFourier)(t) = \sum_{0 \leqslant i}^{\infty} A_i \sin(2\pi i f t + Ph_i) \tag{6.105}$$

of physical time t. Notice that denotators do not represent the physical meaning of an information unit. For example, the instrument names are only symbols, essentially ASCII words, and that the physical nature of the denoted instruments are not included. The same holds for the $Fourier$ form. The transformation from "symbolic" to physical or technological meaning relates to the paratextual signification processes described in Section 18.4.

RegDen-11. To denote an *envelope function* on the unit interval, we may use the real vector space $C^0[0,1]$ of continuous functions on the real unit interval $[0,1]$ and set

$$Envelope \underset{Id}{\longrightarrow} \mathbf{Simple}(C^0[0,1]). \tag{6.106}$$

For $Envelope = V$, $PhysOnset = e$, and non-vanishing duration $PhysDuration = d$ we obtain the deformed envelope

$$V_{e,d}(t) = \begin{cases} V((t-e)/d) & \text{for } t \in [e, e+d], \\ 0 & \text{else} \end{cases} \tag{6.107}$$

which vanishes outside $[e, e+d]$ and has its support between e and $e+d$ if V has its support between 0 and 1, see Figure 6.10.

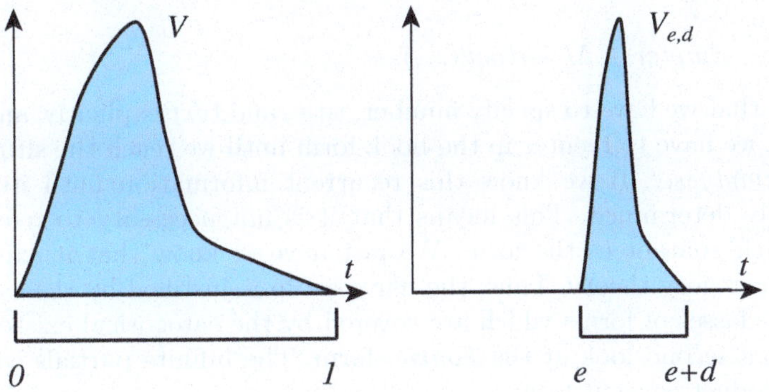

Fig. 6.10. Affine deformation of an envelope denotator V on the unit interval. The deformed envelope $V_{e,d}$ "starts" at onset e and "ends" at onset $e+d$.

RegDen-12. *Combining elementary data* such as onset, duration, Fourier synthesis, and envelope, we can introduce the corresponding physical sound form by

$$\begin{aligned} FourierSound &\underset{Id}{\longrightarrow} \mathbf{Limit}(\mathbf{D}), \\ \mathbf{D} &= (PhysOnset, PhysDuration, Fourier, Envelope). \end{aligned} \tag{6.108}$$

A denotator

$$myFourierSound = (e, d, myFourier, V) \tag{6.109}$$

is associated with the physical sound function

$$sound(myFourierSound) = V_{e,d} \cdot sd(myFourier). \tag{6.110}$$

6.7 Circular Denotators

Summary. This second set of denotators is devoted to important constructs of musical objects which are the foundation of powerful musicological concepts. They make use of circular definitions which do reuse forms or denotators on different recurrent coordinator levels.

– Σ –

Already for the example of Fourier denotators, it was felt that there was a sort of superfluous conceptualization. The inductive repetition of partials was not quite the most economic way to think of forms. In this section we shall take care of a more efficient way of conceptualization: circular forms and denotators. More precisely we have

Definition 7 *A* circular *denotator is a denotator of a circular form.*

CircDen-1. Recall the *book* form discussed in (6.56). An M-addressed denotator $myBook$ of this form looks as follows:

$$\begin{aligned}
&myBook : M \rightsquigarrow book(myNumber, myTitle, myText, myChapters),\\
&myNumber : M \rightsquigarrow No.(N : M \to \mathbb{Z}),\\
&myTitle : M \rightsquigarrow title(Tt : M \to \mathbb{Z}\langle ASCII\rangle),\\
&myText : M \rightsquigarrow text(Tx : M \to \mathbb{Z}\langle ASCII\rangle),\\
&myChapters : M \rightsquigarrow chapters(chapter_1, \ldots chapter_k),\\
&chapter_1 : M \rightsquigarrow book(\ldots),\\
&\ldots\\
&chapter_k : M \rightsquigarrow book(\ldots).
\end{aligned}$$

This means that we have to specify number, title, and text explicitly, and that for each collection of chapters, we have to reenter in the book form until we reach the situation where the chapters build the *empty set.* If we know this recurrent information until its empty roots, the book is completely determined. This means that it is not necessary to refer explicitly to the space functor which guarantees the form. We just have to know that it exists. And this is what we know from category theory. Thus, the naive setup is justified by the categorical background for determined classes of forms which are covered by the categorical existence theorems.

CircDen-2. Let us have a second look at the *Fourier* form. The infinite partials can also be defined by the following product construction:

$$\begin{aligned}
&CcFourier \underset{f:Fu \xrightarrow{\sim} P\times Fu}{\longrightarrow} \mathbf{Limit}(Partial, CcFourier), \text{ where}\\
&Fu = Fun(CcFourier), P = Fun(Partial), \text{ and}\\
&Partial \underset{Id}{\longrightarrow} \mathbf{Simple}(\mathbb{R}^3).
\end{aligned} \tag{6.111}$$

We know from 6.47 that such functors exist. Denotators $myCcFourier$ of form $CcFourier$ are identified by sequences

$$myCcFourier = ((Partial(A_i, f_i, Ph_i))_{0\leqslant i}, \tag{6.112}$$

and this can be interpreted by the Fourier expansion

$$ccsd(myCcFourier) = \sum_{0\leqslant i}^{\infty} A_i \sin(2\pi f_i + Ph_i). \tag{6.113}$$

For the special values $f_i = if$ we get back the classical case. Of course, the numerical information is the same as for the non-circular concept of the Fourier decomposition, but conceptually, the circular form is much simpler—once we have settled the functor existence question.

CircDen-3. We want to look at the *FM synthesis* (see Appendix A.1.2.2 for historical and technical discussion) which looks much more complicated than the Fourier situation if defined without circular forms. In fact, it essentially consists of a sinoidal "carrier" function which in its argument adds other sinoidal "modulator" functions which in turn refer to other modulator functions, etc. The point is that the modulator typology is not fixed and may vary from stage to stage. We may, however, introduce directed graphs (so-called "algorithms" (A.1.2.2)) but the spirit is more obscured than evidenced by such a construction.
Let us first define the FM form and then make a picture of the form. The form is called *FM-Object* and looks as follows:

$$FM\text{-}Object \underset{f:F \xrightarrow{\sim} Fin(FK) \rightarrowtail \Omega^{FK}}{\longrightarrow} \mathbf{Power}(Knot) \text{ with}$$
$$F = Fun(FM\text{-}Object), FK = Fun(Knot) \text{ and the product form}$$
$$Knot \underset{Id}{\longrightarrow} \mathbf{Limit}(Partial, FM\text{-}Object). \tag{6.114}$$

We need to solve the functor equation $F \xrightarrow{\sim} Fin(P \times F)$, but according to Proposition 115 in Appendix G, a solution F always exists.
For each *FM-Object* denotator *myFMObject*, its *graph* $\Gamma(myFMObject)$ is defined as follows: Draw a vertex for each knot, and then an arrow from each knot vertex to all the knot vertexes of its modulator, etc. by induction (including circularity!).

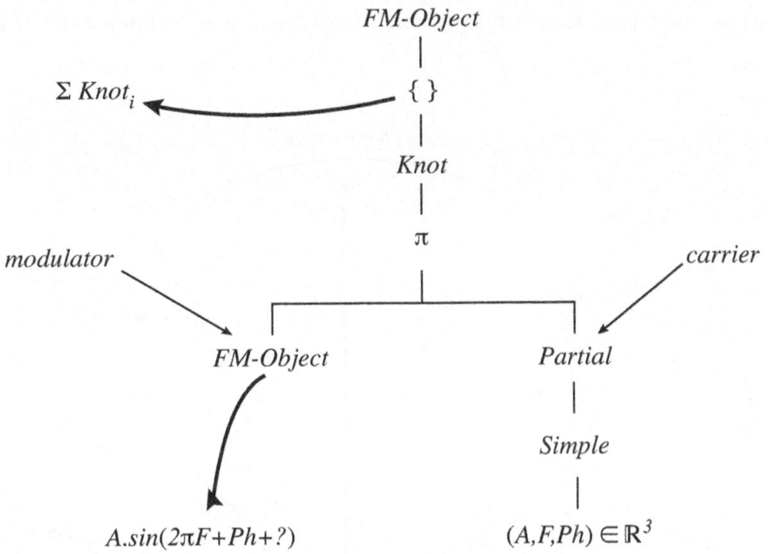

Fig. 6.11. The graphical representation of an *FM-Object* denotator.

What does this mean for the sound synthesis function? Intuitively (see Figure 6.11), an *FM-Object* denotator *myFMObject* is a finite set of knots $Knot_i$ which form the basic carrier of the FM sound. Each knot has it partial denotator which defines the sinoidal function of the knot. In addition, each knot has its modulator which is another FM sound obtained like the basic carrier, etc. Starting from $FMsound(\varnothing) = 0$, the sound function is recursively defined as follows:

$$FMsound(myFMObject)(t) = \sum_i^n A_i \sin(2\pi F_i t + Ph_i + FMsound(myModulator_i)(t)) \quad (6.115)$$

where $myModulator_i$ is the *FM-Object* factor of $Knot_i$.

Recursion here means that after a finite number of downward steps from *FM-Object* factor to *FM-Object* factor, we end up with the empty factor, and the sound function is well-defined. However, under appropriate convergence conditions, infinite descent may also yield well-defined sound functions.

CircDen-4. In music, it is common to consider sound events which share a specific grouping behavior, for example when dealing with arpeggios or trills. We want to deal with this phenomenon in defining *Makro* forms. These forms follow the same schema as discussed for *FM-Object* forms. We only have to replace the partial by an event of a particular type. Put generically, let *Basic* be a form which describes a sound event type, for example the piano specific event type $Basic = Onset \oplus Pitch \oplus Loudness \oplus Duration$ introduced in (6.78). We then set

$$Makro_{Basic} \underset{f:F \xrightarrow{\sim} Fin(FK) \rightarrowtail \Omega^{FK}}{\longrightarrow} \mathbf{Power}(Knot_{Basic})$$

with $F = Fun(Makro_{Basic}), FK = Fun(Knot_{Basic})$

and the product form

$$Knot_{Basic} \underset{Id}{\longrightarrow} \mathbf{Limit}(Makro_{Basic}, Basic). \quad (6.116)$$

How must we interpret this construction to get 'real' events back from this grouping structure?

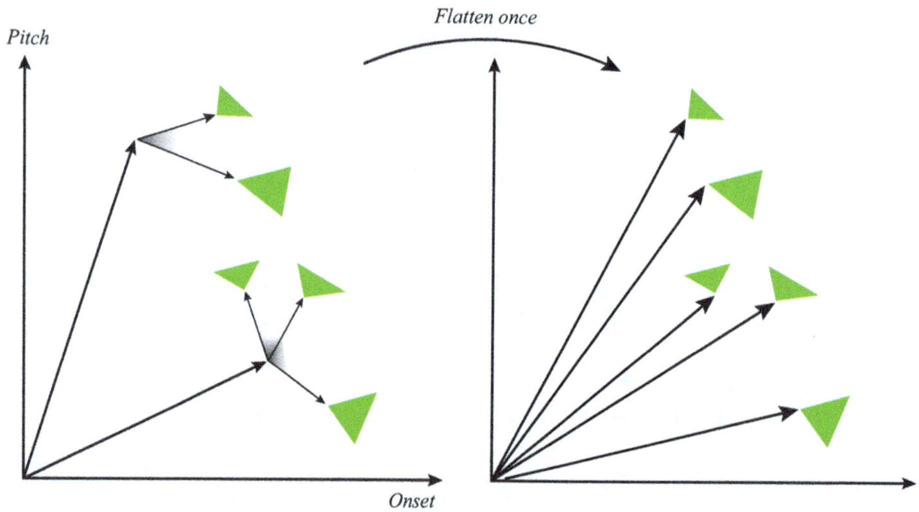

Fig. 6.12. The flattening of a macro-event in the mental onset-pitch domain.

Suppose that we have a module structure on the space of the *Basic* form (e.g. a vector space $\mathbb{R}^4$ of the piano events). To begin with, let a denotator $Knot_{Basic} = D$ of form $Knot_{Basic}$ have coordinates $(myEvent, \emptyset)$ with the denotator *myEvent* in the event space, e.g. in $\mathbb{R}^4$ in the above piano form. We introduce a *flattening* operation and define the $Makro_{Basic}$-formed denotator $Flatten(D) = \{D\}$. For a denotator $Knot_{Basic} = D = (myEvent, myMakro)$ with non-empty

Makro coordinate $myMakro = \{Knot_1, \dots Knot_r\}$ and knot events resp. Makro coordinates $myEvent_i$ resp. $myMakro_i$ for $Knot_i$, we set

$$Flatten(D) = \{(myEvent + myEvent_i, myMakro_i), i = 1, \dots r\}. \qquad (6.117)$$

The general definition of the flattening operation for a denotator $Makro_{Basic} = myMakro = \{D_1, \dots D_k\}$ is

$$Flatten(myMakro) = \bigcup_{i=1,\dots k} Flatten(D_i). \qquad (6.118)$$

If a Makro event *myMakro* ends up with empty Makro coordinates after a finite descent, the flattening operation also ends up, i.e. becomes stationary after an n-fold concatenation:

$$\begin{aligned} &Flatten^n(myMakro) = \\ &Flatten^{n+1}(myMakro) = \dots \\ &Flatten^{\infty}(myMakro). \end{aligned}$$

Since each element of $Flatten^{\infty}(myMakro)$ has empty Makro coordinate, we may view this as a finite set of *Basic*-formed events, i.e., a *local composition* in the terminology to be introduced later in Chapter 7. We shall come back to this construction in Section 21.3 and Section 39.2.5; refer to Figure 6.12 for a first intuition about Makro events.

6.8 Ordering on Forms and Denotators

Summary. In a number of important foundation modules, (linear and partial) orderings on systems of denotators can be defined. These orderings induce canonical (linear and partial) orderings on denotators, founded on these types of modules. The option is fundamental in defining receptive navigation tools. These tools are applied to representation and retrieval problems of general musical objects.

– Σ –

In this section, we want to look for generalizations of the naive order relation among forms and denotators. As in the "naive discussion" in Section 6.1.3, we consider exclusively regular forms here. This means that we are allowed to define order relation recursively on the level[40] of forms.

Recall from Section 6.1.3 that naive forms were ordered lexicographically as three-letter words, i.e., if two forms $F_1 = Nm_1 \longrightarrow Tp_1(Co_1)$ and $F_2 = Nm_2 \longrightarrow Tp_2(Co_2)$ were given, we had $F_1 < F_2$ iff for the first position in the "words" $Tp_1Co_1Nm_1$ and $Tp_2Co_2Nm_2$ where the letters differ, the letter of the first form is smaller than the corresponding letter of the second. We understood that the coordinator was itself presented as a word $C_1C_2 \dots C_m$ of forms in the compound case and the respective module name in the simple case. Functorial forms

$$F = Nm \underset{Id}{\longrightarrow} Tp(Co)$$

are a bit more complicated. We want to define orderings on forms relating to their names in a generative way. Recall that we require that form names be a key to forms, i.e., different forms must bear different names. Since names are strings of characters, they are linearly ordered, and we may order forms according to their names. But we do not want to do this in a completely arbitrary way. The form names should reflect the recursively given orderings of the form components: type, identifier, and coordinator. This is the general policy behind the following procedure. *So whenever we are given already ordered forms, the ordering must be read from the forms' names.*

[40] See Definition 5 in Section 6.5.

In this spirit, we would like to interpret them as four-letter words $TpCoIdNm$ with the lexicographic ordering[41] as above. The naming policy then means that if ever the three first letters define an ordering among two forms, then the form names should also follow that ordering. However, if the first three letters are fixed, synymous naming is of course still allowed. Let us inspect the different orderings on the four positions of $TpCoIdNm$. The name Nm poses no problem, we have the lexicographic ordering on usual $ASCII$ strings. The types are ordered by

$$\mathbf{Simple} < \mathbf{Syn} < \mathbf{Limit} < \mathbf{Colimit} < \mathbf{Power}, \tag{6.119}$$

similar to the ordering defined in (6.12). The difference is that for **Simple** type, we have all the modules as coordinators, whereas in the naive case, we only had four modules $\mathbf{CHR}, BIT, \mathbb{Z}$, and $\mathbb{R}$, giving rise to the simple types STRING, BOOLE, INTEGER, and FLOAT. Since we now have all the modules of **Mod** as coordinators of simple forms, we need an order relation $<_{\mathbf{Mod}}$ on these modules, more precisely: on all morphisms of this category, thereby englobing the modules as identities. This is supposed to extend the "naive" ordering, i.e.

$$\mathbb{Z}_2 <_{\mathbf{Mod}} \mathbb{Z} <_{\mathbf{Mod}} \mathbb{Z}\langle ASCII\rangle <_{\mathbf{Mod}} \mathbb{R}. \tag{6.120}$$

In other words: *The simple forms are ordered according to their coordinator modules.* We suppose to be given such an ordering $<_{\mathbf{Mod}}$ which extends the naive ordering. This is part of the basic, purely module-theoretic "level of reality"; see Section 6.8.1 for such an ordering.

Next, we want to look at the identifier, supposing by construction that we have fixed a form type Tp and a coordinator Co. This means that the form's identifier is a monomorphism $f : Fu \rightarrowtail Uf$, where $Fu = Fun(F)$ and the frame space $Uf = Uni(Tp, Co)$ is the universal functor construction associated with Tp and Co. If we have two forms of fixed form type Tp and coordinator Co, we have two monomorphisms $f_1 : Fu_1 \rightarrowtail Uf$ and $f_2 : Fu_2 \rightarrowtail Uf$ and these are canonically compared by the subobject relation[42], i.e. we set $Id_1 < Id_2$ iff the subobjects verify $f_1 < f_2$, i.e. iff there is a (necessarily mono) morphism $g : Fu_1 \rightarrowtail Fu_2$ such that $f_2 \cdot g = f_1$. Observe, however, that the latter relation is neither antisymmetric nor total in general, it is only a partial ordering[43]. The naive case appears here as related to the identity identifier. When generating new form names, one should however respect the identifier orderings. This is always possible since the form genealogy is a question of finitely many forms, and therefore, we may suppose that identifier ordering is carried over to name ordering.

We are left with the ordering of the coordinator in case it is not a module (compound type). If the fixed type is either **Syn** or **Power**, the coordinator is a form of smaller level, and we are done by recursion. We are thus left with a diagram coordinator $\mathbf{D}$. The naive setup was concerned with discrete diagrams where the vertex family $V(\mathbf{D}) = (C_1, \ldots C_m)$ was finite and linearly ordered by its indices. Comparing diagrams in the naive case was reduced to the lexicographic comparison of words $C_1 \ldots C_m$ of forms, where the ordering among the forms C_i was supposed to be given by recursion. To generalize that situation, we suppose that

- by recursion, the vertex forms are already linearly ordered by their names—let us index them for notational comfort by C_i.

We are therefore able to compare two such diagrams $\mathbf{D}_1$ and $\mathbf{D}_2$ with respect to their vertex forms $C_{1,i}$ and $C_{2,j}$: We build the words $C_{1,0}C_{1,1} \ldots$ and $C_{2,0}C_{2,1} \ldots$ and compare them lexicographically. Upon this ordering, we further differentiate according to the involved diagram arrows. We may thus suppose that $V(\mathbf{D}_1) = V(\mathbf{D}_2) = (C_i)_{0 \leqslant i}$.

We further suppose that the set of morphisms $f : C_i \to C_j$ is linearly ordered for all vertex indices i and j. This suffices to terminate the construction. Fix a pair i, j of vertex indexes. We are going to build a word with the morphisms $f : C_i \to C_j$ of our diagrams. To begin with, sort these words by the number of their arrows, and therefore suppose that the arrow number $s(i, j)$ is the same for both diagrams. Let $mor^d_{i,j} = f^d_{i,j,1} f^d_{i,j,2} \ldots$ be the word of diagram morphisms $f^d_{i,j,s} : C_i \to C_j, s = 1, \ldots s(i, j)$ in the

[41] Concerning lexicographic orderings, see Appendix C, Example 94.
[42] See Appendix G.
[43] It is reflexive and transitive, see Appendix C.2

given linear morphism ordering of diagram $\mathbf{D}_d, d = 1, 2$. We may now arrange the words $mor^1_{i,j}, mor^1_{i,j}$ in their lexicographic order. Since the couples i, j of vertex indices are ordered lexicographically, we can build a word for each diagram $\mathbf{D}_d, d = 1, 2$ whose letters are the $mor^d_{i,j}$, in the given double index ordering: $mor^d = mor^d_{0,0} mor^d_{0,1} \ldots$. It is now clear how to compare the words mor^1 and mor^2 lexicographically, and we are done with the ordering of forms.

Let us now look at possible orderings among denotators. In view of the components

$$N(D) : A(D) \rightsquigarrow F(D)(CT(D))$$

of a denotator D, we have a similar situation as with forms. We again order denotators lexicographically according to the four-letter wording $F(D)A(D)CT(D)N(D)$. Since by the preceding, forms, addresses, i.e., modules, and names are already ordered, we are left with the question of ordering coordinates.

According to the lexicographic ordering, we may thus look at two denotators D_1 and D_2 with identical form F and address M and with coordinates Co_1 and Co_2, both elements of the functor value set $M@Fun(F)$. Since the latter functor is identified to a subfunctor of the universal construction $U = U(Type, Coordinator)$ of the form, we can restrict our question to orderings on value sets of such universal construction functors. We have to investigate the situations $M@N$, $M@Fun(\mathbf{Syn}(C))$, $M@Lim(\mathbf{D})$, $M@Colim(\mathbf{D})$, and $M@\Omega^{Fun(C)}$ for a module N, a diagram $\mathbf{D}$, or a coordinator form C.

In case of simple type, $M@N$ bears an ordering by hypothesis on the category **Mod**. For **Syn** type, we have $M@Fun(\mathbf{Syn}(C)) = M@Fun(C)$, and we are done by recursion. For the power type $\Omega^{Fun(C)}$, we have

$$M@\Omega^{Fun(C)} = Sub(@M \times Fun(C))$$

which is naturally provided with the ordering among subobjects as already discussed above. However, this ordering is not related to the recursively given ordering on (the value sets of) $Fun(C)$. For the frequently used subfunctor $Fin(Fun(C)) \rightarrowtail \Omega^{Fun(C)}$, the subobject ordering can be refined to an ordering which is also linear in case the ordering on $Fun(C)$ is. In fact, by Proposition 77 in Appendix C.2, every set $A@Fin(Fun(C))$ is ordered by a partial/linear ordering $\preceq$ induced from the partial/linear ordering on $A@Fun(C)$. The ordering $\preceq$ refines the inclusion ordering among finite subsets, in other words, the natural inclusion $Fin(Fun(C)) \rightarrowtail \Omega^{Fun(C)}$ with the subobject orderings admits a refinement on $Fin(Fun(C))$ which is linear if the ordering on $Fun(C)$ is. Whether this ordering on $Fin(Fun(C))$ can be extended to a partial ordering on all of $\Omega^{Fun(C)}$ that includes the subobject ordering, is an open question. If the contrary is not stressed we shall always use the $\preceq$ ordering on $Fin(H)$ for a functor $H \in \mathbf{Mod}^@$ whose value sets are ordered.

For **Limit** type, we know from Appendix G.2.1 that

$$M@Lim(\mathbf{D}) \subset \prod_i M@Fun(C_i)$$

with the above notation of the diagram's vertexes C_i. The functor values $M@Fun(C_i)$ are ordered by recursion, and we may consider the lexicographic ordering on the product, restricted to the limit subset, under the hypothesis that the index set is also linearly ordered, see Example 94 in Appendix C.

We are left with the **Colimit** type which is less comfortable to handle than the other types. In fact, we have a surjection

$$\coprod_i M@Fun(C_i) \twoheadrightarrow M@Colim(\mathbf{D})$$

defined by the equivalence relation that is generated by the diagram's arrows. The left set carries a canonical ordering as described in the naive discussion in Section 6.1.3. We define it by use of the linear ordering among the coordinator forms of $\mathbf{D}$, and by the recursively given orderings on the functor sets $M@Fun(C_i)$. If we are to carry over this data to the colimit, we have to turn the above projection into an order-preserving map, i.e. we define the ordering on the colimit set as the partial ordering generated by the image relation of the coproduct's ordering. But observe that this will—in general—be something much larger than the image relation since the latter is not transitive, in general! This terminates the construction of orderings among denotators as far as this is possible.

We should make a final comment on the sense of ordering forms and denotators. This is not so much a purely mathematical exercise. After all, mathematicians know that any set can be well-ordered, at least in acceptance of certain common axioms (see Appendix C.1). The point is much more a constructive approach which lends itself to data base management systems and the related successive growth of denotator arsenals—as a function of a dynamically expanding knowledge base. This perspective stresses the "natural" ordering and not an artificial one which may be built exclusively upon the axiom of choice. The only presence of choice in our context is the historical appearance of concepts, i.e. of forms and denotators, in the development of theory and empirical material, but see Chapter 18 for an extensive study of "what is the case" and what is not. This is taken into account on the level of the ordering within the fundamental category **Mod** of modules and affine morphisms. In this sense we should also understand the following set of applications.

6.8.1 Concretizations and Applications

Summary. We give concrete order relations and applications to musical situations.

– Σ –

Modules and Morphisms. We start this section with a discussion of possible order relations on the category **Mod** of modules and their affine morphisms. Modules in **Mod** appear in two functions: as coordinators of simple forms and as addresses of denotators. When building compound forms and denotators on these forms, we may apply existing orderings on the already used modules for the compound structures. Only if a new module intervenes in the knowledge extension process is it necessary to insert it relatively to the already ordered module set. In other words, the recursive construction of partial orderings is also dynamic. Before we really need a module, it is not necessary to care about its order position. The only essential point here is that one should not reorder modules after dynamical extension: Once an order relation $M <_{\mathbf{Mod}} N$ is defined, it should persist for ever. Otherwise, reusable retrieval on database management systems break down.

For the purpose of ordering modules, we want to observe the following principle:

Principle 1 *Order* first *modules from a skeleton*[44] *of* **Mod**, *and* then *the objects within each isomorphism class. Then order modules lexicographically according to the skeleton and the member of the isomorphism class, i.e. as an ordered coproduct of ordered isomorphism classes.*

For example, we start the skeleton by the few examples of hitherto basic modules and order them as follows[45]:

$$\begin{aligned}\emptyset <_{\mathbf{Mod}} 0_{\mathbb{Z}}(=\mathbb{Z}_1) <_{\mathbf{Mod}} \mathbb{Z}_2 \ldots <_{\mathbf{Mod}} \mathbb{Z}_n <_{\mathbf{Mod}} \mathbb{Z}_{n+1} \ldots \\ <_{\mathbf{Mod}} \mathbb{Z} <_{\mathbf{Mod}} \mathbb{Z}\langle ASCII\rangle \\ <_{\mathbf{Mod}} \mathbb{Q} <_{\mathbf{Mod}} \mathbb{Q}^2 \ldots <_{\mathbf{Mod}} \mathbb{Q}^n <_{\mathbf{Mod}} \mathbb{Q}^{n+1} \ldots \\ <_{\mathbf{Mod}} \mathbb{R} <_{\mathbf{Mod}} \mathbb{R}^2 \ldots <_{\mathbf{Mod}} \mathbb{R}^n <_{\mathbf{Mod}} \mathbb{R}^{n+1} \ldots \end{aligned} \quad (6.121)$$

To give an example of the ordering within an isomorphism class, we look at a fixed positive power $\mathbb{R}^n$. As usual in linear algebra, the vectors of this module are viewed as columns, i.e. $n \times 1$ matrices. We thus write $n = n \times 1$ and identify $\mathbb{R}^n$ with $\mathbb{R}^{n\times 1}$. In this spirit, we arrange the factorizations $n = r \times s$ of n by increasing second factors $1 < s_1 < s_2 < \ldots n$ and obtain an ordering among the following matrix $\mathbb{R}$-modules isomorphic to $\mathbb{R}^n$, inserted between (classes of) modules $\mathbb{R}^{n-1}$ and $\mathbb{R}^{n+1}$:

[44] See Appendix G, Example 117.

[45] In this sequence, the modules $\mathbb{Z}_n$ are viewed as $\mathbb{Z}$-modules, i.e. we should write more correctly $(\mathbb{Z}_n)_{[\mathbb{Z}]}$, but we avoid this clumsy notation whenever possible.

$$\mathbb{R}^{1\times(n-1)} <_{\mathbf{Mod}} \mathbb{R}^{n\times 1} <_{\mathbf{Mod}} \mathbb{R}^{r_1\times s_1} <_{\mathbf{Mod}} \ldots \mathbb{R}^{1\times n} <_{\mathbf{Mod}} \mathbb{R}^{(n+1)\times 1}. \quad (6.122)$$

Before further discussing orderings on forms we have to look at orderings on individual modules since orderings on diagrams depend on orderings on sets of morphisms between functors, and this includes ordering sets of affine homomorphisms between modules which are also, under certain conditions, modules[46].

Ordering on individual modules is quite problematic since nothing functorial can be expected. We have to select specific modules and give them *individual* orderings (without regard to compatibility with affine transformations such as automorphisms). However, a minimum of universality should be observed since one should not produce more incompatibilities than necessary. For example, the identification of forms related to direct sums of modules in point 5 of Section 6.4.1 should be compatible with order relations[47]. Suppose therefore that we have a product form $F \times G$ of two simple forms $F \underset{Id}{\longrightarrow} \mathbf{Simple}(M)$ and $G \underset{Id}{\longrightarrow} \mathbf{Simple}(N)$ over modules M, N with respective orderings[48] $A@ <_M, A@ <_N$ at an address A. Then, according to the preceding discussion, A-addressed denotators $D : A \rightsquigarrow F \times G(Co)$ of $F \times G$ with $Co \in A@M \times A@N$ are ordered lexicographically following $A@ <_M$ and then $A@ <_N$. Hence, if these denotators are identified with A-addressed denotators in $F \oplus G$, we should order the direct sum $A@(M \oplus N)$ by the lexicographic ordering of its summands under the canonical bijection $A@M \oplus N \xrightarrow{\sim} A@M \times A@N$. We therefore have to take the lexicographic ordering from the summands $A@M, A@N$ whenever we construct an ordering on a direct sum $A@(M \oplus N)$. The common case is zero address, and then, $0@M$ identifies to M. So we have to take care of orderings on modules M and their direct sums first. To begin with, we have to order coefficient rings[49]. We have already listed some important rings in (6.121) and will now define orderings on these prototypes:

- $\mathbb{Z}_0$ and $\mathbb{Z}_{p^k} = \{\overline{0}, \ldots \overline{p^k - 1}\}$, $p =$ prime and $k = 1, 2, 3, \ldots : \overline{i} < \overline{j}$ iff $i < j$.
- For general n, let $n = p_1^{k_1} \cdots p_r^{k_r}$ be the (unique) prime decomposition with increasing prime factors and non-zero exponents (see Appendix E.2). Then we have the Sylow decomposition (Appendix C.3.4) and a canonical isomorphism $\mathbb{Z}_n \xrightarrow{\sim} \mathbb{Z}_{p_1^{k_1}} \oplus \ldots \mathbb{Z}_{p_r^{k_r}}$. On this direct sum, we take the lexicographic ordering based on the previous definition of the orderings on the direct factors.
- $\mathbb{Z}$: usual ordering.
- $\mathbb{Z}\langle ASCII\rangle$: Take the lexicographic ordering on the basis set W of $ASCII$-words, then the lexicographic ordering as defined in Appendix C, Example 94, via the identification[50] of $\mathbb{Z}$-modules $\mathbb{Z}\langle ASCII\rangle = \mathbb{Z}^{(W)}$.
- $\mathbb{Q}$: usual ordering.
- $\mathbb{R}$: usual ordering.

Direct Sums. Generalizing the above case with $\mathbb{Z}\langle ASCII\rangle$, we suppose given a family of indices I, well-ordered by $<$, as well as for each index $i \in I$ a module M_i with an ordering $<_i$. Then the direct sum $\bigoplus_I M_i$ is ordered lexicographically according to the index ordering. This includes the finite direct sum R^n of n copies of an ordered ring, $R = \mathbb{Z}_n, \mathbb{Z}, \mathbb{Q}, \mathbb{R}$ for example. Again we should pay attention to the fact that isomorphic modules may not have isomorphic orderings. The latter are strictly related to the identity and construction of specific modules.

Matrix Modules. In (6.122), we have ordered isomorphic copies of $\mathbb{R}^n$ for different factorizations $n = r \times s$ as matrix modules $\mathbb{R}^{r\times s}$. Here, we have to order these copies themselves, and this will be done in the canonical way: Identify $\mathbb{R}^{r\times s}$ canonically with $(\mathbb{R}^r)^s$, order the column summands $\mathbb{R}^r$ and then the total sum as defined above for direct sums. The same ordering procedure works for any matrix module over a ring R which is provided with an ordering, instead of $\mathbb{R}$.

---

[46] see Appendix E for more details on this somewhat delicate subject!

[47] Observe, however, that switching from **Limit** to **Simple** type changes the order position of the underlying form.

[48] The following symbolism $A@ <_F$ for an ordering on a value set $A@F$of a functor F in $\mathbf{Mod}^{@}$ does not mean that the ordering is functorial, i.e. the functor does not necessarily evaluate on the *category* of ordered sets.

[49] Mathematicians should pay attention not to confuse the orderings in this section with the ordering concept for ordered rings.

[50] See Appendix E.

Ordering on Morphisms. If we are given two forms F and G with functors which are represented by modules, $Fu(F) = @M$ and $Fu(G) = @N$, then the morphisms $F \to G$ correspond canonically[51] to affine morphisms $e^n \cdot f : M \to N$. In particular, if M and N are the vector spaces $M = \mathbb{R}^r$ and $N = \mathbb{R}^s$, we canonically identify $M@N = e^N \cdot Lin(M, N)$ with $\mathbb{R}^{r \times s} \oplus \mathbb{R}^s$, ordered as a direct sum of the ordered module $\mathbb{R}^{r \times s}$ with the ordered matrix space $\mathbb{R}^s$. The same construction is valid for any free modules $M = R^r$ and $N = R^s$ over a ring R which is provided with an ordering, instead of $\mathbb{R}$.

With the previous constructions we are in a state of ordering diagrams of forms having functors which are represented in free modules over rings which already carry an ordering. And the last construction yields orderings on functor values $A@M$ at address A if both, A and M, are free over the coefficient ring which is provided with an ordering.

This being true, it is a good exercise to apply the mathematical constructions to concrete musical examples and to give them interpretations in terms of musical meaning.

Example 3 Ordering $\mathbb{Z}_n$.
Consider the form $PiMod_n$ of n-pitch classes as defined in (6.41). Its A-addressed denotators live in the set $A@\mathbb{Z}_n$ and we have the ordering of $\mathbb{Z}_n$ on the zero-addressed denotators. For $n = 12$, we have the prime decomposition $12 = 2^2 3$ and therefore inherit the lexicographic ordering of $\mathbb{Z}_4 \oplus \mathbb{Z}_3$ via the canonical bijection

$$\mathbb{Z}_{12} \xrightarrow{\sim} \mathbb{Z}_4 \oplus \mathbb{Z}_3 : x \mod 12 \mapsto (x \mod 4, x \mod 3).$$

This gives the following ordering on zero-addressed denotators

$\mathbb{Z}_4 \oplus \mathbb{Z}_3$:	(0,0)	(0,1)	(0,2)	(1,0)	(1,1)	(1,2)	(2,0)	(2,1)	(2,2)	(3,0)	(3,1)	(3,2)
$\mathbb{Z}_{12}$:	0 <	4 <	8 <	9 <	1 <	5 <	6 <	10 <	2 <	3 <	7 <	11

which starts from 0 with two major third steps, then jumps by one semitone and repeats the major third steps from there etc. If played in notes from middle C on, this look like four arpeggiated augmented triads[52], as shown in Figure 6.13.

Fig. 6.13. The ordering on 12-pitch classes according to the standard procedure described in this section.

If instead we look at self-addressed denotators of $PiMod_n$, we have to consider the elements (b, a) of the ordered module $\mathbb{Z}_n \oplus \mathbb{Z}_n$ which parametrize the affine morphisms $e^a \cdot b$. In case of $n = 12$, we have the self-addressed tones studied in *[802]*. The ordering thus starts with the so-called *constant* self-addressed tones $e^0 \cdot 0 < e^4 \cdot 0 < \ldots e^{11} \cdot 0$ and ends with the *inversion* tones $e^0 \cdot 11 < e^4 \cdot 11 < \ldots e^{11} \cdot 11$, the last being the inversion between B and C, see Section 7.2.1.

Example 4 Ordering Note Groups from a "Chord Perspective".
Intuitively speaking, we may look at (finite) sets of notes on a score and try to order them for lexical retrieval or/and for listing them in a data base management system. But we may also look at these objects from a harmonic or metrical point of view. This means that we do not only consider them as sets of tone events, but with respect to their harmonic or metric content. Let us develop on this scope by use of forms, orderings on

[51] By Yoneda's Lemma, see Appendix G.2.
[52] See Section 11.3.5 for chord classification.

forms and the corresponding denotator systems. For the sake of transparency we only look at two parameters, onset and pitch. We need the three corresponding forms introduced in Section 6.6:

$$\begin{aligned}
&Onset \underset{Id}{\longrightarrow} \mathbf{Simple}(\mathbb{R}),\\
&Pitch \underset{Id}{\longrightarrow} \mathbf{Simple}(\mathbb{R}),\\
&Onset \oplus Pitch \underset{Id}{\longrightarrow} \mathbf{Simple}(\mathbb{R}^2).
\end{aligned}$$

Corresponding to first and second projection $p_1 : \mathbb{R}^2 \to \mathbb{R}$ and $p_2 : \mathbb{R}^2 \to \mathbb{R}$, we have two morphisms of forms

$$p_1 : Onset \oplus Pitch \to Onset \text{ and} \tag{6.123}$$
$$p_2 : Onset \oplus Pitch \to Pitch. \tag{6.124}$$

Recall that $Fun(Onset) = Fun(Pitch) = @\mathbb{R}$ and $Fun(Onset \oplus Pitch) = @\mathbb{R}^2$ and consider the forms of finite sets in the three preceding forms:

$$Fin(Onset) \underset{Fin(@\mathbb{R}) \rightarrowtail \Omega^{@\mathbb{R}}}{\longrightarrow} \mathbf{Power}(Onset), \tag{6.125}$$
$$Fin(Pitch) \underset{Fin(@\mathbb{R}) \rightarrowtail \Omega^{@\mathbb{R}}}{\longrightarrow} \mathbf{Power}(Pitch), \tag{6.126}$$
$$Fin(Onset \oplus Pitch) \underset{Fin(@\mathbb{R}^2) \rightarrowtail \Omega^{@\mathbb{R}^2}}{\longrightarrow} \mathbf{Power}(Onset \oplus Pitch). \tag{6.127}$$

Corresponding to the above projections and by Lemma 101 in Appendix G.2.1, we have two projections

$$Fin(p_1) : Fin(Onset \oplus Pitch) \to Fin(Onset), \tag{6.128}$$
$$Fin(p_1) : Fin(Onset \oplus Pitch) \to Fin(Pitch), \tag{6.129}$$

and corresponding diagrams $\mathbf{D}_{p_1}$ and $\mathbf{D}_{p_2}$. We then have limit forms defined by these diagrams:

$$L(\mathbf{D}_{p_1}) \underset{Id}{\longrightarrow} \mathbf{Limit}(\mathbf{D}_{p_1}), \tag{6.130}$$
$$L(\mathbf{D}_{p_2}) \underset{Id}{\longrightarrow} \mathbf{Limit}(\mathbf{D}_{p_2}). \tag{6.131}$$

Let us consider order relations of denotators in this context. According to the general theory, we first need an ordering on the diagram's vertexes and arrows. Our diagram scheme is just two points and one connecting arrow. We order the points as follows: $\underset{0}{\bullet} \leftarrow \underset{1}{\bullet}$, i.e. we have $C_0^1 = Fin(Onset), C_1^1 = Fin(Onset \oplus Pitch)$ in $\mathbf{D}_{p_1}$ and $C_0^2 = Fin(Pitch), C_1^2 = Fin(Onset \oplus Pitch)$ in $\mathbf{D}_{p_2}$. Therefore, the vertex words satisfy $C_0^1 C_1^1 < C_0^2 C_1^2$ since forms verify $Onset < Pitch$; in fact, the latter differ only by their names, and then it is clear.

This implies that $L(\mathbf{D}_{p_1}) < L(\mathbf{D}_{p_2})$, and therefore denotators on $L(\mathbf{D}_{p_1})$ precede denotators on $L(\mathbf{D}_{p_2})$. For the first form $L(\mathbf{D}_{p_1})$, an A-addressed denotator $D \in A@L(\mathbf{D}_{p_1})$ looks like this: We have $D = (D_0, D_1)$ with these specifications:

$$D_0 \in A@Fin(@\mathbb{R}) = Fin(A@\mathbb{R}), \tag{6.132}$$
$$D_1 \in A@Fin(@\mathbb{R}^2) = Fin(A@\mathbb{R}^2), \tag{6.133}$$
$$A@Fin(p_1)(D_1) = D_0. \tag{6.134}$$

Correspondingly, in the second form $L(\mathbf{D}_{p_2})$, a denotator $D = (D_0, D_1)$ stays in the same mathematical space but verifies the relation

$$A@Fin(p_2)(D_1) = D_0. \tag{6.135}$$

To understand the meaning of such denotators, let us—as usual—start with zero-addressed denotators and see Figure 6.14 for an illustration. In this case, the denotator $D \in 0_{\mathbb{Z}}@L(\mathbf{D}_{p_1})$ consists of two finite sets $D_0 \subset \mathbb{R}$,

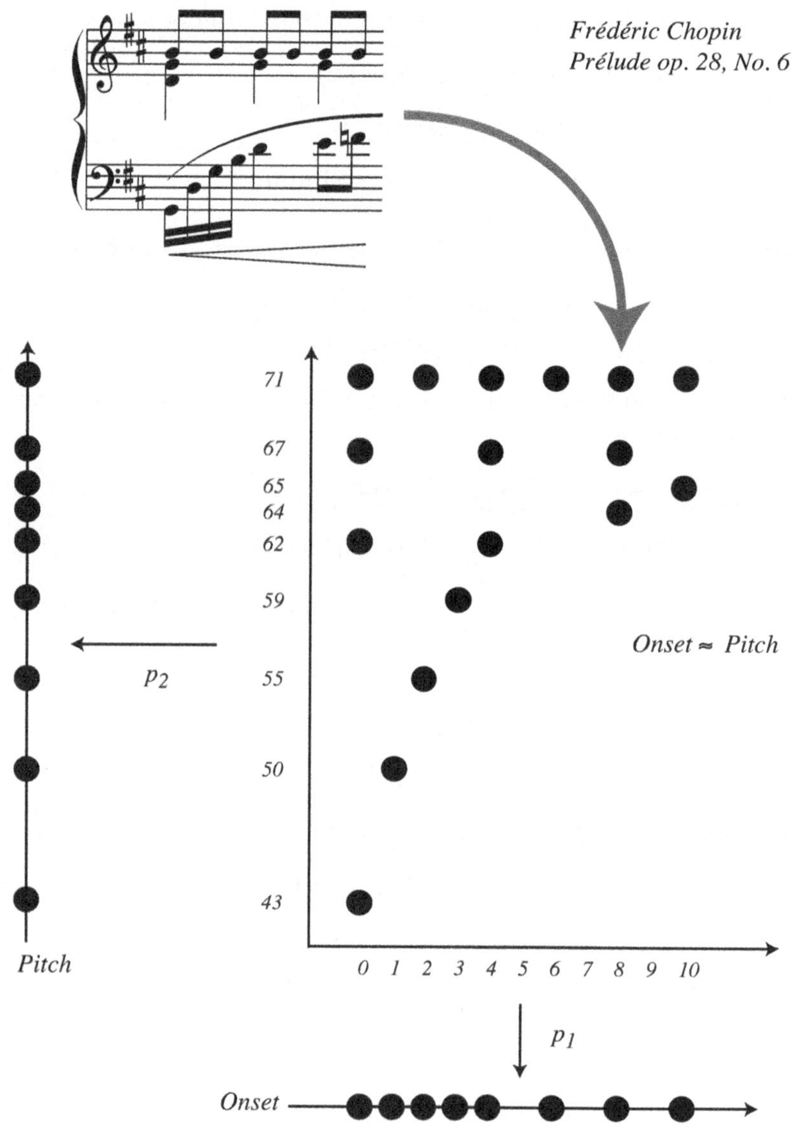

Fig. 6.14. A bar of Frédéric Chopin's Prélude op. 28, No. 6 is projected to an *Onset*$\oplus$*Pitch*-formed denotator in the plane. Its two one-dimensional projections onto the *Onset* and *Pitch* forms reveal metrical/rhythmical and harmonic aspects of the plane denotator, respectively.

$D_1 \subset \mathbb{R}^2$ such that D_1 projects onto D_0 when we forget about the *Pitch* coordinates. Correspondingly, a denotator $D \in 0_{\mathbb{Z}}@L(\mathbf{D}_{p_2})$ consists of two finite sets $D_0 \subset \mathbb{R}$, $D_1 \subset \mathbb{R}^2$ such that D_1 projects onto D_0 when we forget about the *Onset* coordinates.

We have therefore two denotators from one subset D_1 in the onset-pitch plane. The first gives information about the relation between the onset configuration of D_1 via first projection $p_1(D_1)$ to the onset axis. The second tells about the pitch configuration of D_1 via second projection $p_2(D_1)$ to the pitch axis. Roughly speaking, the first deals with metrical or rhythmical aspects of D_1, whereas the second deals with its harmonic aspects. Our order principles imply that metrical or rhythmical aspects precede harmonic aspects if ever we are going to allocate such denotators in a library or in a data base system. So we are left with comparison of denotators within one fixed form.

Consider now two A-addressed denotators $D^i = (D^i_0, D^i_1)$, $i = 1, 2$ of form $L(\mathbf{D}_{p_1})$ (for the second projection form $L(\mathbf{D}_{p_2})$, we have analogous explications). According to the general discussion of **Limit** and **Power** orderings, we have $D^1 < D^2$ iff either $D^1_1 \prec D^2_1$ or $D^1_1 = D^2_1$ and $D^1_2 \prec D^2_2$. Each of these orderings $\prec$ happens to be the type defined according to Proposition 77 in Appendix C.2. This means that we first look at the ordering among metrical/rhythmic aspects of denotators, and after they have been identified, we turn to the tones behind these aspects. Recall that $D^1_1 \prec D^2_1$ means that on the onset axis either $D^1_1 \subset D^2_1$ or $max(D^1_1 - D^2_1) < max(D^2_1 - D^1_1)$.

There may be many different configurations with the same metrical/rhythmic aspect. What does it mean that $D^1 < D^2$ when their metrical/rhythmic aspects coincide? We then have $D^1_1 = D^2_1$ and $D^1_2 \prec D^2_2$, where the latter is induced by the lexicographic ordering on $A@\mathbb{R}^2$. This means that for any $x = (x_1, x_2) \in (A@\mathbb{R})^2$ which stays in $D^1_1 - D^2_1$, there is $y = (y_1, y_2)$ in $D^2_1 - D^1_1$ such that either its onsets coincide and pitches verify $x_2 < y_2$, or onsets verify $x_1 < y_1$.

6.9 Concept Surgery and Denotator Semantics

Summary. Denotators can be used as dynamic tools for concept production. This shows that denotator semantics must be understood as an open-ended specification.

– Σ –

Denotators are very useful for formally repertorizing ethnomusicological data and facts. The possibility to apply character strings as coordinates opens the way to incomplete semantics. For example, it is not necessary to misrepresent or over-interpret the pitch data of a note in an ethnological context in favor of the European lexemata of pitch when denotating an Indian melody. However, when building a pitch model for that context, it is still possible to map the character strings to numbers, such as European pitch classes, MIDI numbers or whatever may be relevant to the given scientific project[53]. Whereas a European score symbol for dynamics is recorded as a character string coordinates by *'mf'*, *'pp'*, the user may interpret these coordinates by numerical evaluation for performance purposes. But until that realization, the denotators keep their full potential of "European ethnological context of notation". This is no luxury for ethnomusicological research, quite the contrary: Without such an open denotator system, the retrieval of ethnomusicological data would fail from the beginning since the original information would be distorted in an irreversible way if the record was meant to be a reliable first order source.

This procedure may be generalized to the end of substituting character strings by entire denotators in order to 'fill up symbolic strings' with more explicit denotators. This substitution process is understood as yielding a 'decoding' method for 'deepening meaning' of signs. Let us explicate the technical aspects. To begin with, one is looking for forms that express the above 'symbolic' representation of semantically open data. Suppose that the symbols we are interested in stem from the set **CHR** of *ASCII*-character strings. We want to consider European dynamic signs with a minimum of superfluous information, more precisely, with an intrinsic character of a non-resolved semantics. This is given by circular forms! Let [**CHR**] be the constant functor[54] of **CHR** and $@\mathbb{R}$ the Yoneda functor of the *ring* $\mathbb{R}$, applied to the coefficient ring of a given address A_R, i.e. $A_R @ @\mathbb{R} = R @ \mathbb{R}$. We consider the form

$$'EuroDyn' \underset{Id_{@\mathbb{R}\times[\mathbf{CHR}]}}{\longrightarrow} Syn(EuroDyn). \tag{6.136}$$

Next, let us consider the technical aspect of modeling this symbolic status on more concrete parameters. We still stick to our *EuroDyn* example to make the general idea clear. A more concrete parametrization of

[53] This is precisely what the RUBATO analysis and performance platform makes possible in the case of European score denotators such as dynamics *mf*, *pp* etc., see Section 41.5 for details.

[54] See Example 123 in Section G.1.3.

loudness would be the simple form $Loudness \underset{Id}{\longrightarrow} \mathbf{Simple}(\mathbb{R})$ (6.69). Giving the symbolic denotators of form *EuroDyn* a more concrete meaning in terms of form *Loudness* means defining a "meaning" or "modeling" morphism of forms $m : EuroDyn \to Loudness$. This means defining a map $A@m : R \circledcirc \mathbb{R} \times \mathbf{CHR} \to A@\mathbb{R}$ which is functorial in address A over the ring R, i.e., for any address change $c : B \to A$ over the scalar restriction $\gamma : S \to R$, we must have a commutative diagram

$$\begin{array}{ccc} R \circledcirc \mathbb{R} \times \mathbf{CHR} & \xrightarrow{A@m} & A@\mathbb{R} \\ {\scriptstyle \gamma\circledcirc\mathbb{R}\times Id}\downarrow & & \downarrow{\scriptstyle c@\mathbb{R}} \\ S \circledcirc \mathbb{R} \times \mathbf{CHR} & \xrightarrow{B@m} & B@\mathbb{R} \end{array} \tag{6.137}$$

of set maps. For example, such a morphism can be defined if we know just what happens on the zero address. Since we have a unique address change $! : B \to 0_S$ of S-modules, we get the map $B@m$ by commutativity of the diagram

$$\begin{array}{ccc} S \circledcirc \mathbb{R} \times \mathbf{CHR} & \xrightarrow{0_S@m} & 0_S@\mathbb{R} \\ {\scriptstyle =}\downarrow & & \downarrow{\scriptstyle !@\mathbb{R}} \\ S \circledcirc \mathbb{R} \times \mathbf{CHR} & \xrightarrow{B@m} & B@\mathbb{R} \end{array} \tag{6.138}$$

and therefore, any modeling transformation is determined by its data on the zero addresses. But we have an isomorphism of functors $\circledcirc\mathbb{R} \times [\mathbb{R}] \xrightarrow{\sim} 0_?@\mathbb{R}$ on the category **Rings**, the latter evaluating to $0_S@\mathbb{R}$ at ring S. But then, the morphism m, restricted to the zero addresses defines a morphism between set-valued functors on **Rings**. Since the identity is the only endomorphism of $\mathbb{R}$, we easily see that m is defined by the product $Id_{\circledcirc\mathbb{R}} \times m_0$ for a set map $m_0 : \mathbf{CHR} \to \mathbb{R}$. In other words, we have this result:

Proposition 1 *The modeling morphisms $m : EuroDyn \to Loudness$ are in one-to-one correspondence with the set maps $m_0 : \mathbf{CHR} \to \mathbb{R}$, defined by commutative diagram* (6.138).

The set maps in Proposition 1 are precisely what one would define in an ethnomusicological modeling of dynamics. The functorial extension is for free in this case. Conceptually, what happens is that we have mapped a circular denotator into a regular one and thereby deepened the semantic short circuit of circularity into regular meaning in terms of real numbers.

This example shows that though it is necessary to work with incomplete concepts such as *EuroDyn*, we should also be capable of making their signification more precise *on demand* in the language of denotators. We want to complete this technique in order to have the generic procedure at hand. What follows will be called *concept surgery* since it deals with replacement of given forms by "better" ones. We want to start with the general situation of a given corpus of forms. All these forms are constructed by use of more or less large coordinators which are diagrams of forms if they are compound, i.e. built upon other forms. Suppose now that a particular form *ArchaicForm* is connected to a "better" form *ModelForm* by means of a "modeling" morphism, i.e. we are given a morphism $model : ArchaicForm \to ModelForm$. Denote by $\mathbf{D}_{model}$ the diagram defined by this morphism. Define a new form.

$$BetterForm \underset{Id}{\longrightarrow} \mathbf{Limit}(\mathbf{D}_{model}) \tag{6.139}$$

which displays two projections

$$\begin{aligned} p_{archaic} &: BetterForm \to ArchaicForm, \\ p_{model} &: BetterForm \to ModelForm, \end{aligned} \tag{6.140}$$

related by $model \cdot p_{archaic} = p_{model}$. We then replace every morphism of forms $g : G \to ArchaicForm$ by the universal morphism $Better^g : G \to BetterForm$ defined by the pair $(g, model \cdot g)$ of morphisms into $\mathbf{D}_{model}$:

$$\begin{array}{ccc} G & \xrightarrow{g} & ArchaicForm \\ {\scriptstyle Better^g}\downarrow & & \downarrow{\scriptstyle model} \\ BetterForm & \xrightarrow{p_{model}} & ModelForm \end{array} \tag{6.141}$$

Similarly, we replace every morphism $g : ArchaicForm \to G$ of forms by its composition $Better_g = g{\cdot}p_{archaic}$, and we obtain a commutative (!) diagram

$$\begin{array}{ccc} H & \xrightarrow{h} & ArchaicForm \\ {\scriptstyle Better^h}\downarrow & & \downarrow{\scriptstyle g} \\ BetterForm & \xrightarrow{Better_g} & G \end{array} \tag{6.142}$$

of "bad and better" morphisms of forms for any morphisms $g : ArchaicForm \to G$ and $h : H \to ArchaicForm$, i.e. $g \cdot h = Better_g \cdot Better^h$ which means that the "deviation" of morphisms to $BetterForm$ does not disturb any of the given factorizations through the old $ArchaicForm$. In this surgical intervention, we have excised the $ArchaicForm$ from all its appearances in other forms, and replaced its role by that of a new $BetterForm$. So this surgery makes it possible to improve the whole concept body of forms and denotators once one form has found an improved version. Set more formally, we may now replace every form $OldForm$ where $ArchaicForm$ is involved somewhere in its recursive tree by a new form

$$model \pitchfork OldForm, \tag{6.143}$$

defined by the surgically inserted $BetterForm$ from the $model$ morphism.

Despite this semantic surgery, the denotator system has its very strict limits and requires a supersystem of connotation to grasp deeper layers of meaning. Already with the above example of ethnomusicological signification it became evident that the denotator system is only a signifier surface pointing at interpretative, performative, and emotional, resp. social, meaning.

Moreover, denotators are not sensible to what really exists in music, as a member of the repertoire of a given culture, for example. In fact, this is an essential fact about music: In the real world, denotators are far from realized in their complete potentiality. Beethoven's Fifth Symphony is not just one of a large repertoire of given denotators! Its existence has far-reaching consequences on the theory and practice of music and its meaning. This is the reason why denotators must be deepened to yield this existential aspect underlying music. We shall develop this subject in full detail in Chapter 18.

Part III

Local Theory

7

Local Compositions

Ich kann ein lokales Objekt, eine lokale Struktur nicht abstrakt *schaffen;*
so elementar ich sie auch immer denke: ich bin bereits gezwungen,
sie zu kodieren, um sie niederschreiben zu können
– im Grunde fällt der Lokalstruktur auch
eine alphabetische Rolle zu (...)
Pierre Boulez [138, II, p.66]

Summary. Local compositions are introduced as elementary objects of music. They derive from powerset denotators. It is shown that all denotators may be transformed into local compositions. A special type of local compositions can be defined from a fixed set of denotators of a given "ambient space". With these so-called *objective* local compositions the problem of universal construction of new concepts from given ones cannot be solved. This imposes a deeper theory of non-objective *functorial* local compositions. The basic vocabulary as well as an introductory list of common local compositions—such as scales, ordinary and fractal chords, meters, rhythms, and motives—are presented. The chapter concludes with a discussion of tangent objects in local theory, a concept framework which leads to alterations and related results by Mason and Mazzola.

– Σ –

The previous chapter was devoted to the universal concept and navigation formalism of forms, denotators and orders among these objects. The attentive reader could observe that the category of forms has been defined, but no categories of denotators were considered. More down-to-earth speaking, our attention to *relations among denotators* has not been set up, denotators are—until now—entities which are only related by orders and not by more in-depth relations which take care of the precise positions of denotators within their space functors.

In this chapter, we want to set up the *objects* of the category *Loc* of local compositions, i.e. of the type of denotators which are the core of musical conceptualization. The next chapter will be devoted to a thorough discussion of the morphisms among local compositions. It is however not straightforward in which generality such objects should be established. There are two directions of generalization which must be observed. The first one relates to the dichotomy of *fixed versus variable addresses* and has been introduced previously. The second relates to the dichotomy of *local versus global structure.* This chapter will only deal with the local aspect and introduce and justify the global one in chapter 13.

7.1 The Objects of Local Theory

Summary. After a heuristic introduction, objective and functorial local compositions are introduced.

– Σ –

G. Mazzola, *The Topos of Music I: Theory*, Computational Music Science,
https://doi.org/10.1007/978-3-319-64364-9_7

According to Arnold Schönberg [948], tones are "the material of music". In this spirit, modules and the form space functors deduced by universal constructions are the mathematical spaces where this material is surveyed; we did already introduce the basic role of modules in Section 6.4.1. But it should be recognized that single tones do not constitute what one would recognize as a musical unit. Musical material needs an elementary structuring in order to build elements of music. What Pierre Boulez refers to in this chapter's header is an alphabet of music, built from local, encoded entities. We agree with Boulez' encoding requirement, and we contend that positioning musical material in the shape of a denotator-point within its form space functor at a determined address does meet encoding. But it is not true that a singular tone is the only elementary unit which one will encounter. Besides singular 'points', triadic chords, intervals, motives or rhythms can also be 'elementary' *as* musical units. We cannot prescribe that elementary objects, i.e. letters of the musical alphabet be reduced to single events. For example, it is known from harmony of the 18th century that single notes were only harmonically significant as material parts of triads.

On the other hand, we will certainly require from an elementary system of sounds that it is viewed in a common context which enables us to determine the relations of the constitutive sounds. In the language of gestalt psychology [293] one could restate our project in that "musical gestalts" should be defined which are elementary in the sense that no relations among their material points remain indeterminate. The gestalt's parts, the single sounds, have to be linked in a 'rigid' way.

We illustrate this rigidity requirement by an example. Look at the 12-tempered tuning in the Euler module pitch space, i.e. at zero-addressed denotators $p : 0 \rightsquigarrow EulerModule(p/12 \cdot o)$, $p \in \mathbb{Z}$. It represents pitch c by $p = 0$, $c\sharp$ by $p = 1$, etc. In this context, a triad is determined by three pitch numbers u, v, w. We then consider not explicitly selected triples (u, v, w) but only the following relations among u, v, and w:

$$|u - v| = 4, |v - w| = 3, \text{ and } u = 0. \tag{7.1}$$

Then there are four solutions

$$\begin{aligned} (u, v, w) = &(0, 4, 1) && \text{corresponding to} && (c, e, c\sharp) \\ &(0, 4, 7) && \text{corresponding to} && (c, e, g) \\ &(0, -4, -1) && \text{corresponding to} && (c, A\flat, B) \\ &(0, -4, -7) && \text{corresponding to} && (c, A\flat, F) \end{aligned} \tag{7.2}$$

with triads which verify our conditions (7.1), see also Figure 7.1. By these relations, the triad is not uniquely determined. Intuitively speaking, it looks like a hinged mechanical system which consists of two rigid rods (u, v) and (v, w) whose ends are tied by hinge v whereas hinge u is fixed at 0. Such a configuration would *not* be termed "elementary". It is composed of different loosely related parts.

Evidently, it would be difficult if not impossible to relate the concept of an elementary musical gestalt to relations as above in the general case. Therefore, we shall restrict the concept of such a structure to the really 'indecomposable' configuration of a *local composition* and introduce to non-elementary structures later.

Definition 8 *A* local composition *is a denotator $D : A \rightsquigarrow F(x)$ whose form F is of* **Power** *type. The coordinator S of form F is called the* ambient space *of D; the coordinates*[1] *x is called its* support. *The local composition D is called* objective *iff $x \in 2^{Fun(S)}$ i.e. iff there is $X \subset A@Fun(S)$, with $x = \hat{X}$, in this case X is also called a support of D. The* cardinality *$card(D)$ of an objective denotator D is the cardinality*[2] *of its coordinates set X. In particular, an objective denotator D is called* empty, finite, infinite, *iff its cardinality is zero, finite, or infinite. Non-objective local compositions are also called* functorial.

[1] Recall that "coordinates" is a plural in singular mode and as a such covers the muliplicity of coordinate values within such an object. However, sometimes we also use the common language if the strict terminology sounds too far-out.

[2] See Appendix C.2 for the concept of cardinality.

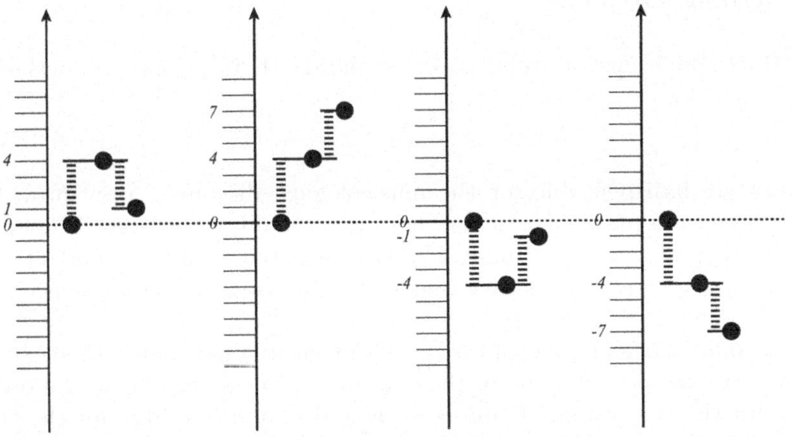

Fig. 7.1. Example of a non-elementary pitch configuration in 12-tempered pitch (parametrized by integers) with non-fixed pitches. One only asks that the first pitch be zero whereas the first stays four semitones from the first, and the third stays three semitones from the second. We then obtain four solutions as shown in the quasi-mechanical 'hinge' system.

Up to names, a local composition is a subfunctor $x \hookrightarrow @A \times Fun(S)$; and an objective local composition is a subset $X \hookrightarrow A@Fun(S)$. A bit more sloppily, we often write $D : x \hookrightarrow @A \times S$ or $D : X \hookrightarrow A@S$ resp. $D = (x, @A \times S)$ and $D = (X, A@S)$.

If we are given a form F, any denotator D of this form can be wrapped as a singleton $\{D\}$ of the form named Loc^F and defined by

$$Loc^F \underset{Id_{\Omega^{Fun(F)}}}{\longrightarrow} \mathbf{Power}(F)$$

We shall always apply this wrapping formalism if talking of a general denotator D of form F as being "wrapped as local composition" or simply "wrapped"; we then consider the objective local composition $\{D\}^\frown$ in Loc^F. Evidently, the construction of Loc^F is a kind of "generic" since any local composition of ambient space F can be derived from Loc^F by synonymy and subfunctors, but we shall not systematically exploit this rigid approach.

Before entering into the very generality of functorial local compositions, we should like to get acquainted with the objective local compositions and try to understand the yoga of this concept. At first sight the semantic potential of a simple structure such as $D = (X, A@S)$ seems to be quite poor. This impression is faulty and stems from the fact that we have packed a lot of information in the preparation of this concept. Recall that this includes:

- selection of the ambient space and address
- the recursively hidden module structures to the ambient space
- the relations among the elements ξ of the support X.

Further, the classical Ehrenfels criterion of "invariance under transpositions" [293] seems to be lacking. We shall introduce this requirement *in extenso*, but not in the much too narrow frame of simple pitch transposition. In fact, the latter is a very elementary special case of general symmetries which are embedded in the concept of morphisms of denotators, to be introduced in Chapter 8. Under this general perspective, Ehrenfels' criterion is but one of a large number of possibilities to define symmetry invariance for the gestalt concept. Not least is it historical development of music which does force an opening of the symmetry invariance. Fixing a symmetry (such as transposition) would cast the concept of gestalt in a too rigid framework, but see Sections 11.7.2 and 18.1.1 for more in-depth discussions of the historical dimension of music, its science, and time-invariant laws of music.

7.2 First Local Music Objects

Summary. We introduce and discuss concepts of scales, chords, rhythms, motives in the language of objective local compositions.

– Σ –

Musical resources are indispensable for the musicological discourse, they provide us with the basic vocabulary of music where concepts such as scale, chord etc. should be declared. Whoever tries to establish a standard of basic musicological concepts is unavoidably confronted with the ambiguity of the existing usage and with the problem of arbitrarily normative and materially overrestrictive or even dogmatic "standard language".

But ambiguity of musicological concepts is not so much a defect but rather expression of a complex situation. Should we—for example—include in the concept of "scale" any temporal order of its tones? Is it then c, d, e, f, g, a, b for the "ascending" C-major scale and b, a, g, f, e, d, c for the "descending" variant? And which "c" do we intend? Any "c" or 'the' "c"? How long does such a tone last? What is its loudness, tuning, etc. And if you play in C-major, does this mean that all tones are contained in the C-major scale? And do they even have to be played in the order indicated by the ascending or descending scale?—We immediately recognize that it is easier to tell what a scale is not than what it is.

We do take into account this ambiguity by usage of a variety of forms and denotators and therefore local compositions (later also global ones). And this is done in full consciousness of the variability of topographic perspectives (discussed in Chapter 2). This provides us with the possibility to build successively operative concept fields from precise aspects and to realize inner coherence by precise relations. For the backing principle of "Yoneda philosophy", roughly stating that knowledge comes through the sum of aspects or perspectives, we refer to Chapter 9. The price for this gain in conceptual profoundness and universality is a certain slowness of ascent from elementary to more complex concept of practical usage. This procedure has, however, the methodological advantage to exhibit a general scheme of musicological conceptualization by which the reader can easily build new concepts according to individual needs. This scheme will be fully developed in the frame of the global theory (Chapter 13). It should be stressed that this scheme is in perfect harmony with the knowledge and data base requirements alluded to in Section 2.5 and Chapter 5 and realized more down-to-earth in Chapter 40.

Let us now introduce some basic concepts which are elementary in the sense that they pertain to the language level of objective local compositions. Momently, these concepts are presented in an unsystematic way; the concept of paradigmatic classification (Chapter 10) will lead to a more systematic treatment in Chapter 11.

7.2.1 Chords and Scales

Summary. This section defines pitch-related local compositions, such as chords and scales as local compositions and gives examples from music and music theory.

– Σ –

7.2.1.1 Chords

Although there are many ways to view a configuration of tones as being a chord, the essential is that intuitively it is a finite collection of pitches. Duration, onset, sound color etc. play a secondary role. We therefore make this definition:

Definition 9 *A* chord *is a finite local composition with EulerModule as ambient space, more precisely*[3], *if we abbreviate F= Fun(EulerModule), it is a denotator of form*

[3] Precision in fact regards the naming, nothing more.

$$EulerChord \underset{Fin(F)\rightarrowtail\Omega^F}{\longrightarrow} \mathbf{Power}(EulerModule) \tag{7.3}$$

If $Cr : A \rightsquigarrow EulerChord(Cr)$ *is a chord at address* A*, we say that it is an* n-chord *if its cardinality* $card(Cr) = n$.

Observe that—as always with denotators—the address is relevant, and therefore we have a 0-chord for each address! To distinguish chords from their instantiation in time, one adds onset to pitch and sets

Definition 10 *A* chord event *is a denotator of form*

$$EulerChordEvent \underset{Id}{\longrightarrow} \mathbf{Limit}(Onset, EulerChord) \tag{7.4}$$

where limit means product of the two factor coordinators.

Thus, a chord event is not a local composition, but:

Exercise 1 *Redefine chord events as being special local compositions with ambient space* $Onset \oplus EulerModule$*, see (6.69) for the Onset form.*

Often, pitches of chords are considered modulo a pitch period p. Usually, this period is the octave $o : 0_{\mathbb{Z}} \rightsquigarrow EulerModule(o)$. More generally, in order to include Wyschnegradsky's non-octave-periodic constructions *[284]*, Bernhard Stopper's tuning *[1016]* or contemporary jazz composition principles *[701]*, period p may be any non-zero denotator $p : 0_{\mathbb{Z}} \rightsquigarrow EulerModule(p)$. One then works in the quotient Euler module[4] $\mathbb{Q}^3/\mathbb{Z} \cdot p$ and then defines as follows:

Definition 11 *Let* p *be a period in the Euler module. The* p-class chord *is a finite local composition with ambient space*

$$p\text{-}EulerClass \underset{Id}{\longrightarrow} \mathbf{Simple}(\mathbb{Q}^3/\mathbb{Z} \cdot p), \tag{7.5}$$

more precisely, setting $F = Fun(p\text{-}EulerClass)$*, it is a denotator of form*

$$p\text{-}ClassChord \underset{Fin(F)\rightarrowtail\Omega^F}{\longrightarrow} \mathbf{Power}(p\text{-}EulerClass). \tag{7.6}$$

The octave period o is particularly common, and we have this representation of a direct sum of two modules:

$$\mathbb{Q}^3/\mathbb{Z} \cdot o \xrightarrow{\sim} \mathbb{Q} \cdot o/\mathbb{Z} \cdot o \oplus (\mathbb{Q} \cdot q \oplus \mathbb{Q} \cdot t) \tag{7.7}$$

where the first is the circle of rational cosets modulo o, and the second is the Euler plane. Hence the points $x : A \rightsquigarrow p\text{-}EulerClass(x)$ of a o-class chord consist of two components $x = (x_{temp}, x_{just})$ which live in the corresponding space forms:

$$x_{temp} : A \rightsquigarrow p\text{-}TempClass(x_{temp}), \text{ with form}$$
$$o\text{-}TempClass \underset{Id}{\longrightarrow} \mathbf{Simple}(\mathbb{Q} \cdot o/\mathbb{Z} \cdot o), \tag{7.8}$$
$$x_{just} : A \rightsquigarrow EulerPlane(x_{just}), \text{ with form}$$
$$EulerPlane \underset{Id}{\longrightarrow} \mathbf{Simple}(\mathbb{Q} \cdot q \oplus \mathbb{Q} \cdot t). \tag{7.9}$$

Accordingly, if a o-class chord has all its points in the 'tempered' ambient space $o\text{-}TempClass$ (resp. in the 'just' ambient space $EulerPlane$) we say that it is tempered (resp. just). If there is no danger of confusion we shall omit the octave period in this context and simply speak of *pitch classes* or *class chords* when we deal with points or chords in ambient space $o\text{-}EulerClass$.

[4] To be precise, the underlying module is a $\mathbb{Z}$-module, i.e. we are taking $\mathbb{Q}^3_{[\mathbb{Z}]}/\mathbb{Z} \cdot p$.

On the (octave-)tempered ambient space *TempClass*, one often considers pitch classes in a regular subdivision of these pitch classes, defined by a positive natural number w. We then look at the tempered ambient subspace

$$w\text{-}PitchClass \underset{Id}{\longrightarrow} \mathbf{Simple}(\mathbb{Z} \cdot \frac{1}{w} \cdot o/\mathbb{Z} \cdot o) \tag{7.10}$$

of *TempClass*. When representing its circle module coordinator as a circle, the submodule $\mathbb{Z} \cdot \frac{1}{w} \cdot o/\mathbb{Z} \cdot o \xrightarrow{\sim} \mathbb{Z}_w$ appears as a regular w-sided polygon, and a tempered class chord is called *w-tempered* if its points are contained in the ambient space *w-PitchClass*. Of course, the ambient space *w-PitchClass* is isomorphic to the form $PiMod_w$ introduced in (6.41). The difference here is that we have embedded the pitch classes in the general Euler module context. We shall often identify the underlying modules and the associated forms.

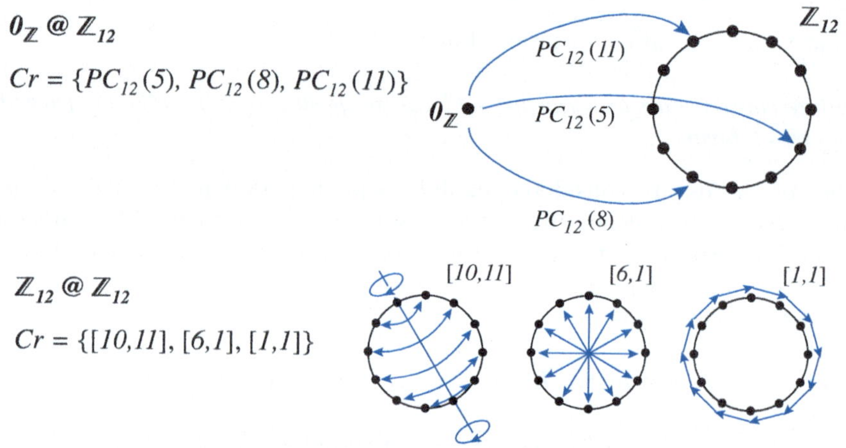

Fig. 7.2. Above, a zero-addressed 12-tempered class 3-chord, below a self-addressed 12-tempered class 3-chord.

Whereas there are infinitely many (coordinate sets for) just class chords, there are only 2^w (resp. $\binom{w}{n}$) different (coordinate sets for) w-tempered class chords (resp. n-element class chords). For every couple Cr_1, Cr_2 of A-addressed chords or class chords, we can build their Boolean combinations: *union* $Cr_1 \cup Cr_2$, *intersection* $Cr_1 \cap Cr_2$, and *difference* $Cr_1 - Cr_2$. For a w-tempered class chord Cr, one may also build its *complementary chord* $Cr^\frown = \chi_w - Cr$, i.e. the difference from the *w-chromatic* class chord χ_w of support $\mathbb{Z}_w$.

We conclude with a remark on different addresses for chords. With the identification from (6.41) and notation from Chapter 6, RegDen-9, a 0-addressed w-tempered class n-chord can be written as a set

$$Cr = \{PC_w(r_1), \dots PC_w(r_n)\},$$

and a self-addressed w-tempered class n-chord can be written as set

$$Cr = \{[a_1, b_1], \dots [a_n, b_n])\}$$

with $r_n, a_n, b_n \in \mathbb{Z}_w$. Figure 7.2 shows two class chords with different addresses.

7.2.1.2 Scales

This concept regards special configurations in pitch spaces. We shall work in the *EulerModule* form as an ambient space. To define a scale, one has to fix a periodicity in order to express the repetition of scale tones under a particular period. For the same reasons as explained for chords in example 7.2.1, we select a non-zero period denotator $p : 0 \rightsquigarrow EulerModule(p)$. We have a morphism

$$mod_p : EulerModule \twoheadrightarrow p\text{-}EulerClass \tag{7.11}$$

which is induced by the canonical projection $\pi : \mathbb{Q}^3 \twoheadrightarrow \mathbb{Q}^3/\mathbb{Z} \cdot p$. This associates with every objective local composition $L : A \hookrightarrow A@EulerModule$ its pointwise defined projection $mod_p(L) : A \hookrightarrow A@p\text{-}EulerClass$. We then define

Definition 12 *Given a period p, a* p-scale *is a non-empty local composition S of ambient space EulerModule, and such that*

(i) *its projection $mod_p(S)$ is finite, i.e. a p-class chord,*
(ii) *$S = e^p S$, i.e. S is p-periodic.*

The associated p-class chord is called the scale's chord.

Conversely, the scale S may be recovered from its p-class chord by inverse image:

$$S = mod_p^{-1}(mod_p(S)),$$

see also Figure 7.3. But it is not true in general that each non-empty p-class chord Cr defines a scale. In fact, for a critical address such as $A = \mathbb{Z}_{12}$, there is no point $x : \mathbb{Z}_{12} \rightsquigarrow EulerModule(x)$ which projects to the canonical injection $\xi : \mathbb{Z}_{12} \rightsquigarrow o\text{-}EulerClass(\xi)$ in the tempered part. One may rightly define a form which

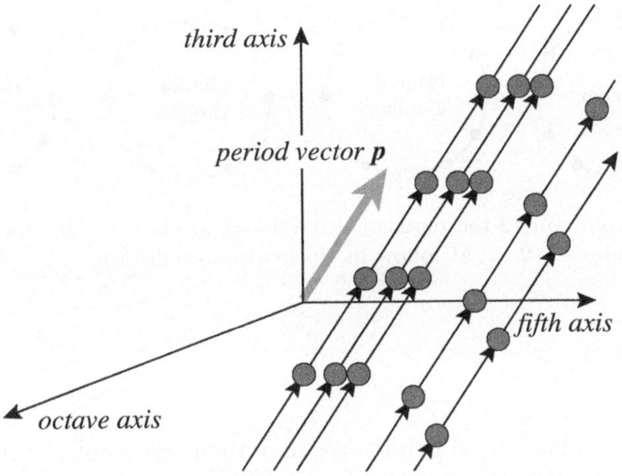

Fig. 7.3. A scale with period p in Euler space.

comprises exactly the scales as its denotators. We have

Lemma 1 *Let $F = Fun(EulerModule)$. Then, for addresses A, the map*

$$A \mapsto Sc(A) = \{S \in A@2^F \mid S \text{ is a p-scale}\} \tag{7.12}$$

defines a subfunctor of 2^F.

Exercise 2 *Give a proof of Lemma 1.*

With Lemma 1 we can introduce the scale form

$$p\text{-}Scale \underset{Sc \rightarrowtail \Omega^F}{\longrightarrow} \mathbf{Power}(EulerModule) \tag{7.13}$$

and we have a morphism $scalemod_p : p\text{-}Scale \to p\text{-}ClassChord$ of forms. According to the above, this morphism allows us to recover all p-scales by inverse images of non-empty p-class chords. We therefore stick to the latter in the following discussion of common scales. Again, these scales will refer to the octave period. In practice, we encounter two main types of o-scales, w-tempered and just scales.

7.2.1.3 *w*-Tempered Scales

Let w be a positive natural number. A scale, represented by its class chord S, is said to be *w-tempered* iff the scale's class chord S is w-tempered. This also means that the scale is contained in the *w-tempered scale space*

$$w\text{-}TemperedScale \underset{Id}{\longrightarrow} \mathbf{Simple}(\mathbb{Z} \cdot \frac{1}{w} \cdot o).$$

In this case, a zero-addressed scale can be represented as a subsets of this polygon, see Figure 7.4 for common scales in 12-tempered tuning.

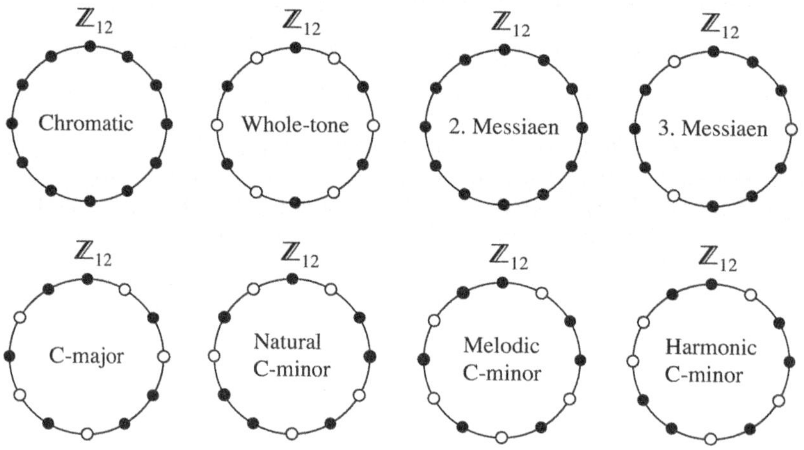

Fig. 7.4. Some of the most common 12-tempered scales, related to pitch $c = \overline{0}$ which is shown as the highest point of the polygon; the other numbers $\overline{1}, \overline{2}, \ldots \overline{11}$ follow in clock-wise orientation.

7.2.1.4 Just Scales

A scale is said to be *just* if its class chord is just. We may then represent the scale by a class chord living in the *EulerPlane* ambient space. When representing zero-addressed pitches on the *EulerPlane*, the horizontal axis is the fifth axis, and the vertical one is the third axis. We show three common scales via their class chords in Figure 7.5.

This gives a first impression of how different tempered and just tuning look from the geometric point of view. This insight becomes even more dramatic if we visualize the 12-tempered and just chromatic scale pitches within one octave, see Figure 7.6. Whereas the 12-tempered configuration looks like a pearl chain, the just configuration seems to be completely random. We shall learn in Section 8.2 that this is a wrong impression. It can be shown that both, the 12-tempered and the just octave configurations, admit a distinguished inner symmetry each, and that this symmetry is intimately related with the theory of consonances and dissonances (see part VII).

7.2.2 Local Meters and Local Rhythms

Summary. This section deals with onset time specific local compositions, such as local meters and local rhythms.

– Σ –

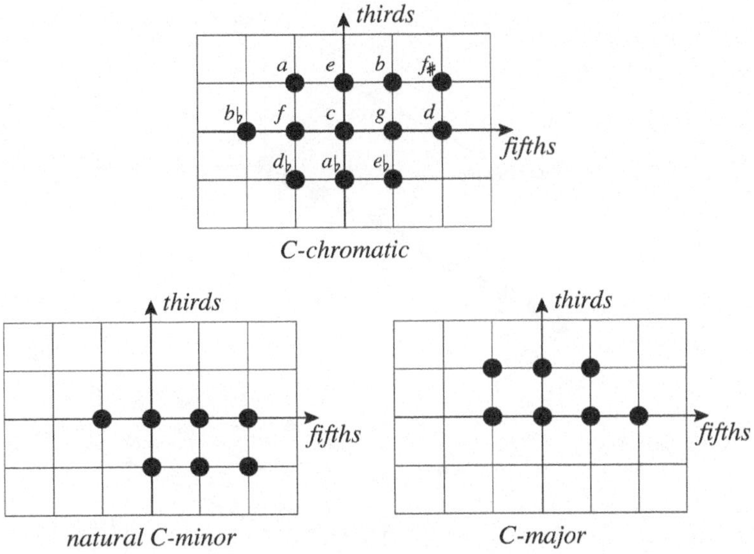

Fig. 7.5. The just C-major, natural C-minor, and C-chromatic scale, the latter according to Martin Vogel [1089], as represented by their class chords in the *EulerPlane*. Observe that natural C-minor contains another pitch class for $b_\flat$ than C-chromatic. The difference is the class of the syntonic comma, see (6.34).

Before discussing the precise concepts we should recall that meters and rhythms are among the most fluffy concepts in musicology. This is partly due to the extremely complex phenomenon of time-based regularities, but it is also due to a remarkable (and not forgivable) lack of corresponding theory in European musicology. This lack is not only due to the traditional preponderance of harmonic considerations since the Pythagorean school was founded, it is also an effect of the incapability of musicological theorists to construe complex concepts from simple ones in order to grasp complex phenomena. There is a strange belief that either a concept encompasses the intended phenomenon—or it cannot and *should not* be defined[5]! We come back to this special issue in Section 10.4.

We start this concept framework with the elementary concept of a local meter, see also Figure 7.7 for an illustration.

Definition 13 *Given an address A, let $p : 0_{\mathbb{Z}} \rightsquigarrow Duration(p)$ and $o : A \rightsquigarrow Onset(o)$ be a positive duration period and an onset origin, respectively. Let $-\infty \leqslant a \leqslant b \leqslant \infty$ be two extended integers. Then a (A-addressed)* local meter *is an objective local composition $L \hookrightarrow A@Onset$ of ambient space Onset defined as the p-periodic interval*

$$L = \{e^{tp} \cdot o \mid t \in [a, b]\} \tag{7.14}$$

of A-addressed onsets. The difference $b - a$ is called the length *of L; in case of positive length, the uniquely determined duration p is called the* period *of L.*

In particular, if we are given a time signature $TimeSig(x, y, e)$ (6.92), two canonical local meters are associated:

- The *barline meter* is defined by *barline* period $Duration(x/y)$, origin $Onset(e)$, and interval numbers $[a, b] = [-\infty, \infty]$.
- The *beat meter* is defined by *beat* period $Duration(1/y)$, origin $Onset(e)$, and interval numbers $[a, b] = [-\infty, \infty]$.

[5] See the problem of defining a motif in [892, p.11/12, footnote], for example.

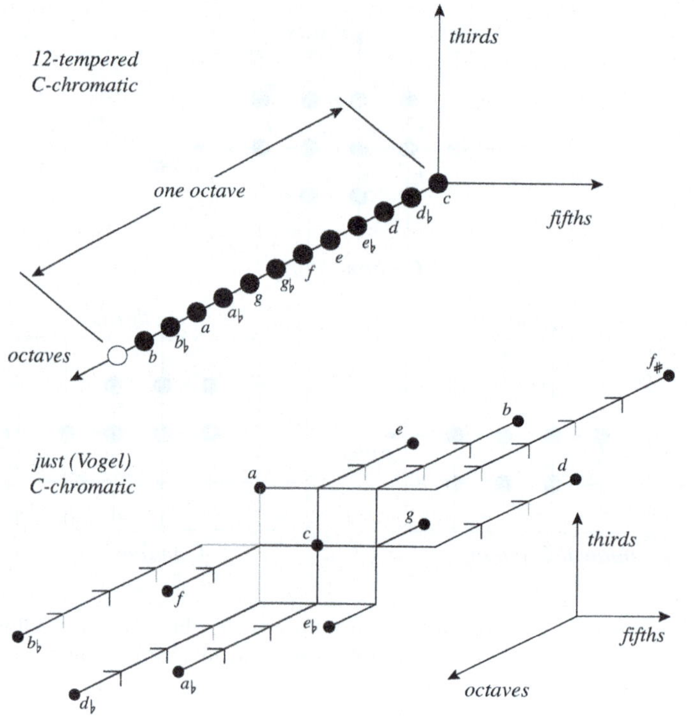

Fig. 7.6. One octave of the 12-tempered chromatic C-scale (above) and the just (Vogel) chromatic scale (below) The images seem to be very different but share an important symmetry each.

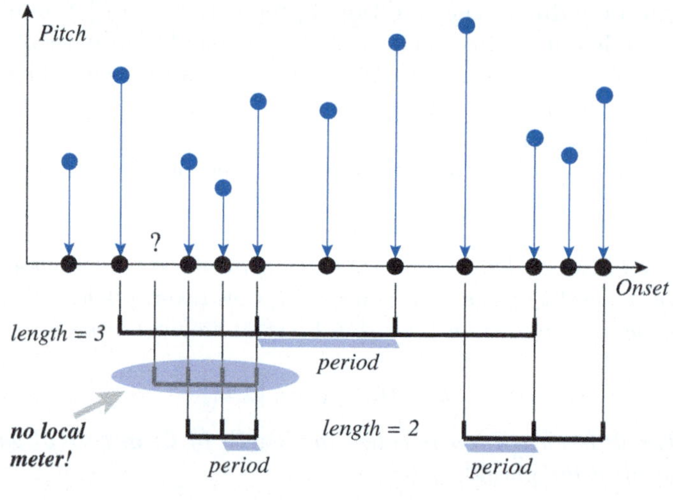

Fig. 7.7. Three local meters as appearing in a local composition in pitch-onset space. Observe that we will not consider local meters within the given local composition if their onsets are not realized as projections of events to the onset axis.

Observe that the barline meter is independent of the time signature representation of the fraction x/y whereas the beat meter isn't. Later, we shall introduce finite length meters for bars and beats to take care of the limits of concrete scores.

From the music(ologic)al point of view, a local meter is the *minimum of time regularity* that can be observed: a simple, periodic repetition of onsets. In the understanding of Hugo Riemann [904], metrical structures deal with onset weights. We may thus view the concept of a local meter as a basic account of the weight of selected onsets, the selection being driven by a period of regular distribution over time. This approach has a psychological interpretation insofar as it suggests that elementary perception and/or expression of temporal stress or "weighting" relies on a repeated appearance of the same object in regular distances.

There are several generalizations and complexifications of this conceptual germ which build a viable tool to understand metrical and rhythmical phenomena. The *second* one deals with "globalization" issues, which means that local meters are only the elements of a compound time structure. We shall deal with global meters in Section 13.4.3, more in depth in Chapter 21, and more concretely in Section 41.1 which deals with software for metrical analysis.

The *first* generalization concerns local rhythms and will be dealt with now. A local meter is only concerned with periodicity on the onset axis. But normal sounds do have lots of additional parameters to make them "audible". And these parameters are responsible for producing rich shapes which are beyond pure onset structures. In [904], Riemann describes rhythm as being related to a periodicity of higher level which means that a melodic movement may involve short onset differences and durations but can nonetheless produce a "rhythmical" period of larger extent. We want to give this idea a first conceptualization, a more general and powerful one will be added when morphisms between local compositions are available. To do so, observe that for a denotator $D : A \rightsquigarrow (Onset \times F)(D)$ in a product space of $Onset$ with a form F and a period $p : 0_{\mathbb{Z}} \rightsquigarrow Duration(p)$, we have a shift $e^p \cdot D$ of D by e^p as above. In fact, since $A@(Onset \times F) \xrightarrow{\sim} A@Onset \times A@F$, we may let e^p act on the onset component of D, leaving the F-component invariant, and this is it. Carrying over this action point by point to objective local compositions with ambient space $Onset \times F$, we obtain a shift operation on such local compositions.

Definition 14 *Given an address A, let $p : 0_{\mathbb{Z}} \rightsquigarrow Duration(p)$ be a positive duration period and $-\infty \leqslant a \leqslant b \leqslant \infty$ two extended integers. Given a local composition germ $\hookrightarrow A@(Onset \times Para)$ in a product ambient space of onset and some additional "parameter" space Para. Then a (A-addressed)* local rhythm *is a local composition R of ambient space $Onset \times Para$ of shape*

$$R = \bigcup_{t=a}^{b} e^{tp} \cdot germ. \tag{7.15}$$

Observe that now, period and length are no longer uniquely determined from the datum of R alone. Intuitively speaking, a local rhythm is a local meter whose recurrent unit has been blown up to a whole set of points with additional parameters, see also Figure 7.8.

Exercise 3 *Give an exact description of the space Para in Figure 7.8.*

Exercise 4 *Define a metronome as a local rhythm with space $Para = Loudness$.*

Evidently, the purely local approach to meter and rhythm is unsatisfactory since it is felt that real phenomena of "time grouping" are much more involved. Whereas we shall extensively develop the global perspectives and prove that this satisfies many of the present urges, it is already now possible to construct local "polyrhythms" from given collections of local rhythms, see Section 7.4.

7.2.3 Motives

Summary. Motives are introduced as a type of local compositions which share (among others) pitch and onset time in their coordinators.

– Σ –

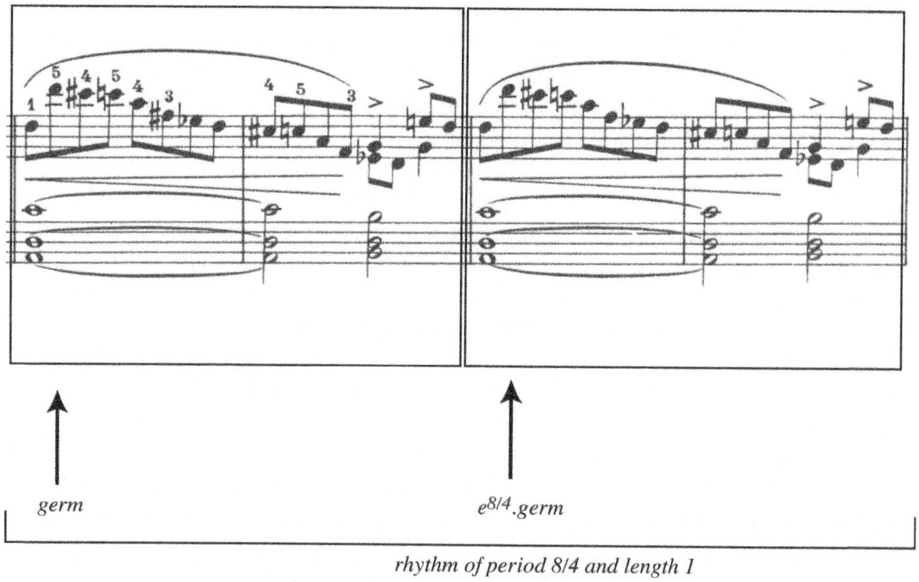

Fig. 7.8. A local rhythm Beethoven's op.106, Allegro, bars Nr.75 to Nr.78. The parameters which are added to onset are numerous, including duration, loudness, legati, crescendi, and accents. The germ consists of the entire bars Nr. 75-76, period is $2 = 8/4$, and length is 1.

In a rather generic setting we could define a motif as a local composition whose ambient space projects to a product of onset and pitch. Let us, however, be more precise in order to make the ideas transparent:

Definition 15 *If A is an address, a* (A-addressed) motif *M is an objective local composition* $M \hookrightarrow A@F$ *whose ambient space* $F = Onset \oplus Pitch \times Para$, *for any form Para of additional "parameters", and such that for any pair* m, n *of different elements in M, projections* $m_{Onset} = p_{Onset}(m)$, $n_{Onset}(n) = p_{Onset}(n)$ *to the onset space are different;* $p_{Onset} : F \to Onset$ *is the canonical projection. If the cardinality of a motif is c we say that it is a c-motif.*

To see examples of motives, refer to Figure 7.9. Notice that this concept is very restrictive in that it does not allow for motives which contain chords or several contrapuntal voices, say. Notice also that in contrast to local meters and rhythms, motives are not attributed by "local". The reason is that naming tradition views meters and rhythms as being something complex whereas motives are conceived as being "cells" of melodic structures. Whenever we want to make the situation unambiguous, the attribute "local" is inserted.

Let us shortly comment on the address question here. Naively, we always think of zero address. So it is useful to study a less elementary address situation. If we take $A = \mathbb{Z}$, a motif at this address means a set of denotators $D : \mathbb{Z}{\rightsquigarrow}(D_{Onset}, D_{Pitch}, D_{Para})$ from which the onset coordinate is an affine homomorphism $D_{Onset} : \mathbb{Z} \to \mathbb{R}$. As in Section 6.2.1, this denotator D_{Onset} can be symbolized by an arrow $ON_D \Rightarrow OFF_D$ in *Onset* space, see also Figure 7.10. The motif condition of separation of motif elements by their onset projection then means that for any two denotators D, E of the motif, the arrows $ON_D \Rightarrow OFF_D$ and $ON_E \Rightarrow OFF_E$ are different, i.e. either their *'ON'* values or their *'OFF'* values are different.

Musically speaking such a motif consists of 'intervals' of zero-addressed denotators such that no two such intervals intervene twice with same *'ON'* and *'OFF'* values. A prototypical case is a two-voice counterpoint with cantus firmus and discantus, the latter being added simultaneously to each note of the cantus firmus voice. In this case, a $\mathbb{Z}$-addressed denotator D consists of the ON-denotator of the cantus firmus and the OFF-denotator of the discantus voice. They differ in pitch but not in onset coordinates. Hence $ON_D \Rightarrow OFF_D$ is the 'zero'-length arrow $ON_D \Rightarrow ON_D$, and a motif just distinguishes its points by the ON_D-value.

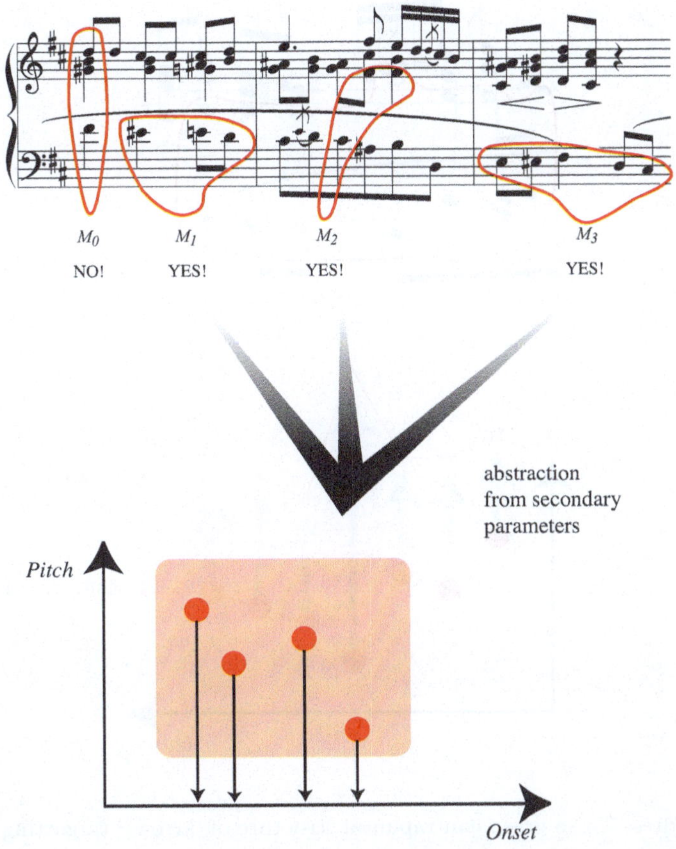

Fig. 7.9. Three examples of motives: M_1, M_2, M_3, and one local composition, M_0, which is not a motif. Though these local compositions live in spaces with possibly many parameters, their relevant 'shape' for motivic analysis is centered around onset and pitch.

7.3 Functorial Local Compositions

Summary. The generalized conceptual framework of non-objective, proper functorial local compositions is presented. It settles problems of universal constructions, such as fiber products of local compositions, which do not always exist in the frame of objective local compositions.

– Σ –

Recall from Section 6.3.1 that an objective denotator $D \hookrightarrow A@F$ at address A in ambient space F is identified with its functorial 'version' $\hat{D} \hookrightarrow @A \times F$. Conversely, we may associate to each functorial local composition $S \hookrightarrow @A \times F$ at address A and in ambient space F its *objective trace* $\check{S}$.

Definition 16 *Let $S \hookrightarrow @A \times F$ be a local composition at address A and in ambient space F. Its* objective trace *is defined as the following objective local composition:*

$$\check{S} = \{s \in A@F \mid (Id_A, s) \in A@S\} \tag{7.16}$$

at address A and in ambient space F. The associated objectivized composition qua functor *will be defined and denoted by*

$$S^{\diamond} = \widehat{(\check{S})} \tag{7.17}$$

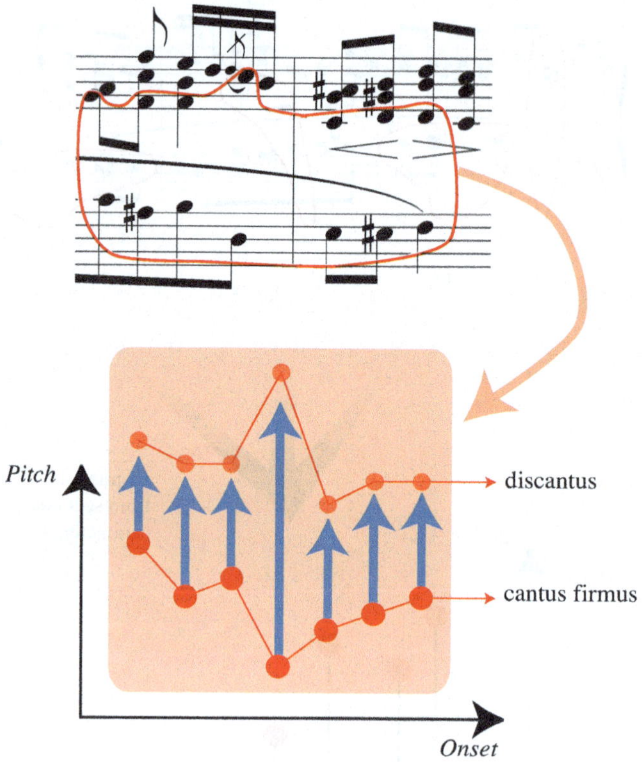

Fig. 7.10. A motif at address $\mathbb{Z}$ can grasp contrapuntal structure of 'arrows' targetting from cantus firmus notes to discantus notes.

The following is evident:

Sorite 1 *Let $S \hookrightarrow @A \times F$ be a local composition at address A and in ambient space F. Then $S^{\diamond} \subset S$, and $S = S^{\diamond}$ iff S is objective.*

We now want to make more explicit pictures of functorial local compositions in order to understand the specific difference to objective local compositions. To this end, let $f : B \to A$ be an address change and let $S \hookrightarrow @A \times F$ be a local composition at address A and in ambient space F. Then we set

$$f@S = \{s \in B@S | \ s = (f, t)\}, \tag{7.18}$$

and call this set the *f-slice of S* whence

$$B@S = \coprod_{f:B \to A} f@S, \tag{7.19}$$

see Figure 7.11 for the visual representation which we shall adopt henceforth. In particular, the notation of Section 6.2.3 give

$$f@\hat{T} = \{f\} \times T \cdot f \tag{7.20}$$

for an objective local composition T at address A and address change $f : B \to A$. It is now evident that in general, local compositions are not objective.

For example, the full local composition $S = @A \times F$ gives the cartesian product $A@S = A@A \times A@F$ at address A, i.e. $f@S = \{f\} \times A@F$ for all $f : A \to A$ whereas $f@S^{\diamond} = \{f\} \times A@F \cdot f$. Take a simple form

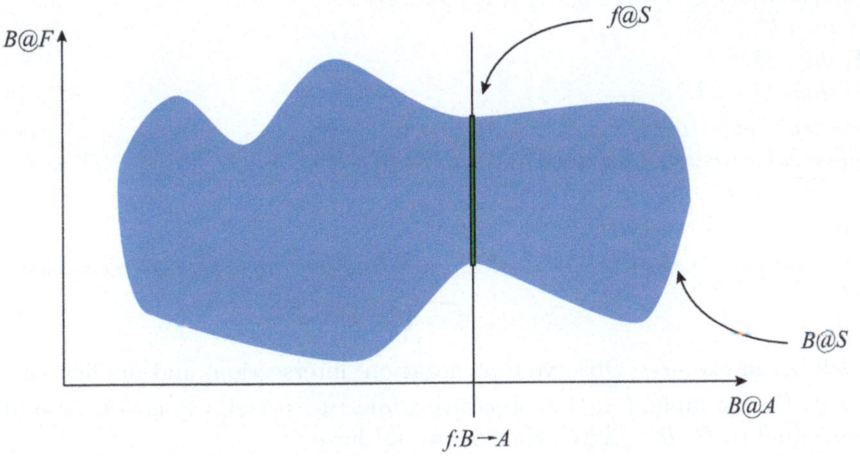

Fig. 7.11. Graphical representation of a functorial local composition S at address A, as evaluated at address B with its 'slice' $f@S$ at address change $f : B \to A$.

of functor $@M$ and the zero morphism $z : A \to A$. Then we have $z@S^{\diamond} = \{z\} \times A@M \cdot z = \{z\} \times e^M \cdot z$, i.e. the constant morphisms in $A@M$ which is much less then the full set $A@M$ of affine morphisms. Since full local compositions play an essential role in the theory of coordinate functions for musical parameters of global compositions—the deep question of realization of abstract compositional ideas, see Chapters 15 and 19—we cannot refrain from non-objective constructs.

Other examples of limitations for objective local compositions will appear in questions of universal constructions such as fiber products, see Section 8.3.2. Since fiber products are fundamental tools of global theory and also in all questions of Grothendieck topologies, this is one further motivation to transgress the frame of objective local compositions which a decade ago were the core of mathematical music theory [682].

7.4 First Elements of Local Theory

Summary. The elementary concepts in local theory are discussed: cardinality, associated module of a local composition, dimension, general position of points, sub-compositions, Boolean constructs, products, coproducts, projections and injections.

– Σ –

In the following discourse we deal with many different local compositions at a time and with their recombinations under specific operations. From the perspective of rigorous denotator theory, one should always specify the names of forms where these denotators belong, and not only the names of the ambient spaces of these denotators. However, for the description of our recombination tools, this would be an 'overdressed' formalism. Rather would this hide what we are doing than clarify anything. Hence we shall not specify these names and just name denotators and ambient spaces. However, for a thoroughly unambiguous discourse, e.g. within an implementation of the theory in data base management systems (viz. the implementation for the RUBATO software described in Chapter 40), the strict formalism must be followed. Also it is necessary in a conceptually full-fledged musicological context where names are essential.

Heyting Constructions. We fix an address A and an ambient space F. It makes sense to invoke the Heyting algebra $Sub(@A \times F)$ of subfunctors[6] of $@A \times F$. On the other hand, we have the Boolean algebra $2^{A@F}$ of subsets of $A@F$. Here are some facts which relate the two algebras:

[6] See Appendix G.5.1, Proposition 125.

Lemma 2 *With the preceding notation, for two objective local compositions $U, V \in 2^{A@F}$ and two functorial local compositions $D, E \in Sub(@A \times F)$ we have:*
(i) If $U \subset V$ then $\hat{U} \subset \hat{V}$.
(ii) If $D \subset E$ then $\check{D} \subset \check{E}$.
(iii) If $D \subset E$ then $D^\diamond \subset E^\diamond$.
(iv) Union preserves objectivity, i.e. $\widehat{U \cup V} = \hat{U} \vee \hat{V}$.
(v) Union preserves associated objective local compositions, i.e. $(D \vee E)^\diamond = D^\diamond \vee E^\diamond$.
(vi) We have $\widehat{U \cap V} \subset \hat{U} \wedge \hat{V}$.
(vii) For $A = 0$, we have $\widehat{-U} \subset \neg\hat{U}$.
(viii) For $A = 0$, and with respect to the Boolean and Heyting implications, respectively, we have

$$\widehat{U \Rightarrow V} \subseteq \hat{U} \Rightarrow \hat{V}.$$

The proof is left as an exercise. Observe that negation, intersection, and implication behave in a rather complicated way. For example, $\widehat{(-\varnothing)}$ is objective while $\neg\hat{\varnothing} = @A \times F$ is not. Also, if $\hat{U} \wedge (A@F - U)^\wedge$ were objective, equal to $\hat{R}$, $R \subset A@F$, then we would have

$$Id_A@\hat{R} = Id_A@\hat{U} \wedge (A@F - U)^\wedge = Id_A@\hat{U} \cap Id_A@(A@F - U)^\wedge,$$

in other words

$$R = R \cdot Id_A = U \cdot Id_A \cap (A@F - U) \cdot Id_A = U \cap (A@F - U) = \varnothing.$$

But $\hat{U} \wedge (A@F - U)^\wedge$ is clearly not empty in general. This also shows that the inclusion in point (vi) above is strict in general. This is a very special case of a fiber product which is not objective, as announced above.

Products, Coproducts, Projections and Restrictions. If we are given two local compositions $S \hookrightarrow @A \times F$ and $T \hookrightarrow @B \times G$, the product functor[7] $S \times T$ can be identified with a local composition in the following way. Since limits and colimits of set-valued functors in $\mathbf{Mod}^@$ are calculated as corresponding constructs in **Ens**, for each address A, $S \times T$ is a subfunctor of $(@A \times F) \times (@B \times G)$. But we have a canonical isomorphism[8] of functors $@A \times @B \xrightarrow{\sim} @(A \times B)$. Therefore, $S \times T$ is isomorphic to a local composition in $@(A \times B) \times F \times G$. Observe that we take the new ambient space with the name $'F \times G'$, called the *product ambient space*, but observe that the new address is also the product $A \times B$ of the old ones. We shall henceforth take this local composition when speaking of the *product* $S \times T$ of local compositions. More concretely, if $f : C \to A \times B$ is an address change, the f-slice of $S \times T$ is calculated as follows. Let $f_A : C \to A$ and $f_B : C \to B$ be the two components of f. Then we have an isomorphism of sets

$$f@(S \times T) \xrightarrow{\sim} f_A@S \times f_B@T \tag{7.21}$$

which is functorial in the variable address C. This formula also defines two projections $p_S : S \times T \to S$ and $p_T : S \times T \to T$ induced by the first projections $p_A : A \times B \to A$, $p_F : F \times G \to F$ and the second projections $p_B : A \times B \to B$, $p_G : F \times G \to G$. In other words, we have two commutative diagrams of functors

$$\begin{array}{ccc} S \times T & \xrightarrow{\text{subfunctor}} & @(A \times B) \times F \times G \\ {\scriptstyle p_S}\downarrow & & \downarrow{\scriptstyle @p_A \times p_F} \\ S & \xrightarrow{\text{subfunctor}} & @A \times F \end{array} \tag{7.22}$$

$$\begin{array}{ccc} S \times T & \xrightarrow{\text{subfunctor}} & @(A \times B) \times F \times G \\ {\scriptstyle p_T}\downarrow & & \downarrow{\scriptstyle @p_B \times p_G} \\ T & \xrightarrow{\text{subfunctor}} & @B \times G \end{array} \tag{7.23}$$

[7] See Appendix G.2.1.
[8] See Section E.3.8: take $M = 0$ and $C = 0$ in Theorem 62.

which yield a model for general morphisms between local composition to be introduced in Chapter 8.

Coproducts of local compositions S and T can be built in case they pertain to the same address $A = B$, i.e. $S \hookrightarrow @A \times F$ and $T \hookrightarrow @A \times G$. The coproduct functor[9] $S \amalg T$ is a subfunctor of $@A \times F \amalg @B \times G$ which in turn is canonically isomorphic to $@A \times (F \amalg G)$; as with the product we identify $S \amalg T$ with the corresponding local composition of $@A \times (F \amalg G)$, including the name $'F \amalg G'$ of the *coproduct ambient space*. We also have $f@(S \amalg T) \xrightarrow{\sim} f@S \amalg f@T$, an equation which is functorial in the domain C of address change $f : C \to A$ and induced by the canonical injections $i_F : F \to F \amalg G$, $i_G : G \to F \amalg G$. We again deduce two commutative diagrams of functors:

$$\begin{array}{ccc} S & \xrightarrow{\text{subfunctor}} & @A \times F \\ {\scriptstyle i_S}\downarrow & & \downarrow{\scriptstyle Id \times i_F} \\ S \amalg T & \xrightarrow{\text{subfunctor}} & @A \times (F \amalg G) \end{array} \tag{7.24}$$

$$\begin{array}{ccc} T & \xrightarrow{\text{subfunctor}} & @A \times G \\ {\scriptstyle i_T}\downarrow & & \downarrow{\scriptstyle Id \times i_G} \\ S \amalg T & \xrightarrow{\text{subfunctor}} & @A \times (F \amalg G) \end{array} \tag{7.25}$$

which also anticipate the concept of morphisms between local compositions.

Cardinality. The cardinality of a local composition S at address A is defined by $card(Id_A@S) = card(\check{S})$, and S is called finite, infinite, etc. iff its cardinality is so.

Module of a commutative local composition. In general, the space functors of local compositions do not carry algebraic structures. But in the special and important case of simple ambient space F with $Fun(F) \xrightarrow{\sim} @M$ for a module M over a commutative ring R, we can say much more about objective local compositions S which are addressed at R-module A and their objective trace $\check{S}$ is contained in the subset $A@_RM$ of R-affine[10] morphisms $f : A \to M$. Such objective local compositions are called *commutative* (*over* R if this specification is required).
Let us fix a commutative local composition S over ring R, address A, and ambient space $@M$. Identify S with its objective trace $\check{S}$. We then have a subset $S \subset A@_RM$ of the R-module[11] $A@_RM$.

Lemma 3 *Let $(S, A@_RM)$ be a non-empty commutative local composition over ring R and let $x, y \in S$ be two elements of S (more precisely: of $\check{S}$). Then the sub-R-modules $\langle d_x \rangle$ and $\langle d_y \rangle$ of $A@_RM$ generated by the sets $d_x = \{s - x|\ s \in S\}$ and $d_y = \{s - y|\ s \in S\}$ of differences to x and to y, respectively, are equal. Further, the non-zero elements of d_x form a basis of $\langle d_x \rangle$ iff the non-zero elements of d_y do so.*

Proof. We have $s - y = (s - x) - (y - x)$, so $\langle d_y \rangle \subset \langle d_x \rangle$, and a symmetric argument shows the other inclusion. Suppose that $\{s - x|\ s \in S, s \neq x\}$ is a set of linearly independent vectors, and let $\sum_{s \in S - \{y\}} \lambda_s(s - y) = 0$ be a vanishing linear combination. We then have

$$\begin{aligned} 0 &= \sum_{s \in S - \{y\}} \lambda_s(s - y) = \sum_{s \in S - \{y\}} \lambda_s((s - x) - (y - x)) \\ &= -\lambda_x(y - x) + \sum_{s \in S - \{x,y\}} \lambda_s(s - x) - (\sum_{s \in S - \{x,y\}} \lambda_s)(y - x). \end{aligned}$$

By linear independence of the elements in $\{s - x|\ s \in S, s \neq x\}$, we have $\lambda_s = 0$ for $s \in S - \{x, y\}$, and therefore also $\lambda_x = 0$. The symmetric argument yields the other implication. QED.

[9] See Appendix G.2.1.

[10] See Section E.3 for notation.

[11] See sorite 14 in Appendix E.3.

Definition 17 *For a commutative local composition $(S, A@_R M)$ over ring R, an R-module $R.S$ is defined by*

$$R.S = \langle d_x \rangle \text{ with } d_x = \{s - x|\ s \in S\} \tag{7.26}$$

for any element x of S if S is non-empty, otherwise we set $R.\varnothing = \varnothing$. By Lemma 3, this module[12] *is well-defined. It is called* the module of S. *We say that S is* generating *iff $R.S = A@_R M$, and that S is* embedded *iff $S \subset R.S$.*
If for a non-empty S, $d_x - \{0\}$ forms a basis of $R.S$ we say that the points of S are in general position[13]. *Again, by Lemma 3, this statement is independent of the chosen element x.*

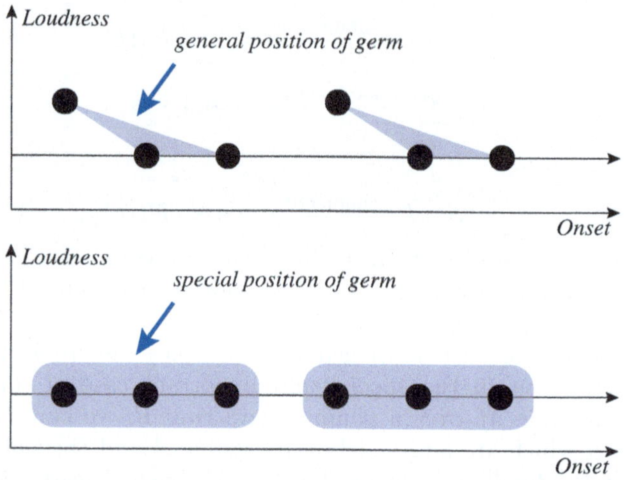

Fig. 7.12. Below, we have a germ of a local rhythm in *Onset* ⊕ *Loudness* space. Since the germ's elements are in special position, in fact on a horizontal line, the germ is no more retractable from the local rhythm. Above, we see the new germ with first event in higher loudness, and thus giving general position and evidence of the germ, together with the rhythm's period.

Sorite 2 *If S is a finite commutative local composition, its points are in general position iff its module is free of rank $card(S) - 1$.*

Proof. The condition is necessary by definition. Conversely, the characterization of projective finitely-generated modules in Appendix F.2, Theorem 66, guarantees that the points are in general position if the module $R.S$ is free of rank $card(S) - 1$. QED.
Local commutative compositions with points in general position play a fundamental role in classification theory and for understanding performance strategies. We should therefore give one preliminary illustration to make evident the musical meaning of general position. Refer to Figure 7.12 for the following comments. Let us consider the zero-addressed local three-element composition $S = \{(0, 10), (1, 10), (2, 10)\}$ in ambient space *Onset* ⊕ *Loudness*. Evidently, $\mathbb{R}.S = \mathbb{R}.(1, 0)$, and the points of S are not in general position but aligned on a horizontal straight line of height 10. Musically, we have to play a regularly spaced sequence of three notes of equal loudness. If we use this local composition as a germ for a local rhythm with period 3, this yields a uniform sequence of events. However, if we want to stress the period and play the first event of this germ louder than the others, at value 18 say, we obtain a germ

[12] Recall that in the category **Mod** the empty module is admitted, see Appendix E.3.
[13] Including the zero-dimensional case $d_x = \{0\}$ and $R.S = 0$.

$S^* = \{(0,18),(1,10),(2,10)\}$. Here, we have $\mathbb{R}.S = \mathbb{R}.(1,0) \oplus \mathbb{R}.(1,-8)$, and this is a general position situation. Musically, putting the germ events into general position makes a hidden (purely mental) idea—in this case: the periodicity and the germ—evident, i.e., audible.

7.5 Alterations Are Tangents

Summary. By use of dual numbers, alterations of tones in pitch or any other coordinate are interpreted as 'tangent' objects. Associated transformations between tangents and base-points are introduced.

– Σ –

The technique of altered notes is more than a notational formalism. For example, take a piano note denoted by "e-sharp". On a well-tempered piano, the key it denotes is just f, and no deeper reason exists to write "e-sharp" instead of "f" to define a particular key and pitch. But for variable temperament such as is present on string instruments, this cannot be done. Whatever the concrete meaning of such alterations, they do point to a refined situation: We have to keep in mind two notes: the origin and the amount of shift from the origin, to speak geometrically.

From Appendix E.3, (E.5), we know that any restriction $r : R \to S$ of commutative scalars gives rise to a functor $S \otimes_R ? : \mathbf{Mod}_R \to \mathbf{Mod}_S : N \mapsto S \otimes_R N$ which carries an R-affine transformation $f = e^k \cdot f_0 : N \to K$ to the S-affine $S \otimes_R f = e^{1 \otimes k} \cdot S \otimes_R f_0$. To describe alterations we consider the dual number restrictions

$$i : R \to R[\varepsilon] : r \mapsto r + 0\epsilon$$
$$p : R[\varepsilon] \to R : a + b\varepsilon \mapsto a$$

and set $M[\varepsilon] = R[\varepsilon] \otimes_R M$ and $f[\varepsilon] = R[\varepsilon] \otimes_R f$ for corresponding functor values on objects and morphisms in $\mathbf{Mod}_R$. Recall from Appendix E.2.2 that over R, $M[\varepsilon] = M + \varepsilon M \xrightarrow{\sim} M^2$ such that scalar multiplication on dual numbers is given by $\varepsilon(a,b) = (0,a)$.

In this section we concentrate on (not necessarily objective) A-addressed local compositions in simple ambient spaces $@M$, M modules over commutative rings. Let $T \hookrightarrow @A \times @M$ be such an object for the ring R. Call T_ε the image of T under the embedding $Id \times @i : @A \times @M \rightarrowtail @A \times @M[\varepsilon]$ such that we obtain a commutative diagram

$$\begin{array}{ccc} T & \xrightarrow{\text{subfunctor}} & @A \times @M \\ {\scriptstyle i_T}\downarrow & & \downarrow{\scriptstyle Id \times @i} \\ T_\varepsilon & \xrightarrow{\text{subfunctor}} & @A \times @M[\varepsilon] \end{array} \tag{7.27}$$

of functors. Conversely, let $T \hookrightarrow @A \times @M[\varepsilon]$, and call T_{red} the image of T under the projection $Id \times @p : @A \times @M[\varepsilon] \twoheadrightarrow @A \times @M$. Then we have a corresponding commutative diagram

$$\begin{array}{ccc} T & \xrightarrow{\text{subfunctor}} & @A \times @M[\varepsilon] \\ {\scriptstyle p_T}\downarrow & & \downarrow{\scriptstyle Id \times @p} \\ T_{red} & \xrightarrow{\text{subfunctor}} & @A \times @M \end{array} \tag{7.28}$$

of functors; both diagrams remind us of the previous diagrams which we have encountered in (7.22) and (7.23). Evidently, there is need for more systematic treatment of such relations among local compositions, this will be done in Chapter 8.

If we suppose that the elements of a local composition with ambient space $@M$ are musically reasonable entities, this is no longer the case with *dual ambient space* $@M[\varepsilon]$. We view this artifact as follows. If $x : A \to M[\varepsilon]$ is a denotator in this ambient space, it can be written as $x = x_{red} + \varepsilon x_{alt}$ with two components $x_{red}, x_{alt} : A \to M$. Musically speaking, the *reduced* component x_{red} is the meaningful musical event whereas the *alteration* component x_{alt} symbolizes a shift from the reduced value. This suggests that we view alterations as tangents[14]. If, for example, $M = Pitch$, we may view x as being a 'real' pitch x_{red} plus an 'pitch alteration quantity' x_{alt}. For instance, if pitch is codified such that a semitone corresponds to real number 1, and 60 to middle C, then the zero-addressed $x = 60 - \varepsilon 2$ codifies $C_{\flat\flat}$.

To recover 'real music events' from this formalism, consider two R-linear projections $\alpha_+, \alpha_- : M[\varepsilon] \to M$, defined by $\alpha_+(x) = x_{red} + x_{alt}$ and $\alpha_-(x) = x_{red} - x_{alt}$, respectively. Clearly, in the above example, α_+ gives the altered pitch number $58 = \alpha_+(60 - \varepsilon 2)$ from the dual representation, whereas α_- gives 62.

Definition 18 *Transformation α_+ and the corresponding natural transformations $@\alpha_+$ is called* sweeping orientation*; the transformation α_- and the corresponding natural transformation $@\alpha_-$ is called* hanging orientation.

These concepts will become central in the counterpoint theory of part VII. Again, we have corresponding commutative diagrams on A-addressed local compositions T in dual ambient space $M[\varepsilon]$:

$$\begin{array}{ccc} T & \xrightarrow{\text{subfunctor}} & @A \times @M[\varepsilon] \\ \downarrow & & \downarrow{\scriptstyle Id \times @\alpha_\pm} \\ \alpha_\pm T & \xrightarrow{\text{subfunctor}} & @A \times @M \end{array} \tag{7.29}$$

with $\alpha_\pm T$ being the image of T under the 'ambient morphism' $Id \times @\alpha_\pm$. The consideration of both, positive and negative alterations is a well known—though not very frequent—tool to produce two alterated pitches from one 'abstract' datum, i.e. viewing simultaneously pitch 65 as $\alpha_-(66 + \varepsilon 1)$, and pitch 67 as $\alpha_+(66 + \varepsilon 1)$, a so-called "di-alteration".

7.5.1 The Theorem of Mason-Mazzola

Summary. The use of alterated pitch is a standard argument in music analysis to associate neighboring tones. The theorem of Mason deals with the richness of possible alterations within 12-chromatic pitch classes between determined scales; Mason's theorem is generalized to n-chromatic classes. Musicologically, the theorem gives information about the significance or insignificance of alteration arguments in traditional analysis.

– Σ –

In classical music theory, alterations are very frequent. Most chords are loaded with alteration signs, thereby reinterpreting given pitch as an alteration of a different 'pitch origin'. This is often inevitable since in real scores, very rarely do we find the pitch arsenal within a determined diatonic scale. But it is nevertheless "clear" that one essentially stays in a given tonality, as represented by the basic diatonic scale. The problem of such an approach is that reinterpreting any given set of pitch values as being the result of a specific alteration process provokes a huge amount of polysemy. In the worst case, one could argue that this technique may be abused to 'demonstrate' anything by construction of appropriate alterations. The following results deal with this problem and give a general answer to the question of how broad an alteration process can become in terms of ambiguities. Our results generalize the theorem of Robert M. Mason [660] concerning alteration with 12-tempered pitch classes.

In the following discourse we shall only deal with commutative local compositions in ambient spaces $@M$ of modules M over a ring R and the associated dual version. *Commutative local compositions of ambient space $M[\varepsilon]$ are also called* alterations.

Definition 19 *Let T be an A-addressed alteration of ambient space $M[\varepsilon]$, and let $0 \neq m \in A@_R M$. Then T is called m-elementary iff it contains one element $k = k_{red} + \varepsilon m$ such that $T - \{k\} = (T - \{k\})_{red}$ and $T - \{k\} \cap \{\alpha_+(k), p(k)\} = \emptyset$.*

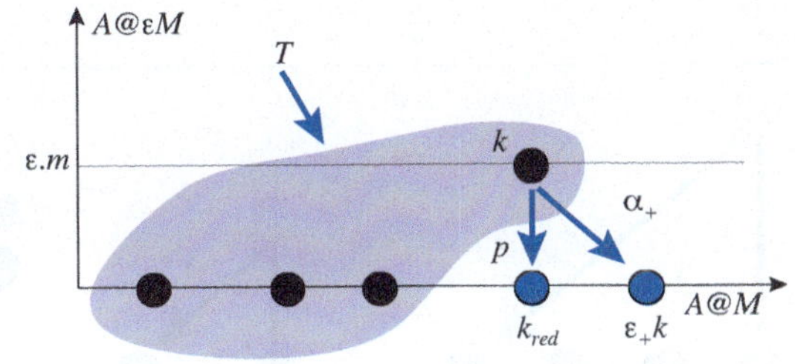

Fig. 7.13. An m-elementary alteration T controls a shift between its reduced and its alterated images.

This means that there is a bijection $T_{red} \xrightarrow{\sim} \alpha_+ T$ which is induced by two bijective projections $T \xrightarrow{\sim} T_{red}$, and $T \xrightarrow{\sim} \alpha_+ T$, see also Figure 7.13. This bijection consists in an m-shift of one element $p(k)$ to $\alpha_+(k) = m + p(k)$, whereas the other elements remain fixed. In particular, if $M = R$ and $m = e^{\pm 1}.0$ (constant shift), we say that T is *elementary*.

Definition 20 *If E and F are two local compositions such that there is an elementary alteration T over ring R with $E = T_{red}$ and $F = \alpha_+ T$, we write $E \rightsquigarrow F$ and say that F* is an elementary shift of *E. If there is a sequence $E_0 = E \rightsquigarrow E_1, E_1 \rightsquigarrow E_2, \ldots E_{i-1} \rightsquigarrow E_i = F$ of elementary shifts, we again write $E \rightsquigarrow F$ and say that F* is a shift of *E.*

If we have an elementary shift $E \rightsquigarrow F$, we also have an *associated bijection* $E \xrightarrow{\sim} F$. The following is then immediate[15]:

Sorite 3 *If there is a shift $E \rightsquigarrow F$, there is a unique alteration T in $A@R + \varepsilon e^{\mathbb{Z}1_R}.0$ such that the bijection $E \xrightarrow{\sim} F$ associated with the shift is defined by two bijections $T \xrightarrow{\sim} T_{red} = E$ and $T \xrightarrow{\sim} \alpha_+ T = F$.*

The following results which were solicited by Mason's investigation [660] show that the existence of shifts is a strong condition on couples of *equipollent*[16] local compositions.

Theorem 1 *Let n be a positive natural number, and let E, F be two equipollent zero-addressed local compositions in $\mathbb{Z}_n$ with cardinality m. Then there are two non-negative integers k, l with $k + l = n - m$ and a shift $E \rightsquigarrow F$ which is induced by an alteration $K \subset \mathbb{Z}_n + \varepsilon[-l, k]$, where $[-l, k] = \{-l, -l+1, \ldots k-1, k\}$. We can even choose K as being defined by a sequence of* elementary *shifts $E_i \rightsquigarrow E_{i+1}$ which all satisfy the condition imposed on K.*

See Figure 7.14 for an illustration. From the proof of Theorem 1 we will also obtain Mason's result:

Proposition 2 (Mason, [660]) *For $n = 12$ and $7 \leqslant m$, we can choose $|k - l| \leqslant 1$ in the above theorem.*

For classical European music, the special case $m = 7$ yields $k = 3, l = 2$ or $k = 2, l = 3$. This means that any scale with seven-element class chord F can be shifted to any other equipollent scale with class chord E by an alteration K whose elements have at most three sharps and two flats or at most two sharps and three flats. In particular, any class chord Ch with $card(Ch) \leqslant 7$ is the shift of another equipollent class chord in any fixed diatonic scale. This fact is very important to judge musicological argumentations built upon alterations. It could happen that reinterpreting a chord as a shift of another, distinguished, chord (such as

---
14 In algebraic geometry, points in dual modules $M[\varepsilon]$ are effectively identified with Zariski tangents, see Appendix F.4.1 and [775, p.25].

15 Here, the summand $e^{\mathbb{Z}1_R}.0$ relates to the constant shift morphisms $e^{t.1_R}.0 : A \to R, t \in \mathbb{Z}$.

16 Equipollence means that they have equal cardinality.

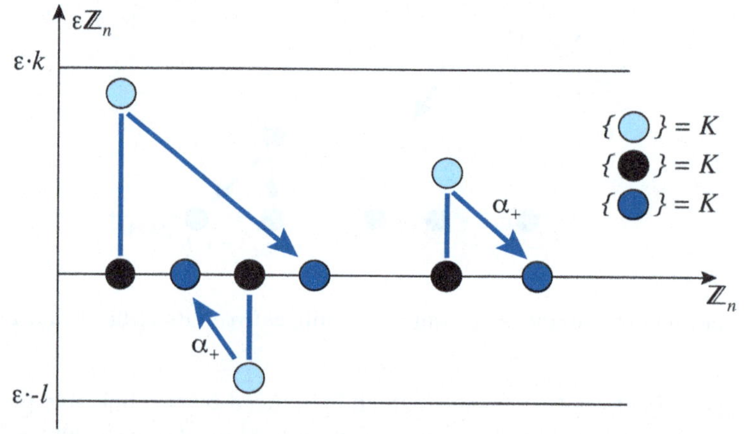

Fig. 7.14. The condition imposed on on alteration K in Theorem 1.

one sitting in a particular scale as above) is always possible and therefore does not mean anything to the actual statement!

For the proof of Theorem 1 we need some notation. Let k, l be as in the theorem, and let $x \in \mathbb{Z}_n$. Write $[l|x|k] = e^x \cdot [-l, k]$ for the x-shift of the interval $[-l, k]$ as defined in the theorem. If k, l are clear we also write $x_+ = x + k$ and $x_- = x - l$, whence $[l|x|k] = [x_-, x_+]$. Since $k + l = n - m < n$, $[l|x|k]$ has exactly $n - m + 1$ elements, and it makes sense to say that an element $y \in [l|x|k]$ is to the right or to the left of x according to whether y appears before of after x in the sequence

$$x_-, x_- + 1, \dots x, \dots x_+ - 1, x_+.$$

Consider now the 'second' $\sigma = \{y, y+1\}$. If y stays to the left of x, we write $x_-|\sigma|x$, if $y+1$ stays to the right of x we write $x|\sigma|x_+$.

Lemma 4 *Let $E \rightsquigarrow F$ be a shift whose alteration K verifies the theorem's conditions with k, l and let $s \in E$.*

(i) *If σ is a 'second' such that $\alpha_+ p^{-1} s = s_-$ and $s_-|\sigma|s$, then there is no $t \in E$ such that $\alpha_+ p^{-1} t = t_+$ and $t|\sigma|t_+$.*
(ii) *If σ is a 'second' such that $\alpha_+ p^{-1} s = s_+$ and $s|\sigma|s_+$, then there is no $t \in E$ such that $\alpha_+ p^{-1} t = t_-$ and $t_-|\sigma|t$.*

Proof of lemma. The proof scheme is the same for both cases, and we may restrict to case (i). Look at the subset $[t_-, t_+] \times [s_-, s_+]$ of $\mathbb{Z}_n \times \mathbb{Z}_n$ and refer to Figure 7.15. The diagonal of $\mathbb{Z}_n \times \mathbb{Z}_n$ is denoted by Δ. If we had a t, which is forbidden by statement (i) of the lemma, then evidently $(t, s) \notin \Delta$. Under elementary shifts the couple (t, s) is shifted one unit in horizontal or vertical direction within $[t_-, t_+] \times [s_-, s_+]$, but it cannot hit Δ. Since according to our premises, the couple (t, s) stays to the left of Δ, it cannot reach the bottom right edge (t_+, s_-). QED.

Proof of theorem. We make an induction on $n - m$ and observe that the cases $n - m = 0, 1, 2, 3, 4$ are easily verified under the stronger hypothesis $|k - l| \leqslant 1$. So suppose $5 \leqslant n - m$. Suppose first that $E \cup F \subsetneq \mathbb{Z}_n$, and then WLOG[17] we can suppose that $z = \overline{n-1} \notin E \cup F$. We may therefore suppose that $E, F \subset \mathbb{Z}_{n-1}$ by contracting the 'second' $\{\overline{n-1}, \overline{0}\}$ to point $\overline{0} \in \mathbb{Z}$ and creating the 'second' $\sigma = \{\overline{n-2}, \overline{0}\}$ in $\mathbb{Z}_{n-1}$. By induction hypothesis, there are non-negative integers k', l' with $k' + l' = n - m - 1$ and a sequence $E_0 = E \rightsquigarrow E'_1, E'_1 \rightsquigarrow E'_2, \dots E'_{i-1} \rightsquigarrow E'_i = F$ of elementary shifts in $\mathbb{Z}_{n-1}$ which shift every element $s \in E$ within $[s_-, s_+]$. According to Lemma 4, we cannot have simultaneously a shift of $s \in E$ to s_- and a shift

[17] Mathematicians' short form of "without loss of generality'.

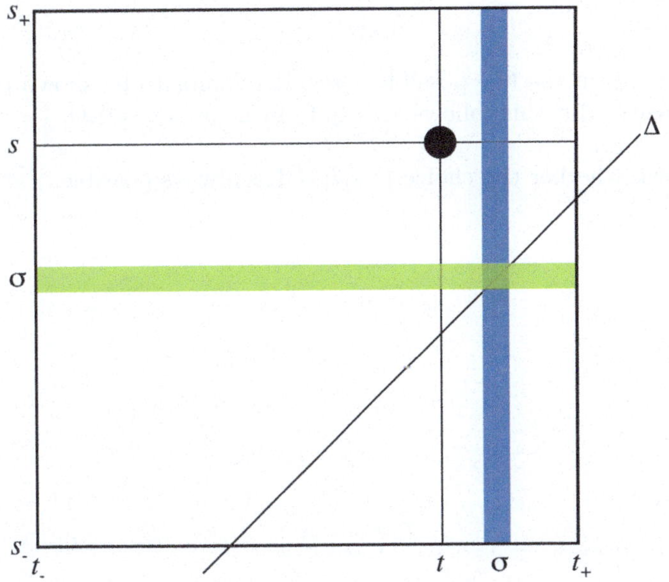

Fig. 7.15. The product set $[t_-, t_+] \times [s_-, s_+]$ and the relative position of (t, s) with respect to the diagonal Δ and the 'second' diagonal (σ, σ).

of $t \in E$ to t_+ with $s_-|\sigma|s$ and $t|\sigma|t_+$. At worst, either some s are shifted to s_- with $s_-|\sigma|s$ or some t are shifted to t_+ with $t|\sigma|t_+$. After reinsertion of z we have a required shift with $k = k', l = l' + 1$ in the first case; in the second case we have a shift with $k = k' + 1, l = l'$, and we are done. The case $E \cup F = \mathbb{Z}_{n-1}$ is a bit more involved. Take a point $z \in \mathbb{Z}_n$ where $z - 1 \in F$ and $z \in E - F$. Suppose that here, we have $z, z + 1, \ldots z' \in E$ whereas $z' + 1 \notin E$.

To begin with, we shift $z, z + 1, \ldots z'$ by one unit to the right and obtain $\tilde{E}$. Let $\sigma = \{z - 1, z + 1\}$ be the 'second' in $\mathbb{Z}_{n-1}$ as above from eliminating z. By induction hypothesis, there are $0 \leqslant k', l'$ with $k' + l' = n - m - 1$ and a sequence of elementary shifts starting from $\tilde{E}$ and ending in F, which shift the elements $s \in \tilde{E}$ within $[l'|s|k']$. The point here is to understand what happens to the shift quantities if we reinsert z 'in the middle of' σ. If $s \in \tilde{E}$, we denote the points s_-, s_+ with respect to k', l'. There are six cases:

1. Point z is unshifted and neither $s_-|\sigma|s$ nor $s|\sigma|s_+$.
2. Point z is shifted and neither $s_-|\sigma|s$ nor $s|\sigma|s_+$.
3. Point z is unshifted and $s_-|\sigma|s$.
4. Point z is unshifted and $s|\sigma|s_+$.
5. Point z is shifted and $s_-|\sigma|s$.
6. Point z is shifted and $s|\sigma|s_+$.

The last possibility cannot happen since $s - (z + 1) + k' \leqslant m - 1 + n - m - 1 = n - 2$, and therefore, σ cannot be to the right of s. We shall do the following: Whenever an element s is shifted to jump from $z - 1$ to $z + 1$ in σ or conversely within one of the existing elementary shifts, we shall replace this elementary shift by the succession $z - 1 \rightsquigarrow z \rightsquigarrow z + 1$ or $z + 1 \rightsquigarrow z \rightsquigarrow z - 1$ of two elementary shifts. So we are left with the question of what happens to the possible total shift intervals. *After reinsertion* of z there is no change in the shift intervals in cases 1. and 2. since z does not intervene in those intervals. Set $\tau = \{z, z + 1\}$. In case 3. we get $(s - (l' + 1))|\tau|s$, in case 4. we have $s|\tau|(s + (k' + 1))$, and in case 5. we have $(s - (l' + 1))|\tau|s$. If we go back to the *original* positions, we have these total shift intervals

1. $[l'|s|k']$
2. $[l' - 1|s|k' + 1]$
3. $[l' + 1|s|k']$

4. $[l'|s|k'+1]$
5. $[l'|s|k']$

for (unshifted) elements $s \in E$ in the five possible cases. It remains to be shown that case 3. versus cases 2. and 4. are mutually exclusive. But this follows rightly from Lemma 4. QED.

Remark 2 It is not known whether the choice $|k-l| \leqslant 1$ is always possible.

8

Symmetries and Morphisms

Hoffentlich habe ich diejenigen beschwichtigt,
die ein gewisses Unbehagen vor dem Worte Symmetrie,
das zu phantasievollen Spekulationen Anlaß gegeben hat,
nicht verbergen können, die darin Formalismus, Schematisums sehen und es
am liebsten ganz aus der 'göttlichen' Vernunft verbannen möchten.
Wolfgang Graeser [387]

Summary. After having introduced the objects of local theory, their relations are discussed and formalized. Thereby, symmetry is a key concept—however in its modern version which by far exceeds traditional axial or rotational symmetries. The core process of symmetry leads to structure-conserving transformations which we rebuild in the context of local compositions and the underlying forms.

– Σ –

The basically 'rigid' structure of a local composition as it was discussed in Section 7.1 requires an additional ingredient of conceptualization to be able to compare such 'rigid' objects. Such a concept of comparison must then conserve the interior correspondences among parts of the local compositions, it must be a concept of "morphological" comparison which is able to grasp these correspondences as expressions of underlying gestalts.

Such a type of comparative paradigm leads to the concept of *symmetry* as it is used in its modern form [1129] and may be rephrased[1] more generically as follows:

Definition 21 *Symmetry is correspondence of parts as an expression of a whole.*

On one hand, this definition by far exceeds simple axial symmetry, on the other it recognizes the semantical function of symmetry and is no longer limited to formal aspects. The modern concept of symmetry is in fact a semiotical one: Symmetry bears the character of a sign. The *form* (the significant) of a symmetry is a mathematical transformation which specifies the correspondence, whereas the function (signification) of a symmetry consists of a production mode of meaning (the significate) from the relation between meanings (significates) of the form's parts (expressions) which are transformed into each other under the given transformation. For example, the dynamical symmetry of a ropedancer produces an equilibrium of his body which prevents him from crashing down and risking his life. The physical symmetry of forces means life to the ropedancer.

We insist on replacing the word "Gleichheit" (equality) in Wille's definition by "correspondence" in the above definition since there is no a priori replication under correspondence, and this is essential to avoid the widespread belief that symmetry is an 'equalizer' concept. The concept of symmetry does in fact *not* imply equality of symmetrically related parts, only correspondence! In a symmetry of moral qualities, for example, good and evil are not equalized, but they do correspond.

[1] Rudolf Wille's definition in [1129, p.444] is this: "Symmetrie ist Gleichheit von Teilen als Ausdruck eines Ganzen."

G. Mazzola, *The Topos of Music I: Theory*, Computational Music Science,
https://doi.org/10.1007/978-3-319-64364-9_8

We also feel obliged to contradict the stubborn simplistic argument against symmetries in fine arts, literature, and music which debates that symmetries would cast the artistic freedom of shaping highly sophisticated works of human expression into a trivial or even fascistoid Procrustean bed and thereby fail to explain the very essence of artistic expression. The point is that

- First, *symmetries are not the only instance* of artistic expression, local-global strategies are at least as important (see also Chapter 13.1 for this subject).
- Second, symmetries often appear as broken or *hidden symmetries* which means that they are not immediately visible, and only appear on the phenomenological surface when investigated with powerful tools. In elementary particle physics, this fact is well known and—hélas—responsible for exorbitant costs of particle accelerators: Fundamental laws of physics are virtually always driven by hidden symmetries which are only visible under extremely high energies.
- Third, the stratification of symmetries into form and content initiates *complex relations of form and content*. For example, Kepler's second law states that a planet sweeps out area at a constant rate[2]. Hence it is not the distance between a planet and the sun which remains constant but the area which is swept out by the straight line connecting planet and sun. The visible symmetry of a "perfect" circle motion is replaced by a more abstract symmetry where "distance" is replaced by "area per time unit". The meaning of planetary circle motion is lifted to a more involved physical concept, area per time unit relates to angular momentum[3].

As a final argument for using non-trivial symmetries in music, we should admit that deep musical facts cannot be understood by use of trivial tools, this would be as absurd as to develop physical theories without infinitesimal calculus. If the great composers and musicians participate in a sphere of high or even divine sophistication it is only logical to approach their works by use of adequate, equally skilled tools of human insight. We therefore hope that the reader will not reject conceptually involved investigations because of a conflicting belief that there should be easy ways to understand the genius in music.

8.1 Symmetries in Music

Summary. This introductory section reviews the multiple and sometimes controversial presence of symmetry in music. We give a short historical account of the quasi-simultaneous appearance of explicit symmetry paradigms in music and mathematics in the works of Schönberg, Graeser, and Noether around 1928. The discussion is complemented by reference to Jakobson's poetic function which reveals a dominant role of symmetries in every poetic effort.

– Σ –

A classical "counterexample" against symmetries in music is the retrograde of a melody. In music psychology it is traditionally argued that the retrograde (see example 7 below) is not recognized (heard) as being essentially the "same" underlying melody. Such a claim fails to grasp the very concept of symmetry which in this case only refers to a *privileged relation* between the original and the retrograde version. The question would rather be whether the listener feels a privileged relation between melody and its retrograde, as compared to any random melody.

This example reveals the essential role of the *topographic* approach to symmetry in music in the sense of Section 2.4. Questioning the role of symmetries in music must always begin with the topographic initial "Where?", otherwise the effort is futile. Whatever its vagueness, the problem whether symmetries can be "heard" is above all an esthesic one and has—to our knowledge—not been investigated systematically in music psychology. On the other hand, from the neutral and poietic point of view, symmetries play a prominent role. We come back to this aspect in Chapter 11. Let us just remark here that first, symmetries have been applied (among many others) from Johann Sebastian Bach in fugal composition to Pierre Boulez in serial

[2] See [617, p.524].

[3] It is the law $\mathbf{x} \wedge \mathbf{p} = const.$ of conservation of the angular momentum but see [617, p.523]

techniques. Following music theorist Wolfgang Graeser we may further ask whether the incessant structural fascination of Bach's music cannot be explained from a network of locally present symmetries. Second, neutral signification of symmetries is explicitly given in Arthur von Oettingen's [806] and in Sigfrid Karg-Elert's [521] modulation theories. Nowadays, symmetries are widely used in musical composition, for instance in the various transformation tools of sequencers, and explicitly in the author's *presto*®(see Chapter 50) and in Opcode's MAX or in music software with fractal[4] composition tools as realized by Mesias Maiguashca [461] or by Peter Stone on his ingenious "Symbolic Composer" software, see Figure 8.1.

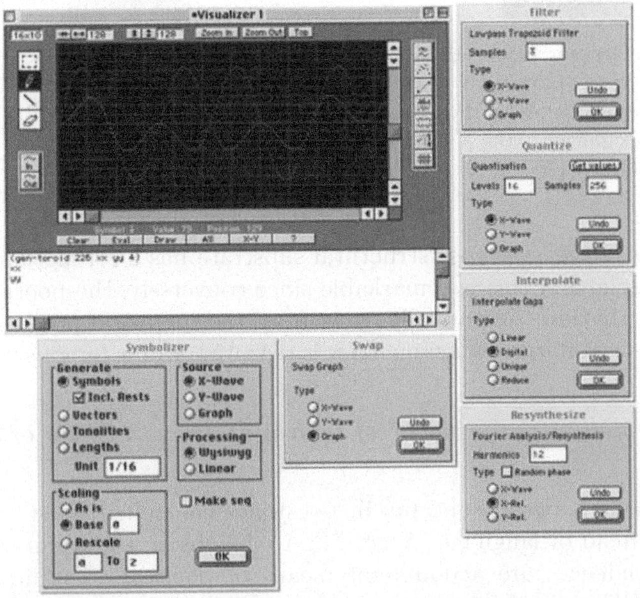

Fig. 8.1. Peter Stone's "Symbolic Composer" [1015] software includes a large number of composition operators, in particular for fractal techniques.

From the historical point of view there is an interesting coincidence of musicological and mathematical development regarding symmetries. Around 1924–1929, the paradigm of symmetries grew into its first order prominence in music as well as in mathematics. In 1928, Arnold Schönberg's (Figure 8.3) "Orchestervariationen" op. 31 had its premiere, realizing the thoroughly symmetric dodecaphonic composition principle (see example 10 below). It is very likely the first time in the history of western music that an entire mathematical group orbit (the orbit of a given series under the full contrapuntal affine group generated by inversion, retrograde, and transposition (see Section 8.1.1, example 12, for precise definitions) of a kernel motif (the composition's fundamental series) was *consciously* designed to define a motivic paradigm. In music analysis, it was the aforementioned young Wolfgang Graeser (Figure 8.2) who in 1924 applied symmetries with great success to Johann Sebastian Bach's "Kunst der Fuge" [387]. Here is his summary:

> *Die Eigenschaft der Symmetrie spielt in der Musik eine so ungeheure Rolle, daß sie verdient, an erster Stelle betrachtet zu werden. Wir werden in der "Kunst der Fuge" ihre fast uneingeschränkte Herrschaft besonders deutlich erkennen.*

In modern abstract algebra, it was Emmy Noether who accomplished a persuasive theory of general symmetries, i.e., the theory of group representations in vector spaces, in the paper "Hyperkomplexe Größen und Darstellungstheorie", written in 1929 [794]. Needless to say that symmetries play a crucial role in contemporary mathematics and therefore in mathematical physics.

[4] Observe that fractals are generated by symmetries in the sense of iterated application of one fixed transformation.

Fig. 8.2. Wolfgang Graeser (1906–1928) was the first music theorist to apply systematically symmetry groups to music analysis. Permission from Zentralbibliothek Zürich, Mus NL 162 : G (Nachlass Wolfgang Graeser).

Fig. 8.3. Arnold Schönberg (1874–1951) was the first composer who consciously applied full orbits of symmetry groups to musical composition. Permission from ÖNB/Wien, 162.542 B

Let us shortly digress on poesy whose structural substrate has a strong musical character. Here, symmetries play a prominent role, and this is remarkable since conversely, the poetical moment in music—as it is mediated by meter and rhythm—historically grew from the linguistic poetical form, and since by these facts the apparently unpoetical nature of symmetries is defeated. In his famous statement, Roman Jakobson has defined poeticity [500]:

> *The poetical function projects the principle of equivalence from the axis of selection to the axis of combination.*

This means that associations between concepts in poesy are controlled by strict correspondence within a rhythmically segmented thread of language. According to Jakobson, the symmetries which appear in these configurations of correspondences are a dominant means of poetical structuration [504]. In its study of classics such as Friedrich Hölderlin [503], Dante Alighieri [441] or François Villon [632], modern poetology has indeed discovered an astonishing density and even dominance of symmetrical constructions; we come back to Jakobson's poetical function in Section 11.6.1.—Also in traditional European musicological analysis, symmetries have been recognized with growing success [786], see Section 11.5.2 for an extensive discussion in symmetry-oriented musicological analysis as cultivated in the American tradition of "musical set theory".

8.1.1 Elementary Examples

Summary. Before starting the general theory of morphisms between local compositions, which encompasses a large portion of what can be thought of being a symmetry phenomenon in music, we should include a number of elementary examples of symmetries in music.

– Σ –

A word of caution: In the course of these examples we use words such as symmetry, transformation etc. in an informal way. Only in Section 8.2 will we introduce the corresponding technical terms.

Example 5 Transposition.

The symmetry of transposition is fundamental for thinking in terms of pitch classes. More precisely, we are given a "translation vector" $t \in \mathbb{Q}^3$ which defines a translation transformation $e^t : \mathbb{Q}^3 \to \mathbb{Q}^3$ and therefore a transformation of functors $@e^t : @\mathbb{Q}^3 \to @\mathbb{Q}^3$ which transforms any EulerModule-denotator $x : A \rightsquigarrow EulerModule(x)$ into its t-translate $e^t(x) : A \rightsquigarrow EulerModule(e^t \cdot x)$. If x is even a zero-addressed point in the *EulerModule* space, it may be identified with an element $x \in \mathbb{Q}^3$, and we have $e^t \cdot x = t + x$. Applying all translates $e^{no}, n \in \mathbb{Z}$ of integer multiples of the octave o to x yields

$$mod_o(x) : 0 \rightsquigarrow o\text{-}EulerClass(e^{\mathbb{Z}o} \cdot x),$$

the *octave class* of x. For such an octave class $mod_o(x)$ and an octave class k-chord

$$Ch : 0 \rightsquigarrow o\text{-}ClassChord(c_1, c_2, \ldots c_k),$$

we can consider the translate $e^{mod_o(x)} \cdot Ch = \{e^{mod_o(x)} \cdot c_1, e^{mod_o(x)} \cdot c_2, \ldots e^{mod_o(x)} \cdot c_k\}$. If for example Ch is the 12-tempered C-major scale $C_{maj} = \{\overline{0}, \overline{2}, \overline{4}, \overline{5}, \overline{7}, \overline{9}, \overline{11}\}$, the translate $e^{mod_o(11)} \cdot C_{maj}$ is equal to the 12-tempered B-major scale $B_{maj} = \{\overline{11}, \overline{1}, \overline{3}, \overline{4}, \overline{6}, \overline{8}, \overline{10}\}$, whereas in just temperament, the just B-major scale (chord) is obtained from the just C-major scale (chord) via a translation $e^{q+t} = e^q \cdot e^t$ of a fifth (class[5]) q plus a major third (class) t.

Example 6 Repetition, da capo, canon, rhythm.

Translation e^t on the simple space *Onset* of onsets means something very different from the same mathematical transformation on pitch domain. We have four different common meanings in this case:

- If we view the onset axis as being a time line, translation of a local composition G in an ambient space $S = Onset \oplus Para$ by e^t, i.e., the translation $e^t \cdot G$ in the sense of rhythm theory as introduced in Section 7.2.2, means playing G a second time, repeating it later in the course of world time. This seemingly trivial fact is the case if we *replay a composition* on a media such as tape, LP, CD, etc. In the technically ideal case, every replay is a translation on the physical time axis. We have interiorized this symmetry to such a degree that we do not even realize it as being essentially different from identity; it is often and erroneously felt that any such reproduction is the piece itself; this is a point where Sergiu Celibidache [944] was right.
- We can also repeat parts of a composition *within* the composition in the sense of a *da capo*.
- A special case of a repetition within one and the same composition is the duplication of a melody in the *canon* construction.
- Finally, we may view the looping, i.e., periodic repetition of a given "percussive pattern" as a generation principle for *rhythmical* structures, this was already introduced in Section 7.2.2.

Example 7 Inversion, retrograde, retrograde inversion.

These are classical symmetries in composition and theory. Let us have a closer look at the highly instructive analysis of these concepts and their multiple meaning and corresponding mathematical interpretations.

Intuitively, *inversion*[6] is a reflection of the pitch axis at a fixed pitch. However, its visual shape depends on the selected pitch space. Everyone of the hitherto discussed pitch spaces *Ptch* were of simple form:

Pitch has the $\mathbb{R}$-vector space coordinator $\mathbb{R}$,

MathPitch has the $\mathbb{Q}$-vector space coordinator $\mathbb{R}_{[\mathbb{Q}]}$,

EulerModule has the $\mathbb{Q}$-vector space coordinator $\mathbb{Q} \cdot o \oplus \mathbb{Q} \cdot q \oplus \mathbb{Q} \cdot t = \mathbb{Q}^3$,

EulerPlane has the $\mathbb{Q}$-vector space coordinator $\mathbb{Q} \cdot q \oplus \mathbb{Q} \cdot t = \mathbb{Q}^2$,

w-TemperedScale, the w-tempered scale from origin 0,

has the $\mathbb{Z}$-module coordinator $\mathbb{Z} \cdot \frac{1}{w} \cdot o$, and

w-PitchClass has the $\mathbb{Z}$-module coordinator $\mathbb{Z} \cdot \frac{1}{w} \cdot o / \mathbb{Z} \cdot o \xrightarrow{\sim} \mathbb{Z}_w$.

---

[5] Since the fifth and third coordinates are not touched by factorization through the octave in just classes of the Euler plane (see Equation (7.7)) we can refrain from adding the class attribute here.

[6] Pay attention to distinguish the symmetry of inversion from chord inversion which just means moving some of their notes one octave up or down.

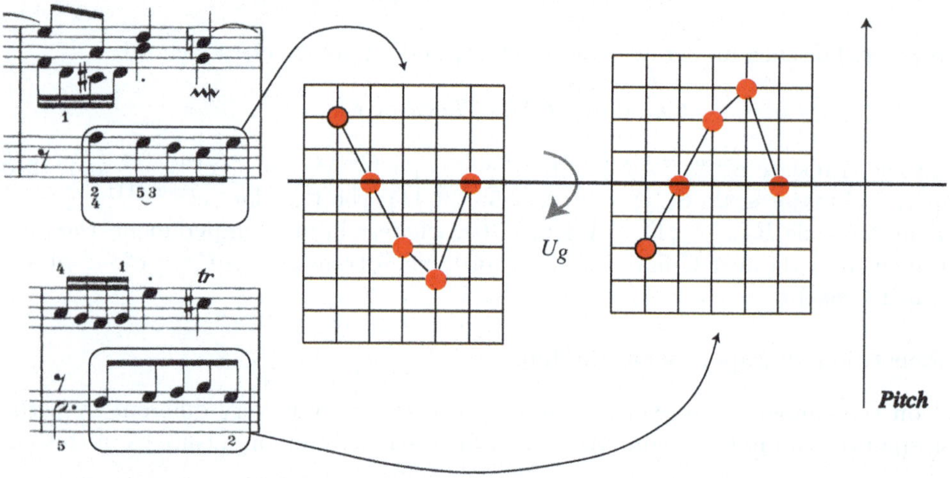

Fig. 8.4. This example from Fuga 6, D minor, in Johann Sebastian Bach's "Wohltemperiertes Klavier" I, BWV 851, shows an inversion at g in 12-*TemperedScale*. The symbol U_g is written in the traditional notation, i.e., the fixpoint is indicated. For fixpointless inversions between pitch g and $g\sharp$, for example, that notation would yield. The grid shows semitone steps and eighth notes onset units. The motivic note groups shown here (above: bar 29, below: bar 33) represent the characteristic motif of this fugue.

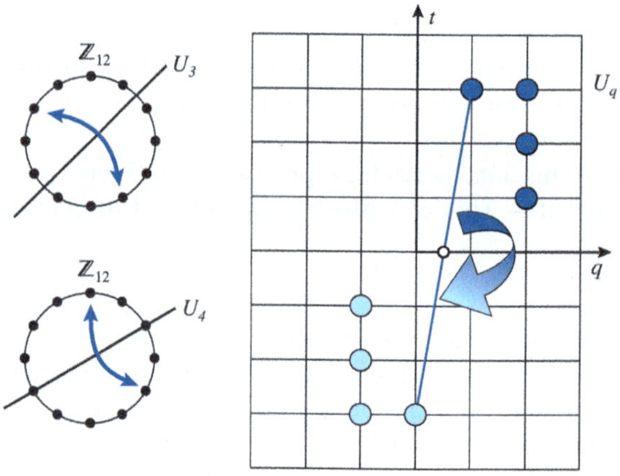

Fig. 8.5. To the left, we show two inversions U_3 and U_4 on zero-addressed denotators in 12-*PitchClass*. The first has no fixpoint and is also denotated by $U_{1/2}$ or $U_{c\sharp/d}$ in traditional notation since its virtual fixpoints are between $c\sharp$ and d as well as between g and $g\sharp$. The second inversion is U_4 has fixpoint $2 = \frac{4}{2}$ and would traditionally be denoted by $U_d = U_{g\sharp}$. To the right we have inversion U_q on the *EulerPlane* space, showing zero-addressed denotators again for simplicity. It leaves invariant the middle $q/2$ between origin and q. This symmetry can be viewed as a 180°-rotation around the fixpoint. If the inversion's shift vector p has integer coordinates with respect to the grid's basis q, t, the fixpoint will stay either on one vertex of the integer grid, or in the middle of one mesh side (as in the case of this figure), or in the middle of one mesh.

Then inversion is an affine symmetry of shape $U_s = e^s \cdot -1$ which sends a denotator $D : A \rightsquigarrow Ptch(x)$ to $U_s D : A \rightsquigarrow Ptch(U_s(x) = s - x)$. This is inversion at the origin of the coordinator module, followed by a shift e^s. The inversion has a fixpoint $x = U_s(x)$ iff with the unique decomposition $x = e^t \cdot x_0$ into translation and linear factor, we have $2t = s$ and $2x_0 = 0$. For real or rational coefficients in the first four forms this yields a unique solution $t = \frac{1}{2}s$ and $x_0 = 0$, whereas in case of w-*TemperedScale*, we have at most one solution, and there is one iff s is an even multiple of $\frac{1}{w}$; otherwise, the inversion is a reflection in the middle between two adjacent points $\eta\frac{1}{w}o, (\eta+1)\frac{1}{w}o$. For form *w-PitchClass*, we are dealing with affine di-homomorphisms in $x \in A@\mathbb{Z}_w$, and we are looking for numbers $s \in 2\mathbb{Z}_w$ and di-linear homomorphisms $x_0 : A \to \mathbb{Z}_w$ with $2im(x_0) = 0$. If w is odd, 2 has an inverse in $\mathbb{Z}_w$, and fixpoints are given by $t = \frac{1}{2}s$ and the zero-di-homomorphism $x_0 = 0$. If $w = 2q$ is even, the condition $s \in 2\mathbb{Z}_w$ is not automatic, whereas the di-linear factor must map into the subgroup $q\mathbb{Z}_w \xrightarrow{\sim} \mathbb{Z}_2$ and is parametrized by $Lin(A, \mathbb{Z}_2)$. Figures 8.4 and 8.5 show the picture of an inversion in the pitch space 12-*TemperedScale* and in pitch class spaces for 12-tempered and just tuning.

Exercise 5 *Calculate all fixpoints of an inversion U_s for self-addressed denotators*

$$D : \mathbb{Z}_{12} \rightsquigarrow 12\text{-}PitchClass(x).$$

Intuitively, the *retrograde* symmetry is a reflection of time's run down. Since several attributes are related to time, this idea is not quite clear. Time intervenes in onset, duration, sound color, and envelope of a sound. If we try to invert time naively as it is done by inversion of a tape, then obviously, something different from musicological "retrograde" happens. So let us have a more detailed look at the retrograde phenomenon.

To begin with, onset is concerned—independently of whether we consider mental or physical time. Let us select the mental onset space *Onset* with simple coordinator $\mathbb{R}$. On this space, retrograde symmetry means the complete analogy to pitch inversion and can be written[7] as $k_s = e^s \cdot -1$ for a shift number s. Whereas U_s has the same formula and acts on pitch, k_s acts on onset. However, this one-dimensional symmetry is somewhat strange if we look at its effect on two events at times *onset* and *offset* = *onset* + *onsetdistance* which are *onsetdistance* apart. If we apply k_s to such a couple $\{onset, offset\}$, we obtain $k_s\{onset, offset\} = \{k_s(onset), k_s(offset)\} = \{s - offset, s - onset\}$, and the "retrograde" of the second event will start *onsetdistance* before the "retrograde" of the first event, see bottom part of Figure 8.6. Such a symmetry may be interesting if we disregard durations, such as in drum patterns. However, if the duration of the second event at time *offset* is longer than *onsetdistance*, the transformed couple will have a disturbing effect because $k_s(offset)$ will not end when $k_s(offset)$ starts. In fact, this transformation is not what usually happens when a score is played from the right to the left in reversed order.

In order to include the role of duration in defining retrograde symmetry, we have to work in the two-dimensional space $OD = Onset \oplus Duration$ with coordinator $\mathbb{R}^2$ as defined in formulas (6.39) ff. and Section 6.6. So we have to take two events $OD(onset, duration_1)$ and $OD(offset, duration_2)$, see left upper part in Figure 8.6. We first perform k_s on their onset parameters and obtain an intermediate transformation $OD(k_s(onset), duration_1)$ and $OD(k_s(offset), duration_2)$ as shown in the lower part of Figure 8.6. To avoid overlaps due to durations, we go one step further and shift the intermediate onset by the event's duration to the left and get the events $OD(k_s(onset) - duration_1, duration_1)$ and $OD(k_s(offset) - duration_2, duration_2)$ as shown in the upper right part of Figure 8.6. Ths latter is a horizontal transvection or 'shearing' operation[8]. Putting all together we have defined a duration-sensitive retrograde which is given by the affine map

$$K_s : \mathbb{Q}^2 \to \mathbb{Q}^2 : (o, d) \mapsto (s - o - d, d)$$

which means

$$K_s = e^{(s,0)} \cdot \begin{pmatrix} -1 & -1 \\ 0 & 1 \end{pmatrix} \tag{8.1}$$

and extends functorially to any A-addressed denotator $D : A \rightsquigarrow OD(D : A \to \mathbb{Q}^2)$ if we set $K_s(D) : A \rightsquigarrow OD(K_s \cdot D)$. Examples of this duration-sensitive symmetry are abundant in classical European literature, see Figure 8.7. Observe that sound colors are not altered by this symmetry. We do obtain what is required in a fugue for *piano*, for example!

[7] The letter "k" stands for German "Krebs".

[8] See Appendix E.3.4

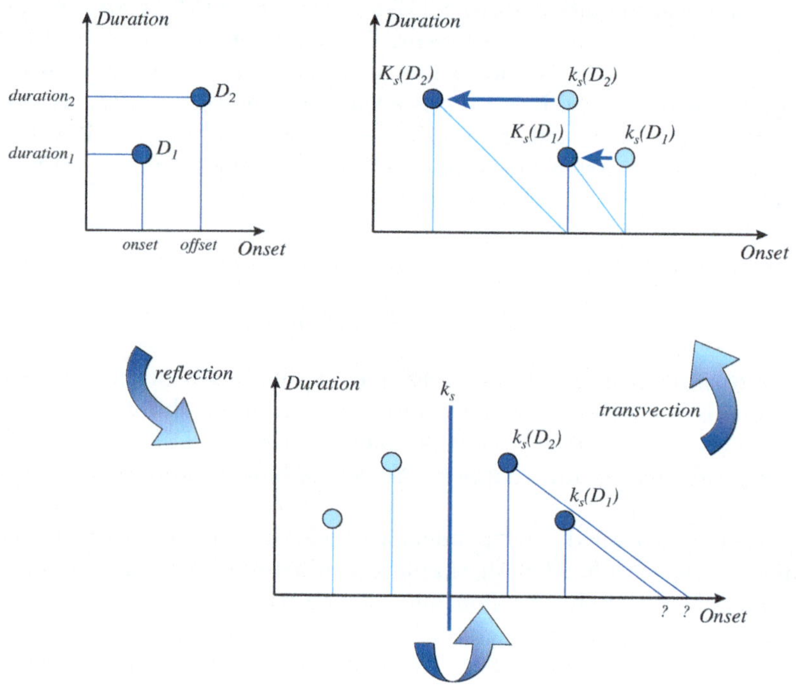

Fig. 8.6. In contrast to pitch inversion, usual retrograde concerns onset as well as duration. It is composed from a horizontal reflection (resulting in the lower configuration) and a horizontal transvection (shown in the right upper part).

Finally, *retrograde inversion* $KU_{s,t}$ is defined as composition of a retrograde symmetry K_s and a pitch inversion U_t. It involves three parameters: pitch, onset, and duration, and we may consider the typical three-dimensional space[9] $OPD = Onset \oplus Pitch \oplus Duration$ with coordinator $\mathbb{Q}^3$. We then obtain the formula

$$KU_{s,t} = e^{(s,t,0)} \cdot \begin{pmatrix} -1 & 0 & -1 \\ 0 & -1 & 0 \\ 0 & 0 & 1 \end{pmatrix} \tag{8.2}$$

which means $K_s \cdot U_t = U_t \cdot K_s$ if these symmetries are canonically lifted to the three-dimensional space OPD. More practically interpreted, $KU_{s,t}$ is a rotation of 180° on the score plane. It is reported that Beethoven, in one of those then famous competitions among pianists, forced his contrahent nicknamed "Tremolo Steibelt" to leave the salon steamed up. Beethoven had just turned around 180° Steibelt's string quartet score and played that symmetric transformation of the composition—evidently turning it into garbage (in those days it very probably *was* garbage).

Example 8 Sound transformations.

Applying symmetries to abstract sound descriptions as they occur in score notation, and which we have dealt with above, is completely different from applying symmetries directly to sound events on the physical level or reality. Let us consider this latter situation to review some details of what happens when a tape is reversed, to name a typical operation. This type of transformation on the sound level is a standard processing tool for analog and digital audio sampling technology.

[9] We choose this sequence of parameters instead of POD, for example, since one normally groups the basis parameters and then adds the pianola parameters.

Fig. 8.7. J.S. Bach: retrograde canon from "Musikalisches Opfer", BWV 1079, [55] (with kind permission of Bärenreiter Publishers).

We stick to the sound description in Fourier synthesis space

$$FourierSound \underset{Id}{\longrightarrow} Limit(PhysOnset, PhysDuration, Fourier, Envelope)$$

introduced in (6.108). The space contains denotators $myFourierSound = (e, d, myFourier, V)$ which give rise to physical sound function

$$p = sound(myFourierSound).$$

For all triples $\beta, \sigma \neq 0, \mu$ of real numbers, we obtain a new time function $p_{\beta\sigma\mu}$ which at physical time t evaluates to

$$p_{\beta\sigma\mu}(t) = \beta p(\sigma t + \mu). \tag{8.3}$$

For $\sigma = -1$, we obtain a reversed run though time and therefore the well-known tape reversion. Let us look at the effect of this symmetry type on the underlying Fourier denotators. Let the Fourier denotator be given by $myFourier = (f, (A_i, Ph_i)_{0 \leqslant i})$. Then the transformed sound function evaluates to

$$\begin{aligned} p_{\beta\sigma\mu}(t) =& \beta V(((\sigma t + \mu) - e)/d) \cdot \sum_{0 \leqslant i}^{\infty} A_i \sin(2\pi i f(\sigma t + \mu) + Ph_i) \\ =& V((t + \frac{\mu - e}{\sigma})/\frac{d}{\sigma}) \cdot \sum_{0 \leqslant i}^{\infty} (\beta A_i) \sin(2\pi i (f\sigma) t + (2\pi i f \mu + Ph_i)), \end{aligned}$$

and we have the affine transformation

$$Q_{\beta\sigma\mu} : (e, d, (f, (A_i, Ph_i)_{0 \leqslant i}, V) \mapsto (\frac{\mu - e}{\sigma}, \frac{d}{\sigma}, (f\sigma, (\beta A_i, 2\pi i f \mu + Ph_i)_{0 \leqslant i}, V) \tag{8.4}$$

which relates the symmetry on the sound level to the underlying parametrization. Let us make this a bit more transparent for later constructions. If we interpret the sound functions $p, p_{\beta\sigma\mu}$ as being zero-addressed denotators of form

$$Sound \underset{Id}{\longrightarrow} \mathbf{Simple}(C^0_{cp})$$

where the coordinator C^0_{cp} is the real vector space of functions $f : \mathbb{R} \to \mathbb{R}$ which are continuous and have compact support[10]. With this, the map

$$sound : myFourierSound \mapsto sound(myFourierSound)$$

commutes with the above symmetries, i.e., we have

$$sound(Q_{\beta\sigma\mu}(myFourierSound)) = sound(myFourierSound)_{\beta\sigma\mu} \qquad (8.5)$$

which means that *sound* is an equivariant map[11] for the two symmetry actions $Q_{\beta\sigma\mu}$ and $?_{\beta\sigma\mu}$. We also see that the envelope is left invariant by $Q_{\beta\sigma\mu}$ which means that it can even be taken as an envelope normed to maximal value 1, say, which is the situation in various sound synthesizer drivers. The loudness change is solely attributed to the Fourier sum expression. The prize for this invariance is in particular that duration may become negative for negative σ. This is the case where the sample will run in reversed direction, see also Figure 8.8.

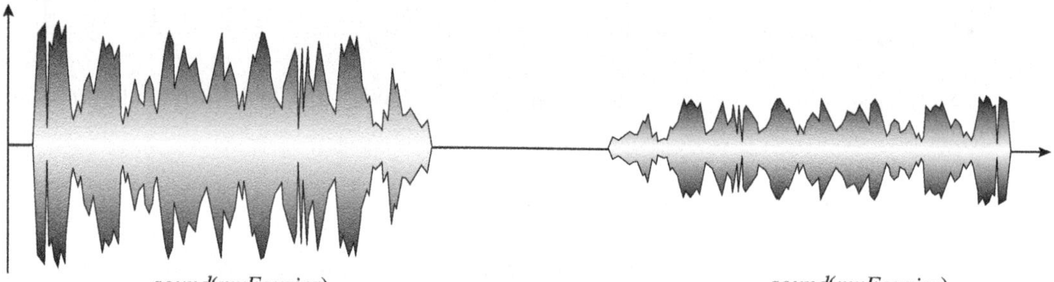

Fig. 8.8. A sound sample and its transform by a symmetry $Q_{\beta\sigma\mu}$, showing time inversion.

Example 9 The major-minor problem.

The major-minor problem is one of the classical schisms in occidental music theory. The problem is to understand the relation between the "major" and "minor" paradigms. We are not going to shed philosophical light on this conceptual field but will concentrate on the concrete and most common context of just temperament where this discussion is incessantly virulent. Historically and theoretically, it was Gioseffo Zarlino [1151] who initiated the subject by the observation that there is a *dualism* between major and minor. In fact, he realized that in just tuning, the major chord $EulerChord(0, t - 2 \cdot o, q - o)$ becomes a minor chord $EulerChord(0, -t + o, q - o)$ after a symmetry of *EulerModule*.

Exercise 6 *Describe all affine endomorphisms f of EulerModule which transform the major chord into the minor chord, i.e.,*

$$f(EulerChord(0, t - 2 \cdot o, q - o)) = EulerChord(0, -t + o, q - o).$$

To discuss this subject in simpler terms, let us look at the situation in the *EulerPlane* space. Here, we have the inversion symmetry U_q which converts the major class chord $\{c, e, g\}$ into the minor class chord $\{c, e_\flat, g\}$, more precisely, $U_q(c) = g, U_q(e) = e_\flat$,and $U_q(g) = c$. This symmetry extends to a transformation of the C-major scale into the C-minor scale:

$$U_q(C\text{-}major) = C\text{-}minor,$$

[10] The support $supp(f)$ is the topological closure of the set $\{x \in \mathbb{R} |\ f(x) \neq 0\}$.
[11] See Appendix C.3.1, example 100 for this concept.

but see Figure 7.5 to evidence this fact. This fact was used by Arthur von Oettingen [806] and by Hugo Riemann [901] to deduce the minor cadence from dualism. Under the "dualism" U_q the major cadence[12] sequence $I = \{c, e, g\}, IV = \{f, a, c\}, V = \{g, b, d\}, I = \{c, e, g\}$ of major triadic degrees is transformed into the sequence $I = \{c, e_\flat, g\}, V = \{g, b_\flat, d\}, IV = \{f, a_\flat, c\}, I = \{c, e_\flat, g\}$ of minor triadic degrees. This contradicts the function of dominant and subdominant in Riemann's harmony in that the fourth degree in minor should have a subdominant function whereas the fifth degree in minor should have dominant character. Carl Dahlhaus [231] has interpreted this fact as contributing to the erosion of the concept of a harmonic function[13].

The theory of dualism is confronted with the theory of *turbidity*[14] which was prominently forwarded by Paul Hindemith [467]. It states that the minor triad is derived from the major triad by a "turbidating" lowering shift of the major third ($c - e$ in $\{c, e, g\}$, for example) to the minor third ($c - e_\flat$ in $\{c, e_\flat, g\}$). This view is charged by the emotional connotation of gloom as being conveyed by the minor triad.

Already Johann Wolfgang von Goethe did not accept this psychologically charged theory which had its roots long before Hindemith. We should however ask whether turbidity has nothing to do with symmetry and must therefore be structurally inferior to dualism as being founded on a purely emotional level. We want to show that this is not true: The theory of turbidity is based on symmetry just as dualism is. The key to this unexpected symmetry lies in the inner symmetry of the major scale itself. Let us consider the C-major scale (scale chord, to be precise) to fix the ideas, see Figure 8.9.

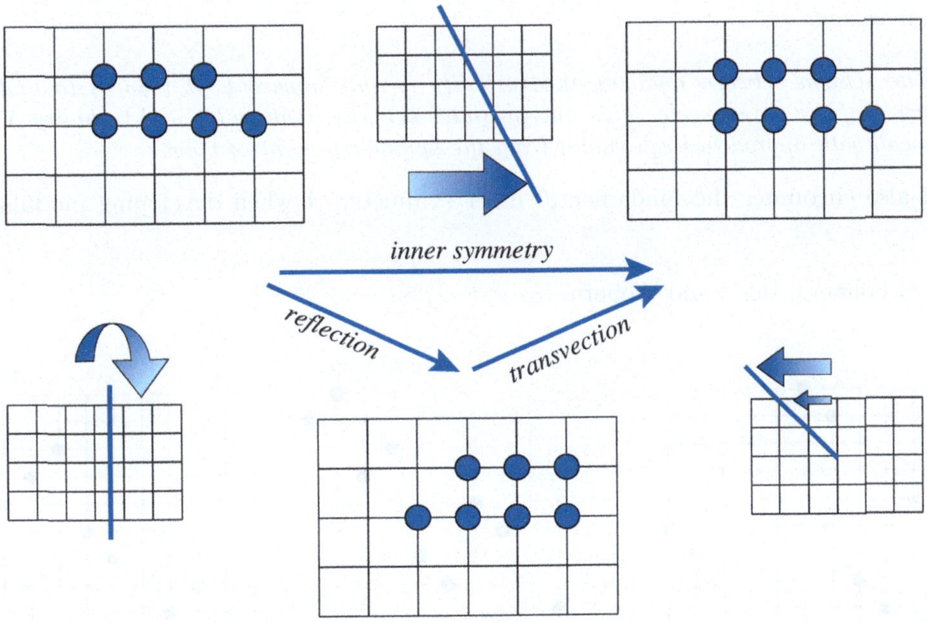

Fig. 8.9. The inner symmetry of a major scale in just tuning. This horizontal reflection at a skew axis is composed of a horizontal reflection at a vertical axis through $q/2$, followed by a horizontal transvection of 45°, and is the "tonal inversion" at the scale's third from the tonic on the keyboard (e in C-major).

The inner symmetry of C-major runs as follows: It is a reflection of the *EulerPlane* space at the vertical line through $q/2$. Then we add a shearing by 45° in direction $-q$. We thus obtain the symmetry

[12] We shall discuss the concept of cadence in more detail in Chapter 26.

[13] We come back to Riemann's function theory in Section 25.3.

[14] German: "Trübung".

$$A = e^q \cdot \begin{pmatrix} -1 & -1 \\ 0 & 1 \end{pmatrix} \tag{8.6}$$

with $A(C\text{-}major) = C\text{-}major$ where e remains fixed whereas we have exchanges $c \Leftrightarrow g$, $f \Leftrightarrow d$, and $a \Leftrightarrow b$. If we execute this symmetry on a keyboard, it turns out that we obtain a reflection of white keys (C-major) at key e. The black keys are so to speak eliminated. In music theory such a reflection is called *tonal inversion* since the transformation is performed within the given "tonality" if we think of the white keys as being equidistant (which they are not). To distinguish the terms, the above inversion U is termed *real inversion.* We shall see in connection to the formalism of alterations (Section 50.3, example 58) how real and tonal inversion relate to each other.

Besides identity, there is only the above A as an "inner" symmetry[15] of C-major. The interesting fact about this non-trivial symmetry is that this one mediates between the theories of dualism and turbidity! In fact, if U_q defines the dualism, then $A \cdot U_q$ evidently defines another symmetry from C-major to C-minor. And this does the following:

$$\begin{array}{llll} U_q \cdot A(c) = c & U_q \cdot A(g) = g & U_q \cdot A(f) = f & U_q \cdot A(d) = d \\ U_q \cdot A(e) = e_\flat & U_q \cdot A(a) = a_\flat & U_q \cdot A(b) = b_\flat . & \end{array} \tag{8.7}$$

We recognize that $A \cdot U_q$ lowers exactly the three tones e, a, b by a semitone, and *this is the process which turbidity theory describes.* This makes evident that turbidity is just the other symmetry $A \cdot U_q$ from major to minor, and we have learned that the unique inner symmetry A of major mediates between dualism and turbidity. Summarizing:

Scholion 1 *The schism between dualism and turbidity is only apparent. It reduces to understanding the inner symmetry A of the major scale. This one mediates between dualism U_q and turbidity $A \cdot U_q$ so that we are allowed to call both approaches equivalent from the symmetry point of view.*

We shall also encounter the fundamental inner symmetry A when developing modulation theory in Chapter 80.

Example 10 Schönberg, Berg, and Webern.

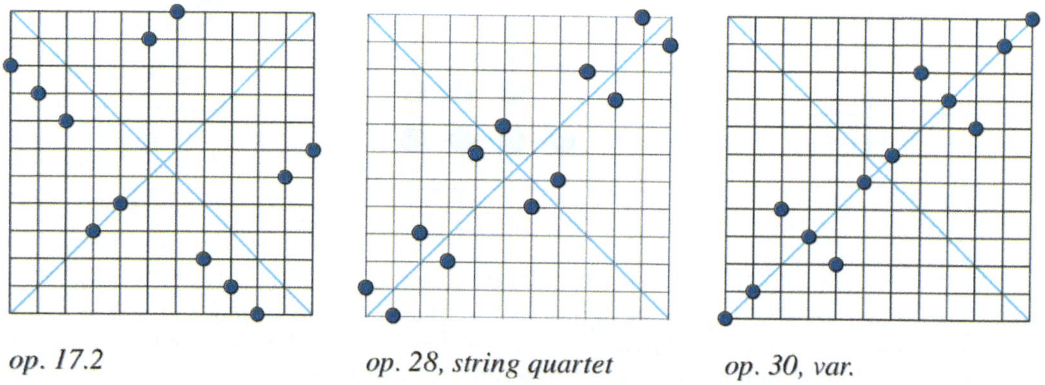

Fig. 8.10. Three dodecaphonic series by Webern, each showing diagonal or/and codiagonal symmetry.

In the framework of dodecaphonic composition, symmetries play a fundamental role. They define the paradigm associated with a *dodecaphonic series*[16]. This means that they generate a field of twelve-note

[15] We shall make this statement precise in Section 8.2.

[16] from German: "Zwölftonreihe".

series which are deduced from a basic form of the series. To make this compositional framework more precise, we invoke the ambient space $PiMod_{12}$ from (6.43) and look for local compositions which are motives in this ambient space:

Definition 22 *Let k, n be positive naturals. A (k, n)-*serial motif *is a $\mathbb{Z}^{k-1}$-addressed denotator*

$$Ser : \mathbb{Z}^{k-1} \rightsquigarrow PiMod_n(Ser_0, Ser_1, \dots Ser_{k-1})$$

with coordinates[17] $Ser_i = Ser(e_i)$. *If these coordinates are pairwise distinct, the serial motif is called a* (k, n)-series. *An* (n, n)-series *is called* n-phonic series, *in particular a* dodecaphonic *series for* $n = 12$.

The set of (k, n)-serial motifs is denoted by $SERM_{k,n}$, and we abbreviate $SERM_{n,n} = SERM_n$. The subset of (k, n)-series is denoted by $SER_{k,n}$, and we abbreviate $SER_{n,n} = SER_n$.

See Figure 8.10 for historic examples. The coordinate index i is viewed as being an 'abstract' onset time line which controls the compositional process. In general, this axis will not be present in a straightforward form within a concrete score. A general rule in twelve-note composition is that within a local realization of the basic series in the particular score, a note with larger onset index should not appear earlier than a note with smaller index, it may however occur simultaneously in a chord. We discuss this issue in more detail in Section 11.5.1.

The genealogical principle in twelve-note music starts with a particular dodecaphonic series Ser—called basic series[18]—and then creates 48 transformed versions according to a complete group of symmetries. In our generic setup, this group is the direct product group $D_{k,n} = \langle rev_k \rangle \times TI_n$ where $TI_n = e^{\mathbb{Z}_n} \rtimes \langle -1 \rangle$, and where rev_k is the 'retrograde' address involution defined by $rev_k(e_i) = e_{k-1-i}$.

This group consists of transformations known from classical counterpoint: transpositions e^s, pitch inversions U_s, onset retrograde rev_k, and their combinations[19]. A typical element $g = (h, f)$ of the group acts on a serial motif Ser by $(g.Ser)_i = f(Ser_{h(i)})$. The 48 versions are defined by the set

$$D_{12,12} \cdot Ser = \{c.Ser |\ c \in D_{12,12}\} \tag{8.8}$$

where, however, several elements may coincide. According to the general theory of group actions (see Appendix C.3.1, Example 100), the set SER_{12} of all twelve-note series splits into a disjoint union of twelve-note orbits, also called "paradigms"

$$SER_{12} = \coprod_r D_{12,12} \cdot Ser^r \tag{8.9}$$

running over a set $\{Ser^r\}$ of representative series.

The musical meaning of symmetries in Schönberg's, Webern's and Berg's twelve-note composition is a poietical and mental principle, it gives you a procedure to fabricate a composition. Joseph Rufer has exposed this approach in detail in [919]; the dominant role of symmetries is made explicit. Semiotically, the symmetries from $D_{12,12}$ have the task to build a paradigmatic field around the basic series. There is an unsolved problem in this approach: It is not clear how far the poietical approach in Schönberg's method may be transferred and evidenced on the neutral and/or esthesic levels. As a matter of fact, the historically problematic and poor reception of dodecaphonism shows that a poietical paradigm cannot only be practised, its protagonists should also take care of mediating it for the sake of public's recognition and acceptance on the esthesic level.

Example 11 Messiaen modi and non-invertible rhythms.

[17] This abuse of language with respect to denotator formalism is fairly acceptable.

[18] German: "Grundgestalt".

[19] To simplify notation, we denote U_s and k_{11} without adding the identity components in the respective other coordinate.

Essentially, Olivier Messiaen's composition technique [733] intends to treat all parameters of time and pitch equally. We shall come back to the systematic aspects of this method in Section 10.2. Here we merely make evident the structural context and the role of symmetries.

In the dimension of pitch, Messiaen introduced *modes with limited transpositions*. These are special scales in $\mathbb{Z}_{12}$, i.e., in 12-tempered tuning. Messiaen's list includes seven modes which are however not selected according to any evident logic. We come back to this selection problem when dealing with local classification in Section 11.3.5. The explicit selection criterium which is recommended and felt by Messiaen is limited transposition. This means that there is a non-zero pitch class $t \in \mathbb{Z}_{12}$ such that a Messiaen mode $M \hookrightarrow 0@PiMod(12)$ remains invariant under the transposition by t: $e^t \cdot M = M$. For a given mode M, the smallest candidate for such an invariant transposition is $t = 1, 2, 3, 4, 6$. Further, if M has this limited transposition, its complement $M^\frown$ shares this property, and we may concentrate on M having $6 \leqslant card(M)$. finally, the full chromatic "mother" scale $M = \mathbb{Z}_{12}$ can be omitted as a trivial case. From the classification theory in Section 11.3.5 we learn that—up to transposition—there are ten such scales, among which Messiaen for obscure reasons omits $M_9 = \{c, c_\sharp, d_\sharp, g, a\}$ as well as its inversion M_{10}, and $M_8 = \{c, c_\sharp, e, f, g_\sharp, a\}$. Here are the remaining seven scales $M_i, i = 1, \ldots 7$ as listed by Messiaen:

1. M_1 = whole-tone scale = $\{c, d, e, f_\sharp, g_\sharp, b_\flat\}$
2. M_2 = (diminished seventh chord)$^\frown = \{c, c_\sharp, d_\sharp, e, f_\sharp, g, a, b_\flat\}$
3. M_3 = (augmented triad)$^\frown = \{c, c_\sharp, d, e, f, f_\sharp, g_\sharp, a, b_\flat\}$
4. $M_4 = \{e, f, b_\flat, b\}^\frown = \{c, c_\sharp, d, d_\sharp, f_\sharp, g, g_\sharp, a\}$
5. $M_5 = \{c, c_\sharp, d, f_\sharp, g, g_\sharp\}$
6. $M_6 = \{d_\sharp, f, a, b\}^\frown = \{c, c_\sharp, d, e, f_\sharp, g, g_\sharp, b_\flat\}$
7. M_7 = (tritone)$^\frown = \{c, c_\sharp, d, d_\sharp, e, f_\sharp, g, g_\sharp, a, b_\flat\}$

Whereas Messiaen's modes are not problematic to define, the way Messiaen uses them to introduce non-invertible rhythms is quite disquieting. Let us anticipate the discussion by the original text [733]:

> *Diese Modi realisieren im Vertikalen (Transposition), was die nicht-umkehrbaren Rhythmen im Horizontalen (Umkehrung) realisieren. Tatsächlich können diese Modi nicht über eine gewisse Anzahl von Transpositionen hinaus transponiert werden, ohne daß man wieder in dieselben Töne hineingerät (enharmonisch gesprochen); ebenso können diese Rhythmen nicht rückläufig gelesen werden, ohne daß man genau dieselbe Anordnung der Werte wiederfindet wie in der Grundform. Diese Modi können nicht transponiert werden, weil sie sich — ohne Polytonalität — in der modalen Atmosphäre mehrerer Tonarten zugleich bewegen und in sich keine Transpositionen enthalten; diese Rhythmen können nicht umgekehrt werden, weil sie in sich selbst keine Umkehrungen enthalten. Diese Modi lassen sich in symmetrische Gruppen teilen; diese Rhythmen auch, mit dem Unterschied, dass die Symmetrie der rhythmischen Gruppen eine rückläufige ist. Endlich hat jede Gruppe dieser Modi jeweils ihren letzten Ton gemeinsam mit dem ersten Ton der folgenden Gruppe; und die Gruppen dieses Rhythmus umrahmen einen für beide gemeinsamen Zentralwert. Die Analogie ist also vollkommen.*

The last sentence is as typical for mathematically oriented phrasings by musicians as it is erroneous. The analogy is anything but perfect. Let us start with the underlying modules for modes and rhythms, respectively. For modes, we clearly have pitch classes referring to coordinator $\mathbb{Z}_{12}$. But to deal with rhythms, we cannot stick to local meters $L \hookrightarrow 0@Onset$ of period p since durations play an important role with Messiaen. Further, it is not sufficient to consider exclusively durations since we have to include pauses in Messiaen's rhythms, and this means that we deal with rhythms in the technical sense of Definition 14 in Section 7.2.2 where the parameter space *Para* includes duration as well as possibly some additional characterization to differentiate pauses from non-pauses. The least we can say is that we have $Para = Duration \oplus Para_0$, and therefore, the coordinator where Messiaen's rhythms are defined is $Onset \oplus Duration \oplus Para_0$. Now, not only the rhythm concept but also the concept of symmetry includes onset and duration. In fact, non-invertibility is defined in terms of onset *and* duration. There is not the least reason to invoke "perfect analogy" here. The retrograde which Messiaen refers to in [733] is precisely the one which we have denoted by K_s in formula (8.1): It is the symmetry of "reading from right to left what is normally read from left to right". And this symmetry is very different from the simple transposition as we know from the preceding discussion. Under these conditions, non-invertibility means this: Messiaen limits himself to indicate just one *germ* of a rhythm, and to ask from *germ* that it be symmetric under a retrograde symmetry, i.e., $germ = K_s \cdot germ$.

Apart from the fact that like transposition, retrograde transformation is also a symmetry in the mathematical sense (to be discussed in detail in Section 8.2), no such analogy as suggested by Messiaen can be discovered. Albeit such an analogy would be feasible, Messiaen seems satisfied with the fact that there is symmetry at all, a fact which is responsible for establishing criteria of compositional esthetics. This does nothing else than reconfirm the dominant role of symmetries in Messiaen's approach.

Example 12 Serial techniques: Eimert, Stockhausen, Kagel, and Boulez.

As instances of the compositional process, symmetries have been considered by exponents of the serial school. Serialism is derived from the idea that Schönberg's series should be extended to all sound parameters. As a point of departure for this approach we have to consider Messiaen's construction of non-invertible rhythms as a codex for durations in analogy to his scales (see Example 11). However, serialists start from treating all parameters equally, in other words: to elaborate Messiaen's analogy into a symmetry of the parameters' roles. Another example is Webern's symmetrization of the roles of onset and pitch in his twelve-note series. Figure 8.10 shows three such series; each provides *exchange of pitch and onset as an inner symmetry.*

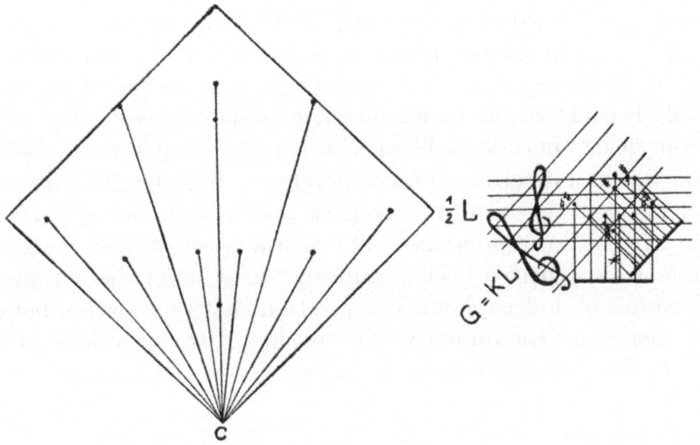

Fig. 8.11. (From [298], with kind permission of Universal Edition, Vienna, ©1964.) Example of a serial transformation described by Eimert in [298]. It results from a counter-clockwise rotation by 45°, followed by dilatation by $\sqrt{2}$.

The idea of a parameter exchange was adopted by Herbert Eimert, Maurizio Kagel, Karlheinz Stockhausen, Pierre Boulez, and other composers. It has been extended to the degree of admitting more general geometric transformations on parameter spaces in order to generate new sound configurations from given ones. For instance [298], such serial transformations include rotation by 90° or rotation by 45°, followed by dilatation by $\sqrt{2}$, see Figure 8.11. This more general type of symmetries leads to the general concept of morphisms of denotators to be introduced in Section 8.2. Examples of compositions using such methods include Kagel's "Transición II", Boulez' "Structures pour piano", or Stockhausen's "Kontra-Punkte". The musical meaning of symmetries is again of poietical and mental topography and serves—like dodecaphonism—the generation of a structurally specified paradigmatism of musical gestalts wherein hitherto separate musical parameters become freely interchangeable.

Example 13 Non-invertible symmetries.

There is one point which we must make clear before leaving the level of examples. In Definition 21 of symmetry as given at the top of this chapter, correspondence of parts is mentioned. But it is not specified what should be the nature of this correspondence. In particular, it is not clear whether such a correspondence should be one-to-one. Could it be that in special situations several parts correspond to one single part? From traditional mathematical terminology, symmetry transformations are understood as being one-to-one. But this is not mandatory.

Let us in fact give two easy examples of symmetries which do not involve one-to-one correspondences. The first one is projection. In twelve-note composition, it is allowed to play a selected part P of the fundamental series Ser or of its transforms $c(Ser)$ as a chord, i.e., simultaneously. Mathematically, we have a subset $P \subset c(Ser)$, and we consider its projection $\pi_2(P) \subset Pitch$ which is a chord. Clearly, no two elements of P are identified under π_2, but we can no longer tell the temporal order of the original elements. The correspondence $p \mapsto \pi_2(p)$ for $p \in P$ does not conserve temporal order. But the dodecaphonic paradigm precisely means expressing the compositional parts (such as chords) as instances of a whole, and this is rightly done by projection in this case: the projection $\pi_2(P)$ is in correspondence with P (and indirectly with Ser) as an expression of a unifying principle.

The second example is the nearly invisible symmetry of projection mod_o from the pitch space $EulerModule$ onto the classical octave pitch class space $o - EulerClass$ (7.11). In this context, a chord

$$Cr : A \rightsquigarrow EulerChord(Cr)$$

projects onto a class chord $mod_o(Cr)$ in $o - classChord$, but elements of the original chord which are multiples of the octave apart are identified under mod_o. Here, symmetry acts as an identifier and is far from one-to-one. Nonetheless, the correspondence under mod_o creates a semantic added value since on the side of $o - classChord$, harmony furnishes structural insight which is not visible on the original chord. In fact, harmonic progressions only become visible under such a correspondence.

This short discussion yields another problem which underlies symmetry theory: the codification of a symmetry transformation within a determined correspondence. It is not the same to write down an explicit correspondence among elements of two local compositions and to write down a mathematical transformation rule which is responsible for that correspondence. This is a delicate subject since it makes visible interface problems between what is a constructive background and what is on the surface of a given composition. We already saw in the example of dodecaphonic composition that the relation between compositional parts on the score and poietic genealogy (preliminary and subsidiary constructions by use of symmetries) is of fundamental importance to the communication of the composers message. We come back to this issue in Chapter 48.

8.2 Morphisms of Local Compositions

Summary. This section introduces the formal definition of morphisms between local compositions. The definition is motivated by representative examples from music as discussed in Section 8.1.1.

– Σ –

From the representative examples in Section 8.1.1 we have to draw three essential construction basics for general morphisms between local compositions (and whatever will be derived from this nucleus):

- As a realization of the general idea of symmetry, a morphism has to deal with a correspondence between the denotator 'points' which are involved in local compositions.
- The shape of this correspondence must rely on a generic type of structural transformation which operates on the 'background' of the local composition's ambient space.
- The background transformation is only subsidiary and should not be confused with foreground correspondence which really identifies the morphism. There may exist many different background transformations which yield the same correspondence, the essential is only that we find at least one of them which induces the visible correspondence.

We shall now give the formal definition of a morphism between local compositions on two levels of abstraction: first for objective local compositions, second on proper functorial local compositions. We shall then establish the connection between the two levels and show that everything fits well. In a last step, we shall show how to integrate morphisms among completely general denotators in this concept framework.

Definition 23 *(Morphisms for objective local compositions) Let $K \hookrightarrow A@F$ and $L \hookrightarrow B@G$ be two objective local compositions*[20] *at addresses A and B and ambient spaces*[21] *F and G, respectively. A* morphism *from K to L is a quadruple (K, L, f, α) composed of K, L, an address change $\alpha : A \to B$ and a set map $f : K \to L$, such that there exists a morphism of forms $h : F \to G$ with $\alpha.f = A@h|_K$, i.e., the diagram*

$$\begin{array}{ccc} K & \longrightarrow & A@F \\ {\scriptstyle \alpha.f}\downarrow & & \downarrow{\scriptstyle A@h} \\ L\alpha & \longrightarrow & A@G \end{array} \tag{8.10}$$

commutes ($L\alpha$ being the image of L under the functorial map $B@G \to A@G$, also denoted by α if no confusion is possible). This morphism is denoted by a fraction

$$f/\alpha : K \to L, \tag{8.11}$$

if the underlying ambient spaces are clear. Any morphism h of forms which gives rise to f/α as above is called an underlying symmetry *of this morphism.*

If the morphism's address change denominator α is the identity $1 = Id_A$, we also write f instead of $f/1$.

Exercise 7 *Review the examples of symmetries in Section 8.1.1 and give interpretations in terms of the above definition. Explicate all possible underlying symmetries.*

Lemma 5 *If $f/\alpha : K \to L$ and $g/\beta : L \to M$ are two morphisms of objective local compositions, then so is $g/\beta \cdot f/\alpha = gf/\beta\alpha : K \to M$. This morphism is called the* composition *of f/α and g/β.*

Proof. Let $K \hookrightarrow A@F$, $L \hookrightarrow B@G$, and $M \hookrightarrow C@H$ be the complete data, and suppose that $\alpha.f$ is induced by $h : F \to G$, whereas $\beta.g$ is induced by $k : G \to H$. To begin with, naturality of k with respect to α defines a set map $\alpha(\beta.g)$ which is uniquely determined by $\beta.g$ and by α and makes the diagram

$$\begin{array}{ccc} L & \longrightarrow & L\alpha \\ {\scriptstyle \beta.g}\downarrow & & \downarrow{\scriptstyle \alpha(\beta.g)} \\ M\beta & \longrightarrow & M\beta\alpha \end{array} \tag{8.12}$$

commute. We also have these commutative diagrams

$$\begin{array}{ccc} K & \longrightarrow & A@F \\ {\scriptstyle \alpha.f}\downarrow & & \downarrow{\scriptstyle A@h} \\ L\alpha & \longrightarrow & A@G \end{array}$$

$$\begin{array}{ccc} L\alpha & \longrightarrow & A@G \\ {\scriptstyle \alpha(\beta.g)}\downarrow & & \downarrow{\scriptstyle A@k} \\ M\beta\alpha & \longrightarrow & A@H \end{array}$$

[20] It is understood that the notation K stands for $K \hookrightarrow A@F$ and has to be explicated if ever necessary.

[21] With their functors being extracted for the functorial calculations, i.e., we use the usual shorthand $A@F$ to denote $A@Fun(F)$.

and hence the composed set map $\beta\alpha.gf = \alpha(\beta.g)\alpha.f : K \to L\alpha \to M\beta\alpha$ which is induced by $A@h \cdot A@k = A@h \cdot k$, QED.

The following is straightforward from the above definitions:

Sorite 4 *(Existence of identity, associativity)*

(i) *For an A-addressed objective local composition K, the identity* $1_K = 1/1 = Id_K/Id_A : K \to K$ *is a morphism.*
(ii) *If* $u/\alpha : K \to L$, *we have* $1_L \cdot u/\alpha = u/\alpha = u/\alpha \cdot 1_K$.
(iii) *If* $u/\alpha : K \to L$, $v/\beta : L \to M$, *and* $w/\gamma : M \to N$ *are three morphisms, then we have associativity of composition:* $(w/\gamma \cdot v/\beta) \cdot u/\alpha = w/\gamma \cdot (v/\beta \cdot u/\alpha) = w/\gamma \cdot v/\beta \cdot u/\alpha$.

For simple ambient spaces and zero-addressed local compositions, the definition of a morphism between objective local compositions boils down to the classical definition of morphisms of local compositions in [682]. In fact, these local compositions can be viewed as subsets of their coordinator modules, and the underlying morphisms by the Yoneda Lemma (Appendix G.2) are precisely the affine homomorphisms between these coordinator modules inducing set maps in the classical definition.

Let us next look at morphisms in the functorial setup.

Definition 24 *(Morphisms for functorial local compositions) Let* A, B *be two addresses, and let* $K \hookrightarrow @A \times F$ *and* $L \hookrightarrow @B \times G$ *be two (functorial) local compositions. A morphism from K to L is a couple* (f, α) *where* $\alpha : A \to B$ *is an address change, and where* $f : K \to L$ *is a natural transformation such that there exists a morphism of forms* $h : F \to G$ *such that the diagram*

$$\begin{array}{ccc} K & \longrightarrow & @A \times F \\ {\scriptstyle f}\downarrow & & \downarrow{\scriptstyle @\alpha \times h} \\ L & \longrightarrow & @B \times G \end{array} \tag{8.13}$$

commutes. As with objective local compositions such a morphism is denoted by $f/\alpha : K \to L$.

Exercise 8 *Equivalently, we could ask for a natural transformation* $f_\alpha : K \to L\alpha$ *where* $L\alpha$ *is the fiber product*[22] *subfunctor of* $@A \times G$ *defined by the pullback diagram*

$$\begin{array}{ccc} L\alpha & \longrightarrow & @A \times G \\ \downarrow & & \downarrow{\scriptstyle @\alpha \times Id_G} \\ L & \longrightarrow & @B \times G \end{array}$$

and where we ask for an underlying $h : F \to G$ *such that*

$$\begin{array}{ccc} K & \longrightarrow & @A \times H \\ {\scriptstyle f_\alpha}\downarrow & & \downarrow{\scriptstyle @1_A \times h} \\ L\alpha & \longrightarrow & @A \times G \end{array}$$

commutes.

This latter definition reminds us more of the previous definition for objective local compositions. However, it is less elegant than the one we gave in Definition 24; this is already evident from the definition of composition of morphisms:

Definition 25 *If* $f/\alpha : K \to L$ *and* $g/\beta : L \to M$ *are two morphisms of functorial local compositions, then their composition* $g/\beta \cdot f/\alpha : K \to M$ *is defined by* $g/\beta \cdot f/\alpha = gf/\beta\alpha$.

[22] See Appendix G.2.1 for fiber products or pullbacks.

This time, the fact that the composition is also a morphism is immediate. We have a similar sorite to the above one for objective local compositions:

Sorite 5 *(Existence of identity, associativity, and canonical factorization for functorial local compositions) The statements of Sorite 4 are literally true if we replace local compositions and morphisms by their functorial homonyms. Moreover, we have a factorization $f/\alpha = 1/\alpha \cdot f_\alpha/1$ via the fiber product composition $L\alpha$ and the morphism f_α defined in the previous exercise.*

Of course this terminological duplication is not accidental and we can in fact establish a complete embedding of the objective setup in the functorial one as follows. We already know from Section 7.3 that every objective local composition $K \hookrightarrow A@F$ has its functorial counterpart $\hat{K}$ with slice $f@\hat{K} = \{f\} \times K \cdot f$ for any address change $f : X \to A$. Suppose that we are given a morphism $f/\alpha : K \to L$ of objective local compositions with $L \hookrightarrow B@G$ and an underlying $h : F \to G$. The obvious claim is that this same h and α induce a commutative diagram

$$\begin{array}{ccc} \hat{K} & \longrightarrow & @A \times F \\ \downarrow & & \downarrow {\scriptstyle @\alpha \times h} \\ \hat{L} & \longrightarrow & @B \times G \end{array}$$

of functors, i.e., that the embedding of $\hat{K}$, followed by the product $@\alpha \times h$, factorizes (necessarily uniquely) through $\hat{L}$.

So let us calculate the image of an element $(g, k\cdot g) \in g@\hat{K}$ for $g : X \to A$. We have $X@\alpha \times X@h(g, k\cdot g) = (\alpha g, X@h(k \cdot g)) = (\alpha g, (A@h(k)) \cdot g)$. But we know that $A@h(k) \in L\alpha$, i.e., $A@h(k) = l \cdot \alpha$. Therefore $(A@h(k)) \cdot g = l \cdot \alpha g$ and therefore $X@\alpha \times X@h(g, k \cdot g) = (\alpha g, l \cdot \alpha g)$, i.e., this is an element of $\alpha g@\hat{L}$, a subset of $X@\hat{L}$. This functorial morphism $\hat{K} \to \hat{L}$ is denoted by $\hat{f}$. It remains to be shown that $\hat{f}$ is well-defined, i.e., depends only on f and not on h. In fact, the evaluation of $\hat{f}$ at Id_A sends $Id_A@\hat{K}$ to $\alpha@\hat{L}$ and yields $\alpha.f$ on the second coordinates of these sets. Further, for any address change $x : X \to A$, we have a commutative diagram

$$\begin{array}{ccc} Id_A@\hat{K} & \xrightarrow{\text{surjection}} & x@\hat{K} \\ {\scriptstyle \alpha.f}\downarrow & & \downarrow{\scriptstyle \hat{f}} \\ \alpha@\hat{L} & \xrightarrow{\text{surjection}} & \alpha x@\hat{L} \end{array} \tag{8.14}$$

with surjective horizontal arrows. Therefore the right vertical arrow is uniquely determined by f, and we have defined a map

$$\hat{?} : f/\alpha \mapsto \hat{f}/\alpha \tag{8.15}$$

which transforms an objective morphism into a functorial one.

Exercise 9 *For objective local compositions, there is no factorization $f/\alpha = 1/\alpha.f_\alpha/1$ into a pure address change morphism and a morphism with fixed address. We shall see later (Proposition 3 in Section 8.3.2), that the map $\hat{?}$ defines a full functor from objective morphisms to functorial ones, but it is not faithful, although it reflects isomorphisms. Therefore, the factorization of the image $\hat{f}/\alpha$ does not, in general, imply a corresponding factorization on the objective side. However, if α is an isomorphism, one has the factorization $f/\alpha = \alpha^{-1}/\alpha.f_\alpha/1$, where $f_\alpha/1 : K \to L.\alpha$ and $\alpha^{-1}/\alpha : L.\alpha \to L$. Prove this statement.*

Before turning to the overall discussion of categories of local compositions we should show how to integrate more morphisms for general denotators. As already introduced in special situations (see also Example 4 in Section 6.8) a denotator $D : Address \rightsquigarrow Form(Coordinates)$ could be wrapped in the following local composition: We first set up a new **Power**-typed form

$$Form^{Form} \underset{Id}{\longrightarrow} \mathbf{Power}(Form) \tag{8.16}$$

and then define the singleton objective local composition

$$\{D\} : Address \rightsquigarrow Form^{Form}(D). \tag{8.17}$$

What does it mean to have a morphism of two such singleton local compositions? We are just given two elements $D_1 \in A@Form_1$ and $D_2 \in B@Form_2$ and a form morphism $h : Form_1 \to Form_2$ as well as an address change $\alpha : A \to B$ such that we have the equation[23] $hD_1 = D_2\alpha$. We now recognize that this is a standard concept in category theory: the category of elements[24] for set-valued functors. In that context, we have $h = Id_F$ for the underlying functor $F = Fun(Form) : \mathbf{Mod}^{opp} \to \mathbf{Ens}$, and the category of elements is $\int_{\mathbf{Mod}} F$. In other words, we have considered denotators as "elements" and morphisms between such "elements" D_1, D_2 as couples of a unique set map of singletons $! : \{D_1\} \to \{D_2\}$ which is induced by a functorial morphism of underlying spaces, and an address change α. We shall denote this type of morphisms of singleton local compositions by $!/\alpha$ and also write $!/\alpha : D_1 \to D_2$ to ease notation. This means that the category of local compositions generalizes the category of elements of each of its forms.

8.3 Categories of Local Compositions

Summary. The formal definition of the category *ObLoc* of *objective* local compositions and morphisms is given and discussed with respect to its musicological relevance. A semantic model of general symmetries over the n-dimensional integer module $\mathbb{Z}^n$ is discussed.

– Σ –

After having prepared all necessary prerequisites, we are ready to define the category of objective local compositions:

Definition 26 *The category ObLoc of objective local compositions has as its object collection* $_0ObLoc$ *the set of all objective local compositions as declared in Definition 8. They are usually denoted by their names (often capital letters* L, K, M_1*, etc.) if no ambiguities are likely. The set* $_1ObLoc$ *of morphisms is the disjoint union of the sets* $ObLoc(L_1, L_2)$ *of morphisms* $f/\alpha : L_1 \to L_2$ *as declared in Definition 23. The composition of morphisms is that declared in Lemma 5. The identity morphism of an objective local composition L is the morphism* 1_L *as declared in point* (i) *of Sorite 4.*

Mathematically, there is no further point in this definition, once its ingredients have been introduced. However, there is one delicate remark to be made as to the musical signification of the entire construct. The remark is that the categorical framework not only suggests general ambient spaces, viz. modules for **Simple** Type, but also that any 'wild' morphisms may intervene, and that any concatenation, i.e., composition[25] of morphisms is allowed—in fact a critical musical and musicological point.

Let us therefore digress on a semiotic paradigm for general symmetries, i.e., affine transformations, on 'prototypical' modules $\mathbb{Z}^n$ which underly morphisms of objective local compositions in such ambient spaces. Of course we know that underlying symmetries are only the background transformations for visible correspondences on the supports of local compositions. But we have to ask whether symmetries—as a structural potential for *possible* morphisms—could be musically significant—independently of the possible points which could be moved around between such spaces. This question becomes of primordial importance not later than when local compositions are transformed by automatic processes as they are implemented in music software, such as in *presto*®, see Chapter 50. For this case we should have a certain theoretical guarantee for the musical meaning of what is being done.

---

[23] We are a bit sloppy in identifying the denotator names with their coordinates and also omitting the address in the natural transformation h, but this makes notation less clumsy.

[24] See Appendix G.2.1.

[25] We should apologize for maintaining the mathematical terminology of "composition" of morphisms; we believe however that no confusion will arise thereby. In delicate contexts we switch to the synonym "concatenation" to make things more distinct.

Therefore we are going to discuss a semantical paradigm for symmetries between modules which play an important role in practice. Very often practice deals with gauged and then digitized parameters such that we may suppose that they are integers. We may therefore work on $\mathbb{Z}^n$ and consider symmetries $f : \mathbb{Z}^n \to \mathbb{Z}^m$. Further any such symmetry can be interpreted as an endomorphism of a large power $\mathbb{Z}^N$, for example by viewing both spaces as direct summands of a sufficiently large superspace, i.e., we may suppose $n = m$ without loss of generality. In this case, a symmetry $e^t \cdot P : \mathbb{Z}^n \to \mathbb{Z}^n$ is described by $n + n^2$ integers via vector $t = (t_1, \ldots t_n)$ and matrix $P = (P_{ij})_{1 \leqslant i,j \leqslant n}$. This manifold of symmetries can be described in a very reduced language by the following theorem which is proven in Appendix E.3.5 and E.3.6, and which we restate in terms that fit with the musical context:

Theorem 2 *Consider the set*

$$\mathfrak{MusGen} = \{T, D_m(m \in \mathbb{N}), K, S, P_s(s = 2, \ldots n)\}$$

of affine endomorphisms on $\mathbb{Z}^n$ *which consists of the following elements:*

1. $T = e^t, t = (0, 1, 0, \ldots 0)$, *the translation in the direction of the second coordinate axis.*
2. *If* $m \in \mathbb{N}$, *let* $D_m = (d_{ij})$ *be the m-fold dilatation in the direction of the first coordinate axis. This means* $d_{11} = m, d_{22} = \ldots d_{nn} = 1$, *and* $d_{ij} = 0$ *else.*
3. $K = D_{-1}$, *dilatation by* -1 *or reflection in the first coordinate.*
4. $S = (d_{ij})$ *the transvection or "shearing" of the second coordinate in the direction of the first coordinate axis, i.e.,* $d_{11} = d_{12} = d_{22} = \ldots d_{nn} = 1, d_{ij} = 0$ *else.*
5. $P_s = (d_{ij})$, $s = 2, \ldots n$, *parameter exchange of first and sth coordinates, i.e.,* $d_{ij} = 0$ *for* $ij = 11, ss$, *and for* $i \neq j$ *and* $ij \neq 1s, s1$, *whereas* $d_{1s} = d_{s1} = d_{33} = \ldots d_{nn}$.

Then every affine endomorphism on $\mathbb{Z}^n$ *can be written as a concatenation (composition) of some of the elements of* $\mathfrak{MusGen}$ *(including repetitions of the same element).*

Affine automorphisms (i.e. the invertible symmetries) on $\mathbb{Z}^n$ *can be written as a concatenation of elements of* $\mathfrak{MusGen} - \{D_m, 0 \leqslant m\}$, *i.e., no m-fold dilatations for* $0 \leqslant m$.

In particular, each of the generators from $\mathfrak{MusGen}$ is concerned with affecting at most two coordinates. Except generators of type parameter exchange $P_s, 3 \leqslant s$, they affect only the first two coordinates!

The following *concatenation principle* yields a semantical interpretation of the preceding result. This principle can obviously not be demonstrated in full generality (this would be a test of an infinite number of musical situations—God beware!), but it states, in its very generality, a cornerstone of generic musical semantics:

Principle 2 *Concatenation of two musically meaningful symmetries or morphisms yields a musically meaningful result.*

Before discussing this principle in Section 8.3.1, let us apply it to Theorem 2. Accordingly, every symmetry $f : \mathbb{Z}^n \to \mathbb{Z}^n$ can be written as a product

$$f = e_1 \cdot e_2 \cdot \ldots e_s$$

all factors being taken from $\mathfrak{MusGen}$. In order to turn this purely mathematical fact into a musicological one, we have to view the modules $\mathbb{Z}_n$ as a coordinator for a reasonably chosen form space. Let us take a direct sum space of shape $Onset|_{\mathbb{Z}} \oplus w\text{-}Pitch \oplus Para$ where $w\text{-}Pitch$ is the w-tempered scale space $w\text{-}TemperedScale$ (see Section 7.2.1.3) with canonical identifier $@\mathbb{Z} \xrightarrow{\sim} @\mathbb{Z} \cdot \frac{1}{w} \cdot o$, and the space $Para$ has coordinator $\mathbb{Z}^{n-2}$ for any 'accessory' parameters such as loudness, duration, whatever. But it should be stressed that this selection is a function of the selection of the special generator set $\mathfrak{MusGen}$, and that there are many other semantical interpretations of symmetries. The subject here is not coverage of all possible meanings but evidence of the very existence of musical meaning behind such a priori constructs from mathematics. On this space, the generating symmetries from $\mathfrak{MusGen}$ have the following musical meaning:

1. Translation T means **transposition** (Example 5 in Section 8.1.1) by one pitch unit upwards in w-tempered pitch space, a classical musical process.
2. Reflection D_{-1} means **retrograde** of onsets (Example 7 in Section 8.1.1), a classical technique in counterpoint.
3. Shearing S in onset direction means **arpeggio** since the notes of a chord instance at a determined onset are played one after another from low to high pitch.
4. Dilatations D_m are classical **augmentations** in onset directions: distances of onsets are augmented by factor m.
5. Operations P_s of **parameter exchange** are related to classical compositional techniques or established musical thinking. concerning exchange of onset and pitch we had already seen Messiaen's—unfortunately not very consistent—exchange principle (Example 11 in Section 8.1.1). Already in the middle ages (to which Messiaen refers [329]) we can observe a symmetry and exchangeability of treatment of modes in time and pitch. Here, P_2 is not applied on a given compositional material but as structural 'isomorphism of meaning', in fact a semiotical symmetry. Further we should recall the P_2-symmetry in twelve-note series from Example 10 in Section 8.1.1, as well as rotation by 90° which can be viewed as concatenation of retrograde and parameter exchange.
   Parameter exchange of onset with other parameters in P_3, P_4 etc. is less investigated by composers and theorists. Serial technique makes use of such symmetries in some cases [1092]; a systematic application can however only be traced in nowadays computer-assisted compositional environments, see Chapter 50.

Let us conclude by a remark on the group $\overrightarrow{GL}(\mathbb{Z}^n)$ of invertible symmetries on $\mathbb{Z}^n$ which is generated by transposition T, retrograde K, arpeggio S, and parameter exchange $P_s, s = 2, 3, \ldots n$. By such symmetries, local compositions are always transformed in a 'reversible' way. All relations among their tones are preserved without loss. In other words, every local A-addressed composition K in ambient space $Onset|_\mathbb{Z} \oplus w\text{-}Pitch \oplus Para$ is embedded in its *orbit* $\overrightarrow{GL}(\mathbb{Z}^n) \cdot K = \{F(K)|\ F \in \overrightarrow{GL}(\mathbb{Z}^n)\}$ which we call K's *general affine paradigm*. This concept, together with specific selections of (parametrized families of) subgroups of the full symmetry group $\overrightarrow{GL}(\mathbb{Z}^n)$ will lead to a far-reaching conceptualization of paradigmatic phenomena in music and music history, see Section 10.2.

8.3.1 Commenting on the Concatenation Principle

Summary. We give some musicological comments on the concatenation principle by means of two examples.

– Σ –

A. *Counterpoint.* In classical counterpoint, one considers translations, retrogrades, inversions, and retrograde inversions. This collection meets the requirement that concatenation (composition) of two of its members is musically meaningful. But there is more: Musically, the retrograde inversion (see Example 7) KU isn't understood as a 180°-rotation, this does not make any sense in classical terms of European music theory. If Beethoven rightly operated on Steibelt's score by such a rotation, this was nothing more than destruction of its very intention, rotation was—in those days—simply ridiculous. The only explanatory access to this symmetry is its decomposition *qua* concatenation of retrograde K and inversion U! In this interpretation, retrograde inversion is composed by a reasonable action in time and one in pitch, there is no genuine theory and semantics in two-dimensional space $Onset \oplus Pitch$. The concatenation principle precisely generalizes this classical mechanism of musical semantics.

B. *Dodecaphonism* (see also [919], [1092]). Concerning semantics of symmetries on $\mathbb{Z}^2$, twelve-note composition offers an interesting example. Schönberg's exegete Joseph Rufer comments on the esthetic principles of dodecaphonism [919]:

> *Die beiden an der Entstehung einer musikalischen Form vornehmlich beteiligten Gestaltungsprinzipien sind: die* Wiederholung *und die* Variation. *Wiederholung ist das Anfangsstadium, Variation und Entwicklung die hohere Entwicklungsstufe musikalischer Formtechnik.*

The musical space where these transformations take place is described as follows by Schönberg [949], see also Figure 8.12:

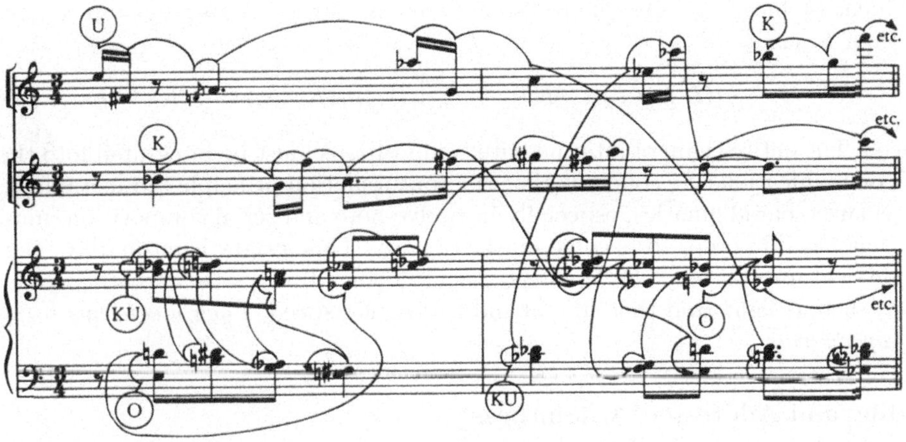

Fig. 8.12. (With kind permission of Reclam Publishers.) This graphics from [1092] shows complex combinations of the fundamental series and its variations in twelve-note technique.

Der zwei- oder mehrdimensionale Raum, in dem musikalische Gedanken dargestellt werden, ist eine Einheit. Obwohl die Elemente dieser Gedanken dem Auge und dem Ohr getrennt und unabhängig voneinander erscheinen, enthüllen sie ihre wahre Bedeutung nur durch ihr Zusammenwirken, so wie kein Wort allein, ohne Beziehung zu anderen Worten, einen Gedanken ausdrücken kann. Alles was an irgend einer Stelle dieses musikalischen Raumes geschieht, hat mehr als nur lokale Wirkung. Es wirkt nicht nur in seiner eigenen Ebene, sondern auch in allen Richtungen und Flächen und zeigt seinen Einfluß selbst an entfernten Stellen.

We pick up from this very modern text that Schönberg thinks at once in two dimensions and not twice in one dimension. This is quite a progress with respect to classical counterpoint. With this example not only semantic relevance of symmetry groups is exemplified, one also learns to understand the communicative problem implied by dodecaphonism. It seems that dodecaphonists had realized that symmetry groups could unfold their presence to the listener only 'subcutaneusly'. Neither the "ideal" listener nor diligent analysis are in a state to decode the massive ambiguities and intertwinements in twelve-note composition. This compositional process is strictly poietic. From this we have to abandon hope to transmit the message more than in a diffuse emotional sphere. The fact that the hoped-for emotional echo has mainly failed does not prove conservative auditory habits (though this is true to a certain degree) but rather a deficiency of compositional strategy.

Let us make this more explicit. The contrapuntal symmetries which de facto embed the basic series in its paradigm is instantiated by symmetries which as musical thoughts exist on the poietical and—at best—neutral level (the latter via analytical activities). However, the temporal course of a composition does not allow the listener to reflect symmetry relations. Here the composer has to bring symmetry relations into audible shape, for instance by means of compositional parts which make these symmetries evident, as inner symmetries, say. The sonata principle has been so successful since its basic components—exposition, development, and recapitulation—build a thoroughly communicative concept. Explicating a musical thought within a musical composition, i.e., on the neutral level, is no compromise to the listener but a genuine ingredient of whatever art claims to communicate.

The concatenation principle attributes musical meaning to a concatenation of symmetries or morphisms if the factors share such a meaning. This does not provide us with a justification automatism for twelve-note composition. In fact, if the factors $e_1, \ldots e_m$ are immediately evident to the listener, their product $e = e_1, \ldots e_m$ does not automatically bear musical evidence, it could as well be hidden to the listener. The turning point from construction to audible evidence is the evidence of concatenation, and, even more prominently, evidence of the underlying factors since they are not at all unique. It is as if you were listening to a spoken sentence in a foreign language but you do not know its word grouping. Somebody has to make the grouping evident by means of prosodic tools like stress or pauses. A step-by-step construction of a local composition L from K via

$$K \mapsto e_1(K) \mapsto e_2(e_1(K)), \ldots e_{n-1}(\ldots e_1(K) \ldots) \mapsto e(K) = L$$

must be evidenced. The entire chain of intermediate compositions must be integrated into the composition to become part of communication. We conclude that the concatenation principle is a metamusical thought. The correspondent semantic paradigmatics, especially in twelve-note and serial composition, must be turned into neutral musical 'material' in order to become musical reality. This insight is intimately related to Einstein's principle that physical information—for instance on a reference frame's time—can only be transmitted by physical support, such as light, and therefore at most at light's speed. *Each media has to unfold (communication) in its proper terms.*

8.3.2 Embedding and Addressed Adjointness

Summary. The category *ObLoc* of objective local compositions is embedded in a category *Loc* of functorial local compositions which admits sufficiently general universal constructions. It is shown that the embedding of *ObLoc* in *Loc* is full and reflects isomorphisms.

– Σ –

After discussion of some topics pertaining to objective local compositions, we are ready to define the category of local compositions:

Definition 27 *The category Loc of local compositions has as its object the collection $_0Loc$ the set of all local compositions as declared in Definition 8 of Chapter 7. They are usually denoted by their names (often capital letters L, K, M_1, etc.) if no ambiguities are likely. The set $_1Loc$ of morphisms is the disjoint union of the sets $Loc(L_1, L_2)$ of morphisms $f/\alpha : L_1 \to L_2$ as declared in Definition 24 of this chapter. The composition of morphisms was declared in Definition 25 of this chapter. The identity morphism of a local composition L is the morphism 1_L as declared in point* (i) *of Sorite 5.*

From Definition 8 and formula (8.15) we know that objects and morphisms of *ObLoc* are mapped into objects and morphisms of *Loc* under a map $\hat{?}$. Recall from Section 7.3 and Definition 16 that for a functorial local composition K at address A, the objective trace $\check{K}$ is given by the set $Id_A@K$. We now have this embedding proposition:

Proposition 3 *The map*

$$\hat{?} : ObLoc \to Loc \tag{8.18}$$

is a full functor which reflects isomorphisms.

Proof. Functoriality of $\hat{?}$: Let $(K, A@F), (L, B@G), (M, C@H)$ be three objective local compositions, $\alpha : A \to B, \beta : B \to C$ two address changes, and $f/\alpha : K \to L, g/\beta : L \to M$ two morphisms of objective local compositions. We have to show that $\hat{g}/\beta \cdot \hat{f}/\alpha = \widehat{gf}/\beta\alpha$, i.e., $\hat{g} \cdot \hat{f} = \widehat{gf}$. Let $q : X \to A$ be a point at address A, and look at the composed map on the q-slice $q@\hat{K} = \{q\} \times K \cdot q$. Take $(q, k \cdot q) \in q@\hat{K}$, then

$$\begin{aligned}
\hat{g}(\hat{f}(q, k \cdot q)) &= \hat{g}((\alpha q, f(k) \cdot \alpha q)) \\
&= \hat{g}((\alpha q, l \cdot \alpha q)) \text{ for an } l \in L, \\
&= (\beta\alpha q, g(l) \cdot \beta\alpha q) \\
&= (\beta\alpha q, g(f(k)) \cdot \beta\alpha q) \\
&= \widehat{gf}(q, k \cdot q),
\end{aligned}$$

and functoriality is established. To see that the functor reflects isomorphisms, look at the evaluation of $\hat{f}/\alpha$ at $Id_A@\hat{K}$: We have $\hat{f}(Id_A, k) = (\alpha, f(k).\alpha)$ which uniquely determines $\alpha.f$ and α. But if α is an isomorphism, so f is also determined. So we are done if we can show that the functor is full. Take any morphism $f/\alpha : \hat{K} \to \hat{L}$. Its evaluation $f/\alpha : Id_A@\hat{K} \to \alpha@\hat{L}$ defines a commutative diagram

$$\begin{array}{ccc} K & \longrightarrow & A@F \\ {\scriptstyle s}\downarrow & & \downarrow{\scriptstyle A@h} \\ L.\alpha & \longrightarrow & A@G \end{array} \tag{8.19}$$

of set maps where the right vertical arrow is the natural transformation h underlying $\alpha.f$. Take any $r : K \to L$ such that $s = \alpha.r$. We contend that $f/\alpha = \hat{r}/\alpha$. Let $(q, k \cdot q) \in q@\hat{K}$. Then

$$\begin{aligned} f/\alpha(q, k \cdot q) =& (\alpha q, h(k \cdot q)) \\ =& (\alpha q, h(k) \cdot q), \text{ by functoriality of } h, \\ =& (\alpha q, s(k) \cdot q) \\ =& (\alpha q, r(k).\alpha q) \\ =& \hat{r}/\alpha(q, k \cdot q), \end{aligned}$$

and we are done with the claims. QED.

In other words, the objective local compositions can be classified up to isomorphism in their proper category *ObLoc* or in *Loc*, it does not matter. Further, every local composition at address A can be 'frozen' to the 'identity point' Id_A and thereby reduced to its objective trace at A.

So the extension to functorial local compositions is seamless and can be 'traced back' to the objective framework. The deeper reason why we have built the functorial point of view upon a universe of objective local compositions is a double one. First, from the geometric and logical point of view, one needs to compare local compositions by use of universal constructions such as limits, or colimits. These not only have generic interest from category theory but are essential to model logical derivates of given ensembles of sounds and other musical objects. We shall see in Chapter 18 that predicate logic on the topos theoretic level needs substantially universal constructs. But there is also a more musicological reason for our extension policy: Musical objects are not only local in their spatial nature, we shall see in the global concept frame work of part IV that gluing local objects to global patchworks is genuine musicology. However, the mathematical description of global objects has to make use of fiber products of local compositions, and it is precisely this type of universal construction which fails within the purely objective setup.

Although we can build the objective trace $\check{K}$ for any object in *Loc*, this construction does not carry over to morphisms. To achieve such a goal, we have to restrict to a fixed address. Let us therefore define addressed comma categories as follows. Fix an address A. When talking of $@A$ as of a local composition we identify it with the functor—in fact the objective local composition—$@A \times @0_0$, where 0_0 is the zero module over the zero ring, i.e., the terminal[26] object of **Mod**. We have $X@(A \times @0_0) = X@A \times \{0\} \xrightarrow{\sim} X@A$. With this convention, we denote by $ObLoc_{@A}$ and $Loc_{@A}$ the comma categories over the local composition $@A$ where the structural address change is the identity on the address, i.e., the unique morphism $!/Id_A = !/1 : K \to @A$. On $Loc_{@A}$, we have an objective trace functor

$$\check{?}_{@A} : Loc_{@A} \to ObLoc_{@A}$$

which carries K to $\check{K}$ and $f/Id_A : K \to L$ to $\check{f}/Id_A : \check{K} \to \check{L}$ via $Id_A@f : Id_A@K \to Id_A@L$. This gives us an evident restriction of the embedding functor and the addressed objective trace functor

$$ObLoc_{@A} \underset{\check{?}_{@A}}{\overset{\hat{?}_{@A}}{\rightleftarrows}} Loc_{@A} \tag{8.20}$$

since the basis object $@A$ and the structural morphisms $!/Id_A$ remain fixed under both functors.

[26] See Appendix G.2.1.

Proposition 4 *The addressed restriction functor $\check{?}_{@A}$ is a left inverse to the addressed embedding $\hat{?}_{@A}$*

Exercise 10 *Give a proof of the preceding proposition.*

We terminate this section with a short discussion of adjointness [27] of the embedding functor $\hat{?}_{@A}$ and the objective trace functor $\check{?}_{@A}$, a subject which is central for the topos-logical approach, see for example [639]. We have adjointness for fixed address, i.e., 'addressed' adjointness:

Proposition 5 *The morphisms $\hat{?}_{@A}$ and $\check{?}_{@A}$ build an adjoint pair $\hat{?}_{@A} \dashv \check{?}_{@A}$.*

Proof. By Definition 16 in Section 7.3 and Proposition 3, we have an isomorphism

$$Hom_{ObLoc_{@A}}(K, \check{L}) \xrightarrow{\sim} Hom_{Loc_{@A}}(\hat{K}, L^{\diamond})$$

and by the injection $L^{\diamond} \rightarrowtail L$ from Sorite 1, we have an injection

$$Hom_{Loc_{@A}}(\hat{K}, L^{\diamond}) \rightarrowtail Hom_{Loc_{@A}}(\hat{K}, L).$$

So all we have to show is that every morphism $f/1 : \hat{K} \to L$ over A factorizes through $L^{\diamond}$. Consider the effect of a morphism $f/1 : \hat{K} \to L$ at a point $x : X \to A$, i.e., the application $x@f : x@\hat{K} \to x@L$. By definition, $Id_A@L = Id_A@L^{\diamond}$. So we have the commutative diagram

$$\begin{array}{ccc} Id_A@\hat{K} & \xrightarrow{Id_A@f} & Id_A@L^{\diamond} \\ x\downarrow & & \downarrow x \\ x@\hat{K} & \xrightarrow{x@f} & x@L \end{array} \tag{8.21}$$

But the very definition of the embedding $\hat{?}$ (see (6.16)) implies that the image of $x : Id_A@L^{\diamond} \to x@L$ stays in $x@L^{\diamond}$. By the surjectivity of the left vertical x-arrow, the image of $x@f$ also lives in $x@L^{\diamond}$, QED.

8.3.3 Universal Constructions on Local Compositions

Summary. We show the finite completeness of *Loc* and its interpretation in musical terms.

– Σ –

It is well-known [639] that $\mathbf{Mod}^@$ is a topos, and in particular admits universal constructions such as finite limits, colimits and a power object. Since the denotator space functors are in $\mathbf{Mod}^@$, there is a good chance that local compositions too admit such universal constructions. We want to investigate the specific construction in this section. Recall from Appendix G.2.1 that in $\mathbf{Mod}^@$, limits and colimits are "calculated pointwise".

Theorem 3 *The category Loc has fiber products.*

Proof. Given three local compositions $K \hookrightarrow @A \times F, L \hookrightarrow @B \times G, M \hookrightarrow @C \times H$ and two morphisms $f/\alpha : K \to M,\ g/\beta : L \to M$, we are looking for a local composition $P \hookrightarrow @D \times R$ and two morphisms $q/\kappa : P \to K,\ p/\lambda : P \to L$ such that we have a pullback diagram

$$\begin{array}{ccc} P & \xrightarrow{p/\lambda} & L \\ q/\kappa\downarrow & & \downarrow g/\beta \\ K & \xrightarrow{f/\alpha} & M \end{array} \tag{8.22}$$

[27] See Appendix G.2.1 for the concept of adjointness.

of local compositions. From Appendix E.3.8 we know that the underlying address changes have a fiber product

$$\begin{array}{ccc} D & \xrightarrow{\lambda} & B \\ {\scriptstyle\kappa}\downarrow & & \downarrow{\scriptstyle\beta} \\ A & \xrightarrow{\alpha} & C \end{array} \tag{8.23}$$

of modules and di-affine morphisms in **Mod**, i.e., $D = A \times_C B$. Further, suppose that f, g are induced by the functor morphisms $h : F \to H, k : G \to H$. Then we have two cartesian diagrams in $\mathbf{Mod}^@$:

$$\begin{array}{ccc} R & \xrightarrow{s} & G \\ {\scriptstyle r}\downarrow & & \downarrow{\scriptstyle k} \\ F & \xrightarrow{h} & H \end{array} \tag{8.24}$$

with $R = F \times_H G$ and

$$\begin{array}{ccc} \tilde{P} & \xrightarrow{\tilde{p}} & L \\ {\scriptstyle \tilde{q}}\downarrow & & \downarrow{\scriptstyle g} \\ K & \xrightarrow{f} & M \end{array} \tag{8.25}$$

with $\tilde{P} = K \times_M L$ *as set-valued functors*. But then, the product diagram

$$\begin{array}{ccc} @(A \times_C B) \times F \times_H G & \xrightarrow{@\lambda \times s} & @B \times G \\ {\scriptstyle @\kappa \times r}\downarrow & & \downarrow{\scriptstyle @\beta \times k} \\ @A \times F & \xrightarrow{@\alpha \times h} & @C \times H \end{array} \tag{8.26}$$

is also cartesian. Since the cartesian diagram (8.25) has its functors K, L, M and their morphisms f, g as subfunctors of the corresponding ambient space functors in diagram (8.26), the fiber product $\tilde{P}$ clearly injects into $@(A \times_C B) \times F \times_H G$, and we can replace $\tilde{P}$ by its image $P = im(\tilde{P})$, and the morphisms $\tilde{q}, \tilde{p}$ by the corresponding morphisms q, p from P, without losing the fiber product properties of the functor $\tilde{P}$. In other words, we have a commutative diagram of local compositions with ambient spaces from diagram (8.26). We now have to show that this diagram is cartesian. To this end, give any commutative diagram

$$\begin{array}{ccc} T & \xrightarrow{v/\nu} & L \\ {\scriptstyle u/\tau}\downarrow & & \downarrow{\scriptstyle g/\beta} \\ K & \xrightarrow{f/\alpha} & M \end{array} \tag{8.27}$$

of local compositions with $T \hookrightarrow @X \times W$, where u/τ is induced by $m : W \to F$, and where v/ν by $n : W \to G$. Then we have a uniquely determined functor morphism $t : T \to P$ and a uniquely determined di-affine morphism $\rho : X \to D$ such that $q \cdot t = u, p \cdot t = v$ and $\kappa \cdot \rho = \tau, \lambda \cdot \rho = \nu$. So if t/ρ is a morphism of local compositions, it is uniquely determined and we are done.

If $z : W \to R$ is the universal arrow induced by r, s, we have a space functor morphism $@\rho \times z : @X \times W \to @D \times R$, and this one evidently induces t, QED.

Corollary 1 *The category Loc is finitely complete, i.e., has finite limits.*

Proof. We use the fact (see Appendix G.2.2) that a category has finite limits iff it has fiber products and a terminal object. Let 0_0 be the terminal object (see Appendix G.2.1) in **Mod**. The local composition $1_{Loc} = @0_0 \times @0_0$ then is terminal since any A-addressed local composition K has a unique morphism $!/! : K \to 1_{Loc}$. Together with Theorem 3 we can apply the criterion for finite limits, QED.

Corollary 2 *For a morphism $f/\alpha : K \to L$ in Loc the following statements are equivalent:*

(i) *f/α is mono.*
(ii) *f and α are mono.*
(iii) *$X@f$ is injective for every address X and α is diinjective.*

Proof. It is well-known (see Appendix G.2.2) that f/α is mono iff in the cartesian product

$$\begin{array}{ccc} \Delta & \xrightarrow{f_2/\alpha_2} & K \\ {\scriptstyle f_1/\alpha_1}\downarrow & & \downarrow{\scriptstyle f/\alpha} \\ K & \xrightarrow{f/\alpha} & L \end{array} \tag{8.28}$$

both projections $f_1/\alpha_1, f_2/\alpha_2$ coincide and are isomorphisms. But clearly, this is equivalent to $f_1 = f_2$ being an isomorphism, and $\alpha_1 = \alpha_2$ being an isomorphism, and this in turn means that f and α are mono. Therefore (i) and (ii) are equivalent. Clearly, (iii) implies (ii). Conversely, we know from Appendix G.2.2 that a natural transformation f in $\mathbf{Mod}^@$ is mono iff its evaluation at every address X is injective. By Proposition 98 in Appendix E, the required characterization of mono diaffine morphisms is given, QED.

Remark. The dual situation is less simple, in fact, general fiber sums do not exist in *Loc*. This implies that we cannot glue together arbitrary local compositions to a new local composition. In other words, 'global compositions' as a result of gluing together local compositions present a proper extension of the local concept framework. This is the *mathematical* reason for the entire global theory to be developed later in part IV.

Let us now evaluate the finite completeness of *Loc* in more concrete terms and with regard to the subcategory of objective local compositions. To begin with, we should discuss an example where the fiber product of objective local compositions is a proper functorial local composition.

To understand this phenomenon we want to construct a fiber product which does not fulfill the following evident requirement for objective local compositions: For any objective local compositions K at address A, and for any X-valued point $x : X \to A$, the map $x : Id_A@K \to x@K$ is surjective. We shall in fact exhibit a pair $f/\alpha : K \to M,\ g/\beta : L \to M$ of objective local compositions for address changes $\alpha : A \to C,\ \beta : B \to C$ such that the fiber product map $u : Id_{A\times_C B}@(K \times_M L) \to u@(K \times_M L)$ is not surjective for a selected address change $u : W \to A \times_C B$.

Take $A = B = C = \mathbb{Z}_\mathbb{Z}$ and take the zero point $x = 0 : 0_\mathbb{Z} \to A$. Consider two non-empty objective local compositions $K, L \subset A@Pitch$ in the pitch form with ambient space $F = @\mathbb{R}$. This means that we take sets of intervals $i = i_0 \Rightarrow i_1$, corresponding to morphism $e^{i_0} \cdot (i_1 - i_0)$ as in Section 6.2.1. Suppose that $K \cap L = \emptyset$, more precisely that $Kx = Lx$, i.e., that the base points of the intervals in K, L are the same, but the interval arrow heads i_1 never coincide. Take $M = A@F$, and look at the natural inclusions $K \hookrightarrow M,\ L \hookrightarrow M$. Then the evaluation of the fiber product $T = \hat{K} \times_{\hat{M}} \hat{L}$ gives $Id_A@T = \emptyset$, and $x@T \cong K$ so that the morphism $x : Id_A@T \to x@T$ cannot be surjective and the fiber product T not objective. In other words, the objective perspective at address A yields no common points whereas the zero-addressed functorial perspective at x shows common base points, see also Figure 8.13.

8.3.4 The Address Question

Summary. We make the address concept explicit: Variation of addresses in music and musicology. We approach categories as point spaces which are sorted and structured by addresses.

– Σ –

We should now review the entire universe of denotators, objective and functorial local compositions and the related address navigation in order to understand the overall 'geography' of addresses.

From an overall point of view, we have built a comprehensive category *Loc* of local compositions. It extends the full subcategory *ObLoc* of objective local composition which are essentially sets of denotators of address-fixed "ambient space". The latter include the *singletons* which are the very general denotators

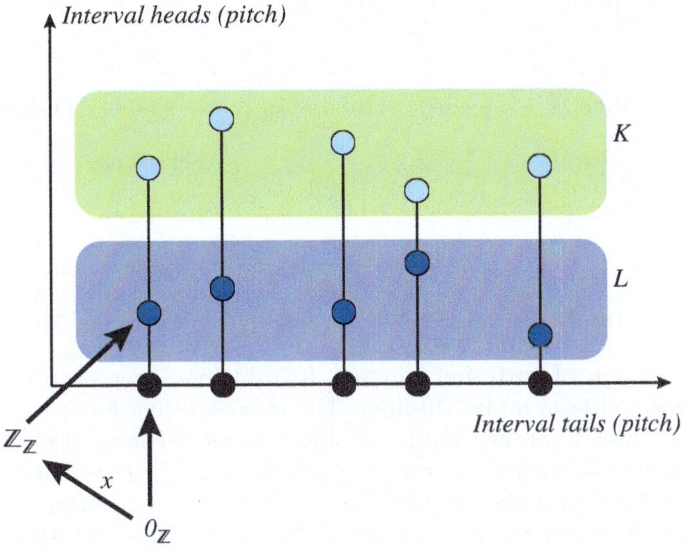

Fig. 8.13. Two objective local compositions K, L in pitch space M at address $\mathbb{Z}$ with identical 'perspective' at the zero-addressed point $x = 0 : 0_{\mathbb{Z}} \to \mathbb{Z}$. The functorial fiber product $\hat{K} \times_{\hat{M}} \hat{L}$ identifies these common 'base points' whereas the objective fiber product (in fact the intersection) is empty.

and take care of the classical construction of "categories of elements". Let us denote this latter category by *SinLoc*. So we have the chain

$$SinLoc \hookrightarrow ObLoc \hookrightarrow Loc \tag{8.29}$$

of full subcategories.

On the other hand, every denotator D has an address $Ad(D)$, i.e., a module object in **Mod**. This defines a map

$$Ad : Loc \to \mathbf{Mod} : D \mapsto Ad(D) \tag{8.30}$$

which is indeed a functor, the *address functor*, since morphisms f/α of local compositions include address change α. Within this framework, we have different fibers which are of interest in determined contexts. For example, if we consider the full subcategory $End(A) \hookrightarrow \mathbf{Mod}$ whose only object is A, and its 'identity' subcategory Id_A with the identity of A as its only arrow, then we have two corresponding fiber categories

$$\begin{array}{ccccc}
SinLoc_{@A} & \longrightarrow & SinLoc_{End(A)} & \longrightarrow & SinLoc \\
\downarrow & & \downarrow & & \downarrow{\scriptstyle \text{subcategory}} \\
ObLoc_{@A} & \longrightarrow & ObLoc_{End(A)} & \longrightarrow & ObLoc \\
\downarrow & & \downarrow & & \downarrow{\scriptstyle \text{full and reflexive}} \\
Loc_{@A} & \longrightarrow & Loc_{End(A)} & \longrightarrow & Loc \\
\downarrow & & \downarrow & & \downarrow{\scriptstyle Ad} \\
Id_A & \xrightarrow[\text{subcategory}]{} & End(A) & \xrightarrow[\text{subcategory}]{} & \mathbf{Mod}
\end{array} \tag{8.31}$$

defined by this evident system of cartesian squares. In principle, every subcategory of **Mod** defines an associated theory of musical objects on its *Ad*-fiber. The indicated fixations of A are only the most obvious constructs. But there is more: If we are given an objective local composition $K \subset A@F$ in the A-fiber $ObLoc_{@A}$, the passage to its associated functorial 'copy' $\hat{K} \hookrightarrow @A \times F$ is indeed an extension of the fixed address to variable address perspectives. The evaluation of $\hat{K}$ at a morphism $x : X \to A$ is a kind of 'K's view as seen from address X under the perspective x'. What is seen is the 'slice' $x@\hat{K}$, the fiber of $X@\hat{K}$ at point x (see also Figure 7.11 for this interpretation). We shall see in Chapter 9 that the Yoneda Lemma essentially instantiates this 'philosophy of perspectives': We look at an address A from all other addresses and try to understand what's happing at A by collecting all these relative views.

But the address question is not only a matter of relativization of perspectives, it also deals with ontology. Let us ask for the existentiality of a local composition. Viewed as a functor it is not one naive set but an entire connected collection of sets, viz all its evaluations at addresses X. Even an objective local composition is solicited by an underlying space functor whose evaluation at the composition's address produces the foreground set structure. The evaluation at a determined address generates a set which is specifically related to this address. The functor can only be 'seen' under such an evaluation.

But what part of a functor can be seen at a determined address is not only mathematically different, it is also ontologically different. For instance, in our example from Figure 8.13, we have $\mathbb{Z}$-valued points which reflect concepts around intervals in pitch space. Musicologically this perspective traces an ontology of different flavor compared to the poor reduction to base points under the zero address. Intervals are not just an ordered couple of zero-valued points! For instance, they have a radically different perceptual quality in music psychology, esthetics and in compositional contexts. In this spirit we can make the point as follows:

Principle 3 *The fibration of Loc and its subcategories via the address functor constitutes a 'bundle of ontologies', one for each address—or even for each subcategory of* **Mod**. *Address change therefore is an expression of an ontological shift and thereby constitutes a fundamental reconstruction tool of musical reality from a bundle of addressed ontologies.*

8.3.5 Categories of Commutative Local Compositions

Summary. This section deals with the structure of the practically very important category of commutative (objective) local compositions.

– Σ –

Definition 28 *Let A be an address which is an R-module for a commutative ring R. Then we denote by $ComLoc_A$ the subcategory of $ObLoc_{@A}$ whose objects are commutative, A-addressed local compositions*[28] *as*

[28] Recall that we often suppress form names and simply write $(L, A@M)$ or $(L, A@_R F)$ for a simple ambient space F with Functor $Fun(F) = @_R M$ for such an object since the names do not intervene for classification (= determination of isomorphism classes in a category). They however play a role for identification, but this is a straightforward aspect which may be suppressed for the majority of mathematically oriented considerations.

defined in Section 7.4. Its morphisms are all morphisms $f = f/1$ of $ObLoc_{@A}$, which means that no address changes are admitted. Hence a morphism $f : K \to L$ is a set map f such that there exists an R-affine morphism $h : M \to N$ of the respective coordinator modules such that the diagram

$$\begin{array}{ccc} K & \longrightarrow & A@_RM \\ {\scriptstyle f}\downarrow & & \downarrow{\scriptstyle A@_Rh} \\ L & \longrightarrow & A@_RN \end{array} \tag{8.32}$$

commutes[29].

Lemma 6 *With the preceding notations, if a morphism $f : K \to L$ in $ComLoc_A$ has underlying morphism $h : M \to N$ of the supporting coordinator modules, and if h_0 is the R-linear part of h, then the R-linear application*

$$R.f : R.K \to R.L : x \mapsto h_0.x \tag{8.33}$$

is well-defined, it only depends on f. The application $R.f$ is called the linear map associated with f. *The association $f \mapsto R.f$ is functorial, i.e., $R.1_K = 1_{R.K}$ and $R.(g \cdot f) = R.g \cdot R.f$. We therefore have a covariant functor*

$$R : ComLoc_A \to \mathbf{Mod}_R \tag{8.34}$$

called the module functor.

Proof. Once the first assertion is proven, it is clear that $f \mapsto R.f$ is functorial. If K is empty, the first assertion is clear since, by definition, $R.\varnothing = \varnothing$. But if $k_0 \in K$, we have $x = k - k_0$, and $h_0.x = h_0.k - h_0.k_0 = h.k - h.k_0 = f(k) - f(k_0)$, so h does not really matter, and the result lives in $R.L$. QED.

Consider the following subcategories:

- The full subcategory $ComLoc_A^{emb}$ of $ComLoc_A$ whose objects are the embedded commutative local compositions.
- The full subcategory $ComLoc_A^{gen}$ of $ComLoc_A^{emb}$ whose objects are the generating commutative local compositions.
- The full subcategory $ComLoc_A^{in}$ of $ComLoc_A^{emb}$ whose objects have injective R-modules as coordinators.

Fact 3 *For an address A, there is a faithful functor*

$$gen : ComLoc_A^{emb} \to ComLoc_0^{gen}$$

defined by $gen(K, A@_RM) = (K, 0_R@R.K)$ and $gen(f : K \to L) = R@f$ according to Appendix E, Lemma 93.

Sorite 6 *For zero address $A = 0_R$, a local composition $(K, 0_R@M)$ may be identified with (K, M). With this notation, we have the following results:*

(i) The embedding $ComLoc_0^{emb} \rightarrowtail ComLoc_0$ is an equivalence of categories.
(ii) Let $ingen : ComLoc_0^{in} \to ComLoc_0^{gen}$ be the functor which maps (K, M) to $(K, R.K)$ and a morphism $f : (K, M) \to (L, N)$ to $R@f$ as defined in the above fact 3 for embedded local commutative compositions. Then ingen is an equivalence of categories.
(iii) Let (K, M) and (L, M) be two objects in $ComLoc_0$ such that their coordinator has finite length[30] *$l(M)$. Then each isomorphism $f : K \xrightarrow{\sim} L$ is induced by an invertible symmetry $F \in \overrightarrow{GL}(M)$, i.e., the orbit $\overrightarrow{GL}(M) \cdot K$ describes all local compositions with coordinator M which are isomorphic to K.*

[29] The notation $A@_Rh$ with index R means that di-morphisms are restricted to the identity of the coefficient ring R.
[30] See Appendix E.2.4

Proof. (i) If K is an object of $ComLoc_0$, and if $k_0 \in K$, we have an isomorphism $e^{-k_0} : K \xrightarrow{\sim} e^{-k_0}(K) = L$, and L is evidently embedded. Since the embedding of embedded objects is fully-faithful, we are done. (ii) Since every generating (K, M) is the image of $(K, I(M))$ for the injective envelope[31] $I(M)$ of M, *ingen* is surjective. Since $R@f$ is an extension of f to the compositions's modules, *ingen* is faithful. Let $f : K \to L$ be a morphism in $ComLoc_0^{gen}$. Then any underlying symmetry $F : R.K \to R.L$ extends to a symmetry $H : I(R.K) \to I(R.L)$, i.e., the diagram of modules and R-affine morphisms

$$\begin{array}{ccc} R.K & \xrightarrow{i_K} & I(R.K) \\ {\scriptstyle F}\downarrow & & \downarrow{\scriptstyle H} \\ R.K & \xrightarrow{i_L} & I(R.L) \end{array} \tag{8.35}$$

commutes. In fact, we may suppose without loss of generality that F is linear, and then, by injectivity of $I(R.L)$, the map $i_L \cdot F$ factorizes through injection i_K, and *ingen* is full. Therefore *ingen* is fully faithful and surjective, i.e., an equivalence.

(iii) Clearly, we may suppose that K and L are embedded and that the isomorphism $f : K \xrightarrow{\sim} L$ and its inverse $f^{-1} : L \xrightarrow{\sim} K$ are induced by linear symmetries G and H, in particular $f = R.f|_K$. According to Fitting's lemma (Appendix E.2.4, Lemma 88), there are a natural number n and two direct decompositions $M = S_1 \oplus T_1 = S_2 \oplus T_2$ such that:

1. $RK \subset S_1, R.L \subset S_2$,
2. $(HG)^n|_{S_1} \in GL(S_1), (GH)^n|_{S_2} \in GL(S_2)$,
3. $T_1 = Ker(HG)^n, T_2 = Ker(GH)^n$.

Since $(GH)^n$ kills T_2, $G(HG)^n|_{S_1}$ maps S_1 into S_2, and by a symmetric argument, $H(GH)^n|_{S_2}$ maps S_2 into S_1. But the composition $H(GH)^n|_{S_2} \cdot G(HG)^n|_{S_1} = (HG)^{2n+1}|_{S_1}$ has its nth power in $GL(S_1)$, so it is also in $GL(S_1)$. By a symmetric argument on S_2 we conclude that $G(HG)^n|_{S_1} : S_1 \xrightarrow{\sim} S_2$; also $G(HG)^n|_K = f$. Since $l(M) < \infty$, the theorem of Krull–Remak–Schmidt (Appendix E.2.5) implies that $T_1 \xrightarrow{\sim} T_2$, and we obtain an invertible extension of $G(HG)^n|_{S_1}$ to all of M, QED.

Corollary 3 *If R is a semi-simple commutative ring, i.e., a finite direct product of commutative fields (see Appendix E.2), then the categories $ComLoc_0$ and $ComLoc_0^{gen}$ are equivalent.*

Proof. Since in this case, every module is injective, we have $ComLoc_0^{in} = ComLoc_0^{emb}$, and the corollary follows from Sorite 6, (i) and (ii).

Statement (iii) of Sorite 6 can be restated in the following way. For an A-addressed commutative local composition $(K, A@_RM)$ in $ComLoc_A$, we denote by $Sym(K)$ the *symmetry group of* K of all invertible symmetries $h \in \overrightarrow{GL}(M)$ which induce an automorphism of K. By definition, we have a group homomorphism

$$r : Sym(K) \to Aut(K) : h \mapsto h|K \tag{8.36}$$

into the automorphism group of K. This is neither surjective nor injective in general, but *for a ambient module M of finite length, the homomorphism is surjective*, by Sorite 6, (iii), and in this case, if we denote $Triv(K) = Ker(r)$ the group of those symmetries of K which act trivially (as identity) on K, we have

$$Sym(K)/Triv(K) \xrightarrow{\sim} Aut(K) \tag{8.37}$$

if K's ambient module M is of finite length. But be aware that the isomorphism class of $Sym(K)$ is *not* an invariant of the class of K while the automorphism group $Aut(K)$ clearly is.

[31] See Appendix E.4.3.

9

Yoneda Perspectives

Kant m'apprit
qu'il n'y a point de nombres,
et qu'il faut faire les nombres
chaque fois qu'il faut les penser.
Alain [18]

Summary. This chapter reviews and completes the inherent paradigm change from "object-based" to functorial mathematics which was initiated by category theory, and completed in the celebrated Yoneda Lemma. Beyond an apparant technical innocence, this lemma introduced a revolution in understanding structures of general types, and transcending pure mathematics. Its implications touch general hermeneutics as well as esthetics of art, and—quite paradoxically!—principles of "object-oriented" programming.

– Σ –

At first sight Nobuo Yoneda's Lemma[1] is a technical tool of category theory. In algebraic geometry as it was developed by Alexander Grothendieck in his Paris School, the lemma serves as a background to rebuild the very concept of a point as an elementary structure of generic geometry. But the deeper impact of this lemma is a radical review of the nature of human conceptualization. This is what we should keep in mind when tracing the role of what is called the "Yoneda philosophy" since its introduction in mathematical music theory [674], [677].

The drama behind the Yoneda revolution in mathematical sciences is that it proposes a thorough geometrization of human concepts—without limitations. And it does this in an approach of infinite recursive descent: a point is a point is a point... there is no lowest level, and this means that there is no basic level of concepts; thinking in this style is bottomless and creates existence from the very fact of performing logically consistent mental processes of conceptualization. This is a modern form of Paul Finsler's principle [321]:

Principle 4 *Mathematical existence of an object means that it can be thought of in a logically consistent*[2] *way.*

It resembles Murray Gell-Mann's "Eleventh Commandment" which states that whatever is not forbidden is mandatory, it has to happen somewhere and sometimes. To Finsler (see Figure 9.1), mathematics was not the usual formal game with empty symbols[3], to him, mathematical objects were entities that existed

[1] The contribution of Yoneda to this lemma is not completely clear, but the name has been commonly accepted.

[2] To Finsler logic was not a formal game but "absolute logic" which reaches as far as the three fundamental laws: A is identical to A, A and Non-A are mutually exclusive and a third possibility is excluded.

[3] He used to call the formal games "paper science", and we adhere to this verdict. There is no substitute for really performing thoughts; no arsenal of strings, no memory storage device can replace mental activity since the latter is semantic. It is not important which organism does the job—a computer could as well—but it is not legitimate to "delegate thoughts to philosophers", which mathematicians often prefer for shear comfort of passing to their beloved game of empty forms, see [561], for example.

G. Mazzola, *The Topos of Music I: Theory*, Computational Music Science,
https://doi.org/10.1007/978-3-319-64364-9_9

like physical objects, though in a more "platonic" mental reality. Existence of such "logical" objects meant consistence with absolute logic, and this was not only the usual concept framework: Finsler was extremely concerned with circular definitions as an important device for creating basic objects of mathematics. To him, the definition of a circular set $M = \{M\}$ is acceptable as long as it does not create logical inconsistencies.

Usually, we are not aware that *conceptual circularity is unavoidable.* For instance, the set concept is circular: Set theory is defined by a fundamental concept "set" and a binary element relation "$\in$" among instances of the set concept. However, to bring set theory to existence, we need examples of sets which are generated by special definitions or definition schemes. To do so, we have the well-known identification of a set S: It is a reference to the elements of S, i.e. those sets X such that $X \in S$. This is a circular concept: We ask "What is S?" And you answer: "It is the collection of those *sets* X which are elements of S." Your set S is only identified via a pointer to a well-defined collection of instances of the *same* concept. The only set which (in the usual setting, see C.1) exists from scratch is the empty set $\emptyset$: It is defined by the statement that "for every set X, $X \notin \emptyset$", or shorter: "S refers to no element". The empty set is the pure reduction to the concept. It is the reference to all instances of the concept, together with the overall negation of the element relation.

To make this evident, we could as well restart set theory with a new name for the basic concept: Call the new objects "newset", and define a "newset" theory with everything except the name unaltered. We could then start with the "newemptyset" $\emptyset_{new}$ defined by "for every newset X, $X \notin \emptyset_{new}$". From the perspective of mathematical model theory, we would say that the newset theory is just another model for set theory. But this is not the point since we are not dealing with formal languages which are mapped to some models in a given reality: The question is about the very concept building: What do we think when we think of mental objects named "sets"? The point then is that we think of self-referential, circularly defined objects and that, while doing so, we have to investigate a universe of such objects which can be built without logical inconsistency. The set $\emptyset$ is such an object for the concept type "set". And $\emptyset_{new}$ is such an object for the concept type "newset".

This is exactly the situation of denotators and their forms: Up to form names, the denotators are the same—one species lives in the form named "set", the other in the form named "newset". These spaces are isomorphic, the difference is reduced to their names. The first form is $set \underset{Id}{\longrightarrow}$ **Power**(set), the second is $newset \underset{Id}{\longrightarrow}$ **Power**$(newset)$, and the **Power** type is nothing but the element relation in its category-theoretic transfiguration. These reflections make clear that we understand concepts as being pointers, i.e.

Fig. 9.1. The mathematician Paul Finsler (1894–1970) was a pioneer of circular and non-formal concept design. Photo from the author's private collection.

structured recursive references with possibly circular recurrence. Quite radically, there is nothing behind concepts except the paths and ramification modes leading down to their referenced 'coordinate concepts'. This is a thorough geometrization of the concept structure: Concepts are points that point to other points...

Now this is very much the same idea that you learn from Yoneda's Lemma: It tells mathematicians that all they can understand about mathematical objects is from the way they point at other objects, from these to still others, and so forth. Replacing a "real" object X by its functor $@X$ means forgetting about the "point as a such" and being exclusively concerned with its reference to other points. Of course, it is not obvious that this methodology gives you back all you need to know about an object, viz its isomorphism class. But this is Yoneda's kick: It is indeed sufficient to know about the pointer chains. And this in turn means:

Thesis 1 *Concepts are nothing more than this recurrent pointing. Their semantics is nothing more than this thorough instantiation of the coordinates of your concepts.*

We should however not enter a more detailed discussion after this bunch of provocations of mathematical catechism[4] without stressing that Yoneda's basic epistemological insight is fundamentally the same as it is recognized in the theory of fine arts and of musical performance. We shall discuss this topic in Section 9.4 below and hope that the reader may now feel and then understand that the previous remarks about concept architectures are only apparently limited to mathematics.

9.1 Morphisms Are Points

Summary. Reviewing Euclid: points, addresses, pointers, and arrows. Why morphisms are elementary.

– Σ –

In the first book of his elements [455], Euclid defines point, line, and surface. Definition one reads: "A point is that which has no part." Definition three: "A line is breadthless length." So a line has a part: length, but no part called "breadth". And definition five reads: "A surface is that which has length and breadth only." In contrast to a point, a line has an attribute and a surface has two of them. This does not mean that a point is nothing, and that two lines are equal if their lengths are. Points, lines, and surfaces have an identity, but points have no further specification of their identity whereas lines and surfaces have[5].

So we are confronted with the identification of points. What can be the identification criterion if the intended point object has no part or attribute which can be used to make its identity more explicit? In analytical geometry, a point would have its coordinates as parts, and we were done, but here, the concept is irreducible. Why is it imaginable, and at least intuitively acceptable, that a point object could have an identity and nothing else? Evidently, the identification act seems a *secure* process which is not further problematic. In the etymology of the word "point", we indeed find a strong argument for the acceptance of the said identification mechanism: "Point" stems from latin "punctum", the past participle of "pungere", to prick. The latter is defined by the circular image "to pierce slightly with a sharp point"[6]. The important fact is that a point is a result of a movement with an arrow-like, acute (sharpened) instrument, its identification is established by a pointing action—the English expression "to point at" is essential: a point is what we point at. It is the result of a pointing gesture which in German reads "zeigen", to show.

4 We should remark that Finsler's point of view was passionately attacked by the establishment of "paper math" logicians [322]. I have participated in the last Finsler seminar before his death, and I admit that his anti-formalist point of view never failed, he had a unique instinct in unveiling hidden formalia.

5 We are not discussing synthetic geometry here where points as well as lines are formal objects of an axiomatic system, but Euclid's attempt to introduce meaningful objects via attributes. The fact that Euclid did virtually not use his fundamental definitions in his axiomatic system of geometry is not our argument.

6 This is Webster's explanation. The explanation of the German Brockhaus is not less circular: "Punkt" refers to "Geometrie"/"Raum", and there, "Raum" is explained as "Menge von Dingen, die *Punkte* genannt werden".

Thesis 2 *Identification of what has no parts is that pointing gesture.*

In Euclid's identificatory approach, two particularities were unspecified: The pointing subject and the variety of identification types. First, the pointing subject was uniquely there to 'throw the identification arrows', it was hidden as a subsidiary instance behind the targeted points, had to hold a quiver filled with identification arrows, nothing else. The second lacuna is the variety of arrows that is offered in Euclid's geometry. There is only the 'point' type of arrows. No other irreducible targets are considered.

For a long time, the problem of identification arrows was not recognized. From the naive Platonic point of view, mathematical objects were cast somewhere in a Platonic topos and one had to recognize them, to describe their eternal properties. The turning point came through a complex, but fascinating, review of the theory of polynomial equations which eventually ended in modern algebraic geometry. In the vein of analytical geometry as initiated by Descartes, a point p is a compound concept, it is identified via its numerical coordinates: $p = (x, y, z)$. In the classical setting, the coordinates are 'given' numbers, there is no question of inventing whatsoever; either we have the coordinates or we don't. But the algebraic approach suggested a radical review of this view. The point p can be seen as a solution of three polynomial equations $X - x = 0, Y - y = 0, Z - z = 0$. The three linear polynomials $X - x, Y - y, Z - z$ intervene in the following restatement: We look at the ring homomorphism[7]

$$ev_p : \mathbb{R}[X, Y, Z] \to \mathbb{R} : f(X, Y, Z) \mapsto f(x, y, z) \tag{9.1}$$

on the ring of polynomials $f(X, Y, Z)$ with real coefficients in the variables X, Y, Z. The point p is then associated with the kernel of ev_p which is the (maximal) ideal[8] $(X - x, Y - y, Z - z)$ generated by the three polynomials $X - x, Y - y, Z - z$. So the coordinates of p are recovered by generators of $Ker(ev_p)$.

This means that we can restate a point p by the evaluation homomorphism ev_p. This seemingly complicated restatement has a far-reaching advantage. We may now look at homomorphisms $e : \mathbb{R}[X, Y, Z] \to B$ with codomain *any* commutative ring B and identify it with a kind of 'generalized point'. This point has coordinates $x = e(X), y = e(Y), z = e(Z)$, but these values are no longer visible in the naive context of real coordinates. Let us see why this is completely natural! Take the canonical factorization homomorphisms

$$e : \mathbb{R}[X, Y, Z] \to \mathbb{R}[X, Y, Z]/(X, Y, Z^2 + 1)$$

with $x = X \pmod{X, Y, Z^2+1}, y = Y \pmod{X, Y, Z^2+1}, z = Z \pmod{X, Y, Z^2+1}$. Thus we have $x = y = 0$ and $z^2 = -1$. Clearly the third equation cannot be fulfilled on $\mathbb{R}$. In other words, this 'generalized point' e is something very exotic! We have not found a classical coordinate triple, but created a new one *uniquely through the homomorphism* e. What is the key argument here? *The introduction of 'points' via homomorphisms allows the solution of equations which were not solvable in the previous real-valued coordinate context.* So the creation of such generalized points identifies with the construction of solutions of algebraic equations.

In this setup, finally, the Euclidean pointing subject and the variety of arrows are instantiated. In algebraic geometry, a homomorphism $d : A \to B$ of rings is read as an arrow with reversed direction $Spec(d) : Spec(B) \to Spec(A)$ of associated geometric objects[9], the spectra $Spec(R)$ of rings R. To us the only point is that algebraic geometry views our homomorphism e as a morphism $Spec(e) : Spec(B) \to Spec(\mathbb{R}[X, Y, Z])$ which in the spirit of our general denotator terminology starts at 'address' B instead of the usual address $\mathbb{R}$ of real coordinates. So the Euclidean subject is instantiated by this address selection whereas the Euclidean 'arrow variety' is precisely the set of arrows from $Spec(B)$ to $Spec(\mathbb{R}[X, Y, Z])$. In algebraic geometry, these arrows are called "B-valued points of $\mathbb{R}[X, Y, Z]$", see Figure 9.2.

In our above case, the quotient ring $B = \mathbb{R}[X, Y, Z]/(X, Y, Z^2 + 1)$ is isomorphic to the domain $\mathbb{C}$ of complex numbers via the identity on the reals and $z \mapsto i = \sqrt{-1}$. And this is a crucial situation in the development of modern mathematics: Construction of solution spaces for algebraic equations! The complex numbers are the solution space whose points are built around the 'square root of -1 point'. In other words,

[7] See Appendix D.1.

[8] See Appendix D.1.

[9] See Appendix F.2; the essence is that such an object is really controlled by its functor $\mathbf{Rings} \to \mathbf{Ens} : B \mapsto \mathbf{Rings}(R, B)$ on the 'addresses' B.

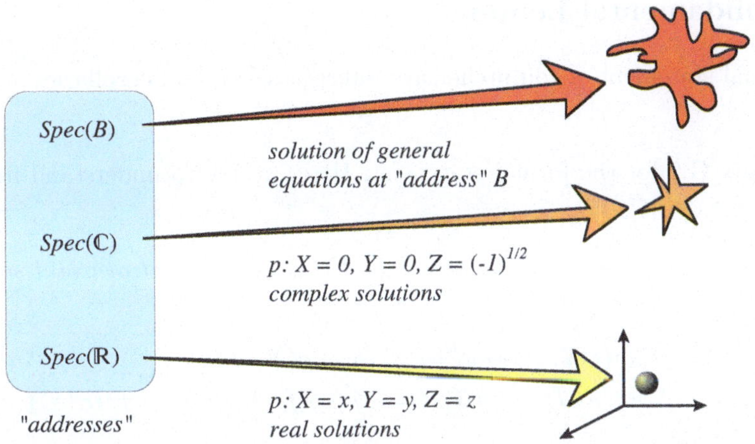

Fig. 9.2. In modern algebraic geometry, points are arrows of spectra, corresponding to ring homomorphisms in reversed direction. Real-valued points can be drawn in usual three-space by three coordinates x, y, z. Complex points define a different space which is no more visible on the naive level and is built upon the solution of equation $Z^2+1=0$. General rings B in the role of 'point addresses' yield solutions for general algebraic equations.

the relativization of the point's subject domain adds new points to the usual ones from naive real geometry. At this point it also becomes evident that the view of addresses as ontological parameters such as was described in principle 3 is completely canonical for mathematical ontology: New numbers provide new ontologies of mathematical objects. *The reality of real numbers is only one of a huge universe of ontological addresses.* In electromagnetic theory and engineering, this is standard: things such as imaginary currents are common practice.

It is not the place here to pursue in depth the deep consequences of this change of paradigm in mathematics which Alexander Grothendieck introduced around 1963, we should however notice that the very concept of a space was relativized according to the address variable. In fact, a Grothendieck topology (see Appendix G.4 and our more specific coverage of the subject in chapter 19) on a category C is a collection of point sets where points are *arrows* on all possible addresses in C with special properties. Unlike classical topology, the Grothendieck concept works on the point concept which we have introduced above: Points are arrows in categories.

This is completed in the 1963 revolutionary approach to mathematics of William Lawvere in joint work with Myles Tierny [585]: He started his successful foundation program to lay the basis of mathematics not on set theory but on category theory. From now on, the elementary objects are no longer sets as given by their element points, but arrows which replace the old points and are now parametrized by their 'domain addresses'. What then is left in this new paradigm from the old pointing gesture? Exactly what was the essence in concept construction: You have a domain address for your arrows and for any arrow $f : X \to Y$, if its codomain Y is the domain of another arrow $g : Y \to Z$, you can concatenate the arrows to $X \to Y \to Z$ and end up with a composed pointing gesture $g \cdot f : X \to Z$. This axiomatics of category theory reduces to the very nature of 'pointing at': the only thing which can be done in the universe of arrows is to put them together if ever possible.

Thesis 3 *This is semantics, reduced to its very essence. There is no more than pointing and pursuing the arrows' path as long as there is a way out...*[10]

[10] The reduction of semantics to 'pure arrows' in cognitive science is approached in a remarkable work by Daniel Dennett [260].

9.2 Yoneda's Fundamental Lemma

Summary. The formal statement, a comprehensive interpretation, and corollaries.

– Σ –

See also Appendix G.2 for the formal statement. Here, we try to understand its contents. This is the statement:

Lemma 7 *Let C be a category, $X \in {}_0C$ an object, and $F \in {}_0C^{@}$ a contravariant set-valued functor on C. Then the applications*

$$\iota : Hom(@X, F) \to X@F : \quad g \mapsto g(X)(Id_X), \tag{9.2}$$

$$\eta : X@F \to Hom(@X, F) : \quad \gamma \mapsto (f : Y \to X \mapsto f@F(\gamma)) \tag{9.3}$$

are inverse to each other. In particular, if $Y \in {}_0C$ is another object in C, we have a bijection

$$\eta : X@Y \to Hom(@X, @Y) : f \mapsto @f, \tag{9.4}$$

i.e. the Yoneda functor $@ : C \to C^{@} : X \mapsto @X$ is fully faithful.

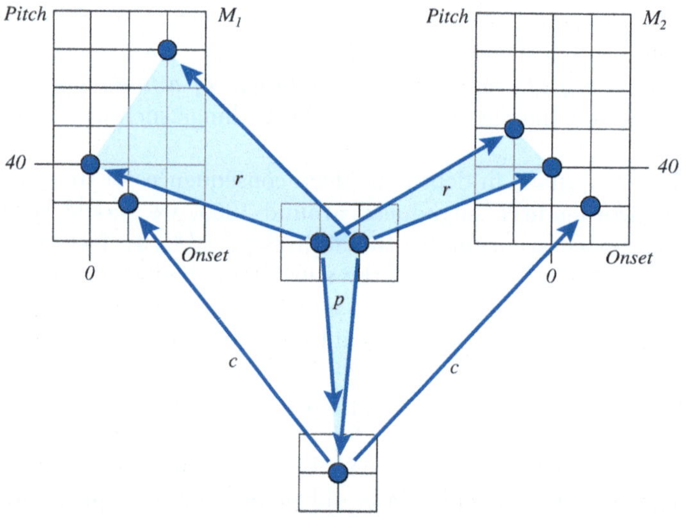

Fig. 9.3. Two three-element motives M_1, M_2 and points addressed in the singleton and in a two-element motif. On this level, M_1 and M_2 look the same: For both, M_1 and M_2, we have the three arrows c from the singleton and three arrows factorizing through the singleton from the two-element motif, as well as six injective arrows r from the two-element motif.

We want to give a comprehensible interpretation of the last statement concerning the Yoneda functor. To begin with, what does it mean to replace an object X by its functor $@X$? We are indeed creating an overall perspective to X since $@X$ means selecting all possible 'addresses' Y in C and then pointing at[11] X via all possible Y-addressed arrows $f : Y \to X$. We also call such arrows f 'perspectives of X viewed from Y'. Let us make an example to learn what such perspectives can teach us about X, and why this terminology is adequate, see also figures 9.3 and 9.4. To this end, take the category $ComLoc_{0_{\mathbb{R}}}$ of commutative zero-addressed local compositions over the reals. With the notation of Section 8.3.5, sorite

[11] This is the reason for selecting @ as a symbol for the arrows $f : Y \to X$ in $Y@X$

6, take two three-element motives $(M_1, Onset \oplus Pitch), (M_2, Onset \oplus Pitch)$ with each three elements: $M_1 = \{(0,40),(1,39),(2,43)\}, M_2 = \{(-1,41),(0,40),(1,39)\}$. Let us look at different 'addresses' from which we look at our motives. The first address is the singleton motif S. The arrows $c : S \to M_i$ correspond to the elements in motives M_i. Since we have three of them in each target motif, the arrows are in one-to-one correspondence. So, intuitively, M_1 and M_2 look the same when viewed from S.

Observe that if we were dealing with set theory, the singleton perspective would already do: The perspectives from the singleton just define the cardinality of a set, and this means fixing the isomorphism class of the set! But we are not in set theory and want to vary perspectives. The next standpoint of perspective is a typical two-element motif as shown in Figure 9.3. We have two types of arrows r here: First those which factorize through the singleton: $r = c \cdot p$. Then we have injective arrows hitting two different points in the target motives. These can be any set-theoretically defined maps, and we get six arrows which again, unfortunately, do not yield any visible difference between the motives.

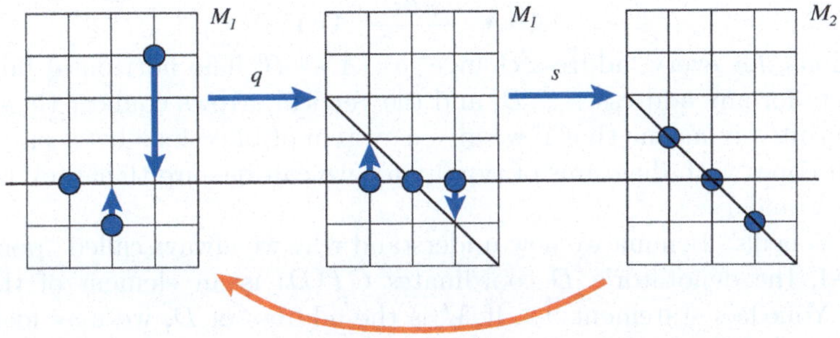

Fig. 9.4. Whereas M_1 can be pointed-at from M_2, there is no *bijective* M_2-valued point at M_1: In its totality, M_1 is literally 'invisible' to M_2.

So let us try a reciprocal confrontation of these motives, see Figure 9.3! We have a bijective morphism $p \cdot s : M_1 \to M_2$ through the projection p onto the first coordinate and the shearing s in pitch direction (transvection). But in the converse direction, every morphism must leave the collinear elements of M_2 collinear. But the points in M_1 are not collinear and we have no bijection. In other words, you can never see all of M_1 from M_2. This breaks the symmetry of roles between the candidates.

Exercise 11 *Another argument would have been to inspect the module functor introduced in Section 8.3.5, Definition 28. Use the fact that a bijection in this situation must be an isomorphism and the fact that a functor maps isomorphisms to isomorphisms.*

From this easy example from motif theory we learn that replacing an object X (a motif in our case) by its functor $@X$ does not yield a full picture of the object at any one of its arrow perspectives. If we restrict to the singleton perspective, we only see the set-theoretic aspect, but not the full affine geometry of the category of local compositions. However, a sufficiently general variation of the perspective can reveal subtler structures of arrow point sets.

In a more intuitive setup[12], we may look at 2D and 3D objects and perspective maps from 3D objects X to 2D objects Y which occur in painting and photography. If we view such maps as arrows (in an appropriate category) $p : Y \to X$, we may ask about the information on the 3D object X which is obtained through its photographs Y under the viewpoint p. For instance: Is it possible to reconstruct X from a bunch of photographs $Y_1, \ldots Y_n$ without knowledge of the perspectives $p_i : Y_i \to X$? A historically famous example for this situation is the discovery by Edward Whymper of an easy route to the Matterhorn's top. The

[12] It can be made rigorous by use of projective geometry, but this is beyond our concern.

common view from Zermatt showed no evident path, every possibility seemed too steep to be attacked by mountain climbers of those early days. However, a change of the 'dogmatic' perspective unveiled the solution to Whymper: Viewed from the eastern perspective of the Theodul glacier, the northern crest appeared as much less steep than they believed from the common 'knowledge' down in Zermatt. The moral of this lesson is that, whereas a single 'address' may blur a lot of a particular object, variation of address and perspective may afford a more complete view of our object.

With this in mind, we now have the crucial statement of Yoneda's Lemma: Passing from the objects to their functors yields not only a rich system of perspectives but the entire information we need to tell whether objects are isomorphic or not. In fact, by bijection (9.4), $\xi : @X \xrightarrow{\sim} @Y$ is an isomorphism of functors iff ξ stems from an isomorphism $x : X \xrightarrow{\sim} Y$, $\xi = \eta(x)$. Mathematically, the isomorphism ξ means that we are given a commutative diagram

$$\begin{array}{ccc} B@X & \xrightarrow{B@\xi} & B@Y \\ {\scriptstyle g@X}\downarrow & & \downarrow{\scriptstyle g@Y} \\ A@X & \xrightarrow{A@\xi} & A@Y \end{array} \tag{9.5}$$

with horizontal bijections, for every 'address change' $g : A \to B$. The horizontal bijections are correspondences among point sets for any addresses A, B, and the vertical arrows connect these bijections of variable address points. This intuitively means that if we give a system of bijections between points or 'perspectives' at objects X, Y, which copes with diagrams of type (9.5), we can be sure that they indeed generate a 'real' isomorphism between X and Y.

As a corollary of Yoneda's Lemma we now understand why we always called "points in a space functor" the denotators. Indeed, the denotator's D coordinates $CT(D)$ is an element of the denotator's functor $F(D) \in {}_0\mathbf{Mod}^@$. By Yoneda's statement 9.4, if M is the address of D, we may identify $CT(D)$ with the arrow $\eta(CT(D)) : @M \to F(D)$. And therefore, we have this statement:

Corollary 4 *The coordinates of denotator D is a point of the denotator's space $F(D)$ with values in the denotator's address $@M$.*

Notice, however, that we are not given all possible 'addresses' to be able to calculate $F(D)$ from its denotators since our addresses $@M$ are only the *representable* functors, see Appendix G.2.

9.3 The Yoneda Philosophy

Summary. The Yoneda revolution in mathematics: functors instead of objects. "The functorial point of view is the geometric one". Perspective and truth, objectivity and dialog.

– Σ –

As this is not a book on mathematics, we cannot explain all the proper mathematical consequences of the Yoneda Lemma. But it is essential to understand its impact on musicology as it is exposed in the present viewpoint based upon functors and points in the form of spaces and denotators.

We shall call the following body of insights the "Yoneda philosophy" because it is more than a technical approach, it is a paradigm of far-reaching power which reshapes not only mathematics but also concept architectures and in particular software engineering as well as methodologies in fine and other arts.

This philosophy is built around the restatement of the elementary character of Euclid's points in terms of arrows: points are addressed perspectives, and on each address, we have an ontology in the sense that the point space of this address is a specific layer of spatiality.

We further know from Yoneda's isomorphism (9.4) that classification of objects is equivalent to classification of their functors, i.e., we can discover everything about an object via its point spaces on variable addresses: The object is recognized through variation of perspectives. Knowledge is the result of a network of (ontological) perspectives.

Therefore, all we know from an object is determined via its behavior (its functorial point system). Whatever the 'real ontology' of an object, it is completely manageable on the behavioral level.

In particular, an object may also be defined by means of its behavior! In mathematical terms, an object can be first defined by its functorial properties, and then, if required, retrieved among 'real' objects. But even if a functorial definition fails corresponding to a real object, i.e. if the functor $F \in C^{@}$ is not representable (see Appendix G.2), it can be shown (see Appendix G.2.1, Proposition 109) that it is the colimit of a diagram of representable functors $F_\iota = @X_\iota$. In other words: In the colimit, all behavioral definitions are reachable via 'objective' definitions, i.e., representable functors[13].

Points are not only mathematical details, they are also technical realizations of concept constructs, this is the philosophy of denotators. Since denotators are basically defined by recursion to ramification instances (via universal constructs such as limits etc.), building concept points is also a concatenation of points qua arrows.

In particular, circular constructs in concept architectures are not only possible, they are even necessary on a basic building level since rooted axiomatics can not yield all the conceptual foundations which are needed to think the concepts—instead of just playing around without knowing what instance is really doing the hardware of thinking.

The Yoneda embedding as functorial review of objects yields a bundle of addressed ontologies which are by no means abstract: "The functorial point of view *is* the geometric one!"[14] This profiled statement by Alexander Grothendieck's pupil Peter Gabriel claims that geometry cannot terminate on the limits of everyday's poor intuition and experience. Of course, such a statement is more general if we are not only looking at instances of geometric provenance. It essentially states that the behavioral reality *is* the essential one! In music this has another interpretation which we shall review in Section 9.4.

In order to understand an object via its functor, there is a basic research agenda to pursue:

- Exhibit a minimal selection of 'ontological addresses and perspectives' which are sufficient to classify an object (in which case we say that they are "classifying").
- Understand what are "equations", and what their "solution" means on a particular address.

In traditional architecture it is known that three mutually orthogonal projections are classifying. We have seen that in the theory of local compositions, singleton perspectives are not classifying. We shall see later, in the global theory, to what degree local compositions are classifying.

Understanding "equations" and "solutions" really is a deep problem: What does a particular address unveil about the object in question? What problem does a particular 'geometric' space perspective address? This agenda item drives us towards the approach of *viewing the Yoneda philosophy as a fundamental concern of understanding things in spaces which are created to make problems evident.*

9.4 Understanding Fine and Other Arts

9.4.1 Painting and Music

Summary. Understanding painting and music is a synthesis of perspective variations. The year 1954. Comparison of Adorno's, Bätschmann's, and Valéry's esthetic principles in view of the Yoneda philosophy. Interpretation as an identification process.

– Σ –

In the fine arts, above all in painting, the consciousness of a particular view has always been part of the art because of its technical requirement to map a higher reality onto a canvas or another plane surface. With the development of the central perspective in the early renaissance, the explicit point of view became

[13] In mathematics it has become common to use this technique to introduce objects since often, the very object is less important than its behavior!

[14] "Der funktorielle Gesichtspunkt *ist* der geometrische!" [350]

even more prominent in that the divine position "sub specie aeternitatis" was replaced by an explicit human place. It is about the same time when in music too, the passage from the representation of metaphysical harmonies as conveyed from Pythagorean tradition to the individual, human composition, the musical work as we understand it in modern times, took place. *The musical work is the discovery of the central perspective in music.*

It is a commonplace to conceive the artist as the first observer or listener. The work is never neutral and independent of its perception, interpretation and understanding. It is virtually never created without its message to an esthetic instance. And it is always clear that the multiplicity of views and interpretations is either desired or at least accepted by the artist.

But it was much less evident and acceptable to the artists and even less to their religious and political patrons that a work of fine art cannot be understood except through a complex and long-term process of variation of the perspective. Perspective variation was—on the contrary—a well-protected tabou of presentation. A perspectivic view was accepted, but not as a possibility, rather as a mandatory presentation of dignity, symmetry and devotion. The selection of a determined perspective, be it in painting or sacral or political ceremonies, was an expression of a weltanschauung or dogma for the humble addressee[15].

On one hand, the prescription of a fixed authoritative perspective was explication of a rigid ontology: "This is the world, this is, how we all should see and accept it!" On the other, this would obscure the relative position of the work of art in a variety of possible alternatives. You cannot really understand if you cannot see the special choice among a variety of—possibly worse or questionable—alternatives. We refer to our computer-aided graphical analysis of Raffael's "School of Athens" which is documented in [671] and which was highly controversial because the computer simulation admitted free perspective variation and thereby destroyed the hitherto dogmatic frontal view of the monumental fresco in the Vatican's sacred rooms, see also figures 9.5 and 9.6.

Fig. 9.5. Original frontal perspective of Raffael's "School of Athens" in a computer-aided simulation. The simulation includes 58 human figures and architectural essentials.

Fig. 9.6. Perspective from the right side of Raffael's "School of Athens" in a computer-aided simulation. It reveals hidden symmetries of the work of art, see [671] for details.

Whereas in sculpture, walking around the work which is positioned in free space is mandatory to understand it[16], perspective variation is practically forbidden in musical compositions of classical European tradition. To the contrary: Following the composer's indications has for a long time been a must and was even a severe criterion of success in the European tradition of music critique. We come back to this subject in Chapter 46. But there is one important point which broke this music tradition, initiated a new approach to understanding musical composition, and which must be seen in the context of Yoneda's Lemma: It was Theodor Wiesengrund Adorno who recapitulated the question of identity of a musical work with respect to

[15] I remember that a computer-aided graphical analysis of Raffael's *School of Athens* during the Darmstadt symmetry exposition [671] was vehemently attacked by conservative historians because the computer simulation included arbitrary variation of the original frontal central perspective. This freedom of manipulation was felt as a serious disrespect against Rome's Sancta Ecclesia.

[16] Here, Yoneda's statement that the object is understood via the totality of its perspectivic views—its functor—is completely obvious.

performance. According to Adorno [7] we have to recognize this fact: "Die Idee der Interpretation gehört zur Musik selber und ist ihr nicht adkzidenziell." Performance is a substantial part of a composition's identity. This means that the work of art is not fully existential before or without performance. Of course, this addendum is a critical one since *now, performance with all its variations would no longer provide us with a unique perspective but with an a priori infinite bunch of alternatives which share the composition's essence.* We could still argue that the best one is the only admitted one, but this is no longer the unique one, we must inspect a variety before selecting the optimal alternative.

In poetry, Adorno's insight was anticipated by Paul Valéry's reflections on literature which culminated in the famous dictum: "C'est l'exécution du poeme qui est le poème." According to Hans Robert Jauss [506], Valéry's approach was a rupture with the classical tradition that viewed a work of literature as an autonomous unit, "ens causa sui" which appeared as one and the same object at any time and to everybody. Jauss concludes that a work of art only unveils its substance in the course of its historical life. This is quite the same as to say—with Adorno—that the literary work realizes its identity in the light of a time chain of performance perspectives.

It seems that Adorno and Valéry have uniquely stressed the exterior performance activity in music and poetry, but their concern is much deeper. We should keep in mind that performance is strongly tied to its rhetoric function as a means to express understanding, and in this respect, performance is not only a perspective of action but instantiation of understanding, of interpretation of given structures. In other words, the Adorno and Valéry approach is classificatory on the level of functorial representation of works of art.

In the hermeneutics of fine arts, Oskar Bätschmann has explicitly thematized the multiplicity of interpretative esthesic perspectives in his systematic treatise [90]. Bätschmann not only presents a picture of multiple interpretations where the artist's is only the very first, he also argues for an *interesthetic dialog* in a community of argumentation[17]. This is nothing less than the requirement of a coherent network of single interpretations, in fact the functorial interconnectivity of the collection of address ontologies, to use our terminology from Yoneda philosophy as developed in Section 9.3. There is, however, a critical point in Bätschmann's system. He does accept mutual contradiction within a set of interpretations which—each in itself—are consistent in argumentation and method. It is not necessary to reject the total set in this case, nor is it of any particular quality if an interpretation is materially complete and consistent. We cannot accept such a "tolerance" which in fact denies any overall system of comprehension. Rather should the fact of contradiction within a system of interpretations (and not within the proper art work where contradictions are common) be a source of refined research.

It is a remarkable fact that Adorno's approach was virtually simultaneous with Yoneda's, namely around 1954–1956. This is exactly a hundred years after the equally simultaneous emergence of the local/global paradigm as proposed by Hanslick and Riemann. Let us summarize that interpretation in the sense that functorial point perspectives distributed among a variety of ontologies is not a subsidiary task of understanding objects of the fine arts or of mathematics or even conceptual constructs, it is the very identification process which is solicited in so doing. Conversely, it should not be argued that we already know everything about classical works of art, be this Bach's, Scriabin's, Bartok's compositions, Raphael's, van Gogh's, Picasso's paintings, or Villon's, Poe's, Benn's poetry. The complete identification is not yet, and will for a long time not be, settled as long as the full functorial poly-ontology of point arrows is not developed to maturity.

9.4.2 The Art of Object-Oriented Programming

Summary. In contrast to its name, object-oriented programming is essentially a categorical access to programming. Characteristics, such as encapsulation, inheritance, methods, and class or instance variables do realize what the Yoneda Lemma suggests: to replace program entities by the behavior they can show under determined conditions, i.e., identification by behavior.

– Σ –

[17] "Argumentationsgemeinschaft", [90, p.160].

Concluding this chapter, we feel obliged to complete the circle from concept frameworks through mathematics and philosophy of fine arts to the core of our knowledge society and its cultural impact: computer science. We would in particular like to discuss the most advanced constructive paradigm of theoretical informatics: object-oriented programming.

Without explaining technicalities which are of no interest in this chapter[18] we can characterize object-oriented programming languages as being built on programming units called objects. These have an identity (in fact a pointer to their data structures) and are accessible through so-called methods which can be performed in a messaging action. All you know about objects is through their response upon messaging to their methods. For instance, if an object describes a point on the graphical interface, the point's coordinates are not automatically accessible. A priori, they pertain to the object's privacy and must be accessed via a special method. This is the encapsulation principle. So an object has an interior structure which is hidden to the program's context, you can only understand the object by its response to method messaging. Methods can be formalized in the language of categories [700]. There, roughly speaking, a method is an arrow from the given object's space functor (technically speaking: its class) to a second object's space functor which is responsible for the method's output[19].

Object-oriented programming is also an application of the point arrow approach to semantics which we discussed in the context of concept architectures: As such the identity of an object is fairly insignificant, it is just an arrow to an address where you can access further data. In terms of denotator theory, the identity is the denotator's arrow $@A \to Fun(Form)$, it will only give you further information upon request of coordinate arrows which—in this programming paradigm—correspond to the object's methods. It is remarkable that the implementation of the denotator theory in the Java-based RUBATO®software environment (see Chapter 40) has realized the full functorial point of view exposed in this chapter.

[18] But they will intervene in the chapters concerning design and implementation of software, see parts X and XIII

[19] The dot-notation for method actions on objects in the Java language is quite near to this category-theoretic formalization.

10

Paradigmatic Classification

The poetic function projects
the principle of equivalence
from the axis of selection
to the axis of combination.
Roman Jakobson [500]

Summary. Paradigmatic classification deals with formation of classes of objects which belong to specific paradigms. Here, paradigms are fields of equivalence or association. In a more general setting than in mathematics, equivalence is *not necessarily* understood as being a transitive relation. We give motivation of the paradigm concept from musicology, semiotics and poetology, and mathematics. Our taxonomy yields two types of paradigms: by transformations and by similarity—however, in practice, they often appear in mixed form. In a mathematical perspective, the first type is covered by group theory, the second by topology. This means that fuzzy concepts in the humanities are not a priori useless, they can be incorporated into exact reasoning by means of a refined paradigmatic reconstruction.

– Σ –

This chapter is a preliminary discourse on local classification theory. Since Chapters 11 and 12 will have more technical flavor, we want to give an overview of the principles of local classification and the corresponding references to sciences which are sensitive to this type of classification. We do not deal with lexical classification here; this was already discussed in Section 6.8 dealing with orders on denotators. Concept taxonomy following classical Porphyrean trees and similar constructs are completely absorbed in that discussion. Rather are we concerned with paradigmata, i.e. "fields of equivalence", whatever this may signify—we have to review the differentiations of this subject. Its essence is very clear: In most human activities, from concept building and abstraction to concrete gestural actions, identity is not easy to get under control. Rather is this onto-logical cornerstone a necessary evil which in virtually all cases is overridden by grouping of large sets of objects whose identity is much less relevant than some common features, be they precise or only fuzzy. We are going to discuss the actual types of fuzziness and their genealogy in musicology, linguistics, and mathematics.

We should however stress that classification is highly controversial in musicology for three reasons:

- It is commonly believed that classifying musical objects on whatever level is contrary to the individual expressivity of compositions as they have been cultivated since the renaissance, but see our discussion in Section 9.4.
- Classification is misunderstood as a purely bureaucratic activity of list compilation.
- Due to a catastrophical lack of technical tools, traditional musicology has only rarely been able to control the variety of their objects. A disdain of detailed technical work which is psychologically comprehensible must scientifically be blamed for a major scientific retardation even with respect to other humanities such as linguistics.

G. Mazzola, *The Topos of Music I: Theory*, Computational Music Science,
https://doi.org/10.1007/978-3-319-64364-9_10

We shall vaporize these mystifications in the following chapters, the second yet seems already partly settled from our previous discussion on the Yoneda philosophy: classification in the sense of a determination of isomorphism classes is nothing else than the overall task of understanding an object as we have identified it in the functorial approach in the spirit of Adorno, Valéry, and Bätschmann in Chapter 9.

10.1 Paradigmata in Musicology, Linguistics, and Mathematics

Summary. Transformation and similarity in music: Neumes, melodic contours, variations, and counterpoint. De Saussure's paradigmatic and syntactic axes. Characteristic combination of transformation and similarity in poetics. Group theory and topology: the poetics of mathematics.

– Σ –

The etymology of paradigm is Greek παρὰδειγμα: example, model, that thing side by side which you show or point at. Already this Greek root has a double meaning for the relation of that thing: On one hand it is "side by side", on the other, you "point at" it. The first suggests a space where neigborhood makes sense, the second is just a connecting action from me to that thing. It is however open whether the object is in my neighborhood because I am pointing at it or vice versa: whether I am pointing at something because it is in my neighborhood anyway. This ambiguity is after all not a defect, it shows a multiple specification which is essential to the word's explanatory power.

The difference between "side by side" and "point at" as explanatory roots is that in the first case, the model object is situated in a spatial neigborhood without my intervention, all I do is to pick it from a multitude of available objects in that neighborhood. We could also say that the model object and its neighboring specimens are taken into account because they are similar to the point where I am standing, so that we can better understand our point when recognizing similar points. We call this meaning the *topological* one.

In the second case, the model object is "side by side" because I have shown it in a pointing gesture. Its neighborhood to my own position is not given from an a priori ambient space but by a construction of this arrow whose head is my model object. The explanatory power here results from a transformation of my position to the target position, i.e. the model is a transformed object issued from my operation upon the given proper position. In this case, the resulting model object is not random as in the topological case, it is a well-defined functional value. The examples can help understanding of our point because they are related under a specific operation. We call this meaning the *transformational* one.

Fact 4 *Both topological and transformational paradigms help understanding of the proper position by the formation of classes of equivalent objects: the "example set" of what is to be explained.*

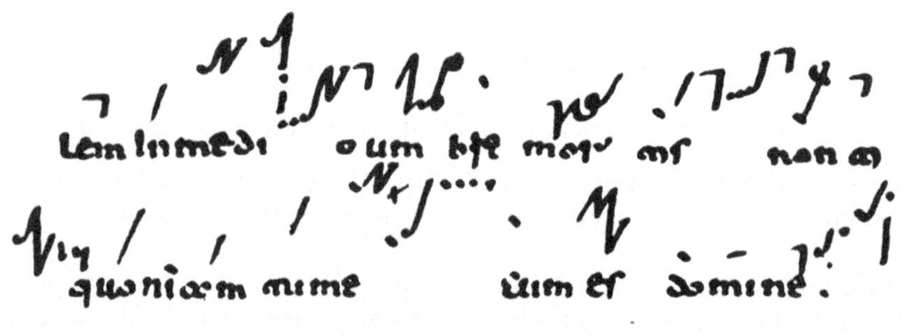

Fig. 10.1. Neumes from the Beneventan writing (Rome, Bibl. Vat. Ms. lat. 10673; 11th century).

Let us first see how both these paradigmatic explications arise in **music**. The topological type means that we are given a musical object which is similar to the model object. This similarity approach in comparison of musical objects is widespread, we shall deal with it in several contexts and can restrict here to a characteristic summary. To begin with, before written notation of European music was developed, the cheironomic practice of hand gestures served as a paradigmatic sign system to indicate melodic movements in choral singing. This gestural system was then abstracted in the neume system with its different writing styles and developmental completion, tracing articulation, accentuation and diastematic movement in Gregorian choral, see figure 10.1. The word neume is rooted in Greek $\nu\varepsilon\tilde{\upsilon}\mu\alpha$, hint, and by this very etymology alludes to a movement of the voice that is only defined in the vague similarity paradigm. This rooting of melodic conceptualization is not only historic, it is—strangely enough—still a fundamental attitude towards melody: Music theory has not yet been able to really define a musical motif or melody apart from a priori vague concept fields. It is understood that the precise identity of a melody is less important than its—however ill-defined—similarity class which may be termed "contour" to indicate some type of topological paradigmatics[1].

After a long period of hidden or even patent refusal to work out explicit definitions of terms such as (melodic) "motif" [2], the vagueness of the contour concept has been attacked by what lends itself to the imprecise: fuzzy theory [870]. We shall come back to this approach but should remark here that fuzzy relations are a poor imitation of what general topology, and even theory of pseudometric spaces, could grasp much better. The interesting point is that the fashionable expression "fuzzy" was necessary to introduce instances of topological reasoning to musicology, the imprecise as a scientific category was the real thing which musicologists seemed to wait for.

The consciousness of similarity is also present in variational methods. Variation does not mean to apply a determined transformation but to move around in a given topological neigborhood. In this sense a tonal variant of a melody or harmony, from major to minor tonality, say, is just a restatement of a structure with neighboring values; the theory of turbidity (see our discussion in example 9 of Section 8.1.1) is indeed thought in this topological spirit.

A huge field of extremely topological paradigmatics opens in the subject of sound colors. However, the decision of whether sound S_1 is more similar to a reference sound than sound S_2 is very difficult and in fact points at one of the most complex problems in sound synthesis and analysis as well as associated topologies. We shall discuss it more carefully in Section 12.3.2.

In contrast, the contrapuntal technique of fugue and canon constructions typically recurs to the second meaning of transformational paradigmatics: The construction of a *comes* answer from the *dux'* thematic germ makes use (among others) of contrapuntal symmetries, such as inversion, retrograde, retrograde inversion, or augmentation. These secondary instantiations of the original object are transformations and not similar objects. On the contrary: It is not true that these "models" are topologically similar to the original. A retrograde theme may look very *dissimilar* from the original; we have already explained this fact in the course of the general symmetry discussion in Section 8.1. This is also completely logical from the construction guideline of a fugue or canon: Answering to the dux melody is a "mental reflection", a correspondence that cannot be similar at random. Note that, in general, an answer need not be a faithful image of the original, it can as well be a poor shadow. But it has to be a functional consequence and not just something vaguely deformed.

Also in the vein of the transformational paradigm is the classification of chords according to transposition and/or inversion: A chord B is a model of a given chord A if it is deduced from A by a transposition and/or pitch inversion. On a more basic level, the formation of pitch classes also creates a pitch paradigm by transpositions of multiples of an octave. Clearly, pitch classes or chord classes are far from similar in the topological sense, but they are essential in the understanding of chord or pitch concepts and structures. Classification is often based upon transformational paradigms. For example, harmony is based upon these classifications in the pitch domain; without comparison of "equivalent" chords, harmony would be an im-

[1] Contour theory is an interesting object for the problematic building of valuable paradigmatic concepts in musicology.

[2] See our discussion of Rudolph Reti's theory of motives in Chapter 22.8.

possible task. So much the more is it remarkable that the entire Riemann program of functional harmony [231] which was designed to include the entire zoo of chord classes did not incite musicologists to write down complete lists of chord classes once for all.

In structuralist **linguistics**, paradigmatic relations intervene in the basic dichotomic system, as sketched by Ferdinand de Saussure in his famous "Cours de Linguistique" [933]. To Saussure, language is spanned within a twofold dimensionality: the syntagmatic and the paradigmatic axes, see the introductory Section 2.3.4. The paradigmatic axis covers relations between language units, but these relations are not present in the textual syntagm, they occur on a more abstract level "in absentia", see figure 10.2.

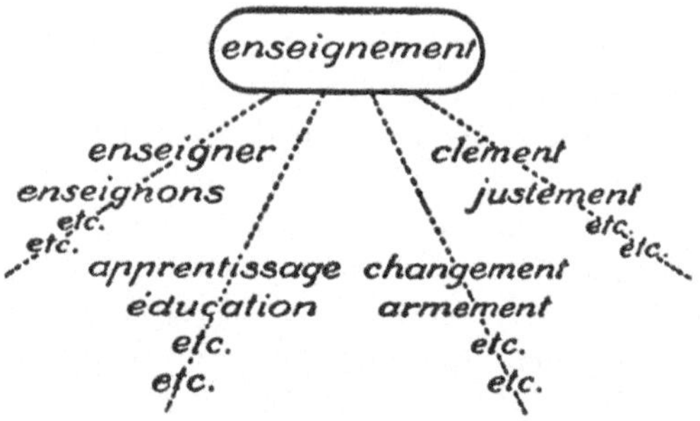

Fig. 10.2. A visualization of the paradigm of the word "enseignement" in Saussure's "Cours de Linguistique". Observe that the paradigmatic relation is defined on different levels: phonological, grammatical, semantical, etc.

What Saussure had called an associative field has been reviewed in different theories by subsequent research. The differentiation among these theories also reflects the double meaning of "paradigm". In a more transformational spirit[3], relations among instances of one paradigm are termed "in opposition" and *suggest* a binary operationality which of course does not cover the general phenomenology, as Roland Barthes has rightly criticized in [86]. Both situations do occur: Grammatical paradigmatics is transformational in the sense that the inflection of a word, for example, obeys the transformational paradigm whereas phonological vicinity in "minimal pairs" [17] such as "vine"/"fine", for example, is topological.

The classificatory function of paradigmatic activity is perhaps best understood in the linguistic context. Understanding a word as a member of a language is virtually never a question of abstract and isolated identification. Understanding means identifying not the individuum but its role, its position in the overall language environment. This position can be realized in two ways: topological deformation and transformation. In each, you learn rules of the language game which determine what you can do to an instance of the game without altering its essence.

In **mathematics**, paradigmatic processes and methods are well known but not in the semiotic interpretation, rather it is a basic mathematical activity of classification to boil down the overwhelming multitude of individual mathematical objects and structures. Topology, originally termed "analysis situs", deals with invariants of geometric objects under shape deformation: What is preserved if we stretch a rubber band? What is the common property of all kinds of holes? Transformations are commonly dealt with in group theory: We want to describe the nature and essence of systematically performing transformations upon a set of objects: Is it the concrete mechanism or are we given a general context which is present each time we proceed to a transformational activity? Is it the case that there are no plane ornaments with fivefold rotational symmetry? Why are there exactly five platonic, i.e. perfectly regular bodies?

Before the advent of these paradigmatic theories mathematics was a thoroughly sober, anti-poetical science. It was about geometric, numerical and functional quantification. There are still uninformed persons—often in the humanities—who believe that mathematics is this type of mechanical quantification business. But it should be recognized that paradigmatic theories have not only introduced a strong qualitative profile of mathematical concerns, it is also true that linguistic poetology has strongly profited from the group-theoretic

[3] Observe that we in no sense allude to Noam Chomsky's transformational theory here.

approach, for example in Jakobson's identification of the poeticity as a projection from the paradigmatic axis to the syntagmatic axis [500]. In fact, this projection introduces group theory in poetology since the functional connection of paradigmatics to syntagmatics uses syntactical symmetries, typically realized in rhyme constructs, to accumulate paradigmatic connectivity. Rhyme periodicity, the most common syntactical symmetry transformation, acts on the association of rhyming units, be it on the phonological or on the semantic level. Such an entanglement of mathematical and poetical roots induces a reversal of perspective: introduction of poetical aspects in mathematics. Even among conservative mathematicians it is commonly felt that works of art such as Cornelis Maurits Escher's or Johann Sebastian Bach's or Dante Alighieri's approaches are a kind of incarnation of mathematics in the fine arts, music, and literature, and *that this is a substantial correlation*[4]. The recent impact of fractal theory and quasi-crystal theory on mathematical rationales of esthetics has confirmed Hermann Weyl's symmetry-oriented approach to the arts [1113] as well as René Thom's topological catastrophe theory of cognitive processes [1052].

Fact 5 *The poetical character of qualitative theories in mathematics is a restitution of a common root of artistic and scientific thinking in the tradition of renaissance universalism as issued from medieval constructivism*[5] *which goes back to the Pythagorean tradition of operationalized thinking in music and mathematics.*

Fig. 10.3. Detail from Maurits Cornelis Escher's *Cycle* (©2017 The M.C. Escher Company-The Netherlands. All rights reserved. `www.mcescher.com`). The generating cube is positioned to the left middle, then iterated according to a plane ornament symmetry, the ornament is then deformed to organic shapes, ending up with the garden gnome figuration.

As may be expected, the double meaning of the word paradigm is often realized in combined form. The model object is then either a topological deformation of a transformed specimen or vice versa: a deformed version of a transformed specimen. For example, a contrapuntal answer may be the inversion of the leading theme followed by a tonal alteration to a new scale. We shall deal with this combined paradigmatics in Chapter 22 on motif gestalts. A very intuitive application of mixed paradigmatics is realized in Escher's metamorphosis works, such as shown in figure 10.3

[4] This has been popularized in [474].
[5] See [945] and [706] for this background.

10.2 Transformation

Summary. Transformation as a transfer of association to generative rules. Regular and degenerate rules: symmetries, morphisms, specializations, alterations, and fractals.

– Σ –

There is a fundamental and far-reaching difference between topological and transformational paradigmatics. The former not only refers to an ambient space topology but also remains much more passive than the latter. You simply have to recognize the model object as compared to the original object, and the rest is a plain verification of similarity. Everything is already given by inherent structures, the only 'activity' which is left to the observer is a cognitive act. Of course this one need not be completely evident, the psychology of topological pattern recognition is not trivial. But once a general acquaintance with the underlying ambient space topology is present, nothing has to be added beyond perception of what is there.

Transformational paradigmatics is much more complex. Not only are we asked to retrieve the actual arsenal of admitted transformations and therefore recall a non-automatic competence, the level of understanding is also quite different from the topological one. In fact, a transformation is a process which unfolds on a metalevel, the objects which are acted upon will only be results of this activity, the activity as such pertains to a hidden operating system. So, whenever transformational paradigmatic association is required, not only Saussure's "absence" is felt (in contrast to the syntagmatic presence of manifest contiguity), but also the highly demanding activation of our transformational competence. In other words:

Fact 6 *On the level of transformational paradigmatics, the simple association of manifest objects is transmuted to a complex construction activity of the subject who is concerned with the paradigm, and therefore the phenomenon is degraded to a secondary level which can only be understood by a competence on the generative rules rather then their effects.*

There are several degrees of complexity of generative rules as related to their effect upon objects. In the simplest case of an axial symmetry, the relation is evident and the operation as such is not difficult to retrace. Also the reflected object is a faithful image, double reflection restores the starting point. As a transformation becomes irreversible it is important to know the underlying transformation, its 'objective trace' no longer enables the transformation's reconstruction. For example, the projection of a part of a dodecaphonic series onto a chord is a "degenerate map" and destroys the temporal order of the series: the chord is a very special 'shadow' of the original series. This is why the morphisms have become the core of structure theories, such as category theory and its crystallizations in the theory of local composition as presented and commented on in Chapters 8 and 9. In the theory of alterations which we introduced in Section 7.5, transformations which are associated with alterations virtually disappear in the relative local compositions, altered local compositions are nearly topological deformations. The extremal situation of degeneracy is fractal theory. Here, the naive visibility of symmetry transformations has completely disappeared or at least been disguised in a fascinating esthetics. A transformation of type $f_c : \mathbb{C} \to \mathbb{C} : z \mapsto z^2 + c$ is used to distinguish those complex numbers x which go to infinity with successive iteration $x, f_c(x), f_c^2(x), \ldots f_c^n(x)$ for $n \to \infty$ from those x which have all their iterations $f_c^n(x)$ in a finite region of the complex numbers. So we are not looking at a repeated shift of a point, i.e. $x, a+x, a+a+x, \ldots a+\ldots a+x, \ldots$ but of an operation: $Id, f = f \cdot Id, f^2 = f \cdot f, f^3 = f \cdot f^2, \ldots, f^{n+1} = f \cdot f^n, \ldots$. The set of points which approach infinity is separated from the rest by a boundary known as the Julia set[6], and which shows a breathtaking beauty, see figure 10.4. This trace of the hidden transformation—an iterated application of a fixed function—shows no simple evidence of the transformation; on the contrary, *without its generating transformation rule* the knowledge of the Julia set per se is completely mysterious.

It may seem that the entire discussion of transformational paradigms is mainly an analytical affair, but it is equally concerned with composition and poiesis, as the attentive reader will have noticed. Paradigmatic methods in counterpoint and serial composition are transformational techniques which *create* ways of

[6] The set is named after French mathematician Gaston Julia [513].

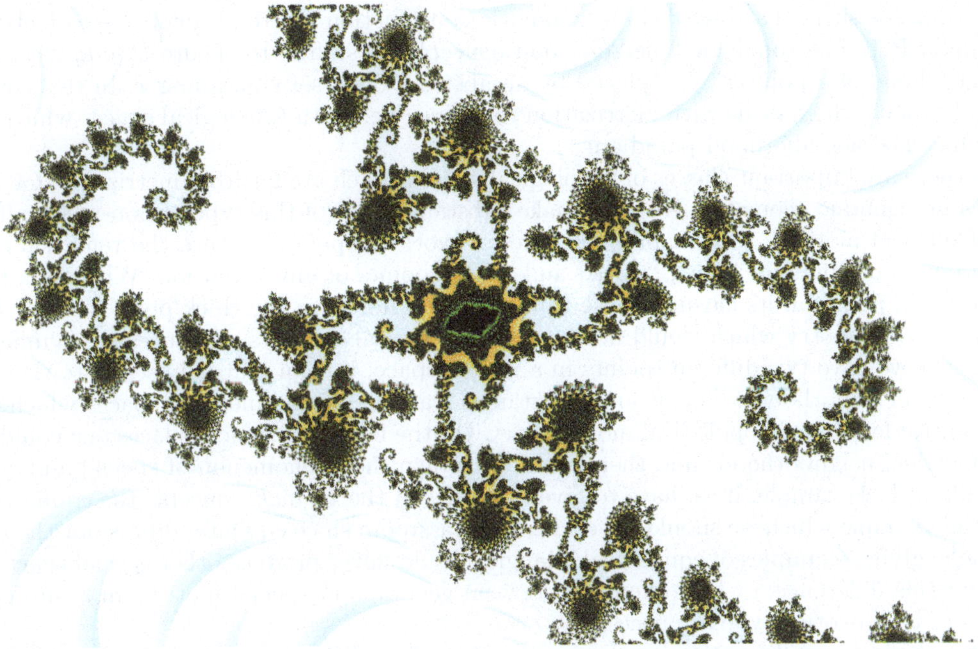

Fig. 10.4. The Julia set, of which a detail is shown here, is a trace of a transformation $f_c = z^2 + c$ on the complex plane which completely disappears in favor of the visible result. Powers of f_c produce chains of associated points which tend to infinity or stay in a finite region, a paradigm of highly complex overall appearance.

understanding the musical ideas as paradigmatic germs of an organic universe of sounds; Anton Webern's compositional techniques following Goethe's organic principles realize these techniques, see Section 17.2 for the corresponding discussion. In view of this, a composer has to pay especial attention to the mentioned vanishing degree of transformations because the communicative evidence of a composition depends strongly on the emergence of the operational metalevel *within* the unfolding composition material, see Section 48.1 for this subject.

10.3 Similarity

Summary. Similarity as a quantification of comparison. Metrical and topological similarity. Quantities and qualities: What is a parametrization?

– Σ –

The scientific goal of comparison is a difficult when the specific difference of the comparanda has to be objectivized. Whereas exhibiting logical *differentiae specificae* is an easy task, comparison of objects which pertain to floating paradigms is much more delicate. In psychometrics this problem is classically tackled by numerical quantification of similarity, for example by Charles Osgood's *semantic differentials* [810]. The same spirit controls the theory of fuzzy sets [72], or fuzzy relations in musical contour theory [870], where the classical logical differentiae specificae are refined by a distance measure of fuzzy validation.

This approach is completely natural since experimental verification is primarily concerned with numerical measurement, and metrical distance functions are the right thing to invoke when such quantification is required. However, music is not a branch of psychometrics, metrical similarity is a very special realization of the topological paradigm. In fact, the original concern of paradigmatic neigborhood was to understand what it means to stay in the vicinity of some given point, and *no allusion to metrical measurement* of vicinity was implicit. In mathematical topology the generic concept of "vicinity" or "neigborhood" has been analyzed, and this "analysis situs" has revealed an axiomatic construction where no reference to metrical data is left, see Appendix H.1. The topological neigborhood concept has shifted to a pure *quality* where the naive "epsilon neigborhood of a point x" is replaced by an abstract open set containing x. In this context, there is a theoretical branch which deals with metrization theorems, i.e., with topological spaces which can be deduced from the metrical neighborhood paradigm.

But there are important classes of topological spaces which are far from metrizable, for example Zariski topologies in algebraic geometry. We shall make substantial use of this type of non-metrizable spaces in the theory of musical motives, see Chapter 22. These "exotic" topologies share the remarkable property that there are points of different "size", "thick" and "thin" points of any extension. Whereas a thick point may have many thin points in its closure, these thin points do not have the thick point in their closures. This is an important asymmetry which would not happen in metrical contexts: distances are symmetric quantities, and whenever we have two different points in a metric space, we can mutually separate them by sufficiently small disjoint neighborhoods. This is an important advantage of non-metrical spaces which can be used in, and is a central feature of, topological motif theory. On the basis of metrical spaces one could never arrive at non-symmetrical neighborhoods and therefore never grasp the phenomenon of special and generic examples of a paradigm! For example, if we have to give examples of the *triangle* concept, the equilateral triangle is a very special example which we should never draw in classroom since equilaterality is not the typical property of a triangle; giving examples of equilateral triangles would never produce other special types of triangles. To understand this difference, or the difference between generic and special motives in a musical composition, topologies of the "exotic" type are necessary.

So topological paradigmatics is a huge extension of traditional metrical similarity. It nevertheless remains within the field of objectivized description of paradigmatic vicinity, it is a scientific refinement of the common language concept of comparison. The characteristic difference lies in the renouncement from plain quantification. Topology is the qualitative theory of neighborhoods, this is a piece of mathematics without a priori quantification which helps building powerful concepts beyond measurement. We should however notice that the absence of immanent quantification has its prize: Classification of topological spaces is a hard program which is described in the so-called *algebraic topology*, see Appendix H for more details.

The common denominator of topological paradigmatics is to distribute comparable objects in a space which allows the objects

- attributes of spatial location which share the property of
- being objective, explicit, adequate, and highly differentiated.

So the idea is to create spatial parameters for paradigmatic fields, parameters which include but go far beyond numerical quantizations. Parametrization is an auxiliary measurement system but not necessarily by numbers, rather by precise concepts of spatial, i.e. topological character. This is the decisive difference between old-style and new-style precision.

10.4 Fuzzy Concepts in the Humanities

Summary. In the humanities, fuzzy concepts are a frequent phenomenon. Paradigmatic reconstruction of such concepts by use of precise similarity and/or transformation concepts can help in validating amphibological discourses instead of rejecting them.

– Σ –

There are two objections to precise terminology in the humanities:

- historically and culturally generated polysemy,
- uncontrolled paradigmatics.

Polysemy has been a major objection to scientific discourse beyond what Hermann Hesse called "feuilleton science" [463]. It is contended that polysemy is in contradiction to precision. But there is no logical argument against a precise description of polysemy. A well-known example of an exact theory of polysemy is the mathematical theory of algebraic equations. In general, a solution X of an equation $f(x) = 0$ is a sign having multiple significates. The description of the variety of solutions, i.e. the collection of all valid semioses of the signifier "solution X of $f(x) = 0$", is a successful branch of mathematics, viz algebraic geometry. The reason for this success is that the signification mechanism specified in the equation $f(x) = 0$ is explicit and precise. *Hence it is not the problem to eliminate polysemy, but to analyze it in a powerful language.* The theory of global compositions developed in part IV is an example of adequate scientific formalization of polysemy for the analysis of musical texts. Quite radically, the neutral identification of a musical work (see Section 2.2.2) is precisely the organized ensemble of polysemies on the analytical level.

The more serious obstruction to scientific precision in the humanities is a fuzzy concept in a non-technical sense, i.e. **uncontrolled paradigmatics**. These are typically produced as follows: Initially, a specific research yields a field of conceptual variants which seems to point to a coherent phenomenon. But any attempt to build a well-defined paradigma from the vague concept field fails. Usually this happens for one or more of three reasons:

- blurred consciousness of the layer of reality or—more generally—the topographic location of the bunch of phenomena, i.e. a smeared topographical analysis,
- taking blurred reflections for blurred facts;
- defectuous concept tools for building well-defined paradigmata.

Whereas the first two reasons are simple defects in analytical mastery, detection of the third is more delicate. In the argumentation the defectuous concept tools are not recognized. Rather does the discourse either seek a deep mystery in the phenomenon itself, or the addressee is persuaded that an uncontrollable infinity of cases and variants simply rules out any conceptual effort. In short: the complexity of the phenomenon transcends human intellectual power.

We shall discuss this latter attitude at length in Section 22.8 on Rudolph Reti's concept sketches dealing with musical motives, themes and melodies. Here, we want to give a short illustration of how failure of paradigmatic tools can create defectous results in the humanities: In the course of an exhibition project on the Darmstadt Mathildenhöhe [671] and with respect to the "School of Athens", Raphael's famous fresco in the Vatican, the question concerning the significance of the star being constructed by the Bramante figure in the fresco's lower right was raised. To date it had been assumed that the star represented a star of David [317], [1077]. However, this semantic was never examined with respect to the precise star shape. A star of David must comprise two regular triangles. But the central perspective in which this masterpiece was demonstrably painted permits more precise conclusions on the form of the geometric configurations in the painting.

In collaboration with the Institute for Architecture of the Technische Hochschule Darmstadt and the Fraunhofer-Gesellschaft computer graphics working group, a computer graphics reconstruction was made, including the 20 most important among the total of 58 human figures, as well as important geometric elements (steps, blocks, spheres, Bramante star etc.). Visualization of the object configurations from all perspectives with the aid of computer graphics software permitted new insights due to visualization of previously concealed features. It came to light that the star does not comprise equilateral but right-angled triangles which configure in relation to the lateral faces of Platonic bodies and thus also create a link to Plato, a central figure of the composition. At the same time it turned out that the points of these triangles are set with amazing precision in relation to the base-points of the principal human figures in the work [670]. Thus a new immanent semantic has become apparent in the fresco by means of precise geometric analysis, and in sharp contradiction to previous interpretations.

Result 1 *The moral of this anecdote is that the paradigmatic tools for analyzing the deformed version of the star in the spatial paradigm of central perspective were not available in the traditional interpretations, and that only rigorous elaboration of these tools by use of adequate computer software and geometric background theory led to a reliable understanding of the star paradigm in the given context.*

11

Orbits

2 230 741 522 540 743 033 415 296 821 609 381 912
The number of isomorphism classes (orbits) of
72-element motives in $\mathbb{Z}_{12}^2$.
Harald Fripertinger [342]

100 000 000 000
The average number of stars in a galaxis.
Hubert Reeves [883]

Summary. This chapter deals with groups of symmetries, their action and orbits as musicological and mathematical concepts. Elementary local compositions—chords, self-addressed chords, and motives are classified under group actions. Enumeration theory of orbits of local compositions in finite $\mathbb{Z}$-modules—including traditional pitch class sets and motives—is presented and discussed for its implications towards a "Big Science" in music. Follows a discussion of group-theoretical methods in composition and theory, including a review of the American tradition and recent developments.

– Σ –

11.1 Gestalt and Symmetry Groups

Summary. Group orbits and the gestalt concept. Orbits are tools of conceptual abstraction.

– Σ –

Transformational paradigmatics is addressed in Christian von Ehrenfels' concept of a *gestalt* as proposed in [293]. The concept is characterized by *transposability* and *super-summativi-ty*. Ehrenfels's famous example of a melodic gestalt explains these attributes: If a melody is transposed in pitch, we still recognize the melody, and—according to Ehrenfels—this also demonstrates that the melody "is more than the sum of its parts", the single notes. We come back to this second point in Chapter 12. Here we concentrate on the first point. The statement is that although the melody, as an individual local composition, say, changes after transposition, something remains invariant, and this is an attribute of the melody's gestalt. In Ehrenfels' example, this invariant is a transformational paradigm under translation symmetries in the pitch domain (see example 5 in Section 8.1.1) and could as well be extended to temporal translation (see example 6 in Section 8.1.1). The very concept of gestalt refers to the entire set of transpositions of a given melody object. The melody's gestalt is something which is identically the same in all the melody's transpositions. But then, it is obvious to identify the set of all transpositions $e^t(M)$ of the melody M with an attribute of this gestalt of M. This means that the orbit set $\{e^t(M)|\ t \in \text{ambient space of } M\}$ is an attribute of the gestalt of M.

This motivates the conclusion that building classes of local compositions under actions of determined groups of symmetries is a conceptual abstraction tool for constructing the gestalt concept. Again, as discussed

G. Mazzola, *The Topos of Music I: Theory*, Computational Music Science,
https://doi.org/10.1007/978-3-319-64364-9_11

in Chapter 10, we recognize paradigmatic classification as a classical subject of the humanities, here: cognitive science in music and music psychology. The concrete selection of the acting symmetry group is not the concern of our calculations, these must be justified by other criteria. But the a priori calculations of orbits is a technical prerequisite without which no effective cognitive science can emerge. Ehrenfels' approach can only be a prototypical one, since the question of which "transpositions", i.e., symmetries, acting on a melody are really responsible for the cognitive construction of a gestalt cannot be answered ante rem, before cognitive experiments are performed to answer empirically to the empirical question in cognitive science.

Principle 5 *So mathematical music theory has to furnish the orbits for all possible—or at least all reasonable—applications, be it for cognitive science, for compositional purposes, or for musicological analysis.*

11.2 The Framework for Local Classification

Summary. We delimit the categories of local compositions where one presently has significant classification results.

– Σ –

In this chapter we make the general hypothesis that all coefficient rings R of modules are *commutative*. We also restrict to commutative local compositions, i.e., those having R-modules as simple spaces and identify them whenever reasonable with their objective trace (see also Section 7.4 and Section 8.3.5 for these local compositions and their categories). So our commutative local compositions are addressed at an R-module; call the category of these commutative local compositions, together with R-affine module morphisms $ComLoc_R$. In terms of the general denotator theory of form semiotics exposed in G.5.3, we are working over the topos $\mathbf{Mod}_R^@$ instead of the larger topos $\mathbf{Mod}^@$. However, address changes including different commutative rings may occur. But such exceptional situations will be expressely mentioned.

We should also notice a canonical technique of 'address killing' which occurs in the following situation: We are given a functor $F \in {}_0\mathbf{Mod}_R^@$ and an address $B \in {}_0\mathbf{Mod}_R$. Using the affine tensor product $\boxtimes$ as defined in Appendix E, we define a new contravariant, set-valued functor on R-modules $B\tilde{@}F \in {}_0\mathbf{Mod}_R^@$ as follows. Let $X, Y \in {}_0\mathbf{Mod}_R$ and $f : X \to Y \in {}_1\mathbf{Mod}_R$. Then

$$X@B\tilde{@}F = X \boxtimes B@F, \tag{11.1}$$

$$f@B\tilde{@}F = f \boxtimes Id_B@F. \tag{11.2}$$

By the equation $0_R \boxtimes B \xrightarrow{\sim} B$ (Appendix E, sorite 15) this functor has the property that $0_R@B\tilde{@}F \xrightarrow{\sim} B@F$, i.e., it gives us back the B-addressed points of F as zero-addressed points. We therefore call it the *B-address killer of F*. For $F = @_R M$, $M \in {}_0\mathbf{Mod}_R$, we get the well-known identification $X@B\tilde{@}M \xrightarrow{\sim} (X \boxtimes B)@_R M \xrightarrow{\sim} X@_R(B@_R M)$, see Appendix E, Lemma 90 and Proposition 95.

Finally, we shall exclusively deal with finite, non-empty commutative local compositions here.

11.3 Orbits of Elementary Structures

Summary. This section deals with different common types of local compositions and their classifications.

– Σ –

In complex categories such as *Loc*, *ComLoc*, and other categories as displayed in diagram 8.31, classification is a multi-threaded task. At present we are far from a complete classification for three reasons: First, classification is far from settled in the module categories which underlie local compositions. Second, as we shall see, even if module category had solved the classification problem the algebro-geometric classification problem would be open, and the latter is essential for local compositions. Third, even if we had a theoretical solution of a classification problem, it would be an unsolved problem to find algorithmic solutions for concrete cases. Already the strict enumeration of isomorphism classes is, as we shall see in Section 11.4, an interesting problem of computer algebra.

So the following sections are only the state of the art and not complete classification. However, there is a considerable number of cases where complete classification is feasible, even on the algorithmic level. This means that musicology disposes of important complete lists of representatives of isomorphism classes, in particular for zero-addressed and self-addressed chords and rhythms, and of small cardinality motives. We have added such lists in the appendices and the working musicologist may consult such lists if she or he does not want to plug in these technicalities. However, we repeat it, classification is much more than the naked list of representatives. If ever any random local composition is given, it has to be attributed to one of the list's representatives, and this requires understanding the candidate in depth! So what we suggest is at least a passive lecture on the following classification discourse in order to trace the conceptual path and to memorize it in case deeper investigations should occur.

11.3.1 Classification Techniques

The first step in any classification program is to turn the given objects into points of a space where isomorphism classes are controllable by canonical parameters. To begin with, let us recapitulate the isomorphisms we will deal with. We fix an address $B \in {}_0Mod_R$ and an ambient module $M \in {}_0Mod_R$; we look at local compositions $K \subset B@_RM$ of positive cardinality $card(K) = n+1$. We further concentrate on generating local compositions, however admitting all endomorphisms of B as address changes, i.e., we extend from the category $ComLoc^{gen}_B$ as defined in Section 8.3.5 to $ComLoc^{gen}_{End(B)}$ as discussed in diagram 8.31, but with fixed coefficient ring R. Let $K \subset B@_RM, L \subset B@_RN$ be two objects of $ComLoc^{gen}_{End(B)}$. An isomorphism $f/\alpha : K \xrightarrow{\sim} L$ in this category is given by a diagram

$$\begin{array}{ccc} K & \longrightarrow & B@_RM \\ {\scriptstyle \alpha.f}\downarrow & & \downarrow{\scriptstyle B@_Rh} \\ L.\alpha & \longrightarrow & B@_RN \end{array} \tag{11.3}$$

with α an automorphism of B and an underlying morphism $h \in M@_RN$ of ambient spaces[1]. This means that we have an associated linear isomorphism $R.(\alpha.f) : R.K \xrightarrow{\sim} R.L.\alpha$. But since both K, L are generating, we are given an automorphism $B@_Rh$ which sends the constant elements $0_R@_RM = M$ bijectively onto the constant elements $0_R@_RN = N$, and this is h, i.e., h is an affine isomorphism. This implies that we may select one representative of every (linear) isomorphism class of $\mathbf{Mod}_R$ and then concentrate on the local compositions within one fixed such module M as ambient space. So we are dealing with the simultaneous action of the groups $\overrightarrow{GL}(B)$ from the right and $\overrightarrow{GL}(M)$ from the left on the set $ComLoc^{gen,M}_{End(B)}$ of objects of $ComLoc^{gen}_{End(B)}$ with ambient space M. If $h \in \overrightarrow{GL}(M)$, $\alpha \in \overrightarrow{GL}(B)$, and if K is such an object, the operation is

$$K \mapsto h.K.\alpha. \tag{11.4}$$

We next transform these local compositions into singleton denotators, i.e., points in ambient spaces, as follows. We view K as a zero-addressed local composition in ambient space $B\tilde{@}M$. But then, this is also defined by a singleton (denotator) $K : R^n \rightsquigarrow B\tilde{@}M(k)$ in the following way: Write the elements of K as a sequence $k = (k_0, \ldots k_n)$ and observe that $R^n@B\tilde{@}M = R^n \boxtimes B@_RM \xrightarrow{\sim} R^n@_R(B@_RM) \xrightarrow{\sim} (B@_RM)^{n+1} = (0_R@B\tilde{@}M)^{n+1}$. So k is a zero-addressed point of $(B\tilde{@}M)^{n+1}$ which is identified with an injective R^n-addressed denotator, i.e., an R^n-addressed generating[2] point in $B\tilde{@}M$ with pairwise different coordinates; we denote this point set by $R^n\dot{@}B\tilde{@}M$. Identify the symmetric group $\mathfrak{S}_{n+1}$ with the group of affine automorphisms of R^n which permute the canonical affine basis $(e_0 = 0, e_1 = (1,0,\ldots 0), \ldots e_n =$

[1] And we have a canonical linear isomorphism $R.L \xrightarrow{\sim} R.(L.\alpha)$ since the action $x \mapsto x.\alpha$ is linear and α is auto.

[2] i.e., k defines a generating local composition.

$(0, \ldots 0, 1))$ of R^n. Then the address-change orbit $k.\mathfrak{S}_{n+1}$ of the denotator k corresponds 1-1 to K since it abolishes the arbitrary choice of indices of K-elements. So the local $(n+1)$-element compositions K correspond to the elements of the orbit space $R^n \dot{@} B \tilde{@} M / \mathfrak{S}_{n+1}$.

Putting these actions together, we have a left action of $\overrightarrow{GL}(M)$, and two commuting right actions, one by $\mathfrak{S}_{n+1}$ and one by $\overrightarrow{GL}(B)$. This gives the following theorem:

Theorem 4 *With the above notation, let* $ComLoClass^{gen,M}_{n+1,End(B)}$ *be the set of isomorphism classes of* $(n+1)$*-element generating commutative local compositions in* $ComLoc^{gen,M}_{End(B)}$*, and consider the set*

$$DenOrb(R, n, B, M) = \overrightarrow{GL}(M) \backslash R^n \dot{@} B \tilde{@} M / \overrightarrow{GL}(B) \times \mathfrak{S}_{n+1}$$

of denotator orbits. Then we have a canonical bijection

$$ComLoClass^{gen,M}_{n+1,End(B)} \xrightarrow{\sim} DenOrb(R, n, B, M). \tag{11.5}$$

11.3.2 The Local Classification Theorem

Summary. Classifying orbits of group actions on such standard morphisms. The geometric framework.

– Σ –

The next step is concerned with geometric parameters for the denotator orbit space $DenOrb(R, n, B, M)$.

Definition 29 *With the preceding notation, let*

$$\begin{aligned} X(R, n, B, M) = \{(V, W)|\ & V \subset W \text{ submodules of } R^n \text{ with } e_i - e_j \notin V, \\ & \text{all } 0 \leqslant i < j \leqslant n, \text{ and there are } R\text{-linear isomorphisms} \\ & \text{such that diagram 11.6 commutes.}\} \end{aligned}$$

$$\begin{array}{ccc} W/V & \xrightarrow{iso.} & 0 @_R M \\ {\scriptstyle can.}\downarrow & & \downarrow{\scriptstyle can.} \\ R^n/V & \xrightarrow{iso.} & B @_R M \end{array} \tag{11.6}$$

and define a right action of the symmetric group $\mathfrak{S}_{n+1}$*,*

$$X(R, n, B, M) \times \mathfrak{S}_{n+1} \to X(R, n, B, M) : ((V, W), \sigma) \mapsto (\sigma_0^{-1} V, \sigma_0^{-1} W) \tag{11.7}$$

where σ_0 *is the linear part of* σ.

We have a projection of the denotator orbit space into the orbit space $X(R, n, B, M)/\mathfrak{S}_{n+1}$ in the following sense. For $k \in R^n \dot{@} B \tilde{@} M$, we have an element

$$drap(k) = (Ker(k_0), k_0^{-1}(0_R @_R M)) \tag{11.8}$$

of $X(R, n, B, M)$.

Before proving the next proposition we need a technical lemma,

Lemma 8 *With the above notation, the linear part* $(\alpha @ g)_0$ *of the affine map* $(\alpha @ g) : B @_R M \to B @_R M$ *is* $\alpha @ g_0$.

Proof. If $e^\gamma \cdot g_0$ is the decomposition of g, and if $f \in B @_R M$, then $\alpha @ g(f) = e^\gamma \cdot g_0 \cdot f \cdot \alpha$ whereas $\alpha @ g_0(f) = g_0 \cdot f \cdot \alpha$ is linear in f and $\alpha @ g(0) = e^\gamma$. In other words, $\alpha @ g(f) = (e^{e^\gamma} \cdot (\alpha @ g_0))(f)$, whence the claim.

Proposition 6 *With the above notation, there is a (necessarily unique) surjective map dr such that the diagram*

$$\begin{array}{ccc} R^n \dot{@} B \tilde{@} M & \xrightarrow{p} & DenOrb(R,n,B,M) \\ {\scriptstyle drap}\downarrow & & \downarrow{\scriptstyle dr} \\ X(R,n,B,M) & \xrightarrow{q} & X(R,n,B,M)/\mathfrak{S}_{n+1} \end{array} \tag{11.9}$$

with canonical horizontal surjections commutes. The map dr is bijective if B is the zero-address.

Proof. Suppose we are given two denotators $k, l \in R^n \dot{@} B \tilde{@} M$. Let us first see that dr is well-defined, i.e., that if they have same orbit in $DenOrb(R,n,B,M)$, then their images $drap(k) \cdot \mathfrak{S}_{n+1}$ and $drap(l) \cdot \mathfrak{S}_{n+1}$ coincide. If $p(k) = p(l)$, there is a triple $\sigma \in \mathfrak{S}_{n+1}, \alpha \in \overrightarrow{GL}(B), g \in \overrightarrow{GL}(M)$ such that the diagram

$$\begin{array}{ccc} R^n & \xrightarrow{k} & B@_R M \\ {\scriptstyle \sigma}\downarrow & & \downarrow{\scriptstyle \alpha@g} \\ R^n & \xrightarrow{l} & B@_R M \end{array} \tag{11.10}$$

commutes. Therefore, the corresponding diagram

$$\begin{array}{ccc} R^n & \xrightarrow{k_0} & B@_R M \\ {\scriptstyle \sigma_0}\downarrow & & \downarrow{\scriptstyle \alpha@g_0} \\ R^n & \xrightarrow{l_0} & B@_R M \end{array} \tag{11.11}$$

of linear parts according to Lemma 8 also commutes. Since the right arrow of the latter diagram is iso, the kernels of k_0, l_0 verify $\sigma_0(Ker(k_0)) = Ker(l_0)$. Further, the subspace $0_R@_R M$ of constant maps in $B@_R M$ is invariant under $\alpha@g_0$, and therefore, $\sigma_0(k_0^{-1}(0_R@_R M) = l_0^{-1}(0_R@_R M)$. This means that $q(drap(k)) = q(drap(l))$. Surjectivity of dr follows since $drap$ and q are evidently surjective.

Last, we want to see that dr is injective for $B = 0$. In this case, $B@_R M \xrightarrow{\sim} M$, $drap(k)$ specializes to $(Ker(k_0), R^n)$, and the action of $\overrightarrow{GL}(B) = 1$ becomes trivial. Therefore, if for $\sigma \in \mathfrak{S}_{n+1}$, we have $\sigma_0^{-1} \cdot drap(k) = drap(l)$, σ_0 induces a linear automorphism of M by the quotient automorphism from σ_0. So k and l have the same orbit in $DenOrb(R,n,B,M)$. QED.

For the rest of this section, we suppose that B is the zero-address[3], and we want to present algebro-geometric classification results in this case.

For the zero-address we have several simplifications of the above formalism. To begin with, consider the short exact sequence

$$0 \longrightarrow M \xrightarrow{\Delta} M^{n+1} \xrightarrow{d} M^n \longrightarrow 0 \tag{11.12}$$

of R-modules and linear maps with the diagonal embedding $\Delta(m) = (m, m, \ldots m)$ and the difference formula $d(m_0, m_1, \ldots m_n) = (m_1 - m_0, \ldots m_n - m_0)$. We have a right linear action of $\mathfrak{S}_{n+1}$ on M^{n+1} defined by $(m_0, m_1, \ldots m_n) \cdot \sigma = (m_{\sigma 0}, m_{\sigma 1}, \ldots m_{\sigma n})$ which leaves $\Delta(M)$ invariant and therefore induces a linear action on the quotient M^n. We also have the left diagonal actions of $\overrightarrow{GL}(M)$ on M^{n+1} and of $GL(M)$ on M^n. Further:

Lemma 9 *The projection d is equivariant with respect to the above actions of the symmetric group $\mathfrak{S}_{n+1}$ and the canonical group homomorphism $\overrightarrow{GL}(M) \to GL(M)$.*

[3] It is an open problem to find a canonical generalization of the *drap* map for general address B such that dr becomes a bijection.

Proof. The first claim is true by construction. For the second, let $m_. \in M^{n+1}$ and $e^m \cdot g \in \overrightarrow{GL}(M)$. Then $d(e^u \cdot g(m_.)) = g(d(m_.))$ since u cancels out by the difference homomorphism d.

Let $M^{\overline{n}}$ be the subset of M^n consisting of the n-tuples with pairwise different *and non-vanishing* coordinates and such that these coordinates generate the module M. Set $M^{\underline{n+1}} = d^{-1}M^{\overline{n}}$. Clearly, both sets are invariant under the above actions, and d projects $M^{\underline{n+1}}$ onto $M^{\overline{n}}$. Moreover:

Lemma 10 *The canonical map of orbits*

$$\overrightarrow{GL}(M)\backslash M^{\underline{n+1}}/\mathfrak{S}_{n+1} \to GL(M)\backslash M^{\overline{n}}/\mathfrak{S}_{n+1} \tag{11.13}$$

is a bijection. The left orbit set (and therefore also the right one) identifies to the set of isomorphism classes of $(n+1)$-element generating local compositions in M.

Proof. The last claim is clear by construction. The map of orbits is also surjective since $d : M^{\underline{n+1}} \to M^{\overline{n}}$ is surjective. Let $m_., n_. \in M^{\underline{n+1}}$ such that there are $g \in GL(M), \sigma \in \mathfrak{S}_{n+1}$ with $g \cdot d(m_.) \cdot \sigma = d(n_.)$. Then for all indices $i = 1, \dots n$, we have $n_i - n_0 = g(m_{\sigma i} - m_{\sigma 0}) = g(m_{\sigma i}) - g(m_{\sigma 0})$ and hence the affine equation

$$n_i = (e^{n_0 - g(m_{\sigma 0})} \cdot g)(m_{\sigma i}),$$

and this shows that the orbits of $m_.$ and $n_.$ in the left orbit set coincide. QED.

Therefore we may observe the linear action of $GL(M)$ and of $\mathfrak{S}_{n+1}$ on $M^{\overline{n}}$. Clearly, by the canonical isomorphism $M^n \xrightarrow{\sim} Lin_R(R^n, M)$, this action identifies to the left-right actions of these groups on $Lin_R(R^n, M)$, i.e., the canonical left action of $GL(M)$ and the right action of $\mathfrak{S}_{n+1}$ on R^n via a group homomorphism $D : \mathfrak{S}_{n+1} \to GL_n(R)$ which on transpositions $(1, i)$ is given by the matrix

$$D(1,i) = \begin{pmatrix} 1 & & & & 0 & \\ 0 & 1 & & & & \\ \dots & \dots & 1 & & & \\ -1 & -1 & \dots & -1 & \dots & -1 \\ & & & & & \\ & & 0 & & & 1 \end{pmatrix} \tag{11.14}$$

with -1 entries overriding the $n \times n$ identity matrix on row $i-1$.

Through this identification $M^{\overline{n}}$ identifies to a stable subset $\overline{Lin}_R(R^n, M)$ of $Lin_R(R^n, M)$, and this yields the following zero-address version of Theorem 4. This reduction from affine to linear structures is only possible for the zero-address.

Theorem 5 *There is a canonical bijection*

$$ComLoClass^{gen,M}_{n+1,O_R} \xrightarrow{\sim} GL(M)\backslash\overline{Lin}_R(R^n, M)/\mathfrak{S}_{n+1}.$$

We now proceed to the linear analogy of the flag space $X(R, n, B, M)$ for zero-address. We set

$$X_n(R) = \{V|\ V \text{ a submodule of } R^n \text{ with } e_i, e_i - e_j \notin V, \text{ all } 1 \leqslant i < j \leqslant n\}, \tag{11.15}$$

$$X_n^M(R) = \{V|\ V \in X_n(R) \text{ and there is an } R\text{-linear isomorphism } R^n/V \xrightarrow{\sim} M\}. \tag{11.16}$$

Clearly, the symmetric group $\mathfrak{S}_{n+1}$ acts from the right on both, $X_n(R)$ and $X_n^M(R)$, via $V \cdot \sigma = D(\sigma^{-1}) \cdot (V)$, an action which we recognize as being the linear analog to the above action on $X(R, n, B, M)$. Let $\mathcal{M}$ be a system of representatives of isomorphism classes in $\mathbf{Mod}_R$. Then we have a disjoint union

$$X_n(R) = \coprod_{M \in \mathcal{M}} X_n^M(R) \tag{11.17}$$

(most of these co-factors are empty, of course) and its corresponding orbit space union

$$X_n(R)/\mathfrak{S}_{n+1} = \coprod_{M \in \mathcal{M}} X_n^M(R)/\mathfrak{S}_{n+1}, \tag{11.18}$$

Lemma 11 *The orbit set $X_n^M(R)/\mathfrak{S}_{n+1}$ is in canonical bijection with the class set*

$$ComLoClass^{gen,M}_{n+1,O_R}.$$

Proof. We know from Theorem 5 that this class set is in bijection with the orbit set

$$GL(M)\backslash\overline{Lin}_R(R^n, M)/\mathfrak{S}_{n+1}.$$

In the diagram

$$\begin{array}{ccc} \overline{Lin}_R(R^n, M) & \xrightarrow{Ker} & X_n^M(R) \\ q\downarrow & & \downarrow p \\ GL(M)\backslash\overline{Lin}_R(R^n, M)/\mathfrak{S}_{n+1} & \xrightarrow{f} & X_n^M(R)/\mathfrak{S}_{n+1} \end{array} \tag{11.19}$$

with canonical vertical arrows and the upper horizontal arrow $R^n \xrightarrow{g} M \mapsto Ker(g)$, there is a unique bijection f which makes the diagram commute. In fact, the kernel of g does not change under an automorphism of M, and we may factorize through the $GL(M)$-orbits. Since the horizontal map Ker is $\mathfrak{S}_{n+1}$-equivariant, the factorization through f is defined, and f is surjective since both, Ker and p, are. Finally, if two R^n-addressed denotators $k, l : R^n \to M$ in $\overline{Lin}_R(R^n, M)$ have kernels which differ by a permutation, then their quotients which identify to M are related by an automorphism of M and therefore, $q(k) = q(l)$, so f is mono. QED.

Let $ComLoClass^{gen}_{n+1,O_R}$ be the set of all isomorphism classes (any ambient R-module M allowed). Then the above lemma gives

Theorem 6 (Classification of n+1-element, generating, zero-addressed local compositions over ring R) *There is a canonical bijection*

$$X_n(R)/\mathfrak{S}_{n+1} \xrightarrow{\sim} ComLoClass^{gen}_{n+1,O_R}.$$

We therefore may concentrate on the left space and look to situations which may lead to geometric classification spaces, i.e., where the classes appear as points on appropriate geometric spaces.

Before delving into techniques of algebraic geometry we should try to understand the geometric aspect of Theorem 6 and its musical interpretation. Let us concentrate on the intuitive situation of local compositions over the rationals. Observe that for $\mathbb{Q}$, every such composition is isomorphic to a generating one[4] and we may forget about this specification and even concentrate on ambient spaces of coordinator $\mathbb{Q}^r$.

To begin with, replacing a generating local $(n+1)$-element composition $(K, \mathbb{Q}^r)$ by a $\mathbb{Q}^n$-addressed denotator $k : \mathbb{Q}^n \to \mathbb{Q}^r$ means that we really have an instance of a more general approach which is based on standard local compositions (see Figure 11.1):

Definition 30 *Given the ring R and a natural number n, we denote by Δ_n the zero-addressed commutative local composition of cardinality $n+1$ in ambient space $R^{\oplus n}$ consisting of the canonical affine base, i.e.,*

$$\Delta_n = \{e_i|\ 0 \leqslant i \leqslant N\} \tag{11.20}$$

with the standard notation $e_0 = 0$ and $e_i = (\delta_{ij})_{1 \leqslant j \leqslant N}$ for $i > 0$ and the identification of $R^{\oplus n}$ with $0_R@R^{\oplus n}$. The local composition Δ_n is called standard simplex composition of dimension n.

With this in mind, the denotator $k : \mathbb{Q}^n \to \mathbb{Q}^r$ corresponds 1-1 to the morphism $k_\Delta : \Delta_n \to \mathbb{Q}^r$ with generating image, and K identifies to the orbit $k_\Delta \cdot \mathfrak{S}_{n+1}$. So we have abstracted from K and concentrated

[4] Simply because every local composition (K, M) over a field F is isomorphic to $(K, F.K)$.

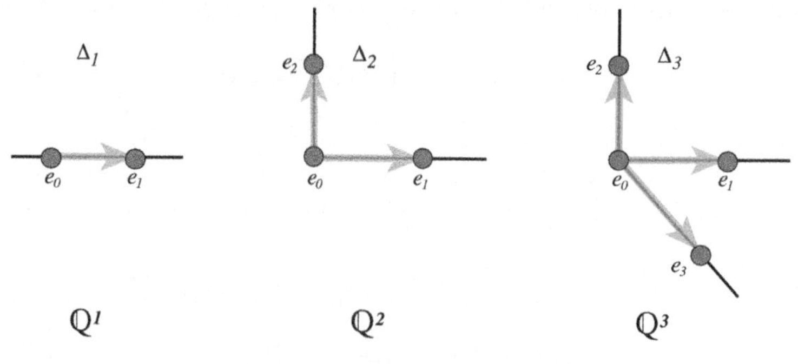

Fig. 11.1. Local standard simplex compositions over $\mathbb{Q}$ of dimension 1, 2, and 3.

on *one distinguished local composition*: the standard simplex composition Δ_n. The variability of K has been transferred to that of $k : \Delta_n \to \mathbb{Q}^r$, and finally, by the *drap* function, that of k has been reduced to the kernel $V = Ker(k_0)$. But what does this restatement mean? We have a finite orbit $V \cdot \mathfrak{S}_{n+1}$ and therefore[5], there is a *single* submodule $W \subset \mathbb{Q}^r$ which is a linear complement for all elements $V \cdot \sigma$ of this orbit: $\mathbb{Q}^n = V \cdot \sigma \oplus W$, *all* $\sigma \in \mathfrak{S}_{n+1}$. By construction, projection $p_{V,\sigma} : V \cdot \sigma \oplus W \to W$ gives us K as the bijective image of the standard composition, i.e., $K = p_{V,\sigma}(\Delta_n)$, see Figure 11.2.

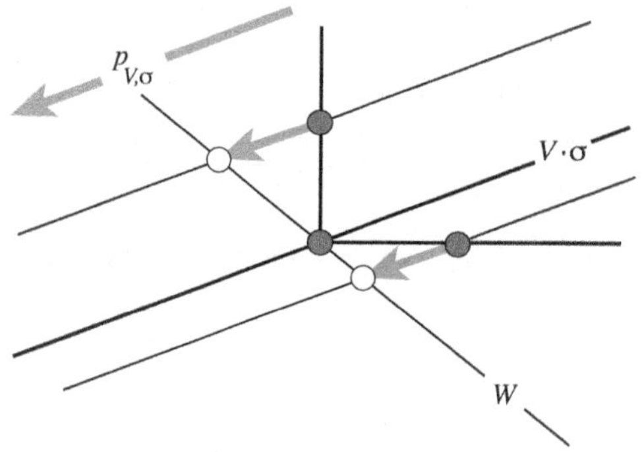

Fig. 11.2. Each kernel space $p_{V,\sigma}$ that is complementary to a given subspace W in $\mathbb{Q}^n$ gives rise to a specific projection of the standard composition Δ_n whose image defines a special instance of our ensemble of local compositions. This may be viewed as a variable coordinate system and leads to musical interpretation of the classification technique.

In other words: the ensemble of local $n+1$-element local compositions K with $dim(K) = r$ is transformed into *one single* "standard representative" Δ_n, together with an ensemble of projections $p_{V,\sigma}$ indexed by V and the symmetric group, and defined by the variable "coordinate system" $\mathbb{Q}^n = V \cdot \sigma \oplus W$ which works for all orbit members. This means that we deal with one "generic" local composition whose points are *in general position*[6] and then project this generic specimen onto specialized images defined by *selection of special coordinate functions*. We shall come back to this classification approach which also solves the problem for global classification in Section 15.1.1. Musically speaking, this process means to give an "abstract, generic composition" Δ_n and to find a set of coordinates to realize a special variant of this generic data. Ontologically, this passage can be viewed as a realization of a symbolic composition in special, musically

[5] See Appendix F.5. Observe also that the orbit $V \cdot \mathfrak{S}_{n+1}$ is contained in the $\mathfrak{S}_{n+1}$-invariant set of the $n-r$-dimensional subspaces which are complementary to all $W \cdot \sigma$. This is a stable open set in the corresponding Grassmannian variety (Appendix F.5).

[6] See Appendix H.1.1.

meaningful coordinates. We shall work out this now somewhat artificially blown up interpretation in 15.1.1, but want to mention it already now since this sheds light on the entire classification business.

For the time being, the more important point is that the passage to kernel spaces via *drap* turns the seemingly dispersed appearance of local compositions in their ensemble into a geometric situation because subspaces can easily be compared with each other by continuous movement. This means that we may now look for class-invariant geometric relations among local compositions, for example linear incidence among their points. In other words, the present efforts aim at *a geometrically sophisticated understanding of local compositions*, a scope which allows a very refined comparison of musical structures—such as motives or chords—in concrete compositions; applications of this view are given in Section 11.6.2.

To construct a geometric parametrization of classes, we restrict to local compositions (K, M) which are *locally free of rank* r, i.e., their module $R.K$ is locally free of rank r (see Appendix F.2, Theorem 158). Denote by $\mathcal{M}_r$ the representative subsystem of modules in $\mathcal{M}$ which are locally free of rank $r \in \mathbb{N}$ and set $X_n^r(R) = \coprod_{M \in \mathcal{M}_r} X_n^M(R)$. In other words,

$$X_n^r(R) \subset \{V \subset R^n |\ R^n/V = \text{ locally free of rank } r\},$$

but the latter is by definition[7] the evaluation $Grass_{n,r}(R)$ of the Grassmann functor

$$Grass_{n,r} : \mathbf{ComRings} \to \mathbf{Ens} \tag{11.21}$$

at ring R. More precisely, if $x \in R^n$, let$V_{n,x}^r(R) = \{V |\ V \in Grass_{n,r}(R), x \in V\}$. Further, if $f : R \to S$ is a ring homomorphism, the functorial map

$$Grass_{n,r}(f) : Grass_{n,r}(R) \to Grass_{n,r}(S) : V \mapsto S \otimes V \tag{11.22}$$

restricts to

$$V_{n,x}^r(f) : V_{n,x}^r(R) \to V_{n,1\otimes x}^r(S).$$

In fact, we know from Appendix E.4, Lemma 94 that $V \in Grass_{n,r}(R)$ is in $V_{n,x}^r(R)$ iff $\wedge^r x = 0$ for the direct summand $V \subset R^n$. But scalar extension commutes with formation of exterior powers ([145, ch.III, §7, no.5, Prop. 8]) and so $\wedge^r 1 \otimes x = 0$ with respect to the direct summand $S \otimes V$ of S^n. In particular, setting $e_{ij} = e_i - e_j$ for $i \neq j$ and $e_{ii} = e_i$ for the canonical basis vectors e_i in $\mathbb{Z}^n$, we have subfunctors $V_{n,e_{ij}}^r$ of $Grass_{n,r}$, defined by $V_{n,e_{ij}}^r(R) = V_{n,1\otimes e_{ij}}^r(R)$. The above action of the symmetric group $\mathfrak{S}_{n+1}$ permutes the subfunctors $V_{n,e_{ij}}^r$, and the complement of their union

$$V_n^r(R) = \bigcup_{1 \leqslant i \leqslant j \leqslant n} V_{n,e_{ij}}^r(R)$$

gives us the $\mathfrak{S}_{n+1}$-invariant set

$$X_n^r(R) = Grass_{n,r}(R) - V_n^r(R).$$

In contrast to the $V_{n,e_{ij}}^r(R)$, to $V_n^r(R)$, and to $Grass_{n,r}(R)$, $X_n^r(R)$ is not functorial in R.

Exercise 12 *Prove this last claim with* $R = \mathbb{Z}$.

We need to define a subfunctor $\tilde{X}_n^r(R) \subset Grass_{n,r}(R)$ which is disjoint from $V_n^r(R)$ and "essentially" gives us $X_n^r(R)$. The idea is that $V_n^r(R)$ is representable by a closed subscheme of the Grassmannian, and that we want to find a functorial expression $\tilde{X}_n^r(R) \subset X_n^r(R)$ of its open complement. We can show the following:

Lemma 12 *The subfunctors* $V_{n,e_{ij}}^r(R) \to Grass_{n,r}(R)$ *are representable by closed subschemes* $V_{n,e_{ij}} \to Grass_{n,r}$. *Each of the open complement schemes* $Grass_{n,r} - V_{n,e_{ij}}^r$ *is covered by the open subfunctor* $\tilde{X}_{n,e_{ij}}^r(R) \subset Grass_{n,r}(R)$ *consisting of the direct factors* $V \subset R^n$ *such that* $\wedge^r e_{ij} : \bigwedge^r V \to \bigwedge^{r+1} R^n$ *splits (has a right inverse).*

[7] See Section F.5 for the theory of Grassmann schemes.

Proof. Let $i_. = i_1, i_2, \ldots i_r$ be an increasing subsequence $1 \leqslant i_1 < i_2 < \ldots i_r \leqslant n$. We know from Appendix F.5 that the functor $Grass_{n,r}(R)$ is covered by the affine open subfunctors $Grass_{n,r,i_.}(R)$ of those direct factors $V \subset R^n$ such that $R^{i_.}$ projects isomorphically onto R^n/V. If $i'_.$ denotes the complementary index sequence of $i_.$, then $Grass_{n,r,i_.}(R)$ identifies to the set of graphs Γf of linear maps $f \in Lin_R(R^{i'_.}, R^{i_.})$, i.e., to the $n \times (n-r)$ matrices with columns $(e_i, f(e_i))^t$, i in $i'_.$. Restricted to such an open affine subfunctor, $V^r_{n,e_{kl}}(R)$ identifies to the set of $n \times (n-r+1)$-matrices $M_{f,kl}$ with $n-r$ columns $(e_i, f(e_i))^t$ for $j = 1, \ldots n-r$ and an $(n-r+1)^{th}$ column e_{kl} such that all $(r+1) \times (n-r+1)$-determinants of $M_{f,kl}$ vanish. This is a closed condition defined by the ideal which is generated by these determinants. We then look for the intersection of $Grass_{n,r,i_.}(R)$ with the subfunctor $\tilde{X}^r_{n,e_{ij}}(R)$. To see that this is an open subfunctor, observe that an R-valued point

$$M_{f,kl} \in Grass_{n,r,i_.}(R) \cap \tilde{X}^r_{n,e_{ij}}(R) \tag{11.23}$$

is characterized by the property that the $(n-r+1) \times (n-r+1)$-determinants $d_b, b = 1, \ldots \binom{n}{n-r+1}$ of $M_{f,kl}$ define the first column of an invertible $\binom{n}{n-r+1} \times \binom{n}{n-r+1}$-matrix over R. This set in affine $\binom{n}{n-r+1}$-space is open by Proposition 108 in Appendix F.5, and the determinants are polynomial functions of the coordinates of Γf, so the intersection $Grass_{n,r,i_.} \cap \tilde{X}^r_{n,e_{ij}}$ is an open subscheme of $Grass_{n,r,i_.}$. To see that the open complements $Grass_{n,r} - V^r_{n,e_{ij}}$ are covered by $\tilde{X}^r_{n,e_{ij}}$, observe that if R is a field, the intersections (11.23) coincide with the complements of the sets $V^r_{n,e_{kl}}(R)$ since non-vanishing of the determinants creates a basis vector by the exchange theorem for vector spaces. QED.

Lemma 13 *The open subscheme*

$$\tilde{X}^r_n = \bigcap_{1 \leqslant i \leqslant j \leqslant n} \tilde{X}^r_{n,e_{ij}} \tag{11.24}$$

is invariant under the given action of $\mathfrak{S}_{n+1}$, and every orbit in $\tilde{X}^r_n$ is contained in an affine neighborhood. We have $\tilde{X}^r_n(R) \subset X^r_n(R)$, and if R is semi-simple we have $\tilde{X}^r_n(R) = X^r_n(R)$.

Proof. Since the action of the symmetric group permutes the elements e_{ij}, it permutes the members of the intersection (11.24). Further, since the Grassmannian is projective, every finite orbit in $\tilde{X}^r_n$ is contained in an affine neighborhood by [396, ch.II, 4.5.4]. The inclusion $\tilde{X}^r_n(R) \subset X^r_n(R)$ follows from construction. The last statement is clear for a field, and hence for a semi-simple ring which, by Wedderburn's theorem is a finite direct product of fields (see Appendix E.2.3, Theorem 56). QED.

Theorem 7 *With the above notation, there is a quotient scheme*[8] *$\tilde{X}^r_n/\mathfrak{S}_{n+1}$, i.e., there is an exact sequence of schemes*

$$\tilde{X}^r_n \times \mathfrak{S}_{n+1} \underset{\mu}{\overset{pr_1}{\rightrightarrows}} \tilde{X}^r_n \to \tilde{X}^r_n/\mathfrak{S}_{n+1}$$

with group action μ and first projection pr_1. Its R-valued points are the orbits of $\tilde{X}^r_n(R)$ and of $X^r_n(R)$ in case R is a semi-simple ring.

Proof. The exactness property follows from Lemma 13 and [259, ch.III, §2, no.6.1]. QED.

The translation of the defining property for the scheme $\tilde{X}^r_n$ is that for such a generating local composition (K, M),

$$\left\{ \begin{array}{l} \text{all module homomorphisms} \\ .(k-l) : R \to M : r \mapsto r(k-l) \\ \text{for any two different elements } k, l \in K \\ \text{are split injections.} \end{array} \right. \tag{11.25}$$

Definition 31 *If a zero-addressed commutative local composition has the splitting property* (11.25) *we say that it is split.*

[8] In the sense of [259, ch.III, §2, no.1.3], see Theorem 67 in Appendix F.6.

Exercise 13 *Verify that the splitting property (11.25) is characteristic for membership in* $\tilde{X}^r_n$.

Corollary 5 *The set* $ComLoClass^{gen,lf,sp}_{n+1,0_R}$ *of isomorphism classes of zero-addressed, local, split, locally free, and generating* $(n+1)$*-element compositions with any* R*-module as ambient space is in canonical bijection to the set of* R*-valued points of the scheme*

$$Cl_n = \coprod_{1 \leqslant r \leqslant n} \tilde{X}^r_n / \mathfrak{S}_{n+1}.$$

If R *is a semi-simple ring, the set* $ComLoClass_{n+1,0_R}$ *of isomorphism classes of zero-addressed, local (not necessarily generating, locally free, or split)* $(n+1)$*-element compositions with any* R*-module as ambient space is in canonical bijection to the set of* R*-valued points of the scheme* Cl_n.

Let us have a closer look at the set $ComLoClass^{gen,lf,sp}_{n+1,0_R}$ in a special case which is useful in practice. If the coefficient ring R is self-injective[9], the condition that the linear homomorphism $.(k-l) : R \to M : r \mapsto r(k-l)$ is split boils down to the vanishing of the annulator of $k-l$: $Ann(k-l) = 0$. In fact, we have $Ker(.(k-l)) = Ann(k-l)$ and in that case, $R(k-l) \xrightarrow{\sim} R$ means by Appendix E.4.3, Proposition 100, that $R(k-l)$ is a direct factor of any module which contains it.

11.3.3 The Finite Case

Summary. Recursive classification on finite modules via socles and geometric classification.

$-\Sigma-$

Although a general geometric classification of local compositions in the above sense is not accomplished, we may use geometric classification to settle local compositions in finite ambient spaces, a situation which is particularly important in music theory. We shall make this more explicit in Sections 11.3.5, 11.3.7, and 11.3.8.

More precisely, we will consider zero-addressed local compositions $(K, \mathbb{Z}^n_N)$ for $1 < N, 0 < n$. Since $\mathbb{Z}_N$ is self-injective by Appendix E, Proposition 108, we may apply sorite 6 in Section 8.3.5 to classify such local compositions: Since the categories $ComLoc^{in}_{0_R}$ and $ComLoc^{gen}_{0_R}$ are equivalent, we may concentrate on the latter.

If $N = p_1^{n_1} \cdot \ldots p_r^{n_r}$ is the prime factorization (see Appendix D.2), we have $\mathbb{Z}_N \xrightarrow{\sim} \mathbb{Z}_{p_1^{n_1}} \times \ldots \mathbb{Z}_{p_r^{n_r}}$. So we first have to deal with direct products of rings and corresponding categories of local compositions. Recall the following from Appendix D.1.1, definition 143. Let $R = S \times T$ be a direct product ring, M an R-module. Denote by $x = x_S + x_T$ the decomposition of element $x \in M$ into its S-component $x_S = 1_S x$ and T-component $x_T = 1_T x$, and by $M_S = 1_S M$, $M_T = 1_T M$ the corresponding decomposition of M such that $M = M_S \oplus M_T$ is a direct sum of an S- and of a T-module. For an address A in $\mathbf{Mod}_R$, we also have a corresponding canonical decomposition $A@_R M \xrightarrow{\sim} A_S@_S M_S \oplus A_T@_T M_T$ which carries over to local compositions in these spaces. With any object $(K, A@_R M) \in ComLoc_A$, we then associate the pair

$$\times(K, A@_R M) = ((K_S, A_S@_S M_S), (K_T, A_T@_T M_T)) \in ComLoc_{A_S} \times ComLoc_{A_T}.$$

For a morphism $f : (K, A@_R M) \to (L, A@_R N)$ in $ComLoc_A$ we have an obvious decomposition into two morphisms $f_S : (K_S, A_S@_S M_S) \to (L_S, A_S@_S N_S)$ and $f_T : (K_T, A_T@_T M_T) \to (L_T, A_T@_T N_T)$ and therefore a map $f \mapsto \times f = (f_S, f_T)$.

Lemma 14 *The above map*

$$\times : ComLoc_A \to ComLoc_{A_S} \times ComLoc_{A_T}$$

is a functor, has well-defined restrictions

[9] See Appendix E.4.3.

$$\times : ComLoc^{emb}_A \to ComLoc^{emb}_{A_S} \times ComLoc^{emb}_{A_T},$$
$$\times : ComLoc^{gen}_A \to ComLoc^{gen}_{A_S} \times ComLoc^{gen}_{A_T},$$
$$\times : ComLoc^{in}_A \to ComLoc^{in}_{A_S} \times ComLoc^{in}_{A_T},$$

and all these functors are surjective on the objects.

We leave the proof as an exercise. Observe that for a pair

$$((K_1, A_S@_SM_S), (K_2, A_T@_TM_T)) \in {}_0ComLoc_{A_S} \times {}_0ComLoc_{A_T},$$

there is a unique maximal element $((K_1 \times K_2, A@_RM)) \in \times^{-1}(K_1, K_2)$. So the fiber consists of all local subcompositions $U \subset K_1 \times K_2$ with $U_S = K_1, U_T = K_2$. We call them *K_1-K_2-projecting*. Suppose that classification is settled for the factors and that we are given a pair K_1, K_2 of representatives of each class. Then classification of the set $K_1 \curlywedge K_2$ of K_1-K_2-projecting local compositions amounts to the calculation of the orbit set

$$Aut(K_1) \times Aut(K_2)\backslash K_1 \curlywedge K_2$$

under the obvious left action of $Aut(K_1) \times Aut(K_2)$. Observe that the latter groups are subsets of the symmetric groups of the underlying sets K_1 and K_2. If these are finite, the problem boils down to finite product group actions on finite set products, a subject of pure combinatorics, see Section 11.4.1 on this subject.

Let us now return to the more substantial problem of classifying the factor objects, i.e., objects in the category $ComLoc^{gen}_{0_{\mathbb{Z}_{p^n}}}$. We want to settle this problem by means of a recursive procedure that runs on the set of orbits of the set

$$U_r(R) = \{V|\ V \text{ a submodule of } R^r\} \tag{11.26}$$

under the action of a subgroup $G \subset \mathfrak{S}_{r+1}$ which extends the known linear action on $X_r(R)$ defined in 11.17. Recursion runs on the cardinality $card(\mathbb{Z}^r_{p^n})$ of our free rank r module over the ground ring $R = \mathbb{Z}_{p^n}$. Let $V \in U_r(R)$.

1. Let $soc(R^r) \subset V$ (see Appendix E.2.4 for the socle concept). Since the socle is invariant under any linear action, we may as well calculate the orbit of $V/soc(R^r)$ in $U_r(R/soc(R))$ which is recursively earlier.
2. Let $soc(V) \subsetneq soc(R^r)$. If $soc(V) = V$, then $V \in U_r(R/Rad(R))$, and we are in the case of the characteristic p prime field $R/Rad(R) = \mathbb{Z}_p$ (see Appendix E.2.4 for the radical concept). This is the geometric situation which is settled in Theorem 7.
3. Let $soc(V) \subsetneq soc(R^r)$ and $soc(V) \subsetneq V$. Since R^r is injective, there is an injective envelope $I \subset R^r$ of V. By Appendix E.4.3, Proposition 100, it is a direct factor of R^r, and since R is local, it is free, i.e., $I \xrightarrow{\sim} R^s$, $s < r$. This direct factor of R^r is an element of the Grassmannian $Grass_{r,r-s}(R)$ and can be classified by the geometric method which is standard for Grassmannians. Let $G = (\mathfrak{S}_{r+1})_I \subset \mathfrak{S}_{r+1}$ be the isotropy group (see Appendix C.3.1) of I. Then we have $V \in U_s(R)$, and we have to calculate its G-orbit for smaller cardinality. This completes the recursion process.

This process settles many of the practically important cases as we shall see in a moment. But we have to stress that we do not control the general geometric classification with address changes—a fortiori we do not control the deeper musicological consequences of this general geometric setup. We shall however give several partial answers in the following sections, and in Section 11.6, the latter concerning the esthetic aspect of classification.

11.3.4 Dimension

Summary. We discuss the problematic concept of dimension of local compositions.

– Σ –

We already learned in Section 6.4.1, formula (6.26), that the pitch space is not the seemingly one-dimensional real space but the subtler construction of Euler module which in the common setting (octave, fifth, third) has three dimensions over the rationals although it injects into the real line. Music theory thinks of octave, fifth, and third as being three independent musical directions, and restriction to rational scalars does the job in the mathematical sense.

This is even more delicate for progressive abstraction as it happens for pitch and onset classes modulo some positive periods. And this has a dramatic impact on the musical consequences of classification. To fix the ideas, look at a zero-addressed motif (7.2.3) $(M, Onset|\mathbb{Z} \oplus Pitch|\mathbb{Z})$ with integer coordinates. In this case, one agrees that the dimension of the underlying space is the rank 2 of a space isomorphic to $\mathbb{Z}^2$. But if we proceed to an identification of integer onsets modulo m and integer pitch modulo n—a completely common practice, we recall it—, then the residual motif $\overline{M} \subset OnPiMod_{m,n}$ (defined in (6.44)) shares a more complex dimensionality. To understand what happens, we compare the residual motives for $m, n = 12, 12$ (typical to dodecaphonism) with those for $m, n = 5, 12$. So in both cases, pitch is taken modulo octave, but in the first case, onset is taken modulo 12 and therefore, rhythmic phenomena are understood to share period 12 whereas in the second case, the period has to be 5.

There are $\binom{144}{3} = 487'344$ zero-addressed three-element motive residues in $OnPiMod_{12,12}$ whereas there are only $\binom{60}{3} = 34'220$ three-element motive residues in $OnPiMod_{5,12}$. However, anticipating classification of motive residues in these ambient spaces (see Section 11.3.8, and Appendix O), we have 26 isomorphism classes in case $m, n = 12, 12$ against 45 classes for $m, n = 5, 12$ which at first glance is very surprising.

The surprise vanishes if we look at the Sylow groups of these modules. We have $\mathbb{Z}_{12}^2 \xrightarrow{\sim} \mathbb{Z}_4^2 \times \mathbb{Z}_3^2$ whereas $\mathbb{Z}_5 \times \mathbb{Z}_{12} \xrightarrow{\sim} \mathbb{Z}_5 \times \mathbb{Z}_4 \times \mathbb{Z}_3$. The first space has a 2- and a 3-Sylow group, the second has a 2-, a 3-, and a 5-Sylow group. This corresponds to the general affine groups

$$\overrightarrow{GL}(\mathbb{Z}_{12}^2) \xrightarrow{\sim} \overrightarrow{GL}(\mathbb{Z}_4^2) \times \overrightarrow{GL}(\mathbb{Z}_3^2),$$
$$\overrightarrow{GL}(\mathbb{Z}_5 \times \mathbb{Z}_{12}) \xrightarrow{\sim} \overrightarrow{GL}(\mathbb{Z}_4) \times \overrightarrow{GL}(\mathbb{Z}_3) \times \overrightarrow{GL}(\mathbb{Z}_5).$$

The orders are $card(\overrightarrow{GL}(\mathbb{Z}_{12}^2)) = 663'552$, and $card(\overrightarrow{GL}(\mathbb{Z}_5 \times \mathbb{Z}_{12})) = 960$ (see Appendix C.3.5 for the calculation). We have ≈ 14.2 more motives in $OnPiMod_{12,12}$ than in $OnPiMod_{5,12}$, but the corresponding operating group $\overrightarrow{GL}(\mathbb{Z}_{12}^2)$ is ≈ 691.2 times larger than $\overrightarrow{GL}(\mathbb{Z}_5 \times \mathbb{Z}_{12})$. This is due to the larger number of Sylow groups in the second module and therefore produces ≈ 1.7 times more isomorphism classes for the 5-periodic motive residues.

This difference has a remarkable interpretation in terms of music theory and even music practice. If we want to classify motives modulo octave and in onset periods dividing 12, e.eg in 2/4, 3/4, 4/4, 6/8 or 12/8 time signatures, this is ≈ 1.7 times easier than to do so modulo octave and in onset period 5, i.e., in 5/8 time signature. This means that our cognitive effort to identify motive classes is nearly double for 5/8 time signature compared to the other situations.

It is of course not known—but to find out would be an important experiment in cognitive psychology— whether and to what degree humans do perform classification tasks in the above sense. There is however strong evidence from practice that this could be the case in light of the above results: Among jazz musicians it is well known how difficult it is to improvise in 5/8 time signature compared to 3/4, 4/4, or 6/8. For example, coherent, bar-exceeding improvisation in Desmond's *Take Five* is much more difficult than improvisation in a waltz such as Sherman's *Chim Chim Cheree*. So the increase in difficulty is not due to odd against even time signature, but the fact that 5 and 12 are relatively prime, whereas 3 and 12 are not.

These observations suggest that cognitive dimension in music is rather related to Sylow groups than to the more obvious length of a finite ambient space (which would be 6 vs. 3 in 12-periodic vs. 5-periodic rhythms). The systematic problem behind these observations is to investigate a general understanding of what is structural "independence" among musical attributes and how this one could be related to cognitive "independence". Structural/cognitive dimension would be a measure for the amount of structural/cognitive independence. *The systematic investigation of dimension as a degree of freedom in a musical system is undoubtedly a central issue for the description of global strategies in musical composition and cognition.*

11.3.5 Chords

Summary. Classification of zero-addressed chords in $PiMod_{12}$, their symmetry groups, automorphism groups, and conjugacy classes of endomorphisms in 12-chromatic pitch class space $\mathbb{Z}_{12}$.

– Σ –

We shall not discuss historic questions of this basic classification issue here, but refer to Section 11.5.2 on the American musical set theory tradition for more detailed remarks. But we should stress right here that, besides this American tradition, no systematic research on classification of chords has ever been undertaken. European classical musicology and music theory has only dealt with some important chords (such as minor, major, diminished, and augmented triads) and has not dealt with classification in the sense of determination of isomorphism classes under any non-trivial group beyond transposition. Even in recent times [73], so-called chord dictionaries are published which are anything but complete. This does not only testify to a fragmentary practice but much more (as a foundation of practice) a fragmentary *theory*. The complete list of isomorphism classes of zero-addressed chords in $PiMod_{12}$, together with additional determinants and class invariants, is accessible in Appendix N. This list was first discussed in a series of University lectures [669] in 1981, and published in [670]. In [802], more information concerning conjugacy classes of endomorphisms has been added.

As the chord class list can be calculated by a computer program, we concentrate on the description of the information which specifies the isomorphism classes. The list has five columns:

- The first column denotes the *class number*. The numbering first reflects the cardinality, starting with the special case of the full, 12-element chord, then the one-element class, then the two-element classes, and so forth until the six-element classes. Classes with more than six elements, seven,..., eleven, and the zero-element, the empty class, are not listed explicitly, since they appear as complements of the explicitly listed classes. There are, however, exceptions to this rule: There are six-element classes which are isomorphic to their complements. These classes have a star exponent with their class number, for example number 82*. So when counting the isomorphism classes, those 18 classes with star count only once, all others count double. This gives us a total of 157 non-empty classes.
  The numbering for fixed cardinality is as follows: We order chords according to the powerset order defined in 6.8, but with two (more historically motivated) differences: The order on $\mathbb{Z}_{12}$ is the clock order $\bar{0} < \bar{1} < \bar{2} < \dots \bar{11}$, and we take the minimal elements x_{min}, y_{min} of the differences $X - Y$ and $Y - X$ instead of the maximal ones to define the ordering. This gives the lexicographic ordering of subsets of $\mathbb{Z}_{12}$ where a bullet is prior to a circle, as drawn in column two of the list. For example, we have

  $$\bullet\bullet\circ\bullet\circ\circ\circ\circ\circ\circ\circ\circ < \bullet\circ\bullet\bullet\circ\circ\circ\circ\circ\circ\circ\circ$$

  According to this order, the first representatives of an orbit are taken, and the numbering of the classes for given cardinality is the ordering among the first representatives.
- The second column shows the first representatives as discussed in the previous item, indicated by sets of bullets. The complements are shown in circles. In particular, the classes of the complements (circle representatives) are ordered according to the order of the bullet representatives.
- The third column shows the symmetry group of $Sym(K)$ of the representative K, as defined in 8.3.5. Although, in general, this group is not an invariant of the class, if we stick to local compositions in $\mathbb{Z}_{12}$, and if we recall that all isomorphisms of local compositions in this ambient space are induced by affine isomorphisms, the symmetry group is unique up to conjugation.
- The conjugation class of the symmetry group from column three is indicated in the fourth column, symbolized by numbers as explained in the list's preamble.
- The fifth column shows the pair $a|b$ where a is the number of conjugacy classes of endomorphisms modulo automorphisms of the given local composition, and b is the corresponding number for the complement.

Contemplation of this list shows that the following data nearly defines a complete set of invariants:

1. Cardinality

2. Conjugacy class of the symmetry group
3. Pair of numbers $a|b$ of numbers of conjugacy classes of endomorphisms
4. Autocomplementarity, i.e., the star exponent in the class number

The only exceptions are four pairs of classes which are not distinguished:

- 18, 19
- 40, 41
- 64*, 68*
- 75*, 82*

We shall learn in Chapter 30 on counterpoint theory that the classes of the last two couples can also be separated by the span and diameter invariants, so essentially, only the first two couples are indistinguishable with the above data.

We cannot discuss the richness of the chord classification here, this will, however be addressed in more specialized and musically motivated chapters. We should instead make a final remark on the musicological range of the above list. Evidently, the pitch parameter was only relevant from the denotator perspective here. Mathematically, we have simultaneously classified all zero-addressed local compositions which implement the same ambient space, for example local compositions of form $OnMod_{12}$ as defined in formula (6.41), i.e., local compositions in 12-cyclic metrical onsets. According to definition 14 in Section 7.2.2, *we have classified 12-periodic local rhythms in the (reduced) onset domain.*

11.3.6 Empirical Harmonic Vocabularies

Summary. The systematic classification is compared to some empirical material.

– Σ –

Unfortunately, to this day no systematic repertorization of chords in classical literature has been realized. We shall see in Chapter 41.3 that this project is also due to some conceptual difficulties concerning the very definition of what can be identified as a chord in a given score. It is, however, not a serious problem and this repertorization should absolutely be realized in order to know the empirical distribution of chord types. The following exposition—initiated by Thomas Noll in [799]—is a first sketch of this project, but it already shows some remarkable results.

The material in [799] was taken from Nikolai Garbusow's work [386] on harmony. The difference of Grabusow's chord notation to the original works he refers to is not relevant to our objective since we only look at pitch classes in $PiMod_{12}$, including enharmonic identification. Our examples split into three lists.

The first list (from [386, Chapter IV]) refers to 85 examples which boil down to 23 chord classes. Except for class 78 which has a top number of 23 conjugacy classes of endomorphisms, these (22) classes are shown in Figure 11.3.

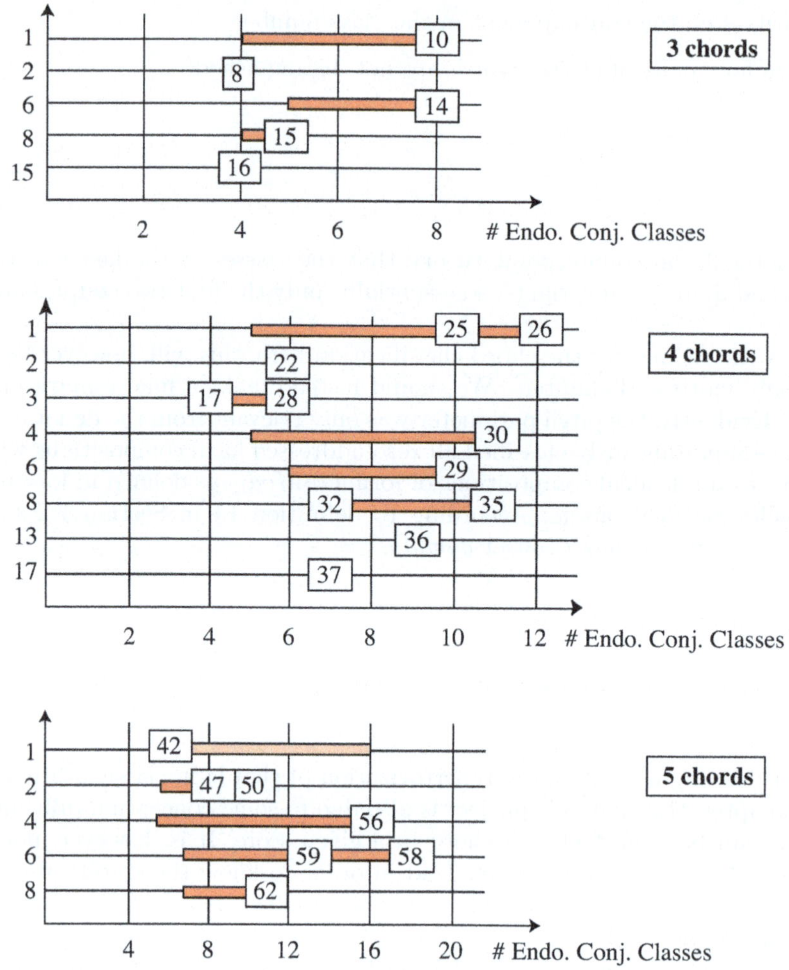

Fig. 11.3. The 22 chord classes from Garbusow's list from [386, Chapter IV] in their distribution among cardinalities 3,4,5, symmetry group conjugacy classes (ordinate axis), and cardinality of endomorphism conjugacy classes. The total range of the latter cardinalities is shown as a dark bar. We recognize that except for class 42, all classes range among the top numbers, i.e., have large numbers of endomorphisms.

The second list refers to 28 examples from Alexander Scriabin's sonatas 6 to 10. These examples boil down to 15 classes, one of which is a 4-chord of class 29 with maximal cardinality of endomorphism conjugacy classes. The other 14 classes are all 5-, 6-, and 7-chords and are shown in Figure 11.4.

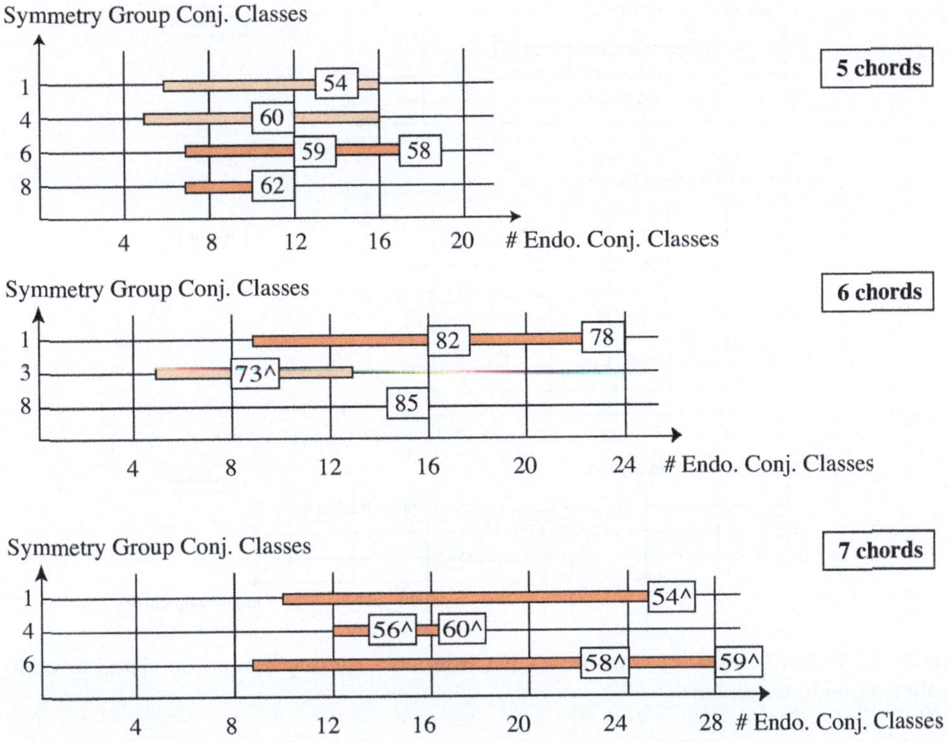

Fig. 11.4. The 14 chord classes of Garbusow's list from Scriabin's sonatas 6 to 10. Again, many classes have large numbers of endomorphisms.

Table 11.1: Examples from 20th Century Composers

Debussy: "Ibérie"	62, 56, 80^, 47^
Debussy: "L'isle joyeuse"	42^, 80^, 62, 78, 88
Ravel: "Ma Mère l'Oye"	40^, 85, 53^, 62
Prokofiev: "Sarcasmes"	(8), 62, 58, 78, 85^, 86
Prokofiev: "Le sacre du printemps"	54^, 17^, 82, 54^, 70
Schönberg: "Erwartung"	20^, 45^, 54^, 33^, 62^, 36^, 65^, 12^, 10^, 26^, 26^, (2^)
Casella: "Sonatine"	40^, 54^, 39^, (1), 10^, 4^

The third list refers to 45 examples from 20th century composers, table 11.1 shows the class list, together with the composers. Our third "statistics" of classes is shown in figures 11.5 and 11.6.

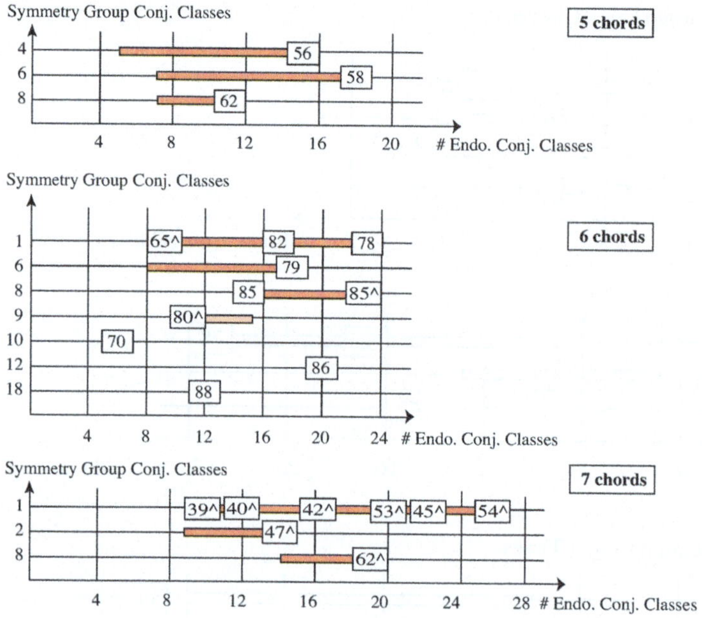

Fig. 11.5. The 5-, 6-, 7-chord classes of Garbusow's list from 20th century composers listed in table 11.1. We again have large numbers of endomorphisms.

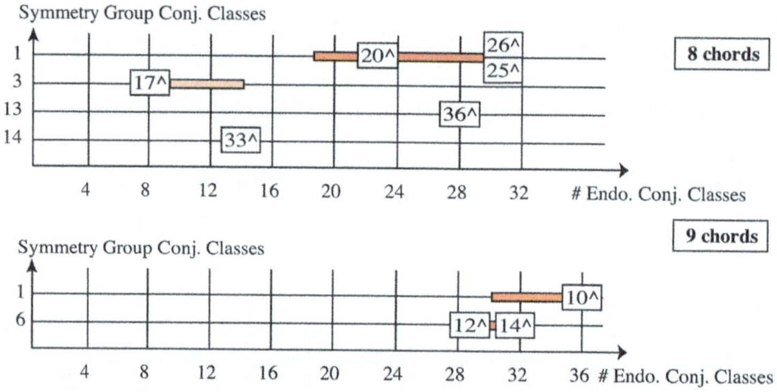

Fig. 11.6. The 8, 9 chord classes of Garbusow's list from 20th century composers listed in table 11.1. We again have large numbers of endomorphisms.

From the figures 11.3, 11.4, 11.5, and 11.6, we see that *there is a strong tendency in this empirical repertoire to be placed where the numbers of conjugacy classes of endomorphisms for a given symmetry group within a determined cardinality is high.* This observation is an empirical hint to the approach introduced in [799] which interprets endomorphisms as generalized tones. In our language this is, of course, the situation of $\mathbb{Z}_{12}$-addressed chords in $PiMod_{12}$ which will be treated in Section 11.3.7. This approach also suggests that harmony could be based to the differentiated richness of $\mathbb{Z}_{12}$-addressed chords in $PiMod_{12}$ which are canonically associated with zero-addressed chords.

11.3.7 Self-addressed Chords

Summary. We discuss classification of self-addressed local compositions.

– Σ –

The reflections from the previous Section 11.3.6 suggest that endomorphisms of ambient spaces could play a major role in harmony. This idea was pursued in [802] and will be taken up in our part VI on harmony, in particular Chapters 24 through 26. In this section we want to discuss classification of these endomorphisms.

More precisely, we have learned that sets of affine endomorphisms $F = e^t \cdot s : \mathbb{Z}_{12} \to \mathbb{Z}_{12}$ (modulo certain automorphisms) are relevant to important chords in European tradition. In order to restate this context we introduce a type of denotators in simple forms, the self-addressed denotators[10]:

Definition 32 *If F is a simple form with coordinator module M, a denotator $S : M \rightsquigarrow F(s)$ is called* self-addressed *(at address M). An objective local composition is called self-addressed if its elements are. A functorial local composition is self-addressed (at M) if it is a subfunctor of the product functor $@M \times F$.*

Observe the conceptual difference between endomorphisms of ambient spaces, local compositions and self-addressed denotators or local compositions. For local compositions, address change makes sense, but for endomorphisms this does not make sense.—So we want to discuss classification of finite local compositions $K \subset M@M$. As before, we look for a classification for self-addressed local compositions at the fixed address M. Quite generally, given two A-addressed local compositions $K \subset A@M, L \subset A@M$, let

$$f/\alpha : K \xrightarrow{\sim} L$$

be an isomorphism with underlying endomorphism $F = e^m \cdot F_0$ of M. If M is indecomposable of finite length, and if $1 < card(K)$, we know from Fitting's lemma 88 in Appendix E that F must also be an isomorphism. In fact, looking at the associated linear isomorphism $R.\alpha f : R.K \xrightarrow{\sim} R.(L.\alpha)$, we conclude that F_0 is an isomorphism. If $1 = card(K)$, it can only be isomorphic to L under a non-invertible endomorphism if it is a constant, since the linear part of the isomorphism on M must be nilpotent. But then, clearly, it can be shifted to the constant L by a translation isomorphism. Since the denominator α is an isomorphism by definition, we are then left with the calculation of orbits under the left and right action of $\overrightarrow{GL}(M)$ on $M@M$. Now, if we are given a ring R which is of finite length and which is a product of local rings, classification of self-addressed compositions at R also boils down to the above orbit calculation since we are then essentially dealing with each local factor of R which is indecomposable and therefore endomorphisms are induced by automorphisms. This is what is required when classifying self-addressed local compositions at address $\mathbb{Z}_{12}$, for example, or at address $\mathbb{Z}_{12}[\varepsilon]$ for counterpoint theory.

In the following discussion we first concentrate on local rings R of finite length. We are given local compositions $K \subset R@R$ of finite cardinality n. The group action is the following. Let $e^s \cdot t \in R@R$ be a self-addressed point which we identify with the row vector $(s,t) \in R^2$. If $(u,v) \in R \times R^\times$ and $(k,l) \in R \times R^\times$ represent an automorphism $e^u \cdot v$ and a base change $e^k \cdot l$ of R, respectively, their action $e^u \cdot v \cdot e^s \cdot t \cdot e^k \cdot l$ is represented by $(u + v.s + v.t.k, v.t.l)$. So, since the elements u,k are arbitrary and v,l are arbitrarily invertible, our action is also described[11] by

$$G^u_{v,k,l} = e^{(u,0)} \cdot G_{v,k,l} : (s,t) \mapsto e^{(u,0)}(G_{v,k,l}(s,t)) = (u + v.s + k.t, l.t)$$

where $e^{(u,0)} \cdot G_{v,k,l}$ is in the affine triangular group $\overrightarrow{T}_2(R)$, i.e., by definition of $\overrightarrow{T}_2(R)$, it has linear part

$$G_{v,k,l} = \begin{pmatrix} v & k \\ 0 & l \end{pmatrix}$$

[10] In [802], such objects were called "fractal" because of evident self-similarity structure.

[11] We make the common abuse of language and write a matrix multiplication $A.(x,y)$, meaning $(A.(x,y)^t)^t$ with transposed rows.

in the group $T_2(R)$ of upper triangular 2×2-matrices over R (with invertible diagonal coefficients). Taking up the methods of Section 11.3.2, fix a positive natural number n and look at the short exact sequence

$$0 \longrightarrow R \xrightarrow{\Delta_1} R^{2n} \xrightarrow{d_1} R^{2n-1} \longrightarrow 0 \tag{11.27}$$

of R-modules defined as follows. We denote the $2n$-tuples in the middle module by

$$(s_1, s_2, \dots s_n; t_1, t_2, \dots t_n)$$

and the $(2n-1)$-tuples in the right module by $(s_2, \dots s_n; t_1, t_2, \dots t_n)$. Then the sequence is defined by

$$\Delta_1(u) = (u, u, \dots u; 0, 0, \dots 0) \text{ and}$$
$$d_1(s_1, s_2, \dots s_n; t_1, t_2, \dots t_n) = (\sigma_2, \dots \sigma_n; \tau_1, \tau_2, \dots \tau_n)$$

where

$$\sigma_i = s_i - s_1, 1 < i \leqslant n,$$
$$\tau_1 = t_1,$$
$$\tau_i = t_i - t_1, 1 < i \leqslant n.$$

If we let the symmetric group $\mathfrak{S}_n$ act linearly from the right on R^{2n} by

$$(s_1, s_2, \dots s_n; t_1, t_2, \dots t_n) \cdot \pi = (s_{\pi 1}, s_{\pi 2}, \dots s_{\pi n}; t_{\pi 1}, t_{\pi 2}, \dots t_{\pi n})$$

for $\pi \in \mathfrak{S}_n$, we get local compositions from sequences $(s.; t.)$ via orbits under $\mathfrak{S}_n$ (and as before, also restricting to the sequences with n pairwise different couples (s_i, t_i). Again, the diagonal image $Im(\Delta_1) = R.\delta_1$, with $\delta_1 = (1.; 0.)$, is invariant under this action and therefore, we have an induced action on the cokernel R^{2n-1}. We have the diagonal left action of the affine triangular group $\overrightarrow{T}_2(R)$. Let $G = G^u_{v,k,l}$. Then $G.(s_1, s_2, \dots s_n; t_1, t_2, \dots t_n)$ leaves the diagonal image $R.\delta_1$ invariant and commutes with the action of the symmetric group. Moreover, the translational part of the action of $\overrightarrow{T}_2(R)$ has orbits which include $R.\delta_1$-cosets , i.e., it induces a linear action of the upper triangular group $T_2(R)$ on the cokernel R^{2n-1}. If $G_{v,k,l} \in T_2(R)$, we have

$$G_{v,k,l}(\sigma_2, \dots \sigma_n; \tau_1, \tau_2, \dots \tau_n) = (v.\sigma_2 + k.\tau_2, \dots v.\sigma_n + k.\tau_n; l.\tau_1, l.\tau_2, \dots l.\tau_n), \tag{11.28}$$

and this means the following: First, the action of the invertible elements $l \in R^\times$ generates orbits $R^\times.\tau.$ on the second half R^n of R^{2n-1} which, according to Appendix F, Lemma 98, is equivalent to considering monogenous subspaces $R.\tau.$ of R^n. Moreover, the contribution of the τ-component to the first half is only a function of these monogenous spaces. We therefore have a first invariant: the monogenous subspaces $R.\tau. \subset R^n$. When fixing one of these spaces, V, say, and denote the projection pr_2V onto the second to last coordinates, we have the induced action of $v \in R^\times$ on the quotient space R^{n-1}/pr_2V. This again gives monogenous subspaces as invariants. Of course, only those generators $(\sigma_2, \dots \sigma_n; \tau_1, \tau_2, \dots \tau_n)$ with

$$(\sigma_i, \tau_i) \neq (0, 0), 1 < i \text{ and} \tag{11.29}$$
$$(\sigma_i, \tau_i) \neq (\sigma_j, \tau_j), 1 < i < j \tag{11.30}$$

correspond to local compositions. Summarizing, we have these invariants of the left action of the triangular affine group:

$$U^1_n = \{(V, W) |\ V \subset R^n, W \subset R^n/pr_2V \text{ both monogenous}\}. \tag{11.31}$$

On this space, the symmetric group acts from the right and we are in a similar situation as for the finite classification discussed in Section 11.3.3. More precisely, we first have to classify the monogenous subspaces of R^n under the action of $\mathfrak{S}_n$. Again, let us suppose that R is self-injective. Then we may proceed by literal copy of the recursive algorithm from 11.3.3. If we fix a representative V from this classification, and if $G_V \subset \mathfrak{S}_n$ is its isotropy group, we are left with the action of this group on monogenous spaces $W \subset R^{n-1}/pr_2V$, and recursion can go on until the algorithm stops. This may be summarized in

Theorem 8 *If R is a local, self-injective commutative ring of finite length and maximal ideal m, then there is an algorithm which classifies self-addressed local compositions of cardinality n in R, and the invariants from this algorithm are parametrized by S-rational points of quotient schemes from actions of finite subgroups of $\mathfrak{S}_n$ on Grassmann schemes, where the occurring rings $S = R/m^r$ are quotients of the ground ring R.*

In particular, if we have a finite ring $\mathbb{Z}_N$, or an infinitesimal extension $R[\varepsilon]$ of a local, self-injective commutative ring R of finite length, this one has the same property, and we therefore have classified the self-addressed local compositions in $\mathbb{Z}_N[\varepsilon]$ which will be used in counterpoint, see Chapter 29.

11.3.8 Motives

Summary. Classification of motives of small cardinality.

– Σ –

The classification techniques which we discussed in the previous sections will now be applied to describe explicitly classification of small motives. The following case of three-element motives is far from general classification but it gives a complete description of the algorithms and, in contrast to the more abstract performance of enumeration theory which is discussed in 11.4, yields explicit lists of representatives of isomorphism classes. We shall also need this kind of list in Sections 11.6.2, 11.6.3, so it is useful to pay them a short moment of attention. The list which we shall calculate now is shown in Appendix O.3.

We now classify the three-element commutative, zero-addressed local compositions K in $OnPiMod_{12,12}$, i.e., mathematically speaking, $K \subset \mathbb{Z}_{12}^2$ with $card(K) = 3$. But we know that $\mathbb{Z}_{12}^2$ is injective, and therefore we know from 11.3.3 that we may concentrate on three-element objects in $ComLoc^{gen}_{3,0_{\mathbb{Z}_{12}}}$. Conversely, according to the theory which yields Theorem 6, any such object K has a subspace $V \subset \mathbb{Z}_{12}^2$ such that $\mathbb{Z}_{12}^2/V \xrightarrow{\sim} \mathbb{Z}_{12}.K$. But then, the module $\mathbb{Z}_{12}.K$, as a quotient of $\mathbb{Z}_{12}^2$ is also a submodule[12] of $\mathbb{Z}_{12}^2$, and therefore, K is already realized in $\mathbb{Z}_{12}^2$.

So we are left with the calculation of $X_2(\mathbb{Z}_{12})/\mathfrak{S}_3$, and we may start by the calculation of orbits of the larger space $U_2(\mathbb{Z}_{12})$ under a subgroup of $\mathfrak{S}_3$. Let us first describe the representation

$$D^{opp} : \mathfrak{S}_3 \to GL_2(\mathbb{Z}_{12})^{opp} : \pi \mapsto D^{opp}(\pi) = D(\pi^{-1})$$

of the right group action in terms of 2×2-matrices over $\mathbb{Z}_{12}$:

$$D^{opp}(Id) = \begin{pmatrix} 1 & 0 \\ 0 & 1 \end{pmatrix} \qquad D^{opp}(Id)(x, y) = (x, y) \tag{11.32}$$

$$D^{opp}(1\ 2) = \begin{pmatrix} -1 & -1 \\ 0 & 1 \end{pmatrix} \qquad D^{opp}(1\ 2)(x, y) = (-x - y, y) \tag{11.33}$$

$$D^{opp}(1\ 3) = \begin{pmatrix} 1 & 0 \\ -1 & -1 \end{pmatrix} \qquad D^{opp}(1\ 3)(x, y) = (x, -x - y) \tag{11.34}$$

$$D^{opp}(2\ 3) = \begin{pmatrix} 0 & 1 \\ 1 & 0 \end{pmatrix} \qquad D^{opp}(2\ 3)(x, y) = (y, x) \tag{11.35}$$

$$D^{opp}(1\ 2\ 3) = \begin{pmatrix} -1 & -1 \\ 1 & 0 \end{pmatrix} \qquad D^{opp}(1\ 2\ 3)(x, y) = (-x - y, x) \tag{11.36}$$

$$D^{opp}(1\ 3\ 2) = \begin{pmatrix} 0 & 1 \\ -1 & -1 \end{pmatrix} \qquad D^{opp}(1\ 3\ 2)(x, y) = (y, -x - y) \tag{11.37}$$

Let us now calculate the $\mathfrak{S}_3$-orbits of subspaces V of $\mathbb{Z}_4^2$.

[12] This follows from the main theorem on finitely generated abelian groups, see Section C.3.4.2.

V is semi-simple. It therefore is a $\mathbb{Z}_2$-vector space, i.e., $V \subset 2\mathbb{Z}_4^2$. So we have $V = 0$ for $dim(V) = 0$, $V = 2\mathbb{Z}_4^2$ for $dim(V) = 2$, and the $\mathfrak{S}_3$-orbit of the one-dimensional space $V = \mathbb{Z}_4.(2,0)$ is the set of all one-dimensional spaces in $2\mathbb{Z}_4^2$. So we have three orbits of semi-simple submodules.

Suppose $2.V \neq 0$. Then the injective envelope W of V in $\mathbb{Z}_4^2$ must be either free of rank one or rank two by Proposition 102 in Appendix E.

In the rank two case, $W = \mathbb{Z}_4^2$, and we claim that in this case, $2\mathbb{Z}_4^2 \subset V$. In fact, otherwise, there is $X \subset W$ with $X \xrightarrow{\sim} \mathbb{Z}_2$ and $X \cap V = 0$ and W cannot be the injective envelope of V. But then, we are in the semi-simple situation since the socle is invariant and we have to look for the orbits of subspaces in $\mathbb{Z}_4^2/2\mathbb{Z}_4^2$. The zero space has already been recognized above. Hence we have the two cases $V = \mathbb{Z}_4 \times 2\mathbb{Z}_4$ and $V = \mathbb{Z}_4^2$.

In the rank one case, we have $V = \mathbb{Z}_4.x \xrightarrow{\sim} \mathbb{Z}_4$ since otherwise, V would be semi-simple. According to the group action table above, we have two orbits, one for $x = (1,0)$, one for $x = (1,2)$.

Summarizing, we have these subspace orbit representatives, including their isotropy groups which we need later:

Representative	Isotropy Group	
0	$\mathfrak{S}_3$	semi-simple
$2\mathbb{Z}_4^2$	$\mathfrak{S}_3$	semi-simple
$\mathbb{Z}_4(2,0)$	$\langle D^{opp}(1,2)\rangle$	semi-simple
$\mathbb{Z}_4(1,0)$	$\langle D^{opp}(1,2)\rangle$	injective
$\mathbb{Z}_4(1,2)$	$\langle D^{opp}(1,2)\rangle$	injective
$\mathbb{Z}_4^2$	$\mathfrak{S}_3$	injective
$\mathbb{Z}_4 \times 2\mathbb{Z}_4$	$\langle D^{opp}(1,2)\rangle$	

We have the following orbits of subspaces of $\mathbb{Z}_3^2$ under the two isotropy groups of the above list:

Isotropy Group	Representatives
$\mathfrak{S}_3$	$0, \mathbb{Z}_3^2, \mathbb{Z}_3.(1,0), \mathbb{Z}_3.(1,1)$
$\langle D^{opp}(1,2)\rangle$	$0, \mathbb{Z}_3^2, \mathbb{Z}_3.(1,0), \mathbb{Z}_3.(1,1), \mathbb{Z}_3.(0,1)$

From this repertory, we have to select those products $V_4 \times V_3 \subset \mathbb{Z}_4^2 \times \mathbb{Z}_3^2$ which do not contain the vectors $e_1 = ((1,0),(1,0)), e_2 = ((0,1),(0,1))$, and $e_1 - e_2$. This gives us the list of kernel spaces in column three of table O.3, and we are done with the subspaces.

Exercise 14 *Recapitulate the general algorithm described in 11.3.3 along the lines of the previous example.*

Exercise 15 *Calculate the representatives of motives in that table from the spaces. Recall that these kernel spaces $Ker(f)$ stem from canonical linear maps $f : \mathbb{Z}_{12}^2 \to \mathbb{Z}_{12}.K$ which is associated with the motif K.*

Table O.3 also contains two more invariants, the *class weight* in column four, and the *volume* in column five. Whereas the class weight will be defined and discussed later in Section 14.5, the volume of a local composition is defined as follows:

Definition 33 *Let $K \subset R^n$ be a commutative local composition in the free R-module of rank n. Then the image of the n-fold exterior power $\bigwedge^n R.K$ in $\bigwedge^n R^n$ is an ideal $Vol(K)$ of R. The ideal $Vol(K)$ is called the* volume *of K.*

If the ring R is of finite length, two isomorphic local compositions $K \subset R^n, L \subset R^n$ are related by an invertible symmetry F, according to sorite 6 in Section 8.3.5. This symmetry's linear part yields an automorphism of the nth exterior power $\bigwedge^n R^n \xrightarrow{\sim}_R R$, and this leaves the ideal invariant. Therefore:

Proposition 7 *With the above notation, and if R is of finite length, the volume is an invariant among all local compositions in the same ambient space R^n.*

In particular, if every ideal in R is principal, and if R is a product of local rings of finite length, the generator of $Vol(K)$ is unique up to invertible factors in R. By abuse of language, we also call this generator the volume of K. This number is included in the last column of table O.3.

In [1017], Hans Straub has classified of four-element motives in $OnPiMod_{12,12}$ by use of a software written in C. This work made use of the methods introduced in the previous sections and confirmed the list of Egmont Köhler [540] which was communicated without any specification of the used methods. It also completed the redundant list in [670] which applied the present methods, but without computer support.

The procedure in [1017] first uses classification of two-element motives to build a reasonably small set of motives which includes all isomorphism classes. From this classification, which is listed in table O.1, we can fix two of the four tones to one of the five representatives of two-element motives of our table. The other two tones are free, and we obtain a total list $List_4$ of $5 \cdot 142 \cdot 141/2 = 50055$ motives to be investigated. The elements of this list are ordered by an index: $List_4 = \{K_i | \ i = 1, \dots 50055\}$. The program picks each of the list's elements in this order, decides whether its class has already been found and then adds it to the found class representatives or else throws it away and picks the next element.

To decide upon class membership, the program first calculates the volume and class weight for a selected motif $K_i \in List_4$. If this data is not yet present in the list of class representatives, we have found a new representative and proceed to the next element K_{i+1}. Else, if volume and class weight is already present, we concentrate on those representatives for the following comparison: For each four-element motif $K_i \in List_4$, the program calculates a set $G(i)$ of generators of the kernel $Ker(f_i)$ of the linear map $f_i : \mathbb{Z}_{12}^2 \to \mathbb{Z}_{12}^2$ which is canonically associated with K_i in our theory, since the elements of K_i are listed in a determined order (ordered representation is automatic in computer programs).

Then the program calculates the orbit of $G(i)$ under the action of the symmetric group $\mathfrak{S}_3$ and checks whether an element of this orbit is included in the kernel $Ker(f_j)$ of a motif $K_j \in List_4$ which has already been selected as a representative and which has the same volume and class weight as K_i. If this is the case, the orbit $Ker(f_i) \cdot \mathfrak{S}_3$ contains $Ker(f_j)$ and we can proceed to the next candidate, otherwise, K_i is a new representative. By the preselection via volume and class weight, the orbit calculations happen less frequently and the program terminates quickly.

This program also settles the question of finding the class of any given motif and therefore can and has been used for experimental purposes. In [1017], three- and four-element motif classification was used to compare melodies. The selection includes: Johann Sebastian Bach/Gigue Nr. 32, Johann Sebastian Bach/Sarabande Nr. 52, John Lord[13]/Sarabande, Wolfgang Amadeus Mozart/KV 449/T.1-8. These Melodies were stratified into motives built from three- and four-element groups of successive notes. The classification of these motives shows rather strong commonalities for three-element motif classes:

Composer/Piece	Number of Occurring Classes	Most Frequent Class (Nr.)
J.S. Bach/Gigue Nr. 32	13	10
J.S. Bach/Sarabande Nr. 52	12	10
J. Lord/Sarabande	15	10
W. A. Mozart/KV 449	14	10

The pieces differ significantly for the refined four-element motif stratification:

[13] John Lord is the organist of the rock group "Deep Purple".

Composer/Piece	Number of Occurring Classes	Most Frequent Class (Nr.)
J.S. Bach/Gigue Nr. 32	52	19
J.S. Bach/Sarabande Nr. 52	66	14
J. Lord/Sarabande	75	15
W. A. Mozart/KV 449	65	23

We see that there are major differences in the latter data. However, as it is pointed out in [1017], the second number is not very reliable since the orbit cardinalities should be included in these figures, i.e., frequency should be divided by orbit cardinality. This type of analysis should be investigated systematically and with statistical skill. Without doubt, it yields a refined melody classification method which can be used for stylistic classification. We shall perform a detailed and poetological melodic analysis built on three-element motives in Section 11.6.2.

11.4 Enumeration Theory

Summary. Enumeration theory is a quantitative approach to classification of local compositions via actions of permutation groups. We review Harald Fripertinger's work on this subject.

– Σ –

Enumeration theory deals with counting of types of structures, i.e., orbits of finite group actions on finite sets. These sets represent particular species of objects. In musical enumeration theory, these structures are chords, partitions of pitch class sets, motives, twelve-tone rows and the like. The groups consist of selected systems of musically interesting transformations, such as transposition or inversion. Evidently, counting orbits is much less than listing them, or even describing them as rational points of schemes or similar classifying spaces of geometric nature. Why is there any interest to know such numbers—for example the incredible number of

$$2\ 230\ 741\ 522\ 540\ 743\ 033\ 415\ 296\ 821\ 609\ 381\ 912 = 2.23\ldots \cdot 10^{36}$$

isomorphism classes of 72-element motives in $\mathbb{Z}_{12}^2$? We shall deal with the more philosophical aspect and the question concerning big science in musicology in Section 11.4.2. Here, we can anticipate that first, a good part of the enumeration work automatically yields or eases explicit lists of representatives, and second, knowing the class numbers is a strong prerequisite to apply probabilistic and statistical reasonment for classification of large music repertories. We gave a small account to this direction in the final words of Section 11.3.8.

In musical enumeration theory, scholars often deal with enumeration of local and global compositions without clear conceptual distinction. We only deal with *local enumeration theory* here and postpone global enumeration to Section 16.2.

11.4.1 Pólya and de Bruijn Theory

Summary. Pólya and de Bruijn theory yields a crucial tools for enumeration of classes of local compositions in the case of finite abelian groups as ambient modules.

– Σ –

This section mainly refers to Harald Fripertinger's work [341, 342, 343, 344]. It rests on combinatorics of finite group actions on finite sets[14]. In order to apply the Pólya theory to the enumeration of isomorphism classes of zero-addressed local compositions $K \subset \mathbb{Z}_n$, the set K is identified with its characteristic function[15]

[14] For another approach to enumeration theory of local compositions in finite abelian groups, see [435].
[15] See Appendix C.1, Lemma 74.

$\chi_K : \mathbb{Z}_n \to 2$ with the identification $2 = \{0, 1\}$. So the set of zero-addressed local compositions $K \subset \mathbb{Z}_n$ is identified with the powerset $2^{\mathbb{Z}_n}$. If a group $G \subset \overrightarrow{GL}(\mathbb{Z}_n)$ of affine automorphisms acts on the set of zero-addressed local compositions $K \subset \mathbb{Z}_n$, this action is reflected on the powerset via the obvious right action $(\chi_K, g) \mapsto \chi_K \cdot g$. To count the G-orbits in $2^{\mathbb{Z}_n}$, we introduce weight functions $w : 2 \to R$ with values in a ring R which express special sets of local compositions.

Let us shortly digress on the general Pólya context to understand the ideas. One looks at powersets F^P of finite sets F, P and canonical right group actions of permutation groups $G \subset \mathfrak{S}_P$ on F^P via $(f, g) \mapsto f \cdot g$ as above for $P = \mathbb{Z}_n, F = 2$. A *Pólya weight function* is a function $w : F \to R$ with values in a ring R which contains the field of rational numbers. This induces a *product weight function*

$$\Pi_w : F^P \to R : f \mapsto \prod_{p \in P} w(f(p)). \tag{11.38}$$

Clearly, the product weight is invariant on a G-orbit and we may define the product weight on the orbit space via

$$\Pi_w : F^P/G \to R : f \cdot G \mapsto \Pi_w(f). \tag{11.39}$$

For our above example, we can take the ring $R = \mathbb{Q}[x]$ of rational polynomials over the indeterminate x and the weight

$$w(i) = \begin{cases} 1 & \text{if } i = 0, \\ x & \text{if } i = 1. \end{cases} \tag{11.40}$$

Whence the product weight $\Pi_w(\chi_K) = x^{card(K)}$. The *configuration counting series* $C(G, F, P, w)$ of this weight is defined by

$$C(G, F, P, w) = \sum_{\omega \in F^P/G} \Pi_w(\omega) \tag{11.41}$$

and we see that *this polynomial in* $\mathbb{Q}[x]$ *has the number of G-orbits of local, zero-addressed k-element compositions* $K \subset \mathbb{Z}_n$ *as its* x^k*-coefficient* in the above case $P = \mathbb{Z}_n, F = 2$.

The Pólya theory offers tools to calculate the configuration counting series. The central object of the theory is the *cycle index*, a rational polynomial which depends on the subgroup $G \subset \mathfrak{S}_P$. It gives a formal account of the cycles[16] of the elements of G in P. Then:

Definition 34 *With the above notation, the* cycle index *of the permutation group G is the rational polynomial*

$$Z(G) = card(G)^{-1} \sum_{g \in G} X^{cyc(g)}. \tag{11.42}$$

For several common permutation groups, the cycle index is known, see Appendix C.3.6 for the explicit formulas. The relation between the configuration counting series and the cycle index is set up by the main theorems of Pólya's enumeration theory:

Theorem 9 *With the above notation, we have*

$$C(G, F, P, w) = Z(G)\Big(\sum_{y \in F} w(y), \sum_{y \in F} w(y)^2, \ldots \sum_{y \in F} w(y)^d\Big). \tag{11.43}$$

Corollary 6 *With the above notation, if the weight $w = 1$ is the constant weight with value 1, $C(G, F, P, 1) = card(F^P/G)$ is the number of orbits and this evaluates to*

$$card(F^P/G) = Z(G)(card(F), card(F), \ldots card(F)). \tag{11.44}$$

In particular, if we have the above action of a group $G \subset \overrightarrow{GL}(\mathbb{Z}_n)$ on the set of zero-addressed local compositions $K \subset \mathbb{Z}_n$, we have the orbit number

$$card(2^{\mathbb{Z}_n}/G) = Z(G)(2, 2, \ldots 2). \tag{11.45}$$

[16] See Appendix C.3.6, definition 141 for the concepts of cycles and the cycle type $cyc(g)$ of a permutation g.

These results yield formulas for counting the number of classes of zero-addressed local compositions with respect to isomorphism types which are common in chord classification. This is a straightforward application of cycle index formulas for these groups.

For the translation group $T_n = e^{\mathbb{Z}_n}$, and for the group of translations and inversions $TI_n = e^{\mathbb{Z}_n} \rtimes \langle -1 \rangle$, the number of G-orbits of zero-addressed local compositions in $\mathbb{Z}_n$ is calculated from the coefficient of the k-th power x^k in Theorem 9 and from the formulas for $Z(T_n)$ and for $Z(TI_n)$. We have

Proposition 8 [343] *The number of T_n-orbits of zero-addressed k-element local compositions in $\mathbb{Z}_n$ is*

$$\frac{1}{n} \sum_{j|gcd(n,k)} \varphi(j) \binom{n/j}{k/j}, \tag{11.46}$$

and the number of TI_n-orbits of zero-addressed k-element local compositions in $\mathbb{Z}_n$ is

$$\begin{cases} \frac{1}{2n}\left(\sum_{j|gcd(n,k)} \varphi(j)\binom{n/j}{k/j} + n\binom{(n-1)/2}{[k/2]}\right) & n \text{ odd,} \\ \frac{1}{2n}\left(\sum_{j|gcd(n,k)} \varphi(j)\binom{n/j}{k/j} + n\binom{n/2}{k/2}\right) & n,\ k \text{ even,} \\ \frac{1}{2n}\left(\sum_{j|gcd(n,k)} \varphi(j)\binom{n/j}{k/j} + n\binom{(n-2)/2}{[k/2]}\right) & n \text{ even, } k \text{ odd,} \end{cases} \tag{11.47}$$

where φ is Euler's φ function[17]*, and where $[t]$ denotes t's integer part*[18].

For the special case $n = 12$, the cycle index $Z(\overrightarrow{GL}(\mathbb{Z}_{12}))$ and Theorem 9 yield the $\overrightarrow{GL}(\mathbb{Z}_{12})$-orbits [343], i.e., the isomorphism classes, and we obtain this orbit number list:

$k =$	0	1	2	3	4	5	6	7	8	9	10	11	12
T_{12}	1	1	6	19	43	66	80	66	43	19	6	1	1
TI_{12}	1	1	6	12	29	38	50	38	29	12	6	1	1
$\overrightarrow{GL}(\mathbb{Z}_{12})$	1	1	5	9	21	25	34	25	21	9	5	1	1

11.4.1.1 Enumeration of Interspace Structures

For special groups of affine automorphisms of $\mathbb{Z}_n$, *interspace structures*[19] of zero-addressed local compositions in $\mathbb{Z}_n$ are useful; they realize a special instance associated with our above construction of difference structures, for example in short exact sequence (11.12). For a non-empty local composition $K \subset \mathbb{Z}_n$, write $(K) = (k_1, k_2, \ldots k_t)$ for the ordered sequence (also called the *sequence of* K) of the $t = card(K)$ elements of K in the canonical linear order $0 < 1 < 2 < \ldots n-1$ on the canonical representatives of $\mathbb{Z}_n$. The *interspace sequence of* K is defined to be the image $\delta(K)$ of the sequence (K) under the linear endomorphism

$$\delta : \mathbb{Z}_n^t \to \mathbb{Z}_n^t : (x_1, \ldots x_t) \mapsto (x_2 - x_1, \ldots x_t - x_{t-1}, x_1 - x_t) \tag{11.48}$$

with the evident kernel $Ker(\delta) = \Delta\mathbb{Z}_n$.

Exercise 16 [343] *Consider the subset $IS_{t,n} \subset [1, n-1]^t$ of those sequences of positive integer t-tuples $(i_1, i_2, \ldots i_t)$ such that $i_1 + i_2 + \ldots i_t = n$. Then the canonical map $IS_{t,n} \to \mathbb{Z}_n^t$ is a bijection onto the set of* interspace sequences *of t-element local compositions in $\mathbb{Z}_n$.*

[17] See Appendix, C.3.4.1.

[18] The largest integer $\leqslant t$.

[19] Terminology of Fripertinger [343].

We next consider three linear automorphisms of $\mathbb{Z}_n^t$:

$$\varphi : \mathbb{Z}_n^t \xrightarrow{\sim} \mathbb{Z}_n^t : (x_1, \dots x_t) \mapsto (x_t, x_1, \dots x_{t-1})$$
$$\psi : \mathbb{Z}_n^t \xrightarrow{\sim} \mathbb{Z}_n^t : (x_1, \dots x_t) \mapsto (x_t, x_{t-1}, \dots x_1)$$
$$\rho : \mathbb{Z}_n^t \xrightarrow{\sim} \mathbb{Z}_n^t : (x_1, \dots x_t) \mapsto (-x_t, -x_{t-1}, \dots - x_1)$$

and observe that $\delta \cdot \varphi = \varphi \cdot \delta$ and $\varphi \cdot \delta \cdot \rho = \psi \cdot \delta$. The *interspace structure* $[K]$ of K is defined as the orbit of $\delta(K)$ under the dihedral group $\langle \varphi, \psi \rangle \xrightarrow{\sim} TI_t$. We then have

Lemma 15 [343] *If $K, L \subset \mathbb{Z}_n$ are non-empty local compositions, their orbits under the group TI_n are equal iff $[K] = [L]$.*

Proof. Clearly, the translation $e^1(K)$ and the inversion $-K$ have the same interspace structure as K. Conversely, if K, L are such that $\delta(K) = \varphi \cdot \delta(L)$, then $(K) - \varphi(L) \in Ker(\delta) = \Delta\mathbb{Z}_n$, and we conclude that K is a translation of L. If $\delta(K) = \psi \cdot \delta(L)$, the above equation gives $\delta(K) = \psi \cdot \delta(L) = \varphi \cdot \delta \cdot \rho(L)$, and by the preceding case, the TI_n-orbits of K and $-L$ coincide, so K and L have same TI_n-orbit. QED.

Together with exercise 16, Lemma 15 yields the following technique for enumerating classes of local compositions which have special interspace configurations. If $x, y_1, y_2, \dots y_n$ are indeterminates, consider the weight function

$$w : [1, n-1] \to \mathbb{Q}[x, y_1, y_2, \dots y_n] : i \mapsto x^i y_i \tag{11.49}$$

which is defined on the interval $F = [1, n-1]$ of positive integers. Therefore, with the formalism introduced in formula 11.38, we have the product weight

$$\Pi_w : F^t \to \mathbb{Q}[x, y_1, y_2, \dots y_n] : g \mapsto x^{\sum g(i)} \prod y_{g(i)}.$$

Since we can identify the set of interspace sequences of t-element local compositions in $\mathbb{Z}_n$ with $IS_{t,n}$, the orbits of such local compositions are identified with TI_n-orbits of sequences g in $IS_{t,n}$ under the same action. Then the above weight function is invariant under the action of the above dihedral group $\langle \varphi, \psi \rangle \xrightarrow{\sim} TI_t$, and since the set $IS_{t,n}$ is characterized by the sum of coordinates being n, we have a weight function

$$W : TI_t \backslash IS_{t,n} \to \mathbb{Q}[x, y_1, y_2, \dots y_n] : g \mapsto x^n \prod y_{g(i)} \tag{11.50}$$

where the indices of the y give information about the intervals between successive elements in these local compositions. We then have

Theorem 10 *With the above notation, the number of interspace structures of t-element local compositions in $\mathbb{Z}_n$ is given via the x^n-coefficient of the cycle index evaluation*

$$Z(TI_t)(\sum_{i=1}^{n-1} x^i y_i, \sum_{i=1}^{n-1} x^{2i} y_i^2, \dots \sum_{i=1}^{n-1} x^{ti} y_i^t), \tag{11.51}$$

i.e., the coefficients of the monomials in the indeterminates y_i of the x^n-coefficient are the numbers of interspace structures of given interval distribution.

Example 14 For example, if we look for interspace structures of local 3-element compositions in $\mathbb{Z}_{12}$, the coefficient of x^{12} is

$$y_1^2 y_{10} + y_1(y_2 y_9 + y_3 y_8 + y_4 y_7 + y_5 y_6) + y_2^2 y_8$$
$$+ y_2(y_3 y_7 + y_4 y_6 + y_5^2) + y_3^2 y_6 + y_3 y_4 y_5 + y_4^3$$

and we can calculate, for example, those interspace structures with all intervals being at least k units by setting $y_1 = y_2 = \dots y_{k-1} = 0$ and $y_i = 1$ else. For $k = 2$ we get seven such structures.

11.4.1.2 Enumeration of Series

Recall from definition 22 in Section 8.1.1 that a n-phonic series is a denotator

$$Ser : \mathbb{Z}^{n-1} \rightsquigarrow PiMod_n(Ser_0, \dots Ser_{n-1})$$

with pairwise distinct coordinates. As we know from the discussion of n-phonic series following definition 22, the group $D_{n,n}$ acts on the set SER_n of n-phonic series. Enumeration theory yields this number of orbits:

Proposition 9 *For* $3 \leqslant n$, *the number of orbits in* $D_{n,n} \backslash SER_n$ *is as follows. Set*

$$n!! = \begin{cases} n(n-2)(n-4) \cdot \ldots 2 & \text{if } n \equiv 0 \mod 2, \\ n(n-2)(n-4) \cdot \ldots 1 & \text{else.} \end{cases} \tag{11.52}$$

Then the numbers are

$$\begin{cases} \frac{1}{4}((n-1)! + (n-2)!!(\frac{n}{2}+1)) & \text{if } n \equiv 0 \mod 2, \\ \frac{1}{4}((n-1)! + (n-1)!!) & \text{else.} \end{cases} \tag{11.53}$$

A proof can be found in [341].

Example 15 For example, the number of $D_{12,12}$-orbits of a dodecaphonic series is 9 985 920. So there is still some uncovered material for dodecaphonic compositions.

More generally, the Bruijn extension of Pólya's enumeration theory yields formulas for orbits of (k, n)-series [343, Satz 2.2.5]. We omit the details here and just reproduce the particularly interesting list of orbits for $n = 12$:

k	number of orbits of $(k, 12)$-series
2	6
3	30
4	275
5	2 000
6	14 060
7	83 280
8	416 880
9	1 663 680
10	4 993 440
11	9 980 160
12	(dodecaphonic series) 9 985 920

To deal with all-interval series, we invoke the context of definition 22 in Section 8.1.1. For natural numbers $0 < k \leqslant n$, we have the set $SERM_{k,n}$ of (k, n)-serial motives which is an additive abelian group by pointwise addition, i.e., $(Ser + Ser')_i = Ser_i + Ser'_i$ for $Ser, Ser' \in SERM_{k,n}$. We have a projection homomorphism

$$D : SERM_{k+1,n} \rightarrow SERM_{k,n} \tag{11.54}$$

defined by $D(Ser)_i = Ser_{i+1} - Ser_i$, $0 \leqslant i < k$, call $D(Ser)$ the *derived serial motif.* This projection has a right inverse injection

$$I : SERM_{k,n} \rightarrow SERM_{k+1,n} \tag{11.55}$$

defined by $I(Ser)_0 = 0$ and $I(Ser)_i = Ser_0 + \dots Ser_{i-1}, 0 < i \leqslant k$, i.e., $D \cdot I = Id_{SERM_{k,n}}$. Call $I(Ser)$ the *integrated serial motif.*

By definition, the set of *all-interval n-phonic series* is the intersection set $ALLSER_n$ in the cartesian diagram

$$\begin{array}{ccc} ALLSER_n & \longrightarrow & SER_n \\ \downarrow & & \downarrow \\ D^{-1}SER_{(n-1,n)} & \longrightarrow & SERM_n \\ {\scriptstyle D}\downarrow & & \downarrow{\scriptstyle D} \\ SER_{(n-1,n)} & \longrightarrow & SERM_{(n-1,n)} \end{array}$$

of sets of serial motives with all horizontal arrows being inclusions. Consider the left action of the group $D_{n,n}$ on $SERM_n$. Then, SER_n is invariant, and for any affine automorphism $g : \mathbb{Z}_n \xrightarrow{\sim} \mathbb{Z}_n$, and the reversion operation rev, we have

$$D \cdot g = g_0 \cdot D,\ D \cdot rev_n = -rev_{n-1} \cdot D. \tag{11.56}$$

Therefore $D^{-1}SER_{(n-1,n)}$ is also invariant under $D_{n,n}$ and even under the larger group where we take all affine pitch automorphisms instead of transpositions and inversions only. So the set $ALLSER_n$ of all-interval series is invariant under $D_{n,n}$ and under the larger group G_n generated by the reversion and all affine pitch automorphisms.

In [343], these properties of all-interval series, as well as specific weight function (as considered above) are used to calculate the size of the orbit spaces of all-interval series under the actions of different, musically indicated subgroups $H \subseteq G_n$. In particular, for $n = 12$, we have

Proposition 10 [343, Satz 2.3.5] $card(G_{12}\backslash ALLSER_{12}) = 267$.

11.4.1.3 Enumeration of Motives

We have discussed the classification techniques for zero-addressed local compositions $(K, \mathbb{Z}_N^n)$ in Section 11.3.3, and more concretely of local compositions K in $OnPiMod_{12,12}$ in Section 11.3.8. Here, we deal with purely enumerative aspects for local compositions K in $OnPiMod_{n,m}$, as discussed in [343]. We are again in the situation exposed in Theorem 9, however, this time, $F = 2, G = \mathbb{Z}_n \oplus \mathbb{Z}_m$, and the weight is again given by formula (11.40). Therefore,

Theorem 11 [343, Satz 2.5.2] *The number of orbits of zero-addressed local compositions K in $OnPiMod_{n,m}$ under a group H of affine automorphisms of $\mathbb{Z}_n \oplus \mathbb{Z}_m$ is the coefficient of x^k in the cycle index polynomial*

$$Z(H)(1 + x, 1 + x^2, \dots 1 + x^{nm}).$$

For the special case $n = m = 12$ and $H = \overrightarrow{GL}(\mathbb{Z}_{12}^2)$, the evaluation has been calculated by use of a Turbo Pascal program in [343]. We write down the first few monomials of the cycle index polynomial in Theorem 11:

$$Z(\overrightarrow{GL}(\mathbb{Z}_{12}^2)(1 + x, 1 + x^2, \dots 1 + x^{144}) = x^{144} + x^{143} + 5x^{142} + 26x^{141} + \dots,$$

the entire formula is written down in Appendix C.3.6. It shows that, in particular, the number of 72-element classes is

$$2\ 230\ 741\ 522\ 540\ 743\ 033\ 415\ 296\ 821\ 609\ 381\ 912 = 2.23\ldots \cdot 10^{36},$$

so there are still some motives to be introduced into musical composition!—In [343], the enumeration of some other dodecaphonic structures is calculated. The reader is referred to this work for more details.

11.4.2 Big Science for Big Numbers

Summary. Signification of the exorbitant growth of numbers of classes of local compositions for computational complexity in musical analysis. Complexity and depth in music revisited.

– Σ –

Big Science is the type of science created by physicists since the Los Alamos nuclear bomb project during world war II. Nowadays, it is best illustrated by the research environment of Europe's CERN[20] near Geneva. With a total of roughly 10 000 employees, among others a circular LEP[21], an electricity consumption equivalent to that of the city of Geneva and a financial volume of about one billion CHF [185] in 1995. Big Science is recognized by the following characteristics:

- Scientific language
- Models and theorems
- Experimental paradigms and operationalization
- Universal collaboration and communication
- Adequate laboratories and machines
- Political acceptance and corresponding resources

Why does the exorbitant number of isomorphism classes of local compositions which we have exposed in the previous sections suggest the research context of Big Science? Let us look at the above characteristics and their realization in centers such as CERN, see Figure 11.7.

Fig. 11.7. (© cern) The 27 km long LEP ring on the CERN site, viewed from above. The ring is situated 100 m underground.

Evidently, the first two items are within the reach of mathematical music theorists. The third item is also attainable, at least in principle. We are given an enormous repertoire of musical works and know how to do experiments on this material. For example, one could calculate the distribution of isomorphism classes of chords, or motives, of serial motives, of local meters and so on within a determined material selection. We

[20] CERN = Centre Européen pour la Recherche Nucléaire.

[21] LEP = Large Electron Positron Collider.

have already given small examples of such investigations in 11.3.6 and 11.3.8. And one could then compare this data to the models of harmony etc. in order to verify/falsify corresponding statements, etc. One also disposes of good operationalization techniques, see the annual ICMC[22] meetings, or also the relevant parts and chapters in this book (e.g., parts X, XIII). What is less evolved is universal collaboration, as well as political acceptance and corresponding resources.

However, the negative extremum is the serious lack of adequate laboratories and machines. What is "adequate" in this case? There are several requirements to be fulfilled. To begin with, any serious processing of relevant musical material, a classical sonata with its note size of order 10^4, say, must be based on powerful computers, precisely because of class numbers such as the above number $2.23 \ldots \cdot 10^{36}$ of 72-element motive classes. Second, harmonic, rhythmic, and motivic analysis of such class data requires enormous combinatoric calculations. For example, as we shall see in Section 41.3, the calculation of harmonic paths is proportional to $2^{\text{number of onsets of a composition}}$, and this is practically infinity. Further, navigating on data bases of relevant musical works must be based on powerful visual and auditive representation paradigms, for example 3D-tables, see Figure 11.8, or 3D-caves. See also Chapter 20 for the visualization of musical data.

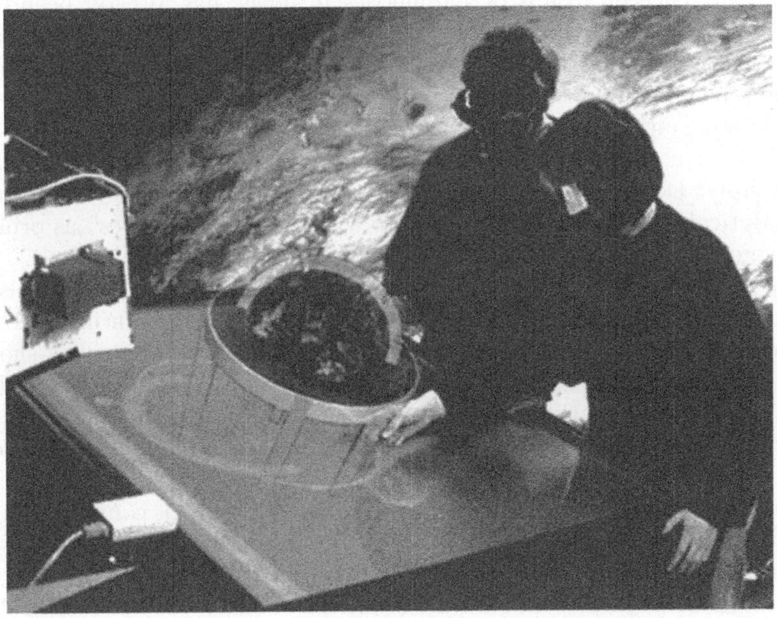

Fig. 11.8. The 3D table [562] as a necessary virtual reality tool for visual navigation in music data bases.

Finally, it is absolutely necessary to have auditory representations of whatever music or analysis data exists in order to control the complex material. All this combines to require a huge apparatus for calculation and representation.

Now, can this be compared to the CERN equipment which deals with experimental necessities deduced from high energy conditions required to unveil hidden symmetries of suspected laws of external nature? But we have already discussed this topic in Chapter 4: Experiments with inner nature require high spiritual energies in a very rational sense, we cannot understand mental event systems having the quality of high-ranked music without adequate analysis, representation, and navigation tools. And these must be set up in technologically sophisticated and collaborative environments. It is not clear and we would not hope that a musicological CERN would be as expensive and monumental as the physicists' display. We would nonetheless like to stress that music is of comparable significance to humans, and this justifies much of such an effort which is not likely to produce destruction devices such as nuclear bombs.

[22] ICMC = International Computer Music Conference.

Whatever the technological apparatus which becomes necessary to control this complex data, it has become evident that depth now is no more a question of empty or censored encapsulation and knowledge hiding. Depth can now completely be recovered in its explicable form of a complex universe which is overwhelming but not disclosed from understanding. The situation is much the same as with the early days of anatomy where opening a human body was prohibited by authorities who feared that detailed and rational access to the subject would annihilate the mumbo-jumbo of Aristotelian and Christian mystification.

11.5 Group-Theoretical Methods in Composition and Theory

Summary. This section reviews the group theoretic methods in composition from Bach to Schönberg, and contemporary serialists. In analysis, approaches from Graeser to Forte and the American set theory school are reviewed.

– Σ –

The wide-spread presence of groups of symmetries in music has already been discussed in Chapters 8 and 10. Here, we want to relate the quantitative and qualitative aspect of orbit spaces to historically traced approaches in composition and analysis. We do, however, refer to further symmetry-related harmonic analysis because this aspect will also be treated in due detail in part VI.

Knowing that paradigmatic classification is a fundamental poetic device that must be taken into account in the analytical perspective, too, what is the specific power of group orbits of musical structures such as motives, chords, rhythms? De facto, once the paradigm and a concrete object have been related either by compositional or analytical efforts, we are thrown into a space of "prototypes" as orbits may be called in a more intuitive language.

So what help is given by moving within a space of prototypes? First of all, this space is a reduction to the paradigmatically essential information, we have abstracted from contingencies which blur the core information. But then, in many cases, as the previous classification discourse has shown, the resulting list is often finite and—for some classical cases such as chord classes—of manageable size. In the tradition of medieval artistic poetics (such as François Villon's poesy [632]) or in the baroque combinatorial rhetoric (such as Athanasius Kircher's classificatory work [945]) such lists of reasonable size are germs of combinatorial approaches to universal classification or artistic construction. This is one essential aspect: Disposing of an alphabet of structural ingredients reduces much of the work to a task of combination or recombination. In other words, synthetical and analytical creativity are projected onto a field of combinatorial activity.

We should nonetheless recognize that lists which turn out to be infinite or so large that even the best computers cannot search them in reasonable time (e.g., the roughly 10^{36} classes of 72-element motives in $\mathbb{Z}_{12}^2$, see Section 11.4.1.3) give a slightly different approach to combinatorics: We have to introduce statistical procedures since systematic combination would fail. So here, aleatoric components cannot be avoided, and this is one aspect which indeed is historically prominent as we shall see.

Whatever the size of combinatorial arsenals, the overall situation is *a massive shift of creative quality*. On one hand, combinatorics gives a feeling of divine power: Everything is reachable, the whole universe at your fingertips. You can play God with the molecular combinatorics of chemical substances from a small number of roughly a hundred atom types. On the other hand, this same "total control" not unexpectedly excites one's boredom with a once-for-all limited alphabet of creativity.

Let us first concentrate on the positive side: complete reachability of classes in an orbit space. To begin with, the specific nature of this space lends itself to an enrichment of semantic depth. More precisely, the classes being points in a particular space type—such as a geometric parameter space for the complete set of class invariants—may be related to each other qua points and therefore induce an additional instrument of poetic construction or analytical investigation. For example, the analysis of historical chord repertories such as discussed in Section 11.3.7 used the size of automorphism groups of classes as a measure for historical distinction between important and marginal chord classes. In other words, the enrichment of having placed classes in a coherent space, and not only in an insignificant list array, yields a deeper principle of unification of the instance of these classes in a given composition. On the one hand, this is a plain added value, on the other, it may also happen that either in composition or in analysis, the space specifics are not or only in part taken into account. So there is no automatic integration of classifying space attributes, but the potential must be observed.

Second, the negative aspect of a well-defined classificatory arsenal is the limitation of the "game's rules" in a rigid framework. We know from music semiotics [703] that a significant difference between the music sign system and the linguistic system is the weak conventional aspect termed "langue" vs. the spoken "parole" in Saussure's dichotomic setting. Vocabularies tend to englobe small portions of space-time in musical cultures. This fact means that the "game's rules" may de facto vary without creating much ado, at least is this a fact without a priori negative consequences. So we observe that each determined combinatorial setup on the basis of a classification arsenal provokes the antithesis of progressive negation of the classification's basics, i.e., the paradigma as such.

In this dialectic dynamics of combinatorics and its negation, of power and creative expansion, history unfolds and lets us recognize a *coupled* syn- and diachronic development of the musical sign system. We shall come back to this overall phenomenon in Section 11.7 after a closer look at representative examples.

11.5.1 Aspects of Serialism

Summary. Extending transformation groups as a developmental parameter: Counterpoint, dodecaphonic composition technique, and genuine serialism.

– Σ –

We have already seen symmetries with Bach, Schönberg and serialists in composition in Section 8.1.1. Here, we reflect their roles with regard to paradigmatic classification via group orbits. In the baroque fugue technique as completed by Bach, the rhetoric scheme of a dialog between dux and comes is not specific enough to define a thematic vocabulary, it only gives empty boxes to be filled with musical structure. But the contrapuntal tradition had prepared a generic procedure of relating voices by a "switch" between counterparts, punctus contra punctum meaning that two opposites are confronted. We shall see in Chapter 29 that this idea can be made completely precise in modeling Fux' contrapuntal rule system. Moreover, the artistic tradition of medieval poesy was highly sensitive towards construction of explicit correspondences between poetic instances: Recall François Villon's poesy [632] which abounds with retrograde, inversion and their combinations in verse and rhyme construction.

In this spirit of combinatorial artistry, the orbit of themes under inversion, retrograde, transposition and their combinations was a perfect completion of the said empty rhetoric scheme. On the one hand, this was however not the only principle of compositional construction. The harmonic level was to be developed into a dominant syntactical force which evolved until its dispersion at the end of the 19th century for reasons we shall discuss in Chapter 23. This led to an apparantly secondary role of the contrapuntal dialog of voices during the harmonic homophony. On the other hand, the construction of themes was not completely explicit in the sense that the interrelation between thematic/motivic and harmonic/contrapuntal constraints had not been made evident as a composed phenomenon.

Only after the dispersion of harmony in the first years of the 20th century was the counterpoint group rediscoverd by Schönberg and his followers. This rediscovery followed Schönberg's declared program of rebuilding the compositional paradigm ab ovo and without any implicit interrelations to harmonic constraints. So the really new message in Schönberg's attempt was *the exclusive basing of local compositional coherence* upon the orbit of a given dodecaphonic series. In contrast to the baroque usage of such orbits, this time, the orbit *defined* the a priori alphabet of local compositions to be distributed over the syntagm of the composition.

So the progress against fugue tradition and the harmonic vocabulary was to make available a proper alphabet for each composition instead of a fixed alphabet being perpetuated along a large number of compositions. The common denominator of the dodecaphonic vocabularies was exactly this meta-vocabulary, i.e., a procedure to build vocabularies, a new one for each composition. Moreover, the corpus of such compositions was classified by a fundamental invariant: the orbit of the governing dodecaphonic series[23]. So the logic of orbits forced a thorough investigation of classes of dodecaphonic series, such as all-interval series and other 'special' types. Under such aspects was it no surprise that Schönberg even patented one of his series.

We know that a dodecaphonic series is a denotator

$$Ser : \mathbb{Z}^{11} \rightsquigarrow PiMod_{12}(Ser_0, Ser_1, \ldots Ser_{11})$$

at the address $\mathbb{Z}^{11}$ which essentially defines a sequence or "row" of pitch values. We could also say that a series is a "discrete curve" with values in a determined parameter space (see Figure 11.9). Then a dodecaphonic series is by no means a generic concept: There is no deeper reason for disclosing other musical parameters, such as duration, loudness, onset, whatever, from such a concept. In other words, if a 12-series is a curve with values in a pitch domain, i.e., form $PiMod_{12}$, then a generic series would be a $\mathbb{Z}^{11}$-addressed denotator with values in a more general form such as $Duration \oplus Pitch \oplus Loudness \oplus \ldots$ or a variant with (partial) modular division $Duration_n \oplus Pitch_m \oplus Loudness \oplus \ldots$.

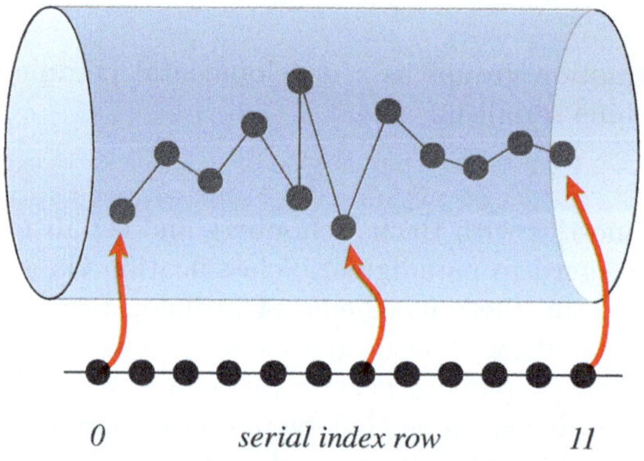

Fig. 11.9. A generic series as viewed by serialists can be interpreted as a discrete curve. It takes values in a parameter space form; here, for example, in the tubular form $Duration \oplus Pitch_m$, for each serial index $0, \ldots 11$.

Asking such a series to fulfill the requirement of unique representation of a pitch, for example, amounts to looking at its projection onto the corresponding form and to making the requirement explicit on this projected parameter space.

In view of this canonical generalization which is immediate from the universal denotator language, the historical realization and handling of serial perspectives has been a poor attempt, and even failure. More precisely, the deficiency has occurred on two levels: Object *description* and control of *transformations* together with the attached orbit spaces.

The first one could be observed when composers were dealing with the question of realizing a given series on the level of a concrete composition. No systematic treatment of admitted series has ever been published. The formal power of composers and musicologists (in serialism often in personal union) was evidently too poor for grasping the complexity of a systematic serialism. So, when reviewing the serial compositions and their interpretations and analyses (see for example [210, 298, 1001, 1092]), we notice a permanent struggle with the fundamental discipline of what parameter should be included, ordered, related to others, and then realized in a concrete composition or its analysis, respectively. The impression is that of a chaotic usage of undigested form and concept fragments from combinatorial and group theory. This chaotic approach is further alimented by an old and strong numerological tradition, such as Fibonacci numbers or the mumbo jumbo of

[23] Of course are there other characteristics than this of dodecaphonic composition technique. They are situated in the global configuration of the alphabet defined by the orbit $D_{12,12} \cdot Ser$ of series Ser (see Section 8.1.1). We shall come back this subject in Chapter 13.

magic squares. It seems that we are participating in a painful process of conceptualization, comparable to the birth of modern score notation.

The second deficiency is a consequence of the first one: Lack of adequate conceptualization makes control of transformations virtually impossible. So there is no evidence for mental control of which transformations of series should be allowed or how they should be formalized, not to mention the question of classification via group orbits. So the chaos of concepts intertwines with the chaos of transformations and results in a wild—euphemistically termed "creative"— conglomerate of operations and listings. We come back to this subject in the introduction of global compositions in Chapter 13. Intuitively, the situation is that of constructing a new musical instrument and having nobody who is capable of playing it in a controlled way, but instead, treating it according to one's inherited capacities.

It should however be emphasized that there is no moral or artistic ground for refuting such a development. It rather turns out to be a typical effect in the history of music: Already it had happened with classical harmony, in the line of Riemann, say, that a fragmentary theory and blurred practice had led to its abandon in the name of (a true or desperate) creative innovation. Schönberg's dodecaphonism was in a strong sense this kind of reaction to fragmentary and deficient harmony. In this spirit of positive redefinition of a crisis, serialism has even been *defined* as a precursor of post-serialism [1001]. History of music is also a history of failures to control results of creative extension.

Though adaptation of classification techniques from mathematics as described above solves some of these deficiencies, the very complexity of explicit control and manipulation of large ensembles of orbits and transformation groups requires more than just conceptual control, it requires auxiliary tools to *operationalize mental handling*. We have already mentioned this perspective in Section 11.4.2. The computer is precisely the tool we are asking for. In Chapter 50, we shall discuss more extensively the computer program *presto*® developed for this type of paradigmatic composition based on structural units and their innumerable transforms (in fact *presto*® allows more than 126 billion affine transformations in each parameter plane).

A typical application of the computer-controlled paradigmatic composition will be discussed in Chapter 51. And we should also mention the impressive Piano concert No.1 by Jan Beran [103] which is the broad instantiation of transformations of a short, but important[24] motif (Figure 11.10) from Beethoven's Hammerklavier sonata op 106.

Fig. 11.10. The fundamental "series" in Beran's Piano Concert Nr. 1: an important motif from Beethoven's Hammerklavier sonata op. 106.

This motif is a local composition of form $OnMod_{71} \oplus PiMod_{71} \oplus LoudMod_{71} \oplus DurMod_{71}$ and is transformed according to the general affine group of $\mathbb{Z}_{71}^4$, a group of cardinality[25]

$$10\ 445\ 260\ 466\ 832\ 483\ 579\ 436\ 191\ 905\ 936\ 640\ 000$$

or $1.0445 \cdot 10^{37}$. This orbit defines the a priori local alphabet of Beran's piano concert. Clearly, no human memory can manage this amount of *a priori* elements in a single orbit.

[24] See the motivic analysis in [670].

[25] See Appendix C.3.5 for the cardinality of general affine groups over finite fields.

11.5.1.1 Serial Strategies in Jan Beran's Piano Concerto Nr.1

In the following, the main structural principles used for the piano concert are described briefly. It should be noted that the equally fundamental 'principle' used for the composition was 'musical intuition'. The structural principles were only used as a tool that helped 'translate' intuition into a large coherent composition, and thus create a 'skeleton' that was a starting point for the actual musical composition.

The technical tool for the effective realization of the composition was the composition application *presto*®. See Chapter 50 for technical details of the *presto*® functionality and sound representation.

The construction consisted of the following steps that were partially carried out jointly and repeatedly, not necessarily in the sequence given here:

Step A: Selection of the basic theme M: The basic theme consists of the upper voice in Beethoven's Hammerklavier sonata op. 106, bars 75 and 76, see Figure 11.10. The note events were put as zero-addressed denotators in general (geometric) position in the four-dimensional space

$$P = Onset \oplus Pitch_{71} \oplus Loudness \oplus Duration,$$

by assigning different values of Loudness and Duration to different notes. (The embedding of *presto*®'s half-tone pitch modulo 71 into a MIDI pitch space is discussed in Chapter 50.)

Step B: Basic pattern in P: A basic pattern (ornament) was created by copying theme M to a large number of starting points in the two-dimensional projection $P_o = Onset \oplus PitchMod_{71}$ of P. The starting points were defined by $(onset, pitch) = ja + lb$ where a and b are two (nonparallel) vectors in P_o and j, l are positive integers in a certain range. (Recall that here all values of pitch are understood modulo 71.) The vectors a and b were chosen such that repetitions of the same pattern occurred only after a very long period.

Step C: Most frequent transformations: A set T of basic transformations (matrices) was defined. These consisted mainly of combinations of slight rotation, stretching and shearing.

Step D: Application of transformations: The pattern created in Step B was varied by local application of transformations, most of them being from set T. The transformations were applied to all two-dimensional projections of P. Of particular interest were transformations of projections that excluded either onset time or pitch. Transforming the pattern led to new musical motives that were used subsequently in the composition.

Step E: Instrumentation: Instruments were assigned (and created) to polygonal areas of the composition such that musically interesting structures in P are emphasized.

Step F: Tempo curve: A detailed tempo curve was defined.

The sequence of the steps was as follows: Steps A through D were done using a piano score only. Steps E and F were done subsequently together with a second application of Step D. Most of the effort was spent with steps D through F, since this is where the actual creative compositional work took place.

11.5.2 The American Tradition

Summary. The purpose of this section is essentially to account for and discuss some crucial contributions of the American tradition to the emergence and proliferation of what Babbitt termed 'professional music theory' [54]. Our account will be split into a genealogical discussion (11.5.2.1) and a contribution to a vocabulary switch (11.5.2.2) between the American set theory (for short: AST) and our present concept environment.

– Σ –

11.5.2.1 Genealogy

The impossibility of giving even a partial (ordered) description of the topics dealt with in the American music-theoretical literature since the 1950s leads us to look for historically and methodologically pertinent 'segmentations' in the domain of contemporary music theory. Perhaps one of the most fruitful approaches is based on the underlying dichotomy between an apparently more compositional attitude (Milton Babbitt) and a radically analytical perspective (Allen Forte) towards music theory. This distinction appears, more or less explicitly, in many studies on the subject, from the important "Perspectives on Contemporary Music Theory" (1972, [134]) to the recent "Music Theory in Concept and Practice" (1997, [69]). Both of these works are divided in two formal parts, meta-theory and methodology/compositional theory and historical and theoretical essays/analytical studies, respectively. The former work points to "the engagement by composers in fundamental music-theoretical explications" ([134, p.vii]), while the latter suggests that "the very fact that Forte is not himself a composer has changed the field of theory considerably" ([69, p.50]).

We may also suggest here that this distinction is not only relevant for an historical discussion on pitch-class set theory, but that it also helps in understanding how this theory successively enlarged its field of applicability thanks to important works by John Rahn [873], Robert Morris [765], and David Lewin ([605, 604]).

As pointed out by George Perle in his comprehensive study on serial and atonal music, "the most important influence of Schönberg's method is not the 12-note idea in itself, but along with it the individual concept of permutation, inversional symmetry and complementation, invariance under transformation, aggregate construction, closed systems, properties of adjacency as compositional determinants (...)" ([831, p.x]). This 'Babbittian' presentation of twelve-tone problematics constitutes perhaps the most appropriate introduction to the composer and theorist Babbitt. It is widely accepted that he first provided "twelve-tone theory with a consistent technical vocabulary" ([831, p.xiv]) and suggested that the relevance and "the force of any 'musical system' was not as universal constraints for all music but as alternative theoretical constructs, rooted in a commonality of shared empirical principles and assumptions validated by tradition, experience, and experiment" ([134, p.ix]).

Reflecting on the problem of "an adequately reconstructed terminology" ([53, p.10]) is a necessary preliminary step for the description of structural characteristics of the twelve-tone system, as Babbitt's four crucial articles make clear [49, 50, 51, 52]. Detailed discussions of Babbitt's terminology also feature in ([49], repr.) and [54]. One cannot emphasize enough that, for Babbitt, a "set" is an ordered collection of pitch classes and it is used as a perfect synonym for row and series. In contrast, the very predecessor of Allen Forte's "pitch-class set" is Babbitt's "source set", a set "considered only in terms of the content of its hexachords, and whose combinatorial characteristics are independent of the ordering imposed on this content" ([49, p.57n]). A synonym for it is the term "collection", first introduced by Lewin in [598] and widely discussed for its analytical pertinence in [600].

Subsequently, a vast body of American literature was devoted to the study of the specific properties of sets and collections, particularly combinatoriality and partitioning [657, 485, 357, 93, 1005, 1006]. Partition problems connected with Babbitt's original idea have also largely proved their relevance to mathematics with their natural embedding into the theory of groups [605, 765]. But even the idea of applying the mathematical concept of group for modeling musical systems can be regarded as one of Babbitt's most fruitful intuitions[26], provided that "the rules of formation and transformation of the twelve-tone system are interpretable as defining a group element (a permutation of order of set numbers) and a group operation (composition of permutations)" ([53, p.20]). This equivalence of structures, first introduced in [50], has important compositional consequences which are "directly derivable from the theorems of finite group theory" ([51], p.8). It is perhaps no exaggeration to see the introduction of groups by Graeser and then Babbitt as the 'Copernican Revolution' of modern music theory, especially if one considers the proliferation of group-theoretical methods applied to music, from Lewin's GIS structure to Clough and Agmon's modern theory of diatonicism. Suggestions for further reading in this area may be found in Rahn's review of Lewin's "Generalized Musical Intervals and Transformation" ([874]).

[26] However anticipated in 1924 by Wolfgang Graeser in his study on Bach's "Kunst der Fuge" [387].

The most important representative of the analytical approach in the American music-theoretical literature is Allen Forte who is the author of a theory of set complexes [332] and of a book primarily devoted to the atonal music of Schönberg, Webern, and Berg in the first 20 years of the 20^{th}century [331]. Forte's main purpose is to "provide a general theoretical framework, with reference to which the processes underlying atonal music may be systematically described" ([331, p.ix]). Forte's starting-point is "firmly analytical, springing from a truly fervent desire to uncover the secrets of an (...) enigmatic repertoire" ([114, p.50]). In his most recent article, Forte particularly emphasized this aspect, stating that "The structure of atonal music (is) above all the study of a musical repertoire rather than a theoretical presentation" ([332, p.83]). However, one of his most striking merits was the introduction of a "consistent terminology for pitch-class collections based on the mathematical properties of the set" ([114, p.49]) together with the elaboration of the "set complex", a topic[27] recently developed by Robert Morris [767]. But it is probably true that "Allen Forte's real success lies in the developments he inspired: beyond his theorization of atonal music, his work convinced many of the interest of a formal study of chromatic space" ([732, p.90]).

An instance of this is Rahn's pedagogically oriented introduction to some problematics common to the atonal and serial repertoire [873]. The book also prepares "its reader for the professional literature in the field" ([873, p.v]) and gives accurate references for further specialized topics including advanced serialism and combinatoriality. Babbitt's "general formative role", together with Forte's terminological heritage are widely recognized in the US-American sphere and the book has become a standard reference for further discussions in set theory. Robert Morris for example developed some "relations between ordering, aggregate completion, and pitch-structure based on groups of operations acting on pitch-classes" ([765], p.xiii).

The following section is dedicated to a detailed discussion of many of the topics dealt with by Morris by means of the theory of local compositions. Morris' recent formalization of his so-called "compositional spaces" provides a new theoretical tool which compensates for some of the weaknesses of his original compositional model. As defined by Morris, "compositional spaces are out-of-time structures from which the more specific and temporally oriented compositional design can be composed" ([766, p.330]). A crucial point here is its intimate connection with Lewin's transformational approach. Like Babbitt and Forte, Lewin's contribution to the field of music theory is both terminological and methodological. His work could also be regarded as one of the clearest attempts to go beyond the music-theoretical dichotomy of compositional vs. analytical procedures. His very first article concerning generalized approaches in describing intervallic collections of notes [598], which is almost in the spirit of Babbitt's compositional attitude, was soon followed by an analytical application of some formal constructions, like the interval function, to Schönberg's hexachordal pieces [600].

As pointed out by Jonathan W. Bernard, the fact that Lewin's function deals with the interval content of a single collection of notes, makes it "the conceptual predecessor of the interval-class vector (...) in Forte's work" ([114, p.44]). Lewin later studied the specific differences between his own "interval function" and other similar theoretical constructions, like Forte's interval vector or Regener's common-note function [601]. Most of the areas developed by Lewin since the sixties find their natural place in his fundamental treatise [605]. Lewin's most important abstract construction is probably the GIS (Generalized Interval System), a more elaborated version of his previous "Formal Interval System" or FIS [603]. The study of such a structure leads to a natural generalization of Forte's Set Theory, also by means of a transformational model that includes transformational graphs, networks, and isographies. Lewin's GIS model is algebraically equivalent to the structure of principal homogeneous G-sets ([1095, p.271]). By taking G to be the additive group of a vector space, one is formally lead to the same family of objects (i.e., affine spaces) that are amply discussed in the present book. An example of the way in which Lewin's theoretical constructions could be described in terms of forms and denotators was considered in Section 6.6, the second small example of 9.

Transformational networks and "their pertinence for music theory and analysis" are largely discussed in Lewin's second book [604], dedicated to Babbitt and Forte. For further applications to Lewin's transformational approach to the problem of voice leading in atonal music see [1019].

[27] See also our discussion of global compositions in Chapter 13, where the natural generalization of this and other global concepts in AST are treated.

We conclude this brief genealogy of the American music-theoretical tradition with the mention of a few noteworthy works on modern diatonic theory. This would be in perfect tune with Babbitt and Lewin's main concepts, for the topic leads to important developments in the field of mathematical music theory. As suggested by Eytan Agmon in his recent comprehensive study in the subject [12], the modern theory of diatonicism originated in some problems first raised by Babbitt [52, 53]. The study of combinatorial properties of generalized well-tempered musical systems [357] leads in a natural way to a formalization of musical structures by means of group-theoretical methods [71] and the elementary theory of numbers [202, 11, 203, 204]. A special issue of JMT[28] confirms the amount of interesting mathematical and musicological problems concerned with diatonic theory and the so-called Neo-Riemannian Theory. Both, by paraphrasing John Rahn, have transformed in the USA the field of music theory into a modern mathematical study.

11.5.2.2 Concepts and Theory A Vocabulary Switch

This section is devoted to an embedding of the Musical Set Theory constructs of *local* structures into the theory of local compositions with their forms and denotators. The global aspects will be dealt with in Section 16.2. We shall comment on the structure and performance of American Music Set Theory (the abbreviation AST also in honor of its characteristic abbreviation overhead) in Section 11.5.2.4. To be consistent with conceptual cross relations, we keep track with Robert Morris' lucid book [765]; for other references, see [180, 873], for example. In fact, Morris' presentation is quite generic, explicit and also recognized among the AST theorists.

The scheme of our proposed embedding is a translation of the AST concepts into forms and denotators, including comments on the semantic background of AST constructs. We do not claim coverage of theorems and definitions, they may be traced back in the AST literature. But a vocabulary switch of the basics is the indispensable minimum for intercultural communication.

A first large class of AST concepts is built around pitch spaces. The basic space form is n-modular pitch, i.e., the form

$$PiMod_n = PiMod_n \underset{Id}{\longrightarrow} \mathbf{Simple}(\mathbb{Z}_n)$$

defined in formula (6.42). This explicitly includes the plain integer pitch module for $n = 0$, i.e., a synonymy to $Pitch|_{\mathbb{Z}}$, but we refrain from this identification since we do not want to connotate too much of the context of $Pitch|_{\mathbb{Z}}$. Morris calls the cases $n \neq 0$ the cyclic cases, whereas $n = 0$ is called the linear case.

If form spaces are deduced from these basic spaces, and if they are related by modular identification with $0 < n$, this always works via the common projection $mod_n : \mathbb{Z}_0 = \mathbb{Z} \to \mathbb{Z}_n$ and its associated natural transformation and form morphism

$$mod_n : PiMod_0 \to PiMod_n$$

which will be assumed to be clear in the present AST context. Very often, AST deals with (k, n)-serial motives or synonymous objects; we shall explicitly include serial motives with integer pitch values, i.e., consider $(k, 0)$-serial motives

$$Ser : \mathbb{Z}^{k-1} \rightsquigarrow PiMod_0(Ser_0, \ldots Ser_{k-1}).$$

C-Spaces. For a positive natural number n, a contour space or *c-space* X of order n has form

$$c\text{-}space_n(X) \underset{Id}{\longrightarrow} \mathbf{Syn}(PiMod_0), \tag{11.57}$$

so that the form name is characterized by the generic "c-space" head and the specification of order and name X. A c-pitch (= cp, plural: cps) of X is a zero-addressed denotator $cp : 0 \rightsquigarrow c\text{-}space_n(X)(x)$ with coordinate pitch numbers $0 \leqslant x < n$. So the order does only affect the possible denotators. The musical background of such a space is encoded in the name "X". In particular, the material nature of the cps is

[28] Spring 1998, Vol. 42/2.

hidden in this name. Contour spaces just parametrize pitch in some generic context where precise pitch is not relevant, but only comparative pitch, and this is numbered by integers from 0 to $n-1$.
A zero-addressed local composition in ambient space $c\text{-}space_n(X)$ (whose elements are c-pitches!) is called a *cpset*[29]. A pcset is always of cardinality at most n.
A *contour* for $c\text{-}space_n(X)$ is defined as a serial motif which takes values in $c\text{-}space_n(X)$, i.e., a denotator

$$Cont : \mathbb{Z}^{k-1} \rightsquigarrow c\text{-}space_n(X)(Cont_0, \ldots Cont_{k-1})$$

whose coordinates are cps, i.e., evaluate to pitch numbers between 0 and $n-1$. The number k is the *length* $l(Cont)$ of $Cont$. We denote the set of these contours by $CONT_{n,k}(X)$, its cardinality is n^k. As with dodecaphonism, the index of a contour is an abstract substitute for onset time. In this sense a contour formalizes an abstract motif structure. We shall deal with the general approach to motif theory in Chapter 22, here, we just describe the AST approach. The AST group operations of contours are those in the Kleinian subgroup K_4 of $D_{k,0}$ (see the discussion of dodecaphonism following Definition 22) generated by the address retrograde $R = rev_k$ and the inversion $I = e^{n-1} \cdot -1$ which flips the interval $[0, n-1]$.
The topological classification paradigm for contours is defined by an "abstract motif" (see Chapter 22), the *comparison matrix* $COM(Cont)$ associated with a contour $Cont$ of length k. $COM(Cont)$ is the integer $k \times k$-matrix with coefficients $COM(Cont)_{i,j} = sign(Cont_{j-1} - Cont_{i-1})$. Clearly, COM is skew-symmetric, i.e.,

$$COM(Cont)^{\tau} = -COM(Cont).$$

The effect of the K_4-action on $CONT_{n,k}(X)$ upon the comparison matrix is this:

$$COM(I \cdot Cont) = -COM(Cont),$$

$$COM(R \cdot Cont) = COM(Cont)^{\rho},$$

the 180-degree rotation ρ of the matrix around its center. We have

$$COM(IR \cdot Cont) = COM(R \cdot Cont)^{\tau} = COM(Cont)^{\rho\tau} = COM(Cont)^{\kappa},$$

the co-diagonal transposition κ. In other words, the COM-map is K_4-equivariant for the above actions of K_4. We have a commutative diagram

$$\begin{array}{ccc} CONT_{n,k}(X) & \xrightarrow{COM} & \mathbb{Z}^{k \times k} \\ {\scriptstyle K_4\backslash}\downarrow & & \downarrow{\scriptstyle K_4\backslash} \\ K_4\backslash CONT_{n,k}(X) & \longrightarrow & K_4\backslash \mathbb{Z}^{k\times k} \end{array} \tag{11.58}$$

and can define the AST *segment classes* or *contour classes* as fibers of the composed map, this is a typical situation in motif theory (see Chapter 22).

P-Spaces. For a more precise pitch information, AST uses the concept of p-space. For positive natural numbers n, M, a pitch space or *p-space* $[M^{1/n}]$ has form

$$[M^{1/n}] \underset{Id}{\longrightarrow} \mathbf{Syn}(PiMod_0), \tag{11.59}$$

where the form name is characterized by the nth root of M which *defines a physical connotation of the pitch numbers.* More precisely, a *p-pitch* (or shorter: a *pitch*) of $[M^{1/n}]$ is a zero-addressed denotator $x : 0 \rightsquigarrow [M^{1/n}](x)$ with coordinate pitch numbers x that stands for a physical frequency $F_0 M^{x/n}$, the case $M = 2, n = 12$ being the 12-tempered tuning starting from F_0.

[29] In the AST, sets are called unordered sets, and sequences are called ordered sets. We do not inherit this unmathematical terminology.

The *ordered p-space interval* $ip < a, b >$ is the integer $a - b$; the *unordered p-space interval* is defined by $ip\{a, b\} = |ip < a, b > |$.
A zero-addressed finite local composition in ambient space $[M^{1/n}]$ is called (unordered) pitch set and abbreviated by *pset*, a serial motif

$$Seg : \mathbb{Z}^{k-1} \rightsquigarrow [M^{1/n}](Seg_0, \ldots Seg_{k-1})$$

is called *pitch segment* and abbreviated by *pseg*; we denote the space of length k pitch segments in $[M^{1/n}]$ by $SEG_k[M^{1/n}]$. *Pitch cycles* are the same as pitch segments, the AST only distinguishes them by different admitted group operations, see below.
The *interval content* of a pset P is the function

$$int : P \times P \to \mathbb{Z} : (x, y) \mapsto ip\{x, y\}$$

which is also written as an equivalence class of matrices according to the permutations of the pset elements.
The *interval succession* of a pitch segment Seg is precisely its derivation $INT(Ser) = D(Ser)$ defined in (11.54), the *iterated interval succession* is denoted by$INT_m(Ser) = D^m(Ser)$.
The *cyclic interval succession* $CINT(Ser)$ is $INT(C(Ser))$, the interval succession of the "cyclic extension"[30] $C(Ser) = (Ser_0, \ldots Ser_{k-1}, Ser_0)$ of Ser; $CINT_m$ is just the iteration $CINT^m$.
In p-space, AST considers *translation* $T_n = e^n$ and *multiplication* $M_n = (n)$ of pitch, and $I = (-1)$ represents *inversion*, whereas general inversions are of shape T_nI. The affine transformations $T_nM_m = e^n \cdot (m)$ on the pitch coordinator $\mathbb{Z}$ define the set of *canonical operators*, i.e., $\mathbb{Z}@\mathbb{Z}$.
Retrogression R on pitch segments is defined as with any serial motif. *Rotation* r_t of a pitch segment Ser is the serial motif $r_t(Ser)$ defined by cyclic permutation address change $i \mapsto i + t \mod k$, i.e., $r_t(Ser)_i = Ser_{i+t}$. We know from (11.54) that INT is invariant under translations, commutes with multiplications and has $INT(R(Ser)) = -R(INT(Ser))$.

Pc-Spaces. The pitch class spaces in AST are thought of as being a modular derivative from pitch spaces, and the standard interpretation of passing from the p-space $[2^{1/12}]$ to a 12-modular pc space makes it useless to introduce a more refined terminology than just "the" n-pc space, i.e.,

$$pc\text{-}space_n \underset{Id}{\longrightarrow} \mathbf{Syn}(PiMod_n)$$

for positive natural number n. A pitch class (abbreviated *pc*) is a zero-addressed denotator

$$x : 0 \rightsquigarrow pc\text{-}space_n(x)$$

in *pc-space*$_n$. The set of all pc is called the *aggregate* and denoted by U_n or just U for the usual $n = 12$.
An *ordered pc interval* between pc x and pc y is $i < x, y >= y - x$; the *unordered pc interval* $i\{x, y\}$ is the smallest nonnegative representative in $\mathbb{Z}_n$ of the differences $i < x, y >, i < y, x >$. AST writes $ic\{k\}$ resp. $ic < k >$ for the set $\{(x, y) \in U^2 |\ i\{x, y\} = k\}$ resp. the set $\{(x, y) \in U^2 |\ i < x, y >= k\}$ and says "$i\{x, y\}$ is $ic\{k\}$" resp. "$i < x, y >$ is $ic < k >$"[31].
An (unordered) pitch class set *pcset* is a local composition in ambient space *pc-space*$_n$. A pitch class segment *pcseg* is a serial motif with coordinates in *pc-space*$_n$.
Operations and TTOs (TTO = Twelve Tone Operators). Operations on *pc-space*$_n$ are, as already with the p-spaces, the affine transformations on the ambient space, i.e., $\mathbb{Z}_n@\mathbb{Z}_n$, mostly used for $n = 12$. Notation is again T_m and M_m for translation and multiplication. The group $\overrightarrow{GL}(\mathbb{Z}_{12})$ is denoted by TTO, and its 48 elements are called *TTO operators*, multiplication M_5 is notated by M. A *set class SC* is an orbit $TTO \cdot A$ of a pcset A.
AST fixes $n = 12$. Given two pcsets A, B the *interval class content vector* is the 7-tuple in square brackets $ICV(A, B) = [ICV(A, B)_0, \ldots ICV(A, B)_6]$ with $ICV(A, B)_0 = card(A \cap B)$, $ICV(A, B)_j =$

[30] This is not AST terminology.
[31] The difference between unordered and ordered interval notation is not AST, but useful.

$\frac{1}{2}card(A \times B \cap ic\{j\})$ for $0 < j < 7$. The *interval vector* is the 12-tuple in square brackets $IV(A, B) = [IV(A, B)_0, \ldots IV(A, B)_{11}]$ with $IV(A, B)_j = card(A \times B \cap ic < j >)$, also called the *multiplicity* of position j and denoted by $Mul(A, B, j)$. The "complement theorem" is the formula

$$MUL(A\hat{}, B\hat{}, n) = MUL(A, B, n) + 12 - (card(A) + card(B))$$

which is a trivial consequence of the facts

$$MUL(X, Y, n) = MUL(Y, X, -n)$$

and

$$MUL(X, Y, n) + MUL(X\hat{}, Y, n) = card(Y).$$

The special case $ICV(A, A)$ is denoted $ICV(A)$. Clearly, the interval class vector of a pcset A is invariant under inversions and translations. The interval vector $IV(A, B)$ is invariant under translations and exchanges its arguments under Inversion.

A pcset A is called *invariant under a TTO K* if it is a symmetry of this local composition. The order of $Sym(A)$ is called the *degree of symmetry* of A. Define[32] $Sym_i(A, B) = \{x = e^t \cdot (i) \in TTO| \ x(A) \subset B\}$ and $sym_i(A, B) = card(Sym_i(A, B))$. Then the *invariance vector* of A is the 8-tuple

$$\begin{aligned} sym(A) =< & sym_1(A, A), sym_{11}(A, A), sym_5(A, A), sym_7(A, A), \\ & sym_1(A, A\hat{}), sym_{11}(A, A\hat{}), sym_5(A, A\hat{}), sym_7(A, A\hat{}) > \end{aligned}$$

which is an invariant of the TTO-orbit of a pcset. The T_n/I *set class* of a pcset is its orbit under TI_{12}. For a given subgroup G of TTO, The *G-prime form* of a pcset is the first representative of its G-orbit in the lexicographic order defined in Section 11.3.5.

Abstract inclusion of pcset A in pcset B means that there is a member in the orbit (SC) $TTO \cdot A$ being included in B, i.e., TTO contains a morphism $f : A \to B$ of local compositions. The number of (categorical) subobjects defined by a TTO $f : A \to B$ is the *"embedding number"*[33] $EMB(A, B)$.

Exercise 17 *Prove that the last assertion is equivalent to Morris' definition "$EMB(A, B)$ = number of pcsets in the SC of A which are contained in a pcset member of the SC of B".*

SC X is the *abstract complement* of SC Y iff they are the orbits of complementary pcsets.

Pcset Similarity. This subject will be dealt with in Chapter 22.

Pc Segments. As with psegs, the mth *interval succession* of a pcseg S is the mth derived serial motif $INT_m(S) = D^m(S)$. The *cyclic interval succession* $CINT(S)$ of a pcseg S is the derived serial motif $INT_m(C(S))$ of its cyclic extension as defined above for psegs; as above, $CINT_m(S)$ is the mth iteration of $CINT(S)$.

Retrograde R and *rotations* and pitch operations from group TTO or the monoid of affine operations are defined mutatis mutandis as above for psegs.

Given two pcsegs X, Y of lengths $l = length(X) \leqslant k = length(Y)$, AST calls X *(literally) included in* Y iff there is an order preserving address change monomorphism

$$\iota : \mathbb{Z}^{l-1} \rightarrowtail \mathbb{Z}^{k-1} : e_s \mapsto e_{\iota(s)}$$

such that $X = Y \cdot \iota$. Abstract inclusion means combining address-change monomorphisms with TTO-induced morphisms f on pitch space. So abstract inclusion is a fractional morphism f/ι of denotators (viewed as singleton local compositions).

A *segment class SGC* of a pcseg S of length k is the orbit $D_{k,12} \cdot S$ (see the discussion of n-phonic series following definition 22). Orbits generated by dodecaphonic series are called *row-classes*.

[32] This is not AST terminology.
[33] Terminology not explicit in AST.

TTOs. Chapter four of Morris' book [765] is entirely devoted to the study of the group $TTO = \overrightarrow{GL}(\mathbb{Z}_{12}^2)$. The justification for considering TTOs and not any general permutations is precisely the affine character ([765, p.125]). The TTO description includes elementary facts about finite groups, the formula for the inverse of a TTO, and cycle representations of the TTO as a permutation group on aggregate U_{12}. The conjugacy classes of subgroups of TTO are discussed and listed ([765, Appendix III]).
Finally, *sets of operations*, i.e., self-addressed local compositions in $pc\text{-}space_{12}$ (subsets of $\mathbb{Z}_{12}@pc\text{-}space_{12}$) consisting of TTOs are considered, together with *strings of operations*, i.e., the serial motives

$$OpStrg : \mathbb{Z}^{k-1} \rightsquigarrow \mathbb{Z}_{12}\tilde{@}pc\text{-}space_{12}(OpStrg_0, \ldots OpStrg_{k-1}),$$

see the technique of address killing defined in Section 11.2. These constructs are used of compositional design, i.e., the abstract configurations of pcs, pcsets, pcsegs, and operators to build background structures for musical compositions. See our discussion in Section 16.3 for more details on global aspects in AST.

11.5.2.3 Software for Musical Set Theory

The thoroughly combinatorial and algorithmic character of AST has led to several implementations of its objects and operations on the level of computer applications, the first ones being already described by Allen Forte [330] in 1970. More recent applications are, for example, the CMAP package developed by Craig Harris together with Alexander Brinkman [443] (see also [180] for further references) and updated by Peter Castine for Macintosh in C++ [180], including MIDI input. His class hierarchy is shown in Figure 11.11 Another implementation in LISP has been presented by John Amuedo [29]. The package includes a computing

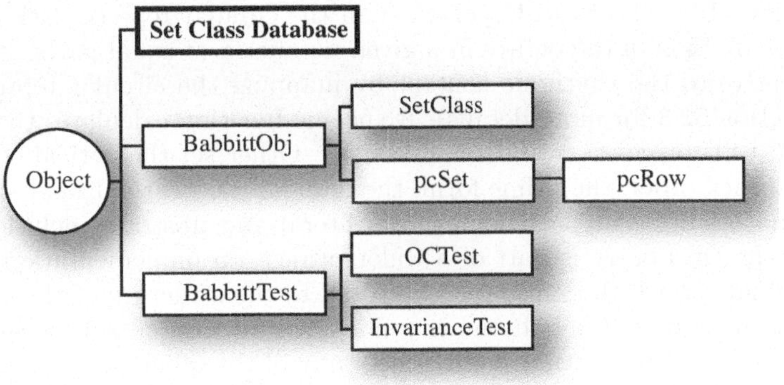

Fig. 11.11. Built on the object-oriented Think Class Library, Castine's principal subclasses are (1) a SetClass-Database class managing database requirements, (2) a BabbitObj class describing the basic objects of AST, and (3) a BabbitTest class which gives a generic framework for operations to be performed on objects.

environment SET-SLAVE for general AST calculations in Common LISP for Macintosh computers.

SET-SLAVE computes several "normal forms" of zero-addressed local compositions c of pitch classes from AST:

- Rahn's normal form NF: This is a standard sequential presentation of c, using a not very systematic cyclic permutation[34] of the ordered sequence of c in $[0, 11]$. (Taking the lexicographically first sequence would be the better solution.)

[34] Take the largest interval of successive pcs on the circle $\mathbb{Z}_{12}$, start the sequence with the right pc with such a maximal interval, and such that this is the lexicographically first among these maximal interval right-hand pcs. Take the smallest pc with these properties.—What a mess!

- Straus' zero normal form ZNF. This is just the transposition-orbits, represented by $e^{-t}NF, t =$ first pc of NF. Amuedo also includes the evident binary codification (BNF) of this normal form via the characteristic function on $\mathbb{Z}_{12}$, and its decimal representation DNF.
- Forte's prime form PF: This is a standard notation of TI_{12}-orbits of zero-addressed local compositions of pitch classes, using the lexicographically first orbit representatives.
- Forte's interval class vector IVC, i.e., the sequence of cardinalities

$$\langle i_1, i_2, i_3, i_4, i_5, i_6 \rangle$$

  with $i_j = card(\{$ unordered intervals (x, y) in c with $i\{x, y\} = j\})$
- Amuedo's decimal normal rotation DNR. This is the refined $ZNF.x$ from ZNF, where the "bass note" is codified by the "rotation index" x for the power of the cyclic permutation ("rotation") $(1, 2, 3, \ldots card(c))$ of the ZNF sequence bringing the "bass note" pc to the first position of the sequence.—This is nothing but the distinction of a particular pitch, stated in terms of inversion of pitches (successive octave increase of a pitch) that yields a new bass note.

Like CMAP, it also computes Forte's "K" and "Kh" inclusion relations (see discussion in Section 16.2 for these relations). Amuedo's thesis also includes software for real-time harmonic analysis. The SET-SLAVE is combined with three interactive MAX programs. This package contains a CHORD-CLASSIFIER, a SCALE-FINDER, and a SCALE-MONITOR for analysis of chords, and analysis resp. visual inspection of scales. See Chapter 25.2 for more details.

Recently a library specifically devoted to American Set Theory has been integrated in Ircam's visual programming language OpenMusic based on CommonLisp/CLOS (see Chapter 52 for a presentation of this computer-aided system for music composition and analysis) by Carlos Agon and Moreno Andreatta. The library, called Dn (from the Dihedral group of order 2n), is a generalization of a previous Lisp library of Janusz Podrazik (http://www.mracpublishing.com) on AST. The originality of the OpenMusic implementation consists in the manipulation of set-theoretical operations in the circular representation, with the possibility of taking into account any division of the octave in a given number n of equal parts. It also allows us to switch from the pitch perspective to the rhythmic content by mapping the circular representation into a 'chord' or into a voice (see Section 52.3 for more details). A specific function calculates the numbers of equivalence classes of k-chords (up to transposition and/or inversion). Other set-theoretical concepts which have been implemented are the normal order, the prime form, the interval vector, the literal and abstract complement, the Z-relation, combinatoriality and partitioning, the literal and abstract inclusion and the sub-complex Kh. We mention that the Dn library is part of a wider project on implementation of algebraic methods in OpenMusic which originated with the Zn library (devoted to the structure of the cyclic group of order n in the classification of chords and of tiling rhythmic canons; see Section 16.2.3 for the presentation of Vuza's model of rhythm) and that will be continued with the study of the affine group and of the symmetric group (in the pitch as well as in the rhythmic domain).

11.5.2.4 Comments

We want to conclude this section with three short comments on the AST, a mathematical, a conceptual, and a model-oriented one.

Mathematically, the AST is a very special achievement. Its concepts are thoroughly out of date from the point of view of 20th century mathematical conceptualization. Even the most standard concepts in group theory are ignored, e.g., the index of a subgroup, or the semi-direct product of translations and multiplications which explain a large portion of the TTO by the corresponding short exact sequence

$$1 \to e^{\mathbb{Z}_{12}} \to TTO \to \mathbb{Z}_{12}^{\times} \to 1.$$

Also the concept of a group action and corresponding elementary facts such as orbit cardinalities in relation to isotropy groups does not appear. Although the theory of categories has been around since the early 1940s and is even recognized by computer scientists, no attempt is visible in AST to deal with morphisms between pcsets, for example. We have seen certain germs of this direction in the definition of abstract subsets, but this is not what leads to a powerful theory of relations between local and/or global musical objects. It would be important to apply the above vocabulary switch in order to adopt the findings of AST to workable mathematical formalism such as it has been used by Fripertinger in Pólya and de Bruijn enumeration theory, for example. It would also be necessary to confront the AST approach with the many other parameters which define musical events, such as onset, duration, loudness, glissando, and crescendo, just to name a few important ones. The AST has never dealt with all these parameters in a global conceptual framework. However, the work of Dan Tudor Vuza or Anatol Vieru does heavily recommend such an extension. We shall deal with these directions in Chapter 21 on metrical and rhythmical global compositions. Nonetheless, many valid lists of isomorphism classes, such as chord classes under translations, translations and inversions, or the full TTO group as well as conjugacy classes of subgroups of TTO have been established. We also have to recognize that the operationalization of musicological concepts has been completely realistic insofar as computer programs and algorithms have been provided.

The conceptual comment must take care of a dramatic need for precise musicological concepts as tools for dodecaphonic analysis and of its theoretical extension in atonal theory. European music theory has only very rarely shown up in this domain, perhaps best in the work of Herbert Eimert and Iannis Xenakis. But the mainstream of after world-war II European musicology had turned towards dialectic mumbo-jumbo and far-out aesthetics and transcendental black-box-theories. So Americans had to start from scratch with precise conceptualization of even the simplest concepts such as pcsets, pcsegs and their classes. Whatever the status of a baby theory the AST concept framework might be, it is an indispensable reset of a rotten conceptualization in musicology where even the most elementary things are blurred.

The question of theoretical modeling is a difficult one. It appears that modeling has predominantly been oriented towards and useful for compositional strategies. Morris' composition designs are a real enrichment to grasp the complex construction of precise sound aggregates when starting from pcsets and similar elementary local compositions. Also the analytical use of the AST language is a considerable one. We are happy that finally, it was possible to simply talk in a precise jargon about analytical problems of atonal and also tonal music (e.g., in Amuedo's work). But there remains a big lack of models in the sense that beyond descriptive tasks, the AST language has very seldom led to musicological modeling. Most theorems of AST are of strictly combinatorial nature. We come back to such achievements in Chapter 25.3.

So we could summarize the AST achievement as a necessary but far from sufficient attempt to escape decadent and impotent European musicology.

11.6 Esthetic Implications of Classification

Summary. Group-theoretical classification is a central issue of poetology. It is used as an esthetical tool in analysis and composition.

– Σ –

This subject is hard to understand from scratch. In many talks and discussions, I have experienced the disbelief that something like abstract group theory, together with group operations and orbit set construction, could have the slightest connection with something as intuitive, emotional or fuzzy-like poesy. This section is not written to replace passionate discussions about the legitimation of abstract algebra in poetology. But we can help in channeling them and focusing attention on the critical points.

To begin with, let us localize the problem on musical topography. As groups pertain to mental reality, we do not discuss psychological or physiological aspects of the phenomenon but only mental perspectives. In turn, these latter may or may not relate to actions on other levels of reality. Perhaps the confusion between mental facts and their psychological or physiological correspondences is one of the important sources of irritation. We shall discuss this subject in extenso in especially Section 30.2.4.

On the mental level, there are three communicative allocations of poetic instances: poiesis, neutral level, and esthesis. The composer may use certain tools to express esthetic categories, and the listener or analyzer may apply appropriate tools to perceive esthetical categories. Esthetic properties may also be defined on the neutral level of the proper work, but esthetic categories emerge on the human consciousness and therefore primarily pertain to the poietic or esthesic level to which we stick from now on.

A strong esthetic category is poetic structure or poeticity. In fact, prosaic esthetics could hardly ever be really esthetic in music, even if we deal with so-called musical prose in the sense of Schönberg or Wagner. They characterize it as an antithesis to metrical regularity[35], but metrics is by far not the only instance of poeticity; musical prose can be perfectly poetic in the sense of the word as obtained by Jakobson's revolutionary approach which we introduce below.

The importance of group theory to poeticity is that it makes the instance of paradigmatic transformation equivalence precise, and this is a core characteristic of poeticity of music. In fact, in the linguistic context, paradigmatic transformations are not so rich and complex as with music. Orbits which are generated via the multidimensional forms of local musical compositions (motives, chords etc.) can only be controlled by explicit group theory. This is the crux and blessing of music. Understanding musical poeticity is far more than a business of evident transformations, and this is what perhaps causes another irritation among those who would prefer that the genealogy of poesy be as straightforward as the evidence of its power upon humans is manifest.

After an introduction to Jakobson's poetic function we want to perform a poetological analysis (Schubert/Stolberg) as well as a poetological (re)construction (Mazzola/Baudelaire). We shall stress the group-theoretical aspect and give a critical comment on this procedure.

11.6.1 Jakobson's Poetic Function

Summary. The turning point of Roman Jakobson's poetic function in poetic analysis. Immanent analysis versus imported knowledge. Construction of poetic structure by use of orbits under symmetry groups.

– Σ –

In his famous paper [500], Roman Jakobson has isolated the poetical function as a central issue of esthetic structure in language. He has also pointed out that the poetic structure is not restricted to poetry, it may be present in mnemotechique or publicity or political rhetorics as well. It is interesting to note on which basic linguistic structures and functions poeticity is located. Jakobson lists six structures: sender, receiver, context, message, contact, and code. He relates these structures to the functions of (in the same order) emotivity, conativity, referentiality, poeticity, phaticity, and metalanguage. So poeticity is the function of language qua message. And poeticity is one important issues of esthetics, together with emotional functionalities, for example.

However, in music, we do not have automatic semantics and therefore, semantical enrichment by techniques of inner-systemic, i.e., autonomous constructions is an essential procedure to elaborate the musical sign system. The message must be strongly constructed in absence of contextual and predefined semantics.

Jakobson defines poeticity—without further empirical deduction—as follows:

Definition 35 *The poetical function projects the principle of equivalence from the axis of selection to the axis of combination.*

The terms in this definition have the following meaning. The "axis of selection" is the paradigmatic axis of language, i.e., the relation among its signs by topological or transformational equivalence as discussed in Chapter 10. The "axis of combination" is the relation among signs in their linear ordering of the syntagm. So we are given two spaces (axes): the abstract ensemble of all possible signs and a concrete string of signs which are juxtaposed in a linguistic message. The first shares the associative relation of paradigmatic fields whereas the second shares the adjacency relation of syntagmatic chains. The principle of equivalence on the paradigmatic axis is association whereas this principle on the syntagmatic axis is relative position on the syntagmatic chain. The first one is a given relation, the second one is not. Apart from immediate adjacency, no a priori relation is given. This enforces the construction of syntagmatic equivalence based on syntagmatic

[35] "...ihr primäres Kennzeichen ist das Abweichen von Normen klassischer musikalischer Metrik...", [907].

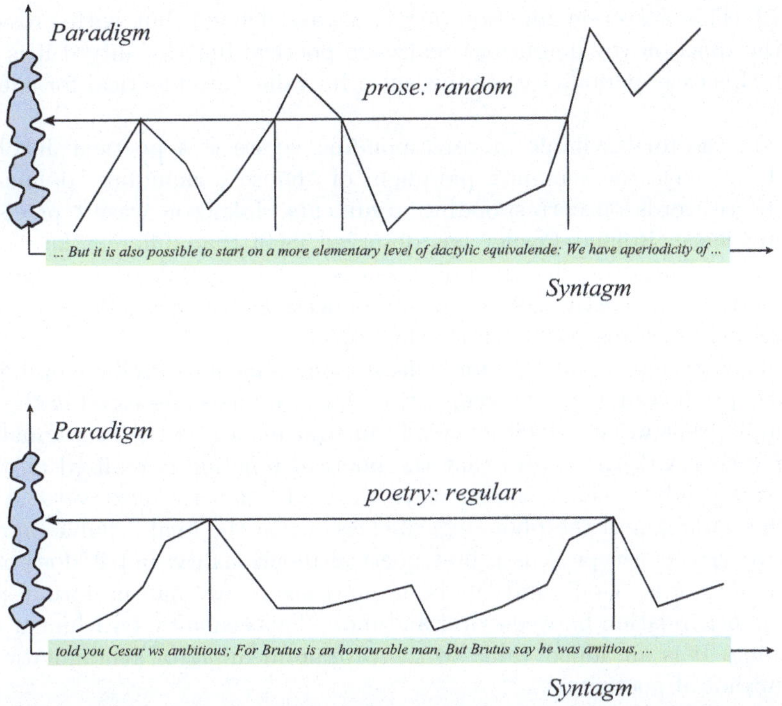

Fig. 11.12. The Jakobson function associates the paradigm with each syntagmatic unit. In case of prose, this function is "random" whereas in case of poetry, this function is poetical, i.e., periodic with respect to a selected equivalence between syntagmatic positions, mostly a periodic relation.

adjacency. The typical example is a periodic distribution of syllables in a verse form, such as a hexameter, the sequence of five dactylic metrical units (/ ⌣⌣) (stressed/unstressed or long/short) plus the ending trochean unit (/ ⌣):

/ ⌣⌣ / ⌣⌣ / ⌣⌣ / ⌣⌣ / ⌣⌣ / ⌣
/ ⌣⌣ / ⌣⌣ / ⌣⌣ / ⌣⌣ / ⌣⌣ / ⌣
...

In such an arrangement of hexametric rows, equivalent positions in syntagm are the syllables on a determined column of the verse matrix. But it is also possible to start on a more elementary level of dactylic equivalence: We have a periodicity of successive triples of syllables, for each dactylic meter one unit, and then equivalence of metrical position means the first, second or third place within each dactylic meter, as may be shown in a more abstract "metrical versification" scheme:

/ ⌣⌣
/ ⌣⌣
/ ⌣⌣
...

Given such a syntagmatic periodicity (just to fix a concrete syntagmatic equivalence relation), the Jakobson projection is related to a special function jak from the syntagmatic axis to the paradigmatic axis. It associates with each syntagmatic unit u_s its paradigm

$$jak(u_s) = \text{paradigm of } u_s,$$

see also Figure 11.12. This Jakobson function *jak* is always defined, but with prosaic texts, it shows no regularity whereas the function (by definition) realizes a poetical functionality if it is periodic with respect to a syntagmatic (in this case: periodicity) equivalence: Its values are identical for equivalent arguments on the syntagmatic axis.

So the rhyme on the final syllable of corresponding verses is a poetical Jakobson function on the phonological level. If we have the semantic paradigm of "house", including "domicile", "home" etc., we would have two of these words on corresponding arguments. Jakobson wasn't precise either on the level of paradigmatic equivalence. Roland Posner has observed [860] that different levels of poeticity must be distinguished, such that besides the "horizontal" poeticity following Jakobson, a "vertical" poeticity must be considered. This vertical perspective relates to the connections between different semiotic aspects of signs when coupled by poetical functions. See [703] for this topic.

Vertical poeticity is an important tool for semantic enrichment, a fundamental feature of poeticity. It relates to the semantic production mode by symmetries which we have discussed in the beginning of Chapter 8. Suppose, for example, that we are given a verse form equivalence on the syntagmatic axis. Suppose that we have alternating verses A,B,A,B,... and that the poetical function is realized on the phonological level by a rhyme at the verse ends, verse A ending with "...God", and the next verse A ending with "mob". This projection of the paradigm of the phonological class /o/ in the final syllable onto the syntagmatically equivalent ends of two verses of type A is a first poetical functionality, but it does more: it correlates the semantic contents of the words "God" and "mob" in a stronger way than in normal prose. We are given a poetic index to a semantic relation between God and mob. This semantic enrichment is a conditio sine qua non for poetic meaning. It is an added value to the prosaic meaning of syntagmatic units. The following three facts give an upshot of poeticity:

1. Poeticity is based on an "unfolding of correspondences in time" or "a tale of symmetries".
2. By means of poetic coupling, semantics is enriched, not only stressed: We obtain an "added value" of meaning.
3. Classification (building equivalence fields) as a prerequisite to perception of equivalence is a central reference of poeticity.

In the following examples, we shall deal with a concrete and precise realization of the Jakobson principle, and we shall do this for the transformational paradigm, mainly in the sense of group actions.

11.6.2 Motivic Analysis: Schubert/Stolberg "Lied auf dem Wasser zu singen..."

Summary. This is a prototypical motivic analysis relating motives to dactylic metrics. It reveals a fundamental regularity in the distribution of three-tone motives in Schubert's composition.

$- \Sigma -$

We want to analyze Leopold Stolberg's poem *"Lied auf dem Wasser zu singen, für meine Agnes"*, written in 1782 and set to music by Franz Schubert 1823, see Figure 11.13 for the vocal score. Science of literature [865, 866] has recognized that in the eighteenth century, no poem realizes symmetry—in the sense of congruence between emotional disposition and esthetic form—so perfectly as these lines. At first sight, the rigorous articulation catches our eye: three strophes of six verses each, ending on identical rhymes, each threefold. Whereas the identical rhyme produces the intended effect of monotony, the dactylic meter achieves another equilibrated movement: the regularity of undulation, again a triple measure.

Whereas in the first strophe, the soul moves rather passively "wie ein Kahn" "auf der Freude sanft schimmernden Wellen", in the second strophe, "atmet die Seele" more actively, namely in a series of four polar word couples: Über/Unter, westlich/östlich, Wipfel/Zweige, Kalmus(säuselt)/Schein (winket). The objects of this world are arranged in a symmetric fashion around something like "meine athmende Seele". The creative power of the poetic Ego incorporates its environment.

What happens to the Ego in the exterior object world, is reflected in the soul of the singer. The way this song accomplishes this identity turns it into a singular artistic achievement [865].

Schubert's setting (op. 72) translates the word's movement into continuous movement of sound. The composer succeeds in establishing a fascinating identity between text and musical transformation, a result which Moritz Bauer [92] has called a masterpiece, one of Schubert's best performances.

After this reference to traditional humanities, we want to start our analysis with the textual poeticity and then try to recognize the corresponding poeticity in music. This is the text:

LIED AUF DEM WASSER ZU SINGEN,
FÜR MEINE AGNES
Mitten im Schimmer der spiegelnden Wellen
Gleitet wie Schwäne der wankende Kahn;
Ach, auf der Freude sanftschimmernden Wellen
Gleitet die Seele dahin wie der Kahn;
Denn von dem Himmel herab auf die Wellen
Tanzet das Abendroth rund um den Kahn.
Über den Wipfeln des westlichen Haines
Winket uns freundlich der röthliche Schein;
Unter den Zweigen des östlichen Haines
Säuselt der Kalmus im röthlichen Schein;
Freude des Himmels und Ruhe des Haines
Athmet die Seel' im erröthenden Schein.
Ach es entschwindet mit thauigem Flügel
Mir auf den wiegenden Wellen die Zeit.
Morgen entschwinde mit schimmerndem Flügel
Wieder wie gestern und heute die Zeit,
Bis ich auf höherem stralendem Flügel
Selber entschwinde der wechselnden Zeit.

We discuss two examples of the Jakobson function. The first relates to the first strophe. We are given the syntagmatic equivalence of feminine and masculine ending syllables of verses 1,3,5, and 2,4,6. On these two classes of positions, we have two phonological paradigms of (1) a trocheic meter / $\smile$ with its two identity paradigms /Wel/ and /len/ and (2) the stressed / with identity paradigm /Kahn/. These two equivalences relate the corresponding verses and their corresponding units: The phonological equivalence /Kahn/ points to the semantic unit of the word "Kahn" (boat).

On verse 2, this Kahn's activity of "gleiten" (sliding) is compared to the same activity of "Schwäne" (swans). On verse 4, the same activity (also identified on the first position of the verse) relates "die Seele" to the said "Kahn". Therefore, two different object categories "Seele" and "Schwäne" are loaded with additional semantics: they slide in the same way as the "Kahn" does. So three object categories: a boat, an animal, and a psychic entity are moving in the same way. This metaphoric intensification sets forth a strong semantic enrichment qua effect of this poeticity.

The second example relates to the second strophe. A phonological poeticity is set up which is isomorphic to that described before: equivalence on verses 1,3,5, and 2,4,6. The turning points of "Heines" and "Scheins", respectively, produce symmetry correspondences of the word couples "Über den Wipfeln"/"Unter den Zweigen", "westlichen"/"östlichen", "(Schein) winket"/"(Kalmus) säuselt". This turns out to be more than a syntagmatic coupling of unrelated objects: Each pairing is a paradigmatic symmetry on the semantic level!

The first is a spatial localization pairing of vertical antipodes relating to the extremal positions on a tree, the second is a pairing of geographic antipodes, the third is a pairing of distinct sensorial modalities: visual-showing against auditive-whispering, the latter pairing again coupled with the other vertical pairing: visual activity is from above ("Über...Winket... der Schein"), auditive activity is from below ("Unter...Säuselt der Kalmus"). This ensemble of polar symmetries presents semantic equivalences. Formally, the pairings are two-element orbits of positions which are arranged by the basic poeticity on the phonological level.

The ensemble's added semantical value is this: We recognize a group of instances being arranged around a central point of rotation which switches the polar positions into one another: west into east and vice versa,

Fig. 11.13. Voice part from Franz Schubert: *Lied zu singen auf dem Wasser, für meine Agnes*, op. 72 [952], with kind permission of Bärenreiter/Henle.

above into below, and vice versa, etc. The poetical function sets up a center of rotation, a force which turns things around, but a force which remains unnamed. However, in the last couple of verses, we are finally informed of the moving force: It is "die Seele" which "athmet" "Freude des Himmels" (from above!) and "Ruhe des Heines" (from below); in other words, the center of rotation is the breathing soul, metaphor for the poetic Ego. This one will become evident in the third strophe. Only there—at the end of a successive and dramatic unvealing process—are we provided with the basic, nominative form of the pronoun "Ich".

Let us now look for *poetical functionality on the musical level.* The Schubert composition is written in 3/4 time, and we may capture all durations in integer multiples of 1/16 notes. So 12/16 time suggests that we introduce onset of form $OnMod_{12}$. Together with octave periodicity, we are led to work in form $OnMod_{12} \oplus PiMod_{12}$. Relating to the dactylic text as discussed above, we want to group melodic parts according to the three dactylus groups in each verse. This gives a grid of 6 times 3 motivic units, corresponding to six verses per strophe and three dactylus meters per verse (the endings are not considered). Each motivic unit will be given its natural structure of a three-element motif M in $OnMod_{12} \oplus PiMod_{12}$. Where we have the repetitive sixteenth note motives (bars 2, 5, for example), we take the three-element motives with eighth note onset distance, conformal with the accentuation. Figure 11.14 shows the grid as well as the isomorphism class number of the three-element motives as listed in the class list of Appendix O.

We take the full affine group $\overrightarrow{GL}(\mathbb{Z}^2_{12})$ as equivalence principle. This is by no means obvious or mandatory, so we should discuss the choice now. The Jakobson function has to be evaluated along the syntagmatic dactylus grid. The values have to be taken in a paradigmatic space which is, contrary to the linguistic situa-

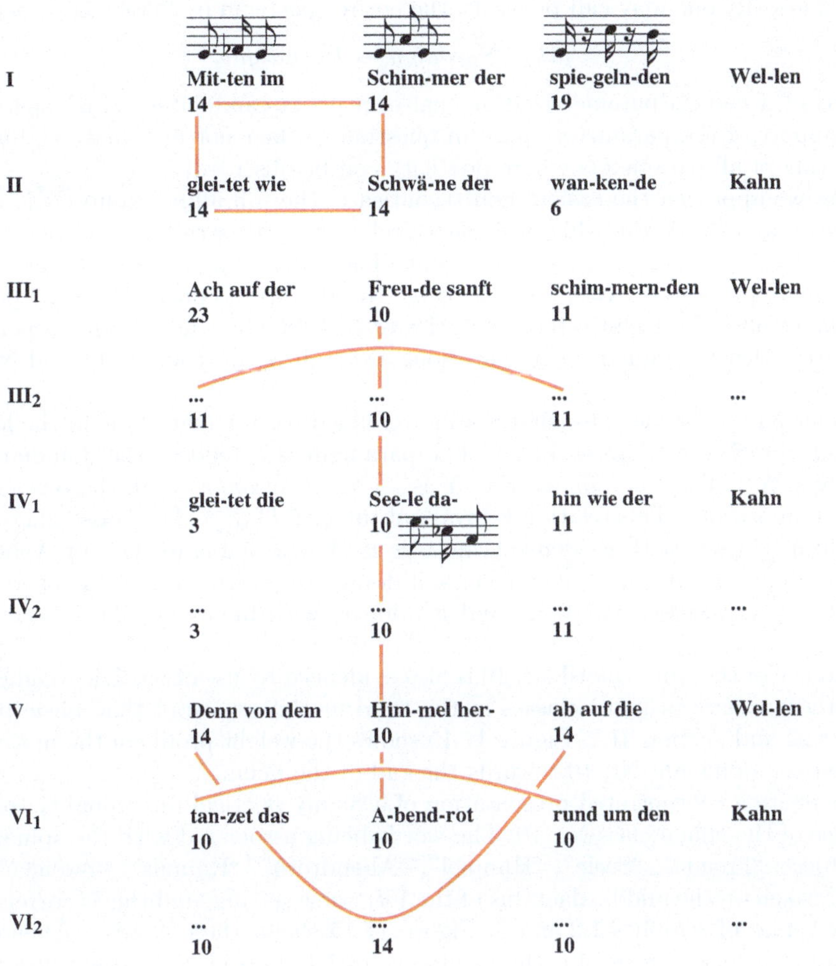

Fig. 11.14. The highly symmetric arrangement of motive classes in the dactylus grid of the voice score in Schubert's setting of Stolberg's poem "Lied auf dem Wasser zu singen, für meine Agnes".

tion, not common or standardized. In music, the idiolect of an individual composer can vary considerably, in fact, it is a highly characteristic feature of a good composer to realize an individual brand of paradigmatic equivalences. The problem is that the analytical work cannot, in this case, rely on known paradigms, and it isn't necessary to copy the poietic standpoint to the esthesic perspective. More precisely, we are not going to judge upon what would be the best analytical choice of paradigmatic equivalence in this piece of music. We just start an experiment (of the mind) and set up a first approach to equivalence.

This forwards an important methodological question: Where should one start on the paradigmatic analysis if no pre-selection is available? Is there the "right" selection against "wrong" alternatives? Globally seen, we have an entire bunch of paradigmata to be applied. However, on the level of group actions there is a hierarchy of orbit size or group size, respectively. In order to decide which group is the best one (if it is any unique one!), we should follow the line of this size hierarchy. One one end, the identity group will very improbably instantiate a poetical functionality (though we have exactly this situation on the phonological level described above). On the other, complete identification (by the full permutation group on $\mathbb{Z}_{12}^2$) of all function values will clearly produce poeticity, but this is a trivial case which is definitely unspecific. So the idea is to start at this end where poeticity is most probable, and then to refine the group choice until non-automatic poeticity emerges.

More systematically one may call poeticity the entire spectrum of Jakobson's poetical functions

$$jak_\pi : Syntagm \to Paradigm_\pi$$

for all paradigms of a certain parametrized arsenal of a priori candidates. This "spectral analysis" would yield a poetic support of the parameter space in question (in fact not one analysis, but a whole variety of analyses), consisting of all parameters where poeticity can be observed.

In this vein, we approach the easiest non-trivial case: the full affine group $\overrightarrow{GL}(\mathbb{Z}_{12}^2)$. Its orbits are 26 in number and we may ask whether this equivalence relation yields structures beyond random functionality on the dactylic grid. Figure 11.14 shows the result. The Jakobson function of this paradigm has highly symmetrical (regular) properties. Above all, the orbit Nr. 10 appears on 10 of the 27 positions, this is 37%, against a probability of 22.7% to hit a representative of this orbit[36]. And it shows a perfect axial symmetry in the dactylic grid. Other symmetric configurations are realized for the Nr. 14 and Nr. 11 orbits, though not globally.

The dominant Nr. 10 "spine" also shares a more refined paradigmatic role in the following sense which will be made more precise in Chapter 12.2.2. The paradigm is extend to the dominance relation between equipollent motives: $X > Y$ (X *dominates* Y or Y is a *specialization* of X) iff there is a surjective morphism $X \to Y$ of local compositions. This relation is invariant under $\overrightarrow{GL}(\mathbb{Z}_{12}^2)$. The Hasse diagram[37] of this relation on the isomorphism classes of three-element motives is shown in Figure O.2 of Appendix O. Under the specialization paradigm, Nr. 10 is the generic class, it dominates every other class. Therefore, the spinal class is also the generic specialization paradigm, and it appears with highly significant overhead above a priori probability.

The importance of the spinal motif Nr. 10 is also evidenced by use of motivic weights, numeric functions which measure the presence of motif classes in the total motivic space of this music piece. Details can be found in Chapter 22 and Section 41.2. Figure 11.15 shows the weight profile of the motives over the dactylic grid, in particular the dominant Nr. 10 towards the end of the poem.

So these facts are a concentrated presentation of a highly poetical functionality together with a motif-weight dominance of the spinal class Nr. 10. The words being associated with the spinal dactylus are nouns of the poetic subject: "Freude", "Seele", "Himmel", "Abendroth", "Kalmus", "Zweigen". The climax of the motivic weight is taken on the middle dactylus of the VI_1 verse setting, and this is correlated with the middle of the three last verses of strophes 1,2,and 3. Figure 11.15 shows these words: "Abendroth", "Seele", and "(Ich) entschwinde", a locus where also the symmetry of Nr. 10 and Nr. 14 becomes perfect. Within this perfection also appears the subject in its explicit form in verse V of the third strophe:

Bis ich auf **hö-** he-rem **stra**-len-dem **Flü**-gel
Sel-ber ent-**schwin**-de der **wech**-seln-den **Zeit**.

This analysis is fairly neutral: It is scarcely esthesic and limits itself to a motive structure which is established by means of a classification list and its extension to the specialization Hasse diagram. In particular, no intention from the side of Schubert has been subtended. Rather is the question emerging from this analysis whether Schubert's motif configuration is an unconscious instantiation of a highly poetic expressivity.

The compactness of this analysis should however not entail that the propedeutical work of classification and specialization data be underestimated or neglected. To understand this analysis, a preliminary understanding of classification is indispensable. Classification in fact has a deep cognitive impact which we do not yet understand: In which way can poetic feeling be generated through group actions and the associated orbits of equivalence? A first answer to this question will be given in Chapter 30.

11.6.3 Composition: Mazzola/Baudelaire "La mort des artistes"

Summary. This section describes the use of the 26 isomorphism classes of three-element motives in the construction of a composition on the basis of Baudelaire's poem.

[36] The total of $\binom{144}{3} = 487\ 344$ motives in $\mathbb{Z}_{12}^2$ contains 110 592 motives of the orbit of Nr. 10, this is $card(\overrightarrow{GL}(\mathbb{Z}_{12}^2))/card(Aut(Nr.10)) = 663\ 552/6 = 110\ 592$.

[37] The minimal diagram whose transitive closure is the given relation, see Appendix C.2, definition 135.

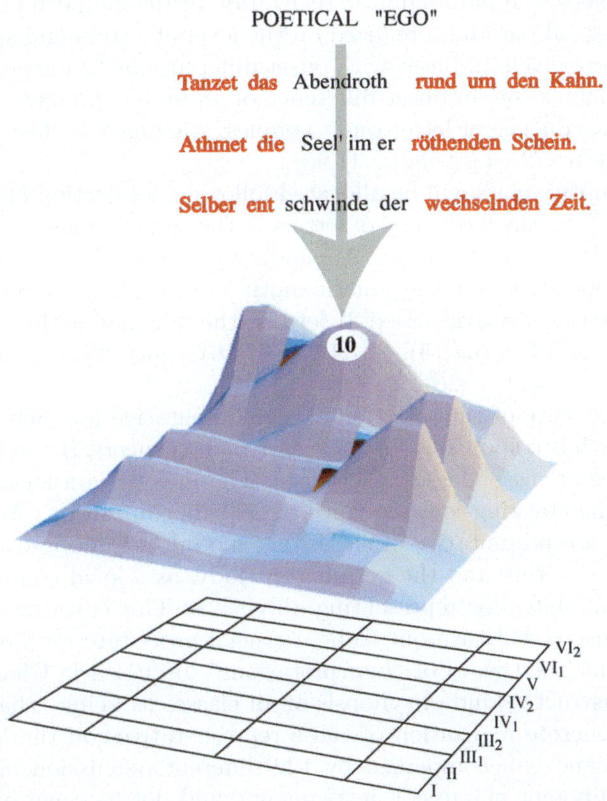

Fig. 11.15. The weight profile of the three-element motive classes as a quantification of the musical paradigm of motives, and related to the corresponding textual instances of the poetic Ego.

– Σ –

This section deals with paradigmatic composition techniques applied in scherzo movement three "Poem of Wind" of the 45 minute concert for piano, percussion and bass "Synthesis", composed and performed on CD [681] in 1990 by the author. The composition was realized on the composition software *presto*® which will be dealt with in Chapter 50.

"Poem of Wind" is a musical transformation of the first and second strophes of Charles Baudelaire's poem "La mort des artistes" from his abysmal "Les fleurs du mal" (in the final version of 1861). These strophes are centered around the poetic Ego whereas the third and fourth strophes are written in the third person. They are four verses each, and follow an embraced A,B,B,A rhyme:

LA MORT DES ARTISTES
Combien faut-il de fois secouer mes grelots
Et baiser ton front bas, morne caricature?
Pour piquer dans le but, de mystique nature,
Combien, ô mon carquois, perdre de javelots?
Nous userons notre âme en de subtils complots,
Et nous démolirons mainte lourde armature,
Avant de contempler la grande Créature
Dont l'infernal désir nous remplit de sanglots!
.....

Compared to the Schubert piece, the paradigmatic realization of the 26 motif classes was much denser here. We started from the phonological paradigm realized on the level of letters and spaces. Each letter and inter-word space within a verse were given an onset with onset difference of 12 integer units, relating to $Onset|\mathbb{Z}$. Each inter-verse space was defined by an onset difference of 48 units, and the inter-strophe onset difference was set to 96. So we have an sequence of letter-onset couples, starting like this: (/C/,0), (/o/,12), (/m/,24), (/b/,36), (/i/,48), (/e/,60), (/n/,72), (/␣/,84),(/f/,96),...

On each such onset, a motive class will be allocated following a bijection between the 26 classes and the 26 letters which is motivated[38] by the frequency of letters in the two strophes, and by the hierarchy of motive classes in the specialization Hasse diagram (see Appendix O, Figure O.2). The most frequent letter is (as expected) /e/, and this is associated with the generic motif Nr. 10. Then, we have letter /n/, corresponding to motif class Nr. 11, etc. For the effectively used 21 letters, the bijection is this: (/e/,10), (/n/,11), (/o/,13), (/r/,12), (/t/,18), (/a/,19), (/s/,17), (/u/,15), (/l/,14), (/i/,16), (/m/,20), (/d/,25), (/c/,2), (/p/,1), (/f/,4), (/q/,3), (/b/,5), (/v/,22), (/g/,23), (/j/,26), (/y/,6).

In the next step, we take care of the selection of a representative for each class. To this end, we make use of a germinal melody which is responsible for the entire piano concert. It was constructed such that it can be covered by representatives of these classes, each once. The melody is a local composition $MelodyGerm$ in $Onset|\mathbb{Z} \oplus Pitch|\mathbb{Z}$ and the covering is a sequence of 26 subcompositions $Motif_t \subset MelodyGerm$ such that each $Motif_t$ reduces to a representative of class Nr. t modulo 12 in onset and pitch.

Figure 11.16 shows the covering and the germinal melody, as a local composition and in conventional notation. This yields the candidates for representing all classes. This covering of a local composition by a system of small sub-compositions will turn out to be a crucial procedure for global structures in music, see the global theory, in particular the theory of "interpretations" discussed in Chapter 16. In our context it is only an auxiliary tool to construct a kind of synopsis of all classes as being "charts" of a global connection.

The next step is the concrete realization of these representatives on the level of instrumental voices. In this piano concert, the piano is accompanied by 122 different percussion instruments[39]. In this setup, the pitches are assigned to different percussion instruments, and not to usual pitch. These instruments are grouped into toms, bells, etc., each group being distributed along an interval of pitch values. For example, we have snares for pitch values 8 to 17. Each motif is given a specific pitch range, i.e., a particular percussion sound so that we may distinguish the motives from their percussion sound range.

Figure 11.17 shows the *presto*® score as elaborated for the original piano score. Vertical axis is *presto*® onset (1-71), horizontal axis is integer onset time. The small quadratic icons denote elements of the local composition. We recognize not only the letter-related onsets of the different motives, but also some ornamentation which was mainly added for echo-type percussion effects. Below the upper rectangular score, we see the tempo curve which determines the micro timing of the piece. This subject will be treated in detail in Section 33.1. Here, we can just remark that the function of tempo—and this is considerable, as one can recognize from the extreme variation of the *presto*® tempo curve—is that of a musical prosody which is parallel to the linguistic one related to the poem's words. So the overall philosophy of this musical construction is a one-to-one transformation of the linguistic text into a musical one, including the prosodic specification by tempo which must be defined here since the *presto*® software is required to furnish a real performance of percussive parts such that the piano can interact as if the synthesizer output would be a human one.

11.7 Mathematical Reflections on Historicity in Music

Summary. Whereas historicity seems abolished from the group-theoretical point of view, it is evidenced that the diachronic line of compositional tools parallels—among others—the size of involved transformation groups. This ideas originate from Jean-Jacques Nattiez' paradigmatic theme.

– Σ –

[38] The bijection is not unique, we just choose a reasonable candidate.

[39] The instruments are taken from Yamaha RX5 and TX802, and Roland R-8M synthesizers. The latter is charged with the sound chips "Jazz", "Contemporary Percussion", and "Mallets".

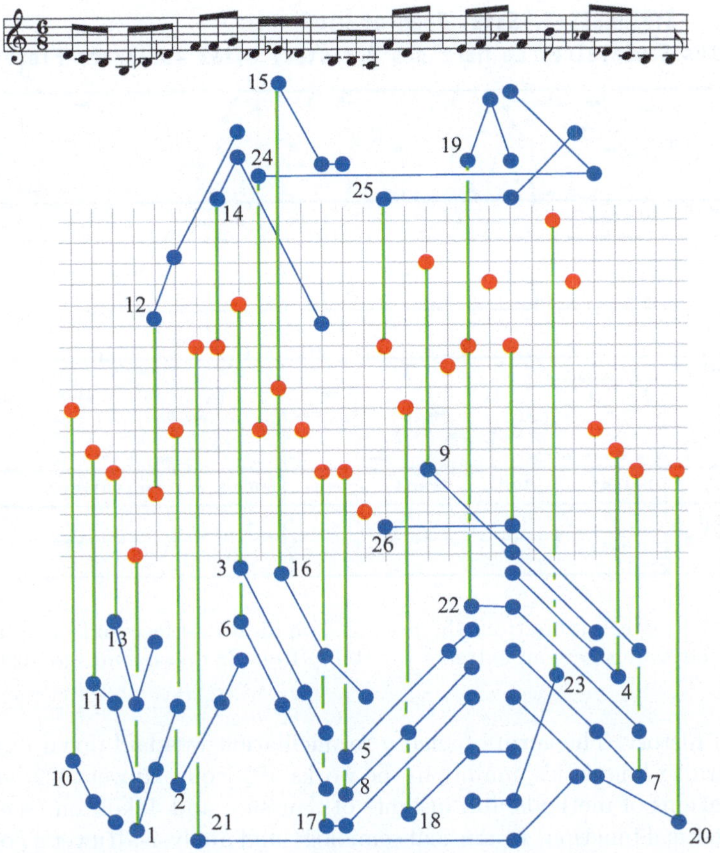

Fig. 11.16. The germinal melody of the *Synthesis* concert for piano, percussion, and bass is a patchwork of 26 three-element motives, each representing one motif class in $OnPiMod_{12,12}$. The melody is shown as a score and as local composition (red-colored points).

The group-theoretical classification theory which we have dealt with in the preceding development is based upon categories of local compositions which are denotator constructs of seemingly exclusive mathematical nature, in other words: Analytical work seems to be reduced to a mathematically oriented systemic process, and as such, apparently contradicts historical perspectives—which are essential to a musical dynamics that parallels the historical dynamics of our physical universe.

This problem arises when we question the program of neutral analysis. Günther Mayer has pointed out this problem in [235]. In fact neutral analysis does not offer tools for historical localization of works. Mathematical structure and diachronic position are independent. In this sense, classification appears to be an ahistorical approach.

We want to discuss this problem in the light of structuralist theory of musical analysis as it has been developed by Ruwet and Nattiez.

11.7.1 Jean-Jacques Nattiez' Paradigmatic Theme

Summary. This section discusses the structure of Nattiez' paradigmatic theme in describing associations in the frame of Jakobson's function.

– Σ –

The above singular "mathematical structure" is misleading. Mathematics offers a wide range of structures, and the possibility to distribute this spectrum on the diachronic axis may initiate a reconciliation

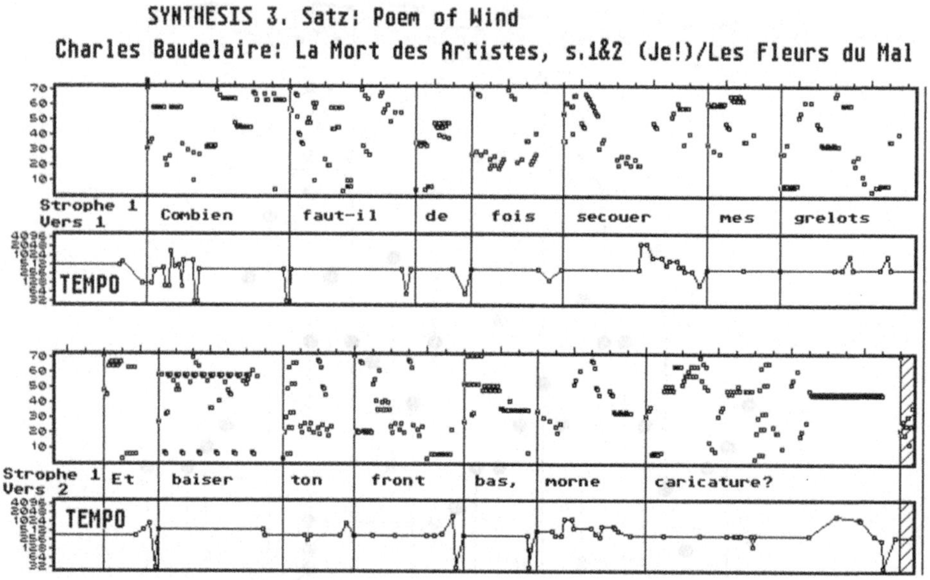

Fig. 11.17. The original *Synthesis* score from the *presto*® graphical interface and piano performance prescriptions, the tempo curve below the score shows an extreme variability from 32 onsets/min. to 4096 onsets/min.

between structure and history. The germ of such a reconciliation was laid down in the investigations of the Paris school of structuralist linguistics, mainly in the works of Nicolas Ruwet [922] and Jean-Jacques Nattiez [782] regarding applications of methods and insights of Saussure and Jakobson (see our previous discussion 11.6.1 of Jakobson's poetical function) to musical semiology and analysis. Ruwet's concept of *neutral analysis* means partitioning the work into a hierarchical system of units which have to be classified up to equivalence among each other and compared to units within other works, in the words of Ruwet [922, p.134]:

> Les diveres unités ont entre elles des rapports d'équivalence de toutes sortes, rapports qui peuvent unir, par exemple, des segments de longeur inégale — tel segment apparaîtra comme une extension, ou comme une contraction, de tel autre — et aussi des segments empiétinant les uns sur les autres.

The totality of classification criteria of a given analysis have been termed *"paradigmatic theme"* by Nattiez. This theme is realized by a concept of equivalence among units, defined by a *determined species* of geometric transformations [782, p.265]:

> Les unités paradigmatiquement associées sont équivalentes d'un point de vue donné (le thème paradigmatique), rarement identiques, et reliées entre elles par des transformations qui décrivent les variants par rapport à des invariants.

The hierarchical system of units is in fact a special case of what we shall call interpretation of a local composition in the framework of global theory, see 13.4. However, in Ruwet's and Nattiez' approach no systematic account is paid to the structure of this system, in particular no classification tools are discussed on this level: the paradigmatic theme is only a local affair.

On this level, equivalence means selecting a special subcategory of local compositions, more specifically, a special subgroup of automorphisms $P_M \subset \overrightarrow{GL}(M)$ for each ambient module of musical parameters. When comparing two hierarchical systems of units under a paradigmatic theme means admitting only the isomorphisms which stem from the system $P = (P_M)_M$ of automorphism groups. For instance, one could ask for a restriction to translations, i.e., $P_M = e^M$. So we could redefine the paradigmatic theme by the above system P, possibly extending the latter to a subcategory of **Mod** which has P as the system of automorphisms.

11.7.2 Groups as a Parameter of Historicity

Summary. This section terminates the chapter with an abstract of the determinants of size in historical growth of paradigmatic transformation groups.

– Σ –

By use of the preceding mathematical rephrasing of a paradigmatic theme P, we may attach P as a diachronic index to a determined analysis. This means that the a priori variability of admitted isomorphisms—which by default is purely mathematical—could be used as a valuation criterion in view of historical localization; the paradigmatic theme becomes a determinant which traces historical dynamics. And this is by no means restricted to esthesic positions, it can be applied symmetrically on the poietical side of a work, a perspective which is equally affected by historical dynamics.

In fact, the presence of a paradigmatic theme in a defined historical moment—be it in esthesic valuation or in poietic construction—is by no means an invariant of the diachronic axis. On the contrary:

Principle 6 *The appearance of certain symmetries as transformations in analytical technique or compositional tools is a secure index of a new musical epoch.*

We could even hypothesize that the paradigmatic theme has been monotonically growing since the last five hundred years of European music history.

12

Topological Specialization

Musical phenomena come to existence
in the constant fluency and motion
of compositional creation.
Therefore any description of them
must finally prove but approximations.
Rudolph Reti [892, p.12]

Summary. In general, transformations will conserve interior relations of objects, but *not* their 'absolute' position, or site, in the ambient space. This aspect is covered by the topological perspective. In its infancy, mathematical topology was in fact called "analysis situs". It deals with the general question of what it means to be in the vicinity of an object. In music, topological considerations are of central importance since slight deformations of objects to neighboring objects are standard identification concepts—though never handled with the necessary care. The point in making these structures precise lies in the sharpening of a fundamental descriptive tool, and in the semantic potential which topology induces. We make explicit the latter topic in a discussion of the problem of topological classification of sounds.

– Σ –

Although in mathematics, general topology is a basic discipline, musicology and also humanities have not yet understood the deep impact of genuinely topological reasoning. Topology has only penetrated the humanities on the metonymous level of metrical reasoning where topology is not at its best.

Genuine topology is a radical antagonist to transformation or metrics. We want to make this plausible in this preliminary chapter, more details will appear in other chapters, for instance in motive theory (Chapter 22) or in inverse performance theory (Chapters 46 and 47). The relatively abstract character of topologies may be a reason for their sparse usage in the humanities, but the abstract character is precisely the power of this approach: It is very helpful in creating concepts which are akin to usual fuzzy situations in the humanities.

The following discussion will in particular evidence the advantage of topologies for the delicate and still unsettled problem of sound classification, and we begin with a classic of shape conceptualization: Ehrenfels' gestalt theory.

12.1 What Ehrenfels Neglected

Summary. Ehrenfels' concept of *gestalt* is characterized by *invariance* under transformations and *super-summativity*. Apart from a marginal remark he did not, however, take into account the question of gestalt stability, i.e., the conservation of the gestalt under more or less small deformations. Without this stability aspect there is no workable concept of gestalt. We discuss the reasons.

G. Mazzola, *The Topos of Music I: Theory*, Computational Music Science,
https://doi.org/10.1007/978-3-319-64364-9_12

– Σ –

In Christian von Ehrenfels' seminal work on gestalt [293] he has characterized gestalt by two attributes: invariance and super-summativity. The former means that the gestalt of an object does not depend upon the identity of the object. We may replace this object by another specimen which is related to the original object by a determined transformation. In Ehrenfels' example of a musical melody, the transposition of the melody by some semitones does affect its identity but not its gestalt. The second attribute, super-summativity is akin to the old Aristotelian principle that the whole is more than its parts. In the musical example this means that the melodic gestalt is more than the enumeration of its tones, in other words: it is related to the relations among its parts rather than to their isolated presence. We do not discuss this characterization here, more will be said in the global theory and in particular in the topological theory of motives, see Chapter 22.

Our present concern is rather that Ehenfels did not give a workable definition of gestalt since he did not specify the problem of gestalt stability. To be clear: we contend that the two well-known Ehrenfels criteria are necessary, but they are not sufficient. Let us discuss this topic on Ehrenfels' example of a musical melody. Suppose that we are given two melodies of equal gestalt, the second being produced from the first by a pitch transposition. Suppose that we have to play these melodies on an ordinary piano. Then, clearly, we will not refuse the gestalt equality of these melodies even if the piano is not tuned in perfect equal temperament. Small deviations of the pitch relations and also small deviations in the temporal reproduction of the second melody against the first one will not bother our gestalt identification.

Now, this invariance against small deformations is neither super-summativity nor transformational invariance. In fact, no difference in the totality of relations is affected. And deformation is not transformation since we do not refer to any transformation rule whatsoever to say that the second melody is 'similar' to the first one. We rather need a concept of distance or neighborhood such that the second melody stays at this small distance or in the neighborhood of the 'prototype'. On the contrary, even the most innocent transformation, such as a transposition, does strongly alter the identity of the melodies constituents: pitch will change dramatically if we transpose by two octaves, for example. Small deformations really have to take care of the original position and small alterations of its coordinates. But the precise quantity of alteration is not relevant, no specification is required—except that it be 'small'. The same observation holds with super-summativity: If gestalt refers to relations among its parts, it is equally true that it is stable under small variations of these relations.

So we may summarize that *gestalt must be preserved under small deformations.* This is not only an argument from cognitive psychology, it is also an argument towards classification in the sense of stable concepts in parameter spaces underlying denotator forms. The point is that not only perception and sensorial processing are subjected to stability requirements. Abstract conceptual processes are also built upon stability phenomena, and this is one of the valid arguments of René Thom's catastrophe theory, see [1052].

12.2 Topology

Summary. This section introduces the topological argument and its formalism. Basically, music involves two radically different topological processes: comparison of *given* neighboring objects, and "degenerative" specialization of an object into a derived object. Both types of topologies are equally important in music and musicology.

– Σ –

We refer to Appendix H.1 for the mathematical definition of a topological space. Here we suppose that the reader has this prerequisite in his/her mind or that he/she is willing to believe that the following discourse is provided with a rigorous mathematical theory. We concentrate on the characteristics of topological argumentation.

Topology deals with the "logic of toposes"[1], it is a conceptual framework for the "analysis situs", the old title for topology in mathematics. The cornerstone of topology is the concept of a neighborhood. We are given a set *Top* of 'points' of whatever nature and want to give an axiomatic account of what it means that we stay in a neighborhood of a selected point x. All we need is a minimum of properties of neighborhoods. Here are our requirements:

Axiom 1 *Every point $x \in Top$ has a non-empty system N_x of subsets $U \subset Top$, called* neighborhoods of x, *such that:*

1. *For every such neighborhood U, we ask that $x \in U$.*
2. *If U is a neighborhood of x, then any larger set $U \subset V$ is also a neighborhood of x.*
3. *If U is a neighborhood of x, there is a special neighborhood V of x such that it also is a neighborhood of every one of its elements $y \in V$, i.e., $V \in N_y$.*
4. *If $U, V \in N_x$, then also $U \cap V$.*

The first requirement is completely natural from common language. The second requirement is not so. It means that if a neighborhood contains the 'very near' points to x, then adding other points to these does not alter this specification; the near ones are still there. In other words, if we say that a gestalt of an object x is conserved under small deformations, and if we agree that a neighborhood of x should *by definition* (!) contain those small deformations, then a larger set a fortiori contains those small deformations and therefore still is a neighborhood. So a neighborhood of x is a set of points 'surrounding' x and containing 'small deformation' points around x. The third axiom is a 'firewall' against too general a situation as suggested in axiom 2: It says that a neighborhood contains a 'core' neighborhood around that point in the sense that this core neighborhood is also a neighborhood of all its elements. This need not be the case for an arbitrary neighborhood: it is not true that any member y of a neighborhood of x has this neighborhood as its *own* neighborhood. But the core neighborhood is stable: everyone of its members contains small deformations in this neighborhood. The fourth requirement is evident from common language.

This suggests that we should consider globally 'stable' sets of points in *Top*: By definition, these are the sets $O \subset Top$ which are neighborhoods of all their elements, the so-called *open sets* of the topological space *Top* defined by the above neighborhood collection. In this language[2], the core neighborhood of x in axiom 3 is an open set (and neighborhood) built around x.

The point of this construction is that it is by no means symmetrical with respect to the points of *Top*. More precisely, if we consider any points x in *Top*, the closure $\{x\}^-$of x is defined as the set of those points z such that x is a member of all their neighborhoods. Intuitively speaking, x is an arbitrarily small deformation such as z. We also say that x *dominates* or *specializes to* z iff $z \in \{x\}^-$, in symbols: $x > z$. In this setup, it is not true *in general* that specialization is a symmetric relation. This means that there may be points x which are arbitrary small deformations of points z, but there are small neighborhoods of x such that z will not appear as a small deformation of x in such neighborhoods. So x is arbitrarily 'near' to z, but not vice versa, a situation which is completely pathological if one thinks of naive neighborhoods in normal life (say a person sitting near you in a crowded subway will have the same feeling that you equally sit near this person).

We insist on this abstract point of view since it shows that by no means can we associate topology with transformational paradigms, it is a paradigmatic framework sui generis! And we shall make full use of it. Let us give one concrete and classical example of such a strange topological space to illustrate the concepts, see figure 12.1. Other examples will be given in the following sections. The example space is the set $Top = Spec(\mathbb{Z}) = \{0, 2, 3, 5, 7, 11, 13, 17, \ldots\}$ consisting of the integer zero and all positive prime numbers, the prime spectrum[3] of $\mathbb{Z}$. We set

[1] Not toposes in the modern technical sense but in the old, set-theoretic understanding. However, the modern topos theory is nothing else but the functorial globalization of this old approach.

[2] Open sets are a common alternative to neighborhoods to write down the axioms of a topology, see Appendix H.1.

[3] See Appendix F.2 for this concept from algebraic geometry.

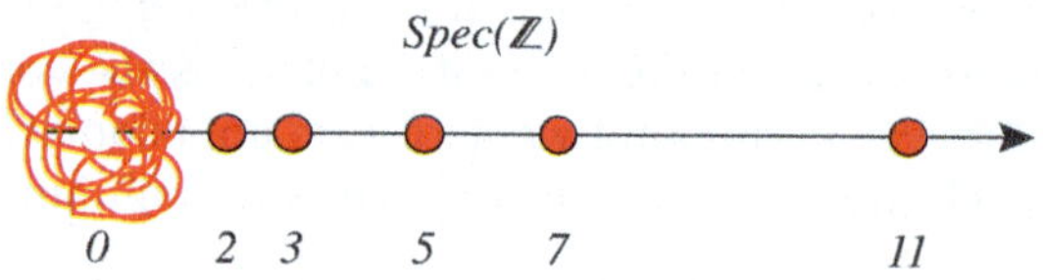

Fig. 12.1. The topology of the prime spectrum of the integers has a unique "generic" point, namely the zero, whereas all other primes are "closed" points.

$$N_x = \begin{cases} x = 0 & \text{all } 0 \in U \subset Spec(\mathbb{Z}) \text{ with } Spec(\mathbb{Z}) - U = \text{ finite,} \\ x \neq 0 & \text{all } U \in N_0 \text{ with } x \in U. \end{cases} \tag{12.1}$$

So we have a slight asymmetry: All primes are dominated by 0, but 0 is dominated by no prime. Further, every prime doesn't dominate any other point: We have

$$\{x\}^- = \begin{cases} Spec(\mathbb{Z}) & \text{if } x = 0, \\ \{x\} & \text{else.} \end{cases} \tag{12.2}$$

One says that a point x in a topological space is *closed* iff $\{x\}^- = \{x\}$. So all primes in our space are closed whereas 0 is not, it dominates all other points and is therefore called *generic*.

Exercise 18 *Prove all these claims concerning the preceding example $Spec(\mathbb{Z})$.*

This example shows that it is very easy, and mathematically standard, to introduce topologies which are far from the common intuition. The zero is a very 'thick' point compared to the prime numbers which are all topologically closed. Of course it is not a question of introducing artificial topologies, and this one is completely natural as a mirror of well-known facts from number theory. We shall introduce such natural topologies on spaces of motives, for example. But here, we should try to make the a priori point: topology is a very powerful tool for creating similarity paradigms without any allusion to naive distance concepts which are well known to psychologists in the wide-spread polarity profiles, for example.

12.2.1 Metrical Comparison

Summary. Metrical comparison is based on geometric distances between objects within their ambient spaces. (Not to be confused with temporal metrics of music!) Metrics are quantifications of similarity and generate a very special—thoroughly intuitive—type of topologies. Essentially, this aspect is a neutral comparison of given objects and does not ask for generative procedures to derive objects from each other. The idea of variation is problematic in this topological paradigm.

– Σ –

In the humanities, above all in psychometrics, metrical comparison arises if a set S of objects is loaded with a non-negative, real-valued distance $d(x, y)$ for every couple of elements x, y in S, measuring something like similarity between the couple's objects. In general, such a quantification of similarity can be arbitrarily wild. For instance, if we measure the distance between two pitches in auditory recognition, this can be a dramatic function; for example we may have $d(x, x) > 0$ or $d(x, y) \neq d(y, x)$ if the pitches x and y are presented enough time apart, and if the presentation time is *not* a parameter of the events. See Appendix B.2 for more details.

Avoiding such pathological—but realistic—phenomena in psychometrics, let us assume that we have a distance function on a set S of denotators such that it defines a metric [4] on S. For example, if S is a local

[4] See Appendix H.1.

composition in an ambient space which is isomorphic to an n-dimensional real vector space V, the usual Euclidean metric on V defines[5] a metric on S. Such a metric canonically defines a topology[6] on S by the neighborhood sets

$$N_x = \{U| \text{ there exists a positive } \epsilon \text{ such that } B_\epsilon x \subset U\} \tag{12.3}$$

where $B_\epsilon x = \{z| \ d(x,z) < \epsilon\}$ is the ϵ-ball (in S) around x.

Example 16 Take the set $S = SERM_{k,0}$ of serial motives of length k with integer values. For such a serial motif Ser we set $\|Ser\|_1 = \frac{1}{k}\sum|Ser_i|$ and then $d(Ser_1, Ser_2) = \|Ser_1 - Ser_2\|_1$. Division by k guarantees that if the series get longer and have only few differences, they will be more similar.

In this setup, if x is in the ϵ-ball neighborhood of y, then the same is true i^{th} exchanged roles, the relation is symmetric. However, the relation is not transitive: In general, $y \in B_\epsilon x$ and $z \in B_\epsilon y$ does not imply $z \in B_\epsilon x$.

Exercise 19 *Check this with the above example.*

But this means that the "paradigmatic ϵ-equivalence" around x defined by being in $B_\epsilon x$ is not a mathematical equivalence relation. It may happen that an object z is in two different ϵ-paradigms, one of x_1, one of a second x_2. This is a trivial mathematical fact, but for musicological conceptualization, it is of dramatic impact: If one tries to define pitch by *perceptual* equivalence via differences ϵ which are just below the famous "smallest just noticeable difference" (jnd), then no pitch concept will emerge since equality will not be an equivalence relation, and this contradicts an elementary requirement of conceptual identification. In other words, there is no such thing as a pitch which can be defined via psychological perception.

So why stick to topologies if one is given metrical functions? What is the added value? The added value is that in general, metrics have an artificial flavor. For example, in the above example, we could have taken another distance function, starting from the Euclidean norm $\|Ser\|_2 = (\sum|Ser_i|^2)^{\frac{1}{2}}$ instead of $\|Ser\|_1$. But the associated topology wouldn't have changed[7]. There are important properties which are intrinsically topological, i.e., they are independent of any chosen metric. For example, the pitches stemming from $(2,3,5)$-just tuning[8] are a dense subset of all pitches: there is no ϵ-paradigm of any pitch without just tuning pitches; this is an important fact for pitch paradigmatics.

Summarizing, the topological paradigms derived from metrics have the symmetry of ϵ-paradigms in common with the transformational, group-theoretical paradigms, but they differ in that transitivity is broken, and therefore, paradigms are not disjoint partitions of the given space. As a rule of thumb, one should observe *that metrics and associated topologies are not good candidates for paradigmatic concept constructions.* However, these topologies are the antagonist of functional relations in paradigms. We have no control of how relata in a given paradigm could have been generated from each other, they are independent, equivalent entities. To put it more topologically: In a topology which is derived from a metric, every point is closed, *dominance is absent.*

Therefore, talking about prototypical objects and variants thereof in this context is a delicate point. Distance does not generate prototypes, nothing is prototypical. The other way round, this says that we lack basic techniques for generating hierarchies among the given objects. A motivic analysis or a sound classification which is built on metrical topologies is completely flat. In particular, memorizing a corpus of objects in this flat world is utterly unattractive, and composition (poietical work) from this perspective will never yield more than an associative chaos. If we are to produce basic directives for compositional strategies, metrical topologies cannot be the motor of quality. We come back to this basic fact in the following discourse on sound classification, Section 12.3.

[5] It is however a non-trivial problem to recalculate a possible ambient space for embedding an arbitrary set S which is provided with a metric *on* S.

[6] See Appendix H.1.

[7] see Appendix I.1.2, Theorem 83 for background.

[8] See Appendix A.2.3.

12.2.2 Specialization Morphisms of Local Compositions

Summary. Specialization is related to generative relations between "dominant" and "dominated" objects. The topologies which are instantiated in this context are radically different from the above topologies, associated with metrical structures. The idea of motivic hierarchies is related to this type of topologies. We discuss the specialization graph of three-element motives in $\mathbb{Z}_{12}^2$, and its signification for the Schubert/Stolberg composition exposed in Section 11.6.2.

– Σ –

To make the ideas concrete and intuitive, we start working in the very small category of zero-addressed finite local compositions K in ambient vector space $M = \mathbb{R}^3$. While deforming such a local composition in the sense of metrical neighborhoods discussed in the previous section, say, by moving each point a bit away from its given position, something may happen which is invisible from the metrical point of view, see figure 12.2, namely the relative position of points. This is a problem which does not arise with affine transformation groups since any set of points on a line will also be colinear after the transformation. This is evidently violated by metrical deformation. Our present concern is something between transformation groups

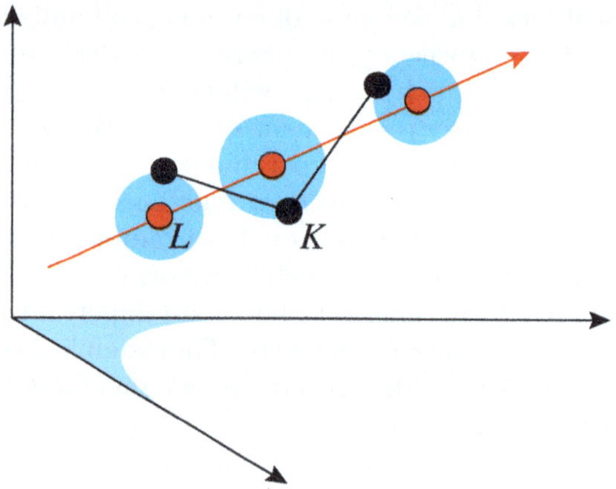

Fig. 12.2. Three (light) points in colinear position may lose this property while performing a very small displacement.

and metrical deformations: Transformations from transformation groups are invertible, and therefore sets of colinear points in local compositions are invariants of the group orbits. Look at the topological situation within the space of affine transformations $e^t \cdot m$ which is isomorphic to $\mathbb{R}^3 \oplus \mathbb{R}^{3\times 3}$, the topology being the ordinary metrical topology. Then the set $\overrightarrow{GL}(\mathbb{R}^3)$ of invertible transformations is open since it identifies to those points $e^t \cdot m$ with nonvanishing determinant, $det(m) \neq 0$. This is a condition which remains true if we change the matrix coefficients of m by a sufficiently small quantity[9]. However, if we approach the boundary of $\overrightarrow{GL}(\mathbb{R}^3)$, we encounter an interesting phenomenon. More concretely, take a curve $m(\lambda)$ of linear transformations

$$m(\lambda) = \begin{pmatrix} 1 & 0 & 0 \\ 0 & 1 & 0 \\ 0 & 0 & \lambda \end{pmatrix}$$

as a function of the real parameter λ. If $\lambda \neq 0$, $m(\lambda) \in \overrightarrow{GL}(\mathbb{R}^3)$. But for vanishing λ, we obtain the projection $m(0) = p_{1,2}$, see figure 12.3. The figure shows that if we apply a transformation $m(\lambda)$ of this curve to a given local composition K, we obtain a one-parameter family or *curve* of local compositions $K(\lambda) = m(\lambda)(K)$

[9] The determinant is a polynomial of these coefficients, and this is a continuous function.

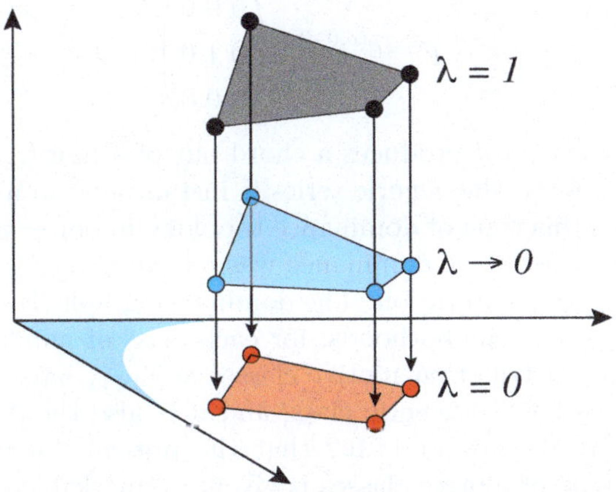

Fig. 12.3. A one-parameter family of local compositions, specializing to the plane projection for value $\gamma = 0$.

which approximates the two-dimensional local composition $K(0)$ for $\lambda \to 0$. Whereas all $K(\lambda)$ give us K back via $K = m(1/\lambda)(K(\lambda))$, we cannot retrace K from $K(0)$. The latter is a proper *specialization* of the general member of the curve $(K(\lambda))_\lambda$ of local compositions.

The point is that we have a curve of morphisms $m(\lambda) : K \to K(\lambda)$ which are isomorphisms, except for the special value $\lambda = 0$, where we have a surjective morphism which is not iso. All colinearity relation of K are preserved since we are dealing with affine transformations, but the limit composition $K(0)$ has strictly[10] more colinearity relations than the other $K(\lambda)$, in fact $dim(\mathbb{R}.K) = 2$, in other words, straight lines connecting pairs of points in K which do not intersect and aren't parallel, now, on $K(0)$, do intersect.

The intuition of a 'dominance' of K over the special position projection $K(0)$ can be turned into a topological dominance relation by the following construction: Let us fix a cardinality n of our local compositions $(K, \mathbb{R}^3)$. Then the set $ObLoc^{\mathbb{R}^3}_{n,0_\mathbb{R}}$ of all these local compositions bears a reflexive partial order relation $K \succ L$ iff there is a surjective (and therefore bijective) morphism $K \twoheadrightarrow L$, and $K \succ L, L \succ K$ is equivalent to K, L being isomorphic. If $[K) = \{L|\ L \succ K\}$ is the closed halfspace above K, we define a neighborhood system on $ObLoc^{\mathbb{R}^3}_{n,0_\mathbb{R}}$ by

$$N_K = \{U|\ [K) \subset U\}, \tag{12.4}$$

and thereby obtain the *dominance topology* on $ObLoc^{\mathbb{R}^3}_{n,0_\mathbb{R}}$. This definition evidently works on much more general spaces, but the idea is best seen on this concrete example.

Exercise 20 *The core neighborhood of K is uniquely determined, namely $[K)$.*

Clearly, the partial order relation $\succ$ is invariant under $\overrightarrow{GL}(\mathbb{R}^3)$ and we may pass to the *quotient dominance topology*[11] on $ObLoClass^{\mathbb{R}^3}_{n,0_\mathbb{R}}$. On both these dominance topologies, the *topological* dominance relation $\succ$ introduced in 12.2 turns out to be identical with the synonymous relation introduced in this section.

The dominance topology is an antagonist to the metrical topologies in that it is highly unsymmetrical: the core neighborhoods of two local compositions are equal iff the local compositions are isomorphic. The point of these topologies is that dominance means existence of degenerate transformations specializing the dominant local composition into the dominated one. Musically, this means that we are given a hierarchy of objects which is defined by 'projection' of generally positioned points onto points which can be thought as living in lower dimensional ambient spaces.

A typical example is the harmonic projection of a motif. If we of think our example space $\mathbb{R}^3$ as parametrizing onset, pitch, and duration, in this order, then projecting into the second axis by

[10] At least for our K having its points in general position as in the figure, i.e., $\mathbb{R}.K = \mathbb{R}^3$

[11] See Appendix H.1.

$$p = e^{(o,0,d)} \cdot \begin{pmatrix} 0\,0\,0 \\ 0\,1\,0 \\ 0\,0\,0 \end{pmatrix}$$

with fixed onset and duration values o, d produces a chord out of a motif. This derivation process is also a standard rule in dodecaphonism when the generic series is instantiated in a score.

We have already alluded to this type of dominance topology in our example 11.6.2 where we worked in the module $\mathbb{Z}_{12,12}$ instead of $\mathbb{R}^3$. There, the dominance was meant as hierarchy among different instances of three-element motives in Schubert's composition. The dominance graph (Hasse diagram shown in figure O.2 of Appendix O.3) visualizes all core neighborhoods: for each class of number x, the set of classes above x are equal to $[x)$. In general, it is not true that motive classes in $\mathbb{Z}_{12,12}$ have a unique dominant top element. Here, the class number 10 stands for the generic class, and it is also the dominant element in the dactylic grid as shown in 11.6.2. We already saw in 11.6.2 that the presence of the generic class is significantly above probability. So the hierarchy of motive classes is given a remarkable presence in Schubert's setting of Stolberg's poem.

We shall develop the dominance topology of motives in Chapter 22. In that discussion, a combination of metrical similarity and dominance topology will be introduced, thereby generating a synthesis of the antagonist views set up in this chapter. So the antagonists are only preliminary stages of a universal topological similarity paradigm.

12.3 The Problem of Sound Classification

Summary. Sound classification is one of the most complex unsolved problems in musicology. The reasons are triple: First, the communicative determinants of sounds are not clear. Second, sound varieties are—independently of selected sound parametrization/representation—completely unclassified objects. Third, the semantic charge of sounds is a substantial constraint for classification; there is no good classification without semantic constraints.

– Σ –

We have positioned this subject within the topological paradigmatics because it is above all a topological problem—besides the other aspects, of communication and semiotics, which are also important but cannot be solved without a thorough reflection on types and tools from paradigmatics.

We start our discussion with an account on the topographic determinants of sound description. This is a delicate discussion since its chronically underpinned relevance has led to serious errors in music psychology and psychophysics, such as the catastrophe of the "Fourier paradigm".

We then turn to the problem of describing varieties of sounds, as a function of the chosen sound representation and parametrization. As may be expected, the shape of such varieties changes dramatically depending on these prerequisites. There is no such thing as a universal approach.

The most irritating problem is the topological classification of sounds with respect to its semiotic potential. In fact, good mathematical candidates for metrical or dominance topologies must be valid with respect to meaning of sounds, meaning on poietical and esthesic, on physiological and psychological levels. This coupling of semiotic constraints in sound classification should be the central concern of any classificatory theory. Purely mathematical games are arbitrary and infinite in number.

12.3.1 Topographic Determinants of Sound Descriptions

Summary. Sound descriptions are not neutral, they depend on the topographic perspective. We discuss the two relevant dimensions: communication and reality.

– Σ –

In the following sections, we give the descriptive level necessary to deal with the semiotic problem which is the third topographic dimension to be treated subsequently in Section 12.3.3.

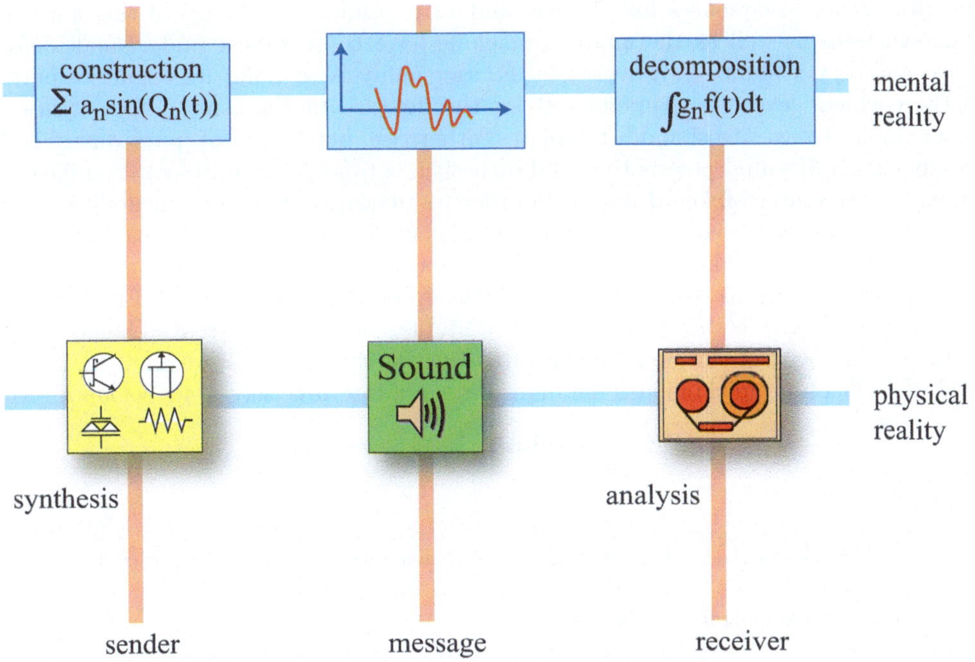

Fig. 12.4. The axes of communication and physical/mental realities of sound production (psychological reality is omitted here). The transition from mental to physical reality is more complex than shown here. The technological codification of the purely mental constructs of mathematics are an intermediate layer. Semiotically speaking, this phenomenon corresponds to a connotation/metalanguage structure.

12.3.1.1 Communication

Summary. Poietic and esthesic descriptions are independent of each other and may intervene for completely different reasons. The necessity of such a distinction is classically misunderstood in the ubiquitous "Fourier paradigm".

– Σ –

As a physical object, sound is produced by a specific synthesis machine and received by another analysis machine which is in a completely independent device environment, see figure 12.4. More concretely, the synthesis machine may be a classical instrument, like a violin, or human voice, or an electronic synthesizer built on analog and/or digital principles. The analysis machine may be a human ear, a microphone and a digital or analog analyzer device built on a specific dynamical system or a digital sample recorder. This diversity on the physical level may be understood in its mirror structures on the mental level.

Suppose we are given a process of Fourier synthesis, i.e., the summation of a series of sinoidal curves as described by a circular Fourier denotator as described in Section 6.7, example 2. The resulting curve is a time function which can be decomposed by a series of functions which are derived from the sinoidal functions via any isometric linear transformation, for example a rotation in a plane spanned by any two of the sinoidal base curves. The resulting parameter representation of the curve message will then have different coordinates which relate to the original ones by an arbitrary isometry, a rotation, say, in our example.

In order to identify these two aspects, one would have to look at the orbits of Fourier coordinates under the group of isometries in our function space. But this is not the common approach. Fourier ideology (above all on the metaphysical level of so-called "pure" sinoidal sounds, maintained by Stockhausen, for example) does not take orbits but the identical Fourier coordinates. Topologically, this point of view is also unstable: Any tiny rotation of the coordinate system will turn pure waves into impure coordinate sets, or conversely: in every neigborhood of a sinoidal function there are an infinite number of impure functions.

This has dramatic consequences for physical and/or physiological mirrors of this mathematical representation. The synthesis as well as the analysis machine have to be robust under small deformations. For example, if the auditory system really works by Fourier analysis to build its cognitive performance (and not only on the cochlear level, see Appendix B for the detailed mechanisms!), then it has to map small deformations of sinoidal waves to sinoidal templates, otherwise, perception of 'pure waves' is illusory.

So communication of sound poiesis to sound esthesis is a topological problem: You have to take care of topologically stable attributes of sound poiesis in order to guarantee its communicability.

A good example of a stable codification of sound phenomena is the phonological system as proposed by Roman Jakobson and Morris Halle in [499] and elaborated to the *Sound Pattern of English* (SPE) system by Halle and Noam Chomsky in [197]. The task of this system is to present the English consonants (e.g., /b/, /m/) and vowels (e.g., /i/, /ε/) as points in a stable coordinate system. The stability has to be a mathematical one and one that can be mapped into acoustic reality.

The SPE is built on a multidimensional digital space. More precisely[12], we set up a form

$$SPE \underset{Id}{\longrightarrow} \mathbf{Colimit}(Consonant, Vovel)$$

with a first cofactor

$$Consonant \underset{Id}{\longrightarrow} \mathbf{Limit}(Syllabic, Consonantal, Sonorant, \ldots Low)$$

consisting of 17 factors of essentially identical form:

$$Syllabic \underset{Id}{\longrightarrow} \mathbf{Simple}(\mathbb{Z}_2), Consonant \underset{Id}{\longrightarrow} \mathbf{Simple}(\mathbb{Z}_2), \text{ etc.}$$

All have the same Bit-type Coordinator $\mathbb{Z}_2$ and differ by name. The values $1, 0$ are traditionally codified by $+, -$. The second cofactor is a six-dimensional product

$$Vowel \underset{Id}{\longrightarrow} \mathbf{Limit}(Syllabic, High, Back, Low, Round, Tense)$$

which has the same type

$$Syllabic \underset{Id}{\longrightarrow} \mathbf{Simple}(\mathbb{Z}_2), \text{ etc.,}$$

of Bit-typed factors. A *phoneme* is a denotator of form SPE, its name is the phoneme's usual name, e.g., the consonant

$$\theta : 0 \rightsquigarrow SPE(-,+,-,-,+,-,-,-,-,-,-,-,+,+,-,-,-)$$

with its coordinates[13] in the cofactor space $Consonant$. This representation is mathematically discrete and the acoustical correspondence to the abstract digital symbols is meant to be produced by a distinctive position/movement of the vocal tract. This implies that any such position is *isolated* with respect to the others in an adequate representation of the vocal tract dynamics. So small displacements do not change the type of physical dynamics in the vocal tract, and there should be no smooth transition between any two of these dimensions. This justifies the digital representation of acoustical dynamics. And it is meant that by identical vocal tract dispositions among humans (at least English speaking individuals) the esthesic interpretation of the symbolism should be unambiguous, i.e., the distinctive nature can be stably replicated on the basis of the receiver's knowledge on personal vocal tract topology.

This latter approach is also interesting because it represents something like a prototype of psychologically oriented topological classification: it is hoped that the continuous space of physical parameters may be split into disjoint regions where discrete values can be attached in a stable way. After all, linguistics has been successful on this path. But music is not linguistics. The variety of musical sounds is far from frozen in a low-dimensional space above $\mathbb{Z}_2$. So communication will have to face dramatic changes on both sides of the message object of neutral sound.

[12] See [17, p.103ff] for details.

[13] Of course, a grouping of some of these coordinator spaces would have allowed for fewer dimensions and more values (2^k per dimension), but the information is unaltered.

12.3.1.2 Reality

Summary. The levels of reality define a strong parameter for sound representations. Psychological, physical, and symbolic descriptions may vary substantially. We discuss the representations inscripted by chronospectra, frequency modulation, wavelet methods, and physical modeling.

– Σ –

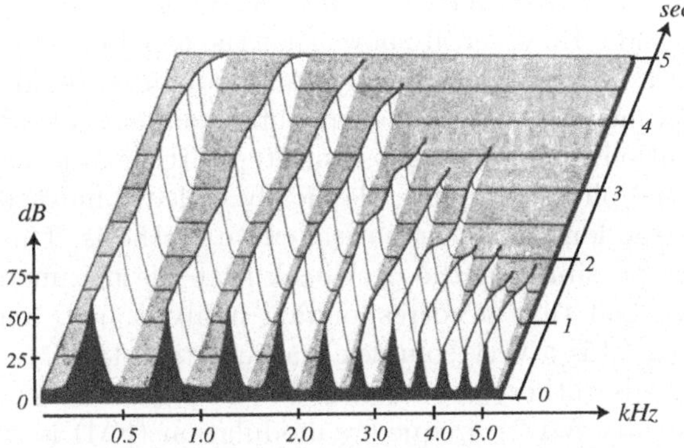

Fig. 12.5. The chronospectrum of a trumpet sound, played decrescendo; the dark sections represents the Fourier spectrum at a given time of the sound's execution.

Sound parameters can be symbolized on the string level of pure names. This is the case for the MIDI[14] code, for example. Here, sound types are accessed via program change numbers and these are just numbers for predefined sounds in synthesizers which understand the MIDI code. There have been attempts to give these numbers a certain standard meaning (General Midi, see [796, p.105], for example), but this has nothing to do with the contents of sounds. However, the attempt is oriented towards grouping 'similar' instrumental sounds (such as piano-like sounds in number 0 to 7) on neighboring places on the axis of natural numbers.

To move deeper into the sound classification problem, one has to open a poietic (construction) or esthesic (decomposition) method beyond pure convention. From classical Fourier analysis on real instrumental sounds, it is standard to model the temporal unfolding of a sound as a function where Fourier components are not constant. Moreover, the amplitude is defined by an envelope curve. We have described a corresponding form $FourierSound$ in Section 6.6, example 12. In order to generalize that form to variable Fourier spectra, we may replace the factor form $Fourier$ by a one-parameter family of such forms $Chrono - Fourier$, parametrized by a variable $\lambda \in [0, 1]$ of the unit interval, exactly like the envelope form $Envelope$ described in Section 6.6, example 11. This defines the form $ChronoFourierSound$ as a generalization of $FourierSound$, see figure 12.5 for an illustration.

It is erroneous to think that this is the end of the classical sound description. In terms of instrumental construction, this is only a basis of general instrumental characteristics. Since we shall deal with this subject in Chapter 54 on parameters for string quartet, we can stick to a short sketch here. For instruments of the violin family, the form $ChronoFourierSound$ is enriched by two classes of parameters: bow application and vibrato.

Bow parameters are split into four groups:

- Bow pressure
- Bow velocity
- Contact point bow to string
- Bow angle.

[14] MIDI = Musical Instrument digital Interface, see [761, 796]

The vibrato effect is given by four numeric parameters:

- Relative delay time from onset
- Modulation frequency (frequency of finger displacement)
- Pitch modulation (extent of finger displacement on string)
- Amplitude modulation (contact point of finger-tip).

We call these eight parameters the *technical parameters* of a violin sound. They are an additional variety of sound construction superimposed on a given chronospectrum.

The complexity of this poietical perspective of physical sound production is in sharp contrast to what can be analyzed on the esthesic side. However, it shows that the topology of such poietic objects can become quite complex, see next Section 12.3.2 for this. The physical modeling technique is related to this situation, it is a strictly poietic construction methodology of instrumental sounds via software modeling of the complete dynamics of instrumental execution. It however goes one step further in that the Fourier chronospectrum is replaced by the simulation of the physical device, classically called a musical instrument. So it models the making of a sound, not the abstraction via mathematical curve synthesis. The high fidelity to natural sounds which physical modeling makes possible has the drawback that the parameter space for sound description becomes increasingly complex. And that a corresponding analysis, i.e., a reconstruction of the defining parameters from the message signal is a very complex mathematical task.

If the Fourier construction is rather easy to decode since the Fourier coefficients can be calculated by classical methods (see Appendix A.1.2), frequency modulation (FM) is rather hard, but less hard than physical modeling. FM sound construction is associated with a form *FM-Object* introduced in Section 6.7, example 3, see Appendix A.1.2.2 for technicalities. It generalizes the classical Fourier construction in that it introduces a hierarchy of sinoidal functions which act as relative modulators in a functional concatenation which can be displayed as a directed graph[15], possibly with cycles. Relating to Fourier construction, there are two important differences in the zoo of these FM-sounds: First, the FM-graph represents a new type of combinatorial variable (with Fourier it was only a discrete set of points). Second, the FM analysis is far from settled: Even if we know about the underlying FM-graph, the FM-coefficients cannot be retraced in the general case, at present, only special graphs show solutions, see Appendix A.1.2.2 for this problem. So FM-synthesis is strongly poietic, though less complex than physical modeling.

Wavelets are a last example in our overview. See Appendix A.1.2.3 for technical details. In principle, the wavelet method is comparable to the Fourier method in that it parametrizes a given time signal function $f(t)$ by a distinguished system of 'coordinate functions'. We have to fix a generic wavelet ψ, for example the Mexican hat function

$$\psi(t) = (1 - t^2)e^{-t^2/2}. \tag{12.5}$$

This furnishes the "wavelet" paradigm prototype of the method. On this function, the affine group $\overrightarrow{GL}(\mathbb{R})$ acts from the left by $\psi_{a,b} = e^b \cdot a \bullet \psi = |a|^{-1/2}\psi \circ (e^b \cdot a)^{-1}$. We then take the orbit $\overrightarrow{GL}(\mathbb{R}) \bullet \psi$ and apply each of the orbit's members as affine deformations of the prototype, thus yielding a system of 'coordinate' wavelets. The coordinate of the test function $f(t)$ with respect to a coordinate wavelet $\psi_{a,b}$ is the scalar product

$$L_\psi f(a,b) = c_\psi \langle f, \psi_{a,b} \rangle_{L^2}$$

(see A.1.2.3). The inversion formula for wavelets says that this representation has an inverse and that it is an isometry, in other words, the wavelet representation of $f(t)$ is faithful and preserves metrical properties among functions. In contrast to the Fourier representation, no fundamental frequency is needed here, but we have to pay the price of two coordinate variables.

However, the wavelet method also must be concerned with questions of topological stability, in particular the reduction to discrete "frames" which represent the continuous variety of parameters. In fact, the theory of *wavelet frames* (see [622, p.79ff]) deals exactly with this problem. It is, however not evident that the

---

[15] In YAMAHA's TX802 synthesizer technology, this graphical structure was called the "algorithm" of the FM synthesis. TX802 offered 32 such algorithms.

'cochlear wavelet' yields a faithful wavelet representation of sound (relating to adequate wavelet frames). And it is absolutely unknown whether on the neocortical level, wavelet analysis is performed.

Let us now turn to more psychological sound representations. In the attempt to transpose the linguistic SPE model to music, several authors have described spaces of sound colors. Among others Wayne Slawson [981] and Fred Lerdahl [594]. The common characteristic of these attempts is a shift from the physiologically founded coordinates of the SPE system to frankly psychological perception coordinates such as "acuteness", "openness", "smallness", "vibrato", "inharmonicity", "brilliance", etc.

All these attributes being on the esthesic side, they can hardly transgress the communicative barrier to poiesis. And they are also very fuzzy as such. For example, Lerdahl's "vibrato" is completely wrong as a one-dimensional attribute, we have seen above that four dimensions are needed for a common description. The universe of sounds (let us talk about the poietic construction universe discussed above) is much too broad to be captured by manifestly reductionist words such as "brilliance". The rationale of Lerdahl's

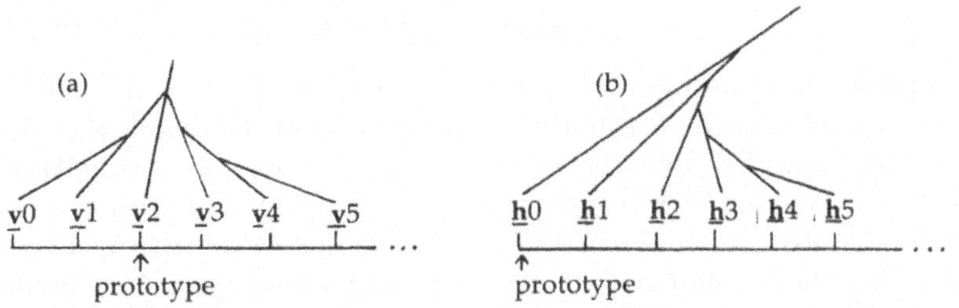

Fig. 12.6. Two hierarchical trees from Lerdahl's attempt to impose a linear order and hierarchy on vibrato and inharmonicity.

construction (see also figure 12.6) is a very limited system of ramification trees from that theory which enforces absurd hierarchies without necessity. An example is the similarity between different variants of a vowel described by prolongational ramifications where no hierarchy is given. The generative theory of tonal music is mathematically primitive since it lacks most elementary tools of topology. The vibrato space is at least four-dimensional, and the vibrato types are much richer than an ordered set which is dominated by a "prototype" (left part of figure 12.6). Even if we accept templates of vibrato, these would be trivially conventional for technical reasons of software preferences and could not claim prototypical roles in musical performance.

Even worse is the linearization of "inharmonicity" (right part of figure 12.6) since there is not the least indication of a one-dimensionality in non-harmonic Fourier components (what would be the theorem to which this fuzzy construct refers?). This perceptual model, which is built on a wrong analysis model, is wrong on every single step, and it is induced by systemic constraints which are completely artificial in this subject. The topological richness would never enforce unique maximal elements since it does not enforce global hierarchies. Topological dominance can have as many maximal elements as necessary. The reduction to hierarchical trees simply destroys facts in favor of graph theory for beginners. Such terrible simplifications from psychology are in dramatic contrast to the complexity of sound description for realistic synthesis and even analysis on the physical (or technological) and physiological level.

We can summarize that the present sound color spaces from psychology are reductionist in their construction, have nothing to do with sound synthesis and analysis whatsoever, have a ridiculously low number (two, three, at most...) of dimensions which are even exceeded by the linguistic system (17 dimensions for consonants in the SPE model), and do not cover any of the fundamental problems in sound description, classification, similarity, dominance, and sound semantics; see also Reinhard Kopiez' review of this situation in music psychology [548].

12.3.2 Varieties of Sounds

Summary. As a function of the chosen representation, the universe of sounds constitutes a more or less difficult variety. We look at possible shapes of such varieties and their meaning for the classification problem.

– Σ –

Already for classical Fourier construction, the infinite-dimensional coefficient space of the amplitude and phase spectrum $(A_i, Ph_i)_{0 \leqslant i}$ is difficult to turn into a reasonable manifold. In fact, the naive Euclidean distance does not express more than local similarity, whereas one is more interested in topological description of 'dominant regions', i.e., locally maximal coefficients in the sense of formants which are also responsible for vowel recognition[16]. This means that one is looking for indices b such that the amplitudes have local peaks at index b, i.e., $A_i < A_j$ for couples of indices i, j with $a \leqslant i < j \leqslant b$ or $b \leqslant j < i \leqslant c$. Such a formant condition defines an open formant set $U_{a,b,c}$, and each[17] formant set is given by the intersection

$$U_{a_1,b_1,c_1} \cap U_{a_2,b_2,c_2} \cap \ldots U_{a_k,b_k,c_k}$$

of all local peak neighborhoods of an increasing sequence of index intervals $a_1 < b_1 < c_1 \leqslant a_2 < b_2 < c_2 \leqslant \ldots a_k < b_k < c_k$. If we agree that such neighborhoods are regions of reasonable sounds, the relation between different such regions is not clear: Possibly, there are no reasonable sounds relating these regions, possibly are there many different paths from one such region to another. So we are left with complex problems regarding subsets of reasonable candidates and the manifolds which are defined by these subsets, see also figure 12.7. It is not even clear whether these manifolds are of a determined dimension since boundary conditions could define closed submanifolds, e.g. by inverse images of functional equations $F(A.) = 0$ or energy constraints (e.g., upper limits $\sum A_i^2 \leqslant const.$). A serious look at these problems is exactly what Lerdahl missed in his

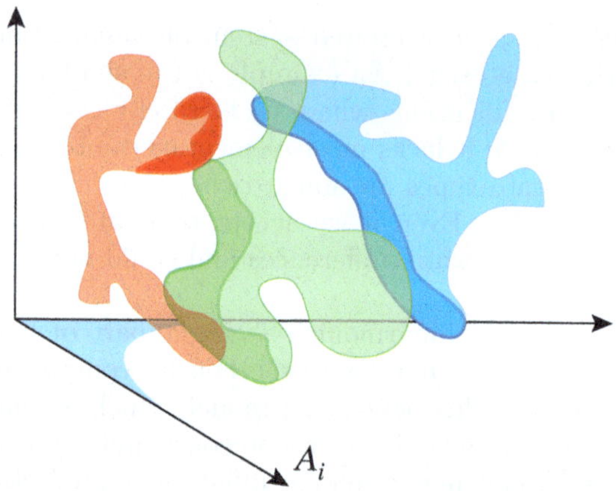

Fig. 12.7. In the infinite-dimensional space of Fourier coefficients, the relevant sounds can define a subset which looks like an inhomogeneous variety, a patchwork of open manifolds and closed connections, defined by algebraic or analytical constraints on energy, spectral formant distribution etc.

work [594] at IRCAM where he applied the CHANT program to deduce his broken sound color space. In fact, CHANT is exactly what would be necessary to make experiments which are based on serious theories of formant manifolds.

What was only defined by the selection of relevant candidates is intrinsically built in for FM objects. We have a double description level of these objects. An *FM-Object*-formed denotator *myFM* has its FM-graph

[16] We stick to discrete Fourier spectra for simplicity

[17] Supposing for simplicity and experience that no two successive amplitudes are equal.

$\Gamma(myFM)$[18] and the partial coefficients $(A_k, F_k, Ph_k) \in \mathbb{R}^3$ for each vertex k of $\Gamma(myFM)$. So the variety of FM-sounds is parametrized by the directed graphs, and is an open set in $\mathbb{R}^{3v_\Gamma}$ for the vertex cardinality v_Γ of the parametrizing graph Γ. This splits the FM-sounds into disjoint sets and is not a workable environment for the totality of FM-sounds since we are not yet able to compare sounds of different graphs.

The point is that graphs are discrete objects which we should embed in a topological space in order to manage deformation of FM-objects. In fact, a graph arrow $a : v_i \to v_j$ in $\Gamma(myFM)$ symbolizes that the partial at v_i acts as one of the modulators on the partial at v_j. The strength of this action is not variable, if it does not act, we simply omit it. So this is a digital reduction which generalizes in a canonical way. Instead of sticking to graphs Γ, we weight their arrows by non-negative real numbers, writing $w(a)$ for the weight of arrow a We therefore consider weighted graphs Γ^w. If we denote 1 the weight having values 1 for all arrows, then the unweighted situation for FM-sound $myFM$ was equivalent to the situation with $\Gamma(myFM)^1$. For a general weight w on $\Gamma(myFM)$, we get the generalized FM-object $myFM^w$ which refers to $\Gamma(myFM)^w$. We then have an obvious projection

$$p : myFM^w \mapsto myFM_w \tag{12.6}$$

which by definition replaces each partial (A_v, F_v, Ph_v) at the tail v of arrow a by the partial $(w(a)A_v, F_v, Ph_v)$ and generates a normal (weight 1) FM-sound from a weighted object. In other words, we have a p-section σ_1 of the set Γ^1 of 1-weighted FM-objects of given graph Γ in the space Γ^- of weighted FM-objects of given graph Γ. The composed map

$$\Gamma^1 \xrightarrow{\sigma_1} \Gamma^- \xrightarrow{p} \Gamma^1$$

is the identity, and the subspace Γ^1 is defined by the closed condition $w = 1$. But on Γ^-, we have the action of the multiplicative group μ^{a_Γ}, the a_Γ-th power of the multiplicative group of the base field $\mathbb{R}$, where a_Γ is the arrow cardinality of Γ: An element $(t_a)_{a=\text{arrow}}$ acts factorwise on the weight w. If $myFM$ is an object in Γ^1, its orbit $\mu^{a_\Gamma} \cdot myFM$ is irreducible in the Zariski topology[19]. In this setup, if we are given an object

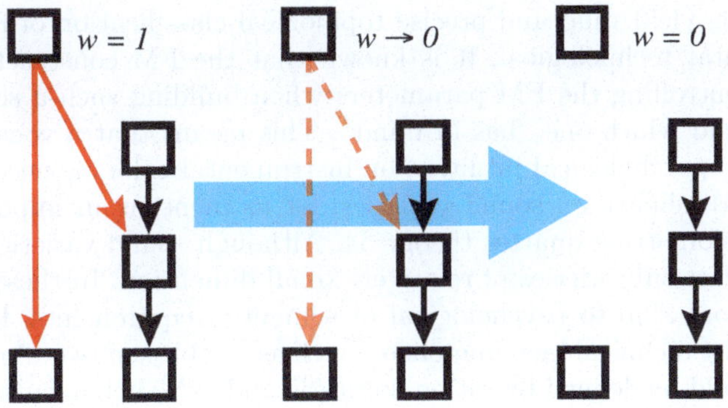

Fig. 12.8. Cutting off arrows means specializing weighted graphs, here for two arrows starting at the same vertex.

$myFM$, we may ask for objects $myFM^*$ which are dominated by the orbit of $myFM$. This means that the generic point of $\mu^{a_\Gamma} \cdot myFM$ dominates the generic point of $\mu^{a_\Gamma} \cdot myFM^*$. Such a specialization is obtained, for example, if we let the weight of arrow a of $\Gamma(myFM)$ vanish. Then, the specialized FM-object has no contribution from this arrow, see figure 12.8. So we can integrate different graphs into a generic graph in such a way that they are defined by cutting adequate subgraphs. This idea was already introduced in the classification of TX802 sounds in the first software prototype MDZ71 of *presto*®, see [682, Chapter 10]. That approach divides the 32 algorithms from TX802 into six *families*, according the number 1-6 of root partials, and then, within each family, a specific graph defines a *species*. Within one species, two graphs Γ_1, Γ_2 are

[18] See Section 6.7, example CircDen-3, for the definition of the FM-graph $\Gamma(myFM)$.

[19] This classical algebro-geometric topology is also given by the prime spectrum of a ring, see Appendix F.2.

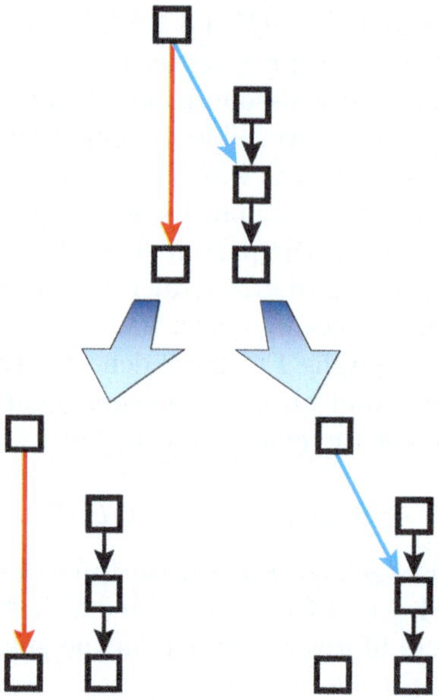

Fig. 12.9. Two graphs (from the TX802 algorithms) are in kindred relation if they are specializations of a dominant third graph and are mutually related by displacement of a particular arrow.

called *kindred* if they are derived from each other by moving one arrow. This means that they are both specializations of a weighted object $\Gamma^w_{1,2}$ in such a way that there are two arrows a_1, a_2 in the latter which yield Γ_1 or Γ_2 if $w(a_1) \to 0$ or $w(a_2) \to 0$, respectively, see figure 12.9.

Though these methods yield valid and precise topological classification of FM-sounds, the perspective is strictly poietic and mental/technological. It is known that the FM construction is far from evident on the psychological level: Controlling the FM parameters when building such a sound is very difficult if one aims at constructing a sound which one "has in mind". This means that a vocabulary switch between the preceding topology and the psychological intuition on instrumental colors is needed.

To complete this short remark on sound varieties, let us mention an important technique which will be applied in the discourse on string quartet theory 54. Although sound varieties are of infinite dimension, it is often reasonable to select subvarieties of relatively small dimension. In these subvarieties, classification may become easier and more akin to psychological or semantic requirements. It should however not be a selection which is dictated by technical incompetence—such as the broken two-dimensional space constructed by Lerdahl in [594]—it should be defined by rational criteria and as a such a variable selection, depending on the context. For example, for questions concerning a fixed orchestration (a string quartet, a vocal ensemble, etc.), selecting a sound subvariety which contains the corresponding instrumental colors will do, but see [594] for details.

12.3.3 Semiotics of Sound Classification

Summary. Sound classification has to cope with constraints of meaning of sound. This concerns above all the topological properties of sound varieties. They are a function of the type of musical compositions and their instrumentations. They also depend on the semantic charge which the composer imposes on his/her message.

– Σ –

Topological classification of sounds is not only a matter of mathematical description of sound varieties on specific levels of reality. The point is that from all possible topologies, metrics, etc., we are interested in those which are able to carry a semantic function. They are good if they reflect meaning of sounds. Meaning on any level: auditory physiology, perceptual category, emotional quality, or compositional/analytical function.

So we are confronted with a complex problem: To map topological sound varieties (varieties of timbral colors, the name is irrelevant) to semantic manifolds—such as emotional landscapes—with their topologies, in such a way that the mapping becomes continuous or even an isomorphism of topological spaces. Or to define metrics on these manifolds such that geodesics correspond to semantically shortest paths. In the example [594] of the string quartet, it is shown that general position of timbral points in the violin family can help to represent contrapuntal and harmonic structures of classical compositions by Joseph Haydn or Luigi Boccherini.

Unfortunately, no significant research has been done to investigate possible maps and associated timbral topologies. The question is highly non-trivial since it is not only a problem of testing topologies, but of defining them in order to settle semantical constraints. Regarding Jakobson's poetical function, the problem is related to the definition of topological paradigms of sounds which turn emotional syntax of musical compositions into poetical syntax. We know the emotional syntax (not so sure, though...), we know the timbral spaces, we know that the emotional syntax is poetically instantiated on the sound colors of musical compositions. But we do not know which timbral topologies and/or metrics turn music into timbral poetics.

12.4 Making the Vague Precise

Summary. Summarizing, topological methods of musical paradigmatics are powerful tools for grasping blurred and ambiguous concepts in musicology. Transcending the evident structural task by far, topological considerations unveil deep problems of communication and semantics of music.

– Σ –

The foregoing reflections are far from systematic. We are at the very beginning of an exact theory of topological classification of musical objects. But we have a good technical basis, and we shall see that in the theory of motives, global metrics, and in the string quartet theory, these techniques can be successfully applied to construct basic concepts, such as "motivic gestalt", "metrical rhythm", and to give good and precise answers to the intriguing question why the string quartet came up to primordial prominence so suddenly in the middle of the eighteenth century.

Topological considerations are not only good technical classification tools, they are also the key to a deeper understanding of many problems of musical communication and—above all—semantics on all levels of reality: mental (composition), physical (auditory perception), and psychological (emotion). It is also clear that the present state of our theory, namely the strictly local point of view, is not sufficient to cope with more refined topological situations. We encountered many situations where the word "manifold" or "variety" was the good pointer, however, this was a more mathematical flavor, and we still lack an intrinsically music-related approach to global views. So we are well prepared to envisage this new challenge: *What are musical manifolds?*

Part IV

Global Theory

13

Global Compositions

... bei der Bestimmung der Großform wirkt
über den ganzen Eingliederungsprozeßhinweg
die gleiche Denkweise, *und zwar*
von der morphologischen Mikrostruktur
bis zur rhetorischen Makrostruktur.
Pierre Boulez [138, II, p.60]

Summary. The categories *ObLoc* and *Loc* are representations of local objects, i.e., of objects that are not composed of proper "parts". However, music mainly deals with compound objects. The adequate concept of a global composition is defined. The corresponding vocabulary of elementary global music objects is described, including ecclesiastical modes, tridadic degrees, meters, rhythms, motives and themes.

– Σ –

This chapter represents a major segmentation point of this book since it introduces the paradigm of global objects into the topos of music. This is a mandatory step as the nature of music is an organically compound one from all topographic points of view. Composition, analysis, interpretation, cognition, understanding, and performance, all of them are immersed in an omnipresent dialectic of analysis into local parts and synthesis of a global whole. However, compared to the investigation of exterior nature, it is very difficult to separate analytical or synthetical activity from the object of investigation. The uncertainty relation in physics, stating that every experimental interaction with nature is subjected to a complementarity between hidden and revealed information, this relation becomes a dominant and even characteristic attribute of music as an expression of human nature.

The investigation of the musical object tends to be an integral part of the object. The composition's germs, the parts of an analysis, the shaping elements of a performance, and the units of cognitive processes, all of them define works and musical systems as multiply varied perspectives of a whole. We know from our previous discussion of Yoneda perspectives in Section 9.4 that the identification of a musical work involves an integration of multiple perspectives. This integration is in particular one of local points of view, a "patchwork" of multiply zoomed micro-views. But the huge variability of such views is not a drawback or defect of music and its theory, on the contrary, is substantial:

Fact 7 *Music is not the definitely ambiguous, it is the definiteness of ambiguity.*

On the level of music performance, this general insight was pronounced by Theodor Wiesengrund Adorno [7]: "Musik interpretieren: Musik machen." Music is only instantiated *via* its interpretation or performance.

A theory which crashes on the ambiguity of its object in misunderstanding it as an indeterminacy cannot, contrary to Carl Dahlhaus' judgment [231], be an adequate one. Rather the problem is to set up an exact theory of ambiguity. This is the methodological background of the theory of global compositions.

G. Mazzola, *The Topos of Music I: Theory*, Computational Music Science,
https://doi.org/10.1007/978-3-319-64364-9_13

There is no way out of making the multiplicity of compositional and interpretative activities a part of the theory's concerns.

The objects of the global theory are mainly global compositions. This concept generalizes that of local compositions in the sense that it takes care of and formalizes ambiguity. In an important special case—where we talk about interpretations—such global objects precisely represent the result of interpretative activity. The possibility of crystallizing selective actions in mathematical objects encompasses their structural comparability, a possibility which will be dealt with in Chapter 14 about morphisms between global objects. This achievement is interesting for musicians and musicologists since a theory of ambiguity also favors mediation between various possible explications of a given work or analysis. Moreover, the precise setup of morphisms qua objects of comparative studies implies an objective (meta)discourse upon comparative studies.

The fact that we were only able to define very elementary objects in the previous theory reflects the musicological fact that ambiguity enters very early in music and its science. Most of the relevant concepts, such as tonality, degree, tonal function, etc., refer to objects which are composed of local ingredients. The following global theory will compensate for this defect. The power of the global approach is, however, proportional to the power of the local components: If we had not prepared a universally valid local concept framework, the global theory would also fail. Recall that we had pushed the local theory to the extreme of its potential and then recognized (regarding the non-existence of general colimits, see 8.3.3) that there is a limit of the local approach which must give rise to a global context.

We should stress that this global approach is not intended as a normative standardization of musicological language, but as a precise methodology of language formation. We also believe that musicology cannot survive in the present status quo where essential competences and problems regarding the intertwining of object and subject are not lifted to a scientific level but still reside in the lowlands of feuilletonistic entertainment. The widespread belief that ambiguity and polysemy in music is a firewall against precise science unmasks its representatives: *There is no ambiguous theory, there are only ambiguous theorists.*

13.1 The Local-Global Dichotomy in Music

Summary. This section motivates the local/global dichotomy in music. Composition, score representation, analysis, and performance do use this dichotomy—virtually without clear-cut conceptualization or even consciousness. Built on the Aristotelian tradition, Ehrenfels' criterion of super-summativity introduces the dichotomy into psychology. The question of a precise description of the difference between the "sum of the parts" and the "whole" in Ehrenfels' criterion is addressed. More generally, it is evidenced that specification of parts in music objects is not accidental for understanding the very nature of music. We discuss the radical difference between the local/global dichotomies in music and mathematics. A short historical account on the simultaneous appearance (1854) of these approaches in musicology and mathematics with the work of Eduard Hanslick and Bernhard Riemann is given.

– Σ –

Let us introduce the subject by the summary of a (non-exhaustive) series of approaches to global structures in music(ology), including Boulez/Webern, Ude/Wieland and Marek, Hoffmann and Kaiser, Graeser and Ruwet/Nattiez, Jackendoff/Lerdahl, Xenakis and other music program designers.

Boulez/Webern. This approach deals with the composer's perspective. Evidently, a musical composition is never created as a local composition. In classical European literature, e.g., for sonatas, string quartets, or symphonies, which consist of 10^4 to 10^5 tone events, it would be absurd to start from such immense local compositions. Rather does the composer start from 'small' local compositions, such as motives, themes, chords, rhythms and similar elements as a basis of the 'creative combinatorics' and then merges these parts or recombinations thereof by use of various transformations and deformations in order to build a compound whole.

A good example is Boulez' example of a dodecaphonic series in [138, I]: He describes a series together with its internal structure[1], i.e., its composition from partial series and their transformations, see Figure 13.1. Boulez' reflections refer to the local composition *Boul* in ambient space $OnPiMod_{0,12}$ of integer onset and

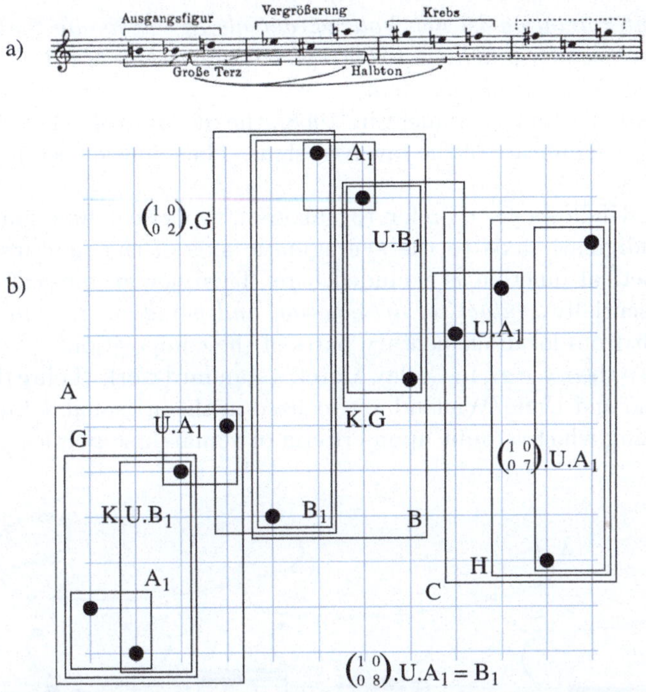

Fig. 13.1. a) Internal structure of a dodecaphonic series following Boulez [138, I], with kind permission of Schott-Verlag; b) representation of the series as a local composition in ambient space $OnPiMod_{0,12}$. Frames are drawn around partial series as suggested by Boulez (part a) of this figure), which are related to each other by symmetry transformations (U = inversion, K = retrograde).

common pitch classes. We see local subcompositions A, A_1, B, G, H of *Boul* and some transformations, drawn as subsets of points which are surrounded by rectangles, defining a total of 15 partial series. These subcompositions are not disjoint in general, they define a rather complex overlapping covering of the given series. We come back to a more systematic and formal discussion of the covering configuration of this example in Section 14.4.

In his discussion of compositional principles with Schönberg, Berg, and Webern, the idea of hierarchies of local compositions is made explicit by Boulez [138, I, p. 86]:

> *Bei Webern findet sich der Keim einer äußerst fruchtbaren Idee, die die Reihe als Einigungsfaktor von* Untergruppen *und* Obergruppen *betrachtet. Im Effekt sind alle isomorphen Figuren einer Grundstruktur von der Tatsache abhängig, daß sie sich immer innerhalb der gleichen Ordnung abwickeln, entsprechend den gegebenen Transpositionen und Umkehrungen; sie integrieren sich in die* Conditio sine qua non *der Zwölftonreihe: die chromatische Totalität. Diese isomorphen Figuren bilden die Basis von privilegierten Mengen, die ihrerseits auf einer höheren Ebene wiederum das vorstellen, was die isomorphen Figuren selbst innerhalb der Reihe bedeuten. Durch Verkettung fügen sich Reihenformen, welche festgelegte Privilegien besitzen, zu einer Ganzheit,*

[1] German: "Binnenstruktur".

einer in gewisser Weise "höheren Reihe" zusammen. Die Grundreihe kann dann als strukturelle Kraft der Vermittlung *zwischen Unter- und Obergruppen betrachtet werden.*

So it is evident that Boulez had learned from the second Viennese school that there is a strong local-global principle in serial composition.

Ude/Wieland and Marek. The art of performance is not accessory, but essential in the constitution of a musical work; this was already pointed out by Adorno [7]:

Musikalische Interpretation ist der Vollzug, der als Synthesis die Sprachähn-lichkeit festhält und zugleich alles einzelne Sprach-ähnliche tilgt. Darum gehört die Idee der Interpretation zur Musik selber und ist ihr nicht akzidentiell.

According to Uhde's and Wieland's comment in [1068], the meaning of "alles einzelne Sprach-ähnliche" is a semantic moment[2], the significate of the involved signs. Thus, music need not be decoded, but requires "imitation" of itself [3].

Semiotically speaking, Uhde seems to refer to Jakobson's poetical function (see Section 11.6.1) as a projection of the paradigmatic axis to the syntagmatic axis of the sign system. A semantic which is independent of the poetical function is excluded here. This view of musical performance as a poetical oriented activity is essentially *syntactical articulation and paradigmatic intertwining.* It establishes an articulated whole, built from local, elementary parts of the composition.

In practical performance theory, e.g., in Ceslav Marek's standard work "Lehre des Klavierspiels" [651], the basic insights of Adorno and Unde/Wieland are realized in the artisanal details. We choose the example of dynamics and phrasing which is built upon criteria of vemarekrse poetics (see Figure 61.1). Here, the

Fig. 13.2. Two grouping proposals by Marek for dynamics and phrasing at the beginning of the C-minor fugue of Bach's *Well-Tempered Piano I*, based upon criteria of verse poetics (from [651], with kind permission of Atlantis-Verlag).

metrical grouping of the melody creates successive parts which induce the shaping in performance by dynamical and phrasing (legato/staccato) prescriptions. This level of grouping into local units is also essential for mnemotechnical purposes and in order to give the fingering strategy a support. In fact, it would be bad piano playing to phrase against the fingering strategy.

Hoffmann and Kaiser. After the foundation of music criticism by Mattheson, Rousseau, or Avison, to name some of the important contributors, it was the merit of Ernst Theodor Amadeus Hoffmann—above all in his famous review of Beethoven's *Fifth symphony* in the Leipzig-based "Allgemeine Musikalische Zeitung" in 1810—of having given music criticism importance as a contribution to the esthesic identification of the work. Since this achievement, music criticism has been cultivated by profiled critics, such as Eduard Hanslick or Joachim Kaiser, and has contributed to the self-estimation of music and to the history of its reception.

It happens only as nonprofessional side-effect that this criticism is eager to celebrate the "uniquely valid" interpretation and performance, or else to lament its vanishing. Although Kaiser may be right in his Beethoven book [517] that whereas it is a rare event in our days when a fundamentally new reading of

[2] Uhde: "meinende Intention".

[3] Uhde: "erheischt Nachahmung ihrer selbst".

the 32 Beethoven sonatas is presented, it is a never-ending objective in this field of classical literature to set forth new approaches to the spiritual torso of these sonatas, approaches which add to the given ones new perspectives, variations of explanatory power. In the media and concert business, the music critics should do exactly the job of commenting on this process and of comparing, questioning and evaluating the singular approaches in the spirit of a work in progress towards understanding an infinite evolution. Now, if the poietic work of the artist is expressed by articulation and correlation of the composition's local parts, then this a fortiori is the work of the critic—only on the level of esthesis. The overall impression of a performance integrates knowledge, prejudice, and personal disposition, which may result in tiny local effects on agogics, dynamics, articulation, and tuning. This is a central effort to open the access of a broader public to the present work.
With the technology of saving works on LP, CD, MC and other media, the articulated listening to music which was founded by music criticism has been enriched by a new aspect: multiply repeated perception. Listening repeatedly to a work on CD changes one's articulation and grouping activity. Every new listening changes or questions the relevant local compositions and their mutual relations, some are eliminated, others are added. Successively, the listener accumulates a *patchwork of elements of comprehension.*

Graeser and Ruwet/Nattiez. On the opposite side of the composers stand the musicologists whose efforts for an adequate analysis are characterized by the need to retrace the composer's thoughts. In this respect it is not astonishing that in the analysis of musical works, organically composed hierarchies are common structures. They start from small local elements, such as chords, degrees, tonalities, tonal functions, voice leading, contrapuntal and harmonic progressions, and end up with large local compositions, such as exposition, development, recapitulation, and coda in the sonata form. This principle of hierarchical organisms was explicitly put into evidence in 1924 by Wolfgang Graeser in his analysis of Bach's *Kunst der Fuge* [387]. He describes a contrapuntal form as follows ([387, p.17]):

Bezeichnen wir die Zusammenfassung irgendwelcher Dinge zu einem Ganzen als eine Menge dieser Dinge und die Dinge selber als Elemente der Menge, so bekommen wir etwa das folgende Bild einer kontrapunktischen Form: eine kontrapunktische Form ist eine Menge von Mengen von Mengen.

Das klingt etwas abstrus, wir wollen aber gleich sehen, was wir uns darunter vorzustellen haben. Bauen wir einmal ein kontrapunktisches Werk auf. Da haben wir zunächst ein Thema. Dies ist eine Zusammenfassung gewisser Töne, also eine Menge, deren Elemente Töne sind. Aus diesem Thema bilden wir eine Durchführung in irgendeiner Form. Immer wird dies Durchführung die Zusammenfassung gewisser Themaeinsätze zu einem Ganzen sein, also eine Menge, deren Elemente Themen sind. Da die Themen selber Mengen von Tönen sind, so ist die Durchführung eine Menge von Mengen. Und eine kontrapunktische Form, ein kontrapunktisches Musikstück ist die Zusammenfassung gewisser Durchführungen zu einem Ganzen, also ein Menge, deren Elemente Mengen von Mengen sind, wir können also sagen: eine Menge von Mengen von Mengen.

The explicit reference to set theory is historically interesting since set theory was a relatively new language in mathematics in 1924. The text is somewhat misleading since it suggests that tones are abstract objects. This is however not the case: Graeser views tones as points in a geometric space, and he also recognizes the role of symmetry transformations, like rotations and reflections, in such a space, transformations which may be applied to alter sets of tones or to compare different tone-sets ([387, p.13]):

Gegenstand der Untersuchungen sind aber nicht die Töne selbst, denn deren spezielle Beschaffenheit spielt gar keine Rolle, sondern die Verknüpfungen und Verbindungen der Töne untereinander.

Es wird uns interessieren, ob wir gewisse Analogien zwischen den Gebilden, die man in der Geometrie aus Punkten aufbaut, und unseren aus Tönen hergestellten erkennen können.

Das wichtigste Grundprinzip der festen Körper, und die Geometrie ist nichts anderes als das Studium der festen Körper, ist die Eigenschaft der Symmetrie.

The Paris school of structuralist linguistics has applied results of semiology after Saussure and Jakobson to musical analysis, above all in the investigations of Nicolas Ruwet [922] and Jean-Jacques Nattiez [782]. The method of *neutral analysis* which was developed by these authors starts from a hierarchical ordering into units and subunits which are a function of the given work. These units are associated with each other by certain equivalence relations:

Les divers unités ont entre elles des rapports d'équivalence de toutes sortes, rapports qui peuvent unir, par exemple, des segments de longueur inégale — tel segment apparaîtra comme une expansion, ou comme une contraction, de tel autre — et aussi des segments empiétinant les uns sur les autres.

The equivalence relations are not arbitrary but realized by specific transformations:

Les unites paradigmatiquement associées sont équivalentes d'un point de vue donné (le thème paradigmatique), rarement identiques, et reliées entre elles par des transformations qui décrivent les variants par rapport a des invariants.

This language resembles the one which Graeser seems to aim at. However the geometric aspect—embedding tones in spaces which admit symmetry transformations—is more radical with Graeser, though less flexible regarding the transformations which may be applied (Graeser limits his approach to transformations of the "rigid" geometry, i.e., isometries).

Jackendoff/Lerdahl. Explicit grouping concepts are described by Ray Jackendoff and Fred Lerdahl in [496]. Grouping is described from an esthesic point of view of music psychology. It deals with portions of notes which are heard as building a unit of hearing. In this approach, a group is always defined by a time interval, i.e., a group consists of strictly time-adjacent notes and cannot be restricted to proper subsets within time-slices which would be defined by voice splitting or parametric splitting of pitch or loudness, for example. Also this grouping is a nearly perfect hierarchy under inclusion: Overlapping neighboring groups may only contain one common onset and are treated as very special situations in this theory. A more general "web of motivic associations" would be beyond the theory *because* it is not hierarchical [496, p.17].

This approach resembles Graeser's concept of a contrapuntal structure which is also a hierarchical grouping of parts, but Graeser's idea was more general insofar as it did not strictly ask for time slices. It is also less paradigmatic than Graeser's, Ruwet's and Nattiez' because no significant statements are made concerning the association of groups under symmetry transformations. From the remarks in [496, p.286] it follows that these authors have no concept of the transformation groups which may be adapted to specific contexts, as it is explicitly proposed by Ruwet and Nattiez under the flag of "paradigmatic theme" (see 11.7.1 for this concept). But it is precisely the psychological claim of grouping—even in its strictly hierarchical appearance of the Jackendoff-Lerdahl theory—as a cognitive basic which undermines the fact of global structures in music.

Schaeffer and Cage. In the last years of the 1940s of the 20th century, the new technology of tape music became a paradigm for local-global constructs. The American music for tape movement was initiated by Otto Luening and Vladimir Ussachevsky and applied by John Cage in the early fifties[4] since his first music for tape *Imaginary landscape.* At the same time, in Paris, Pierre Schaeffer and Pierre Henry initiated the "musique concrète" movement, also based on tape as a flexible medium of syntactical combination and recombination. Pierre Henry produced his first work in musique concrète, *concerto des ambiguités* in 1950.

Tape as a new medium had become a tool for concretely cutting and merging time-slices of music. This was a starting point for an entire group of technological realizations of local-global patchworks. At present, it is largely extended and refined in various software for musical composition, notation, and postproduction as compared to hard disk recording. The local parts are termed tracks, parts, global and local scores, etc. We shall review one of these approaches in Chapter 50 when discussing the composition software *presto*®.

But the tape music movement had not only consequences for music software, it also changed the concept of a score. In particular, Cage realized compositions, such as his *concert for piano and orchestra* (1957),

[4] In the *Project of Music for Magnetic Tape*, together with Morton Feldman, David Tudor, and Christian Wolff.

where the modular structure of the composition, showing a number of relatively autonomous local parts, and an intended ambiguity of identifying such parts, became an explicit and primordial feature of poiesis.

In the light of these rich traces of a local-global paradigm in music, Ehrenfels' approach to gestalt which stresses super-summativity (in fact a warmed-up version of the Aristotelian principle that "the whole is more than the sum of its parts") does not look very original. But it does provoke the question how much more the added value exceeds the sum of the parts. The above examples show that this may be a very complex question, in fact the only interesting point of super-summativity. How is the whole constructed from the simple collection of its parts? When are two wholes different, though having identical parts? How can we compare wholes if we suppose that their parts are comparable?

Fact 8 *As long as no precise structure theory of wholes qua constructs from local parts is available, no real understanding of music is possible. By the arsenal of the preceding examples, this is a problem which touches all levels of the communicative axis. And it is a problem which involves interpretative activity and its innate ambiguities.*

13.1.1 Musical and Mathematical Manifolds

Summary. We shortly discuss the question of how music and mathematics differ in their understanding of local and global structures.

– Σ –

The historical point of creating and pronouncing mathematical and musicological concepts of global gestalts is situated around 1854 when Eduard Hanslick (Figure 13.4) published his famous treatise "Vom Musikalisch Schönen" [438]. Musical content was recognized as being "tönend bewegte Formen". Hanslick added that these forms are by no means elementary but composed in an artistical way, and building a unity within the manifold, as restated by music theorist Hugo Riemann.

Fig. 13.3. Bernhard Riemann introduced global structures (manifolds) as compound mathematical spaces. Sources: See text.

Fig. 13.4. Eduard Hanslick described music as a compound structure, built artistically from parts. With kind permission of ÖNB/Wien, 106.732 C.

In the same year, the mathematician Bernhard Riemann (Figure 13.3[5]) conceived a far reaching generalization of the mathematical concept of space to so-called manifolds [899]. These are understood as being patchworks of locally cartesian charts, similar to geographic atlases. The common ground of both approaches

[5] Source: `http://www.math.uni-hamburg.de/home/grothkopf/fotos/math-ges/` and Archives of the Mathematisches Forschungsinstitut Oberwolfach. With kind permission of Mathematisches Forschungsinstitut Oberwolfach.

is that locally trivial structures can add up to esthetically valid configurations if glued together in a non-trivial way. A simple and well-known example of such a global shape is the Möbius strip. Its fascination stems from gluing together the ends of an ordinary 'belt'-shaped strip after a rotation of half the full circle of one end. For example, in musicology, the Möbius strip is realized by the harmonic strip of triadic degrees within a diatonic scale, see Section 13.2.

The characteristic difference between musical and mathematical manifolds is that musical manifolds are defined with a fixed atlas whereas mathematical manifolds are not tied to fixed atlases. Intuitively speaking, the sphere of our globe, viewed as a mathematical manifold, is the same if we add new small or large charts as long as they are compatible with the given ones. If I add a city map, nobody will complain that the geographic identity of the globe has changed. Mathematically speaking, we may go to the colimit of all atlases. In music, this is completely different: A given covering of a composition by a determined set of charts, such as chords, motives, or periods, is essential to the identification of the composition, two different coverings change the composition qua global structure or gestalt, to use Ehrenfels' concept. The colimit is not allowed, the individual interpretational activity is an integral part of the object's identity.

13.2 What Are Global Compositions?

Summary. This more technical section gives the precise definition of an *objective* global composition. This is a "patchwork" of (objective) local compositions, formally captured by the concept pairing of *charts* and *atlases*, a standard structure in differential and algebraic geometry.

– Σ –

After the preceding propaedeutic reflections, we want to give a precise technical definition of a global composition built upon the objective local compositions as local charts. After these technicalities, we give comments on more intuitive aspects.

Definition 36 *A (global) objective composition is defined by the following data:*

(i) *A set G and a finite, non-empty covering I of G,*
(ii) *an address A,*
(iii) *a family $(K_t, A@F_t)_{t\in T}$ of A-addressed local objective compositions,*
(iv) *a surjection $I_? : T \to I : t \mapsto I_t$,*
(v) *a bijection $\phi_t : K_t \xrightarrow{\sim} I_t$ for each $t \in T$,*
(vi) *for each couple $s, t \in T$ such that $I_s \cap I_t \neq \varnothing$, the induced bijection*

$$\phi_{s,t} := \phi_t^{-1} \circ \phi_s : \phi_s^{-1}(I_s \cap I_t) \to \phi_t^{-1}(I_s \cap I_t)$$

(restricted to the respective domains and codomains) defines an isomorphism

$$\phi_{s,t}/1 : (\phi_s^{-1}(I_s \cap I_t), A@F_s) \xrightarrow{\sim} (\phi_t^{-1}(I_s \cap I_t), A@F_t)$$

of local compositions.

*The data (iii) to (vi) are called an A-*addressed atlas Φ *for the covering I of G. The bijections ϕ_t (or—by abuse of language—the local compositions K_t) are called the* charts of the atlas Φ.

Two A-addressed atlases Φ, Ψ for the covering I of G are called equivalent *iff their disjoint sum*[6] $\Phi \amalg \Psi$ *is an A-addressed atlas for the covering I of G. An A-*addressed objective global composition *is a covering I of G (often abbreviated by G^I) together with an equivalence class of A-addressed atlases for G^I. If no confusion is likely, we may abbreviate the entire data by saying that we are given an objective global composition G, i.e., by just naming the* support set G *of the objective global composition.*

[6] $\Phi \amalg \Psi$ has index set $T_\Phi \amalg T_\Psi$, and the atlas corresponding to the coproduct surjection $T_\Phi \amalg T_\Psi \to I$.

Example 17 Let us start with an illustration of this definition by an example which we have already alluded to: the internal structure of Boulez' dodecaphonic series (see the Boulez/Webern approach in Section 13.1 and Figure 13.1).

We select as the supporting set the support *Boul* of the zero-addressed local composition

$$(Boul, 0@OnPiMod_{0,12})$$

in that example and abbreviate $OP := OnPiMod_{0,12}$. We consider the covering $I = \{A, B, G, \ldots H\}$ by the 15 subsets described in Figure 13.1, and which express the internal structure of the series. These subsets of *Boul* define corresponding zero-addressed local compositions

$$K_1 = (A, 0@OP), K_2 = (B, 0@OP), K_3 = (G, 0@OP), \ldots K_{15} = (H, 0@OP)$$

and the evident atlas Φ via the identities $\phi_1 : K_1 \xrightarrow{\sim} A, \ldots \phi_{15} : K_{15} \xrightarrow{\sim} H$ on the chart supports. Clearly, the gluing condition (vi) is verified.

Evidently, this is not the most economic atlas since many of the local compositions are isomorphic. For instance, we could replace the identity chart $Id : K.G \xrightarrow{\sim} K.G$ by the retrograde isomorphism $K : G \xrightarrow{\sim} K.G$, and therefore introduce a new chart domain $(G, 0@OP)$ of this isomorphism. We can proceed in this way until we obtain a minimal set of eight chart domains

$$(A, 0@OP), (A_1, 0@OP), (B, 0@OP), (B_1, 0@OP)$$
$$(G, 0@OP), (\left(\begin{smallmatrix} 1 & 0 \\ 0 & 2 \end{smallmatrix}\right) \cdot G, 0@OP), (C, 0@OP), (H, 0@OP)$$

together with 15 corresponding isomorphisms ψ_i, defining a second, equivalent atlas Ψ for $Boul^I$. This second atlas makes evident the typology of the internal structure of *Boul*.

Exercise 21 *Show that the relation of atlas equivalence in Definition 36 is in fact an equivalence relation*

13.2.1 The Nerve of an Objective Global Composition

Since the covering G^I is a fixed data of an objective global composition, we should immediately introduce a visualization tool: the nerve of an objective global composition. In the course of the theory, this construction will be extended, but we should look at its most elementary aspect as soon as possible.

Given a global objective composition G^I, the *(abstract) nerve* or simplicial complex $n(G^I)$ of G^I, as well as its *geometric realization* $N(G^I)$ are defined[7]. Recall that $N(G^I)$ is a union of affine simplexes $|\sigma|$ associated with the (abstract) simplexes $\sigma \in n(G^I)$ such that for any two different simplexes $|\sigma|, |\tau|$, their interiors[8] are disjoint, i.e., $|\sigma|^o \cap |\tau|^o = \emptyset$.

Example 18 For our example $Boul^I$, we start with the zero-dimensional skeleton $N_0(Boul^I)$ which is a set of 15 points, the 0-simplexes $|J|$, one for each chart $J \in I$, which we distribute in three-space $\mathbb{R}^3$, see Figure 13.5 a) for this procedure. For the moment, it is not important where to place these points; the only condition is that they be distinct if their charts[9] $J \in I$ are. Now, we look at all couples J, J' with non-empty intersection. Their 0-simplexes $|J|, |J'|$ are connected by a straight line, a 1-simplex $|J, J'|$, see 13.5 b). This visualization idea stems from combinatorial topology, however, musicology has been aware of such a construction. In his treatise on harmony *[949]* Schönberg talks about the harmonic strip[10] between two chords which have one or several tones (pitch classes) in common. If we view chords as being charts in the

[7] See Appendix H.2 for these concepts.

[8] The interior of an affine simplex is the simplex minus its faces. For example, the interior of an affine 1-simplex is the straight line minus the two endpoints.

[9] This is a typical abuse of language: identification of the chart with its codomain. In fact, we are only looking at the codomains in the covering I.

[10] "Harmonisches Band."

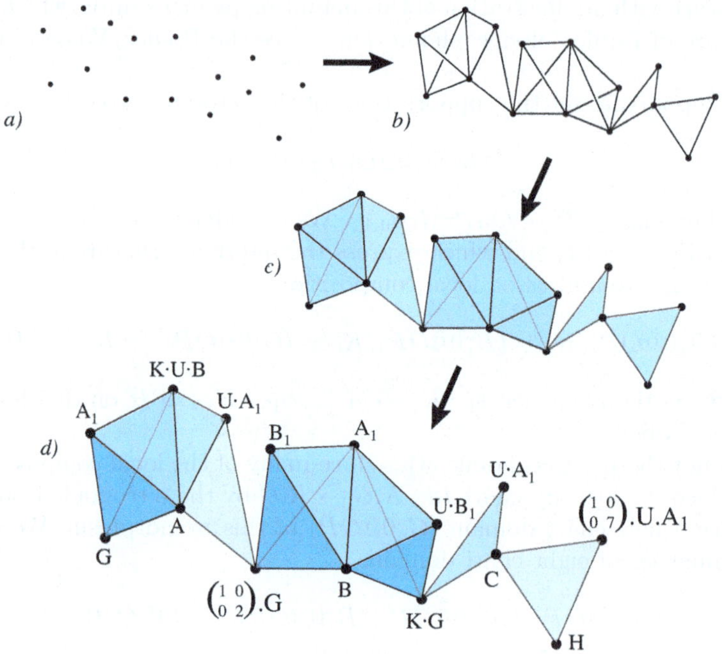

Fig. 13.5. Construction of the nerve of the Boulez series *Boul* for its internal structure defined in Figure 13.1. a) Every chart corresponds to a point. b) Two points are connected by a straight line iff the corresponding charts have non-empty intersection. c) Three different charts having common points are connected by triangular surfaces. d) four different charts with common tones define a full tetrahedron.

framework of harmony (see also Section 13.4.2 and Chapter 80), 1-simplexes are precisely the formalization of Schönberg's harmonic strip.

In combinatorial topology, one goes one step further. There is no deeper reason to stop at 1-simplexes and to proceed with a test for common tones in three or more charts. In our next step, we look at all triples J, J', J'' of mutually distinct charts such that they have common elements. We then add 2-simplexes $|J, J', J''|$, i.e., triangular surfaces with the 0-simplexes $|J|, |J'|, |J''|$ as vertices, see 13.5 c). And the last step consists in joining each group of four distinct 0-simplexes $|J|, |J'|, |J''|, |J'''|$ by a full tetrahedron if the intersection $J \cap J' \cap J'' \cap J'''$ is non-empty, see 13.5 d).

So the internal structure of Boulez' series appears as a complex intertwining of local charts which overlap as shown by the geometric nerve in 13.5 d). Observe that in general, this procedure will not stop in three-dimensional simplexes. Only here, no five charts have common notes.

The shape of the nerve of an objective global composition is a good measure for the complexity of the global configuration. There are two extremal situations of this perspective: On one hand, a nerve may be discrete, i.e., reduced to the 0-dimensional skeleton. This happens if we just draw disjoint groups of notes on a given composition (as principally suggested by Jackendoff/Lerdahl), which is called a *discrete interpretation.* In this case, there are only isolated local objective compositions around, and the connectivity of the global construction is trivial. The extreme case of such a "strategy" happens if we just draw circles around every tone of a composition; this will completely destroy its gestalt, and we essentially boil down everything to counting notes. This is quite silly (though not superfluous for accounting purposes), and we shall call such a global composition a *silly interpretation.* On the other hand, we may just take one single local chart to cover the entire composition, and in this case, we have a huge set of notes which does not observe all the more local similarities or symmetries or other relations. This will be a very difficult, cryptic object, also since classification of large local compositions is quite intricate (see our discussion in Chapter 11), we call it the *indiscrete interpretation.*

Both extrema are no intelligent solutions for human cognition: Good global structures are somewhere in the middle between intractable monsters and insignificant atomized data. But it would be nonsense to set up artificial limits (like Jackendoff/Lerdahl [496]) to the grouping formalism because it is one of the most important objectives to make all grouping perspectives accessible and formally tractable. The breakdown of the Jackendoff/Lerdahl Generative Theory of Tonal Music (GTTM) happens where their grouping levels have to rely on traditional music theory in order to work. The GTTM is built on a competence which is both, formally and semantically, out of the reach of the GTTM, and therefore boils down to an interface between psychology and music theory which does neither solve the formal inconsistency of music theory nor observe the variability of the psychological grouping: a bad glue of two unresolved components.

Often, an objective global composition is not given in advance, but only results from a compatible gluing of local charts. The next theorem describes a necessary and sufficient condition for a system of local compositions to become an atlas for a global composition.

Theorem 12 *Let A be an address, T a set of indexes. Suppose we are given a family of A-addressed objective local compositions*

$$Rel\Phi = ((K_{i,j}, A@F_j))_{i,j\in T} \tag{13.1}$$

of forms F_j, together with a family

$$Iso\Phi = (\phi_{i,j}/1 : (K_{i,j}, A@F_j) \xrightarrow{\sim} (K_{j,i}, A@F_i))_{i,j\in T} \tag{13.2}$$

of isomorphisms in ObLoc whenever $K_{i,j}$ is non-empty. And we assume that each $(K_{i,j}, A@F_j)$ is a subcomposition of the diagonal element $(K_{j,j}, A@F_j)$, i.e., we have inclusion morphisms $\rho_{i,j}/1 : K_{i,j} \subset K_{j,j}$, and that for all indexes, $\phi_{i,j} \circ \phi_{j,i} = Id$, and, more specially, $\phi_{i,i} = Id$. Consider the binary relation $\sim$ on the disjoint union of the sets $K_{j,j}$ defined by $x \sim y$ iff $x \in K_{i,j}$ for some index couple i, j, and $\phi_{i,j}(x) = y$.

Let U be the colimit of the system of set maps $\phi_{i,j}$ and $\rho_{i,j}$. Then the canonical maps $k_j : K_{j,j} \to U$ define an A-addressed objective global composition with atlas $((K_{j,j}, A@F_j))_{j\in T}$ and intersections $(K_{i,j}, A@F_j)$ on the charts, iff the relation $\sim$ is an equivalence relation.

Proof. If the canonical maps define an A-addressed objective global composition with atlas $((K_{j,j}, A@F_j))_{j\in T}$, then the relation $\sim$ is the inverse image of the equivalence relation of equality on U for the surjection $\coprod K_{j,j} \to U$, and the claim follows. If, conversely, the relation $\sim$ is an equivalence relation, then the colimit, which is identified[11] to $U = \coprod K_{j,j}/\sim$, defines injective maps (!) $k_j : K_{j,j} \to U$, and the non-empty intersections $k_i(K_{i,i}) \cap k_j(K_{j,j})$ correspond to the local subcompositions $(K_{i,j}, A@F_j), (K_{j,i}, A@F_i)$ which are isomorphic under the morphisms of the system $Iso\Phi$. QED.

Observe that objective global compositions need not be derived from given local compositions by coverings as in the above Boulez example. The concept is a proper extension to the local framework, and the general construction method is described by the above theorem. We shall develop necessary and sufficient criteria for covering constructs in Section 16.1.

Exercise 22 *Let us define a zero-addressed objective global composition by data corresponding to the above theorem. The index set is $T = \{1, 2, 3\}$. We take a constant space form $F_i = Onset \oplus Pitch$ with functor $@\mathbb{R}^2$. The family $Rel\Phi$ of local compositions is defined as follows: Consider the four (zero-addressed) points $c_1 = (0,0), c_2 = (0,1), c_3 = (1,0), c_4 = (1,1)$ in $\mathbb{R}^2$ and set*

$$\begin{aligned}
K_{i,i} &= (\{c_1, c_2, c_3, c_4\}, 0@\mathbb{R}^2), i = 1,2,3,\\
K_{1,2} = K_{2,1} &= (\{c_3, c_4\}, 0@\mathbb{R}^2),\\
K_{1,3} = K_{3,1} &= (\{c_1, c_2\}, 0@\mathbb{R}^2),\\
K_{2,3} &= (\{c_3, c_4\}, 0@\mathbb{R}^2),\\
K_{3,2} &= (\{c_1, c_2\}, 0@\mathbb{R}^2),\\
K_{i,j} &= \emptyset \text{ else.}
\end{aligned}$$

[11] See Appendix G.2.1.

The family $Iso\Phi$ is given by

$$\phi_{1,2} = Id_{K_{2,1}},$$
$$\phi_{1,3} = Id_{K_{3,1}},$$
$$\phi_{2,3} = \phi_{3,2} = e^{1,1} \cdot \begin{pmatrix} -1 & 0 \\ 0 & -1 \end{pmatrix}.$$

Show that the conditions of the theorem are satisfied and try to draw a picture of the global composition. We shall see in Section 16.1 that this global composition cannot be constructed by a covering of a local composition.

We should observe a hidden subtlety of the global composition context: Global ambient spaces. In fact, every local composition is connected to an ambient space form F. But if we are given an atlas of local compositions, the ambient spaces are no longer uniquely determined. In fact, the construction of a global composition makes chart names exchangeable: it is 'homonymous'. More precisely, we have as many names for a given point x of the support as we have chart names on any of the compatible atlases. So the name of the 'ambient space' of x is an entire collection of local chart names. On the level of local supports, univocal naming would not be a good idea since on the overlaps of two different covering sets, what name should one select? Homonymy is unavoidable for global compositions. One should, however, not refrain from naming or sweep names away in favor of anonymy. Global naming is interchangeable, but not irrelevant:

Definition 37 *The name $Name(x)$ of an element $x \in G$ of an objective global composition G is the set of names of all the local chart forms (in any compatible atlas) which hit this element.*

13.3 Functorial Global Compositions

Summary. Corresponding to functorial local objects, the (functorial) local compositions in the category *Loc*, *global functorial compositions* are introduced as global objects by an atlas of charts, composed of (functorial) local compositions.

– Σ –

The definition of functorial global compositions is the transposition of the objective case with everything made functorial:

Definition 38 *A (global) functorial composition is defined by the following data:*

(i) *A functor $G \in \mathbf{Mod}^@$ and a finite, non-empty, generating set I of subfunctors of G, i.e., $G = \bigcup I$,*
(ii) *an address A,*
(iii) *a family $(K_t, @A \times F_t)_{t \in T}$ of A-addressed functorial local compositions,*
(iv) *a surjection $I_? : T \to I : t \mapsto I_t$,*
(v) *an isomorphism of functors $\phi_t : K_t \xrightarrow{\sim} I_t$ for each $t \in T$,*
(vi) *for each couple $s, t \in T$ such that $I_s \cap I_t \neq \emptyset$, the induced isomorphism*

$$\phi_{s,t} := \phi_t^{-1} \circ \phi_s : \phi_s^{-1}(I_s \cap I_t) \to \phi_t^{-1}(I_s \cap I_t)$$

(restricted to the respective domains and codomains) defines an isomorphism

$$\phi_{s,t}/1 : (\phi_s^{-1}(I_s \cap I_t), @A \times F_s) \xrightarrow{\sim} (\phi_t^{-1}(I_s \cap I_t), @A \times F_t)$$

of functorial local compositions.

The data (iii) to (vi) are called an A-addressed atlas Φ for the covering *I of G. The bijections* ϕ_t *(or—by abuse of language—the local compositions* K_t*) are called the* charts of the atlas Φ.

Two A-addressed atlases Φ, Ψ *for the covering I of G are called* equivalent *iff their disjoint sum*[12] $\Phi \amalg \Psi$ *is an A-addressed atlas for the covering I of G. An A-addressed functorial global composition is a covering I of G (often abbreviated by* G^I*) together with an equivalence class of A-addressed atlases for* G^I*. If no confusion is likely, we may abbreviate the entire data by saying that we are given a global functorial composition G, i.e., by just naming the* support *G of the global composition.*

Example 19 The first and immediate example for this definition is the functorial global composition $\hat{G}$ associated with a given objective global composition G, if the latter has some adequate properties. To construct this object, suppose that we are given an A-addressed objective global composition G with the notation and data of Definition 36, in particular we have an atlas $(K_t, A@F_t)_{t\in T}$ of local compositions.

To begin with, we have a family of functorial local compositions $(\hat{K}_t, @A \times F_t)_{t\in T}$, as well as a system of subcomposition inclusion monomorphisms

$$\hat{\rho}_{s,t} : (\hat{K}_{s,t}, @A \times F_t) \subset (\hat{K}_t, @A \times F_t),$$

defined by the inverse images $K_{s,t} = \phi_t^{-1}(I_s \cap I_t)$ for any two different indexes s, t. We also have induced isomorphisms

$$\hat{\phi}_{s,t}/1 : \hat{K}_{t,s} \xrightarrow{\sim} \hat{K}_{s,t}$$

of functorial local compositions. Let $\hat{G}$ be the colimit functor of the system $\hat{\phi}_{s,t}, \hat{\rho}_{s,t}$ of isomorphisms and inclusions. For an address change $x : B \to A$, we have the formula

$$x@\hat{G} \xrightarrow{\sim} \mathrm{colim}(x@\hat{K}_s, x@\hat{K}_{s,t}) = \mathrm{colim}(K_s.x, K_{s,t}.x) \tag{13.3}$$

on the slices of the colimit. Because of the right colimit expression, the slice $x@\hat{G}$ is also denoted by $G.x$.

In general, $\hat{G}$ is not a global functorial composition, in fact, the canonical morphisms $\hat{K}_t \to \hat{G}$ are not even injective. To obtain a global composition, we suppose that for all address changes $f : B \to A$, the maps

$$A@F_s \to B@F_s$$

are injective for all indexes s. This is the case, for example, if $A = 0_{\mathbb{Z}}$, or if F_s is constant. Then it follows that the system of isomorphisms and inclusions $\phi_{s,t}, \rho_{s,t}$ satisfies the conditions of Theorem 12, and therefore, the canonical morphisms

$$i_t : \hat{K}_t \to \hat{G}$$

are mono, i.e., injective for all addresses. Further, the canonical squares

$$\begin{array}{ccc} \hat{K}_{t,s} & \xrightarrow{\hat{\rho}_{t,s}} & \hat{K}_s \\ {\scriptstyle \hat{\rho}_{s,t}\circ\hat{\phi}_{s,t}}\downarrow & & \downarrow{\scriptstyle i_s} \\ \hat{K}_t & \xrightarrow{i_t} & \hat{G} \end{array} \tag{13.4}$$

are cartesian. Therefore $\hat{G}$ is generated by the isomorphic images $\hat{I}_t$ of $\hat{K}_t$. Furthermore, since the evaluation of the inductive system of functors at the morphism Id_A (the slice at Id_A) yields the original objective configuration, the generators $\hat{I}_t$ are different iff the original covering elements I_t are. So we have obtained a finite covering $\hat{I}$ of $\hat{G}$, an atlas $(\hat{K}_t, @A \times F_t)_{t\in T}$ of functorial local compositions, and a surjection $I_? : T \to \hat{I}$; this completes the construction.

Exercise 23 *Explicate this construction in all the details.*

[12] $\Phi \amalg \Psi$ has index set $T_\Phi \amalg T_\Psi$, and the atlas corresponding to the coproduct surjection $T_\Phi \amalg T_\Psi \to I$.

Exercise 24 *Establish a theorem for functorial compositions which corresponds to Theorem 12.*

Remark 3 Observe that, although the chart morphisms are compatible in the sense of fulfilling the conditions of Theorem 12, no compatibility of the underlying functor morphisms on the ambient spaces is required. This means that we will not be able to extend the gluing procedures to the ambient spaces; in other words: global objects are really global, no common ambient space is available in general!

13.4 Interpretations and the Vocabulary of Global Concepts

Summary. A large set of (objective) global compositions is constructed by "interpretations" of given (objective) local compositions. This construction is a subtlety that tends to escape to common interpretative activities. It is, however, a basic prerequisite for every interpretative activity in music. This is why this type of global construction is called *interpretation.* We give standard examples of such global objects.

$-\Sigma-$

Very often, music analysis and composition explicitly deal with local compositions, but this coarse perspective is not sufficient: implicitly, one specifies local parts of this data and different overlapping relations among such parts. For example, Boulez' series of example 13.1 needs a covering by 15 local parts in order to be fully understood. Many analytical texts (e.g. [496, 782, 922]) also use this kind of local part selection.

As already stressed, this type of chart specification is not temporary, it is a substantial attribute of the analytical work. We shall therefore introduce a special construction method of global compositions which take into account this important technique as an interpretational basic. Here is the general construction method of such global compositions which we therefore call interpretations.

Definition 39 *Let $(K, A@F)$ be an A-addressed objective local composition, and let I be a non-empty covering of the support set K. Then the covering K^I, together with the obvious atlas $(i, A@F)_{i\in I}$, defines an objective global composition, the* interpretation K^I of K *associated with the covering I.*

Let $(K, @A \times F)$ be an A-addressed functorial local composition, and let I be a non-empty covering[13] *of the support functor K. Then the functorial covering K^I, together with the obvious atlas $(i, @A \times F)_{i\in I}$, defines a functorial global composition, the* interpretation K^I of K *associated with the functorial covering I.*

We shall see in Chapter 16 that interpretations define a special case: Global compositions may be far from 'interpretable', i.e., they are not 'isomorphic' to any possible interpretations (we shall make these remarks precise later).

Example 20 A second example of functorial global compositions is associated with interpretations of objective compositions. Let $(K, A@F)$ be an objective local composition, and take an interpretation K^I by an atlas $I = (K_\iota)_\iota$ of subcompositions. Then we have an interpretation $\hat{I} = (\hat{K_\iota})_\iota$ of the local functorial composition $\hat{K}$ since unions and $\hat{?}$ commute by Lemma 2, Section 7.4. But observe that the intersection $\hat{K_{\iota_1}} \wedge \hat{K_{\iota_2}}$ is not objective in general. This is however the case for the zero address $A = 0_{\mathbb{Z}}$, and there, the colimit construction and this one coincide.

13.4.1 Iterated Interpretations

Summary. Interpretation is not a one-step process, it may be iterated on an infinity of levels. We present the formal framework and examples.

$-\Sigma-$

[13] The functor K is generated by the subfunctors in I.

If $(K, A@F)$ is an A-addressed objective local composition, we can give an interpretation by a covering set I of subcompositions $(K_\iota, A@F)$ of K. We then introduce this sequence of forms:

$$G_0 = F,$$
$$G_{i+1} \underset{2^{Fun(G_i)} \rightarrowtail \Omega^{Fun(G_i)}}{\longrightarrow} \mathbf{Power}(G_i), i = 0, \ldots.$$

Then, the interpretation I is an A-addressed objective local composition in G_1 whose union is K, $\bigcup I = K$. Here, the union operator is defined functorially as a transformation $\bigcup_i : G_{i+1} \to G_i, i = 1, 2, \ldots$, with the initial evaluation $\bigcup_1(I) = K$. So we have an infinite diagram

$$\mathbb{D} = (G_1 \overset{\bigcup_1}{\leftarrow} G_2 \overset{\bigcup_2}{\leftarrow} G_3 \overset{\bigcup_3}{\leftarrow} \ldots)$$

and we can define a corresponding limit form

$$Limint(F) \underset{Id}{\longrightarrow} \mathbf{Limit}(\mathbb{D}) \qquad (13.5)$$

whose denotators are local compositions K as above, together with coverings I of K, and coverings J of I etc. This means that we are given an infinite succession of interpretations and interpretations of interpretations etc., in other words, an infinite sequence of interpretations of constantly increasing level of complexity. Of course there are infinitely many such denotators even if we fix an initial sequence of first k positions.

A very classical 'infinite interpretation' is the one induced by the nerve of an interpretation. In fact, if we start as above, i.e., taking the covering I of K, we get an initial sequence $(K, I, \ldots)$. The simplicial complex $n(I)$ is an element of G_3 and can be iterated ad libitum by $n_{t+1}(I) = n(n_t(I))$. This defines a sequence

$$n_\infty(K^I) = (K, I, n(I), n^2(I), \ldots n^t(I), \ldots)$$

which indeed is in $Limint(F)$. In particular, if $x \in K$, the star of x $st(x) = \{U \in I \text{ with } x \in U\}$ is an element of $n(I)$, and the star $st^2(x) = st(st(x))$ of $st(x)$ is an element of $n^3(I)$, etc., $st^i(x) \in n^{1+2i}(I)$.

Suppose that each element $x \in K$ is given a weight w, for example the constant weight $w(x) = 1$ or some weight stemming from an analysis of K as in motivic or metrical theory. Then each covering set $U \in I$ can be given a numerical weight, for example a power $w(U) = (\sum_{x \in U} w(x) - 1)^p$ as in the metrical theory (see Section 21.2) or some other musically motivated evaluation. Then, if $x \in K$, we have a weight $w_1(x) = w(st(x)) = \sum_{U \in st(x)} w(U)$, and then inductively $w_{t+1}(x) = \sum_{u \in st^{t+1}(x)} w(u)$, if we take the above formula $w(u) = (\sum_{x \in u} w(x) - 1)^p$ to calculate recursively the weight of elements of G_i. This associates a power series

$$\lambda_w(x) = 1 + \sum_{1 \leqslant t} \frac{w_t(x)}{t!} T^t.$$

The reader may easily generalize this idea to obtain a numerical evaluation of points in infinite interpretations. This means that infinite interpretations can be boiled down to yield numerical 'coordinates' of points in compositions, and therefore quantitative measures for further processing within performance or analytical contexts.

13.4.2 The Pitch Domain: Chains of Thirds, Ecclesiastical Modes, Triadic and Quaternary Degrees

Summary. Chains of thirds and modes are a classical domain of interpretation. The construction deals with "refinement" of scales qua local compositions. The refinement specifies determined notes or groups of notes which give the scales an interior profile. More elaborate interpretations of scales concern triadic and quaternary degrees for concerns of harmony.

$- \Sigma -$

Recall from Section 7.2.1 that we may identify an octave-periodic scale S by its class chord $S : A \rightsquigarrow$o-Scale$(S_1, \ldots S_k)$. Many common scales and chords are classified by use of the so-called third chain construction. To understand this construction, consider the two standard cases: 12-tempered and just scales.

For the 12-tempered case, we are effectively working in the ambient pitch class space $PiMod_{12}$ with module functor $@\mathbb{Z}_{12}$. Here, we have the major third $M_{12} = e^4$ and the minor third $m_{12} = e^3$ translation. If the pitch class space is identified with its Sylow decomposition: $@\mathbb{Z}_{12} \xrightarrow{\sim} @(\mathbb{Z}_4 \oplus \mathbb{Z}_3) : x \mapsto (x \mod 4, x \mod 3)$, M_{12} identifies to $M_{12} = e^{(0,1)}$, whereas m_{12} identifies to $e^{(-1,0)}$. See right half of Figure 13.6 for the torus representation of $PiMod_{12}$. On the $EulerPlane$ ambient space (see example 7.2.1 in Section 7.2.1), we have the corresponding translations $M_{\text{just}} = e^{(0,1)}$ and $m_{\text{just}} = e^{(1,-1)}$, see left half of Figure 13.6.

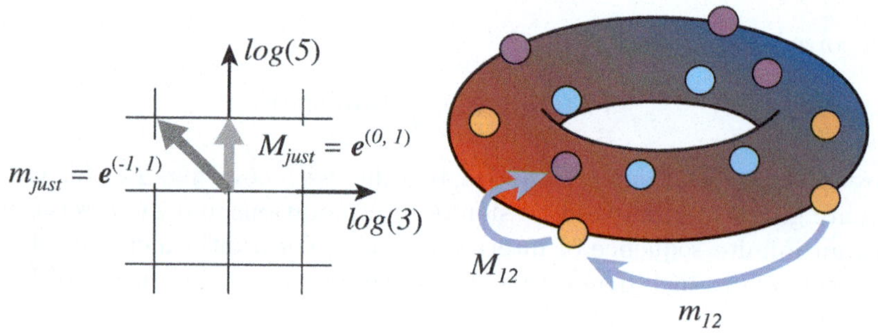

Fig. 13.6. Left: Major and minor third transpositions in just tuning pitch classes. Right: The third transpositions on the third torus representing 12-tempered pitch classes.

With these translations in mind, we can define a third chain:

Definition 40 *A chord $S = \{S_1, \ldots S_k\}$ in $PiMod_{12}$ or in $EulerPlane$, respectively, of cardinality k is called* a third chain *iff the elements of S can be ordered in such a way that $S_{i_{j+1}} = T_j(S_{i_j})$ for all $j = 1, \ldots k-1$ and T_j is either the major third translation M_{12} (resp. M_{just}) or the minor major translation m_{12} or m_{just}, respectively.*

Lemma 16 *Every zero-addressed chord in $PiMod_{12}$ is contained in a third chain.*

If we admit positive and negative translation in the minor third direction m_{just}, we obtain the concept of *weak third chains* in just tuning (see also right half of Figure 13.6).

Lemma 17 *Every zero-addressed chord in $EulerPlane$ is contained in a weak third chain, but not necessarily in a third chain.*

Exercise 25 *Give a proof of the preceding lemmata.*

The set $3Chains$ of all zero-addressed third chains in $PiMod_{12}$ is known and has been used in computer software *presto*®(see Section 25.2.1), as well as in RUBATO®'s HarmoRubette (see Chapter 41.3). See Appendix N.2 for this list. The (objective) interpretation $0@PiMod_{12}^{3Chains}$ of $0@PiMod_{12}$ by the covering $3Chains$ is called the *third chain interpretation* of the pitch class space.

In the 12-tempered case, the set of third chains containing a given zero-addressed chord S with minimal cardinality are called *the minimal third chains of S*, and this set is called the *third chain closure of S*, it is denoted by $3Chain(S)$. This set can grow quite dramatically for non-standard chords. For example, the chromatic 3-chord $S = \{0, 1, 2\}$ has 23 minimal (7-element) third chains, including third chains which start with three of the classical triads: major, minor, and augmented.

In order to describe ecclesiastical modes, we make use of the 12-tempered space $PiMod_{12}$, although other pitch class spaces would also do the job. The point here is less the tuning but the construction of an interpretation from a given scale (chord). Our modeling of ecclesiastical modes will follow the (zero-addressed) scale X = C-major = $\{\overline{0}, \overline{2}, \overline{4}, \overline{5}, \overline{7}, \overline{9}, \overline{11}\}$, the other modes are be derived by evident transposition.

To describe modes, one selects two elements in X: the *tenor* or *recitation tone* t, as well as the *finalis* or *final* f. The first interpretation is the *mode* X_f, i.e., the enrichment of the scale by a determined *tonic* f. This is an interpretation $X_f = X^{I_f}$ of X by an atlas I_f of two charts: $I = \{X, \{f\}\}$, the scale and the tonic singleton. Observe that this is not a full-fledged concept of tonality. To meet this requirement, a mode may and will be enriched by other structural aspects, such as selected degree chords, etc. According to the position of the tonic, the mode has the names as listed in table 13.4.2.A.

Table 13.4.2.A

Nr.	mode and plagal variant	finalis f	tenor t
1.a	Dorian	d	$a = d + 7$
1.p	Hypodorian	d	$f = d + 3$
2.a	Phrygian	e	$c = e + 8$
2.p	Hypophrygian	e	$a = e + 5$
3.a	Lydian	f	$c = f + 7$
3.p	Hypolydian	f	$a = f + 4$
4.a	Mixolydian	g	$d = g + 7$
4.p	Hypomixolydian	g	$c = g + 5$
5.a	Aeolian	a	$e = a + 7$
5.p	Hypoaeolian	a	$c = a + 3$
6.a	Locrian	b	$g = b + 8$
6.p	Hypolorian	b	$d = b + 3$
7.a	Ionian	c	$g = c + 7$
7.p	Hypoionian	c	$e = c + 4$

The next step introduces an ecclesiastical mode[14] as a refined version of the simple mode interpretation: An *ecclesiastical mode* on scale X is the interpretation $X_{f,t} = X^{I_{f,t}}$ with atlas $I_{f,t} = \{X, \{f\}, \{f, t\}\}$. For every final f, there are two modal variants: the *authentic* (.a in table 13.4.2.A) and the *plagal* (.p in table 13.4.2.A), the latter being marked by the prefix "Hypo". For the authentic mode we have either $t = f + 7$ or $t = f + 8$ (sixths), for the plagal one, we have either $t = f + 3$ or $t = f + 4$ or $t = f + 5$ (thirds or the fourth). The modal pair Nr. $6.a, 6.p$ are listed for completeness, however, they are scarcely documented.

The aeolian and ionian modes were only introduced in the 16th century by Glarean, but they are the basis of the modern major-minor system where the tenors have disappeared (and we only consider the modes X_f), the final tone being renamed to *tonic*, such that we now have aeolian =cantus mollis (C_a for scale C), and ionian = cantus durus (C_c for scale C).

In harmony, interpretations are important which associate 3-chords with special harmonic functions. Corresponding interpretations of scales are called triadic interpretations, we shall discuss them in the sequel for the 12-tempered and just tuning. We shall see below that the examples to be discussed here have special automorphism properties. This is why we call them *triadic degree interpretations*.

For 12-tempered pitch classes, we again take the major scales X, as well as the melodic and harmonic minor scale x_m and x_h (see Figure 7.4). Observe however, that the name of such a scale, $X = C$, for example, does not imply the selection of a tonic. The name is only historically loaded and not in our structural setting. We are going to define the triadic degree interpretations for $X = C$, the others being deduced by transposition.

[14] The *ambitus*, i.e., the octave where the modal melody may move, is omitted here because we work in octave classes.

The triadic degree interpretation $C^{(3)}$ is defined by a seven element atlas

$$(3) = \{I_C, II_C, III_C, IV_C, V_C, VI_C, VII_C\}$$

of three-element charts. The analogous notation works for minor scales, i.e., $c_m^{(3)}$ is defined by seven charts $(3_m) = \{I_{c_m}, II_{c_m}, \ldots VII_{c_m}\}$, and $c_h^{(3)}$ by $(3_h) = \{I_{c_h}, II_{c_h}, \ldots VII_{c_h}\}$. If ever the scale is clear, we omit it, writing simply I for the first degree, etc. The precise values for all our charts are shown in table 13.4.2.B. The non-trivial automorphisms of the major and melodic minor scales are shown with the notation U_x = inversion (German "Umkehrung") at pitch x, and $U_{x/x+1}$ = inversion between neighboring pitches $x, x+1$. The arrows in row two are for the alteration shifts in the melodic and harmonic minor scales against the major scale.

Table 13.4.2.B

	C		c_m		c_h	
	●○●○●●○●○●○●		●○●$\overleftarrow{●○}$●○●○●○●		●○●$\overleftarrow{●○}$●○●$\overleftarrow{●○}$○●	
autom.	| $= U_d$		| $= U_g$			
deg.		**tpe.**		**tpe.**		**tpe.**
I	●○○○●○○●○○○○	maj.	●○○●○○○●○○○○	min.	●○○●○○○●○○○○	min.
II	○○●○○●○○○●○○	min.	○○●○○●○○○●○○	min.	○○●○○●○○●○○○	dim.
III	○○○○●○○●○○○●	min.	○○○●○○○●○○○●	aug.	○○○●○○○●○○○●	aug.
IV	●○○○○●○○○●○○	maj.	●○○○○●○○○●○○	maj.	●○○○○●○○●○○○	min.
V	○○●○○○○●○○○●	maj.	○○●○○○○●○○○●	maj.	○○●○○○○●○○○●	maj.
VI	●○○○●○○○○●○○	min.	●○○●○○○○○●○○	dim.	●○○●○○○○●○○○	maj.
VII	○○●○○●○○○○○●	dim.	○○●○○●○○○○○●	dim.	○○●○○●○○○○○●	dim.

Observe that not all major, minor, diminished, or augmented triads in these scales are automatically degrees. For example, the triad $\{a_\flat, b, d\}$ is not a degree in harmonic minor c_h.

Referring to the automorphism groups of these scales, as listed in the chord classification table Appendix N.1, we observe that the degree atlases $(3), (3_m), (3_h)$ are all invariant under the respective automorphism groups. More precisely, we have these degree orbits in the three scales:

$$X : \{I \leftrightarrow VI\}, \{II \leftrightarrow V\}, \{III \leftrightarrow IV\}, \{VII\}$$
$$x_m : \{I \leftrightarrow V\}, \{II \leftrightarrow IV\}, \{III\}, \{VI \leftrightarrow VII\}$$
$$x_h : \{I\}, \{II\}, \{III\}, \{IV\}, \{V\}, \{VI\}, \{VII\}$$

where the third is so since the automorphism group of harmonic minor is trivial, x_h is "rigid". The geometric nerves $N(X^{(3)}), N(x_m^{(3)}), N(x_h^{(3)})$ of each of these global objective zero-addressed compositions are Möbius strips, as shown in Figure 13.7. Following Schönberg's proposal [948], we call it the *harmonic strip*. The harmonic strip has seven simplexes of dimensions zero and two, and fourteen simplexes of dimension one. The 1-skeleton is exactly Schönberg's harmonic strip. The order of the degrees on the strip's boundary is the so-called *fifth sequence*,

$$V \to I \to IV \to VII \to III \to VI \to II \quad (\to V),$$

although this is only a diminished fifth (tritone) for the seventh-fourth passage. In other words, the boundary is connected, unlike with a normal strip where we have two connected components. This is due to the lack of orientation on the harmonic strip: One walk around the entire strip changes your upside to downside, see Figure 13.8.

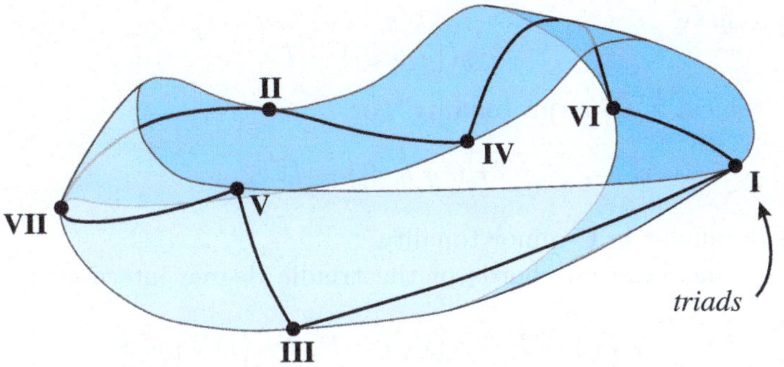

Fig. 13.7. The nerve $N(X^{(3)}), N(x_m^{(3)}), N(x_h^{(3)})$ is a Möbius strip. Following Schönberg's proposal, we call it the harmonic strip.

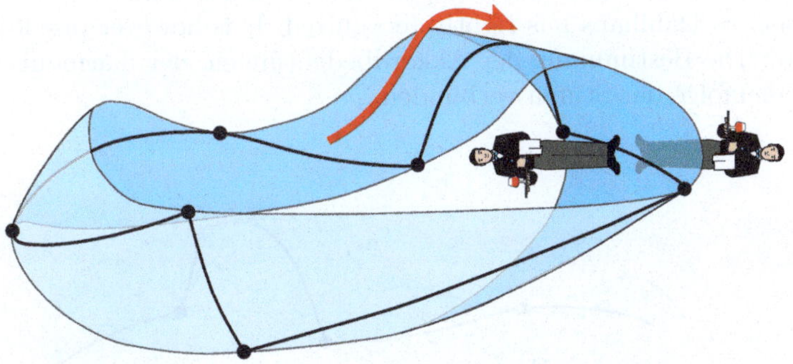

Fig. 13.8. The harmonic strip is not orientable. The mutual position of the boundary changes after a round-trip. This property has consequences for Riemann's harmony.

13.4.2.1 Orientation in Riemann Function Theory

We should add a remark on the failure of Riemann's attempt to build a global function theory of harmony, as was discussed by Carl Dahlhaus in [231]. The central concept of this theory is the function which is to be attached to every (!) possible chord, not only to the chords of common usage. Before starting the discussion, we observe that the problem here is not so much one of tunings, but rather one attached to the non-orientability of the triadic interpretation independently of tuning specifications. Therefore we want to stick to the harmonic strip as discussed above (this nerve will reappear for the triadic interpretation in just tuning which we discuss below).

Riemann's idea was to define "tonality" by use of a function with three possible values: "Tonical" (T), "Dominant" (D), and "Subdominant" (S), which can be attributed to chords. So we have to deal with a function

$$\tau : Ch \to TDS$$

defined on the set Ch of all (zero-addressed objective) chords (in $PiMod_{12}$, to make the ideas precise), with values in the three-element set $TDS = \{T, D, S\}$ of harmonic function values. Moreover, we have to distinguish between two sub-categories of tonal functions: major, and minor, for each pitch class x:

$$^{+}\tau_x : Ch \to TDS \text{ (x-major tonality)}$$
$$^{0}\tau_x : Ch \to TDS \text{ (x-minor tonality)}$$

According the selected tonality, Riemann's program is to attribute a specific value to given chords. For example, one would like to have

$$^{+}\tau_g(\{g,b,d\}) = T$$

which means "$\{g,b,d\}$ is tonical in G-major tonality", or

$$^{0}\tau_c(\{g,b,d\}) = D$$

which means "$\{g,b,d\}$ is dominant in C-minor tonality".

When they define tonality, special chords of the triadic degree interpretation $X^{(3)}$are given special values , i.e.:

$$^{+}\tau_x(I_X) = T,\ ^{+}\tau_x(V_X) = D,\ ^{+}\tau_x(IV_X) = S.$$

However, it is not true that the function concept is not a mathematical one, as Dahlhaus has criticized in [231] relating to an erroneous proposal. Rather is the problem of function theory to extend the values of tonality functions from common first, fifth, and fourth degree triads to any chords in such a way that the harmonic coherence is reflected. This procedure can be termed "musical logic" in the sense of Riemann. This concept is an obscure one, as Dahlhaus has rightly recognized. It is however possible to shed some light on this approach [231, p.96]:"Die Bestimmung der Akkordbedeutungen, der 'harmonischen Logik', ist also mit einer Regel über die Reihenfolge der Stufen verbunden."

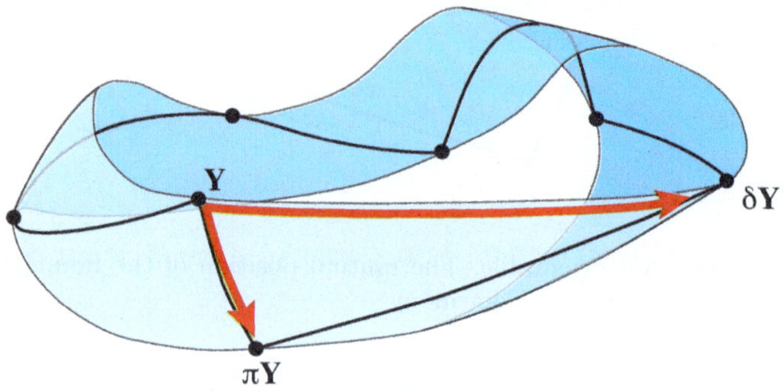

Fig. 13.9. The local orientation on the harmonic strip yields the parallel degree as the one staying in front and to the right when moving in fifth sequence direction such that the strip stays to our right.

According to Dahlhaus, this is the fifth sequence, the fundamental sequence along the harmonic strip's boundary[15]. This sequence lays the basis of the idea of "different" (German:"differente") degrees: the values

$$^{+}\tau(V) = D,\ ^{+}\tau(I) = T,\ ^{+}\tau(IV) = S$$

are different according to the fifth steps $V \to I \to IV$. In order to proceed, the other four triadic degrees must obtain one of the existing values, each.

The second musicological idea relates every degree Y to its "parallel degree" πY. On the harmonic strip, the latter is deduced from the fifth sequence in the following way (see Figure 13.9): Let δY be the successor of Y in the fifth sequence. Then, πY is the third member of the 2-simplex containing Y and δY. If we move along the harmonic strip in such a way that the strip stays to the right when moving from Y to δY, then πY is to the right in front of us. For example, if we move from $Y = I$ to $\delta Y = V$, the parallel degree is $\pi Y = III$. We also have $\pi I = VI$, $\pi IV = II$. The wording "parallel" is also geometrically correct since we look for the degree which is parallel to the present position.

[15] Dahlhaus starts the sequence in degree I, but this is the equivalent to ours since the sequence is cyclic.

The contradiction in function theory comes out from the requirement in function theory that parallel degrees should have equal function values:

$$^{+}\tau(Y) = {}^{+}\tau(\pi Y).$$

In fact, we then must have

$$D = {}^{+}\tau(V) = {}^{+}\tau(\pi V) = {}^{+}\tau(III) = {}^{+}\tau(\pi III) = {}^{+}\tau(I) = T.$$

This contradiction can also be read as follows: When applying the parallel function to every degree on the harmonic strip, the connectedness of the boundardy (=the fifth sequence) leads to the parallelism $\pi\pi Y = \delta Y$ (the "Gegenklang" relation), and therefore the tonal values cannot be different on the vertices of a 2-simplex. The orientation being only a local one, the musicological requirements can only be fulfilled locally and do not glue to yield a global function!

Therefore we cannot follow Dahlhaus [231, p.102] when he says that function theory: "gerade dort versagt, wo auch das Phänomen, das sie erklären soll, ins Vage und Unbestimmte gerät." In reality, the phenomenon is founded in a very precise fact: the non-orientability of the Möbius strip. Any attempt to define a function despite this fact must fail for mathematical reasons.

13.4.2.2 Just Triadic Degree Interpretations

For this situation, recall the just C-major scale in the *EulerPlane* from Section 7.2.1.4. The triadic degrees are not completely clear here, see Figure 13.10. The problem is that one would like to respect certain pitch relations. Except for degree II, VII we obtain major or minor chords. If we want to have a minor chord for degree II which has two tones in common with degree IV, we need to take the tone $d^* = -2q + t$ instead of $d = 2q$, i.e., $d^* - d = Kt$, the third comma—more precisely: its pitch class, see formulas (6.33) for the definitions. In the same sense, the seventh degree must, if we want it to be built from two minor thirds $d - b$ and $f - d$, refer to the second f-variant $f^* = f - Kt$. We then have the alternative degrees

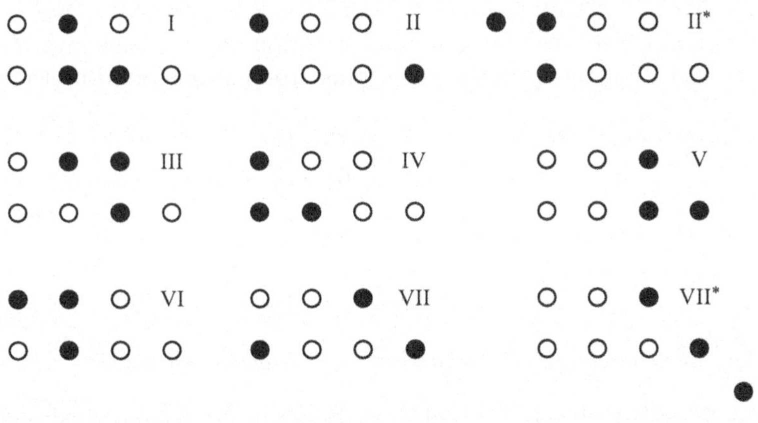

Fig. 13.10. The list of just triadic degrees, together with the variants II^*, VII^* which meet standard requirements of interval distances.

$II^* = \{d^*, f, a\}, VII^* = \{b, d, f^*\}$, and these have an empty intersection, in contrast to the canonical degrees II, VII.

In order to harmonize this irritating situation—which stems from pitch-distance requirements on degrees—one proceeds as follows: We no longer work with pitch classes x but with *third comma classes*

$$]x[= x + \mathbb{Z}Kt.$$

The *justest scales* can then be defined as local compositions by the quotient module $J_{Kt} = \mathbb{Z}^2/\mathbb{Z}Kt$, i.e., with ambient space $@J_{Kt}$, a space which can be identified with the Pythagorean tuning subspace $@\mathbb{Z}q$ in the *EulerPlane*. In fact, we have $\mathbb{Z}^2 = \mathbb{Z}q \oplus \mathbb{Z}Kt$, see also Figure 13.11. With such a construction, we

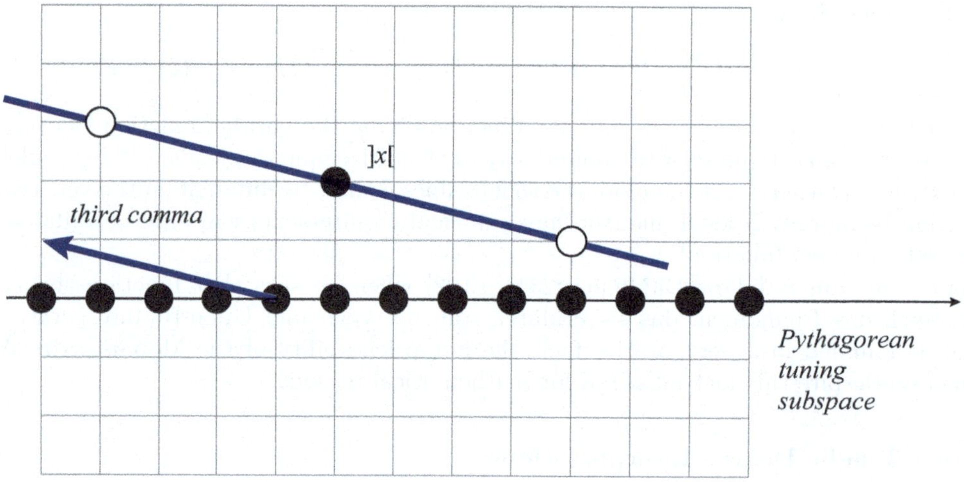

Fig. 13.11. The third (or Pythagorean) comma classes cover all just pitch classes when starting from the Pythagorean subspace. Third comma classes solve the problem of the second and seventh degrees, one is allowed to select any one of the possible representatives within one comma class.

may transport the triadic degrees of just tuning $C^{(3)}$ to C-major in *justest tuning* , and thereby obtain the interpretation $\mathfrak{C}^{(3)}$ with degrees $\mathfrak{I}, \mathfrak{II}, \mathfrak{III}, \mathfrak{IV}, \mathfrak{V}, \mathfrak{VI}, \mathfrak{VII}$ with $\mathfrak{I} =]I[= \{]c[,]e[,]g[\}$, etc. In this setup, ambiguities from just tuning disappear, and the nerve $N(\mathfrak{C}^{(3)})$ is a Möbius strip. The second degree now can be realized by selecting one variant of the third comma equivalent just degrees (tone by tone, not as a whole!).

The atlas of degrees of a just triadic degree interpretation also remains invariant under the automorphism group of the just scale. For the *C-major* scale, this group is generated by the skew reflection

$$A = e^q \cdot \begin{pmatrix} -1 & -1 \\ 0 & -1 \end{pmatrix}$$

(see also example 9 in Section 8.1.1). The orbits of the degrees look as follows:

$$\{I\}, \{II \leftrightarrow VII\}, \{III \leftrightarrow VI\}, \{IV \leftrightarrow V\}.$$

Exercise 26 *Verify the invariance and the orbit structure as described above.*

We shall see however in Chapter 27.1.6 that modulation in just tuning is also easily modeled without necessarily building justest tuning constructs.

13.4.3 Interpreting Time: Global Meters and Rhythms

Summary. Global meters and rhythms as interpretation by atlases of local rhythms and meters. We describe macros for rhythmic germs.

– Σ –

Fig. 13.12. Example of a simultaneous presence of local meters in Beethoven's op.106, Allegro, bars 209-210 [96]. The right hand plays eight quavers while the left hand plays six triplets per meter.

In music, different local metrical structures mostly coexist in the same portion of music, as for example illustrated in Figure 13.12. In this section, we want to give an account of global time constructions deduced from this type of phenomena.

Recall from Section 7.2.2 that a local rhythm is defined as a union of translates of a local 'germ' composition in an ambient space $Onset \times Para$. We would like to include formally the case where the rhythm is a local meter (empty parameter set for space $Para$) on one hand, and where the parameters may vary, on the other. The first requirement is met by the form

$$Rhythm(Para) \underset{Id}{\longrightarrow} \mathbf{Colimit}(Onset, R(Para))$$

with $R(Para) \underset{Id}{\longrightarrow} \mathbf{Limit}(Onset, Para)$. This means that an objective A-addressed local composition Rh in ambient space $Rhythm(Para)$ is a subset of $A@Onset \sqcup A@Onset \times A@Para$ (with corresponding functors of these forms). For any period $p : 0 \rightsquigarrow Duration(p)$, we have the evident translation action e^p on $Rhythm(Para)$, and therefore on the local $Para$-rhythms, in the sense of Section 7.2.2.

Definition 41 *With the above notation, for a rhythm germ*

$$G \subset A@Rhythm(Para)$$

and an interval $[a, b]$ *of extended natural numbers,* a local (A-addressed, objective) $Para$-rhythm *is an A-addressed local composition R defined by*

$$R = \bigcup_{t \in [a,b]} e^{tp} G$$

and denoted by $e^{[a,b]p} G$.

We have an obvious projection endomorphism

$$p_{meter} : Rhythm(Para) \rightarrow Rhythm(Para)$$

which is the identity on cofactor $Onset$ and projects onto the first factor on $R(Para)$. For a local composition L in $Rhythm(Para)$, the projection $p_{meter}(L)$ is called the *associated metrical rhythm*, and the intersection $L \cap Onset$ is called the *metrical component* of L.

A local $Para$-rhythm defined by a singleton germ $G = \{g\}$ is called a *local Para-meter*. Conformal with the known terminology, a local $Para$-meter is again called a local meter if its germ lives in the cofactor $Onset$ and is equal to its *metrical component*; it is called local rhythm if its germ lives in the cofactor $R(Para)$.

A finite local $Para$-meter M with germ $\{g\}$ may be written as $M = e^{[0,b]p}\{g\}$, with non-negative b; we abbreviate this object by $M = [b, p, g]$. For a local $Para$-meter M, g is called the *origin* onset, while

$b = l(M)$ its *length*. For length zero, we shall always assume the period to be also zero; the now uniquely defined period of a local *Para*-meter M is denoted by $p(M)$.

The second requirement is met by suitably generalizing the form *Para*: Suppose that we want to take local rhythms from different parameter spaces

$$Para_1, Para_2, \ldots Para_n.$$

Then we may collect these options to

$$Para \underset{Id}{\longrightarrow} \mathbf{Colimit}(Para_1, Para_2, \ldots Para_n),$$

and the second problem is also solved.

Lemma 18 *The intersection of any two A-addressed local Para-meters R, S is a A-addressed local Para-meter.*

Proof. Let $R = [b, p, g], S = [c, q, h]$ be a representation by germs, periods, and origins. Suppose that $R \cap S$ is not empty (otherwise, the empty germ will do it), and take $x = e^{sp}g = e^{tq}h$. This implies that either g, h are both onsets, or they have the same *Para*-component. WLOG[16], we may assume the first case, i.e., $g, h \in A@\mathbb{R}$ are A-addressed onsets. So g, h have the same linear parts and if their translation parts are $\gamma, \eta \in \mathbb{R}$, respectively, we have the equation $sp + \gamma = tq + \eta$ of real numbers. If there is no other common point, we are done with length zero. Else, if there is another common value $\omega = s^*p + \gamma = t^*q + \eta$, we have $\Delta = (s - s^*)p = (t - t^*)q$. We may suppose that Δ is the smallest possible positive number of this type. Then we claim that the maximal integer interval $a \leqslant b$ such that $e^{[a,b]\Delta}\omega \subset R \cap S$ equals $R \cap S$. Clearly, the interval $[0, 0]$ is in $R \cap S$. So, there are such numbers a, b. Take a maximal interval. It is evidently unique. Take any element $y = s^{**}p + \gamma = t^{**}q + \eta$ in $R \cap S - e^{[a,b]\Delta}\omega$. It cannot lie between two elements of the local meter $e^{[a,b]\Delta}\omega$ since this would contradict minimality of Δ. So the point lies to the right or left of the set $e^{[a,b]\Delta}\omega$. WLOG, suppose it lies to the right. Then we evidently can increase b until we reach y, and this is a contradiction. So the intersection is this local meter. QED.

Lemma 19 *For two local A-addressed Para-meters R, S, if $R \subset S$, then $l(R) \leqslant l(S)$.*

Proof: Exercise.

With this terminology, the interpretation G^I of an objective, A-addressed local composition G in ambient space *Rhythm*(*Para*) is said to be *(Para-)rhythmical* iff every chart is a local (*Para*-)rhythm; it is said to be *(Para-)metrical* iff every chart is a local (*Para*-)meter. These definitions englobe Vuza's concepts of rhythms and of canons developed in his works [1094, 1096, 1097, 1098, 1099].

In general, it is not easy to decide how to interpret a local composition by (*Para*-)rhythms since the decision between large repetition numbers and large germs has no evident rationale. When taking (*Para*-)meters, the situation is easier. We have several basic examples of (*Para*-)metric interpretations of an A-addressed objective local composition X in *Rhythm*(*Para*).

1. Consider the set $Max(X)$ of all maximal local A-addressed *Para*-meters contained in X. Observe that by Lemma 18, the set $MaxMet(X)$ of all finite intersections of elements of $Max(X)$ consists of local A-addressed *Para*-meters which form a base of the so-called *maximal meter topology* on X. We identify it with the corresponding covering and also denote it by the interpretation symbol X^{MaxMet}. Its charts are called the *canonical* (local A-addressed) *Para*-meters of X.
2. Take the atlas of all local A-addressed *Para*-meters $U \subset X$ such that their lengths $l(U)$ are at most equal to a limit L. This interpretation is denoted by $X^{MetLg[L]}$.
3. Select the set of all local A-addressed *Para*-meters $U \subset X$ such that their periods $p(U)$ are at most equal to a limit P. This interpretation is denoted by $X^{MetPer[P]}$.

[16] WLOG = without loss of generality.

4. For any interpretation K^I of a finite local composition K, we may introduce a level function $Lev : I \to \mathbb{N}$ as follows. We define the inverse images $lev_i = lev^{-1}(i)$ by recursion:
   - $lev_o = Max(I)$, the set of (set-theoretically) maximal members of I;
   - $lev_{i+1} = lev_0(I - \bigcup_{k \leqslant i} lev_k)$.

   The charts in lev_j are called *charts of level* j, and we set $I|i = \bigcup_{k \leqslant i} lev_k$. We then evidently have an interpretation $K^{I|i}$, for each i.
   The above example of $MaxMet(X)$ has the atlas $Max(X)$ for its interpretation $X^{I|0}$, i.e., $Max(X) = MaxMet(X)|0$.

With the above notation, the *simplicial metrical weight* $sp(x)$ *of a point* $x \in X$ is defined to be the dimension of the simplex $Sp(x)$ of all maximal (=level zero) canonical *Para*-meters in X^{MaxMet} which contain x. The *level* $lev(x)$ is defined to be the level of the (uniquely defined) smallest canonical *Para*-meter containing x.

Proposition 11 *With the above notation, we have* $lev(x) \leqslant sp(x)$.

Proof. In fact, we have a surjective map of partially ordered sets

$$\bigcap : n(X^{MaxMet|0}) \to X^{MaxMet} : \sigma \mapsto \bigcap \sigma, \tag{13.6}$$

i.e., $\sigma \subset \tau$ implies $\bigcap \tau \subset \bigcap \sigma$, whence the claim. QED.

The open set $U(x) = \bigcap Sp(x)$ is the minimal neighborhood of x. The dominance relations in this topology read as follows:

Lemma 20 *With the above notation, if* x, y *are points in the local composition* X, *and if* $<$ *is the dominance relation*[17] *on the maximal meter topology, then we have*

$$y < x \text{ iff } Sp(y) \subset Sp(x) \text{ iff } U(x) \subset U(y).$$

Proof. We have $y < x$ iff $x \in U(y)$ iff $U(x) \subset U(y)$, whence $Sp(y) \subset Sp(x)$, whence $x \in U(x) \subset U(y)$ and therefore $U(x) \subset U(y)$. QED.

Proposition 12 *The irreducible closed sets of the maximal meter topology are the closures of their points with maximal simplexes in the nerve.*

Proof. Suppose F is an irreducible closed set and x is a point having maximal simplicial metrical weight among the elements of F. If y is any point of F, the minimal neighborhoods $U(x), U(y)$ intersect, so take a z in that intersection. This point dominates x and y. By maximality of the simplicial metrical weight of x, the simplicial metrical weight of z is equal to the weight of x, and Lemma 20 implies that $Sp(x) = Sp(z)$, so x dominates y. The rest is clear. QED.

For example, the closures of points of maximal simplicial weight are irreducible components, but not necessarily vice versa. The irreducible components correspond to the maximal simplexes of the nerve $n(X^{MaxMet})$, and the generic points are the elements of the images of the maximal simplexes under the map $\bigcap$ (13.6). Musically speaking, the generic points are those which participate in maximal collections of local *Para*-meters, and this means that they are present in a maximal collection of regular patterns in onset time. We shall introduce refined metrical weight functions which generalize the above simplicial weight, and take care of the lengths of the local *Para*-meters as well as of the periods and the positions of the points within the maximal local *Para*-meters in Chapter 21.

[17] See Appendix F.2.1.

13.4.3.1 Macros for Rhythmic Germs

We have seen that for general rhythmic germs, there is no unique maximality of germs, and that the metrical special case is built upon the singleton germ set. There are however two approaches in order to circumvent such a problem.

The first deals with address change. Suppose we are given a finite germ G which is an A-addressed local composition which is completely contained either in the sub-form *Onset* or in the sub-form *R(Para)* of *Rhythm(Para)*, and where $Para \xrightarrow{\sim} @PARA$ is simple with a module $PARA$. This means that the germ has coordinates in a module. Then the elements of G are in $A@M$, where M is one of the modules $\mathbb{R}, \mathbb{R} \oplus PARA$, and we suppose that the coefficient ring for A is $\mathbb{R}$. In this case, by the construction of fiber sums of modules (Appendix E.3.8), we have[18] $(A@M)^n \xrightarrow{\sim} A^{\oplus n} \oplus \mathbb{R}^{n-1}@M$ and we can parametrize an n-element germ G by the n-tuple (G) of A-addressed points, and then by the corresponding point $\coprod(G)$ with address $\coprod_n A = (A)^n \oplus \mathbb{R}^{n-1}$.

The second solution to ambiguities in germ definitions uses macros and a flattening operation[19], as discussed in Section 6.7. We start from the hypothesis of the preceding discussion and therefore may have the germ G living in a simple form index$Makro_{Basic}Basic$ of module $M = \mathbb{R} \oplus PARA$ where—for the sake of mathematical simplicity—the first factor parametrizes the *Onset* coordinates. We need the circular form $Knot_{Basic}$:

$$Knot_{Basic} \underset{Id}{\longrightarrow} \mathbf{Limit}(Basic, Makro_{Basic})$$

$$\text{with } Makro_{Basic} \underset{f:F \xrightarrow{\sim} 2^{FK} \rightarrowtail \Omega^{FK}}{\longrightarrow} \mathbf{Power}(Knot_{Basic})$$

$$\text{and } F = Fun(Makro_{Basic}), FK = Fun(Knot_{Basic}).$$

We have the flattening operation on $Knot_{Basic}$ Denotators $D = (b, m)$ defined by

1. $Flatten^{\infty}(D) = \{b\}$ if $m = \emptyset$,
2. $Flatten^{\infty}(D) = \bigcup_i b + Flatten^{\infty}(Knot_i)$, if $m = \{Knot_1, \ldots Knot_k\}$.

Then the idea is to define a macro germ G^* such that the given, 'flat' germ G is the flattened version of the former, $G = Flatten^{\infty}(G^*)$. For example, we may take a germ which contains ornaments as $Makro_{basic}$-formed denotators, and/or drum patterns or the like as 'satellites' of the macro. The translation in time is the same as before, only that it operates exclusively on the basic event in the knot macro, i.e., if e^p is a period translation in time, and if the macro germ is $G^* = (b, m)$, we define $e^p(G^*) = (e^p(b), m)$.

An A-addressed local *Para*-rhythm can be paraphrased by the local compositions in ambient space $Knot_{Basic}$; they are local compositions of the shape $e^{[a,b]p}G^*$, built upon macros G^*. In this language, a global rhythm is an interpretation of a local composition in ambient space $Knot_{Basic}$ by local *Para*-rhythm of macro type. So global macro rhythms can be projected to global rhythms by flattening, and they constitute a hierarchical shaping of the flat rhythmical germs.

We should stress that these reflections are merely 'germs' of a theory of macro-events, and that we are far from having established any systematic treatment of such objects. But there is an evident relation of this approach to the Schenker idea, a perspective which we shall reconsider in Chapter 21 when discussing grouping rules in Jackendoff-Lerdahl's GTTM.

13.4.4 Motivic Interpretations: Melodies and Themes

Summary. Motives are considered as being local ingredients of melodies. We discuss the local/global dichotomy in this subject, and, in particular, the subtle interpretative activity in motivic analysis.

[18] The bijection takes the n-tuple $(g_i = e^{t_i} \cdot g_{i,0})_i$ into $e^{t_1} \cdot g$, where $g(a_1, \ldots) = g_1(a_a)$, $g(0, \ldots a_i, \ldots) = g_{i,0}(a_i)$ for $1 < i$, and $g(0, \ldots 1_{n+j}, \ldots) = t_{j+1} - t_1$ for $j = 1, \ldots n-1$. The injections $\iota_i : A \rightarrowtail \coprod_n A$ are these: $\iota_1(a) = (a, 0, \ldots)$, $\iota_i(a) = (0, \ldots a_i = a, 0, \ldots 1_{n+i-1}, \ldots, 0)$ for $1 < i$.

[19] The present flattening operation is just the infinite one from the previous discussion.

– Σ –

Concepts such as "motif", "melody", "theme" are among the most fuzzy of musicology. There are three main sources of such fuzziness: On the one hand, the very concepts are not clear, neither in the delimitation against each other, nor in the intrinsic attribution. On the other hand, it is not clear how to compare motives, for example, how to decide when two motives are "similar". Thirdly, the concept of a theme or motive is not only a structural one, but includes a *semiotic* perspective. A theme is a subcomposition of a composition which plays a certain role in the entire construction of the composition qua expression of meaning. We shall deal with this aspect in Chapter 21. We should also stress that a theme is not necessarily related to the motif concept, a theme may as well be a harmonic or rhythmic object. Its characteristic property is the semiotic role, not the structure.

In this section, we shortly focus our attention on the problem of elementary and compound motivic structures. We have already introduced the motif concept in Definition 15 (Section 7.2.3) as an objective local composition with onset and pitch coordinators in its ambient space, and such that the onset projection is bijective.

We should therefore define a "motivic" interpretation as being an interpretation of a local composition in a determined motif ambient space (having onset and pitch factors) such that every chart is a motif in this space. In particular:

Definition 42 *A* melody *is defined as a motivic interpretation of a motif.*

In fact, a melody is a local composition with successive tones in time. But it is more: it is understood as a compound object with "characteristic" melodic ingredients, the melody's motives. Now, these units are not neutral data, they depend upon the interpretative, analytical, or poietic interaction. In other words: they are a covering of the motivic tone material of the melody by specific selected submotives.

In traditional musicology (see, e.g., the article "Melodie" in the Riemann Musiklexikon [907, p.554]), a melody is also associated with the Ehrenfels gestalt qualities of super-summativity and transformational invariance. The AST has dealt with this aspect in the contour theory. We shall deal with this ramification in Chapter 22. Gestalt is a central invariant of melody, but it is not the intrinsic concept, it is a *derived* attribute which comes out from a complex and not uniquely defined abstraction process. Every contour or gestalt or shape concept is derived from the motif structure by elaboration of specific motif aspects, such as interval sequences, onset-pitch-interval angles, length ratios, diastematic ambitus, etc.

Before the gestalt aspect comes in, the motivic interpretation which defines the melody upon a motif has a more elementary—but nonetheless dramatic—function which we discuss now. If we cover a motif by characteristic motifs, as expressed by the melody's nerve, this action reflects a disintegration of the original underlying melody motif into several relatively small parts which may intersect in a rather loose way. For example, the covering may split into several disconnected (disjoint) motives, or motives may have just one tone in common. The crucial question is whether we should keep in mind the original melodic motif (the entire set of tones) or whether the splitting activity is really meant as a decomposition in henceforth in- or inter-dependent parts. So this elementary interpretation activity precedes refined considerations: it deals with the initial statement of which are the parts and which is the whole.

But even if we are given two fixed motivic interpretations, i.e., melodies, it is not clear how one should compare two such objects. Gestalt is a concept of equivalence which should be built upon comparison devices for melodies. The only thing which is clear now is that melodies are not neutral objects, they are results of interpretative interventions. We are urged to set up the basic tools for comparison of global compositions, and this is what we shall do in the next chapter.

14

Global Perspectives

The woods are lovely, dark, and deep,
But I have promises to keep,
And miles to go before I sleep,
And miles to go before I sleep.
Robert Frost (1874–1963)

Summary. Global perspectives deal with relations among global music objects. We introduce this subject together with its musical motivation and append the formal definition of morphisms among global compositions. This leads to the categories *ObGlob* of objective and *Glob* of functorial global compositions. We describe the combinatorial aspects of globality and associated functors, as well as corresponding geometric classification tools: nerves and simplicial weights.

– Σ –

14.1 Musical Motivation

Summary. Since global compositions are of musical and musicological interest, requiring their comparison is quasi-automatic. However, this is not in the tradition of musical analysis. We discuss this issue with regard to comparative analysis of melodic themes in a composition, and in the more general context of comparative analysis of musical works. It turns out that the latter is virtually non-existent, at least on the level of a reliable scientific language.

– Σ –

Mathematically, it is straightforward how to extend the categories of local compositions to categories of global compositions. But musicologically, this is not a common situation, on the contrary, The musicological fuzziness of global structures appearing in melodic analysis—which we have made precise via motivic interpretations—makes it impossible to take a step further in the comparative methodology of global structures. In the typical case of a comparison of two melodies, or in the discussion of the motivic-thematic work in, say, Beethoven's compositions, it is intuitively accepted that one deals with smaller and larger units, and that comparison should take into account comparison of parts, but it is not clear how the relations of parts within a first melody (for example) should be taken over to relations among parts of a second melody. Comparison of the gluing of parts is not conceived. So Ehrenfels' requirement of an "added value" to the "sum of parts" against the "whole" is not understood: How much more do we need to get the whole? How can we compare this specific difference. For example, if we have split a local composition into isolated singleton charts (the 'silly' interpretation), the disintregation is extremal: We essentially have to count the tones, and there is no connection left. So the comparison reduces to counting notes, the cardinality of the original local composition. This is the trivial case where the added value is naught. But in general, for example with the

G. Mazzola, *The Topos of Music I: Theory*, Computational Music Science,
https://doi.org/10.1007/978-3-319-64364-9_14

harmonic strip, there is a rich connectivity, in fact the Möbius strip, in the nerve. So the added value is complex: it lies in the single charts (the triads) as well in their intersections (thirds or single notes) and also in the combinatorial configuration. So, how strongly is the harmonic strip related to the original scale? Are there many scales with essentially equal harmonic strips? Same situation for motivic analysis: If we take the main theme of Bach's *Art of Fugue*, is it the 8-tone theme or the 12-tone theme? This still debated question depends on the development of a sophisticated technique of motivic analysis of melodies or themes. And, finally, without answering these questions, a comparison of gestalts is a fortiori impossible.

On the other hand, overall comparative analysis of musical works of course transcends the theory of local or global compositions. In fact, the denotators of musical works, such as European scores, cannot be grasped by compositions (in the technical sense), they deal with more general forms of all possible typologies: limits, colimits, synonymy, and all kinds of simple forms, see the contribution of Mariana Montiel Hernandez [760] concerning a standard form for piano scores. In this chapter we shall introduce only comparative tools (morphisms) for global compositions. The category theory for general denotators is still embryonal.

14.2 Global Morphisms

Summary. This section introduces the technical definition of morphisms between global compositions. Special aspects, e.g., isomorphisms or composition of morphisms, are discussed. The concept of interpretable global compositions is presented and illustrated.

– Σ –

The first definition regards objective global compositions:

Definition 43 *Suppose we are given two objective global compositions G^I at address A and H^J at address B. Then a morphism from G^I to H^J is a triple (f, ι, α) where*

1. *$f : G \to H$ is a set map,*
2. *$\iota : I \to J$ is a set map such that $f(i) \subset \iota(i)$ for all covering sets $i \in I$,*
3. *α is a family $\alpha = (\alpha_i : A \to B)_I$ of address changes, such that*
4. *for any atlases $(K_t, A@E_t)_T$ for G and $(L_s, B@F_s)_S$ for H, if we take the chart isomorphisms $\phi_t : K_t \xrightarrow{\sim} I_t$ and $\psi_s : L_s \xrightarrow{\sim} J_s$ for some pair t, s of indexes which correspond under the map ι (i.e., $\iota(I_t) = J_s$), then the induced maps $f_t : K_t \to L_s$ define morphisms $f_t/\alpha_i : K_t \to L_s$ of objective local compositions.*

This morphism will be denoted by $f/\alpha : G^I \overset{\iota}{\to} H^J$ or, if no misunderstanding is possible, by $f/\alpha : G \to H$.

Observe that this is the precise generalization of the morphism concept from objective local compositions. For functorial global compositions, the definition of morphisms runs correspondingly:

Definition 44 *Suppose we are given two functorial global compositions G^I at address A and H^J at address B. Then a morphism from G^I to H^J is a triple (f, ι, α) where*

1. *$f : G \to H$ is a natural transformation,*
2. *$\iota : I \to J$ is a set map such that we have subfunctor relations $im(f|i) \subset \iota(i)$ for all covering functors $i \in I$,*
3. *α is a family $\alpha = (\alpha_i : A \to B)_I$ of address changes, such that*
4. *for any atlases $(K_t, @A \times E_t)_T$ for G and $(L_s, @B \times F_s)_S$ for H, if we take the chart isomorphisms $\phi_t : K_t \xrightarrow{\sim} I_t$ and $\psi_s : L_s \xrightarrow{\sim} J_s$ for some pair t, s of indexes which correspond under the map ι (i.e., $\iota(I_t) = J_s$), then the induced maps $f_t : K_t \to L_s$ define a morphism f_t/α_i of functorial local compositions.*

This morphism will be also be denoted by $f/\alpha : G^I \overset{\iota}{\to} H^J$ or, if no misunderstanding is possible, by $f/\alpha : G \to H$.

If we have $G^I = H^J$, the identity morphism is the triple (Id_G, Id_I, Id_A). This being the case, defining the composition of morphisms of global compositions is straightforward.

Definition 45 *Given three objective or functorial global compositions G^I, H^J, K^L, a morphism $f\overset{\iota}{/}\alpha : G^I \to H^J$, and a morphism $g\overset{\kappa}{/}\beta : H^J \to K^L$, their composition $h\overset{\zeta}{/}\gamma = g\overset{\kappa}{/}\beta \cdot f\overset{\iota}{/}\alpha : G^I \to K^L$ is defined by the triple $(h = g \cdot f, \zeta = \kappa \cdot \iota, \gamma = \beta \cdot \alpha)$, where $\beta \cdot \alpha$ is the family $(\beta_{\iota(i)} \cdot \alpha_i)_I$.*

Exercise 27 *Show that composition of global morphisms is associative whenever corresponding factors are defined. Show that the identity morphisms are in fact left and right identities in the sense of category theory.*

Definition 46 *The category of objective global compositions ObGlob has the objective global compositions as its object set ${}_0$ObGlob, as defined in definition 36, whereas the set ${}_1$ObGlob of morphisms is defined by definition 43.*

The category of functorial global compositions Glob has the functorial global compositions as its object set ${}_0$Glob, as defined in definition 38 whereas the set ${}_1$Glob of morphisms is defined by definition 44.

It is useful to view functorial global compositions as a system of objective global compositions as follows:

Lemma 21 *Let G^I be a functorial global composition at address A. Suppose that $\phi_\kappa : K_\kappa \xrightarrow{\sim} I_\kappa$ and $\phi_\lambda : K_\lambda \xrightarrow{\sim} I_\lambda$ are any two charts for G^I. Fix any address change $f : B \to A$, and consider the subsets $f@K_\kappa \subset B@K_\kappa$, $f@K_\lambda \subset B@K_\lambda$. Then the intersection of their images under $B@\phi_\kappa$ and $B@\phi_\lambda$, respectively, with $B@I_\kappa \cap B@I_\lambda$ coincide.*

Proof. In fact, the intersections are the same since the induced isomorphism $\phi_{\kappa,\lambda}/\alpha$ on the inverse images of the intersection have denominator $\alpha = Id_A$, and therefore, the parts related to f correspond.

Definition 47 *For every address change $f : B \to A$ and a functorial global composition G^I at address A, we have a B-addressed objective global composition $f@G^I$ which is defined on the subset of $B@G$ which is locally covered by the images $f@I_\kappa$ of the chart subsets $f@K_\kappa$ of any atlas, and which has the atlas defined by the second factors in the charts $f@K_\kappa = \{f\} \times K_{\kappa,f}$, together with the given transition isomorphisms. We denote this covering of $f@G$ by $f@I$, and call the objective global composition the* f-slice *of G^I. The entire construction is well defined by Lemma 21. The identity slice $Id_A@G^I$ is denoted by $(G^I)^\vee$.*

Example 21 Given two interpretations G^I, G^J of an objective local A-addressed composition G, and a *refinement map*, i.e., a map $\iota : I \to J$ with $i \subset \iota(i)$, all $i \in I$, then the identity Id_G and ι induce the *refinement morphism* $Id_A\overset{\iota}{/}Id_A : G^I \to G^J$, also denoted by ι. In particular, any interpretation G^I yields a unique refinement morphism $! : G^I \to G$.

For example, with the notation of 13.4.3, we have a chain of refinements

$$K^{I|0} \to K^{I|1} \to \ldots K^{I|n} \to K^{I|n+1} \to K^I \tag{14.1}$$

of an A-addressed finite objective local composition K.

If the local composition is a major or minor scale scale X as discussed in Section 13.4.2, we may consider the different versions of degrees. We have discussed the triadic interpretation $X^{(3)}$, but this is just one possibility. More generally, set $X = \{x_1, x_2, \ldots x_7\}$ in the order of increasing position on $\mathbb{Z}_{12}$, i.e., also with indexes mod 7. One starts with the singleton interpretation $X^{(1)}$, where $(1) = \{\{x_i\}|\ i = 1, 2, \ldots 7\}$; the third interpretation $X^{(2)}$ is defined by charts of thirds, i.e., $(2) = \{\{x_i, x_{i+2}\}|\ i = 1, 2, \ldots 7\}$; the triadic interpretation is defined by the atlas $(3) = \{\{x_i, x_{i+2}, x_{i+4}\}|\ i = 1, 2, \ldots 7\}$; and the tetradic interpretation $X^{(4)}$ which is important in jazz harmonics is defined by the atlas $(4) = \{\{x_i, x_{i+2}, x_{i+4}, x_{i+6}\}|\ i = 1, 2, \ldots 7\}$.

With this setup, we have 2^7 refinement maps for each succession of the above atlases. More precisely, if Z^i is the degree Z in $X^{(i)}$, we have two embeddings $Z^i \subset W^{i+1}$ depending on whether we take Z^i as

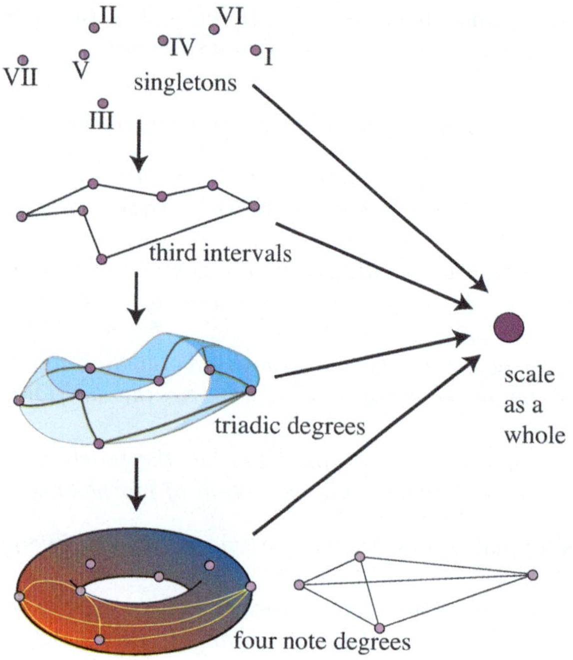

Fig. 14.1. The four interpretations of a major or minor scale, starting with the singletons, up to the tetradic interpretation for jazz harmony. We have successive refinement morphisms, and associated maps between nerves (see Section 14.4 for this association). Every interpretation uniquely maps to the 'uninterpreted' local scale composition shown to the right. Observe that the tetradic interpretation has a nerve which is a full torus, in fact a union of tetrahedra, the intersection configuration around any fixed tone. The harmonic strip of the triadic interpretation is embedded in the tetradic torus as a strip which is 'entwined' along the torus' interior circle. So the absence of orientation of the harmonic strip vanishes on the tetradic jazz harmony.

the lower or higher part within W^{i+1}. The lower embedding is denoted by 0, the higher one by 1, and the refinement map $\delta = (\delta_1, \delta_2, \ldots \delta_7) : (i) \to (i+1)$ denotes the choice of $\delta_i \in BIT = \{0,1\}$ for the ith value. So we have this succession of refinements, ending with the unique morphism to the local scale composition:

$$X^{(1)} \xrightarrow{\delta} X^{(2)} \xrightarrow{\delta} X^{(3)} \xrightarrow{\delta} X^{(4)} \xrightarrow{!} X \qquad (14.2)$$

which has, of course an important musicological meaning: If ever we are given a chart of atlas (i), it may be reinterpreted as being one of two charts of the successive refinements, etc., and this documents the vast ambiguity in elementary harmony of degrees! For the corresponding nerves, see Figure 14.1.

This ambiguity is also dramatic with regard to the geometric configuration of the nerve geometry. The absence of orientation on the harmonic strip vanishes on the tetradic interpretation of jazz. This may not only ease jazz harmony, more radically, it throws the context of jazz harmony one dimension higher: Moving in a three-space such as the torus, is more complex, there is more freedom of choice, more improvisational flexibility—at the cost of unambiguous harmonic syntagmatics.

Example 22 This example refers to the analysis in [670] of the first movement of Beethoven's 'Hammerklavier' sonata Op. 106. The general harmonic principles of this sonata yield a motivic germ $G \subset Onset \times Pitch|_{\mathbb{Z}[1/2]}$ with coefficients in the localization ring $\mathbb{Z}[1/2]$ of fractions with denominators that are powers of 2. This germ is essentially an ascending and then descending chromatic scale of three half-tone steps 'octave' period. The claim of our analysis is that the motivic work (German: "motivisch-thematische Arbeit") relates to interpretations of the *Pitch*-rhythms of type $G(t) = e^{[0,t]6}G$, see also Figure 14.2 for this 'motivic zig-zag'. In [670] it is shown that a large number of motivic charts can in fact be interpreted such as

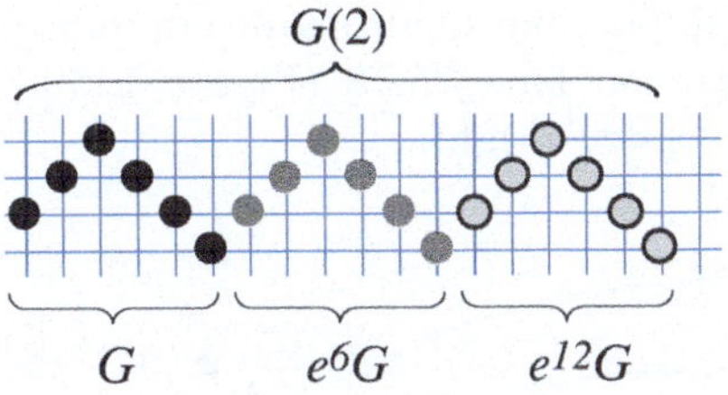

Fig. 14.2. The germ G of motivic work in Beethoven's 'Hammerklavier' sonata Op. 106 is essentially an ascending and then descending chromatic scale of three half-tone steps 'octave' period. Here we show the *Pitch*-rhythm with three copies of the germ.

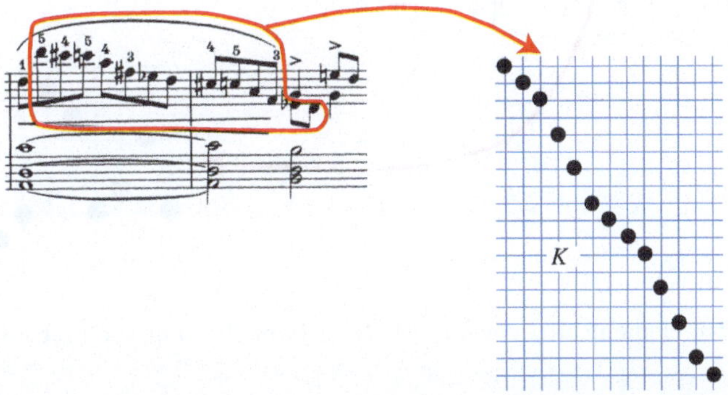

Fig. 14.3. The right-hand melody which appears in bars 75-76, and 77-78 of the first movement in op. 106.

to become isomorphic to interpretations of local subcompositions of $G(t)$. Of course, the problem is not the existence of such interpretations since the silly interpretations always fit in such a framwork. We therefore look for good quality interpretations.

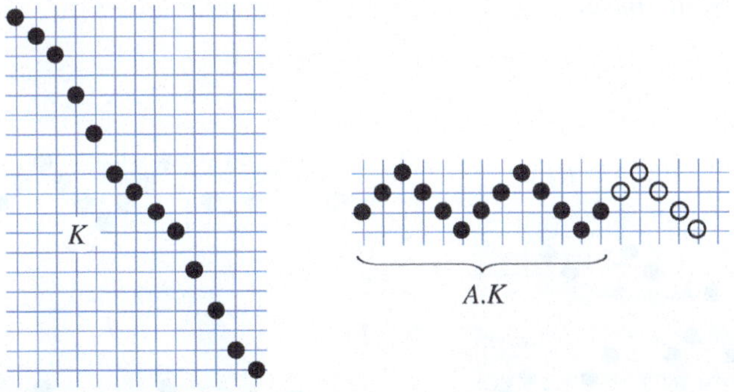

Fig. 14.4. The right-hand melody in bars 75-76 and 77-78 is isomorphic to a local subcomposition of the motivic 'zig-zag' under a vertical arpeggio A.

Definition 48 *Let K^I be an interpretation of a local composition K. The quality of K^I is the pair (s,p), where $s = card(I)$, and p is the cardinality of connected components of $N(K^I)$. Qualities are compared lexicographically: $(s_1,p_1) < (s_2,p_2)$ iff either $s_1 < s_2$ or $s_1 = s_2$ and $p_1 < p_2$. So the best quality is $(1,1)$.*

With this in mind, we look for best possible qualities in our interpretations. To begin with, we look at the right-hand melody which appears in bars 75-78 (Figure 14.3). The (projected) representation K in *Onset* × *Pitch* space is shown to the score's right.

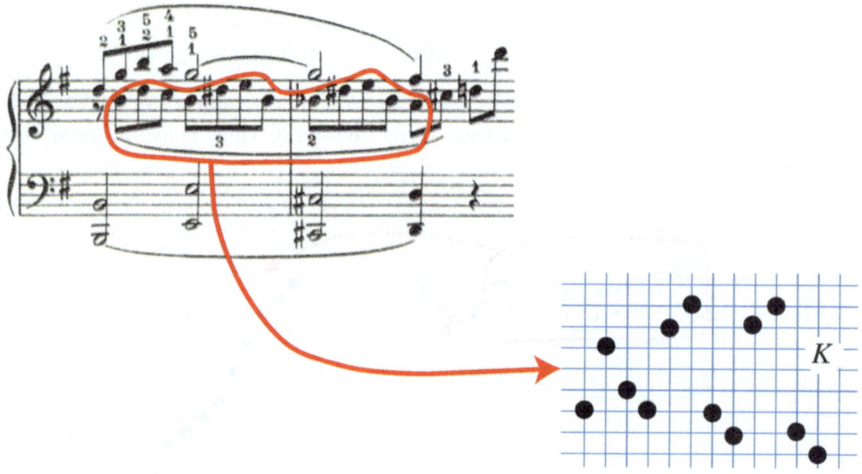

Fig. 14.5. A twelve-element alto-voice motif K in bars 79-80 in the first movement of op. 106.

In Figure 14.4, we see that this melodic part is isomorphic as a local composition (quality $(1,1)$) to a subcomposition of $G(2)$. The isomorphism is given by the matrix

$$A = e^q \begin{pmatrix} 1 & 0 \\ 2 & 1 \end{pmatrix}$$

representing a vertical arpeggio.

The second sample needs a proper interpretation of quality $(3,3)$. The sample, a twelve-element motif, is shown to the left in Figure 14.5. To the right we see the (projected) representation and interpretation K of this motif in *Onset* × *Pitch* space.

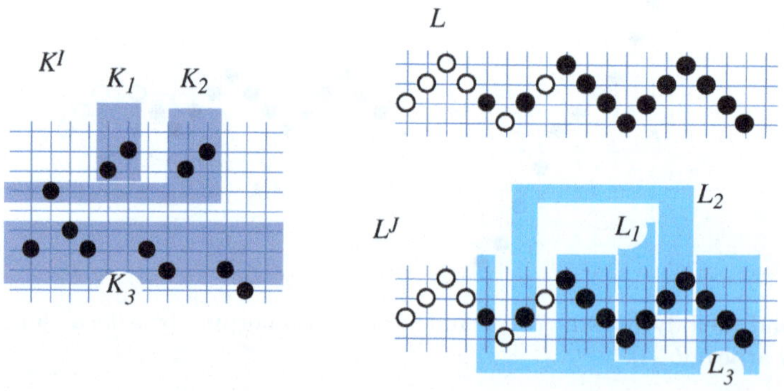

Fig. 14.6. The interpretation of the alto voice in bars 79-80 of quality $(3,3)$.

In Figure 14.6 we see the subcomposition L of $G(2)$, as well as its interpretation L^J by three disjoint charts L_1, L_2, L_3. This interpretation is isomorphic to the interpretation K^I under the following isomorphisms: L_2 and L_3 are translations of K_2 and K_3, respectively. The first chart L_1 equals $B \cdot K_1$ with a transformation

$$B = e^q \cdot \begin{pmatrix} 2 & 9 \\ 0 & -1 \end{pmatrix}$$

where the linear part is an onset dilatation by 2, followed by a ninefold horizontal arpeggio, and the pitch inversion, an isomorphism since its determinant -2 is invertible in $\mathbb{Z}[1/2]$.

Example 23 Take the two following non-isomorphic, zero-addressed three-element motives in $OnPiMod_{12,12}$:

$$K = \{(0,0),(0,1),(1,0)\} \text{ isomorphism class Nr.10,}$$
$$L = \{(0,0),(0,1),(3,11)\} \text{ isomorphism class Nr.13,}$$

and consider the interpretations $K^{(2)}, L^{(2)}$ by the three two-element 'interval' subcompositions. Evidently, every interval is isomorphic to $\{(0,0),(0,1)\}$, and since the intersections are singletons, we have $K^{(2)} \xrightarrow{\sim} L^{(2)}$. So the isomorphism of the interpretations by submaximal charts is no guarantee for the isomorphism of the original local compositions.

Exercise 28 *Consider the following zero-addressed local compositions in* $PiMod_{12}$*:*

$$C = \{0,2,4,5,7,8,9\} \textit{ C-major scale, isomorphism class Nr } 38^\frown,$$
$$X = \{0,2,3,4,7,10,11\} \textit{ isomorphism class Nr } 50^\frown,$$

which are not isomorphic. Take the triadic interpretation $C^{(3)}$ *and the following triadic interpretation of* X*:*

$$I_X = \{11,3,0\}, II_X = \{7,10,2\}, III_X = \{3,0,4\},$$
$$IV_X = \{10,2,11\}, V_X = \{0,4,7\}, VI_X = \{2,11,3\},$$
$$VII_X = \{4,7,10\}$$

and call the associated interpretation $X^{(3)}$*. Show that we have an isomorphism* $f/1 : X^{(3)} \xrightarrow{\sim}_{\iota} C^{(3)}$ *which is defined by*

$$f(0) = 7, f(2) = 9, f(3) = 4, f(4) = 11,$$
$$f(7) = 2, f(10) = 5, f(11) = 0,$$
$$\iota(Y_X) = Y_C, Y = I, \dots VII.$$

Show that this example fails if we do not allow isomorphisms with fifth or fourth transformations, i.e., if only *inversions and transpositions are allowed, then we have* $C \xrightarrow{\sim} X$ *as soon as* $C^{(3)} \xrightarrow{\sim} X^{(3)}$*.*

Exercise 29 *Show that the ecclesiastic modes (Section 13.4.2)* are rigid, *i.e., they have trivial automorphism group:* $Aut(X_{f,t}) = 1$*; it is even true that any two different ecclesiastic modes on the same scale are not isomorphic. This is false for modes* X_f*, in fact, since* $Aut(C) = \langle U_d \rangle$*, we have a permutation of these modes (we take the authentic mode names):*

$U_d(C_d) = C_d$, *dorian* $\leftrightarrow$ *dorian*
$U_d(C_e) = C_c$, *phrygian* $\leftrightarrow$ *ionian*
$U_d(C_f) = C_b$, *lydian* $\leftrightarrow$ *locrian*
$U_d(C_g) = C_a$, *mixolydian* $\leftrightarrow$ *aeolian*

and this means that modes become rigid when transformed into their ecclesiastic castings, a completely traditional effect...

14.3 Local Domains

Summary. The morphisms which are addressed in local compositions separate morphisms of global compositions. We discuss the musical consequences.

– Σ –

The Yoneda Lemma (see our discussion in Section 9) tells us that we may replace any global composition X in *Glob* by its functor $@X \in Glob^{@}$ and we shall not lose any information about its isomorphism class, more precisely, for any two global compositions, X, Y, we have a canonical bijection $Hom(X, Y) \xrightarrow{\sim} Hom(@X, @Y)$. It is a fundamental question in every category to find out whether some determined subclass of arguments of the functor $@X$ is already rich enough to determine X up to isomorphism. In algebraic geometry, it is well known ([395]) that the restriction $@S|_{Aff}$ of the scheme functor $@S$ of a scheme S to the subcategory Aff of the affine schemes is classifying, i.e.,if S, T are two schemes, we have a bijection $Hom(@S, @T) \xrightarrow{\sim} Hom(@S|_{Aff}, @T|_{Aff})$.

We shall now show that we have a weaker result in the theory of global compositions, i.e., the local arguments are 'separating'.

Definition 49 *Let X be a global composition in Glob. By $@_{loc}X$ we denote the restriction of the Yoneda functor $@X$ to the subcategory Loc of local compositions. By definition, we have $Y@_{loc}X = Y@X = Hom(Y, X)$ for any local (functorial) composition Y.*

Since a local composition has just one chart, the map on the atlases for a local domain Y reduces to the indication of one chart I_i or chart index i in the target composition X. Therefore we write

$$f\overset{i}{/}\alpha : Y \to X$$

to indicate elements of $Y@_{loc}X$.

Lemma 22 *If X^I, Y^J are local compositions at addresses A, B, respectively, the canonical restriction map*

$$loc : Hom(@X^I, @Y^J) \to Hom(@_{loc}X^I, @_{loc}Y^J)$$

is injective, i.e., local domains are separating. If $X^I = X$ is a local composition, then loc is a bijection.

Proof. Let $u : @X^I \to @Y^J$ be a natural transformation. We show that u is uniquely determined by its restriction to local arguments. By the Yoneda Lemma we know that $u = @f\overset{\iota}{/}\alpha$ for a morphism $f\overset{\iota}{/}\alpha : X \to Y$. So let us show that $f\overset{\iota}{/}\alpha$ is determined by the restriction $loc(u)$ to local arguments. To this end, consider the canonical embeddings $h_i\overset{i}{/}Id_A : X_i \to X$ of the subfunctors X_i of the covering I of X. We have $u(h_i\overset{i}{/}Id_A) = f\overset{\iota}{/}\alpha \cdot h_i\overset{i}{/}Id_A = f \cdot h_i \overset{\iota(i)}{/} \alpha_i$, where α_i is the address change of the ith covering subfunctor. Therefore, the map $\iota : I \to J$ is determined by the local evaluation at X_i. Also, the address change $\alpha = (\alpha_i)_i$ is determined, and the restriction $f \cdot h_i$ on the covering (X_i) of X determines f. Therefore $f\overset{\iota}{/}\alpha$ and its functorial counterpart $u = @f\overset{\iota}{/}\alpha$ is uniquely determined by the local restrictions.

Let now X be local and let $u : @_{loc}X \to @_{loc}Y^J$ be a natural transformation. We have to extend u to a natural transformation $U : @X \to @Y^J$. Let W^I be a global composition with charts W_t. Then the difference kernel $Ker(\prod_t W_t@X \rightrightarrows \prod_{s,t} W_{s,t}@X)$ of the restriction morphisms $W_{s,t} \to W_t, W_{s,t} \to W_s$ identifies[1] to $W^I@X$. By u, this kernel is mapped into the kernel $Ker(\prod_t W_t@Y^J \rightrightarrows \prod_{s,t} W_{s,t}@Y^J)$ which is a subset of $W^I@Y^J$, and we are done. QED.

[1] For global codomains, the kernel is strictly smaller than the morphism set because of the index map!

Lemma 23 *If X^I, Y^J are local compositions at addresses A, B, respectively, then there is a bijection*

$$l : Hom(@_{loc}X^I, @_{loc}Y^J) \xrightarrow{\sim} \coprod_{\iota: I \to J} \prod_{i \in I} X_i @Y_{\iota(i)}.$$

Proof. In fact, the contravariant functors $@_{loc}X^I, @_{loc}Y^J : Loc \to$ **Ens**, respectively, are isomorphic to $\coprod_{i \in I} @_{loc}X_i, \coprod_{j \in J} @_{loc}Y_j$, respectively. Hence,

$$Hom(@_{loc}X^I, @_{loc}Y^J) \xrightarrow{\sim} \prod_{i \in I} Hom(@_{loc}X_i, @_{loc}Y^J)$$

and by Lemma 22 and Yoneda's Lemma

$$Hom(@_{loc}X_i, @_{loc}Y^J) \xrightarrow{\sim} X_i @_{loc}Y^J \xrightarrow{\sim} \coprod_{j \in J} X_i @Y_j$$

whence our claim follows from the distributivity laws for products and coproducts of set-valued functors. QED.

Lemma 24 *With the above notation a natural transformation $u : @_{loc}X^I \to @_{loc}Y^J$ stems from a morphism $X^I \to Y^J$ of global compositions iff its image*

$$l(u) = (f_i/\alpha_i : X_i \to Y_{\iota(i)})_{i \in I}$$

with index map $\iota : I \to J$ has the property that for every couple $i_1, i_2 \in I$, the transformations f_{i_1}, f_{i_2} coincide on the intersection functor $X_{i_1,i_2} = X_{i_1} \cap X_{i_2}$. By $Hom^{loc}(@_{loc}X^I, @_{loc}Y^J)$, we denote the set of these natural transformations and call them localizable.

Exercise 30 *Prove these statements.*

So we have this proposition:

Proposition 13 *If X^I, Y^J are local compositions at addresses A, B, respectively, then the canonical restriction map*

$$loc : Hom(X^I, Y^J) \to Hom^{loc}(@_{loc}X^I, @_{loc}Y^J)$$

is well defined and bijective.

In other words, we may replace the study of isomorphism classes of global compositions X^I by the study of the isomorphism classes of their local domain functors $@_{loc}X^I$ under localizable natural transformations (which define a non-full subcategory ${}^{loc}Loc^{@}$ on the local domain functors).

The musicological meaning of these results is that the Yoneda philosophy for global musical structures may be restricted to local perspectives, i.e., morphisms on local compositions. With the restriction that the functors of local perspectives only reflect isomorphisms of global compositions if these isomorphisms are localizable.

We are however far from completely understanding the nature of the local domain functors and their category ${}^{loc}Loc^{@}$. But we know that above the local domain argument, the nature of the patchwork of the local aspects is crucial, and this is why we look at the nerve functor now.

14.4 Nerves

Summary. To each global composition is associated a combinatorial structure, its *nerve*, in fact the simplicial complex of the covering of the composition by its charts. This functorial association retrieves important information on the global object.

$-\Sigma-$

Nerves of objective global compositions were introduced in 13.2.1. For functorial global compositions, we have the same procedure. The *abstract nerve* $n(G^I)$ of a global functorial composition is the covering's simplicial complex, i.e., the simplicial complex whose vertexes are the covering subfunctors G_i, and whose simplexes are the (finite) vertex sets with non-empty intersection; the *geometric nerve* $N(G^I)$ is defined as the space $|n(G^I)|$ of $n(G^I)$, but see Appendix H.2.2, example 133. Since I is finite, the geometric nerve can be realized[2] as a polyhedron in $\mathbb{R}^{2m+1}$ if $dim(n(G^I)) \leqslant m$. Clearly, every morphism $f\overset{\iota}{/}\alpha : G^I \to H^J$ of objective or functorial global compositions gives rise to a morphism

$$n(f\overset{\iota}{/}\alpha) : n(G^I) \to n(H^J),$$

and this one generates the continuous map

$$N(f\overset{\iota}{/}\alpha) : N(G^I) \to N(H^J),$$

everything in a functorial way, i.e., we have two functors

$$N : Glob \to \mathbf{Top}, N : ObGlob \to \mathbf{Top}$$

into the category **Top** of topological spaces and continuous maps[3]. In particular, the functor N of geometric nerves generates topological invariants for isomorphism classes of global compositions.

Definition 50 *Two morphisms* $f\overset{\iota}{/}\alpha, g\overset{\kappa}{/}\beta : G^I \to H^J$ *between global compositions are called* mathematically equivalent *iff they coincide on the functors, i.e., iff* $f = g$. *An equivalence class of mathematically equivalent morphisms, i.e., a natural transformation* $f : G \to H$ *which stems from a morphism of global compositions, is called a* mathematical morphism *between global compositions.*

We shall come back to mathematical morphisms in Section 19.1.

Evidently, for two mathematically equivalent morphisms $f\overset{\iota}{/}\alpha, g\overset{\kappa}{/}\beta : G^I \to H^J$, the corresponding simplicial maps are contiguous[4]. So, by Lemma 107 in Appendix H, we have

Proposition 14 *Any two mathematically equivalent morphisms* $f\overset{\iota}{/}\alpha, g\overset{\kappa}{/}\beta : G^I \to H^J$ *induce homotopic maps on the corresponding geometric nerves.*

Exercise 31 *Draw a picture of a homotopy between the equivalent morphisms* $\delta : X^{(2)} \to X^{(3)}$ *from the third interval to the triadic interpretation (yielding the harmonic strip) of a diatonic scale as described in the diagram 14.2.*

Proposition 14 means that the variation of the embedding of a chart—of a refined interpretation, say—in one particular candidate of a larger chart of a coarser interpretation is related to the topological operation of homotopy if we consider the associated geometric nerves. Intuitively, this means that reinterpretation of an embedding is a kind of deformation on the geometric level.

For every address change $f : B \to A$ from the address A of a global functorial composition G^I, we have the f-slice $f@G^I$, an objective global composition. In general, its nerve has less vertexes and these may collapse. Also some simplexes could vanish. However, for every k-simplex $\sigma = \{G_{i_0}, \dots G_{i_k}\}$ of $n(G^I)$, there is an address change $f : B \to A$ such that $f@\cap\sigma = \cap_j f@G_{i_j} \neq \varnothing$. But observe that $\varnothing \neq f@\cap\sigma$ implies $\varnothing \neq f \cdot g@\cap\sigma$ for any address change $g : C \to B$. So take address changes $f_s : B_s \to A$, one for each

[2] See Appendix H.2.2, Theorem 82.
[3] See Appendix H.1.
[4] See Appendix H.2.3.

maximal simplex σ_s, such that all intersections $f_s@\cap\sigma_s$ are non-empty. Take the limit $(g_s : L \to B_s)_s$ of this system of address changes over A (it exists according to Appendix E.3.8). Then we have one unique address change $h = f_s \cdot g_s : L \to A$ with nonvanishing intersections in every simplex of the functorial composition. This means:

Theorem 13 *For every global functorial composition G^I, there is an f-slice $f@G^I$ such that the canonical simplicial map $n(G^I) \to n(f@G^I)$ is well defined and surjective. On the geometric nerves, it appears as a specialization morphism by collapse of particular vertexes.*

For an interpretation X^I of a local objective composition X, we have seen the construction of the global functorial composition $\widehat{X^I}$ associated with the covering of $\hat{X}$ by $\hat{I}$. The nerve $n(\widehat{X^I})$ may be significantly larger than the nerve $n(X^I)$. This can be checked on an adequate slice of $\widehat{X^I}$. We shall discuss such an example relating to the functorial interpretation associated with the harmonic strip in Section 25.3.4, example 61.

14.5 Simplicial Weights

Summary. The structure of a nerve induces weights on the nerve's simplexes with values in isomorphism classes of local compositions. This yields a tool for local and global classification, to be treated in the next chapter. We present the elementary example of simplicial motive weights.

– Σ –

Lemma 25 *Let G^I be an A-addressed objective or functorial global composition, and take a k-simplex $\sigma = \{G_{i_0}, \ldots G_{i_k}\}$ of its nerve. Consider the subobject (set or functor) $\cap\sigma \subset G$ and the chart isomorphism $\phi_{i_j} : K_{i_j} \xrightarrow{\sim} G_{i_j}$ of a G-chart K_{i_j} for the i_jth vertex of our simplex. Make $\cap\sigma \subset G$ into a local composition by the induced isomorphism $\phi_{i_j,\sigma} : K_{i_j,\sigma} \xrightarrow{\sim} \cap\sigma$ on the inverse image $K_{i_j,\sigma} = \phi_{i_j}^{-1}(\cap\sigma)$. Then all the charts defined by any of the simplex vertexes are compatible, i.e., they form an atlas for the (local) composition imposed on the intersection $\cap\sigma$. This local composition (viewed as a global composition with one chart) is again denoted by $\cap\sigma$.*

Proof. The case of a one-dimensional simplex is just the axiom of gluing charts together by transition isomorphisms of local compositions. Clearly, this transition situation restricts to the intersection of any higher number of covering functors. QED.

If we view the abstract nerve $n(G^I)$ as a category by the inclusion morphisms $i_{\sigma,\tau} : \sigma \hookrightarrow \tau$ among its simplexes, we have a contravariant functor

$$\cap : n(G^I) \to Loc : \sigma \mapsto \cap\sigma \tag{14.3}$$

(with the usual identification of local and one-chart global compositions), the *simplicial weight* of G^I.

This functor induces a map

$$iso\cap : n(G^I) \to Loc/\xrightarrow{\sim}$$

from the abstract nerve to the set $Loc/\xrightarrow{\sim}$ of isomorphism classes of local compositions which we call the *class weight* (function) of the global composition G^I.

Summarizing, we have a geometric nerve $N(G^I)$ and a class weight $iso\cap$ for each global composition. If we parametrize the isomorphism classes of local compositions by a determined set of symbols, this means that this information can be visualized by a polyhedron with a class weight attached to each simplex; we call this object the *class nerve* of G^I and denote it by $CN(G^I)$. This is a very intuitive invariant object which we want to illustrate in two examples.

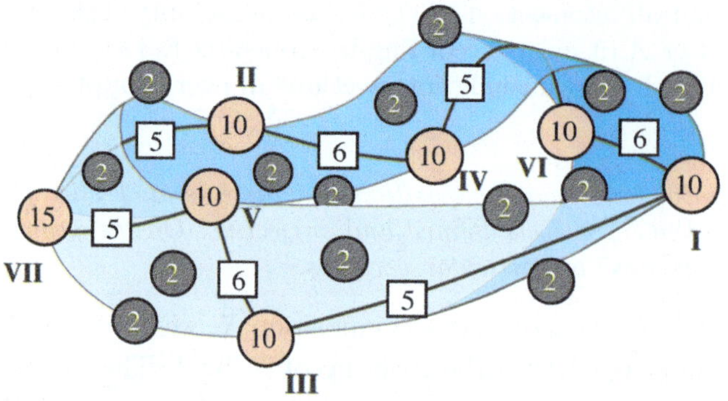

Fig. 14.7. The class nerve of the triadic interpretation of a major scale. The numbers are the class numbers from the classification list in Appendix N.1.

Example 24 We consider the harmonic strip of $X^{(3)}$ of a major scale X. The classes are numbered according to the classification in Appendix N.1. For example, this nerve has seven 2-simplexes with weight number 2 for the unique zero-addressed singleton class in $PiMod_{12}$. The class nerve $CN(X^{(3)})$ is shown in Figure 66.3. We see that the only automorphism of this interpretation must fix the seventh degree. So we have only the exchange of the boundary around the seventh degree, in fact the induced automorphism from the uninterpreted local scale.

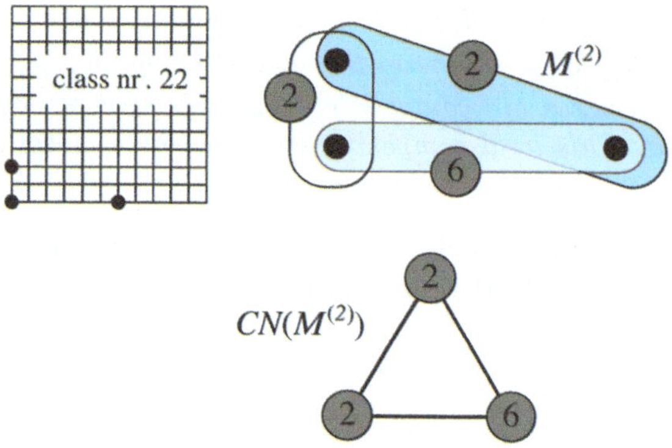

Fig. 14.8. A three-element motif of class number 22 and its class nerve.

Example 25 Consider the motif situation, more precisely, the zero-addressed local three-element compositions M in $OnPiMod_{12,12}$. We look at the 'face' interpretation $M^{(2)}$ by maximal subcompositions, i.e., by the three 2-element 'interval' charts in the motif. Clearly, if two motives are isomorphic, the class nerves are also. So the class nerve yields invariants of the local classification. Figure 66.2 shows a motif of class number 22 in the classification from Appendix O.3, as well as the interpretation, the chart class numbers of the intervals from Appendix N.1, and the class weights $(2, 2, 6)$ on the class nerve. From the table in Appendix O.3, we see that the class nerve is not classifying, but, together with the motif's volume (see Section 11.3.8), the class nerve yields a complete set of class invariants.

For motives with larger cardinality this is, however, no longer true. The four-element motives show counterexamples. The classification list in Appendix O.4 which was calculated by Hans Straub [1017] shows the class weights on the tetrahedra $CN(M^{(3)})$ of the face interpretation of the four-element motives in

$OnPiMod_{12,12}$. The motif classes which are not uniquely determined by volume and class nerve are indicated by a star.

14.6 Categories of Commutative Global Compositions

Summary. The category $ObLoc_A$ of commutative local compositions introduced in Section 8.3.5 has a global counterpart which is discussed here.

– Σ –

We fix an address module A over a commutative ring R. Recall from definition 28 in Section 8.3.5 that the category of commutative local compositions refers to the fixed address module A; the morphisms are of form f/Id_A with f being induced by an R-affine morphism of the Coordinator R-modules. If we consider global objective A-addressed compositions which are glued together by commutative local charts, this means that the transition morphisms are isomorphisms of commutative local compositions. And it means that morphism between commutative global compositions are induced by R-affine morphisms on the ambient modules of the local charts. Denote this category of commutative global composition at address A by $ComGlob_A$.

15

Global Classification

Wir sehen also, daß in der Musik
die Mehrzahl der sogenannten "wissenschaftlichen" Geister
fast so naiv ist wie Monsieur Achras – eine Figur von Jarry –,
der eine Sammlung von Polyedern anlegte.
Nun, es mag dahinstehen, ob in unserem Fall
Polyeder wirklich von unerschöpflichem Interesse sind.
Mir jedenfalls will die Notwendigkeit solcher
pataphysikalischer Spekulationen
nur schwerlich einleuchten.
Pierre Boulez [138, II, p.19]

Summary. Global classification relies on two concepts: affine functions and resolutions of global compositions. These constructs are discussed and exemplified. We derive classifying spaces and compare them to the situation in the Dreiding–Dress–Haegi theory of molecules: The latter are deduced from global compositions by additional structures concerning *orientation*, *distances* and *angles* (bilinear and exterior forms). It is therefore possible to view "molecules" as being global compositions with additional constraints; their musical meaning is discussed.

– Σ –

This chapter deals only with objective (more precisely, commutative) global compositions. This is probably not always necessary from the technical point of view, but presently, no more general theory has been elaborated and we prefer dealing with situations where concrete results are available.

From the musicological point of view, this is one of the most difficult chapters since the relation of classification and musicology, in particular esthetics, is quite implicit. But classification is a deep concern since it reveals the a priori extent of a structural framework and therefore its power as an expressive tool of artistic activity, be it in composition, performance, or understanding. So it is a kind of essence of the overall efforts of the art of music.

15.1 Module Complexes

Summary. Module complexes describe systems of global affine functions in a classical homological language.

– Σ –

For this chapter, we fix a commutative ring R. If we do not stress the contrary, we only consider R-modules and R-affine morphisms between R-modules, in particular, addresses A are R-modules.

Given a global composition G^I, we recall the category $n(G^I)$ of its abstract nerve, see Section 14.5.

G. Mazzola, *The Topos of Music I: Theory*, Computational Music Science,
https://doi.org/10.1007/978-3-319-64364-9_15

Definition 51 *An (R-)module complex over G^I is a covariant functor (a coefficient system, see Appendix H.3)*

$$M : n(G^I) \to {}_R\mathbf{Mod} \tag{15.1}$$

into the category ${}_R\mathbf{Mod}$ of R-modules and R-affine morphisms, where

$$M_{\sigma,\tau} : M(\sigma) \to M(\tau)$$

are transition morphisms of modules for the simplex inclusions (morphisms) $\sigma \subset \tau$.

As usual in sheaf theory, we put $\Gamma M = \lim_{n(G^I)} M(\sigma)$ and call this the set of global sections.

Example 26 Since any morphism of global compositions $\overset{\iota}{f}/\alpha : G^I \to H^J$ yields a natural transformation $n(\overset{\iota}{f}/\alpha) : n(G^I) \to n(H^J)$, every module complex M over H^J induces a module complex on $\overset{\iota}{f}/\alpha \star M$ over G^I, with

$$\overset{\iota}{f}/\alpha \star M(\sigma) = M(n(\overset{\iota}{f}/\alpha)(\sigma)).$$

Example 27 *If M is any R-module, the* constant module complex *of M is the complex with $M(\sigma) = M$ for all simplexes and identity transition. Observe that its global sections are in bijection with the set M^c, if $N(G^I)$ has c connected components.*

15.1.1 Global Affine Functions

Summary. Global affine functions are patchworks of affine functions on charts of atlases. Their role is that of generalized coordinate functions for musical compositions. This approach makes it possible to separate the core process of composing music from its realization in instrumental parameter spaces.

– Σ –

In the sequel of this chapter, we shall work on categories $ComGlob_A$ of global commutative compositions—except when explicitly mentioning the contrary. Suppose that we are given an A-addressed commutative global composition G^I. Let $A@_RR$ be the A-addressed local composition of the full R-module $A@_RR$. For a simplex σ of $n(G^I)$, a morphism $f/Id_A : \cap\sigma \to A@_RR$ may be identified with the supporting map $f : \cap\sigma \to A@_RR$. We call such a morphism an *A-addressed function* on the simplex. The set $n\Gamma(\sigma)$ of these functions is provided with the structure of an R-module by the usual addition and scalar multiplication of function values: $(f+g)(x) = f(x) + g(x), (rf)(x) = rf(x), r \in R$.

Exercise 32 *Show that sums and scalar multiples of A-addressed functions is again an A-addressed function.*

Suppose that we have an inclusion of simplexes $\sigma \subset \tau$ of $n(G^I)$. Then the ambient spaces of the charts of these simplexes are the same, i.e., the inclusion of local compositions $\cap\tau \subset \cap\sigma$ are in bijection with an inclusion of local compositions $K_\tau \subset K_\sigma \subset A@_RN$ for a specific module N. Since an affine function on $\cap\tau$ is the restriction of an affine morphism $A@h : A@_RN \to A@_RR$, f evidently extends to the restriction $A@h|_{K_\sigma}$, and we have proven that the transition morphisms by restriction of affine functions are surjective. The corresponding complex of *affine functions* is denoted by $n\Gamma(G^I)$. The subcomplex $C = C_{G^I}$ of constant functions is defined by $C(\sigma) = \{f \in n\Gamma(G^I)(\sigma), f = \text{ constant on} \cap \sigma\}$. The set of global functions of the function complex is denoted by $\Gamma(G^I)$.

For the following construction, we select one ambient space $A@_RF_\sigma$ for every simplex σ of $n(G^I)$. Take a chart $\cap(\sigma)$ of $\cap\sigma$ in $A@_RF_\sigma$. Then we have the submodule $R.\cap(\sigma)$ of $A@_RF_\sigma$ (see definition 17 in Section 7.4). If we have a morphism $\sigma \subset \tau$, we deduce a R-linear homomorphism $R.\cap(\tau) \to R.\cap(\sigma)$ which induce these homomorphisms

$$Lin_{\sigma,\tau}(M) : Lin_R(R.\cap(\sigma), M) \to Lin_R(R.\cap(\tau), M)$$

on the homomorphism modules for any given module M. This module complex is denoted by $Lin_R(G^I, M)$, although it is only determined up to an isomorphism of module complexes[1] on G^I; it evaluates to $Lin_R(G^I, M)(\sigma) = Lin_R(R.\cap(\sigma), M)$. In the particular case $M = A@_R R$, we abbreviate $Lin_R(G^I, A@_R R) = G^{I*}$ and call this *the module complex of A-addressed forms on* G^I. For an affine function $f : \cap\sigma \to A@_R R$, we have an R-linear homomorphism $R.f : R.\cap(\sigma) \to A@_R R$ (see Lemma 6 in Section 8.3.5), and the map $f \mapsto R.f$ is R-linear and functorial in σ, i.e., we have a morphism of module complexes on G^I:

$$R_{G^I} : n\Gamma(G^I) \to G^{I*} \tag{15.2}$$

If we have a subcomplex situation $N \subset M$ on G^I, we get the quotient complex M/N with values $M/N(\sigma) = M(\sigma)/N(\sigma)$.

Lemma 26 *The kernel of R_{G^I} is the module C_{G^I} of constant functions. Therefore, if N is any module complex with $C \subset N \subset n\Gamma(G^I)$, we have an embedding of module complexes*

$$N/C \rightarrowtail G^{I*} \tag{15.3}$$

Proof. The function f is in the kernel iff its values $R.f(s_i - s_j) = f(s_i) - f(s_j)$ vanish, and this in turn means that f is constant on $\cap\sigma$. QED.

Definition 52 *Let P be a property of modules. If M is a module complex over G^I, it is said to share property P iff all its values $M(\sigma)$ do so. In particular, M is called injective (projective) iff the module $M(\sigma)$ is so for every simplex σ.*

Lemma 27 *If A is the zero address $A = 0_R$, and if $n\Gamma$ is injective (in particular, if R is a field), the morphism R_{G^I} is surjective, and therefore, we have an isomorphism $n\Gamma/C \xrightarrow{\sim} G^{I*}$.*

Proof. Since we have the zero address, any $\cap\sigma$ is isomorphic to an embedded local composition in the same ambient module, and any linear map $l : R.\cap\sigma \to R$ extends to a map on the ambient module. QED.

Let $\overset{\iota}{f}/Id_A : G^I \to H^J$ be a morphism of A-addressed commutative global compositions. Take a simplex σ in $n(G^I)$, and its image σ' under the associated simplicial map. Then, each restricted morphism

$$f|_{\cap\sigma}/Id_A : \cap\sigma \to \cap\sigma'$$

gives rise to a map

$$\overset{\iota}{f}/Id_A \star n\Gamma(H^J)(\sigma) = n\Gamma(H^J)(\sigma') \to n\Gamma(G^I)(\sigma)$$

by right composition with this restricted morphism. Moreover, the map is R-linear. Therefore, if M is any subcomplex of $n\Gamma(H^J)$, its induced complex $\overset{\iota}{f}/Id_A \star M$ is mapped R-linearly onto what is called the *retracted* module complex

$$M|\overset{\iota}{f}/Id_A \subset n\Gamma(G^I). \tag{15.4}$$

In particular, if $M = C_{H^J}$, we have $C_{H^J}|\overset{\iota}{f}/Id_A \subset C_{G^I}$.

If M is any subcomplex of $n\Gamma(H^J)$ which contains the constant complex C_{H^J}, we have the short exact sequence

$$0 \to C_{H^J} \to M \to M/C_{H^J} \to 0$$

and therefore the corresponding short exact sequence

$$0 \to \overset{\iota}{f}/Id_A \star C_{H^J} \to \overset{\iota}{f}/Id_A \star M \to \overset{\iota}{f}/Id_A \star M/C_{H^J} \to 0$$

which projects to an exact sequence on G^I as follows:

[1] By definition of module complexes, this is an isomorphism of functors.

$$\begin{array}{ccccccccc}
0 & \longrightarrow & f\overset{\iota}{/}Id_A \star C_{H^J} & \longrightarrow & f\overset{\iota}{/}Id_A \star M & \longrightarrow & f\overset{\iota}{/}Id_A \star M/C_{H^J} & \longrightarrow & 0 \\
 & & \downarrow & & \downarrow & & \downarrow & & \\
0 & \longrightarrow & C_{H^J}|f\overset{\iota}{/}Id_A & \longrightarrow & M|f\overset{\iota}{/}Id_A & \longrightarrow & M/C_{H^J}|f\overset{\iota}{/}Id_A & \longrightarrow & 0 \\
 & & \downarrow & & \downarrow & & \downarrow & & \\
0 & \longrightarrow & C_{G^I} & \longrightarrow & n\Gamma(G^I) & \longrightarrow & n\Gamma(G^I)/C_{G^I} & \longrightarrow & 0
\end{array}$$

including the definition of $M/C_{H^J}|f\overset{\iota}{/}Id_A$. In particular, for $M = n\Gamma(H^J)$, we have this surjective R-linear homomorphism

$$f\overset{\iota}{/}Id_A \star n\Gamma(H^J)/C_{H^J} \to n\Gamma(H^J)/C_{H^J}|f\overset{\iota}{/}Id_A. \tag{15.5}$$

Finally, by construction of global sections, there is a canonical projective system of R-linear maps

$$\Gamma(H^J) \to f\overset{\iota}{/}Id_A \star n\Gamma(H^J)(\sigma) \to n\Gamma(G^I)(\sigma)$$

and therefore we have a canonical R-linear map

$$\Gamma(f\overset{\iota}{/}Id_A) : \Gamma(H^J) \to \Gamma(G^I)$$

which is a contravariant functor

$$\Gamma : ObGlob_A \to {}_R\mathbf{Mod}. \tag{15.6}$$

15.1.2 Bilinear and Exterior Forms

Summary. Bilinear and exterior forms capture the language of classical geometry of angles, distances, and orientation in Euclidean three-space. This formalism is built on the systems of global functions.

– Σ –

Until now, no considerations regarding angles and similar properties from Euclidean geometry have been made in the theory of local or global compositions. Here is the formal framework for such perspectives.

Suppose that N is a module complex of A-addressed affine functions on G^I which contains the constant complex $C = C_{G^I}$. Consider the module complex B_N defined on simplexes σ by the expression

$$B_N(\sigma) = Lin_R((N(\sigma)/C)^\star, N(\sigma)/C)) \tag{15.7}$$

with $X^\star$ being the R-linear dual of X, and with the evident transition homomorphisms.

With the injection 15.3 in mind we have a canonical R-linear map

$$Lin_R((N(\sigma)/C)^\star, N(\sigma)/C)) \to Lin_R((G^{I*}(\sigma))^\star, G^{I*}(\sigma))$$

and the right side is

$$Lin_R((Lin_R(R.\cap(\sigma), A@_RR))^\star, Lin_R(R.\cap(\sigma), A@_RR))$$

which, according to Appendix E.3.2, Lemma 92 identifies to

$$Lin_R((A@_RR.\cap(\sigma)^\star)^\star, A@_RR.\cap(\sigma)^\star).$$

We now assume that $A = R^n$, $0 \leqslant n$ natural. Then Appendix E.3.2, Proposition 97 yields linear maps

$$d : A@_R R. \cap (\sigma)^\star \to (A@_R R. \cap (\sigma))^\star,$$
$$u : A@_R R. \cap (\sigma) \to (A@_R R. \cap (\sigma)^\star)^\star,$$

and therefore a map

$$Lin_R((G^{I*}(\sigma))^\star, G^{I*}(\sigma)) \to Lin_R(A@_R R. \cap (\sigma), (A@_R R. \cap (\sigma))^\star)$$

which gives an R-linear map

$$\beta_\sigma : B_N(\sigma) \to Lin_R(A@_R R. \cap (\sigma), (A@_R R. \cap (\sigma))^\star). \tag{15.8}$$

The right side describes the bilinear forms on the space of A-addressed points in the module $R. \cap (\sigma)$. We therefore associate a bilinear form $\beta(x)$ with each linear map $x : (N(\sigma)/C)^\star \to N(\sigma)/C$, and in this framework, we also speak of the bilinear form x, even when we cannot instantiate an associated form $\beta(x)$ (which is the case for general address A).

Definition 53 *With the above notation, the pair* (G^{I*}, β) *with a global section* $\beta \in \Gamma B_N$ *on* G^{I*} *is called an* N-formed global composition.

Let $f/\overset{\iota}{I}d_A : G^I \to H^J$ be a morphism of A-addressed commutative global compositions. Let M, N be submodules of $n\Gamma(G^I), n\Gamma(H^J)$, respectively, both containing the constants. Take global sections $\beta \in \Gamma B_M$, $\gamma \in \Gamma B_N$, respectively. Suppose that we have an inclusion of the restriction of N under $f/\overset{\iota}{I}d_A$ in M:

$$N | f/\overset{\iota}{I}d_A \rightarrowtail M \tag{15.9}$$

and therefore an inclusion

$$N/C | f/\overset{\iota}{I}d_A \rightarrowtail M/C$$

of quotients modulo constants. Then we have a canonical map

$$B_f : B_N \to B_M \tag{15.10}$$

over $f/\overset{\iota}{I}d_A$ and therefore one of global sections

$$\Gamma B_f : \Gamma B_N \to \Gamma B_M. \tag{15.11}$$

Definition 54 *With the preceding notation and hypotheses, we say that the morphism* $f/\overset{\iota}{I}d_A$ *is a* morphism of formed compositions $f/\overset{\iota}{I}d_A : (G^I, \beta) \to (H^J, \gamma)$ *if* $\Gamma B_f(\gamma) = \beta$.

Given a submodule complex N of the complex of affine functions on G^I, containing the constants, we have an evident complex $\Lambda^n(N/C)$ of exterior n-forms. If $\Omega \in \Gamma\Lambda^n(N/C)$ is a global exterior n-form, we call (G^I, Ω) an oriented global composition. As with bilinear forms, we have the concept of a morphism $f/\overset{\iota}{I}d_A : (G^I, \Omega) \to (H^J, \Delta)$ of oriented global compositions if $\Gamma\Lambda^n_f(\Delta) = \Omega$, with the evident nth exterior power morphism whenever we have the above inclusion (15.9).

15.1.3 Deviation: Compositions vs. "Molecules"

Summary. Molecule structures, as they are considered in the Dreiding–Dress–Haegi theory, are very similar to representable global compositions. They emerge from the latter by adding bilinear and exterior forms. Geometrically this means that compositions may be "deduced" from molecular structures by abstraction from angles, distances, and orientation. We consider the specific difference of such an abstraction in the musical perspective: What does music gain after adding "molecular" information?

– Σ –

For the coefficient field $R = \mathbb{R}$ of real numbers, and for the zero address, special formed and oriented global compositions may represent molecular structures such as have been considered in [279]. We shall see below (16.4) that classification of formed and oriented global compositions, and therefore of molecular structures, is significantly more difficult than classification of global compositions. So this short deviation is not only an application of the general formalism of global compositions, it also shows that enriching the inner structure of a global composition may cause substantial difficulties in the understanding of the corresponding category.

For this situation, the (zero-addressed) points of G^I are viewed as being momentary positions of atoms of a molecule in $\mathbb{R}^3$. The atom species are parametrized by natural numbers, and their distribution on G^I is defined by a marking application $\alpha : G \to \mathbb{N}$. Moreover, we ask that

- $dim(n\Gamma(\sigma)/C) \leqslant 3$ for all charts σ of G^I,
- G^I is formed by a symmetric bilinear form β which is positive definite on all charts,
- we are given an exterior 3-form Ω which is β-normed, i.e., the Gram identity $\Omega^2 = det(\beta)$ is valid[2].

Definition 55 *A quadruple $(G^I, \alpha, \beta, \Omega)$ with the above properties, i.e., an α-marked, β-formed, and Ω-oriented (zero-addressed) global composition G^I is called a* global molecule. *A* morphism between global molecules *is a morphism of global compositions which respects the additional structures as discussed above. A global molecule which is isomorphic to an interpretation of a local molecule[3] is called interpretable.*

Interpretable molecules are canonically associated with distinguished structures in the theory of Dreiding–Dress–Haegi, the proof may be omitted in this context:

Theorem 14 *The molecular structures which are associated with interpretable molecules are (d, χ)-defined in the sense of* [279].

15.2 The Resolution of a Global Composition

Summary. The second device for classification is the global standard composition, canonically associated with the composition's nerve. Special module complexes on such standard objects allow the reconstruction of the original composition.

– Σ –

This section introduces two methods which are used in classification theory: The first is the global standard composition; it is a free object in the category of global compositions and helps standardize the various special objects via corresponding special module complexes of affine functions. The second method is the construction of quotient compositions from given modules of affine functions.

Combining these methods, we shall classify global compositions by the classification of special module complexes of affine functions in free objects, and by a theorem which tells us how to rebuild compositions as quotients from such special modules.

[2] The Gram form verifies the identity $\Omega(x_1, \ldots x_n)\Omega(y_1, \ldots y_n) = det((\beta(x_i, y_j))_{i,j}$, for all sequences $x_1, \ldots x_n, y_1, \ldots y_n$ of elements in the chart vector space, see also [391, VII,3].

[3] One with a single chart; on the interpretation's charts, bilinear forms and the orientation are induced from the local data.

15.2.1 Global Standard Compositions

Summary. The standard compositions are objects representing compositions with "notes in general position", i.e., their configuration is as 'free' as possible from 'occasional' coincidences. In fact, the standard composition is a geometric realization of the composition's nerve and thus depends only on combinatorial information. There is a natural projection from the standard object onto the generating composition.

– Σ –

We first want to broaden the provisional concept of a standard local composition given in definition 30, Section 11.3.2. Given a module A in ${}_R\mathbf{Mod}$ and a natural number $0 \leqslant n$, we denote by $A^{\sqcup n}$ the $n+1$-fold coproduct $\coprod_{n+1} A$ of A. Recall from its construction (Appendix E.3.8) that there is an isomorphism $A^{\sqcup n} \xrightarrow{\sim} R^n \oplus A^{n+1}$. We denote the canonical basis of R^n by $(e_1, \ldots e_n)$, and for any element $a \in A$ and $0 \leqslant i \leqslant n$, we set $a_i = (0, \ldots a, \ldots 0)$ for the $n+1$-tuple in A^{n+1} having a at its $i+1$-th position and zero else; the zero element is denoted by e_0. We have the inclusion morphisms

$$\sigma_i : A \to A^{\sqcup n} \tag{15.12}$$

for $0 \leqslant i \leqslant n$, with

$$\sigma_i(a) = (e_i, a_i) = \begin{cases} \text{linear if } i = 0, \\ \text{affine if } i > 0. \end{cases} \tag{15.13}$$

This defines a local, A-addressed composition ${}_A\Delta_n \subset A@_R A^{\sqcup n}$ which is called the *A-addressed local standard composition of dimension n*. By construction, it has the following property: If M is any R-module, and if $s. = (s_0, \ldots s_n)$ is any sequence of A-addressed points in M, with associated local composition $S = \{s_0, \ldots s_n\} \subset A@_R M$, then there is exactly one morphism of local compositions

$$(s.) : \ {}_A\Delta_n \to S : \sigma_i \mapsto s_i \text{ for } i = 0, \ldots n. \tag{15.14}$$

This morphism is in fact defined by the universal property of the coproduct and is mediated by the following affine function $f : A^{\sqcup n} \to M$: Write $s_i = e^{t_i} \cdot s_{i,0}$. Then we have

$$\begin{aligned} f(e_0) &= t_0, \\ f(e_i) &= t_i - t_0 \text{ (linear) for } i > 0, \\ f(a_i) &= s_{i,0}(a) \text{ (linear) for } i \geqslant 0, \end{aligned}$$

and the formula $s_i = f \cdot \sigma_i$ is immediate.

Exercise 33 *If A, B are two addresses, sitting in ${}_R Mod$, ${}_S Mod$, respectively, we denote by $Id : {}_A\Delta_n \to {}_B\Delta_n$ the "identity" $\sigma_i \mapsto \sigma_i$. Consider base changes $\alpha : A \to B$. Show that the assignment $A \mapsto {}_A\Delta_n$ and $\alpha \mapsto Id/\alpha : {}_A\Delta_n \to {}_B\Delta_n$ is a functor $\Delta_n : {}_c\mathbf{Mod} \to ObLoc$ on the category ${}_c\mathbf{Mod}$ of modules over commutative rings and diaffine morphisms.*

To define global "free" objects among the A-addressed objective compositions with finite charts, we shall represent the nerve $n(G^I)$ by an isomorphic *standard representative nerve* $n^\star$ induced by a covering of the natural interval $[0, m] = \{0, 1, 2, 3, \ldots m = card(G) - 1\}$ of natural numbers. For $n^\star$, we define the *global standard composition* ${}_A\Delta_{n^\star}$ at address A by the interpretation of the local standard composition ${}_A\Delta_m$ which is given by the present covering of $[0, m]$. We are also given a *standard atlas* of ${}_A\Delta_{n^\star}$. In fact, for any subset $q = \{t_0, \ldots t_c\} \subset [0, m]$ of $c+1$ elements, indexed corresponding to increasing values, we have the canonical injection $i_q : {}_A\Delta_c \to {}_A\Delta_m$ via $\sigma_j \mapsto \sigma_{t_j}$. This defines the standard atlas.

The universal property of this global standard composition reads as follows. Take the category *Covens* of coverings of sets[4], and consider the covariant functor

[4] See Appendix H.2.1, example 132.

$$_ACov_{n^\star} : ObGlob_{@A} \to \mathbf{Ens} : G^I \mapsto Hom_{Covens}(n^\star, (G, n_0(G^I))). \tag{15.15}$$

Then we have this result:

Proposition 15 *The functor $_ACov_{n^\star}$ is representable by the standard global composition $_A\Delta_{n^\star}$, i.e., we have a bijection*

$$Hom_{Covens}(n^\star, (G, n_0(G^I))) \xrightarrow{\sim} Hom_{ObGlob_{@A}}(_A\Delta_{n^\star}, G^I)$$

which is functorial in the A-addressed composition G^I.

The proof is left as an exercise. In particular, if we take the standard nerve $n^\star = n^\star(G^I)$ of the nerve of G^I and then the corresponding 'identity' morphism $Id : n^\star \xrightarrow{\sim} (G, n_o(G^I))$, we obtain a corresponding bijective[5] morphism

$$res_{G^I}/Id_A : \Delta_{G^I} \to G^I \tag{15.16}$$

with the notation $\Delta_{G^I} = {_A\Delta_{n^\star(G^I)}}$, this object and the morphism res_{G^I}/Id_A being called *the resolution of* G^I. Clearly, the associated simplicial morphism $n(res_{G^I}/Id_A) : n^\star(\Delta_{G^I}) \to n^\star(G^I)$ is an isomorphism.

In particular, due to the universal property of the global standard compositions, every morphism $\overset{\iota}{f}/Id_A : G^I \to H^J$ can uniquely be lifted to a corresponding morphism $res_{\overset{\iota}{f}/Id_A}$ of resolutions to make the diagram

$$\begin{array}{ccc} \Delta_{G^I} & \xrightarrow{res_{\overset{\iota}{f}/Id_A}} & \Delta_{H^J} \\ {\scriptstyle res_{G^I}/Id_A}\downarrow & & \downarrow{\scriptstyle res_{H^J}/Id_A} \\ G^I & \xrightarrow{\overset{\iota}{f}/Id_A} & H^J \end{array} \tag{15.17}$$

commute. We therefore have a resolution functor

$$res_{@A} : ObGlob_{@A} \to ObGlob_{@A} \tag{15.18}$$

and a natural transformation

$$\delta_{@A} : res_{@A} \to Id_{ObGlob_{@A}}. \tag{15.19}$$

The following deals with the reconstruction of the category $ObGl_{@A}$ from its subcategory of free objects.

15.2.2 Compositions from Module Complexes

Summary. The projection of the standard composition onto its generating composition canonically induces a module complex of global affine functions on the standard composition. This complex is used to reconstruct the generating composition from the standard composition.

– Σ –

The resolution functor $res_{@A}$ and its associated natural transformation $\delta_{@A}$ give rise to a module complex of affine functions $\Delta n\Gamma(G^I) = n\Gamma(G^I)|res_{G^I}/Id_A$ in $n\Gamma(_A\Delta_{G^I})$, for each global A-addressed composition G^I. Call this complex *the resolution complex of composition* G^I. Moreover, this assignment commutes with the morphism of the resolution functor, i.e., for a morphism $\overset{\iota}{f}/Id_A : G^I \to H^J$ we have a canonical inclusion

$$\Delta n\Gamma(H^J)|\overset{\iota}{f}/Id_A \subset \Delta n\Gamma(G^I) \tag{15.20}$$

of the retracted resolution complex of H^J in the resolution complex of G^I. The next step deals with the reconstruction of G^I from $n\Gamma(G^I)$ and the related question of classification of global compositions by use of the resolution complex which is suggested by the functorial relation (15.20).

[5] It is, however, not an isomorphism in general.

The generic situation from the preceding constructions is that we are given a module complex $M \subset n\Gamma(G^I)$, containing the constants C, and that we would like to construct a kind of "quotient" composition whose affine functions are those of M. We first look at the local situation.

Definition 56 *Let $S \subset A@_R U$ be an A-addressed objective local composition in the R-module U. For a submodule $L \subset \Gamma(S)$ of affine functions on S, the evaluation map $\dot{}: S \to A@_R L^\star$ into the A-valued points of the dual module $L^\star = Lin_R(L, R)$ of L is defined by $\dot{s}(a)(l) = l(s)(a)$.*

The problem is that the evaluation is not a morphism of local compositions in general. We have to investigate sufficient conditions for the existence of a morphism. In the special case which is of interest, we have this guarantee: In fact, let $S = {}_A\Delta_n \subset A@_R A^{\sqcup n}$. Then, the dotted points $\dot{\sigma}_i : A \to L^\star$ define the universal map $H_L : A^{\sqcup n} \to L^\star$, and we have interpreted $\dot{}: {}_A\Delta_n \to A@_R L^\star$ as a morphism of local compositions.

Next, suppose we are given two local compositions $S \subset A@_R U, T \subset A@_R V$ and a morphism

$$f/Id_A : S \to T,$$

together with a module $L_T \subset \Gamma(T)$ whose retract $L_T|f/Id_A$ is included in a module $L_S \subset \Gamma(S)$. We then have a commutative diagram

$$\begin{array}{ccc} S & \xrightarrow{\;\dot{}\;} & A@_R L_S^\star \\ {\scriptstyle f/Id_A}\big\downarrow & & \big\downarrow{\scriptstyle A@|f^\star} \\ T & \xrightarrow{\;\dot{}\;} & A@_R L_T^\star \end{array} \tag{15.21}$$

where $|f^\star$ is the R-dual of the canonical linear map $|f : L_T \to L_S$.

Exercise 34 *Give a proof of the commutativity of diagram (15.21).*

This construction yields a morphism $\dot{f}/Id_A : \dot{S} \to \dot{T}$ of local compositions in the ambient spaces $L_S^\star, L_T^\star$, respectively. With this technique, we may associate a global composition with a module complex $N \subset n\Gamma({}_A\Delta_{n^\star})$ of affine functions in the standard composition ${}_A\Delta_{n^\star}$ of a standard covering $n^\star$.

To this end, suppose we are given such a module complex N of affine functions, and suppose that the restriction transition morphisms $N(\tau) \to N(\sigma)$, $\tau \subset \sigma$, are all surjective. This is certainly the case for retracted function modules from resolution morphisms. For a simplex σ of the nerve $n({}_A\Delta_{n^\star}) \xrightarrow{\sim} n^\star$ we note $c(\sigma) = card(\cap_{q\in\sigma} q) - 1$.

In the following discussion of classification, we shall tacitly assume that all module complexes of affine functions have surjective transition morphisms.

With this, if we apply the construction from diagram (15.21) to the situation where $S = {}_A\Delta_{c(\sigma)} \subset A@A^{\sqcup c(\sigma)}$, and $T = {}_A\Delta_{c(\tau)} \subset A@A^{\sqcup c(\tau)}$ for simplexes $\tau \subset \sigma$ of ${}_A\Delta_{n^\star}$, and with $L_S = N(\sigma), L_T = N(\tau)$, then we have injective vertical arrows in the corresponding commutative diagram

$$\begin{array}{ccc} {}_A\Delta_{c(\sigma)} & \xrightarrow{\;\dot{}\;} & A@_R N(\sigma)^\star \\ {\scriptstyle \text{inclusion}}\big\downarrow & & \big\downarrow{\scriptstyle A@res^\star} \\ {}_A\Delta_{c(\tau)} & \xrightarrow{\;\dot{}\;} & A@_R N(\tau)^\star \end{array} \tag{15.22}$$

where res is the restriction map. We write $\cap\sigma = {}_A\dot{\Delta}_{c(\sigma)}$, and therefore get a surjective morphism of diagrams of local compositions ${}_A\Delta_{c(\sigma)} \to \cap\sigma$ over $n^\star$. Setting ${}_A\Delta_{n^\star}/N = colim_{n^\star} \cap\sigma$, we have a commutative diagram of sets

$$\begin{array}{ccc} {}_A\Delta_{c(\sigma)} & \xrightarrow{\;\dot{}\;} & \cap\sigma \\ \big\downarrow & & \big\downarrow \\ {}_A\Delta_{n^\star} & \xrightarrow{/N = colim\dot{}} & {}_A\Delta_{n^\star}/N \end{array} \tag{15.23}$$

induced by the dot morphisms of local compositions.

Definition 57 *Call a module complex $N \subset n\Gamma(G^I)$ separating, iff for every zero-simplex (chart) $\sigma \in n_0(G^I)$, the dot map $\sigma \to A@_R N(\sigma)^\star$ is injective.*

Intuitively, this means that for any pair of points on any chart, there is a function of N on this chart which separates these points. So, if in the above situation $N \subset n\Gamma({}_A\Delta_{n^\star})$ is separating, the dot maps are all bijective onto the images since the vertical maps in diagram (15.22) are injective. Therefore, the colimit diagram (15.23) has bijective horizontal arrows, and the images $\cap\sigma$ are injected into the limit ${}_A\Delta_{n^\star}/N$. So these images cover the limit and the images of the zero-dimensional simplexes build a canonical atlas of a global A-addressed composition, i.e., the diagram (15.23) becomes a bijective morphism of A-addressed global compositions.

So, if N is separating, we have constructed a canonical global composition and a bijective morphism from the free object to a global composition which is defined by the functions of N.

Definition 58 *We call this composition ${}_A\Delta_{n^\star}/N$ the N-quotient of ${}_A\Delta_{n^\star}$. The morphism ${}_A\Delta_{n^\star} \to {}_A\Delta_{n^\star}/N$ from diagram (15.23) is denoted by $/N$.*

In particular, if $res_{G^I} : {}_A\Delta_{G^I} \to G^I$ is the resolution of the A-addressed composition G^I, we have the resolution complex $\Delta n\Gamma(G^I)$. Clearly, the resolution complex is separating iff the complex of affine functions $n\Gamma(G^I)$ is. So there is a $\Delta n\Gamma(G^I)$-quotient if the complex of affine functions $n\Gamma(G^I)$ is separating, in which case we also say that *G^I is separating.* Of course, this is a property which is invariant under bijective morphisms among A-addressed compositions.

Proposition 16 *If the A-addressed composition G^I admits a projective atlas, i.e., an atlas whose charts have projective R-modules, then G^I is separating.*

Proof. In this case, the canonical bidual map $U \to U^{**}$ is injective for each ambient space U of a projective atlas, see Appendix E.4.2. Now, if $s \neq t$ are two points in the chart $\sigma \subset A@_R U$, there is $a \in A$ with $s(a) \neq t(a)$, and we may take a linear form $h \in U^\star$ such that $h(s(a)) \neq h(t(a))$ which means that the induced affine form $l = A@h|\sigma$ separates s from t. QED.

Proposition 17 *Suppose that G^I is separating. Then we are given a commutative triangle of* covering set *isomorphisms*

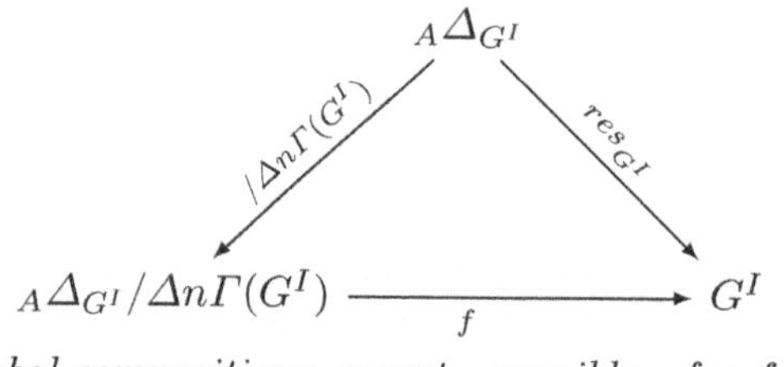

(15.24)

which stem from morphisms of global compositions except—possibly—for f.

To look for conditions when f is a morphism, we may concentrate on the local situation. This means that we are given a separating local composition $S = \{s_0, \ldots s_n\} \subset A@_R U$ of cardinality[6] $n+1$. For the module $\Gamma(S)$ of affine functions $g : S \to A@_R R$ on S, we have the corresponding commutative triangle of set bijections:

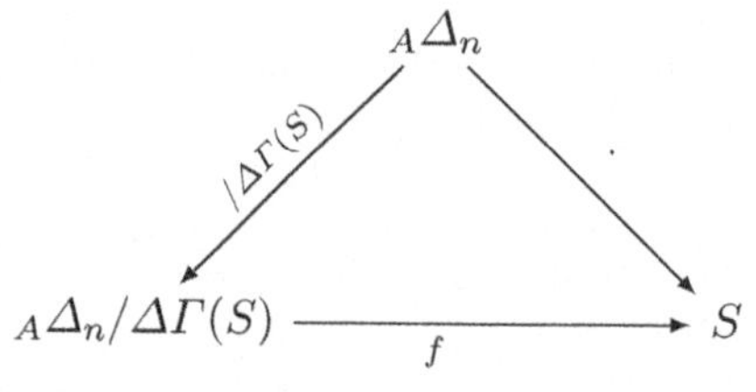

(15.25)

[6] Recall that in this discussion, we are considering global compositions with finite charts.

which becomes a triangle of morphisms of local compositions if we can derive f from an affine morphism $F : \Gamma(S)^\star \to U$ of ambient modules. On the level of ambient modules, we have the following commutative square of affine morphisms:

$$\begin{array}{ccc} A^{\sqcup n} & \xrightarrow{\sqcup \dot{S}} & \Gamma(S)^\star \\ {\scriptstyle \sqcup S}\downarrow & & \downarrow{\scriptstyle q^\star} \\ U & \xrightarrow{b} & (U@_R R)^\star \end{array} \tag{15.26}$$

where the $b(u)(t) = t(u)$ is the extension of the bidual, where $q : U@_R R \to \Gamma(S)$ is the canonical surjection, and where $\sqcup S, \sqcup \dot{S}$ are the universal morphism associated with $S, \dot{S}$, respectively.

Exercise 35 *Verify the commutativity of diagram (15.26) by use of the universal property of the* $(n+1)$*-fold coproduct* $A^{\sqcup n}$.

If U is finitely generated and projective, the bidual $U \to U^{\star\star}$ is an isomorphism (Appendix E.4.2), and the dual $r : (U@_R R)^\star \to U^{\star\star}$ of the inclusion $U^\star \to U@_R R$ defines a right-inverse i of b. Therefore, we are given the commutative triangle

$$\begin{array}{ccccc} & & A^{\sqcup n} & & \\ & {\scriptstyle \sqcup \dot{S}}\swarrow & & \searrow{\scriptstyle \sqcup S} & \\ \Gamma(S)^\star & & \xrightarrow[i \cdot q^\star]{} & & U \end{array} \tag{15.27}$$

which yields the required morphism for f in diagram (15.25). This means that

Theorem 15 *If the composition* G^I *has a finitely generated, projective atlas*[7]*, the factorization diagram of Proposition 17 is a diagram of global compositions.*

Proof. In fact, by Proposition 16, G^I is separating, and by the local diagram (15.27), applied to the charts of this situation, the map f in diagram (15.24) is a morphism of global compositions, QED.

The factorization morphism f has a chance to become an isomorphism if we succeed in constructing an ambient space morphism in the other direction. A sufficient condition for such inverse morphism is that the surjection $q : U@_R R \to \Gamma(S)$ has a section, i.e., a left inverse $p : \Gamma(S) \to U@_R R$. This implies that on the duals, one has $Id_{\Gamma(S)^\star} = q^\star \cdot p^\star$, and we may take $p^\star \cdot b$ to go back from U to $\Gamma(S)^\star$. This condition is certainly satisfied if $\Gamma(S)$ is projective. We say that the global composition *has projective functions* iff its complex of affine functions is projective on the zero simplexes. So we have this result:

Theorem 16 *If* G^I *has a finitely generated projective atlas and projective functions, then the horizontal arrow* f *in the factorization diagram of Proposition 17 is an isomorphism.*

This condition is evidently satisfied if R is semi-simple, i.e., a finite product of commutative fields. This means that in this case, we are able to reconstruct G^I from its retracted affine functions on the resolution. Moreover, in this case, the retracted module can also be recovered from the quotient composition, i.e.,

$$\Delta n\Gamma(G^I) = n\Gamma({}_A\Delta_{G^I}/\Delta n\Gamma(G^I))|/\Delta n\Gamma(G^I) \tag{15.28}$$

so that we are now left with the question of characterizing those module complexes of affine functions in ${}_A\Delta_{n^\star}$ which could give rise to compositions having this free object as their resolution.

Clearly, the first condition on such a module complex is that it is separating. Secondly, we may suppose that it is finitely generated projective, i.e., its modules on the zero-dimensional simplexes are so. The first requirement is in particular the case for the global compositions in Theorem 16 above since the charts have

[7] All charts have finitely generated projective modules.

finitely generated projective ambient modules. The second requirement is the one we had in Theorem 16. Also, under these conditions, the quotient composition has charts $\Gamma(S)^\star$ which are again finitely generated projective and separating. The third and last requirement is obvious: N should contain the constant functions. So we proceed to the analysis of module complexes $N \subset n\Gamma({}_A\Delta_n)$ which are separating, finitely generated projective, and contain the constant functions; call these complexes *representative*.

15.3 Orbits of Module Complexes Are Classifying

Summary. By means of the representation of a composition via its module complex of functions on the purely combinatorial standard composition, we obtain a classification frame for global compositions.

– Σ –

Let us first recall the overall situation of resolutions. For a global composition G^I, we have the functorially associated standard resolution $\Delta_{G^I} = {}_A\Delta_{n^\star}$ with the standard covering $n^\star$ of G^I, and the resolution bijection $res_{G^I} : \Delta_{G^I} \to G^I$, as described in Section 15.2.1. If a global composition H^J is isomorphic to G^I via a morphism $f/Id_A : G^I \underset{\iota}{\to} H^J$, which we now abbreviate by f if no confusion is likely, we have a commutative diagram with horizontal isomorphisms:

$$\begin{array}{ccc} {}_A\Delta_{n^\star} & \xrightarrow{res_f} & {}_A\Delta_{n^\star} \\ {\scriptstyle res_{G^I}}\downarrow & & \downarrow{\scriptstyle res_{H^J}} \\ G^I & \xrightarrow{f} & H^J \end{array} \tag{15.29}$$

By the universal property of the standard compositions, the automorphism group $S_{A,n^\star}$ of the standard composition ${}_A\Delta_{n^\star}$ identifies to a subgroup of the symmetric group $\mathfrak{S}_{m+1}$ of permutations of ${}_A\Delta_{n^\star}$ if the standard covering is defined on the integer interval $[0, m]$ as discussed above. By retraction (see (15.4)), this group acts from the right on the set $Rep_{A,n^\star}$ of representative module complexes on ${}_A\Delta_{n^\star}$

$$ret : Rep_{A,n^\star} \times S_{A,n^\star} \to Rep_{A,n^\star} : (N, g) \mapsto N|g. \tag{15.30}$$

So, if our isomorphic global compositions G^I, H^J have a finitely generated projective atlas and projective functions, their retracted function complexes N_{G^I}, N_{H^J} on the standard composition ${}_A\Delta_{n^\star}$ are representative, and we have $N_{G^I} = N_{H^J}|res_f$.

Conversely, if we are given a representative module complex on ${}_A\Delta_{n^\star}$, we have a bijective morphism

$$r : {}_A\Delta_{n^\star} \to {}_A\Delta_{n^\star}/N \tag{15.31}$$

(see definition 58). Since the charts of the quotient have ambient spaces $N(\sigma)^\star$, the quotient has a finitely generated projective atlas. What are the functions of the quotient? Since the simplicial complexes are isomorphic, it is sufficient to consider the local situation on a chart. So we are given a representative module[8]of affine function $N \subset \Gamma({}_A\Delta_n)$ in a local standard composition. An affine function f on the image $\Delta \subset A@_R N^\star$ is induced by an affine morphism $F : N^\star \to R$, and since N is finitely generated projective, we may write $F = e^t \cdot n^\star$, with the bidual $n^\star$ of an element $n \in N$. So the retraction of the induced function f on σ_i, and on an argument $a \in A$ evaluates to

$$f(\sigma_i)(a) = t + n(\sigma_i)(a) \tag{15.32}$$

and, since the constants are in N, we may absorb the constant t in n, which means that $n\Gamma({}_A\Delta_{n^\star}/N)|r) = N$. Since this retraction is isomorphic to the function module of the quotient, the latter is also representative.

Conversely, if we are given two representative module complexes $M, N \in Rep_{A,n^\star}$, and an automorphism $g \in S_{A,n^\star}$ with $N = M|g$, then we have a commutative diagram with isomorphism on the horizontal arrows

[8] So N is finitely generated, projective, contains the constants, and is separating.

$$\begin{array}{ccc} {}_A\Delta_{n^\star} & \xrightarrow{g} & {}_A\Delta_{n^\star} \\ {\scriptstyle r_N}\downarrow & & \downarrow{\scriptstyle r_M} \\ {}_A\Delta_{n^\star}/N & \longrightarrow & {}_A\Delta_{n^\star}/M \end{array} \tag{15.33}$$

by the factorization of the resolution map $g \cdot r_M$ through the quotient ${}_A\Delta_{n^\star}/N$. So conversely, isomorphic representative module complexes give rise to isomorphic quotients, and we have proven this:

Theorem 17 *The orbit space $Rep_{A,n^\star}/S_{A,n^\star}$ is in bijection with the set of isomorphism classes of A-addressed global compositions with projective functions and finitely generated projective atlases which have a covering complex isomorphic to $n^\star$. This bijection is induced by the retraction of the function module complex to the resolution ${}_A\Delta_{n^\star}$, in one direction, and by the quotient composition on a given representative module complex on ${}_A\Delta_{n^\star}$, in the other.*

In particular, this classification result is valid for the global compositions having as their address a module A over a semi-simple commutative ring R, i.e., a finite direct product of commutative fields (see Appendix E.2.3, Theorem 56).

15.3.1 Combinatorial Group Actions

Summary. We discuss the action of the automorphism group of the standard composition on module complexes since it induces isomorphism classes of global compositions.

– Σ –

By the above classification Theorem 17, we are interested in a more explicit description of the action of the automorphism group $S_{A,n^\star}$ on the set $Rep_{A,n^\star}$ of representative module complexes on the standard composition ${}_A\Delta_{n^\star}$. To begin with, the module complex of affine functions on ${}_A\Delta_{n^\star}$ reads as follows: We recall from Section 15.2.1 that the standard covering $n^\star$ may be given in form of a covering of the integer interval $[0, m]$ by subsets (the zero simplexes) $\sigma = \{t_0, \ldots t_{c(\sigma)}\}$ of $c(\sigma) + 1$ elements, and therefore by the standard atlas injections $i_\sigma : {}_A\Delta_\sigma \to {}_A\Delta_m : \sigma_j \mapsto \sigma_{t_j}$. On each simplex σ of this covering, we have a corresponding local standard composition ${}_A\Delta_{c(\sigma)}$, and its function module is

$$\Gamma({}_A\Delta_{c(\sigma)}) \xrightarrow{\sim} (A@_RR)^{c(\sigma)+1}$$

whereas the constants correspond to the diagonal submodule

$$\Delta(A@_RR) \subset (A@_RR)^{c(\sigma)+1},$$

and our module complex N must be a module

$$\Delta(A@_RR) \subset N(\sigma) \subset (A@_RR)^{c(\sigma)+1}$$

for each simplex σ, finitely generated projective as well as separating in dimension zero. What does it mean that it is separating? It means that for any zero-dimensional simplex σ, and for any pair $i \neq j$ of indices between 0 and $c(\sigma)$, there is an element $n \in N(\sigma)$ such that its coordinates n_i and n_j differ. Moreover, for any two simplexes $\tau \subset \sigma$, we have a commutative diagram

$$\begin{array}{ccccc} \Delta(A@_RR) & \hookrightarrow & N(\tau) & \hookrightarrow & (A@_RR)^{c(\tau)+1} \\ {\scriptstyle Id}\downarrow & & \downarrow{\scriptstyle res_{\tau,\sigma}} & & \downarrow{\scriptstyle pr} \\ \Delta(A@_RR) & \hookrightarrow & N(\sigma) & \hookrightarrow & (A@_RR)^{c(\sigma)+1} \end{array}$$

with surjective vertical arrows induced by the projection to the right side. Further, an automorphism of the underlying global standard composition boils down to a permutation of the interval $[0, m]$ which is compatible with the covering. In particular, this is a finite group action which, on every module $N(\sigma)$ acts by induction from the permutation of components on the supporting direct sum module $(A@_R R)^{c(\sigma)+1}$.

Recall the diagonal embedding and its factorization from formula (11.12) in Section 11.3.2:

$$0 \rightarrowtail A@_R R \overset{\Delta}{\rightarrowtail} (A@_R R)^{c+1} \overset{d}{\twoheadrightarrow} (A@_R R)^c \twoheadrightarrow 0$$

which is equivariant for the given permutation group action. So we have a direct equivariant decomposition $(A@_R R)^{c+1} = \Delta(A@_R R) \oplus (A@_R R)^c$ which carries over to the submodules:

$$N(\sigma) = A@_R R \oplus N(\sigma)_{red}, \text{ with } N(\sigma)_{red} \subset (A@_R R)^c,$$

The action on the reduced factor has been described in formula (11.14) of Section 11.3.2. So the above projection diagram reduces to

$$\begin{array}{ccc} N(\tau)_{red} & \hookrightarrow & (A@_R R)^{c(\tau)} \\ \downarrow{\scriptstyle res_{\tau,\sigma}} & & \downarrow{\scriptstyle pr} \\ N(\sigma)_{red} & \hookrightarrow & (A@_R R)^{c(\sigma)} \end{array}$$

It is known that $N(\sigma)$ is projective of finite type iff $A^\star$ and $N(\sigma)_{red}$ are so (see Appendix E.4.2). So, if we make the general hypothesis that $A^\star$ is projective of finite type, we may concentrate on the reduced part of the module complex, i.e., on a finitely generated projective module complex $N_{red} = N/C$ and on the reduced group action.

15.3.2 Classifying Spaces

Summary. This section is devoted to the classification theorem: There is an algebraic scheme whose rational points represent certain isomorphism classes of global compositions. The scheme is a fine moduli space in the sense of David Mumford.

– Σ –

To begin with, in the local situation of a local standard composition ${}_A\Delta_c$ of dimension c, we consider a (covariant) Grassmann functor $X_r^{A,c}$: **ComRings**$|_R \to$ **Ens** on all commutative R-algebras. Writing $W_{A,c} = (A@_R R)^c$, if $s : R \to S$ is an R-algebra, we have

$$\begin{aligned} X_r^{A,c}(s) &= Grass_{W_{A,c},r}(S) \\ &= \{V \subset S \otimes_R W_{A,c} |\ S \otimes_R W_{A,c}/V = \text{ locally free of rank } r\}. \end{aligned} \tag{15.34}$$

Since[9] $S \otimes_R W_{A,c} \xrightarrow{\sim} ((S \otimes_R A)@_S S)^c$, this functor parametrizes S-modules of affine functions on ${}_{S\otimes_R A}\Delta_c$ which have locally free[10] quotients of rank r.

Next, we have to deal with the separation property. We are in a similar situation as in the discussion of the local classification in Section 11.3.2, where we had to deal with subfunctors of the Grassmannian (formula (11.21)) with respect to a selected point (formula (11.22)). Here, the S-valued ambient module is the scalar extension $S \otimes_R W_{A,c}$. For any pair $1 \leqslant i \leqslant j \leqslant c$, consider the difference projections

[9] In fact, $S \otimes_R (A@_R R) \xrightarrow{\sim} S \otimes_R (A^\star \oplus R) \xrightarrow{\sim} (S \otimes_R A)^\star \oplus S \xrightarrow{\sim} (S \otimes_R A)@_S S$. See [145] for scalar extensions on duals.

[10] The property "locally free" is equivalent to "finitely generated projective", see Appendix F.2, Theorem 66.

$$p_{i,j} : S \otimes_R W_{A,c} \to (S \otimes_R A)@_S S = \begin{cases} pr_i & \text{if } i = j, \\ pr_j - pr_i & \text{if } i \neq j, \end{cases} \tag{15.35}$$

where pr_i is the projection onto the i-th factor.

The separation property means that no difference projection $p_{i,j}$ vanishes on the module $V \in X_r^{A,c}(s)$. For two indices $1 \leqslant i \neq j \leqslant c$, let $\Delta_{A,i,j,c}$ be the submodule of $W_{A,c}$ whose i, j entries are equal, while the other entries vanish. Define these submodules of $W_{A,c}$:

$$V_{i,j} = \begin{cases} (A@_R R)^{i-1} \oplus 0 \oplus (A@_R R)^{c-i} & \text{if } i = j, \\ \Delta_{A,i,j,c} \oplus (A@_R R)^{i-1} \oplus 0 \oplus (A@_R R)^{j-i} \oplus 0 \oplus A@_R R)^{c-j-1} & \text{if } i \neq j. \end{cases}$$

We have these quotients $S \otimes_R W_{A,c}/S \otimes_R V_{i,j} \xrightarrow{\sim} (S \otimes_R A)@_S S$. Let us assume henceforth that A is locally free of rank m. Then we have $S \otimes_R V_{i,j} \in Grass_{W_{A,c},m+1}(S)$. The separation condition for V then means that we do not have $V \subset S \otimes_R V_{i,j}$ for all $1 \leqslant i \leqslant j \leqslant c$. The latter condition defines a subfunctor $X_{r,i,j}^{A,c}$ of $X_r^{A,c}$ which is represented by a closed subscheme $Grass_{r,i,j}^{A,c}$ of the scheme $Grass_r^{A,c}$ over $Spec(R)$. In fact, it is the closed subscheme of the flag scheme $Drap_{r,m+1}^{A,c}$ defined by the $m+1$-codimensional flag component being fixed to $S \otimes_R V_{i,j}$. It is a closed subscheme since a section of structural morphism of the Grassmannian is a closed immersion, see Appendix F.5, Lemma 97 for details.

In other words, we have to take V in the (open) complement scheme $O_r^{A,c}$ of all closed subschemes $Grass_{r,i,j}^{A,c}$ of $Grass_r^{A,c}$. So locally, the separation property defines an open subscheme of the Grassmannian which is also invariant under all permutations of the indexes, an action which is induced by the automorphisms of the underlying standard composition.

So if the zero-dimensional simplexes of the global standard composition are ${}_A\Delta_{c_i}, i = 1, 2, \ldots k$, our module complex is given by an open subscheme $O_{r.}^{A,c.} = \prod_{i=1,2,\ldots k} O_{r_i}^{A,c_i}$ of the fiber product of Grassmannians over $Spec(R)$. Since all combinations of coranks are possible, we should also take the coproduct $O^{A,c.} = \coprod_{r.} O_{r.}^{A,c.}$ of all these open subschemes on which the permutation (automorphism) group of the covering acts. To obtain the effective candidates, we have to consider coincidence on restrictions. So if ρ is any simplex whose vertexes are σ_1, σ_2, we need to know that $N(\sigma_1)|_\rho = N(\sigma_2)|_\rho$. This evidently is a closed, equivariant condition and we have shown that the representative module complexes over the R-algebra S which are of any rank configuration on the respective simplexes define the S-valued points of a locally closed subscheme $C^{A,n^\star}$ of a projective scheme over $Spec(R)$, and since any finite number of points is contained in an open affine subscheme in this situation (projective schemes, see Appendix F.6), we know from Appendix F.6, Theorem 67, that the quotient scheme of orbits of the finite group action of $S_{A,n^\star}$ in the sense of the difference cokernel of the group action μ and the first projection pr_1

$$C^{A,n^\star} \times_{Spec(R)} S_{A,n^\star} \underset{\mu}{\overset{pr_1}{\rightrightarrows}} C^{A,n^\star}$$

exists. This means that we have this result:

Theorem 18 *For an address A which is locally free of rank m over the commutative ring R, there is a subscheme $J^{n^\star}$ of a projective $Spec(R)$-scheme of finite type such that its S-valued points $J^{n^\star}(Spec(S))$ for an R-algebra S are in bijection with the classifying orbits of module complexes N in ${}_{S\otimes_R A}\Delta_{n^\star}$ which are locally free of defined co-ranks on the zero-simplexes of $n^\star$.*

In particular, if the ground ring R is semi-simple, this theorem gives the classification of any global composition which is addressed in a finitely generated R-module A.

16

Classifying Interpretations

Weil die mathematische Methode die Wissenschaft ist,
die zur Zeit die am weitesten entwickelte Methodologie besitzt,
war mir daran gelegen,
sie zum Vorbild zu nehmen,
das uns helfen kann,
unsere gegenwärtigen Schwachstellen
zu beheben.
Pierre Boulez [138, II, p.71]

Summary. This chapter exposes criteria for characterizing interpretable compositions in terms of classifying spaces. This is a central issue since interpretability yields access to instrumental parameters for the physical 'rendering' of a compositional structure. In contrast to general classification, *interpretable* molecular structures are difficult to classify. We also review global enumeration theory as well as global American Set Theory.

– Σ –

We are not going to give further musicological comments and interpretations of the preceding and the following classification results. This will be dealt with in the next chapter. However, it should be kept in mind that this subject is far from pure mathematical exposition; the meaning of the classification techniques for musicology cannot be underestimated. It is the turning point between thinking and making music, between mental construction and physical realization. This is why affine functions are so important: they yield the entire potential of parametrizing music in acoustic realms.

The interpretation of isomorphism classes in terms of points of a scheme also lead to a comparative theory of global compositions: We can now deal with the problem of which compositions are more generic than others, so this entire approach from algebraic geometry opens the subject of germinal vs. derived ideas in musical composition.

In particular, the question of those compositions which can effectively be played on ensembles of musical instruments is embedded in the perspective of algebro-geometric specialization, and this is the view we shall deal with in this chapter.

16.1 Characterization of Interpretable Compositions

Summary. This section presents a condition for a composition to be interpretable. This condition regards the restriction behavior of affine functions which relates to *flasque* sheafs.

– Σ –

G. Mazzola, *The Topos of Music I: Theory*, Computational Music Science,
https://doi.org/10.1007/978-3-319-64364-9_16

Pursuing the path of the last chapter, we again assume that the address A is a module over a commutative ring R, and we consider the category $ComGlob_A$ of A-adressed commutative global compositions. As we were able to classify those commutative global compositions in $ComGlob_A$ with A locally free of a defined rank, and function modules which are locally free of defined coranks on the zero simplexes of their resolutions, by the orbits of the retracted function module complexes, it should also be possible to characterize the interpretable compositions by their function module complexes.

Definition 59 *A module complex on a global composition is called* flasque *iff every section on a zero-simplex stems from a global section.*

For example, if a global composition G^I is interpretable by the covering J of a local composition $K \subset A@_RM$, for an R-module M, then we have a bijective morphism $f/Id_A : G^I \to K$, and the (global) functions of K, by retraction yield all the functions of G^I on their charts (=zero-simplexes), this means:

Lemma 28 *Interpretable global compositions have flasque function module complexes.*

Moreover, if G^I is interpretable by a projective module M, the global functions are separating, this follows from Proposition 16 of Section 15.2.2. If a local composition has ambient module M which is not projective, quite pathological things may happen.

Example 28 Let us study such an example. Take the ring $R = \mathbb{Z}$ of integers, a free module $I = R^d$ of defined rank d, and take a non-zero prime ideal $m = (p)$ in R. Consider the R-module $M = I \oplus R/m$, and any address A. Clearly, we have $M@_RR \xrightarrow{\sim} e^R \cdot I^*$ since all linear forms on R/m vanish. Consider a local composition $L = \{x, y\} \subset A@_RM$ with $x = e^\xi \cdot u$, and $y = e^\eta \cdot u$, where $\xi, \eta \in R/m$ are two different elements, and where $u : A \to I$ is a surjection. Then no affine function on M can separate x from y since their difference is annihilated by any affine function. Moreover, if we consider the interpretation of L by the two singleton charts $\{x\}, \{y\}$, we have free function modules $\Gamma(\{x\}), \Gamma(\{y\}) \xrightarrow{\sim} R \oplus I^*$ of rank $d+1$. We therefore have an interpretation with locally free function modules on charts whereas the global functions are not separating, essentially because the ambient module is not projective.

But we have the following:

Proposition 18 *Take G^I in $ComGlob_A$ with A locally free of rank m, and function modules which are locally free of defined co-ranks on the zero simplexes of their resolutions. Suppose that its function module is flasque, and that the global functions are separating. Then G^I is interpretable.*

Proof. We know from the quotient construction of global compositions from function modules which backs the classification Theorem 18 in Section 15.3.2 that an atlas of G^I is given by the duals $N(\sigma)^*$ of the functions on the given charts σ of G^I. So on the one hand, G is injected into $A@_R\Gamma(G)^*$ with ambient module $\Gamma(G)^*$. On the other, we have a split[1] injection $i_\sigma : N(\sigma)^* \to \Gamma(G)^*$ into the dual of the global function module since the latter is flasque, and since by hypothesis, the function module $N(\sigma)$ and therefore its dual is locally free of defined rank. So the two charts $\sigma \subset A@_R\Gamma(G)^*$, and $\sigma \subset A@_RN(\sigma)^*$ are isomorphic, and we have reconstructed G^I as an interpretation. QED.

In particular, if the ring R is semi-simple, we obtain a complete characterization of interpretable global composition via their function modules in the resolution given in Theorem 18 in Section 15.3.2. But in the general case, a necessary and sufficient condition for module complex classes to yield interpretable global compositions is not known.

Exercise 36 *Let A be the zero address, and R be an infinite field. Take $\mu = max_\sigma(dim(R.\sigma))$ the maximum of dimensions of the modules associated with the charts σ of G^I. Then an interpretation of G^I which is guaranteed in Proposition 18 can be constructed in an ambient vector space of dimension μ.*

[1] The image is a direct summand.

Before terminating this classification discourse, we should add examples of non-interpretable global compositions. We shall discuss a further example of this type in Chapter 17, but with a more semantic orientation. Here, we simply give the examples as facts of the classification discourse.

Example 29 We discuss a non-interpretable global composition G^I in the zero address, and over the field of real numbers, see Figure 16.1. It has six points,

$$G = \{x_1, x_2, x_3, x_4, x_5, x_6\},$$

and three charts for its covering, i.e., $I = \{U_1, U_2, U_3\}$ with

$$U_1 = \{x_1, x_2, x_3, x_4\},$$
$$U_2 = \{x_1, x_2, x_5, x_6\},$$
$$U_3 = \{x_3, x_4, x_5, x_6\}.$$

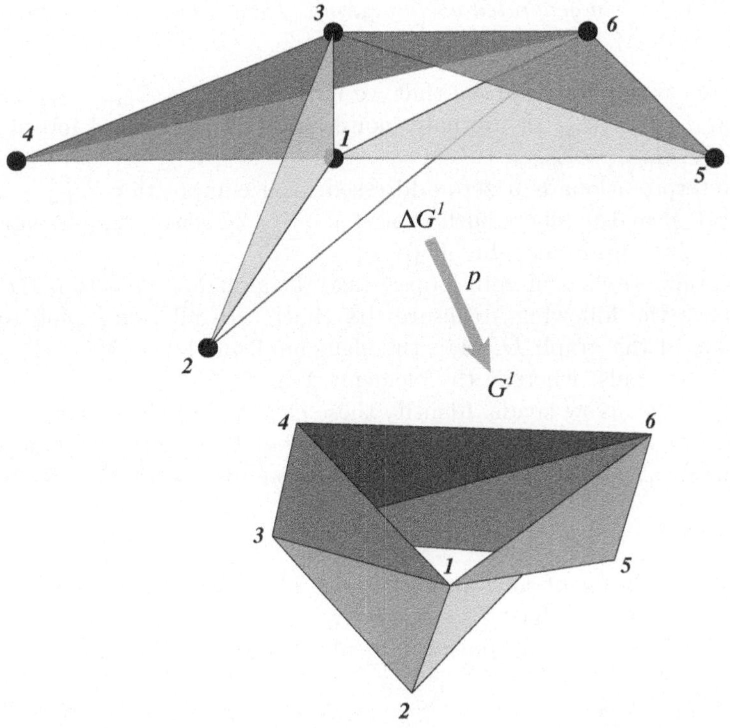

Fig. 16.1. A non-interpretable composition and its resolution.

The three covering charts are all in bijection with a 'square vertex' composition $S = \{a = (0,0), b = (1,0), c = (1,1), d = (0,1)\} \subset 0@_{\mathbb{R}}\mathbb{R}^2$ as follows:

$$U_1 \xrightarrow{\sim} S : x_1 \mapsto a, x_2 \mapsto d, x_3 \mapsto b, x_4 \mapsto c,$$
$$U_2 \xrightarrow{\sim} S : x_1 \mapsto a, x_2 \mapsto d, x_5 \mapsto c, x_6 \mapsto b,$$
$$U_3 \xrightarrow{\sim} S : x_3 \mapsto b, x_4 \mapsto c, x_5 \mapsto a, x_6 \mapsto d,$$

so that the intersections of any two of these charts yield an isomorphism of a two-point composition in $\mathbb{R}^2$. If we visualize this configuration by means of a subdivision of each chart as a union of two triangular surfaces, we obtain a Möbius strip, see Figure 16.1. The resolution of G^I in the same surface representation is also shown. Here, the charts are no longer plane compositions but tetrahedra in three space.

Let us show that G^I is not interpretable. By the above results, it is sufficient to see that the global affine functions on this composition do not separate points. Let $f : G \to \mathbb{R}$ be an affine function (we suppress the zero address and just work in the respective ambient spaces), set $f(x_i) = f_i$ for $i = 1,2,3,4,5,6$ and $f_{i,j} = f_i - f_j$. Since f is affine on each chart, we have on chart U_1: $f_{1,2} = f_{4,3}$, on chart U_2: $f_{1,2} = f_{5,6}$, on chart U_3: $f_{4,3} = f_{6,5}$. Therefore $f_{5,6} = f_{6,5}$, and $f_{5,6} + f_{6,5} = 0$, by definition, so $0 = f_{1,2} = f_{4,3} = f_{6,5}$, and no one of the pairs $(x_1,x_2),(x_3,x_4),(x_5,x_6)$ can be separated by f.

16.1.1 Automorphism Groups of Interpretable Compositions

Summary. We show that every finitely generated abelian group can be represented by the automorphism group of an interpretable composition.

– Σ –

Theorem 19 *Let H be a finitely generated abelian group. Then there is a zero-addressed interpretation G^I over $R = \mathbb{R}$ with $H \xrightarrow{\sim} Aut(G^I)$.*

The proof idea runs as follows. Suppose that we have $H \xrightarrow{\sim} \mathbb{Z}^r \times \mathbb{Z}_{m_1} \times \ldots \mathbb{Z}_{m_k}$. We shall construct a disjoint union $M^r \sqcup M_{m_1} \sqcup \ldots M_{m_k}$ of mutually non-isomorphic connected interpretations M^r, M_{m_i} with $Aut(M^r) \xrightarrow{\sim} \mathbb{Z}^r$, and $Aut(M_{m_i}) \xrightarrow{\sim} \mathbb{Z}_{m_i}$.

The cell of our interpretations is a zero-addressed local composition $L_{\alpha,\beta} = \{0,\alpha,\beta,1\} \subset 0@_{\mathbb{R}}\mathbb{R}$ for any pair $0 < \alpha < \beta < 1$ of real numbers such that $\alpha + \beta \neq 1$. Clearly, $L_{\alpha,\beta}$ is rigid, i.e., $Aut(L_{\alpha,\beta}) = Id$. Moreover, any two $L_{\alpha,\beta}, L_{\gamma,\delta}$ are isomorphic iff $\{\alpha,\beta\} = \{\gamma,\delta\}$.

Given a directed graph D without zero-loops[2], take the global composition $Ex(D, L_{\alpha,\beta})$, *the extension of $L_{\alpha,\beta}$ by D*, which has the following structure: Its charts are all isomorphic to $L_{\alpha,\beta}$, and correspond one-to-one to the arrows of the graph D. Map the elements 0 in the copies of $L_{\alpha,\beta}$ to the graph vertexes corresponding to the arrow tails, whereas the elements 1 in the copies of $L_{\alpha,\beta}$ are mapped to the graph vertexes corresponding to the arrow heads. Identify these elements iff their images under this map coincide. Intuitively, this means that we replace each arrow by a copy of $L_{\alpha,\beta}$, with the 0 on the tail and the 1 on the head. The graph D canonically identifies to the global subcomposition of $Ex(D, L_{\alpha,\beta})$ defined by the arrow tails and heads:

$$D \rightarrowtail Ex(D, L_{\alpha,\beta}) \tag{16.1}$$

What are the automorphisms of such an extension? They are related to an underlying automorphism of the arrow set of D. And on each chart $L_{\alpha,\beta}$ for a given arrow, the chart morphism must be the identity since the charts are rigid. So the automorphism must also conserve heads and tails of arrows, and therefore we have this fact:

Lemma 29 *The automorphism group of $Ex(D, L_{\alpha,\beta})$ is isomorphic to the automorphism group of the directed graph D. No two extensions are isomorphic if their cells aren't.*

Exercise 37 *Give a proof of this lemma.*

Suppose that the graph D has no multiple arrows (independent of their direction). Then such an extension can be realized as an interpretation in $\mathbb{R}^3$. We are now ready to define the interpretation with the required automorphism group. Take the group $\mathbb{Z}^r \times \mathbb{Z}_{m_1} \times \ldots \mathbb{Z}_{m_k}$, and fix $r + k$ mutually different pairs α_i, β_i as indexes of mutually non-isomorphic cells $L_i = L_{\alpha_i,\beta_i}$. Consider the directed graph D_∞ with vertexes $v_i, i \in \mathbb{Z}$ and arrows $a_i : v_i \to v_{i+1}, i \in \mathbb{Z}$. We have $Aut(D_\infty) \xrightarrow{\sim} \mathbb{Z}$, the group of translations of $\mathbb{Z}$. For a

[2] No arrows with identical domain and codomain.

positive integer m, consider the regular polygon graph D_m with vertexes $v_i, i = 0, \ldots m-1$ and arrows $a_i : v_i \to v_{i+1}, i = 0, \ldots m-1$, where we close the polygon by defining $v_m = v_0$. Clearly, $Aut(D_m) \xrightarrow{\sim} \mathbb{Z}_m$. We then consider the extensions $E_i = Ex(D_\infty, L_i), i = 1, \ldots r$, and $E_i = Ex(D_{m_{i-r}}, L_i), i = r+1, \ldots r+k$. No two of these interpretable compositions are isomorphic and we have $Aut(E_i) \xrightarrow{\sim} \mathbb{Z}, i \leqslant r$, as well as $Aut(E_i) \xrightarrow{\sim} \mathbb{Z}_{m_i}, r < i \leqslant r+k$. It is now immediate that the disjoint union

$$G^I = \coprod_{i=1,\ldots r+k} E_i \tag{16.2}$$

does the job, i.e., $Aut(G^I) \xrightarrow{\sim} \mathbb{Z}^r \times \mathbb{Z}_{m_1} \times \ldots \mathbb{Z}_{m_k}$. QED.

16.1.2 A Cohomological Criterion

Summary. Module complexes of define cochain complexes and their cohomology modules. We derive a necessary condition for the vanishing of the first cohomology and for the complex of the module complex of affine functions to be flasque, and hence the interpretability of the composition.

– Σ –

If G^I is a global composition at address A, everything over the commutative ring R, we may define the cochain complex $C^*(G^I)$ as follows, and in congruence with the general cohomology theory of coefficient systems (see appendix H.3.1). Denote by $S_k n(G^I)$ the set of singular k-simplexes of the nerve $n(G^I)$. For each $k \geqslant 0$, we have the module

$$C^k(G^I) = \bigoplus_{s \in S_k n(G^I)} n\Gamma(s) \tag{16.3}$$

of k-cochains of functions. We have a linear differential coboundary map for each k:

$$d_k : C^k(G^I) \to C^{k+1}(G^I) \tag{16.4}$$

where for any singular $k+1$-simplex $s = (i_0, i_1, \ldots i_{k+1})$, we set

$$d_k(f)(s) = \sum_{j=0,1,\ldots k+1} (-1)^j f(s_j) \tag{16.5}$$

where s_j is the face of s after omitting vertex i_j. It is well known that $d_{i+1} \cdot d_i = 0$, and therefore, we have the usual cohomology modules

$$H^k(G^I) = Z^d / B^d \text{ with } Z^d = Ker(d_k), B^d = Im(d_{k-1})$$

for positive k, and $H^0(G^I) = Ker(d_0)$, which evidently identifies to the module $\Gamma(G^I)$ of global sections.

Proposition 19 *Suppose that the A-addressed global composition G^I has a nerve $n(G^I)$ which is a finite acyclic graph. Then $H^1(G^I) = 0$, and $n\Gamma(G^I)$ is flasque.*

Proof. The second statement is obvious by induction on the cardinality of the nerve: There are leaves, i.e., vertexes which are connected to (at most) one other vertex. Take such a leaf, omit it, and extend a function on the rest to the omitted leaf.

For the vanishing of first cohomology, we may suppose that the nerve is connected and take a cochain $f = (f(i,j))_{(i,j) \in S_1 n(G^I)}$ in Z^1. The vanishing of df means that we have

$$f(i_1, i_2) - f(i_0, i_2) + f(i_0, i_1)|_{i_0 \cap i_1 \cap i_2} = 0$$

for any singular simplex i_0, i_1, i_2. In particular, we have $f(i,i) = 0$, and therefore, $f(i,j) + f(j,i) = 0$ for any singular 1-simplex (i,j). We may suppose that there is a tree (a directed acyclic graph) T such that its undirected image $|T|$ is isomorphic to $n(G^I)$. Denote by A_T the arrow set and by V_T the vertex set of T. We then conclude the existence of a 0-chain $g = (g(i))_{i \in I} \in C^0(G^I)$ with $dg = f$ by the following lemma (QED).

Lemma 30 *Let* $f \in \bigoplus_{(i,j)\in A_T} \Gamma(i \cap j)$, *then there exists* $g \in \bigoplus_{i\in V_T} \Gamma(i)$ $(= C^0(G^I))$ *such that* $g(j) - g(i)|_{i\cap j} = f(i,j)$ *for all* $(i,j) \in A_T$.

Proof. Induction on $card(A_T) \geqslant 1$. For $card(A_T) = 1$, suppose that f resides on i, j. Extend $f(i,j)$ to h on i, and choose $g(j) = h, g(i) = 0$.

For the general case, take a $i_0 \in V_T$ which is the head of a non-empty set A' of arrows, and where no arrow tails exist. Consider the subgraph T' of T obtained after removing i_0 and A'. It is a not necessarily connected graph, in fact a forest with possibly several maximal subtrees, i.e., connected components. By induction, there is a cochain $g' \in \bigoplus_{i\in V_{B'}} \Gamma(i)$ such that $g'(j) - g'(i)|_{i\cap j} = f(i,j)$ for all $(i,j) \in A_{T'}$. Since T is a tree, every arrow $(i, i_0) \in A'$ defines exactly one connected component Z_i of T' which contains i.

We may add any global section d to the system $(g'_i)_{i\in A_{Z_i}}$ without altering the differences which yield $g'(i) - g'(j)|_{i\cap j}$. Since $n\Gamma(G^I)$ is flasque, we may therefore choose $g'(i)$ freely; set $g'_i = f(i_0, i)$. The final definition of g runs as follows: Take $g(i)) = g'(i)$ for all vertexes except i_0, and $g(i_0) = 0$. This does the job. QED.

The following examples (which are also exercises) show that we are far from understanding the connection of cohomology of global compositions to the problem of interpretability.

Example 30 Consider the interpretation G^I of the zero-addressed local composition $G \subset \mathbb{R}^2$ by three charts G_1, G_2, G_3 as shown in Figure 16.2 Since every intersection $G_i \cap G_j$ is generating, we have $\Gamma(G^I) \xrightarrow{\sim}$

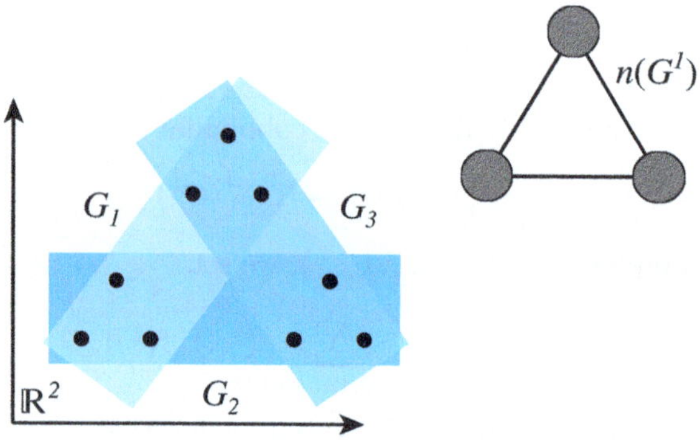

Fig. 16.2. An interpretation with non-vanishing first cohomology.

$\Gamma(G_i) \xrightarrow{\sim} \Gamma(G_i \cap G_j)$. We have $C^0(G^I) \xrightarrow{\sim} (\mathbb{R}^2@\mathbb{R})^3$, $H^0(G^I) \xrightarrow{\sim} \mathbb{R}^3$, therefore $B^1(G^I) \xrightarrow{\sim} \mathbb{R}^6$, and[3] $Z^1(G^I) \xrightarrow{\sim} (\mathbb{R}^2@\mathbb{R})^3 \xrightarrow{\sim} \mathbb{R}^9$, whence $H^1(G^I) \xrightarrow{\sim} \mathbb{R}^3$.

Example 31 This example is the "Möbius" strip composition introduced in Example 29 above. It is not interpretable since the functions are not flasque, however, we have the global functions $\Gamma(G^I) \xrightarrow{\sim} H^0(G^I) \xrightarrow{\sim} \mathbb{R}^3$, and $H^1(G^I) = 0$.

Example 32 Take the zero-addressed global composition over the finite field $R = \mathbb{Z}_3$, defined as a gluing of two copies of $\mathbb{Z}_3 = 0@_{\mathbb{Z}_3}\mathbb{Z}_3$ along two common points, see Figure 16.3 The composition cannot be interpreted since the three points in each card are colinear, but it is not possible to maintain this property in a common ambient $\mathbb{Z}_3$-vector space while distinguishing the two non-common points of the charts!

[3] Please, check!

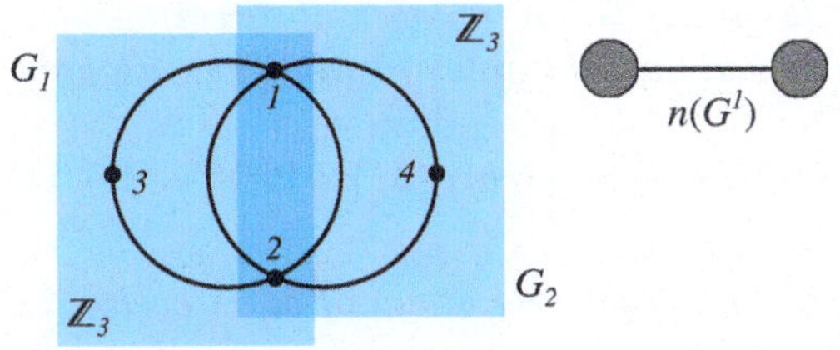

Fig. 16.3. A non-interpretable composition with acyclic nerve and vanishing first cohomology.

16.2 Global Enumeration Theory

Summary. Enumeration of interpretations in special cases, such as tesselations, or canons, has a strong root in the American tradition. Contributions by George Halsey, Edwin Hewitt, David Lewin, Dan Tudor Vuza, and Harald Fripertinger are discussed.

– Σ –

Enumeration theory deals with the calculation of isomorphism classes of interpretations of local compositions in finite abelian groups, i.e., in finite $\mathbb{Z}$-modules. Since these are never projective, the techniques of global classification presented in Chapter 15 do not apply here. Further, to this date, only very special non-trivial coverings have been classified: Global enumeration theory is mostly restricted to zero-dimensional nerves, Fripertinger's recent classification of canons are an exception.

16.2.1 Tesselation

Summary. We review the work by George Halsey and Edwin Hewitt [435] on the tesselation of finite $\mathbb{Z}$-modules (abelian groups).

– Σ –

A finite $\mathbb{Z}$-module M is a zero-addressed local composition, and we may consider its interpretations M^I with mutually disjoint (non-empty) charts, i.e., with discrete nerves. Recall from Sorite 6 that any isomorphism $f : U \xrightarrow{\sim} V$ among two such charts may be extended to an affine automorphism of M. Fix a subgroup $G \subset Aut(M)$. If these charts (U,M) are all isomorphic under G, we call M^I *a G-isotypic tesselation of M*. So this is an interpretation by mutually isomorphic charts which can even be transformed into each other by a transformation in G, in other words, the class nerve $CN(M^I)$ (see Section 14.5) is discrete and has constant weights *under* G. We have the canonical bijective *tesselation morphism* $M^I \to M$ of zero-addressed compositions. The classifications in [435] deals with these tesselation morphisms: A G-isomorphism between such tesselation morphisms $M^I \to M$, and $M^J \to M$ is an element $g \in G$ which defines this commutative diagram

$$\begin{array}{ccc} M^I & \longrightarrow & M \\ g\big\downarrow & & \big\downarrow g \\ M^J & \longrightarrow & M \end{array} \tag{16.6}$$

of global compositions[4]. In [435] a *tesselating chord* is defined as an isomorphism class of tesselation morphisms on M under the group $G = e^M$ of translations on M. For any natural number j, we denote by $\Gamma_t(M, j)$ the number of tesselating chords in M which represent tesselations by charts of cardinality j. We have two formulas for special cases: one for $j = 2$, and one for $j = 3$ if the 3-Sylow group[5] is cyclic.

[4] Instead, we could equivalently consider the interpretation which adds the chart M to the atlas I and then just look for G-isomorphisms of such interpretations of M.

[5] See appendix C.3.4.2.

Theorem 20 [435, Satz 11.10] *Let $m(2)$ be the number of direct cyclic summands in the 2-Sylow group of M, and let c be the number of elements of odd order in M. Then we have*

$$\Gamma_t(M,2) = \frac{1}{2}(card(M) + 2^{m(2)} - c - 1).$$

Theorem 21 [435, Satz 11.15] *If the 3-Sylow S_3 group of M is cyclic, i.e., $S_3 \xrightarrow{\sim} \mathbb{Z}_{3^\lambda}$, and if $card(M/S_3 = d)$, we have*

$$\Gamma_t(M,3) = 1 + \frac{1}{3}(d^2 - 1) + \frac{3}{8}d^2(3^{2(\lambda-1)} - 1).$$

In principle, the number $\Gamma_t(M,j)$ can be calculated by Sands' algorithm for all finite $\mathbb{Z}$-module M which are Hajós groups (this property is related to Hajós' solution of Minkowski's problem; see [435, §12] for further references). For a local composition (W,M), denote $Trans(W)$ for the translation symmetries of W. Then

Definition 60 *A finite abelian group M is called a* Hajós group *if either M is trivial, or for every tesselation $M = \coprod_{v \in V} e^v(U)$ by translates of a subset U, either $Trans(U)$ or $Trans(V)$ is not trivial.*

All finite cyclic Hajós groups have been classified ([435, §12]). They are the groups of order $n = 1, n = p^\alpha, n = p^\alpha q$ for positive natural α, $n = p^2q^2$, and $n = p^2qr, n = pqr, n = pqrs$, for any distinct primes p,q,r,s. In particular, $\mathbb{Z}_{72}$ is the smallest cyclic non-Hajós group.

Based on an algorithm similar to Sands' algorithm, the numbers of tesselation chords have been calculated in [435, §12] for all cyclic groups of order $\leqslant 24$ (they are automatically Hajós by the above). Their list is given in table (12.13) in [435].

16.2.2 Mosaics

Summary. We review the work by Harald Fripertinger [345] on the enumeration of mosaics, i.e., zero-addressed discrete interpretations of finite cyclic groups.

– Σ –

This classification deals with discrete interpretations $\mathbb{Z}_n^I$ of $\mathbb{Z}_n$ by $card(I) = k$ non-empty charts of arbitrary size. So the ambient module is more special than in Section 16.6, but the partition is not necessarily isotypical. Also, in this case the calculation of isomorphism classes is not restricted to translations. Again, we are selecting a subgroup $G \subset Aut(\mathbb{Z}_n) \xrightarrow{\sim} e^{\mathbb{Z}_n} \cdot \mathbb{Z}_n^\times$, and we say that a *k-partition* is the canonical morphism $Part(I) = \mathbb{Z}_n^I \to \mathbb{Z}_n$ of a discrete interpretation $\mathbb{Z}_n^I$ with $card(I) = k$. Isomorphisms among partitions $Part(I)$, $Part(J)$ are defined by commutative diagrams as for tesselations. We ask for $g \in G$ that the diagram

$$\begin{array}{ccc} \mathbb{Z}_n^I & \longrightarrow & \mathbb{Z}_n \\ g\downarrow & & \downarrow g \\ \mathbb{Z}_n^J & \longrightarrow & \mathbb{Z}_n \end{array} \tag{16.7}$$

commutes. With this, an orbit of partitions is called *a G-mosaic*.

According to the general combinatorial Pólya methodology applied by Fripertinger, the calculation of mosaic numbers requires the action of a finite group on a set of objects which represent the partitions. To this end, a partition P is identified to an orbit of a function $p : \mathbb{Z}_n \to [1,n]$ under the left action of the symmetric group $\mathfrak{S}_n$, i.e., P corresponds to an orbit $\mathfrak{S}_n \cdot p$. The reason is that the fibers of the function p define a partition of $\mathbb{Z}_n$, and that we are only interested in the fibers, no matter which values they stem from, whence the action of the symmetric group on the value domain. On the other hand, the isomorphism $g \in G$ acts from the right on p, and we may identify the mosaics with the orbits of the action of the direct product $\mathfrak{S}_{[1,n]} \times G^{opp}$ on the function set $[1,n]^{\mathbb{Z}_n}$. Of course, the set Mos_n^G of G-mosaics in $\mathbb{Z}_n$ is partitioned into the subsets $Mos_{n,k}^G$ of k-element partition mosaics for $k = 1,2,\ldots n$. These subsets correspond to the function sets of functions $p : \mathbb{Z}_n \to [1,n]$ with $card(Im(p)) = k$.

As usual in Pólya theory, we need to calculate the cycle index $Z(G) = Z(G, P)$ as defined in Section 11.4.1, formula (11.42). The notation is related to the second variable $P = \mathbb{Z}_n$, as in that context, and $G \subset \mathfrak{S}_P$ which we omitted in the previous notation of Z since P was fixed. With this index, a rational polynomial in the indeterminates $X_1, \ldots X_n$, with $n = card(P)$, if each variable X_i in $Z(G, P)$ is replaced by a value ξ_i, we write $Z(G, P|X_i = \xi_i)$. Then we have the following enumeration theorem. Write $[\frac{n}{i}]$ for the greatest integer $\leqslant n/i$, and $s_i = i(X_1 + \ldots X_{[\frac{n}{i}]})$.

Theorem 22 [345, Theorem 1] *With this notation, we put*

$$M_k = Z(G, \mathbb{Z}_n|X_i = \frac{\partial}{\partial X_i})Z(\mathfrak{S}_k, [1, k]|X_i = e^{s_i})|X_i = 0 \tag{16.8}$$

for $1 \leqslant k \leqslant n$ (differential operators acting on the second factor, and then being evaluated at zero). Put $M_0 = 0$. Then we have

$$card(Mos^G_{n,k}) = M_k - M_{k-1} \tag{16.9}$$

and therefore

$$card(Mos^G_n) = M_n. \tag{16.10}$$

A calculation by use of the computer program SYMMETRICA in [345] yields this table of mosaics for three important groups:

$k =$	1	2	3	4	5	6	7	8	9	10	11	12
T_{12}	1	179	7254	51075	115100	110462	52376	13299	1873	147	6	1
TI_{12}	1	121	3838	26148	58400	56079	26696	6907	1014	96	6	1
$\overrightarrow{GL}(\mathbb{Z}_{12})$	1	87	2155	13730	30121	28867	13835	3667	571	63	5	1

This yields a total of 351773 T_{12}-mosaics, 179307 TI_{12}-mosaics, and 93103 $\overrightarrow{GL}(\mathbb{Z}_{12})$-mosaics.

This method can be refined to yield finer classifications of mosaics. We summarize one such refinement. If we are given a partition $Part(I)$ of $\mathbb{Z}_n$, not only the cardinality $card(I) = k$ is invariant under isomorphism of the group G, but also the *block type*, i.e., the sequence $\lambda(Part(I)) = \lambda = (\lambda_1, \lambda_2, \ldots \lambda_n)$ where λ_i is the number of charts in I with cardinality i. So we clearly have two linear conditions $\sum_i \lambda_i = k, \sum_i i\lambda_i = n$ for λ.

We again replace sets by functions. Given a type λ, take the set $Bi(\mathbb{Z}_n, [1, n])$ of bijections $f : \mathbb{Z}_n \to [1, n]$, and fix once and for all a partition Λ of $[1, n]$ which is of type λ. Then the inverse image $f_\Lambda = f^{-1}\Lambda$ of Λ under a bijection f defines a partition of $\mathbb{Z}_n$, and two such inverse images f_Λ, g_Λ define the same partition iff $g^{-1} \cdot f$ stabilizes Λ. Let $H_\Lambda \subset \mathfrak{S}_{[1,n]}$ be the stabilizer of Λ. Then the orbit set $H_\Lambda \backslash Bi(\mathbb{Z}_n, [1, n]$ of left action

$$H_\Lambda \times Bi(\mathbb{Z}_n, [1, n]) \to Bi(\mathbb{Z}_n, [1, n]) : (g, f) \mapsto g \cdot f$$

identifies to the set of partitions of type λ. To get the mosaics of this type, we add the well-known right action of G, and we obtain the canonical identification

$$Mos^G_{n,\lambda} \xrightarrow{\sim} H_\Lambda \times G^{opp} \backslash Bi(\mathbb{Z}_n, [1, n])$$

of the set $Mos^G_{n,\lambda}$ of G-mosaics of type λ with the orbits of the two-sided action of G and H_Λ. Using this identification, we have

Theorem 23 [345, Theorem 2] *The number of G-mosaics of type λ is given by the formula (due to de Bruijn)*

$$card(Mos^G_{n,\lambda}) = Z(G, \mathbb{Z}_n|X_i = \frac{\partial}{\partial X_i})Z(H_\Lambda, [1, n]|X_i = iX_i)|X_i = 0. \tag{16.11}$$

Since the cycle index of a wreath product[6] of groups can be deduced from the indexes of its factors, and since the stabilizer is isomorphic to

[6] See appendix C.3.2.

$$\prod_i \mathfrak{S}_{[1,i]} \wr \mathfrak{S}_{[1,\lambda_i]},$$

the formula is controllable. The numbers of T_{12}-mosaics of all types have been calculated by the SYMMETRICA program and yield this table:

λ		λ		λ		λ	
(12)	1	$(1,11)$	1	$(2,10)$	6	$(1^2,10)$	6
$(3,9)$	12	$(1,2,9)$	30	$(1^3,9)$	12	$(4,8)$	29
$(1,3,8)$	85	$(2^2,8)$	84	$(1^2,2,8)$	140	$(1^4,8)$	29
$(5,7)$	38	$(1,4,7)$	170	$(2,3,7)$	340	$(1^2,3,7)$	340
$(1,2^2,7)$	510	$(1^3,2,7)$	340	$(1^5,7)$	38	(6^2)	35
$(1,5,6)$	236	$(2,4,6)$	610	$(1^2,4,6)$	610	$(3^2,6)$	424
$(1,2,3,6)$	2320	$(1^3,3,6)$	781	$(2^3,6)$	645	$(1^2,2^2,6)$	1820
$(1^4,2,6)$	610	$(1^6,6)$	50	$(2,5^2)$	386	$(1^2,5^2)$	386
$(3,4,5)$	1170	$(1,2,4,5)$	3480	$(1^3,4,5)$	1170	$(1,3^2,5)$	2330
$(2^2,3,5)$	3510	$(1^2,2,3,5)$	6960	$(1^4,3,5)$	1170	$(1,2^3,5)$	3500
$(1^3,2^2,5)$	3510	$(1^5,2,5)$	708	$(1^7,5)$	38	(4^3)	297
$(1,3,4^2)$	2915	$(2^2,4^2)$	2347	$(1^2,2,4^2)$	4470	$(1^4,4^2)$	792
$(2,3^2,4)$	5890	$(1^2,3^2,4)$	5890	$(1,2^2,3,4)$	17370	$(1^3,2,3,4)$	11580
$(1^5,3,4)$	1170	$(2^4,4)$	2325	$(1^2,2^3,4)$	8860	$(1^4,2^2,4)$	4463
$(1^6,2,4)$	610	$(1^8,4)$	29	(3^4)	713	$(1,2,3^3)$	7740
$(1^3,3^3)$	2610	$(2^3,3^2)$	6005	$(1^2,2^2,3^2)$	17630	$(1^4,2,3^2)$	5890
$(1^6,3^2)$	424	$(1,2^4,3)$	8725	$(1^3,2^3,3)$	11623	$(1^5,2^2,3)$	3510
$(1^7,2,3)$	340	$(1^9,3)$	12	(2^6)	554	$(1^2,2^5)$	2792
$(1^4,2^4)$	2325	$(1^6,2^3)$	645	$(1^8,2^2)$	84	$(1^{10},2)$	6
(1^{12})	1						

16.2.3 Classifying Rational Rhythms and Canons

Summary. We give a short comment on the classification of rhythms and canons on rational onsets.

– Σ –

Following Vuza's context [1094], we consider rhythms without any further parameters, except the onsets which we also restrict to the rationals. Formally, we work with zero-addressed objective local and global compositions on the space $Onset|_{\mathbb{Q}}$ of rational onsets. In Vuza's theory, a *(periodic) rhythm* is a zero-addressed local rhythm $R = e^{[-\infty,\infty]p}G$ over a finite germ $G \subset 0@Onset|_{\mathbb{Q}}$, and positive period p. Equivalently, R is a zero-addressed objective local composition in $Onset|_{\mathbb{Q}}$ which has a non-zero translation in its automorphism, and which is locally finite. Call *Vuza rhythm* such a local rhythm, and we suppose that it is non-empty since otherwise everything is known. Therefore, the translation automorphisms of R form the group $e^{p\mathbb{Z}}$, $0 < p$ being called *the period* $Per(R)$ of R. The group $Trans$ of rational translations acts on the set R_{Vuza} of Vuza rhythms, and defines the Vuza classes as being the translation isomorphism classes or orbits in $Trans\backslash R_{Vuza}$ of Vuza rhythms. In other words, we are working on the additive group of rationals, i.e., on the $\mathbb{Z}$-module $\mathbb{Q}_{\mathbb{Z}}$.

The determination of Vuza classes runs as follows. Given a Vuza rhythm R, we write $R_u = \frac{1}{Per(R)}e^{-r}R$ the *unified* rhythm with period one which is obtained by contraction after a shift by the minimal non-negative element r of R. We have $Per(R_u) = 1$, and R_u is embedded. Clearly, two Vuza rhythms of fixed period p have the same class iff their unified rhythms do so.

Hence we may concentrate on classifying embedded rhythms with period one modulo translations. Since the $\mathbb{Z}$-module $\mathbb{Z}.R \subset \mathbb{Q}_{\mathbb{Z}}$ is finitely generated, it is monogeneous, and as it contains $\mathbb{Z}$, it must be of shape

$\frac{1}{n_R}\mathbb{Z}$, with uniquely determined positive integer n_R. After a dilatation by the invariant n_R, we are left with the classification of zero-addressed objective local compositions $(R, \mathbb{Z})$ which

1. are generating: $\mathbb{Z}.R = \mathbb{Z}$,
2. have period n_R

under the group of integer translations. But this classification is clearly equivalent to that of the objects in $ObLoc_0^{gen,\mathbb{Z}_n}$ which have trivial translation automorphisms (they are "maximal" in the terminology of Halsey and Hewitt [435]). Since these classes are the classes of $ObLoc_0^{gen,\mathbb{Z}_n}$, minus those which have non-trivial translation groups $H \subset T_n$, and the latter are recursively determined as the classes stemming from $ObLoc_0^{gen,\mathbb{Z}_n/H}$, we have a recursive procedure to enumerate the classes in question. In [435, §6], this has been explicated. The classification itself can be performed along the lines of our exposition for the finite case in Section 11.3.3. We leave it as an exercise to go through that procedure with the translation group instead of the full automorphism group.

Based on this local situation, Vuza [1097] considers "unending rhythmic canons" which are defined as finite, non-empty subsets of a Vuza class, i.e., a non-empty finite set of Vuza rhythms of the same translation class. Equivalently, a Vuza canon is the interpretation C^I of a zero-addressed objective local composition $C \subset 0@Onset|_{\mathbb{Q}}$ by a $Trans$-isotypic covering set I consisting of Vuza rhythms, also called the voices of the canon, although no further parameters are considered; we subtend that the Vuza context must be interpreted as a projection of a Para-rhythm as discussed in Section 13.4.3. Vuza [1097] calls the common chart class the *ground class* $groundclass(C^I)$ *of the canon* C^I, whereas the Vuza class of the union C is called the *resultant class* $resclass(C^I)$ *of the canon.* Two Vuza canons C^I, D^J are called equivalent iff the resulting rhythms are translation equivalent under a translation e^t which is compatible with the respective atlases, i.e., we have a commutative diagram

$$\begin{array}{ccc} C^I & \longrightarrow & C \\ {\scriptstyle e^t}\downarrow & & \downarrow{\scriptstyle e^t} \\ D^J & \longrightarrow & D \end{array} \tag{16.12}$$

as above for tesselations. So this situation is a generalization to non-discrete isotypic coverings, and not necessarily covering the entire ambient module, but this module is $\mathbb{Q}_{\mathbb{Z}}$ instead of a finite abelian group, as discussed by Halsey, Hewitt, and Fripertinger.

As with the local situation of one Vuza rhythm, the *period of a Vuza canon* is a well-defined positive rational number given by $Per(C^I) = Per(groundclass(C^I))$. Two Vuza canons C^I, D^J are equivalent iff they have the same period p and the contracted canons $\frac{1}{p}C^I, \frac{1}{p}D^J$ are equivalent canons of period one, so we may restrict our discussion to C^I, D^J having this special period. Moreover, we may also suppose that the resulting classes are embedded, i.e., $C \subset \mathbb{Z}.C, D \subset \mathbb{Z}.D$. Since these classes are translation-equivalent, we obtain the same module $\mathbb{Z}.C = \mathbb{Z}.D = \frac{1}{n}\mathbb{Z}$, with the invariant *resulting divisor* $n_C = n_D = n$ as already discussed above in the local case. So, if we dilatate this situation by the resulting divisor, we have the generating local compositions $n.C \subset \mathbb{Z}, n.D \subset \mathbb{Z}$ with a period which divides n, and which are covered by local chart compositions which are all (strictly) n-periodic. So modulo this period, we have to look at interpretations c^ι, d^κ of local compositions $c, d \subset \mathbb{Z}_n$ by equipollent translation-isotypic atlases ι, κ which consist of $card(\iota) = card(I)$ (!) not necessarily generating local chart compositions which do not have non-trivial translations in $\mathbb{Z}_n$, i.e., are "translation rigid".

So our classification problem reduces to the classification of "canons" in the cyclic residue groups $\mathbb{Z}_n$. By definition, these are interpretations of generating, zero-addressed objective compositions by translation-isotypic translation rigid charts, but the charts are not necessarily generating. And the classification goes by translations on the ambient space $\mathbb{Z}_n$ which carry over to the interpretations. The classification of such interpretations of local compositions in $\mathbb{Z}_n$ cannot be settled by the known resolution theorems from Chapter 15 since the involved function modules are not projective. However, we have a number of numerical invariants, such as the (translation) class nerve and the cohomology groups. We do not know how far these invariants are away from being classifying. In a recent work, Fripertinger has also classified such canons, but see [346].

A special family of Vuza canons are the *regular complementary canons of maximal category* (for short: RCMC-canons). In [1099], Vuza presents an algorithm which enables the calculation of any tesselation $M = \coprod_{v \in V} e^v(U)$ of a group M by translates of a subset U, where both $Trans((U)$ and $Trans(V)$ are trivial (a non-Hajós group, see Definition 60). In particular, he proves that six is the minimal number of voices of an RCMC-canon and $n_R = 72$ is the shortest period. For a detailed discussion on Vuza's algorithm in the perspective of the theory of non-Hajós groups and the Minkowski conjecture, see [30]. The algorithm has been implemented in OpenMusic by Carlos Agon and Moreno Andreatta, see [33] for a complete list of solutions.

16.3 Global American Set Theory

Summary. American Set Theory has developed a number of "combinatorial" structures which relate to interpretations of pitch class sets. We give a short account to this sector, as it was developed by Forte, Lewin, Morris, and Rahn.

– Σ –

As mentioned in Section 11.5.2.2, part of what we called the American tradition focused on global instead of local musical properties. The local/global dichotomy is already present in Allen Forte's book [331], in which the second part is concerned with the so-called Pitch-Class Set Complexes, i.e., sets of sets associated by virtue of the inclusion relation. Inclusion relations are basically of two type: the K and the Kh relations. By definition, given a pc set class X, called the *nexus*, a pc set class Y is a member of the set complex about X iff Y can contain X or can be contained in X (or the corresponding for the complement of X), with some preliminary conditions on the cardinality of X and Y which are:

1. the inequalities $2 < card(X) < 10$ and $2 < card(Y) < 10$,
2. the inequalities $card(Y) \neq card(X)$ and $card(Y) \neq card(-X)$, where $-X$ means the complement of X.

Two sets that belong to the set complex about a given pc set class X are said to be in the K relation. As pointed out by Forte, "the rule of set-complex membership yields aggregates of considerable size" [331, p.96]. A stronger condition, called the Kh relation, enables one to reduce drastically the number of sets which are in relation of a given set. The new family of sets is called the subcomplex of a given pc set class X. By definition it is the collection of all sets Y such that Y can contain or can be contained in X *and* can contain or can be contained in the complement of X. A preliminary remark consists of the distinction between literal and abstract relations. Literal relations are among unordered collections of pitch classes, whereas abstract relations are among collections of pc sets related by some equivalence relation (usually transposition and/or inversion). In the case of the abstract inclusion, it may happen for example that a pc set class may be included in its complement, a statement that would be absurd in the case of literal inclusion. Forte's K and Kh relations are examples of abstract relations.

We agree with Robert Morris that "aside from reviews of Forte's book, there has been scant theoretical elaboration on the K and Kh relation in the literature" [767, p.175]. In order to give a new interpretation of Forte's set complexes, Morris suggests to represent K and Kh relations as lists of set classes displayed in complementary pairs which are called SC-comp lists [767, p.284]. Before describing the lattice representation of the K and Kh complexes, we follow Morris's discussion of a preliminary relation that he calls the KI relation. By definition two set classes A, B are said to be in the KI relation iff $A \subseteq B$. The main difference compared to Forte's relations consists of the fact that K relates two pairs of set classes whereas KI simply relates two set classes. The KI relation is a partial ordering which can be displayed in a lattice representation called the "KI-inclusion lattice". The following example in Figure 16.4 shows the KI-inclusion lattice about the octatonic collection 8-28 considered as the set class of the pc set $\{0, 1, 3, 4, 6, 7, 9, 10\}$.

Notice that the abstract inclusion relation has been independently theorized by Rumanian composer Anatol Vieru by means of the concept of "modal structure". By definition, a modal structure is an equivalence class of a pc set under simple transposition and without taking inversion into account. In [1086, ch.3] Vieru explains how to generalize the concept of inclusion in the case of modal structures. The greater power of

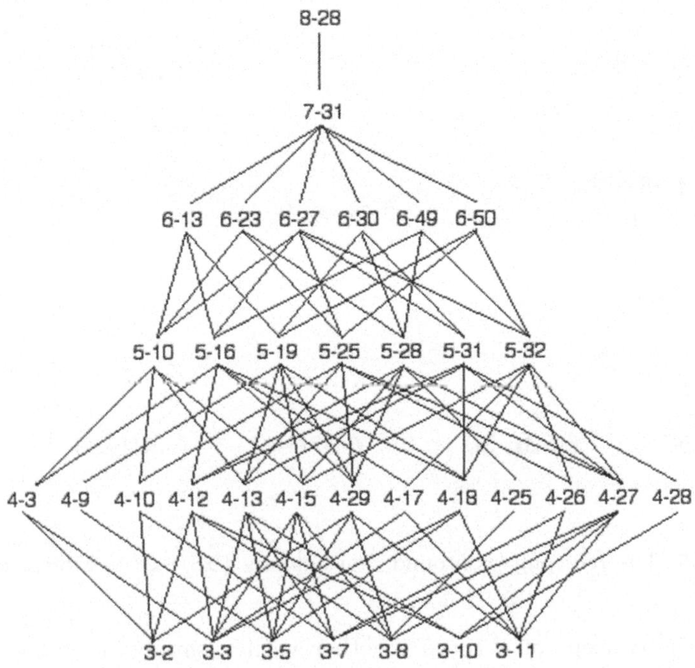

Fig. 16.4. Abstract inclusion lattice for the "octatonic scale".

selectivity of Forte's Kh relation compared to the KI relation is well explained by the concept of SC-comp list we mentioned before. In this representation, each class is listed together with its abstract complement (the two set classes are separated by a slash). In the case of a self-complementary hexachord, e.g., the set class 6-5 corresponding to the pc set $\{0, 1, 2, 3, 6, 7\}$, this hexachord is listed alone. Forte's Kh relation may be reformulated now in terms of SC-comp lists. A couple of (abstract) complementary set classes $A/-A$ is said to be the Kh nexus of the SC-comp list X if for all couples of complementary set classes $Y/-Y$ of X the couple $A/-A$ is in the Kh relation with $Y/-Y$. Note that this new definition drops Forte's original condition by assuming that a couple $Y/-Y$ may be equal to the Kh nexus $A/-A$. The following Figure 16.5 shows the great power of abstraction of the Kh complex about the octatonic collection 8-28/4-28.

For a discussion of some difficulties arising in the analytical application of Morris K and Kh relations as well as of a possible generalization of the SC-list concept in relation to Forte's concepts, see [767, pp.285-288 and pp.299-304]. As mentioned in Section 11.5.2.3, the main properties concerning set complexes have been implemented in Ircam's visual programming language OpenMusic.

We conclude this section by briefly discussing a different aspect of global American theory, which is usually called transformational theory. The theoretical basis of this approach is contained in the second part of David Lewin's book [605] as an alternative approach to the Generalized Interval System (GIS). As suggested by Lewin "Instead of starting with a GIS and deriving certain characteristic transformations therefrom, it is possible to start with a family of characteristic transformations on a musical space and derive a GIS structure therefrom" ([605, xiii]). Transformations produce a global network whose morphology is crucially dependent on the type of transformations which are modeled. Lewin discuss the case of Riemannian networks (i.e., networks involving so-called "Riemann transformations"[7]) and networks making use of serial and inversional transformation. As observed by Vuza ([1095, p.277]), from a mathematical point of view, the GIS structure

[7] As mentioned by Richard Cohn in his recent survey on neo-Riemannian theory [209], Riemann transformations have been introduced in the music theory discipline by David Lewin in [602] and largely discussed in Lewin [605] as something acting on a consonant triad (i.e., a "Klang" after Riemann) in order to produce another "Klang". Examples of Riemann transformations are Lewin's MED function (i.e., the operation by which the transformed triad becomes the mediant of its MED-transform), the DOM operation (i.e., the transposition by 5 semitones)

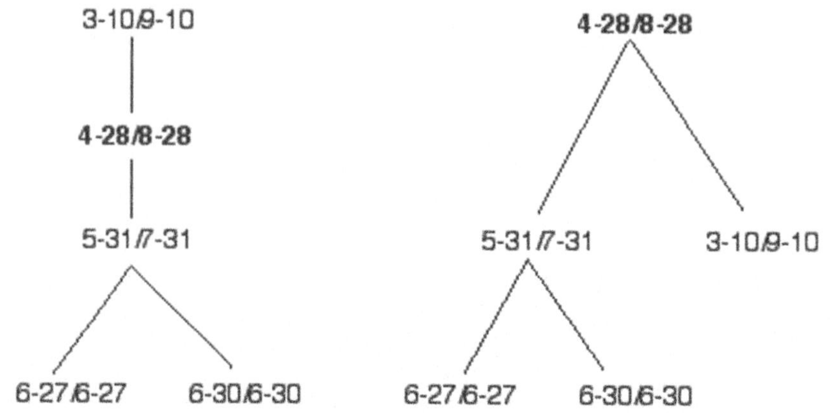

Fig. 16.5. The Kh complex around chord classes 4-28/8-28 in Morris' notation.

cannot be replaced by the concept of a space together with a simply transitive group of operations on it. Nevertheless, the interest in a transformational oriented music analysis has become more and more evident, particularly in relationships with the concept of 'contextual transformations', i.e., transformations which are sensitive to a particular aspect of a given musical context. This approach has been extensively discussed by Lewin in the four analytical essays of *Musical Form and Transformation* [604]. For an example of the musicological relevance of the contextuality of the transformational network approach, see Lewin's analysis of Karlheinz Stockhausen's *Klavierstück III* (pp.16–67), a piece which appears to be generated via a transformational network around the pc-set $\{0, 1, 2, 3, 6\}$. In some sense, as suggested by Robert Morris in his review of Lewin's book, the discovery of various transformational relationships in a piece can be regarded as building an abstract space in which a given piece lives. For a more detailed discussion of the concept of abstract space in composition and improvisation see [766].

16.4 Interpretable "Molecules"

Summary. The question of interpretability of molecular structures is very difficult. We make the problem clear and explain its musical aspect.

$- \Sigma -$

The classification of global molecules in the sense of Section 15.1.3 is evidently based on the classification of global compositions over the real numbers. So the important question concerning interpretable molecules resides in the interpretability of the underlying global compositions, a question which is settled by the criteria of Section 16.1, using the module of affine functions. In order to interpret a global molecule, we need to know how to construct a global space where an underlying local molecule could subsist. So we know that there is a big space where an underlying local composition exists. And we have a bilinear form and an orientation on each chart. At present, we do not know of criteria to decide whether these local Euclidean structures extend to an overall Euclidean space.

For the composer, passing from local Euclidean to global Euclidean structures in his global compositions means that he deals with the question of not only playing a global composition in a space of instrumental

and some so-called "contextual inversions" as the operations taking any Klang into its relative or its parallel major/minor (see also our discussion on harmony in Chapter 25)

parameters, but also of playing the local distances between sound events in this instrumental parameter space. So if the property of interpretability means that an abstract global composition 'idea' can be played at all, the property of interpretability of the molecular Euclidean structures means that the local event configuration shape given by distances and angles (such as pitch or time distances, sound color distances, and similar metric categories) of the abstract 'idea' can also be played in a singular big instrumental parameter space. We are far from having settled this problem in theory or in practice.

17

Esthetics and Classification

La musique est la corporification de
l'intelligence qui est dans les sons.
Józef Marja Hoene-Wroński (1778–1853)

Summary. Contrary to the seemingly bookkeping character of the concept of classification, the subject is deeply tied to esthetics. This is explicated and illustrated with a detailed example.

– Σ –

This final chapter on global classification is a more practical and philosophical one, and it thereby clarifies the theoretical background of a demanding subject which will be dealt with later: performance theory. It stresses the basics of the intriguing problem of performance, viz, the obligation to shape a composition beyond the given score data. Why is this necessary? Isn't the score enough? What is added, why, and how?

In this chapter, we shall not answer all these questions but only one: What is added? We shall try to make this clear on a very abstract level: resolutions of global compositions. But this offers a powerful perspective on the deeper structural rationales for shaping performance.

We hope that we can convey the essential connection between performance and classification: both deal with understanding, the former more on the communicative level, the latter more on the semiotic level. So they are indebted to each other as communication has to control its contents whereas contents should be put in evidence in the making of music.

17.1 Understanding by Resolution: An Illustrative Example

Summary. Before concluding this chapter on classification with a more programmatic and philosophical outlook, the theory and its esthetic implications are illustrated by the detailed analysis of a small two-part composition.

– Σ –

Our example deals with two easy two-voice compositions, and we want to analyze what it means to understand these pieces, each in itself and comparing them to each other: What is their commonality, what is the difference? Are they isomorphic? Are they two expressions of one and the same compositional idea?

To put everything into a completely familiar context without cumbersome mathematical abstracta, we select the ring $\mathbb{R}$ of real numbers and modules over $\mathbb{R}$, i.e., real vector spaces and their affine morphisms. This eases the geometry and gives us an occasion to visualize the classification ideas in ordinary three-space. For the music this means that we may consider real-valued onsets, pitch, duration, loudness, etc., and that we do not consider pitch classes but pitch as such, including microtonal intervals and infinitely subdivided time. Intuitively this means that very 'elastic' parameter conditions are admitted.

G. Mazzola, *The Topos of Music I: Theory*, Computational Music Science,
https://doi.org/10.1007/978-3-319-64364-9_17

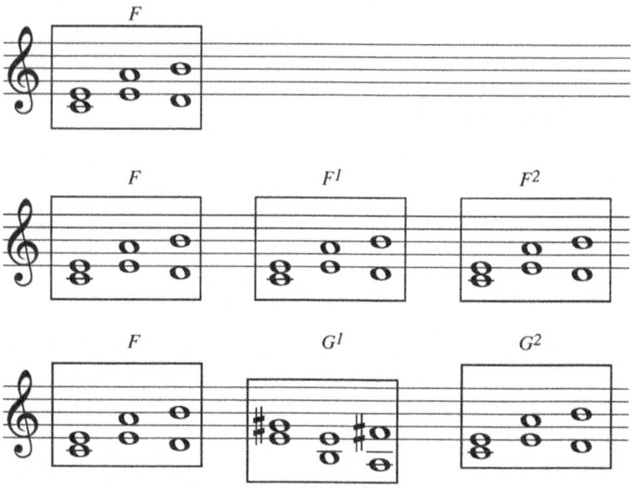

Fig. 17.1. The compositional germ F (top system) gives rise to two compositions X (middle system), and Y (bottom system), by use of contrapuntal replicates: translation and inversion.

The example—composed by a fictitious composer—where we start from, is a two-voice zero-addressed local composition $F \subset 0@OnPi$ from elementary note-against-note counterpoint, see Figure 17.1, top system. Duration and other parameters are of no relevance now and will be added later. Suppose that our composer now extends this germ by two procedures, see Figure 17.1, second and third system. In the first, he unites $F^0 = F$ to successive translates in time: Consider $F^1 = e^{(3,0)}F, F^2 = e^{(3,0)}F^1, \ldots F^{k+1} = e^{(3,0)}F^k$, and then the union $X = \bigcup_{k=0,1,\ldots r} F^k$. In the second, he unites $G^0 = F$ to successive translates in time, but every new translate is also inverted in pitch by I, the inversion at middle e: $G^1 = e^{(3,0)} \cdot IG^0, G^2 = e^{(3,0)} \cdot IG^1, \ldots G^{k+1} = e^{(3,0)} \cdot IG^k$, and we get the union $Y = \bigcup_{k=0,1,\ldots r} G^k$. The composer would like to understand the essential difference between X and Y.

According to the Yoneda philosophy, we first examine the discrete interpretations of X, Y and cannot find any difference since both just enumerate to $6(r+1)$ points. The next interpretation by use of vertical interval 'slices' does not show essential differences since the corresponding interval events in X and Y are just either the same or inverted copies, so these interpretations are isomorphic with each other. In the next refinement, X^C, Y^D, the composer lays the charts of couples of successive contrapuntal intervals, each, see Figure 17.2. So each of the r charts has four events. In X^C as well as in Y^D, the first two charts are the same, they have these event numbers $X_1 = Y_1 = \{1,2,3,4\}, X_2 = Y_2 = \{3,4,5,6\}$. But the third charts are different: $X_3 = \{5,6,1',2'\} \neq Y_3 = \{5,6,1^\star = I(1'), 2^\star = I(2')\}$. After that, things get repeated: Y_4, Y_5 are the inversions of X_4, X_5, respectively, etc.

In the next step, the composer tries to boil down these compositions to their common 'germ' F. In order to do so, he considers two point sets

$$F_X = \{1_X, 2_X, 3_X, 4_X, 5_X, 6_X\}, F_Y = \{1_Y, 2_Y, 3_Y, 4_Y, 5_Y, 6_Y\}$$

consisting of six points each. According to the above contrapuntal interpretations, the composer constructs two global compositions $F_X^\star, F_Y^\star$ on these sets by the following charts for $F_X^\star$ (see left part of Figure 17.3):

$$X_1 \to F_{X,1} = \{1_X, 2_X, 3_X, 4_X\} : i \mapsto i_X$$
$$X_2 \to F_{X,2} = \{3_X, 4_X, 5_X, 6_X\} : i \mapsto i_X$$
$$X_3 \to F_{X,3} = \{5_X, 6_X, 1_X, 2_X\} : 5 \mapsto 5_X, 6 \mapsto 6_X, 1' \mapsto 1_X, 2' \mapsto 2_X$$

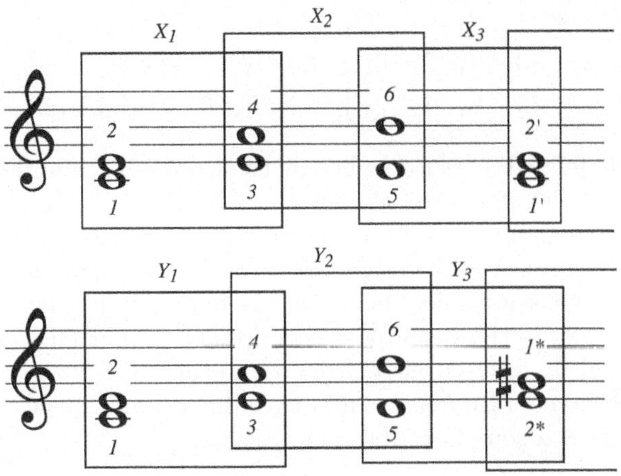

Fig. 17.2. The interpretations shown here refer to charts which represent the contrapuntal sequence of two successive intervals per chart.

which are taken from the interpretation X^C. So this is a global composition which is isomorphic to the interpretation F^A of F by the atlas

$$A = \{\{1,2,3,4\},\{3,4,5,6\},\{5,6,1'2'\}\}.$$

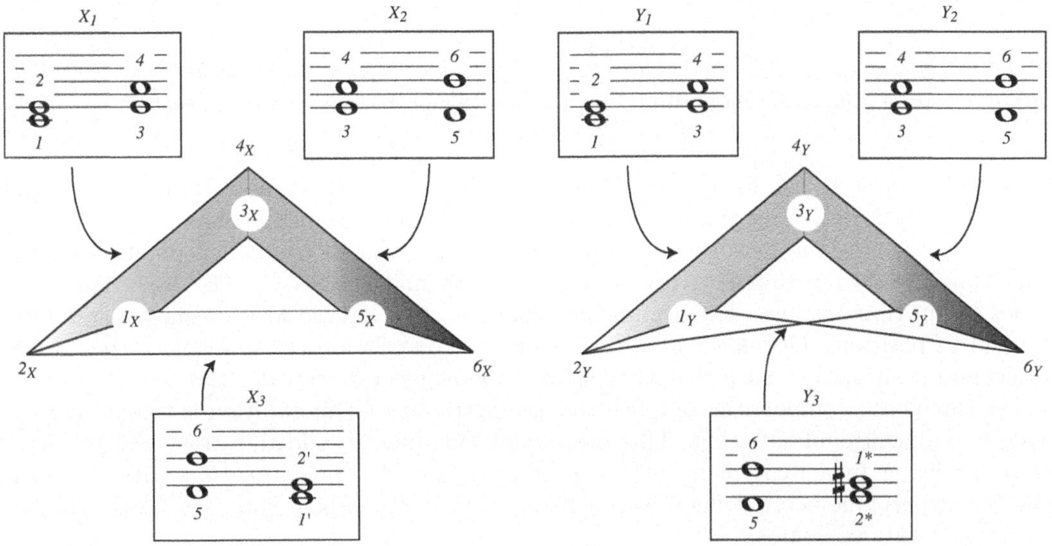

Fig. 17.3. These two global compositions are abstract models of the composition principles which give rise to the contrapuntal interpretations as shown in Figure 17.2. The left one is interpretable whereas the right one is not.

On the other hand, we construct a global composition $F^{\star}_{Y}$ by this atlas (see right part of Figure 17.3):

$$\begin{aligned}
Y_1 &\to F_{Y,1} = \{1_Y, 2_Y, 3_Y, 4_Y\} : i \mapsto i_Y \\
Y_2 &\to F_{Y,2} = \{3_Y, 4_Y, 5_Y, 6_Y\} : i \mapsto i_Y \\
Y_3 &\to F_{Y,3} = \{5_Y, 6_Y, 1_Y, 2_Y\} : 5 \mapsto 5_Y, 6 \mapsto 6_Y, 1^{\star} \mapsto 1_Y, 2^{\star} \mapsto 2_Y
\end{aligned}$$

where the first two charts are defined as before whereas the third is different: The points $1^{\star}, 2^{\star}$ are identified via inversion to the points $1, 2$, therefore we have set $1^{\star} \mapsto 1_Y, 2^{\star} \mapsto 2_Y$ on the third chart.

These global compositions should be viewed as being formalizations of the compositional ideas in X and in Y. Now, although the class nerves $CN(F^{\star}_{X})$ and $CN(F^{\star}_{Y})$ are equal, these compositions are essentially different. Whereas the first is interpretable, the second is not, it is in fact the composition which we had already discussed in Example 29. So these compositions are not isomorphic, and the compositional ideas are essentially different, though subtly differentiated from each other. The first composition can be played in its integrity on one space of musical coordinates, even in two-dimensional space, by construction. The other can never be played in this way. It is a somehow abstract object.

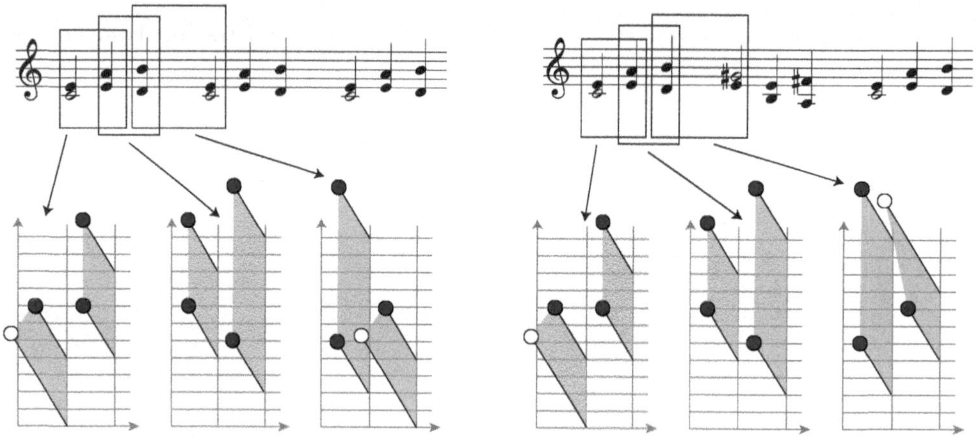

Fig. 17.4. By means of differentiation of durations, the non-isomorphic abstract compositions become isomorphic.

It is a good exercise to ask for the subtle reasons why the second composition cannot be interpreted. The calculation of the function modules which was carried out in Example 29 showed that the impossibility for separating functions is the parallelism of the lines drawn through couples of points in the charts. We have parallelisms $(1_Y, 2_Y) \parallel (3_Y, 4_Y)$, $(3_Y, 4_Y) \parallel (5_Y, 6_Y)$, and $(5_Y, 6_Y) \parallel (1_Y, 2_Y)$. This entails ample dependencies of functional values. The argument to blow up this flat world is to shift points into mutually more independent positions. Geometrically, this means that one should try to displace the four points on a chart to a general position, i.e., such that they span a non-degenerate tetrahedron.

But how should our composer accomplish this geometric task? One solution is to add more parameters to the given two-dimensional situation. This means, for example, to add duration, and to view the given compositions as living in a hyperplane of the product space *OnPi* ⊕ *Duration* with duration as third component, the hyperplane being defined by a constant duration, a whole note, say. This is in fact what we were drawing in the above figures.

To obtain a more general position of the chart events, we just have to alter the relative duration of one event: all notes being set to quarters, the first middle c can be given duration $1/2$. We show this change in Figure 17.4. Now, the three charts are in fact isomorphic to each other: the first to the first, the second to the second, and the third to the third, because any two standard compositions are isomorphic!

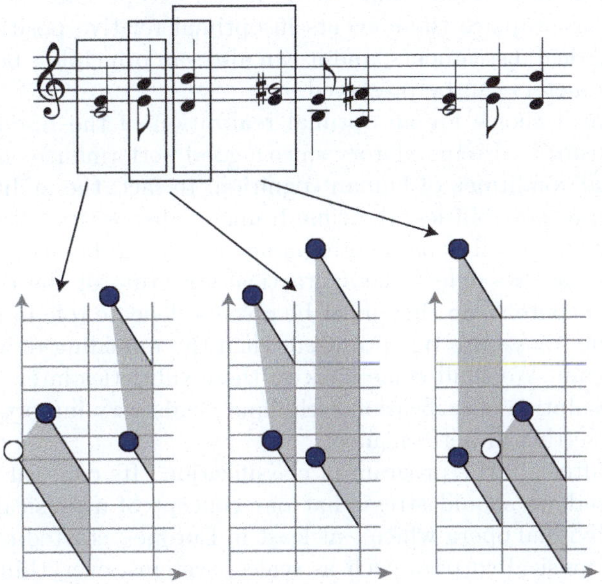

Fig. 17.5. The resolution of the non-interpretable abstract model $F^\star_Y$ is obtained by further differentiation of durations such that each chart becomes a configuration of events in general position: In fact, every chart shows a non-degenerate tetrahedron of points.

The trick simply is to add enough parameters in order to accomplish the goal of being in a state to play the piece in a single big parameter space. But clearly, the new compositions are no longer isomorphic to the old ones, they do, however, project onto the given original samples. More systematically speaking, one could have put any chart configuration into general position, i.e., the second chart, too, by just making an eight note from the second middle e, this is shown in Figure 17.5. This figure of the resolution $\Delta F^\star_Y$ of $F^\star_Y$ shows that we have an evident bijection of note events $\Delta F^\star_Y \to F^\star_Y$ as given in the general theory of resolutions, but the positions of the events in the different charts project from a general to a special position. But where has the special position been hidden in the resolution structure? It is packed in the retract of the function module of the original composition. This retraced module tells the composer, whether and how global coordinate functions for instrumental rendering are at his disposition. We shall shortly come back to this point in the discussion of Varèse's approach to composition.

17.2 Varèse's Program and Yoneda's Lemma

Summary. Implicitly, the part of Yoneda's Lemma dealing with variations of perspectives is of primordial importance in compositional concepts. In Edgar Varèse's programmatic writings, a thoroughly geometric approach to the Yoneda philosophy is sketched. It accomplishes the classical variational principle in composition; we give an overview to this central connection between modern mathematics and music.

$-\Sigma-$

To begin with, let us recapitulate the impact of the resolution of a (commutative) global composition on the esthetics of music. The resolution ΔG^I of a global composition G^I yields points which are in general position in every chart of the atlas. Moreover, the resolution is interpretable, and G^I can be reconstructed from the retracted function module in the resolution.

Since the resolution's nerve is isomorphic to the original nerve, and since the resolution projects bijectively onto the original composition, no note event and no overlapping relation of charts is destroyed in the resolution. So we essentially have the same set of notes, except that they were enriched by a number of parameters which allow us to place these events in optimal relative position. So the resolution can—in principle—be played by physical instruments, and it can also—in principle—be played such that the old idea can be heard since old parameters can be preserved.

However, the freedom of choice for an optimal realization of the resolution (which after all is only determined up to isomorphism) is also mandatory since a good performance on the physical level has to cope with a number of additional conditions of human cognition. In fact, the auditory system, the instrumental skill of an artist, the material possibilities, the time frame at disposition, etc., all these conditions impose serious boundary values on the possible parameter values which can be accepted in a good realization. For example, these parameters have to reflect the syntactical structure of the composition, and not only the resolution's general position context. So time must be given a delicate role in the parametrization of events. And the distinction of events must also be optimized when the unfolding of a performance in time is to be a good communication stream. We shall come back to these subtleties later in the context of performance theory. Nonetheless, the resolution classification technique yields very important necessary conditions for a comprehensible parameter setting in performance.

What could now be, after all, the program of classification? Its core value is that it deals with understanding musical works. And we should stress that our concept of a musical work is not the narrow one which restricts to those individual opera which—at least in Europe—started to emanate in the Renaissance. It includes as well general musical corpora such as scales, systems, everything that can be represented by means of global compositions, and—in the limit—any denotator if we admit the most general topos of this theory.

From the precise parametric description of a work and of its ambiguities, this work appears as a point configuration in a more or less complex space (or form, if you prefer the denotator terminology). However this configuration is already a determinate perspective which shows a multitude of relations among its ingredients. It is the composer's perspective (now including an abstract 'composer' or creator of a general musical structure like a scale). For example, the choice of tonality, instrumentation, tempo, etc., are points of view which may or may not pertain to the composition, this is a question of the epoch of creation. But their character can undoubtedly be subject to variation. Among others, here we do address the question of historical instrumentation for early music.

In order to understand the relations among different parts of a composition, and even to simply recognize them, a change of the given perspective is mandatory. If a never seen object must be inspected, what should we do? You walk around it. This is the most common version of Yoneda's Lemma. The analogy to cartography is straightforward: The natural perspective of the landscape in which we live does not coincide with the perspective which best meets our need for orientation. To reach this goal, we preferably build maps which show the landscape from an infinitely far vertical point.

The same happens to music. You play a piece in slow motion 'from very near', in a zoomed optics, a complex chord is arpeggiated, i.e., viewed from a skew angle, and so forth. This idea of variation of the perspective has also been integrated in the compositional thinking of the 20th century. We want to illustrate this remarkable fact by a citation from Edgar Varèse's comments on his composition "Intégrales" [1081, p.67]:

> *Die Intégrales wurden für eine räumliche Projektion entworfen. (...) Während wir in unserem musikalischen System Klänge anordnen, deren Werte festgelegt sind, suchte ich eine Verwirklichung, bei der die Werte fortwährend im Verhältnis zu einer Konstanten verändert werden. (...)*
>
> *Um dies besser zu begreifen, übertragen wir, da das Auge viel schneller und geübter ist als das Ohr, diese Vorstellungen ins Optische und betrachten die wechselnde Projektion einer geometrischen Figur auf eine Fläche, wobei Figur und Fläche sich beide im Raum bewegen, aber jede nach ihren eigenen Geschwindigkeiten, die veränderlich und verschieden sind, die sich verschieben und rotieren. Die augenblickliche Form der Projektion ist durch die Relation zwischen Figur und Fläche in diesem Augenblick bestimmt. Aber wenn man erlaubt, daß die Figur und die Fläche ihre eigenen Bewegungen haben, ist es möglich, mit der Projektion ein äußerst komplexes und scheinbar unvorhersehbares Bild zu erhalten. Diese Qualitäten können noch vermehrt werden, wenn man die Form der geometrischen*

Figur ebenso wie ihre Geschwindigkeiten variiert. (...) Ich hoffe, innerhalb kurzer Zeit einen Apparat zur Verfügung zu haben, der es erlauben wird, ein räumliches Relief zu geben. Nur des Beweises wegen würde ich daran interessiert sein, die Intégrales eimal so zu realisieren, wie sie ursprünglich konzipiert worden sind.

In 1960 Maurizio Kagel transferred these principles on paper strips and discs of the score for "Transición II" for piano, percussion, and two tapes.

Varese's idea basically is a remake of the *classical variation principle.* Bach's Goldberg Variations (BWV988), Beethoven's Diabelli Variations (op.120), or Webern's Variationen für Klavier (op.27) are compositions in this spirit. The subject is always an artistically interwoven change of perspectives of a theme: the variation in the parts of the theme and their relations.

Principle 7 *Variation as a principle of musical shaping is nothing else than the identification of an idea—such as the theme—as a integral of its perspectives.*

Especially for Webern a composition is a cellular organism, a connected manifold (in the naive sense) of transformations, of ever changing perspectives, of metamorphoses of a single cellular germ (in the sense of Goethe), which in fact is Schönberg's dodecaphonic series.

In front of this historical background the classification problem of global compositions—together with its central process of resolution—appears as a *canonical program.* In particular, the nerve of a resolution, a concept related to that of a "cell complex" from algebraic topology, reminds us of the cellular organism alluded to by Webern. And the projections which Varèse describes in a visionary fashion show a surprisingly similar geometry to the projections of a resolution onto the original composition, projections which are distinguished in that they project a general position onto specializations. Finally, the variation of these projections corresponds to the variation of the modules of affine functions, i.e., the variation of the compositions which are distinguished from each other via their retracted function modules on one and the same resolution.

But the variational principle is not only a compositional strategy, it equally, or even more dramatically, applies to the performance level. Performance—we shall discuss the issue in depth later—deals with a transformation from the mental score space to the physical space of the acoustic realization. But this transformation locally is a deformation of the "rigid" parameter values set out on the score. Why should the artist deform a perfect opus? Wouldn't this be blasphemy or at least a tremendous lack of respect? No, the added value of such a deformation is not a destruction of given structure, it is a subtle change of parametric perspectives which let the auditory still recognize the written relations, but on top of that puts configurations into general position such that their generic, or better: resolved, structure becomes "visible" on the auditory level—to restate it in the wording of Varèse.

Principle 8 *So the* structural *rationale of performance is a strategy of small changes of the composer's perspective to make the resolution of the composition audible and thereby ease understanding of the underlying composition class (in the strict sense of classification).*

18

Predicates

Die Welt ist alles, was der Fall ist.
Ludwig Wittgenstein [1138]

Summary. Denotators are purely mathematical structures which do not specify "what is the case" and what is not. This chapter deals with the existence problem of music-related objects in contrast to mathematical 'fiction'. This amounts to loading mathematical constructs with an additional semiotic signification process in order to express "which denotators are the case". These existence specifications instantiate an interface between mental potentiality and historical actuality. It reveals two fundamentally different existentialities, termed "textual" and "paratextual" signification, respectively. The former involves predicates defined by classical extension over specific denotators, whereas the latter transcends pure extensionality and thus points into domains of open semiosis. Both, textual and paratextual predicates are essential enrichments of mathematical constructs: The platonic ontology is thereby supplemented by a differentiation which cannot be reduced to pure "mentality". The variety of textual predicates follows certain construction rules of logical and geometric nature and is founded in a triply typed set of "atomic" predicates of (a) mathematical, (b) musical, and (c) deictic types.

– Σ –

18.1 What Is the Case: The Existence Problem

Summary. Music and musicology cannot refrain from distinguishing fiction against facticity. There are three reasons for this constraint: (1) historically, works, interpretations, and performances are in a substantial interaction which deals with what is the case; (2) in view of interactive discourse on music, actual perspectives of involved instances are not contingent but proper to the results; (3) the documentation of facts related to music are part of their accessibility *not less* than physical bodies do share and—by Einstein's gravitation theory—even *shape* a site in the universe of space-time.

– Σ –

In mathematics, once a domain of objects has been consistently defined, it is no question that they are all available or 'exist' without further differentiation, respectively. Once prime numbers are defined, their instances are just there, no question of distinguishing explicitly those we have already dealt with from the others[1].

The fundamental difference between mathematical set or category theory and denotator theory is that the 'existence' of denotators with respect to music or musicology is *not* equivalent to their purely

[1] However, for computer mathematics, the prime numbers which have been dealt with are definitely more concrete than the others.

G. Mazzola, *The Topos of Music I: Theory*, Computational Music Science,
https://doi.org/10.1007/978-3-319-64364-9_18

mathematical existence. Rather must we consider a specific type of allocation or instantiation, be it on a computer's memory or in an intellectual framework, such as the composer's mind or a given composition.

The point is that, within a fixed discourse, we do not permit automatic access to denotators. This is a rigorous discipline about what is given and what is only possibly given, a fundamental feature in musical thought or musicological analysis.

For example, in the diachronic evolution of the music system or within a specific material allocation within an information system[2], this may be relevant. In particular, if one deals with musicological analysis, it is essential to make precise the universe of objects one deals with, be it for classification or for ad hoc reference.

Principle 9 *We do not deal with a priori limitation of the available 'material', but with a 'declaration duty' of what we are allowed to refer to.*

Denotators share a kind of mathematical existence. Whether we view them in a set or category-theoretical perspective, they participate in a layer of abstract existence. In the context of music semiotics, it becomes relevant to couple semiotic specification in the sense of Hjelmslev's glossematic [472] with existentiality: Denotation and connotation are ontologically sensitive concepts. They not only reflect ontology but are possibly responsible for its very production. This contrasts musicology with physics: The latter deals with a fairly objective subject to be described by mathematics, whereas the subject of musicology is far more human nature and as such does not only exist but is essentially *created*. Therefore this ontological enrichment has to be dealt with explicitly and in a differentiated way.

In order to differentiate denotator ontology from the musicological one, the denotator system is viewed as the denotative layer of a supersystem whose signs are called (musical) predicates. Here, the denotators play the role of significants, whereas the significates are instances of 'specifically' musicological meaning. This is the model to be developed in the sequel[3].

18.1.1 Merging Systematic and Historical Musicology

Summary. Predicates prepare the field for a reconciliation of systematic and historical musicology.

– Σ –

We cannot refrain from stressing how profoundly the distinction between denotators and predicates hits the very structure of musicology. Following a strong tradition [234], this is commonly divided into systematic and historical musicology. While historical musicology seems to deal with history and historicity of music, systematic musicology seems to deal with musical systems. However, it is well known that history is also history of somewhat, of some systems, whereas systems are also historic in nature. The contrast to physics, for example, is that history of physics is history of the knowledge of one and the same subject: nature. So the physical systems evolve as knowledge bases about one and the same thing. Their coherence is a priori guaranteed by the identity of nature and by the invariance of physical laws in time.

In contrast—and this is the argument which is usually given—music's identity is itself a historic variable since music is a constantly evolving human creation. The argument is however erroneous since music's identity has never been defined or even understood as a time-independent entity. In other words, the diachronic axis of music is essential, but not as a disconnecting instance, on the contrary, coherence of the historical development is part of the overall system. Like language, music is a semiotic system with diachronic and synchronic coordinates and laws which pertain specifically to dia- and synchronic locations. But the study of time-dependent systems is not essentially different from the study of time-independent systems such as

[2] A database, for example.

[3] It is based upon the insight of music semiotics [703] that musicological meaning is a multilayered fact to be successively constructed via Hjelmslev chains of denotator/connotator systems.

physics[4]. In other words: The fact that laws of music are susceptible to be time-dependent is no reason to believe that there are no laws, that there is no all-embracing system beyond a valid chain of stories.

There are several methods to turn this pseudo-dichotomy of historic and systematic musicology into science. One way is to seek for time-invariant laws which model historic development as such in specific areas. Such an approach is undertaken in the theory of increasing paradigmatic groups as a significant parameter for historicity, see Section 11.7.2. This approach resembles the second law of thermodynamics which—for a determined set of physical systems—defines a time arrow towards the future by postulated increase of entropy. This first approach is somewhat naive since it may happen that the dynamics of knowledge acquisition requires a conceptual reengineering. For example, the concept of reference frame in physics had to be revised because of the relativization of time in special relativity. The relativistic time concept had to be attached as a proper attribute to each reference frame, and was no longer an all-embracing, divine quality in the sense of Newton.

Whereas in physics, epistemological events of this size are rather rare, in fact appear as veritable scientific revolutions, *music is in a state of incessant big bang*: Creativity constantly adds to the fundamental data and knowledge. In other words, dynamical concept and theory handling is much more vital to musicology than to physics. Therefore a second method consisting of dynamic concept frameworks and explicit handling of the incessant creative impact is required. Dynamic concept frameworks have been developed in the previous theory of forms and denotators. The impact of incessant big bang creativity is handled by the theory of predicates.

More generally speaking, *historical sciences are likely to be absorbed by dynamical knowledge management* in the following sense. Understanding history means conceptual and data control of the world's diachronic dynamics[5]. Professional data control is no longer feasible without powerful data base management systems, and conceptual control cannot survive if it is not part of a comprehensive concept management system. Otherwise, historic science runs out of self-control or downsizes to well-known variants of ideologies and "weltanschauungen". In this sense, *innovative musicology could play a role of a prototype for future historical science.*

18.2 Textual and Paratextual Semiosis

Summary. Review of Kofi Agawu's and Roman Jakobson's introversive and extroversive semiosis. Truth values and meaning. The problem of open semiosis. Classifying open semiosis: processuality, synchronic and diachronic pointers of competence and tradition. Introducing the formal system of a *predicative semiology.*

– Σ –

To get off the ground with the discourse on musicological meaning, we review Agawu's work on music semiology [10] which builds on the tradition of Jakobson's [500, 501] research in modern poetology. Agawu follows Jakobson in distinguishing introversive vs. extroversive semiosis. Introversive semiosis is production of meaning on the basis of intratextual signs. Agawu calls them the universe of structure. Examples: Schenker's "Ursatz" (beginning/middle/ending), Ratner's model of harmonic functions, and, of course, all elementary signs for metric, rhythmical, motivic, harmonical, etc. structures. Introversive semiosis can be said to be production of textual meaning because the text is the relevant reference level for introversive semiosis.

Extroversive semiosis involves signs which transcend the system of musical signs in the narrow sense of the word. Agawu calls them "the universe of topics". Topics are signs which have a significate beyond the text. The author gives 27 examples within the form of his analysis (reaching from 1770 to 1830, i.e., embracing the first Viennese school):

[4] Even physical laws do not persist eternally, it is argued that in the very beginning of the big bang, physical laws were rightly established; in fact at physical zero time, all laws break down.

[5] Wittgenstein's dictum heading this chapter should then be refined diachronically: "Die Welt ist alles, was mit der Zeit der Fall ist."

alla breve, alla zoppa, amoroso, aria, bourrée, brilliant style, cadenza, Empfindsamkeit, fanfar, French overture, gavotte, hunt style, learned style, Mannheim rocket, march, minuet, musette, ombra, opera buffa, pastoral, recitative, sarabande, Seufzermotiv, singing style, Sturm und Drang, Turkish music.

These signs have a surface which has a regular textual meaning involving particular groups of sound events with a particular structure. But the deeper meaning, in some sense a connotative significate, reaches beyond the text, and understanding it requires historical and/or music(ologic)al competence. It can only be adequately realized by the competence of the listener/musician and his/her idiomatic expertise.

Agawu's access to musical meaning is characterized by a dichotomy between precise textual and denotative semiosis and some kind of black box semiosis referring to an exterior context. This music(ologic)al context is an open system and should be treated as such. With regard to it, there is open ended semiosis dealing with what we call paratextual meaning.

The distinction between textual and paratextual meaning is motivated by the fact that meaning is never formally closed and must therefore be treated in an open way.

When building the predicate concept, one has to make the semiosis that takes place on the denotator level more open with respect to

- diachronicity: the set of predicates may change with time (cultural epochs);
- synchronicity: the set of predicates depends on the spatial (cultural) context;
- incompleteness: meaning may be incomplete, provisional or overtly undetermined.

Let us give an illustration: The diachronic extension of the repertoire of compositions determines the experimental material upon which music analysis has to be executed, tested and developed. It is not time-invariant. For instance, the Tristan chord is far more than just one item within an abstract list of chords. The synchronic analysis of ethnomusicological data is heavily dependent upon the cultural region from where they are applied. The extroversive meaning of data (such as "fermata") may be incomplete (from the point of view of performance) and one has to find a way of handling it formally. We do not yet know 'everything' a sign is supposed to convey, but we have to handle it within its context. In the life of musicians and musicologists, incomplete semiosis is the rule, not the exception!

To control this variety of sign processes it is necessary to set up an adequate system of signification mechanisms.

18.2.1 Textual and Paratextual Signification

Summary. Truth values and meaning; extension vs. intension.

– Σ –

In order to distinguish potential from actual instances of denotators in the spirit of Agawu's universe of structure, it is necessary to be in a position to tell which instances "are the case" and which are not. This is why we stipulate that predicates are related to sets of denotator instances which do exist. For example, if we describe piano notes by special types of denotators, the predicate "piano notes of concerto XY" would cover all these denotators for the notes of the score XY attributed to the piano. We may say that this predicate creates meaning by extension.

Meaning by extension can be distinguished from meaning by intension. This is what Agawu alludes to when introducing topics. It turns out that intensional meaning involves a much richer type of semiosis. The following list enumerates some types of open semiosis (without claim of completeness):

- *Semiosis as a process*—To begin with, semiosis is not a state but a process so that richness of meaning increases or decreases as a function of system time. What could be an intensional meaning at a given moment can be transformed into an explicit extensional meaning after additional information was added, see [257].

- *Synchronic pointers: competence*—Meaning of the intensional type may consist in the reference to another instance which 'knows' more about the music in question. This is a pointer to a competence exterior to the structural data. For example: The expert teacher in piano music knows how a specific articulation sign has to be realized on a grand piano.
- *Diachronic pointers: tradition and progress*—Another pointer type of intensional character is directed towards historical topoi or towards paradigms of progress. As an example, meaning may be anchored in historical style knowledge.

18.3 Textuality

Summary. The category of denotators. The textual predicate system. Predicate expressions as significants. Classification of textual predicates by three dichotomies: arbitrary/motivated, punctual/relational, and objective/morphic. Classification of motivation mechanisms: logic and geometry.

– Σ –

18.3.1 The Category of Denotators

Summary. We first have to introduce a category structure on the set of all denotators.

– Σ –

In Section 8.2, we briefly alluded to an extension of morphisms between local compositions to denotators of more general types. For the following discussion of textual predicates, we need such a framework which admits morphisms between any denotators.

In order to define morphism for denotators of any type, i.e., for the category *Den* of denotators, we start by the following axiom:

Axiom 2 *If x, y are denotators of different form type, the set Den(x, y) of morphisms in Den is empty.*

This means that only typed morphisms are allowed here, in other words: The category of denotators is the coproduct

$$Den = Den_{\mathbf{Simple}} \sqcup Den_{\mathbf{Syn}} \sqcup Den_{\mathbf{Power}} \sqcup Den_{\mathbf{Limit}} \sqcup Den_{\mathbf{Colimit}} \tag{18.1}$$

of the subcategories of types as notated in the indexes. Let us now look at the different types. The type where we already have a good theory is the category $Den_{\mathbf{Power}} = Loc$ of local compositions. Observe that here, the morphisms do not presuppose that the form morphisms which induce the natural transformations f of the respective subfunctors restrict to specific form types, i.e., the coordinator forms may be of any type. This means that if we want to associate a denotator $x : A{\rightsquigarrow}F(\xi)$ of form type t_x with a denotator $y : B{\rightsquigarrow}G(\eta)$ of form type t_y, by a form morphism $h : x \mapsto y.\alpha$, say, then we can do so under their "wrapping" as singleton local compositions, i.e., we have $f/\alpha : \{x\} \to \{y\}$ for $f = h|_{\{x\}}$, as described in Section 8.2. So

Principle 10 *The only way to change type by morphisms is to wrap form morphisms into local compositions.*

Let us now define the subcategory $Den_{\mathbf{Simple}}$. Give two simple denotators $x : A{\rightsquigarrow}F(\xi), y : B{\rightsquigarrow}G(\eta)$, with simple forms $F \underset{Id}{\longrightarrow} \mathbf{Simple}(M), G \underset{Id}{\longrightarrow} \mathbf{Simple}(N)$, we set

$$Den(x, y) = \{(x, y, \alpha), \alpha \in A@B| \text{ there is } h \in M@N \text{ with } h.x = y.\alpha\} \tag{18.2}$$

and we denote such a morphism by $!/\alpha : x \to y$. This definition should be read properly with respect to the underlying identifiers. We have $\xi : A \to Fun(F), \eta : B \to Fun(G)$, a notation which we may use in view of Yoneda's Lemma, by identification of $A@Fun$ with $Hom(@A, Fun)$. The existence of h means that we have a form (i.e., a functor) morphism $\bar{h} : Fun(F) \to Fun(G)$ which is induced by $@h$. Then the equation $h.x = y.\alpha$ means that $\bar{h}.\xi = \eta.\alpha$. The rest is obvious.

Proposition 20 *The subcategory* $Den_{\mathbf{Simple}}$ *has arbitrary fiber products.*

Proof. Suppose we are given three simple forms

$$F \underset{Id}{\longrightarrow} \mathbf{Simple}(M), G \underset{Id}{\longrightarrow} \mathbf{Simple}(N), H \underset{Id}{\longrightarrow} \mathbf{Simple}(L),$$

three respective denotators $x : A{\leadsto}F(\xi), y : B{\leadsto}G(\eta), z : C{\leadsto}G(\zeta)$ and two morphisms

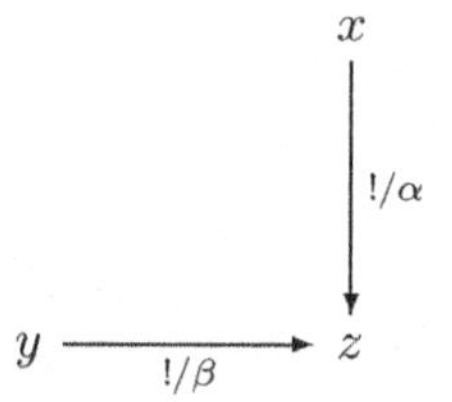

of simple denotators. Suppose that $!/\alpha$ is induced by a morphism $h : M \to L, k : N \to L$. Then we have a fiber product form $F \times_H G : Id{\leadsto}\mathbf{Simple}(M \times_L N)$ with identifier equal to the fiber product of the identifiers of F, G, H under the given form morphisms. Since these form morphisms are induced by two module morphisms h, k, the projections $F \times_H G \to F, F \times_H G \to G$ are also induced by the projections $M \times_L N \to M, M \times_L N \to N$. With this in mind, there is a denotator $x \times_z y : A \times_C B{\leadsto}F \times_H G(\xi \times_\zeta \eta)$ which is universally defined to make this diagram

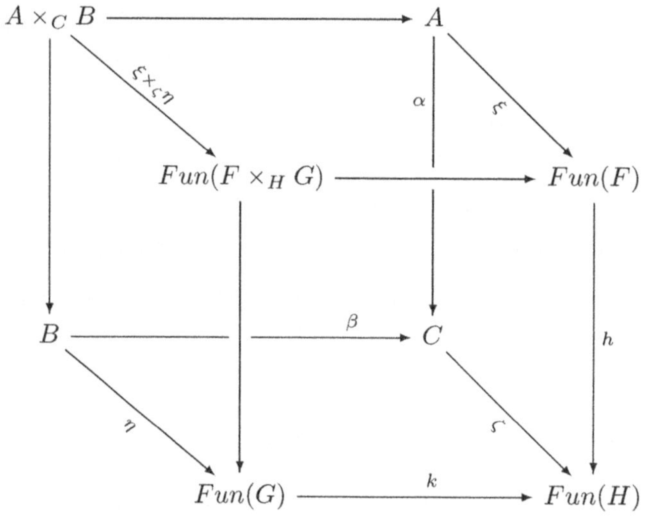

commute, and we are done. QED.

For synonymy, define the *reference denotator* $x^{\sim} : A{\leadsto}F^{\sim}(\xi^{\sim})$ of a denotator $x : A{\leadsto}F(\xi)$ with $F \longrightarrow \mathbf{Syn}(F^{\sim})$ and $\xi^{\sim} = Id(\xi) \in A@Fun(F^{\sim})$. This is essentially the same as x, albeit leaving aside the synonymous 'disguise'. Suppose by recursion that for two synonymy denotators $x : A{\leadsto}F(\xi)$, $y : A{\leadsto}F(\eta)$, we have already defined and symbolized morphisms of referenced denotators by the fractional notation $f/\alpha : x^{\sim} \to y^{\sim}$ with address change $\alpha : A \to B$. Then we identify the morphisms $Den(x, y)$ with the set of triples $\{(x, y, f/\alpha) |\ f/\alpha \in Den(x^{\sim}, y^{\sim})\}$; and we denote such a morphism by the evident symbol $f/\alpha : x \to y$. The composition of such morphisms is evident, and this (sub)category is defined. This subcategory has fiber products iff its referenced counterpieces have. Observe that this is a problematic requirement for circular denotators; in a mathematical discussion, the situation should receive a special treatment.

For the limit type, give two $x : A{\leadsto}F(\xi), y : B{\leadsto}G(\eta)$, with forms $F \underset{Id}{\longrightarrow} \mathbf{Limit}(\mathbb{D}), G \underset{Id}{\longrightarrow} \mathbf{Limit}(\mathbb{E})$. By definition, we have $\xi = (\xi_d)_{d \in |\mathbb{D}|}, \eta = (\eta_e)_{e \in |\mathbb{E}|}$ where $|\mathbb{D}|, |\mathbb{E}|$ denote the corresponding diagram vertex sets. In this case, a morphism $f/\alpha : x \to y$ means the following:

1. an address change $\alpha \in A@B$,

2. the symbol f/α for a family of morphisms $(f_d/\alpha : x_d \to y_{\phi(d)})$ with a map $\phi : |\mathbb{D}| \to |\mathbb{E}|$, and where the denotators $x_d, y_{\phi(d)}$ are reference denotators as above, i.e., $x_d : A \rightsquigarrow F_d(\xi_d)$ for the d-th form F_d of the diagram.

In particular, we suppose that the types of the factor domains and codomains are identical. The composition of two such morphisms is again evident. In this subcategory, fiber products can be constructed as follows: Give a diagram of three limit denotators

$$\begin{array}{ccc} & & x \\ & & \downarrow{\scriptstyle f/\alpha} \\ y & \xrightarrow[g/\beta]{} & z \end{array}$$

with morphisms $f/\alpha : x \to z = (f_d/\alpha : \xi_d \to \zeta_{\phi(d)})$ for a map $\Phi : |\mathbb{D}| \to |\mathbb{F}|$ and $g/\beta : y \to z = (g_e/\beta : \eta_e \to \zeta_{\psi(e)})$ for a map $\Psi : |\mathbb{E}| \to |\mathbb{F}|$. The fiber product has a diagram scheme which is the fiber product set $\mathbb{P} = |\mathbb{D}| \times_{|\mathbb{F}|} |\mathbb{E}|$ (no arrows), together with its canonical projections into the factor diagrams. On a pair (d, e) over the vertex h of $\mathbb{F}$, the diagram evaluates to a denotator $p_{d,e}$ at the address $A \times_C B$ which is given by the cartesian square

$$\begin{array}{ccc} p_{d,e} & \longrightarrow & x_d \\ \downarrow & & \downarrow \\ y_e & \longrightarrow & z_h \end{array} \tag{18.3}$$

and whose existence is supposed by circular or regular recursion. It is easy to verify that this defines a fiber product on $Den_{\mathbf{Limit}}$. For the colimit type, give two $x : A \rightsquigarrow F(\xi), y : B \rightsquigarrow G(\eta)$, with forms $F \underset{Id}{\longrightarrow} \mathbf{Colimit}(\mathbb{D}), G \underset{Id}{\longrightarrow} \mathbf{Colimit}(\mathbb{E})$. Here, we have either $x = y$ where we just take the identity as a morphism, or else it is supposed that the diagrams are both discrete, in which case both, x, y live in coproduct spaces, $x \sim x_d \in A@d, y \sim y_e \in B@e$, say. Then we identify $Den(x, y)$ with the morphism set $Den(x_d, y_e)$ as above. In all other cases, no morphisms are provided in the colimit type. We leave it as an exercise to verify the existence of fiber products for the subcategory $Den_{\mathbf{Colimit}}$.

We stress once more that these are recursive definitions which require a special treatment in case the denotators pertain to circular forms. Observe that, for example, if we have a denotator x in a product form of two simple factors with modules M, N, this form is isomorphic to the simple form associated with the direct sum $M \oplus N$. However, the denotator is not isomorphic to its corresponding denotator x' in the simple form of the direct sum! It only becomes isomorphic if wrapped as a singleton, i.e., $\{x\} \xrightarrow{\sim} \{x'\}$. One could also see this technique as a "type casting" procedure, known from (strongly) typed computer programming languages.

18.3.1.1 Morphisms as Denotators

Summary. For formal programming reasons, it might be useful to transform morphisms into denotators. We look for such techniques.

– Σ –

If $\alpha \in A@_R B$ is an address change over a fixed ring R, and $x : A \rightsquigarrow F(\xi), y : B \rightsquigarrow G(\eta)$ are two denotators, the triple (x, y, α) can be seen as a denotator as follows: We have these sections and projections of addresses: $i_A : A \rightarrowtail A \oplus B, p_A : A \oplus B \twoheadrightarrow A$ and $i_B : B \rightarrowtail A \oplus B, p_B : A \oplus B \twoheadrightarrow B$. We also have the injection $p_A@B : A@B \rightarrowtail A \oplus B@B$. We refer to these maps in the following discussion.

Take the two injections $p_A@F : A@F \rightarrowtail A \oplus B@F$, and $p_B@G : B@G \rightarrowtail A \oplus B@G$. Consider the form $F \times G \times B \underset{Id}{\longrightarrow} \mathbf{Limit}(F, G, @B)$. Then the triple (x, y, α) is identified with a denotator $(x, y, \alpha) : A \oplus B \rightsquigarrow F \times G \times B(\xi, \eta, \alpha)$. This settles the description of simple denotators.

For the local compositions, take the usual morphism situation:

$$\begin{array}{ccc} x & \longrightarrow & @A \times F \\ f \downarrow & & \downarrow @\alpha \times h \\ y & \longrightarrow & @B \times G \end{array} \tag{18.4}$$

which represents the morphism $f/\alpha : x \to y$. We may take the graph $\Gamma_f \subset @A \times F \times @B \times G \xrightarrow{\sim} @A \oplus B \times F \times G$. This is a $A \oplus B$-addressed local composition in the form $F \times G$. Taking the form $exp(F, G) \underset{Id}{\longrightarrow}$ **Power**$(F \times G)$, we have this denotator $\Gamma_{f/\alpha} : A \oplus B \rightsquigarrow exp(F, G)(\gamma)$ and we can couple it with the equally addressed denotator (x, y, α) from above to get a denotator (f/α) in form $exp(F, G) \times F \times G \times B \underset{Id}{\longrightarrow}$ **Limit**$(exp(F, G), F \times G \times B)$:

$$(f/\alpha) : A \oplus B \rightsquigarrow exp(F, G) \times F \times G \times B(\Gamma_{f/\alpha}, (x, y, \alpha))$$

for the given morphism of local compositions.

The case of synonymy is trivial since the morphisms are just "synonymy casts" of already given morphisms.

The limit and colimit situations essentially reduce to a finite list of morphisms or one selected morphism, respectively, and this is settled by the above techniques; we leave it as an exercise to work out the details.

We should add two remarks:

- Morphisms of forms can be restated as new forms. In fact, if $h : F \to G$ is any morphism of forms, look at the form
$$\Gamma_h \underset{\Gamma_h : Fun(F) \rightarrowtail Fun(F) \times Fun(G)}{\longrightarrow} \textbf{Limit}(F, G).$$
The morphism is represented by the identifier of this form. This can be used to represent general limit data by products: If we have a diagram $\mathbb{D} : D \to \mathbf{Mod}^@$, each diagram arrow $f_{d,e,k} : d \to e$ can be wrapped in its graph form $\Gamma_{f_{d,e,k}}$ of limit type (in fact a product of two forms). Of course, one should not believe that wrapping a morphism in an identifier is an essential simplification of real problems, but it shows that morphisms are controlled on the level of identifiers within forms, and that, conversely, controlling forms qua objects means controlling the whole category of forms.
- If two denotators of same address and form have same coordinate ξ but different names "d_1", "d_2", we can shift this difference into the coordinate by adding a simple factor of character strings to the given form, and restating the coordinates as $(d_1, \xi), (d_2, \xi)$ such that now, the coordinates are different and not only the total denotator data.
This discussion makes clear that we may either look at denotators or at the category of denotators, the information is essentially the same: Each morphism yields an object, so that the knowledge about all objects is sufficient to describe knowledge about the whole category *Den*. This is why we shall in fact use the whole category and not only its objects in the following discourse about predicates.

18.3.2 Textual Semiosis

Summary. We introduce the formal system of a textual semiosis, including the question of switching between predicates and denotators.

– Σ –

The textual predicate system is a formal system of semiosis which manages the difference between mathematically possible denotators and denotators which are the case in some technical or more informal data base. The typical application is the storage of denotators on computer data bases for music research including retrieval, communication, and extension of knowledge.

A *textual semiosis (over the category Den)* is a map

$$sig_{Den} : Tex \to Texig(Den)_I \tag{18.5}$$

with a domain $Tex \subset Ex =< UNICODE >$ in the monoid of all $UNICODE$ strings[6], which we call *expressions* here.

To define its codomain, let $Den_\infty = \coprod_{1\leqslant n} Den^n$ be the union of all powers of the arrow set ${}_1Den$, including the identities which represent the denotators as objects. Moreover, we are given a module I of 'truth values' whose role will be discussed in the sequel[7]. This module will only be evoked via its role within a simple form

$$Val(I) \underset{Id}{\longrightarrow} \mathbf{Simple}(I) \tag{18.6}$$

of truth values[8]. This in turn gives rise to what is really needed, i.e., the form

$$TRUTH(I) \underset{Id}{\longrightarrow} \mathbf{Power}(Val(I)) \tag{18.7}$$

whose set of A-addressed *truth denotators* (any address admitted) is denoted by $\mathbb{T}_I^A$, whereas the union of all such sets over the totality of addresses is denoted by $\mathbb{T}_I$ and called the set of truth denotators. In other words, we have a 'fibration' $add : \mathbb{T}_I \to \mathbf{Mod}$ whose fiber $add^{-1}(A)$ at address A is $\mathbb{T}_I^A$.

The codomain $Texig(Den)_I$ is the set $\mathbb{T}_I^{Den_\infty}$ of all *characteristic functions* on the set of all finite length tuples of denotators or morphisms of denotators, and with values in $\mathbb{T}_I$. If $\chi \in Texig(Den)_I$, and if $x \in Den_\infty$, the address of $\chi(x)$ is often related to the addresses which are involved in x. For example, if all these addressed coincide and are equal to A, then $\chi(x)$ should also live in $\mathbb{T}_I^A$, but this is not mandatory and depends upon the construction of χ from previously defined functions.

To understand the truth denotator set, let us first discuss the completely reduced case of zero address and zero truth module, i.e., $A = I = 0_0 = 0$. This is the standard situation of truth values in topos theory: They are just the set of morphisms $1@\Omega$; in our situation of the presheaves over the module category $\mathbf{Mod}$, the final object is representable: $1 = @0$, and by Yoneda, we have $1@\Omega \overset{\sim}{\to} \Omega(0) = Sub(@0)$. Warning: This topos $\mathbf{Mod}^@$ is not Boolean. In fact, a topos of presheaves $\mathbf{Ens}^\mathbf{C}$ is Boolean iff $\mathbf{C}$ is a groupoid, i.e., iff all arrows are isomorphisms (see [639, exercise VI.2,p.343]). In general, we have $A@TRUTH(I) = A@\Omega^{@I} \overset{\sim}{\to} Sub(@(A \times I))$, therefore for the zero situation, a denotator $D \in \mathbb{T}_0^0$ is $D : 0 \rightsquigarrow TRUTH(0)(d)$ and has coordinate $d \subset @0$, a sieve in the zero module over the zero ring. To describe all sieves in the zero module, consider a category $\mathbf{C}$, and the equivalence relation

$$R \sim S \text{ iff } \mathbf{C}(R,S) \text{ and } \mathbf{C}(S,R) \text{ are both non-empty}$$

on $\mathbf{C}$ whose classes are denoted by $[R]$. On $\mathbf{C}/\sim$, we have a directed tree graph[9] structure $B_\mathbf{C}$ where an arrow $[R] \to [S]$ is defined iff there is a morphism $f : R \to S$ for representatives R, S. On any directed tree T, we can build the duplication tree $Dup(T)$ as follows: Take two disjoint copies T_0, T_1 of T and draw one arrow $d_x : x_0 \to x_1$ for each couple (x_0, x_1) of points corresponding to the original point x of the tree T.

Call a subset of a directed graph G *open* iff it contains the domains of all arrows whose codomains it contains. The set $Open(G)$ of all open sets in G defines a topology. In fact, the empty set and G are open, and any union/intersection of opens is open. Therefore, the topological space $Open(G)$ is a Heyting algebra in the usual sense (with intersection, union and implication, see Appendix G.5.1 and [639, p.51]). With this in mind we have the following description of all truth denotators modulo names:

Theorem 24 *The Heyting algebra* $Open(Dup(B_{\mathbf{Rings}}))$ *of open sets of the duplicated tree* $Dup(B_{\mathbf{Rings}})$ *and the Heyting algebra* $Sub(@0)$ *of (zero) truth coordinates are canonically isomorphic.*

[6] $UNICODE$ is an extension of the $ASCII$ set to non-European letters and symbols.

[7] We refrain from introducing non-representable truth value objects here and restrain to modules as there are merely speculative reasons to generalize.

[8] More generally, a truth form may be any form F, and not only a simple one.

[9] This is a directed graph whose undirected associated graph (forget about arrow directions) has no cycles.

Proof. Let $X \subset @0$ be a sieve, and A any address. Then either $A@X = \{!_A\}$ (the unique arrow to the final module) or $A@X = \emptyset$. So such a sieve is defined by the subset $supp(X)$ of addresses where $A@X = \{!_A\}$. Conversely, if any subset S of addresses is given, this corresponds to the support set of a sieve iff it is closed under address changes, i.e., iff, whenever $A \in S$ and there is an address change $B \to A$, then $B \in S$.

Therefore, the support set S of a sieve must be a union of module classes in the above sense, i.e., elements of $B_{\mathbf{Mod}}$. We therefore view S as a subset of vertexes in $B_{\mathbf{Mod}}$ which is open in $B_{\mathbf{Mod}}$.

Now, every non-empty R-module M is equivalent to the zero-module 0_R over the same ring. And for two such modules, there is a morphism iff there is an underlying ring homomorphism. Therefore, the restriction of the module class tree to the non-empty modules is isomorphic to the class tree of the ring category $B_{\mathbf{Rings}}$. Moreover, the empty modules are also classified according to their coefficient rings, and therefore, the class tree of these modules is also isomorphic to the class tree $B_{\mathbf{Rings}}$ of rings. Finally, there is a morphism $\emptyset_R \to 0_R$ whereas there is no morphism in the other direction. Therefore, the tree $B_{\mathbf{Mod}}$ is isomorphic to $Dup(B_{\mathbf{Rings}})$. Moreover, a sieve support set S of module classes in $B_{\mathbf{Mod}}$ must be open, and we are done. QED.

Exercise 38 *Observe that there are open sets in $Dup(B_{\mathbf{Rings}})$ which are not defined by a discrete set of leaves. For example, take the direct product ring $R = \prod_{i=1,2,3,...} \mathbb{Z}_{p_i}$ over all prime fields for increasing primes $p_1 = 2, p_2 = 3, p_3 = 5, \ldots$. Consider the infinite increasing sequence of ideals $I_k = \prod_{k \leqslant i} \mathbb{Z}_{p_i}$ and the corresponding chain of projections*

$$R \to R/I_1 \to R/I_2 \ldots R/I_k \to R/I_{k+1} \ldots$$

which defines non-equivalent rings and whose opening (the smallest open set which contains this sequence) is not defined by a leaf.

Example 33 A first example of a characteristic function is the attribution of one of the two universal values $\top, \bot : 1 \to \Omega$ with $\top = @0, \bot = \neg\top = \emptyset$ for each tuple f of denotators. This may and *will* happen in a completely uncontrollable, mathematically non-foreseeable way, see Section 18.3.3.

Proposition 21 *The coordinate Heyting algebra $0_{\mathbb{Z}}@\Omega$ of truth denotators for address $A = 0_{\mathbb{Z}}$ and truth module 0_0 is isomorphic to the Heyting algebra of sieves $S \subset @\mathbb{Z}$ in the category* **Rings**.

The proof is left as an exercise.

For any truth value module I and address A, and for any local composition $t \subset A@I$, we have the canonical functorialization $\hat{t}$ (see Section 6.2.3), in particular, for every value $\tau \in A@I$, we have $\hat{\tau} = \{\tau\}^\frown$. So every 'set' $t \subset A@I$ of A-addressed 'truth values' in I gives rise to a truth denotator in $\mathbb{T}^A_I$.

Example 34 Take, for example, the value module $S_1 = \mathbb{R}/\mathbb{Z}$, i.e., the real circle group (a $\mathbb{Z}$-module). For each number $\phi \in [0,1]$, we have the local, zero-addressed truth denotator (coordinate) $\phi^\frown = \widehat{[0,\phi[} \in \mathbb{T}^{0_{\mathbb{Z}}}_{S_1}$. This means that we have traced a (naive) fuzzy logic on the unit interval by the interval truth denotators $\phi^\frown$, included in the special values $0^\frown = \emptyset \subset \phi^\frown \subset 1^\frown = \widehat{S_1}$. The upper limit is the largest objective local composition in $@(0_{\mathbb{Z}} \times S_1)$.

Principle 11 *Philosophically, this means that fuzzy logic (and also other variants according to the given value module) is interpreted as a logic of intervallic local compositions in the unit group. This is quite the spirit which we shall now evoke: to view logical values as special local compositions, and to rephrase statements about local compositions as if they were generalized truth values. In music, there is no deeper reason to restrict truth to the final power Ω^1 of the subobject classifier instead of extending the discourse to general powers Ω^I.*

Exercise 39 *For every morphism of truth value modules $h : I \to J$, we have a canonical form morphism*

$$TRUTH(h) : TRUTH(I) \to TRUTH(J)$$

defined by images of subfunctors, and a corresponding map

$$\mathbb{T}(h) : \mathbb{T}_I \to \mathbb{T}_J : d : A \rightsquigarrow TRUTH(I)(\delta) \mapsto h.d : A \rightsquigarrow TRUTH(J)(Im(\delta))$$

on the truth denotators. So we have a corresponding map

$$\text{Texig(Den)}(h) : \text{Texig(Den)}_I \to \text{Texig(Den)}_J$$

which canonically extends to a change of textual semioses, i.e., if

$$sig_{\text{Den}} : \text{Tex} \to \text{Texig(Den)}_I$$

is a given textual semiosis, the composition

$$h.sig_{\text{Den}} = \text{Texig(Den)}(h) \cdot sig_{\text{Den}}$$

defines a new textual semiosis.

More generally, given two textual semioses

$$sig^1_{\text{Den}} : \text{Tex}^1 \to \text{Texig(Den)}_I \text{ and } sig^2_{\text{Den}} : \text{Tex}^2 \to \text{Texig(Den)}_J$$

a morphism $sig^1_{\text{Den}} \to sig^2_{\text{Den}}$ *is a pair* (u, h) *with a set map* $u : \text{Tex}^1 \to \text{Tex}^2$ *on the expressions and a morphism* $h : I \to J$ *on the truth modules, such that the diagram*

$$\begin{array}{ccc} \text{Tex}^1 & \xrightarrow{sig^1_{\text{Den}}} & \text{Texig(Den)}_I \\ u \downarrow & & \downarrow \text{Texig(Den)}(h) \\ \text{Tex}^2 & \xrightarrow{sig^2_{\text{Den}}} & \text{Texig(Den)}_J \end{array} \tag{18.8}$$

commutes. This defines the category of textual semioses *on Den. Verify the details.*

In particular, we may now compare different textual semioses according to variations on the expressive as well as on the truth value module levels.

The construction of the category of textual semioses in exercise 39 will be used when we have to combine characteristic functions which evaluate to different truth domains. This is essential in the whole construction process of musicologically relevant functions and predicates (see below, definition 61). The problem is not a formal one: As soon as one has opened the truth domains from naive logic of "Yes" and "No", combinations of logical statements must deal with a comparison paradigm, otherwise the choice of logic may terminate in a failure of communication between different logical territories.

Very often, a characteristic function is not really relevant except to a small subset of Den_∞, e.g., the zero-addressed local compositions. For the remaining arguments, the function takes value 'false', more precisely, the value is the empty sieve $\perp_A \in A@\Omega^I \xrightarrow{\sim} (A \times I)@\Omega$ at address A, write also $\perp$ if the address is clear or irrelevant. Define the *support* $supp(\chi)$ *of a characteristic function* $\chi \in \text{Texig(Den)}_I$ as the set of tuples $f \in Den_\infty$ such that $\chi(f) \neq \perp$. Whenever the value of a characteristic function is not specified, it is supposed to be 'false' (the address of this $\perp_A$ still being undetermined).

Definition 61 *A* predicate *for the category Den is a pair* (E, sig_{Den}) *of a textual semiosis* sig_{Den} *over* Den, *together with an expression* E *in* sig_{Den}. *The characteristic function* $sig_{\text{Den}}(E)$ *is called the* content *of the predicate, whereas the mapping (the functional relation) of* sig_{Den} *is called its* signification. *The predicate's* extension $ext(E)$ *is the content's support* $supp(sig_{\text{Den}}(E))$.

In a more sloppy language we shall identify a predicate with its expression if the rest (the semiosis in its very meaning!) is clear. Also we will refer to the predicate's textual semiosis when referring to the predicate's truth module etc.

We now introduce the concept of a predicative object (a pre-object), morphism (a pre-morphism), or more generally a tuple of morphisms. The denotators are the fictitious (mathematical) objects and we need those denotators which are supported by predicates. So let E be a predicate of a given semiosis. Then a *predicative object, morphism, tuple* (for this predicate) is a pair, denoted by x/E, where x is a denotator, a morphism or a tuple of morphisms such that $E(x) = sig_{\mathrm{Den}}(E)(x) \neq \perp$, i.e., $x \in ext(E)$. In general, pre-objects and pre-morphisms do not build categories since predicates are completely 'orthogonal' to the underlying mathematical structures. It will however be a special task to delimit classes of predicates which relate to the underlying mathematics. Essentially, pre-objects are introduced to grasp those entities which are the case in the sense of topos-theoretic, and more or less fuzzy sense as explained above. And to guarantee the most unsystematic approach to facticity as it happens to pervade the mud of the humanities.

Summarizing, if we denote the collection of all pre-objects, pre-morphisms and pre-tuples by Den_{∞}/sig, we have a 'forgetful' map

$$sig/ : Den_{\infty}/sig \to Den_{\infty} : x/E \mapsto x$$

whose fiber $sig/^{-1}(x)$ represents the *full facticity* of x with respect to the given semiosis. Setting ${}_0Den_{\infty} = \prod_{1 \leqslant i}({}_0Den)^i$ we complete the general terminology with this definition:

Definition 62 *A predicate E in sig is called*

- *objective iff* $supp(E) \subset {}_0\mathrm{Den}_{\infty}$,
- *morphic iff it is not objective,*
- *punctual iff* $supp(E) \subset \mathrm{Den}$,
- *relational iff it is not punctual.*

The above observation that a morphism of denotators may be restated as denotator means that *we may restate morphic and relational predicates as objective and punctual ones.*

Although time is not mentioned explicitly in this construction, the concrete situation is that the entire textual semiosis is not time-independent. The point is that the construction of new pre-denotators from old ones is not automatic but has to be performed by someone. This is not a serious drawback for the theory, but we should be aware that the setup changes every time a new item is added or removed! The details of this dynamics will become more clear in the following construction discourse.

A final point concerning reality and fiction: finiteness conditions. The realistic situation is that every predicate should only have a finite number of supporting denotators, a computer hard disk can not have infinitely many items. This is a delicate condition if we start producing new pre-denotators from given ones, since evident logical or geometric procedures yield infinite support if they are applied carelessly. So the *axiom of finite facticity support* for predicates is an omnipresent boundary condition to be observed in all the following construction.

18.3.2.1 Predicates as Denotators

Summary. Predicates should be convertible into denotators for the sake of management in information technology, but also because this is an essential method for generating new denotators from given ones. We discuss such techniques.

– Σ –

We suppose that we are given a predicate E, and that it is objective and punctual, a restriction which is not substantial—modulo some restatement acrobatics as described above. By the technique from Section 18.3.1.1, we may further suppose that all the denotators of its support have the same address A, viz the direct product of all the addresses of its denotators. And we may suppose that different denotator names also have different coordinates. The set $supp(E)$ is of course not a denotator, but there is a canonical way to make one: Suppose that we have the list $FormSupport(E) = (F_i)_{i=1,\ldots n}$ of forms pertaining to denotators of $support(E)$. Define this operator[10] on forms:

$$SupportForm(E) \underset{Id}{\longrightarrow} \mathbf{Limit}((P(F_i))_{i=1,\ldots n}) \tag{18.9}$$

with the factor forms

$$P(F_i) \underset{Id}{\longrightarrow} \mathbf{Power}(F_i).$$

This defines a canonical restatement of $supp(E)$, in fact, take $supp(E)_i = supp(E) \cap supp(F_i?)$, where we view $F_i?$ as the predicate with support the denotators of form F_i. Set $Coord_i =$ set of coordinates of $supp(E)_i$. Then we have the denotator

$$den(E) : A \rightsquigarrow SupportForm(E)((Coord_i)_{i=1,\ldots n}). \tag{18.10}$$

As we suppose that the coordinates distinguish denotators, we get a one-to-one representation of the predicate E by a product of objective, A-addressed local compositions.

The usage of this kind of denotators is not harmless! In fact it is important since we want to build denotators by this technique, but we do not want to produce contradictions by circularity, as, for example by Russel's antinomy! However, since the construction involves already given denotators, the new one is just a further construction which uses given supports.

But then, we have to pay attention to avoid conflicts with history! The point is that the construction of $den(E)$ involves the actual status of the predicate E. So, as time passes, E may change its signification (sic!) and we should update the associated denotator. For example, if E is already constructed from other predicates, this may be a cascade of updating duties. So the updating process may even involve circular constructs (referring to E), we will not produce contradictions since the referred to instances of E will be the last one, as with all replacement processes of type "$x = x + 1$" in programming.

We insist on this conflict potential since in the increasingly accelerated cycle of production, documentation, dissemination, and reflection of knowledge, a dramatic interaction between successive cycles may cause such strong intercyclic distortion forces that *knowledge crashes*, since the known and the new may overlap to the degree where the older 'version' cannot even stabilize its contents before the new 'version' intervenes and puts into question its predecessor.

18.3.3 Atomic Predicates

Summary. The criteria for atomicity. Classification of atomic textual predicates: mathematics, primavista, and deixis (shifters).

– Σ –

As announced earlier, predicates are not just invented in the air but germinate from a well-delayed arsenal of elementary or "atomic" predicates. The different types of atomicity are first treated before we deal with the combinatorial chemistry of compound predicates in Section 18.3.4. Evidently, this concept reflects the overall topography of construction in music and its theory. Basically, we distinguish three approaches to music related predicates: 1) those which are given by a fundamentally mathematical reasoning without further relation to music or to other rationales, 2) those which are defined by a purely music(ologic)al reasoning, and 3. those which are just the user's decisions beyond any further foundation, i.e., deictic morphemes in the sense of semiotics (see [703, 2.5.1] for a more semiotic discussion of the present topic). They will be discussed in the sequel.

[10] The index limit n could be a finite of infinite number, but as supports will be finite in every practical situation, the finite case is the default case.

18.3.3.1 Mathematical Predicates

Summary. Why do we need mathematical predicates since denotators already share mathematical existence? Examples of mathematical predicates for the working musician and musicologist.

– Σ –

We insist on distinguishing pure mathematical existence from musical relevance (musical existence) of mathematical properties in the lexical foundation for arbitrary predicates. This may—at first sight—seem unnecessarily restrictive. Why should a mathematical criterion not be unrestrictedly accessible within music and its theory? The reason is that, besides some historical attempts to identify music with mathematics (e.g., the Pythagorean school), mathematical objects are not automatically relevant to music. For example, a Fibonacci sequence does not automatically have musical meaning. Only when it is introduced as a composer's or analyzer's criterion will it gain the status of a predicative instance. Musical meaning consists of predicates which are not free from mathematical criteria, but have to be given a semantic status beyond mathematics. One of the reasons for this duty is that cognitive aspects of music must be examined with respect to their relevance within a given context. For example, the discussion of consonance and dissonance perception should be sensitive towards the mathematical procedures involved in the recognition of sonance classes. As soon as cognitive performance has to be traced in the cortical or subcortical tissue, the question of modules for mathematical tasks becomes primordial.

So mathematical predicates have extensions of denotators defined by mathematical criteria. One looks at mathematically defined properties of denotators, such as onset quantities or Boolean specifications, and then selects those denotators which fulfill these properties. Here fulfillment means attributing to denotators truth values in some set $\mathbb{T}_I^A$ of truth denotators. Let us start with four elementary examples which may be extended and completed in many ways.

1. *Chords.* We want to specify denotators which comprehend chords as sets of simultaneous notes. To this end, sets of notes, for piano say, are defined by the form $NoteGroup \underset{Id}{\longrightarrow}$ **Power**(*Piano-Note*), using the well-known coordinator $Piano - Note$. Then, by definition, the expression "*PianoChord*" will take truth value $\top \in \mathbb{T}_0^0$ precisely on the denotators
$$Ch : A \rightsquigarrow NoteGroup(Note_1, \ldots Note_n)$$
such that all its substance points $Note_1, \ldots Note_n$ have one and the same E-formed onset coordinate $E_1 = E_2 = \ldots E_n$. It takes the false value $\perp$ in all other cases (we take $I = 0_0$ here). Recall that this pre-denotator is denoted by $Ch/PianoChord$.
2. *Melodic motifs.* We again start with the form $NoteGroup$ in the above example, but require now that the onset coordinates E_i be pairwise distinct and define the content of the expression "$Motif$" to evaluate to $\top_A \in \mathbb{T}_0^A$ those denotators
$$Mt : A \rightsquigarrow NoteGroup(Note_1, \ldots Note_n)$$
which verify this condition: $Mt/Motif$, $\perp_A \in \mathbb{T}_0^A$ for all other denotator morphisms with codomain A, and $\perp \in \mathbb{T}_0^0$ in all other cases. So these values also take care of the address for objects and morphisms in *Den*.
3. *N-element groups of notes.* We fix a natural number N and want to select those sets of piano notes which have at most N elements. Denote this predicate by the expression $< N$, thus defining the corresponding pre-denotators by $D/ < N$, the truth values being again taken from $\mathbb{T}_0^0$. This last example shows that expressions may contain variables from the mathematical concept framework, such as N in this case.
4. *Form positions.* Each denotator refers to its form, address, and coordinate. Hence we are also interested in the information which comes from this data. For example, we may look for the form's type and define a characteristic function with values[11] in $\mathbb{T}_{\mathbb{Z}_5}^{A_{[\mathbb{Z}]}}$, with this prescription: We codify the forms by

[11] Recall that the notation $A_{[\mathbb{Z}]}$ means restriction of scalars to the integers, i.e., reduction to the underlying abelian group, see Appendix E.1.1.

$i(\mathbf{Simple}) = 0, i(\mathbf{Syn}) = 1, i(\mathbf{Limit}) = 2, i(\mathbf{Colimit}) = 3, i(\mathbf{Power}) = 4$, and then attribute the value $t(d) \in \mathbb{T}_{\mathbb{Z}_5}^{A[\mathbb{Z}]}$ defined by the constant sieve $i(type(form(d))^\frown$ for denotators d or codomains d of morphisms, and the false value $\bot \in \mathbb{T}_{\mathbb{Z}_5}^{A[\mathbb{Z}]}$ else.

These four examples suggest that there may be a legitimate musical or musicological interest to introduce certain mathematical properties for denotators. Nevertheless, these properties are of a purely mathematical nature and cannot be deduced from other properties by formal arguments; this is why they are arbitrary to the predicate system.

18.3.3.2 Primavista Predicates

Summary. Primavista (= PV) predicates are the main source for score-related predicates. We give a list and discuss some representative PV predicates for classical European notation; further examples of non-European PV predicates.

– Σ –

PV predicates are predicates related to scores. The score concept is however not restricted to the classical European music culture, even if we shall only give a rather complete list of primavista predicates for classical European music. From this list, some selected predicates will be discussed in more detail in order to illustrate the general procedure and to present templates for other predicates.

List of typical primavista predicates from classical European music scores:
Staves, Braces and Systems, Ledger Lines and Octave Signs, Clefs, Stems, Flags and Beams, Rests and Pauses, Ties, Key Signatures, Time Signatures, Accidents, Bar-lines and Repeat Signs, Slurs, Dynamic Marks, Articulation, Ornamentation Signs, Tempo Indications, Arpeggio, Composer, Name of the Composition, Expression, Instruments, Lyrics, Comments, Jazz-Harmony, Gestures, Number Sheets (see [882]). For non-European predicates see [234, Vols. 8 & 9].

Absolute Tempo. The absolute tempo indication can be either a numerical Mälzel sign or by verbal indications. To grasp this information, we take the forms

$$Maelzel \underset{Id}{\longrightarrow} \mathbf{Limit}(Rate, Onset) \tag{18.11}$$

$$Rate \underset{Id}{\longrightarrow} \mathbf{Simple}(\mathbb{Z}) \tag{18.12}$$

$$VerbAbsTpo \underset{Id}{\longrightarrow} Simple(\mathbb{Z} < ASCII >) \tag{18.13}$$

$$AbsTpo \underset{Id}{\longrightarrow} \mathbf{Colimit}(Maelzel, VerbAbsTpo) \tag{18.14}$$

where a denotator $m : 0_{\mathbb{Z}} \rightsquigarrow Maelzel(x, y)$ denotes that the Mälzel sign m is read as "Play x units of duration y!"; the meaning of the $VerbAbsTpo$ is clear, and denotes indications such as "Play *'Lento assai'*!". The total form codifies the alternative between Mälzel and verbal indications.
The next step is the embedding of the absolute tempo information in a named composition and at a precise onset position. We have

$$AbsTpoInComp \underset{Id}{\longrightarrow} \mathbf{Limit}(AbsTpoEvt, NComp) \tag{18.15}$$

$$AbsTpoEvt \underset{Id}{\longrightarrow} \mathbf{Limit}(AbsTpo, Onset) \tag{18.16}$$

$$NComp \underset{Id}{\longrightarrow} \mathbf{Simple}(\mathbb{Z} < ASCII >) \tag{18.17}$$

and understand that the absolute tempo event denoted under the form *AbsTpoEvt* is referred to a composition which is represented by a name. Of course, this could be refined, but the idea is clear from this.

Until now, we have just defined forms which yield the spaces where predicates could play a role. The predicate which is expressed by "Absolute Tempo" then is more than the abstract set of denotators. It should tell us which absolute tempo indications are the case from the point of view of a score reading expert. The trace of such a judgment is the predicate's content. If it is not always very clear from the philological point of view whether there is a determined absolute tempo, we could take the truth value of the characteristic function in $\mathbb{T}^{O_\mathbb{Z}}_{S_1}$ and thereby model the fuzzy approach as explained above. Therefore the value for a denotator $atp : O_\mathbb{Z} \rightsquigarrow AbsTpoInComp(x)$ is set to the fuzzy value $\phi^\frown$ attributed to the expert judgment that atp is the case with "certainty $0 \leqslant \phi \leqslant 1$", and $\perp$ in all other cases.

Jazz CD Reviews. If a jazz researcher wants to investigate the role of certain jazz directions in the historical line, it may be important to refer to a judgment of CD releases in professional jazz media. Let us therefore introduce a PV predicate about the importance of a CD in a determined review. The idea is that the predicate is set up by a jazz expert who judges the importance: Is the CD review saying that the product is important? Yes? No? "Rather yes, but, well, I changed my mind about the contents of that review.", etc. So we want a fuzzy, time-dependent predicate *Importance* on denotators which parametrize jazz CD reviews. Here are the forms:

$$CD \underset{Id}{\longrightarrow} \mathbf{Limit}(Label, Number, Year, Title) \tag{18.18}$$
$$Label \underset{Id}{\longrightarrow} \mathbf{Simple}(\mathbb{Z} < ASCII >) \tag{18.19}$$
$$Number \underset{Id}{\longrightarrow} \mathbf{Simple}(\mathbb{Z}) \tag{18.20}$$
$$Year \underset{Id}{\longrightarrow} \mathbf{Simple}(\mathbb{Z}) \tag{18.21}$$
$$Title \underset{Id}{\longrightarrow} \mathbf{Simple}(\mathbb{Z} < ASCII >) \tag{18.22}$$
$$Review \underset{Id}{\longrightarrow} \mathbf{Limit}(CD, Media, Year, Number) \tag{18.23}$$
$$Media \underset{Id}{\longrightarrow} \mathbf{Simple}(\mathbb{Z} < ASCII >) \tag{18.24}$$

Under the *Importance* predicate, a review denotator rev may take truth values in $\mathbb{T}^{O_\mathbb{Z}}_{\mathbb{Z}^3 \times S_1}$. This means that we attribute the fuzzy value in the module S_1 as above, but we also denote the date (on $\mathbb{Z}^3$) to fix the time-dependence of the judgment which may change even if we concentrate on an individual expert. The rest is clear. Why is this time parameter inserted in the predicate and not in the denotator? Because it is a quality of the predication and not of the review. Of course, one could also have introduced the time parameter in the review form, but then, the predication would be hidden in the denotators instead of the truth values.

Fushi in Noh's Utai Chant The Noh tradition (see [542] for the following facts) includes a sophisticated chant formalism which is titled "Utai", a triply articulated structure. We distinguish between melodic, dynamic, and speech articulation in the Utai. In this example, we shall concentrate on the melodic Utai which is described by a combination of melodic units, the *"fushi"*, see Figure 18.1.

Each fushi (or combination of such) is a prescription to shape the melodic content of one text syllable in duration, pitch, and color. So fushis are paired with syllables which we codify by the 16-BIT UNICODE codification (the Japanese-Chinese-Korean (JCK) character set, in this case). This is encoded in the form

$$JCK \underset{Id}{\longrightarrow} Simple(\mathbb{Z} < JCK >) \tag{18.25}$$

where we use the free $\mathbb{Z}$-monoid algebra over the JCK part of the UNICODE alphabet to encode text. The proper fushi part is built from the three fushi articulations into basic, special, and pitch-change segments. As it is not guaranteed that the special fushi signs are integrated in UNICODE, we better give a plain visual representation of those signs by a sufficiently fine binary encoding, on a $k + l$-Bit basis, say, i.e., in the forms

$$FushiPic \underset{Id}{\longrightarrow} \mathbf{Simple}(\mathbb{Z}_2^k \times \mathbb{Z}_2^l) \tag{18.26}$$

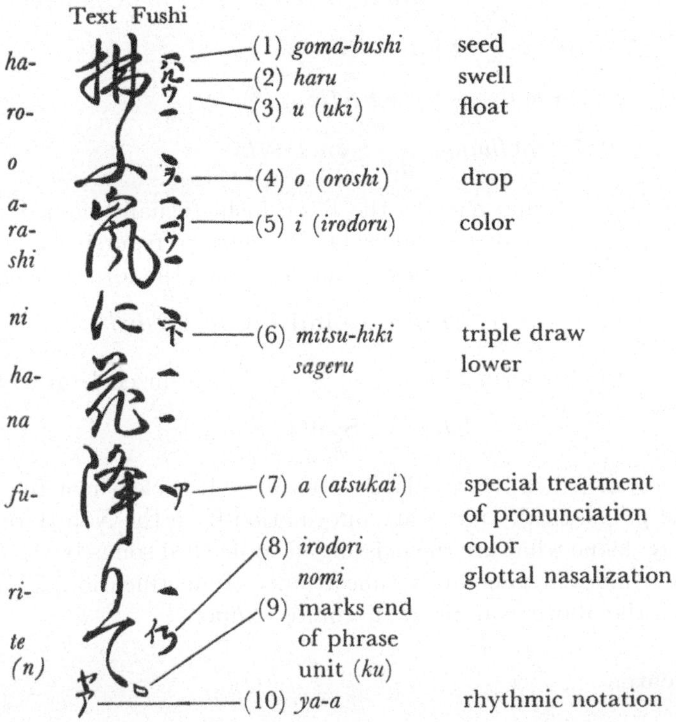

Fig. 18.1. An example of a Noh Utai song text (left column) with the fushi sequence (column to its right). From [542] ©1980, 1983 by Kunio Komparu. Reprinted by arrangement with The Permissions Company, Inc., on behalf of Shambhala Publications Inc., Boulder, Colorado.

with k Bits for the pixel color encoding and l Bits for the pixel number. This enables us to add a name to each fushi sign. However, the fushi part is not just one of these denotators but it has a recursive compound structure since fushis may be enriched by auxiliary sequences (strings or lists) of fushis for pitch and other specifications. For any form F we may consider the circular list form where the entries are of form F:

$$List_F \underset{Id}{\longrightarrow} \mathbf{Limit}(F, ListEntry_F) \tag{18.27}$$

$$ListEntry_F \underset{Id}{\longrightarrow} \mathbf{Colimit}(List_F, Count) \tag{18.28}$$

$$Count \underset{Id}{\longrightarrow} \mathbf{Simple}(\mathbb{Z}) \tag{18.29}$$

Let us then codify the fushi sign strings in form

$$FushiSTRG \underset{Id}{\longrightarrow} \mathbf{Syn}(List_{FushiPic})$$

and then construct the fushi concept as a circular form as follows:

$$FSH \underset{Id}{\longrightarrow} Limit(FushiSTRG, FushiOrnament) \tag{18.30}$$

$$FushiOrnament \underset{Fin}{\longrightarrow} \mathbf{Power}(FSH) \tag{18.31}$$

with the usual finite set identifier. We therefore have a fushi structure which is essentially the same as the $MakroEvent$ structure discussed in 6.7. With this, we may finally complete the fushi structure associated with a syllable:

$$Fushi \underset{Id}{\longrightarrow} \mathbf{Limit}(Basic, Special, PitchChange) \tag{18.32}$$

$$Basic \underset{Id}{\longrightarrow} \mathbf{Syn}(FSH) \tag{18.33}$$

$$Special \underset{Id}{\longrightarrow} \mathbf{Syn}(FSH) \tag{18.34}$$

$$PitchChange \underset{Id}{\longrightarrow} \mathbf{Syn}(FSH) \tag{18.35}$$

where the three synonymous forms refer to the basic fushi (usually named *sugu, hiki, mawashi*), the special fushi (two types, *hashiri*, *yari*, the latter being a class of recursively ornamented trill fushi), and thirdly: pitch-change fushis. Our construction terminates with the form of combination

$$JCKFU \underset{Id}{\longrightarrow} \mathbf{Limit}(JCK, Fushi)$$

which yields the units of the list form $List_{JCKFU}$ of the full fushized Utai text, and we may pose

$$Utai \underset{Id}{\longrightarrow} \mathbf{Syn}(List_{JCKFU}) \tag{18.36}$$

whose denotators are the melodically articulated Utai texts. It is clear how to complete this construction in order to add all the philologically necessary information from the Noh tradition.
In this context, we may define what are the existing melodic Utai songs by *MelodicUtaiSequence*, a PV predicate that takes its values in some fuzzy (uncertainty of identification) $\mathbb{T}^{0z}_{S_1}$ to be ϕ^for $0 \leqslant \phi \leqslant 1$ in the same sense as with the above example on absolute tempo.

18.3.3.3 Shifter Predicates

Summary. Organizing the non-lexical resources: the relevance of shifter predicates in production and reception of music and its science.

– Σ –

Until now, atomic predicates were drawn from lexical data, be they of notational primavista nature related to scores or of mathematical nature. There is another major source for the creation of predicates: deictic criteria. For such a predicate the extension may be different, depending on its user. Whereas in language, lexical signs dominate and determine a fixed ontology, the basic means of semiosis in music (and to some degree also in musicology) are shifters: being and becoming a musical object may be a result of individual decisions. Once a predicate has been introduced, it may be used just like the others; however, its mode of coming into existence will remain important. In accordance with communicative coordinates, shifters are differentiated relative to poietic and esthesic perspectives. The neutral level is omitted here since it excludes the deictic dimension by definition.

Poietic shifters typically come up when the composer decides to specify a determined set of denotator instances which will become the objects of the composition. For example, the choice of a composition's motivic germ $MyMotif : A \rightsquigarrow NoteGroup(N_1, \ldots N_k)$ by a composer does not favor just any possible predicate loaded with the property of being a motif, i.e., verifying a predicate $MyMotif/Motif$; on the contrary, it is this particular motif that was chosen objectively by the composer and it is this choice that gives it a special position among the denotators of form $NoteGroup$ and among the predicates encompassed within the predicate expressed by $Motif$. The expression corresponding to this very individuum could be denoted by $ThisMotivicGerm$, an expression that is evaluated to $\top$ if and only if we deal with this composer's concrete and unique choice of the germ. The predicate $MyMotif/ThisMotivicGerm$ has exactly one affirmative answer and no possibility to be generated within a genuinely lexical environment.

The *esthesic counterpart* of creational shifters is the set of extensions considered by an analytical interest. Often, there is no lexical reason to consider a particular composition or a special chord, there is just scientific curiosity, or interest. In this context, a composition which we may identify by a simple character string denotator S for the composition's name, is marked by !, i.e., we write $S!$ for the fact that there is a special interest in S.

Let us terminate this subject by two remarks and a principle statement:

- Comparison of different truth value domains is eased by the concept of a morphism of textual semioses which is induced by morphisms between corresponding truth value modules. So at least in principle we have a canonical technique for comparing truths as they are modeled in specific and adequate contexts.
- Truth in the topos-theoretic sense is not meant to model some formal languages but to add contents to the poor truth value of absolute logic in order to target towards a possible fusion of logic and beauty—after the marriage of logic and geometry initiated by topos theory.

Principle 12 *After all, it would be wonderful if one day, we could say that something is true* because *it is beautiful. The feeling of such an insight has been around for a long time—let us finally face its substance.*

18.3.4 Logical and Geometric Motivation

Summary. Analysis of the motivation mechanism for textual predicates. Besides the classical logical "recombination" of predicates, music preconizes more geometric methods in order to produce new predicates from given ones. Motivation is the technical counterpart of productive navigation in the EncycloSpace.

– Σ –

We announced that predicates are mainly generated by certain constructions from arbitrary basics such as mathematical, PV, and shifter predicates. Recall that semiotically, constructing the signification of signs by a determined mechanism from given significations is called motivation (see 2.3.4). So we now have to deal with motivation from the three named (arbitrary) basics.

One must, however, point out a major difference between predicative existence and mathematical fiction in the motivation process. The construction (motivation) of new denotators from given ones is a purely mathematical routine process: it simply has to fit in the generic mathematical framework of soundness. But with predicates, generating new denotators which "are the case" from already instantiated facticity is not only an abstract affair: these objects have to be the case for us, they have to be available, on a storage media, in a determined reservoir of shared information, in other words, predicates are the elements of knowledge in the sense of ordered access to information. So accessibility of the predicates' extensions is a serious requirement. Let us look at this condition in the different arbitrary basics.

Start with the arbitrary PV predicates. They mean music(ologic)al facticity. They have extensions somewhere, and this somewhere can be accessed without restrictions, it is not only as if it existed, it is within our reach. So the existing sonatas of Beethoven, the existing Noh Utai, the existing jazz compositions of Duke Ellington are a knowledge base which is strictly antagonistic to possible, but never conceived music works.

Principle 13 *If we are going to extend this level of facticity, we should meet the requirement that the knowledge ontology is conserved and not evaporated into fiction of whatever nature.*

The mathematical predicates are a bit more delicate. We contended above that these are introduced to indicate that there is a mathematical predication which is music(ologic)ally relevant. Nonetheless, such a predicate may have an infinite extent. So it is a fact of music but not necessarily a finite one. The point is that this differentiation from purely mathematical existence has to be combined with the requirement of principle 13. More precisely:

Principle 14 *If a mathematical predicate is conceived, it must be set up to cope with the accessibility requirement of knowledge. This accessibility is guaranteed by an explicit declaration of its characteristic function, and not necessarily by the predicate's extension which could be infinite or simply too large for any human storage device. If such mathematical predicates are used to generate extensions, they must meet the specific storage conditions of the medium, be it a technological or a human one (memory).*

In the case of deictic predicates, the facticity is more a question of declaration discipline: We do not want to deal with uncertainty in the choice of predicate extension instances. Rather would it be important to say that you have to attribute to a determined denotator a fuzzy truth value instead of saying that you do not know whether you want to consider that denotator or not. So this means

Principle 15 *Deictic predicates must be understood as the duty to declare facticity versus the freedom of leaving your decision unsettled. The latter is not part of the predicate system but of the psychology of the decision process in its making. This is not a proper part of the predicate theory which is a precise theory of fuzziness in music and not a properly fuzzy theory.*

Summarizing, we want to construct new predicates from already given ones in the sense of a consistent extension of knowledge ontology:

Principle 16 *The explicit and unrestricted access to information must be declared on the germinal level of arbitrary predicates and conserved in each motivation process.*

After these preliminary reflections on knowledge propagation, we may proceed to the description of motivation mechanisms. Basically, two levels of construction can be distinguished: the logical and the geometric motivation.

Logical motivation means that new predicates are constructed from given ones by use of the *codomains* of characteristic function. In fact, the topos-theoretic truth domains $\mathbb{T}_I^A$ are provided with logical functionality of the associated Heyting algebras, such as conjunction, disjunction, implication, and negation. This type of logical operation extends to universal and existence quantifiers.

Geometric motivation acts on the *domain* or characteristic functions. This means that the topos-theoretic constructs such as fiber products, fiber coproducts, power objects and other universal objects are used to build new denotator objects in Den_∞ on which truth values are applied.

Of course, nothing can prevent us from applying these methods in a not so topos-theoretic manner or to introduce still other motivation methods. This is no problem, the scope of these delimitations is rather to put into evidence the richness of the purely topos-theoretic vein of motivation.

Before giving further comments on these motivation methods, we should present some easy typical examples.

Logics: conjunction, disjunction, implication, and negation. This type is completely straightforward. We suppose given two predicates (named) P, Q which have their truth values in the same domain of truth denotators $\mathbb{T}_I^A$. Then the logical combinations $P \wedge Q$ evaluates as follows. Let $P(f) = p : A{\rightsquigarrow}TRUTH(I)(\pi)$ and $Q(f) = q : A{\rightsquigarrow}TRUTH(I)(\kappa)$ for an object f of Den_∞. Then we set

$$P \wedge Q(f) = p \wedge q : A{\rightsquigarrow}TRUTH(I)(\pi \wedge \kappa)$$

where $\pi \wedge \kappa$ is calculated in the Heyting algebra of the functor $A@Fun(TRUTH(I))$. The analogous construction works for the other logical operations. (If we deal with different truth denotator domains, we have to be provided with a common domain, for example by use of colimits such as fiber sums.)
Despite the evident procedure there is a delicate point in these construction. If we look for the extensions of predicates and suppose that the extensions of the given predicates are traced on some human or man-made data storage or memory devices, it is not clear whether the same may also be the case for the resulting logical combinations. For disjunction and conjunction this works since the extensions are—roughly speaking—unions or intersections of the given extensions. For negation, however, the resulting predicate may obtain an infinite, not controllable extension. In other words, negation of facticity is not necessarily the same type of facticity. In fact, what can be conserved in facticity is not the extension but the operation which takes a decision on every proposed tuple of denotator morphisms.

Logics: Existence and universal quantifiers. In the existence and universal quantifier situation, typically involving a predicate $P(x, y)$ with two free variables x, y, these two logical operators, $\exists_x P(x, y)$, and $\forall_x P(x, y)$, invoke a variable, x, say, which must be run through and checked for truth of the given predicate. In our situation, this check will not be a dichotomic one between true and false, but an 'integration' over all values of $P(x, y)$ as x varies. The statement $\exists_x P(x, y)$ then reads as the maximal possible truth value of all $P(x,y)$ as x varies, so if there is at least one truth value $\top$, then we assert that $\exists_x P(x,y)$ has this value. Analogously, the statement $\forall_x P(x,y)$ reads as the minimal possible value among all $P(x,y)$ as x varies.

To put it more formally, suppose that a predicate P of a textual semiotic evaluates on every n-tuple $f = (f_1, \dots f_k)$ of denotators to a truth denotator $P(f) \in \mathbb{T}_I^A$. Fix an index $1 \leqslant i$ of tuples. Then the quantifiers are defined with respect to the denotator variable at this index. To this end, if $X \subset \mathbb{T}_I^A$, denote by $lub(X)$ its least upper bound whereas $glb(X)$ denotes the greatest lower bound of X in the complete Heyting algebra $\mathbb{T}_I^A$. Further, if $d \in Den, 1 \leqslant i \leqslant k+1$, and if $f = (f_1, \dots f_k) \in Den^k$ is a k-tuple, write $f_{>i}d = (f_1, \dots f_{i-1}, d, f_i, \dots f_k)$. Then the existence quantifier is the predicate $\exists_i P$ which at the k-tuple f evaluates to

$$\exists_i P(f) = \begin{cases} P(f) & \text{if } k+1 < i, \\ lub(\{P(f_{>i}d) | d \in Den\}) & \text{if } 1 \leqslant i \leqslant k+1, \end{cases} \tag{18.37}$$

whereas the universal quantifier at the k-tuple f evaluates to

$$\forall_i P(f) = \begin{cases} P(f) & \text{if } k+1 < i, \\ glb(\{P(f_{>i}d) | d \in Den\}) & \text{if } 1 \leqslant i \leqslant k+1. \end{cases} \tag{18.38}$$

Observe that the calculation of the lub and the glb needs only the set $supp(P)$. So if this one is so small that a computer or human memory can control it, the quantifier predicates can be calculated at a comparably small expense. If this support is not finite, or if it is too large to be stuck on a memory device, the range of the bound ith variable d must be restricted to a controllable predicate or else the predicate P must be joined to a predicate Q such that $supp(P \wedge Q)$ becomes controllable by the available memory. In this situation, we write $\exists_{i,Q}P, \forall_{i,Q}P$.

We cannot solve the details of this problem of facticity here, but we feel obliged to insist on the relevance of its investigation because abstract mathematical reality is not sufficient for predicates to be handled by human technology and culture. So one should in any case point out the delicate turning points.

Geometry: fiber products. Geometric motivation implies predicate construction methods which are built on the domain Den_∞ of characteristic functions. We want to illustrate this for fiber products on denotators, a universal construction which—modulo some circularity traps—is known to exist for all five typed denotator subcategories (see Section 18.3.1).

We take two mathematical predicates having their support on ${}_0Den$ (see Section 18.3.3.1):

$$\text{Chords } Ch/PianoChord, \tag{18.39}$$

$$\text{Melodic motifs } Mt/Motif, \tag{18.40}$$

and a PV predicate $OnBeetSon$ which takes value $\top \in \mathbb{T}_0^0$ exactly for each zero-addressed local composition C consisting exactly of all onsets of the onset-bearing events of one of the 32 Beethoven sonatas (and $\bot$ else), so we are looking for the predicative support objects

$$C/OnBeetSon. \tag{18.41}$$

To relate these three predicates, we consider a mathematical predicate $OnsetProj$ supported on $\mathbb{T}_0^0$ with $\top$ by morphisms of local compositions $f : X \to Y$ such that X has form $NoteGroup$, Y has form $Onset$, and f is the projection morphism retaining the note events' onsets.

To relate the different truth denotator domains, we have to reduce the motif domain $\mathbb{T}_I^A$ to $\mathbb{T}_0^0$, but this is straightforward by the unique projection $A \to 0$, what yields $\mathbb{T}_I^A \to \mathbb{T}_0^0$, and this is what we carry over from the truth values for motifs.

Finally, the geometrically motivated predicate $BeetMotChordFiber$ has its $\top$ support on octuples $f = (X_1, X_2, X_3, X_4, f_1, f_2, f_3, f_4)$ where the first four objects are veritable objects in Den (identity morphisms), and the second four are morphisms of *Den*. The value $\top$ in $\mathbb{T}_0^0$ is taken on all octuples of *cartesian* diagrams (in *ObLoc*)

$$\begin{array}{ccc} X_1 & \xrightarrow{f_1} & X_2 \\ {\scriptstyle f_2}\downarrow & & \downarrow{\scriptstyle f_3} \\ X_3 & \xrightarrow{f_4} & X_4 \end{array} \tag{18.42}$$

such that

$$X_2/Motif, X_3/PianoChord, X_4/OnBeetSon, \tag{18.43}$$
$$f_3/OnsetProj, f_4/OnsetProj \tag{18.44}$$

are all true.
This construction can be decomposed as follows: The predicate *BeetMotChordFiber* on octuples is built as a logical conjunction of several projection predicates. A projection predicate is a geometric motivation as follows: Given an increasing sequence of positive natural numbers $j. = j_1, \ldots j_s$, denote by $\pi_{j.}$ the natural transformation $Den_\infty \to Den_\infty$ which on a k-tuple $f = (f_1, \ldots f_k)$ evaluates to

$$\pi_{j.}(f) = (f_{j_1}, f_{j_2}, f_{j_s}). \tag{18.45}$$

Then the projection jth predicate $P.\pi_{j.}$ deduced from a given predicate P is defined by

$$P.\pi_{j.}(f) = P(\pi_{j.}(f)).$$

This is an easy geometric motivation built upon this projection. We may then consider the mathematical predicate *Cart* of cartesian diagrams which gives $\top$ on octuples $f = (X_1, X_2, X_3, X_4, f_1, f_2, f_3, f_4)$ if they yield a cartesian square 18.42. So our above predicate reads as a logical motivation from projection motivation via

$$\begin{aligned} BeetMotChordFiber(f) &= Cart \wedge Motif.\pi_2 \wedge PianoChord.\pi_3 \\ &\wedge OnBeetSon.\pi_4 \wedge OnsetProj.\pi_7 \\ &\wedge OnsetProj.\pi_8(f) \end{aligned}$$

which means that we have built this predicate by a (logical) conjunctive motivation and a (geometric) projection motivation from mathematical and PV predicates. Strictly speaking, this second construction of *BeetMotChordFiber* is a synonymous predicate: It should have another expression, but its content is the same as the first one's content; in fact, synonymy of predicates is a common feature as in every semiotic system.
Now, the predicate *BeetMotChordFiber* is not quite what we have in mind: We would like to speak about the fiber product as a local composition, and not about the whole bunch of auxiliary structures. So the final construction would be this:

$$\begin{aligned} BeetMotChordFiberObject(X) = \top &\text{ iff there is an octuple } f \\ &\text{with } \pi_1(f) = X \\ &\text{and } BeetMotChordFiber(f) = \top. \end{aligned}$$

Exercise 40 *Try to unify the above existence condition with the logical motivation via the existence quantifier as two special cases of a more general logical existence construction. (Hint: use more general projections.)*

So we finally get a predicate which selects local compositions (in fact chords) consisting of piano notes which share the onset of a note of one motif and of a chord within the onsets of one of Beethoven's sonatas.—But we should, once again, be aware that the verification of the cartesian predicate is a mathematical operation of complex nature. It involves mathematical existence and universal quantifiers for the verification of a universal property. The operationalization of such predicates on the basis of computer implementation can be quite tedious since it always must be solved on a constructivist basis.

18.4 Paratextuality

Summary. Beyond textual semiosis, a type of predicate semiosis can occur which is not limited to extensions over denotators.

– Σ –

Besides textual semiosis, expressions of predicates can also (or exclusively) invoke historical and stylistic competence, for example, and thus point at exterior strata of signification. However, this does not enforce physical reality; it can still live on the mental level of signification. Classification of paratextual signification includes reproductive and productive behavior. Reproductive paratextual predicates include performance, lyrics, expressive, choreographic, and musicological aspects. Productive paratextual predicates can be of lexical nature, such as ornaments or improvisational patterns—or else of shifting character, such as creational and unintentional predicates. For a more systematic treatment of this subject, we have to refer to [703, 2.6].

It is however important to give a short account of the germ of paratextuality as it is preconized by the very nature of truth denotators in the textual semiosis. In fact, we have already learned that the classical logic of extensionality is canonically embedded in the truth denotator domains $\mathbb{T}_I^A$ which means that extensionality and truth evaluation is only a very 'poor' signification codomain. It is natural to view this situation as a special case of a more general validation operator ω process where a validation just gives us back a denotator for every given tuple, i.e., we have a signification process

$$\omega : Den_\infty \to Den$$

which attributes some values to tuples, values which share a proper extratextual semantics, such as 'truth' or 'falsity'.

For example, such an operator is given for the evaluation of certain PV predicates, such as "Fermata": An operator ω would turn this abstract denotator $myFermata$ (which essentially has the coordinates onset and duration) into a tempo curve $\omega(myFermata)$ which may be described by a parameter set for a spline, or by a string which represents a mathematical expression for that tempo curve, etc. This operator process does however not yet produce physical contents, but it gives an explicit representation of such objects in physical terms. So the generalized textuality which evaluates to not necessarily truth-oriented denotators yields a connotation stratification for deeper semantics beyond text of whatever abstraction.

This successive absorption of semantics by denotators is not only a methodological mechanism, it is already a tendency in music history where contents of musical signs have always been progressively split into connotational substrata which in turn have been absorbed by conceptualization and music theory [257].

15.1 Paratextuality

19

Topoi of Music

Das ist wohl schon die Mathematik des "Neuen Zeitalters".
Alexander Grothendieck [397] on "Geometrie der Töne" [682]

Summary. This chapter is a conceptual synthesis of the previous achievements. We show that the overall structure of the category *Glob* of global compositions carries a Grothendieck pretopology via finite covering families. It is well known that such a pretopology generates a Grothendieck topology J and therefore a Lawvere–Tierny topology j on the presheaf topos $Glob^@$. We discuss the associated instances, such as the subobject classifier sheaf Ω, and the subtopos $\mathrm{Sh}(J, Glob)$ of sheaves.

– Σ –

In this chapter we tacitly make use of the topos theory of Grothendieck topoi, refer to Appendix G.4.

19.1 The Grothendieck Topology

Summary. This section introduces the Grothendieck topology on *Glob* via covering families. Musical motivation is given.

– Σ –

The finiteness of common local and global compositions is not favorable to intuition from standard topology, but we have known the application of traditional methods from combinatorial topology and corresponding Čech cohomology for classification purposes of musical compositions in chapter 15. However, this is not the full potential of topological methodology: This one has been developed with great success by Grothendieck in algebraic geometry, and applied—among others—by Pierre Deligne in his celebrated solution of the Weil conjectures, see also [444].

The idea is now fully absorbed by topos theory in the context of sheaf construction from presheaf categories. Here, the yoga of topology boils down to the gluing catechism which gives insights to compatibility of local data without having to deal with the original intuitive neighborhood connotations from Euclidean geometry. Therefore it is a natural idea to integrate the Grothendieck topos construction which generalizes the Čech canon for the analysis of musical composition categories, or even of denotator categories provided they admit fiber products, a mandatory condition for the definition of Grothendieck topologies via covering families (see Appendix G.4). To begin with, we have the following theorem:

Theorem 25 *The category Glob of global compositions is finitely complete.*

Proof. We have to show[1] that *Glob* has fiber products and a final object. The latter is already known from Section 8.3.3 in the local case, but for global compositions, the same object does the job. The former claim

[1] See Appendix G, Proposition 111.

G. Mazzola, *The Topos of Music I: Theory*, Computational Music Science,
https://doi.org/10.1007/978-3-319-64364-9_19

is known for local compositions, and the fiber product of the underlying functors is known to exist. So the proof reduces to a control of the atlas index maps and the gluing of local fiber products. The fiber product of the index sets and their maps defines a system of subfunctors of the fiber product of the functors which underlie the global compositions. The union of these subfunctors is therefore covered by charts which are isomorphic to fiber products of local compositions and which clearly glue together as desired. QED.

Exercise 41 *Show that the fiber product of interpretable compositions is interpretable.*

Despite this positive result, this is not the right perspective to introduce Grothendieck topologies. In fact, consider the fiber product of two charts U_1, U_2 in a global composition G^I. Since their index singleton injects into different indexes of the covering set, the fiber product $U_1 \times_G U_2$ is empty! The reason for this strange fact is that we had—and still have—serious musicological reasons for specifying the index map ι as well as the associated family α of address changes of a morphism $f \overset{\iota}{/} \alpha$. So this is a strong dichotomy between musicological setting of global compositions and the traditional setting with mathematical manifolds, where the atlas is only an existence and not a uniqueness requirement. If we stick to the musicological categories, such as *Loc*, *Glob*, we cannot apply those powerful tools of Grothendieck pretopologies as we would like to do: even the intersection of two charts would be intractable qua fiber product. So the approach for this conceptual environment must be slightly weakened: it is reasonable to forget about the index and address change maps and to retain only the transformation f between sets or functors, respectively. In other words, we consider the categories ${}_\mu Loc$, ${}_\mu Glob$, ${}_\mu ObLoc$, ${}_\mu ObGlob$ as follows (μ for "mathematical"). The objects are those of the old categories *Loc*, *Glob*, *ObLoc*, *ObGlob*. The morphisms are just the maps of the underlying functors (or sets in the objective case). So we have, by definition, surjective "forgetful" functors

$$\mu : \quad Glob \to {}_\mu Glob \tag{19.1}$$
$$\mu : \quad Loc \to {}_\mu Loc \tag{19.2}$$
$$\mu : ObGlob \to {}_\mu ObGlob \tag{19.3}$$
$$\mu : \quad ObLoc \to {}_\mu ObLoc \tag{19.4}$$

(and the other known subcategories) which are all defined by the action $\mu(f \overset{\iota}{/} \alpha : G^I \to H^J) = (G^I, H^J, f : G \to H)$ which we shall also write as $f : G^I \to H^J$, and which are all the identity on the objects. We shall denote the image of $Hom(G^I, H^J)$ by $G^I @ H^J$, the set of mathematical morphisms, as already introduced in definition 50 of Section 14.4, to make the reference to the manifold mathematics evident. So intuitively, a mathematical morphism is just a morphism of compositions where we forget about the underlying chart index and address change information—it is just known that they are there, as in manifold theory.

If we restrict to the subcategories Loc_A over a fixed address and without address changes, then evidently, we have an isomorphism $\mu : Loc_A \xrightarrow{\sim} {}_\mu Loc_A$, whereas the restriction to the global compositions $GlCo_A$ with fixed address A gives us as a fiber of a morphism f all index maps which give rise to f. More precisely:

Proposition 22 *Let $f \overset{\iota}{/} \alpha : G^I \to H^J$ be a morphism of global composition. Let $\kappa : I \to J$ any set map such that the H-chart $V_{\kappa(i)}$ contains $im(U_i)$, for every $i \in I$. Then $f \overset{\kappa}{/} \alpha$ is a morphism.*

Proof. If $im(U_i) \subset V_{\kappa(i)} \cap V_{\iota(i)}$, then we may apply the transition isomorphism between the intersection of the two charts $V_{\kappa(i)}, V_{\iota(i)}$ to induce a morphism of local compositions $U_i \to V_{\kappa(i)}$ with the same address change. QED.

Observe that the general transformation μ does not reflect isomorphisms. In fact, if $G^I \to G$ is a refinement map for a local composition G, as discussed in Example 21, Section 14.2, then the identity $Id_G : G^I \to G$ of the underlying functor G is an isomorphism in ${}_\mu Glob$, but generically the objects are evidently not isomorphic qua global compositions if $card(I) > 1$.

Before introducing a Grothendieck topology on ${}_\mu Glob$, we have to guarantee the existence of fiber products. Here is the theorem:

Theorem 26 *The category $_\mu$Glob of global compositions and mathematical morphisms is finitely complete.*

Proof. Evidently, the final object of *Glob* does also the job for $_\mu$*Glob*. So we are left with the proof that $_\mu$*Glob* has fiber products. We are given a pair of mathematical morphisms $f : G^I \to H^J, g : L^K \to H^J$. As these are natural transformations we may consider the fiber product $G \times_H L$ of functors with its canonical projections to G, L. This functor can be covered by the non-empty fiber products $P_{i,k} = G_i \times_H L_k$ of the chart subfunctors G_i, L_k in $I \times K$. Let $(G_i, L_k) \in I \times K$ be such that $G_i \times_H L_k$ is not empty. Take J-charts $H_{\iota(i)}, H_{\kappa(j)}$ such that f on G_i factorizes through $H_{\iota(i)}$, and such that g on L_k factorizes through $H_{\kappa(k)}$. Then evidently, the image of $P_{i,k}$ in H is contained in $H_{i,k} = H_{\iota(i)} \cap H_{\kappa(j)}$, and we have $G_i \times_H L_k \xrightarrow{\sim} G'_i \times_{H_{i,k}} L'_k$, where G'_i, L'_k are the inverse images of the intersection $H_{i,k}$ in G_i, L_k, respectively. By the existence of fiber products of local compositions, these fiber products are local compositions, and we obtain an evident atlas together with two mathematical morphisms of the fiber product functor (which is a global composition) into the factors. The details are left as an exercise for the reader. QED.

To make the ideas precise, the present approach to Grothendieck topology will be focused[2] on the category $_\mu$*Glob*. We consider the following system of covering families: For a global composition[3] G, the covering basis $K(G)$ consists of all surjective finite families $(f_i : G_i \to G)_i$, i.e., their functorial images generate the functor G. Clearly, isomorphisms are such families, fiber products of such families are such families, and the composition of such families are again finite and surjective. Therefore we have a basis (originally called a pretopology) for a Grothendieck topology. Call the Grothendieck topology J of these covering families the *finite cover topology* on $_\mu$*Glob*.

Example 35 The interpretation G^I of a local composition G by a finite covering I of subcompositions is in $K(G)$.

Example 36 The family of local subcompositions of a defining finite atlas of global composition is a covering family.

Example 37 More generally, if we add all finite intersection compositions as considered in the theory of affine functions on the composition's nerve, we obtain another covering family.

So musicologically, the finite cover topology is the "least common denominator" of all our previous local-global constructs. In other words, all hitherto known concepts and techniques for gluing together finite assemblies of local data in music and music theory are special cases of the overall approach by the finite cover topology. This also applies to the Čech cohomology of the nerve which we have shortly discussed in Section 16.1.2.

Remark 4 The finite cover topology is not subcanonical (this would mean that every representable presheaf $@X$ over $_\mu$*Glob* is a sheaf). The point is that the exactness condition for sheaves only yields a natural transformation from a family of compatible mathematical morphisms, but not (in general) a mathematical morphism. Possibly, a narrower definition of covering families would yield subcanonical topologies.

Example 38 Suppose we are given a predicate P with values in $\mathbb{T}_I^A$ and support on the local compositions, to fix the ideas. Consider this data as a function t on *Loc*. We would like to be able to extend this function to the global compositions. To this end, suppose that this predicate is invariant under isomorphisms of local compositions. As the truth denotator domain is a complete Heyting algebra, we may attribute to a global composition G with charts G_i the truth values $t(G) = sup_i(t(G_i))$ or else $t(G) = inf_i(t(G_i))$ which are well defined by our invariance assumption and do extend the values on local compositions.

[2] Generalizations to other denotator categories are possible but we do not have concrete concepts and leave this subject to future developments.

[3] Pay attention to the different morphism set of mathematical morphisms whereas the objects are unchanged, even if we often omit the covering atlas to ease notation!

19.1.1 Cohomology

Summary. We sketch the topos-theoretic Čech cohomology associated with the finite cover topology J on ${}_{\mu}Glob$.

– Σ –

Following the ideas of Verdier's exposition [42, exposé V], every covering family $f. = (f_i : G_i \to G)_{i \in I}$ induces the following simplicial diagram. Let $n(f.)$ be the simplicial complex of all subsets $S \subset I$ with non-empty fiber products $\prod_G^{i \in S} G_i$. For a natural number $0 \leqslant k$, denote by $S(f.)_k$ the set of singular k-simplexes, i.e., maps $s : \Delta_k \to I$ on the abstract standard k-simplex $\Delta_k = \{0, 1, 2, \ldots k\}$ such that $im(s) \in n(f.)$. For such a singular k-simplex s, we write $\prod_s(f.) = \prod_G^{i=0,\ldots k} G_{s(i)}$ and $\Sigma_k(f.) = \coprod_{s \in S(f.)_k} \prod_s(f.)$.

Every set map $q : \Delta_k \to \Delta_l$ induces a universal map

$$\Sigma(q) : \Sigma_l(f.) \to \Sigma_k(f.) \tag{19.5}$$

which on a cofactor $\prod_s(f.)$ of $\Sigma_l(f.)$ at singular l-simplex s is the arrow

$$\Sigma(q) : \prod_s(f.) \xrightarrow{x} \prod_{s.q}(f.) \longrightarrow \Sigma_k(f.)$$

where x is the universal arrow defined by the respective projections. In particular, we have the arrows $\sigma_i = \Sigma(q_i)$ with $q_i : \Delta_k \to \Delta_{k+1}$ the omission of the ith entry for $0 \leqslant i \leqslant k+1$, i.e., $q_i(j) = j$ if $j < i$, and $q_i(j) = j + 1$ else. Finally, we also consider the augmentation map $\sigma_0 : \Sigma_0(f.) \to G$ given by the coproduct of the original family $f.$ of morphisms into G.

Summarizing, we get the simplicial diagram $\Sigma(f.) =$

$$\ldots \overset{\sigma_0,\ \sigma_1,\ \sigma_2,\ \sigma_3}{\rightrightarrows\rightrightarrows} \Sigma_2(f.) \overset{\sigma_0,\ \sigma_1,\ \sigma_2}{\rightrightarrows\rightarrow} \Sigma_1(f.) \overset{\sigma_0,\ \sigma_1}{\rightrightarrows} \Sigma_0(f.) \xrightarrow{\sigma_0} G$$

of global compositions which is the basis of the Čech cohomology if we succeed in applying contravariant functors with values in abelian groups to this diagram.

Suppose that we are given such a functor F, defined in a category which comprises the objects and arrows of the simplicial diagram $\Sigma(f.)$. Then we have the cochain complex $C^*(f., F) =$

$$\ldots \xleftarrow{d_2} F(\Sigma_2(f.)) \xleftarrow{d_1} F(\Sigma_1(f.)) \xleftarrow{d_0} F(\Sigma_0(f.)) \xleftarrow{d_{-1}} 0 \tag{19.6}$$

with the alternate sums

$$d_k = \Sigma_{i=0,\ldots k+1}(-1)^i F(\sigma_i)$$

of abelian group homomorphisms for indexes $0 \leqslant k$ and $d_{-1} = 0$. This cochain complex has its usual cohomology groups $H^k(f., F) = Ker(d_k)/Im(d_{k-1})$ for $0 \leqslant k$.

Example 39 *Affine functions.* In Section 15.1.1, we have dealt with the category $ComGlob_A$ of commutative global compositions where the address A is a module over a commutative ring R. We now want to look at affine functions on functorial global compositions. More precisely, we work in the following category ${}^{R}Glob$: instead of $\mathbf{Mod}^{@}$, we work in the category $\mathbf{Mod}^{@}_{R}$ of presheaves over R-modules over the commutative ring R. And ${}^{R}Glob$ is the category of global compositions over this context. Clearly, ${}^{R}Glob$ is finitely complete, and so are ${}^{R}Glob_A$, ${}^{R}_{\mu}Glob$, and ${}^{R}_{\mu}Glob_A$.

To define affine functions on a global composition G^I in ${}^{R}_{\mu}Glob_A$, consider the objective composition $\hat{A} = \widehat{A@_R R}$. We claim that the set $Hom_A(G^I, \hat{A})$ of morphisms which leave the address fixed, is canonically provided with an R-module structure. For any commutative local composition $K \subset A@_R M$ with ambient module M, which is an R-module, i.e., $K = R.K$, each f-slice $f@_R \hat{K}$ for $f : B \to A$ is an R-module when identified with its second factor $K.f$. Now, given two morphisms $u/Id_A, v/Id_A : G^I \to \hat{K}$, we can add them or apply a scalar multiplication when restricted to each f-slice:$f@_R G^I \to f@_R\hat{K}$. This defines an R-module structure on $Hom_A(G^I, \hat{K})$. In particular, for $M = R$ and $K = A@_R R$, we get the required module structure. We call the module $Hom_A(G^I, \hat{A})$ the *module of global affine functions on* G^I and denote it by $\Gamma(G^I)$.

We therefore have a representable contravariant functor (15.6)

$$\Gamma : {}^R Glob_A \to {}_R\mathbf{Mod}$$

which associates the R-module $\Gamma(G^I)$ of global affine functions with a given global composition G^I. It is easily seen that this functor is well defined on the mathematical category ${}^R_\mu Glob_A$ and gives rise to a synonymous functor

$$\Gamma : {}^R_\mu Glob_A \to {}_R\mathbf{Mod}$$

entailing cohomology R-modules $H^i(f., \Gamma)$ for any covering family $f.$ in ${}_\mu ObLoc_A$.

- For the default covering family $\iota.$ of charts which define G^I, this gives back the cohomology modules $H^i(\iota., \Gamma) = H^i(\Gamma)$ discussed in 16.1.2.
- Consider next the (mathematical) resolution morphism (15.16)

$$res_{G^I} : \Delta_{G^I} \to G^I$$

and its restriction

$$res_i : {}_A\Delta_i \rightarrowtail \Delta_{G^I} \to G^I \tag{19.7}$$

to the resolution's standard composition chart ${}_A\Delta_i$ for chart $i \in I$ of G^I. This family $res.$ is covering for G^I and we therefore obtain the *resolution cohomology modules* $H^i_\Delta(G^I) = H^i(res., \Gamma)$ for the affine function functor.

Clearly, we have an epimorphism of simplicial diagrams

$$\Sigma_{res} : \Sigma(res.) \twoheadrightarrow \Sigma(\iota.) \tag{19.8}$$

which is induced by the canonical morphisms

$${}_A\Delta_{i_1} \times_G \ldots \times_G {}_A\Delta_{i_k} \twoheadrightarrow U_{i_1} \cap \ldots \cap U_{i_k}$$

for the G^I-charts U_{i_j}. Therefore, the corresponding function complexes are related by an exact sequence

$$0 \longrightarrow C^\star(\iota., \Gamma) \longrightarrow C^\star(res., \Gamma) \longrightarrow C^\star(res., \Gamma)/C^\star(\iota., \Gamma) \longrightarrow 0 \tag{19.9}$$

of differential R-module complex homomorphisms. So by the standard procedures for exact sequences of differential complexes, we have a long exact cohomology sequence

$$\ldots \longrightarrow H^i(G^I) \longrightarrow H^i_\Delta(G^I) \longrightarrow H^i(\Delta/G^I) \xrightarrow{\delta} H^{i+1}(G^I) \longrightarrow \ldots$$

with the cohomology modules $H^i(\Delta/G^I)$ defined by the quotient differential complex $C^\star(res., \Gamma)/C^\star(\iota., \Gamma)$.

Example 40 *Linear forms.* For this example, we work over a fixed R-module A as address, R being a commutative ring. Recall from the discussion of module complexes in Section 15.1 that for a commutative (objective) global composition G^I in $ComGlob_A$ and an R-module V (in particular the already known module $A@_R R$ which yields the classical forms), we have the module complex ${}_V G^I = Hom_R(G^I, V)$ of V-forms on G^I. We want to make this complex into a contravariant functor ${}_\mu Glob_A \to \mathbf{Mod}_R$. On the category $ComLoc_A$, it is the well-known dual functor to the associated module functor $R : ComLoc_A \to \mathbf{Mod}_R$. It sends a local composition $K \subset A@M$ to the module ${}_V K = Hom_R(R.K, V)$. For each global composition G^I in $ComGlob_A$, select an atlas $(K_i \subset A@M_i)_{i \in I}$ and restrictions $(K_\sigma \subset A@M_{i(\sigma)})_{\sigma \in n(G)}$ where $i(\sigma)$ is a selected vertex index of the simplex σ. When taking the limit $\Gamma({}_V G^I) = lim_{\sigma \in n(G^I)}\, {}_V K_\sigma$ for global sections, this is clearly already the limit on the double index restrictions. So a global section $x \in \Gamma({}_VG^I)$ is just a family $(x_i)_I$, $x_i : R.K_i \to V$ such that the restrictions $x_i|R.K_j$ and $x_j|R.K_i$ coincide.

If we are given a morphism $f/Id_A : G^I \overset{\iota}{\to} H^J$, and if we select an atlas $(L_j \subset A@N_j)_{j\in J}$, the global sections are mapped into each other by the universal property of limits via

$$\Gamma({}_VH^J) \to {}_VL_{\iota(i)} \to {}_VK_i$$

for every $i \in I$. Moreover, this linear map is invariant under a change of the index map ι. In fact, if we have a second index map $\kappa : I \to J$, the map

$${}_V\Gamma(H^J) \to {}_VL_{\kappa(i)} \to {}_VK_i$$

factorizes through the intersection of charts:

$${}_V\Gamma(H^J) \to {}_VL_{\iota(i),\kappa(i)} \to {}_VK_i$$

and we are done. Moreover,the functor ${}_V\Gamma(G^I)$ is also natural in the module V, i.e., any module homomorphism $u : V \to W$ entails the evident natural homomorphism ${}_u\Gamma : {}_V\Gamma(G^I) \to {}_W\Gamma(G^I)$. Therefore

Proposition 23 *Given an R-module V, we have a contravariant functor of V-forms*

$${}_V\Gamma : {}_\mu Glob_A \to \mathbf{Mod}_R \tag{19.10}$$

which associates the R-module ${}_V\Gamma(G^I) = \Gamma({}_VG^I)$ of global sections of V-valued linear forms to each global composition G^I in the mathematical category over the fixed address A. The functor ${}_V\Gamma$ is natural in V.

In particular, for the integers as ground ring, and for any covering family $f.$ of a global composition G^I which is addressed in a $\mathbb{Z}$-module A, if we look at the projective $\mathbb{Z}$-module system $\mathbb{Z}/\mathbb{Z}_{l^r}$ and its projective limit, the l-adic integers[4] $\mathbb{Z}_l$, we first have cohomology groups $H^\star(f., {}_{\mathbb{Z}/\mathbb{Z}_{l^r}}\Gamma)$ and then their limit, the *l-adic form cohomology*

$$H_l^\star(G^I) = lim_r H^\star(f., {}_{\mathbb{Z}/\mathbb{Z}_{l^r}}\Gamma) \tag{19.11}$$

on G^I.

19.1.2 Marginalia on Presheaves

Summary. We briefly discuss function presheaves, and the subobject classifier.

– Σ –

This section is a sketch of what should be investigated more carefully and systematically.

19.1.2.1 Function Presheaves

Summary. Function sheaves are a basic instance for global music theory, we review the corresponding discussion of the resolution covering.

– Σ –

The resolution covering is a remarkable example for the finite cover topology since it is not just a covering by subcompositions, it is somewhat analogous to étale topology, the set-theoretic map is even bijective, only the relative position of points has changed. We feel that this presheaf type is crucial for understanding music. Let us make this more explicit—even beyond the resolution covering.

[4] With the usual unusual confusing index notation...

We recall that the resolution was a procedure to lift non-interpretable global compositions to interpretable ones built on point configurations in general position. Now, this approach may be interpreted from still another angle. Whereas classification uses resolutions to calculate the 'minimal perspective of a global composition' which tells you everything of its structural contents, it could also happen that a composer does not want to classify a composition but to view it as a germ of a real composition, a concentrated abstract version of something which will be used only on a different level. So the germinal composition could be a global composition G^I which will only be used via some covering morphism family $f_i : G_i^{I_i} \to G^I$ which the composer uses to grasp particular local aspects of the germ. From this construction, she could also deduce function modules to realize her music in determined parameter spaces. A study of such poietic aspects and applications of general covering families of the finite cover topology is still outstanding, but surely will furnish powerful compositional tools.

19.1.2.2 The Subobject Classifier Ω

Summary. The subobject classifier is a key object in a topos. We have a closer look at its structure and function in the topoi ${}_{\mu}Glob^{@}$ and $\mathrm{Sh}(J, {}_{\mu}Glob)$.

– Σ –

Recall the general relation between presheaves and sheaves. Given the finite cover topology J on ${}_{\mu}Glob^{@}$, we have the subcategory $\mathrm{Sh}(J, {}_{\mu}Glob)$ of J-sheaves and its embedding $i : \mathrm{Sh}(J, {}_{\mu}Glob) \rightarrowtail {}_{\mu}Glob^{@}$. Whereas the subobject classifier Ω of ${}_{\mu}Glob^{@}$ gives the set

$$\Omega(G^I) = \{ \text{ sieves } S \subset @G^I\}$$

of G^I-sieves, the subobject classifier Ω_{Sh} of $\mathrm{Sh}(J, {}_{\mu}Glob)$ evaluates to the subset

$$\Omega_{Sh}(G^I) = \{\text{closed sieves } S \subset @G^I\}$$

of closed G^I-sieves, i.e., the sieves S such that any (mathematical) morphisms $f : H^T \to G^I$ covering[5] S are contained in S. In terms of covering families this means that whenever there is a covering family $f_i : H_i^{T_i} \to H^T$ such that every composite morphism $f.f_i \in S$, then $f \in S$. From a predicative point of view, a closed sieve $S \subset @G^I$ corresponds to a predicate P_S whose support is S, i.e., $P_S(f) = \top$ iff $f \in S$, and $P_S(f) = \perp$ else. The predicate P_S is characterized by

- $P_S(f)$ is invariant among mathematically equivalent arrows,
- invariance under right multiplication, i.e., if $P_S(f) = \top$, then $P_S(f.g) = P_S(f)$ whenever the product makes sense;
- $P_S(f) = \top$ if there exists a covering family $f.$ such that $P_S(f.f_i) = \top$.

There are different approaches to view the sheaves within the presheaves. First, we have the adjointness property: There is a left adjoint $a : {}_{\mu}Glob^{@} \to \mathrm{Sh}(J, {}_{\mu}Glob)$ for the embedding i, called the associated sheaf functor, see Appendix G.4.1: For any presheaf X and sheaf Y, we have a functorial isomorphism $Hom_{{}_{\mu}Glob^{@}}(X, i(Y)) \xrightarrow{\sim} Hom_{\mathrm{Sh}(J, {}_{\mu}Glob)}(a(X), Y)$. It resides on the sheafification operator $P \mapsto P^+$, see Appendix G.4.1 which yields the associated sheaf $\mathbf{a}P = (P^+)^+$. To this property corresponds the construction of the Lawvere–Tierny topology $j : \Omega \to \Omega$ which associates with every sieve $S \subset @G^I$ its closure $\overline{S}$ consisting of all morphisms which cover S. In terms of predicates, this amounts to adding to a predicate all morphisms which fulfill the third property above.

19.2 The Topos of Music: An Overview

Summary. This section summarizes the overall categorization of music with regard to the topos-oriented structures worked out from denotators to Grothendieck topologies.

Fig. 19.1. Alexander Grothendieck. Drawing by Mitty Sheen © 2017.

– Σ –

The overall theory of forms and denotators is a concept framework based on the topos of presheaves over the category **Mod** of modules and of selected subcategories. Special types of denotators: the local compositions (power type) are naturally provided with a local-global paradigm. In fact, they are essentially subobjects of special functors, together with induced morphisms, quite similarly to elementary approaches to affine algebraic varieties and to local differential geometry. This part of the denotator theory—which stresses rather the conceptual construction than the geometric manifold approach—is embedded in the global theory which can be seen as a passage to inductive limits, i.e., 'patchworks' of local objects. This latter construction is essentially different from the algebraic and differential geometry in that the atlas covering is not variable but an intrinsic data from the musicological point of view. The inductive limit over variable atlases is not the point for musicology, it would be a too coarse approach.

However, the patchwork of local structures can be boiled down to a more mathematical manifold theory: the categories of global compositions, together with mathematical morphisms. In this framework, it is possible to introduce natural Grothendieck topology, for example the finite cover topology considered above. This one is also the Grothendieck topology which englobes all the local global approaches which we have performed in interpretation and classification theory.

Undoubtedly, larger denotator subcategories should be accessible by either globalization processes or—at least—by Grothendieck topologies which extend the one sketched above. This is a priori feasible since we have seen that fiber products are also given for a large class of denotators as soon as they are not pathologically circular.

The cohomology theory for Grothendieck topoi, as sketched by Verdier in the legendary SGA seminar, is naturally given via function presheaves, for example. We do not know whether the predicate sheaves can also reveal cohomological spin-offs since it is not evident how truth denotators should be given structures of abelian groups. They rather live in Heyting algebras, and one should first adapt these structures to become candidates for cohomology theory. However, if truth denotators are just singletons in truth modules, they evidently add up to groups. But this perspective is neither within the reach of our experience with concrete examples nor do we have theoretical results which would enforce such a development.

Summarizing, the topos of music first centers around the concept architecture of music objects in the general denotator theory, and therein around the presheaf topos over modules, and secondly, it evolves to

[5] i.e., such that f^*S covers $J(H^T)$.

a universe of local-global perspectives which are readily described by Alexander Grothendieck (see Figure 19.1) with his awe inspiring topologies and their functors and cohomology theories.

The following development of this book will now descend to more concrete investigations, regarding those fields which are more in the tradition of the composers, music theorists, and performers. This does not mean that we can forget about the general framework, on the contrary, it will often not be possible to understand the technical and conceptual procedures without knowing about the topoi which back the more down-to-earth scenery.

20

Visualization Principles

For they shall see eye to eye, when the Lord shall bring again Zion.
ISAIAH 8

Summary. As a compensation to the abstract nature of general topos theory, some principles for visualizing such abstract objects are mandatory, in particular in view of implementations on computerized knowledge bases. We give an account of such principles as they are being applied in graphical interface design.

– Σ –

20.1 Problems

Visual navigation on general databases is a difficult task for three reasons:

- The data structure is not a priori in a geometric shape.
- The geometric shape, if it occurs, is not a priori adapted to human 3D vision.
- Objects may be composed of other objects which in turn are composed, and so on in a recursive way. Visualization then should take care of a recursive architecture.

The first obstacle may happen if we are given a bunch of textual objects. How should we arrange them as if they were points in a geometric space? Or else, if we are given a collection of chords in a score, is it possible to deal with such objects as if they were nicely distributed as points or spheres or whatever geometric objects in an adequate geometric space?

The second obstacle may occur when we have a high-dimensional geometry, for example six-dimensional representation of tones (onset, pitch, loudness, duration, glissando, crescendo). How could we grasp all these dimensions without forgetting some of them by plain projection to 2D or 3D partial representations, such as 2D for onset and pitch?

The third obstacle is a fundamental design problem. If we are able to visualize 2D situations, for onset and pitch, say, this only works if the visualization of these coordinates is already settled. But in general, the coordinate references are not elementary decimal or integer numbers, so how is it possible to set up a recursively stable visualization strategy?

Apart from this agenda, visualization has to make knowledge accessible, i.e., visualization must provide a distinguished knowledge navigation tool. We have already discussed knowledge navigation in Chapter 5. From that discussion, one could draw the following condensed definition of knowledge:

Definition 63 *Knowledge is ordered access to information.*

Accordingly, navigation is intimately related to knowledge. In fact, knowledge involves the access activity. And it obeys ordering principles, this access is not a random walk as it is the case in present internet surfing environments. So the instantiation of knowledge involves navigation on ordered spaces, not just blind

G. Mazzola, *The Topos of Music I: Theory*, Computational Music Science,
https://doi.org/10.1007/978-3-319-64364-9_20

date behavior. For example, in a traditional encyclopedia, the retrieval of a keyword starts with a linear movement along the alphabetic order axis, but then, when one delves into the text, presents a completely chaotic navigation environment: Only some rare linkage arrows tell us where to navigate, the rest must be constructed by semantic lecture, interpretation and then rebuilding of new keywords to deepen the searched concept's understanding. The textual abstraction is produced by an extreme knowledge hiding, whose disclosure requires considerable efforts in supplementary knowledge navigation. Let us put this insight as follows:

Principle 17 *Visual navigation must be built on orderings*[1].

We may now recall the treatment of denotator orderings in Section 6.1.3 and—more technically—in Section 6.8. We have shown there that for quite general addresses, there is a generic recursion process to define linear orderings on denotator systems of any form space. Let us rephrase those results in an informal way:

Fact 9 *Any system of denotators can be linearly ordered according to a generic recursive procedure which is based on linear orderings of address modules. These module orderings can be constructed from canonical orderings of elementary modules, such as ground rings, and basic quotient modules, in a natural way.*

This fact lends itself to visualization: To begin with, in all relevant cases, we only deal with finite, and mostly comprehensible, collections of denotators, of size, say, below 10^6. This means that we can deal with a number of points on the real line. But this is not sufficient, since the abstract order relation is not ideal for representing the underlying 'distances'. If, for example, a number of pitches is represented by pure order relations, this reduces to an equidistant arrangement of points of the real axis and destroys the pitch distances. Of course, it is not wrong to recur to this equidistributed arrangement which we may call the *generic linear visualization*. One must be able to invoke this one if no further information is required. But we have to ask for this:

Principle 18 *A generic construction method for natural distances among successive items in the generic visualization must be developed to generate intuitive visual distance perception.*

Call the visual representation on the basis of principle 18 the *metrical linear visualization*. The ongoing implementation of denotator visualization[2] has already realized this principle on the modules which occur in 'real life'.

The next problem to be settled is to deal with high-dimensionality, more precisely: more than three dimensions in product type denotators. Now, the principle of metrical linear visualization basically solves this problem, but it is too flat, we are able to visualize in 3D and should make use of this human feature. So we have to respect the following *packing principle*:

Principle 19 *In a product type space, the linear visualization should be distributed among a number of packages of coordinate spaces such that at most three spatial visualization axes are needed.*

So, for example, we should pack three real dimensions in the linear visualization of a first axis, another four dimensions in the linear visualization of a second axis. The remaining dimensions should be packed into some different visualization level. Even if we apply the linear visualization principle to these latter dimensions, we cannot, by construction, add more spatial visualization axes. We therefore propose the following *object visualization principle*:

Principle 20 *In the product (more generally speaking: limit) type, the non-spatial—possibly linearly packed groups of—linear dimensions must be attributed to object parameters of multimedia objects.*

[1] Observe that visualization means using the ordered field $\mathbb{R}$, and this means writing—typically—decimal numbers, which in turn amounts to building on the ordering of the natural numbers $0 < 1 < 2 < \ldots 9$.

[2] PhD Work of Stefan Göller at Zürich University.

Such objects can be specified by geometric shapes, such as spheres, cubes, faces, by surface texture, by sonic parameters such as sound, earcons[3], and by gestural parameters, such as handling behavior, like drag and drop response, of more or less physical nature (elasticity, inertia, gravity, magnetism).

Summarizing, we obtain a distribution of high-dimensional denotator visualization among space and object visualization. With this in mind, how does navigation look? You may specify the space/object distribution and associated linear (metrical) visualization according to the previous principles, including selection of adequate multimedia objects. You may move around within such visual spaces, zoom in and out, and interact with the multimedia objects, like listening to tones, feel the response of a piano key, etc.

But other navigation methods are at our reach, let us just name some evident variants:

1. Pick a specific selection of factors or cofactors in products or coproducts; this is the classical "coordinate projection" or "book selection" method.
2. Greeking[4] of deeper recursive information. This means that in the recursive denotator space ramification, one decides to collapse all values and to forget about existing differences. This is very important for first approaches to complex databases.
3. Differentiating between more or less generic visualizations. In a coproduct, one may concentrate on the 3D visualization of one cofactor, leaving the adjacent and more distant cofactors in a decreasingly detailed state of visualization.

20.2 Folding Dimensions

According to the *packing principle* (Principle 19) we need a method to fold dimensions. Taking principles 17 and 18 into account, this method needs to fulfill the following properties:

- order preserving
- injective

Fair enough, it is not *homeomorphic*, as it can't be *injective* **and** *homeomorphic*. Obviously there are many ways to define such a mapping; we choose the following generally valid algorithm. Denote by $p_i : \mathbb{R}^n \to \mathbb{R}$ the ith projection, and by $p_{i\ldots i+k} : \mathbb{R}^n \to \mathbb{R}^{k+1}$ the projection onto the coordinates i through $i+k$.

20.2.1 $\mathbb{R}^2 \to \mathbb{R}$

We are given a finite set $\mathcal{P}$ of points in $\mathbb{R}^2$.

1. Partition $\mathcal{P} = \mathcal{P}_0 \cup \mathcal{P}_1 \cup \ldots \cup \mathcal{P}_{n-1}$ with $\forall i \in \{0, 1, \ldots n-1\}\ \forall x \in \mathcal{P}_i : p_1(x) = t_i = const.$, and $t_i \neq t_j$ if $i \neq j$.
2. Subdivide the x-axis by giving every $\mathcal{P}_i$ a lower and upper bound u_i and o_i:

$$u_i = \begin{cases} 2t_i - o_i & : (i = 0) \wedge (n > 1), \\ \frac{t_i + t_{i-1}}{2} & : (i \geqslant 1) \wedge (n > 1), \\ t_i - \frac{1}{2} & : else; \end{cases}$$

$$o_i = \begin{cases} 2t_i - u_i & : (i = n-1) \wedge (n > 1), \\ \frac{t_{i+1} + t_i}{2} & : (i < n-1) \wedge (n > 1), \\ t_i + \frac{1}{2} & : else, \end{cases}$$

[3] Earcons are icons for the ears.

[4] An operation known from text applications, meaning that a structure is blurred, recalling the English idiom "that sounds Greek to me".

3. Map every $\mathcal{P}_i$ into $]u_i, o_i[$ using a bijective strictly increasing map μ_i on $\mathbb{R}$ (such as of arctan shape, see Figure 20.1).
4. The maps $\mu_i : \mathcal{P}_i \to]u_i, o_i[$ define the map $\mu : \mathcal{P} \to]u_0, o_{n-1}[$.

Fact 10 *Note that the first step in the algorithm needs the set $\mathcal{P}$ to be finite in the first coordinate; therefore μ is* not total *on $\mathbb{R}^2$. This is not a real limitation, as we won't try to visualize infinite sets.*

Lemma 31 *If we impose the lexicographic ordering on $\mathbb{R}^2$, μ is order-preserving and therefore injective.*

Proof. There are two possible cases: If $p_1(x_1) < p_1(x_2)$, then $\mu(x_1) \in]u_i, o_i[$ and $\mu(x_2) \in]u_j, o_j[$ with $i < j$. But then, by construction of the intervals $]u_i, o_i[,]u_j, o_j[$, $]u_i, o_i[<]u_j, o_j[$, hence $\mu(x_1) < \mu(x_2)$. Else, if $p_1(x_1) = p_1(x_2)$ and $p_2(x_1) < p_2(x_2)$, then on these points, $\mu = \mu_i$ for an index i, and $\mu_i(x_1) < \mu_i(x_2)$ since μ_i is strictly increasing, QED.

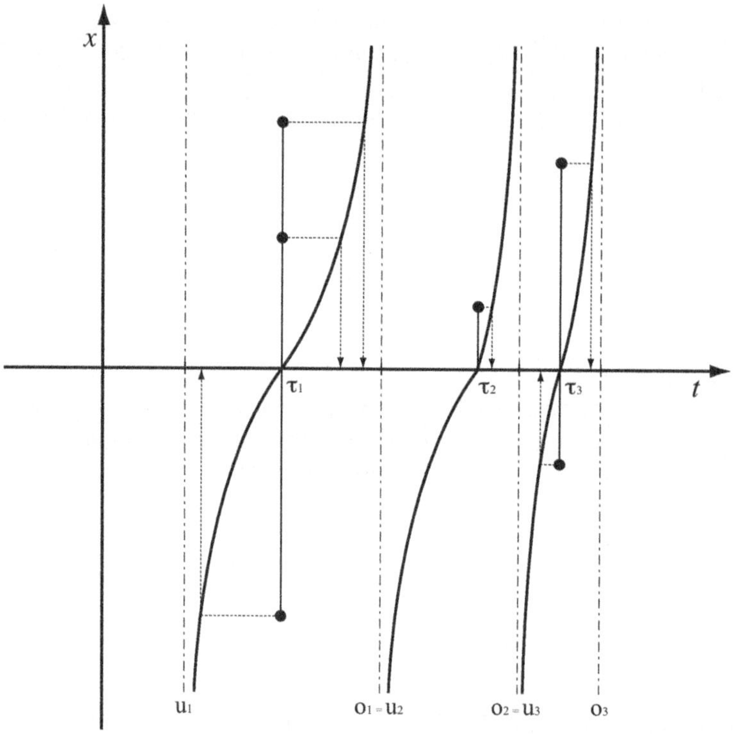

Fig. 20.1. The two-dimensional folding algorithm μ.

20.2.2 $\mathbb{R}^n \to \mathbb{R}$

With $\mu : \mathbb{R}^2 \to \mathbb{R}$ we can define $\hat{\mu} : \mathbb{R}^n \to \mathbb{R}$ inductively:

$$\hat{\mu}(x) = \begin{cases} \mu(x) & : n = 2, \\ \mu(p_1(x)\ ,\ \hat{\mu}(p_{2\ldots n}(x))) & : n > 2. \end{cases}$$

Lemma 32 *$\hat{\mu}$ is order-preserving and hence injective.*

Proof. The map $x \mapsto (p_1(x), \hat{\mu}(p_{2\ldots n}(x))$ is order-preserving by induction, and μ is so by Lemma 31, QED. Note that $p_1(\mathcal{P})$ becomes the coarse factor, whereas $p_{2\ldots n}(\mathcal{P}))$ the fine one.

20.2.3 An Explicit Construction of μ with Special Values.

What remains undefined so far is the *bijective proper increasing mapping* from step 3 in 20.2.1. We define this map with two special real values *low* and *high* in mind, which will be explained in the following Section 20.3. We define the following (partial) maps τ and ϕ on $\mathbb{R}^5, \mathbb{R}^{10}$, respectively, see Figure 20.2.

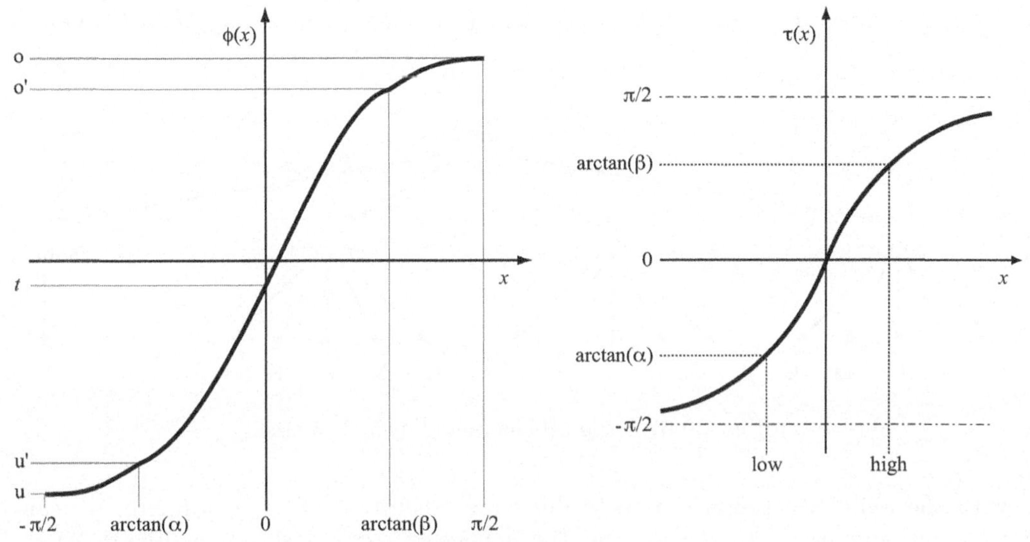

Fig. 20.2. The maps τ and ϕ.

$$\tau(\alpha, \beta, h, l, x) = \tan^{-1}\left(\frac{\beta - \alpha}{h - l}(x - l) + \alpha\right),$$

$$\phi(o, o', u, u', t, \alpha, \beta, h, l, x) = \begin{cases} \frac{u'-u}{\frac{\pi}{2}+\tan^{-1}(\alpha)} x + \frac{\pi}{2} \quad +u & : \quad -\frac{\pi}{2} < x \leqslant \tan^{-1}(\alpha) \\ \frac{t-u'}{-\tan^{-1}(\alpha)} x - \tan^{-1}(\alpha) + u' & : \tan^{-1}(\alpha) < x \leqslant 0 \\ \frac{o'-t}{\tan^{-1}(\beta)} x \quad +t & : \quad 0 < x \leqslant \tan^{-1}(\beta) \\ \frac{o-o'}{\frac{\pi}{2}-\tan^{-1}(\beta)} x - \tan^{-1}(\beta) + o' & : \tan^{-1}(\beta) < x \leqslant \frac{\pi}{2} \end{cases}$$

The composed map on $x \in \mathbb{R}^2$

$$\begin{aligned} &\phi_\tau(o_i, o_i', u_i, u_i', \alpha, \beta, high, low)(x) = \\ &\phi(o_i, o_i', u_i, u_i', p_1(x), \alpha, \beta, high, low, \tau(\alpha, \beta, high, low, p_2(x))) \end{aligned}$$

maps

$$\begin{aligned} p_2(x) &= low \mapsto u_i' \\ p_2(x) &= high \mapsto o_i' \\ p_2(x) &= -\infty \mapsto u_i \\ p_2(x) &= +\infty \mapsto o_i \end{aligned}$$

which is useful for our visualization. Note that in the variable $p_2(x)$, ϕ_T is *continuous* at $\tan^{-1}(\alpha)$, 0, $\tan^{-1}(\beta)$ but not *differentiable*.

20.3 Folding Denotators

With $\hat{\mu}$ defined, we can visualize finite sets in real vector spaces of any dimension in 3D. But what we really want, is to visualize *denotators*. For the time being, we exclusively address *zero-addressed* denotators. Recall from Chapter 6 that a denotator can be either simple—with its coordinate in a module—or compound, such as a limit, a colimit or a powerset of 'lower-level' denotators, to speak in the naive denotator terminology. A typical example is shown in Figure 20.3, where A to L are again denotators. If we provide every simple

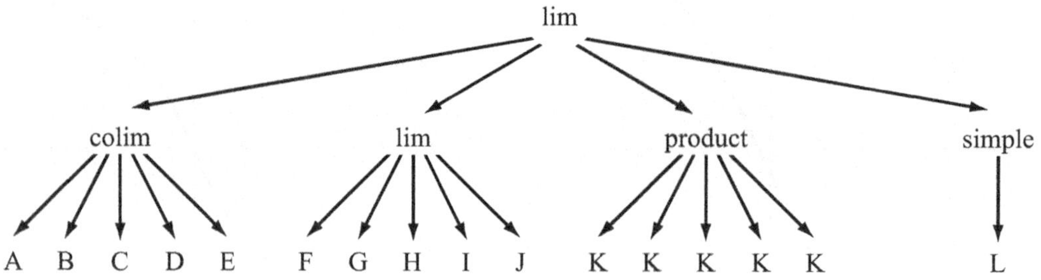

Fig. 20.3. A typical compound denotator tree.

denotator with the ability to produce a real-valued representation of its coordinate, we can map every denotator to $\mathbb{R}$ by applying $\hat{\mu}$ all the way up the denotator tree, as shown in Figure 20.4. The simple

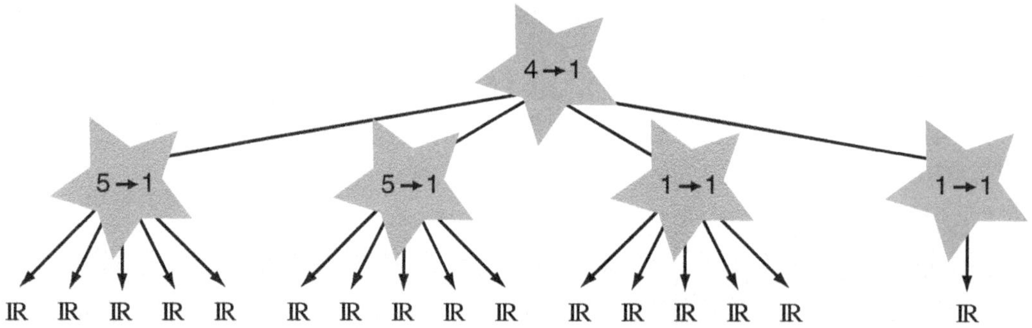

Fig. 20.4. Mapping every denotator to $\mathbb{R}$ by applying $\hat{\mu}$ all the way up the denotator tree

denotator gets this real value from the underlying module; the module also provides us with two more values *low* and *high*, representing *natural* borders such as lowest and highest note on the piano, which is useful for the subsequent visualization. But how can we apply $\hat{\mu}$ to a limit- or colimit-denotator?

20.3.1 Folding Limits

When we forget about the limit's diagram, a limit is nothing but a product. Assuming that we already know (by induction) how to fold its factors, we just apply $\hat{\mu}$ to the limit as if it were of type $\mathbb{R}^n$. Later in the visualization we can add the diagram's information by use of color, transparency, etc. By this we visualize a denotator as shown in Figure 20.5 as a set of points in the ordinary n-space as shown in Figure 20.6.

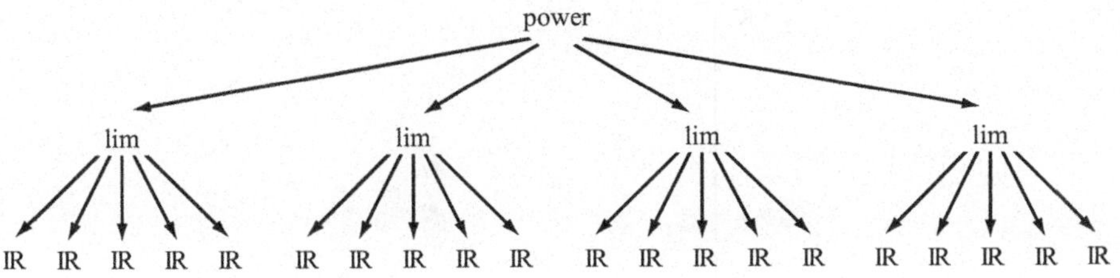

Fig. 20.5. Folding limit denotators.

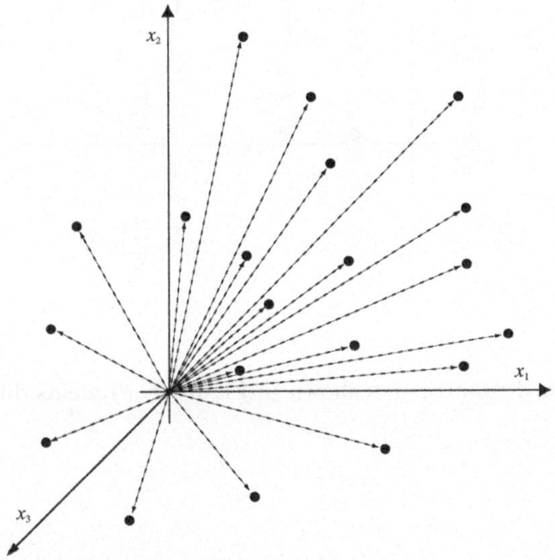

Fig. 20.6. Visualization of a denotator as shown in Figure 20.5 by a set of points in an ordinary n-space.

20.3.2 Folding Colimits

The folding of colimits is something different. While the semantic of a limit is some kind of a vector space with some additional constraints, the notion of a colimit is the one of a library. A point in a colimit space is like a book or a set of books from a library while the cofactors mean different types of books and the diagram controls which books you may take out together. Therefore we decided to fold a set of colimit denotators (like in Figure 20.8) in the following way:

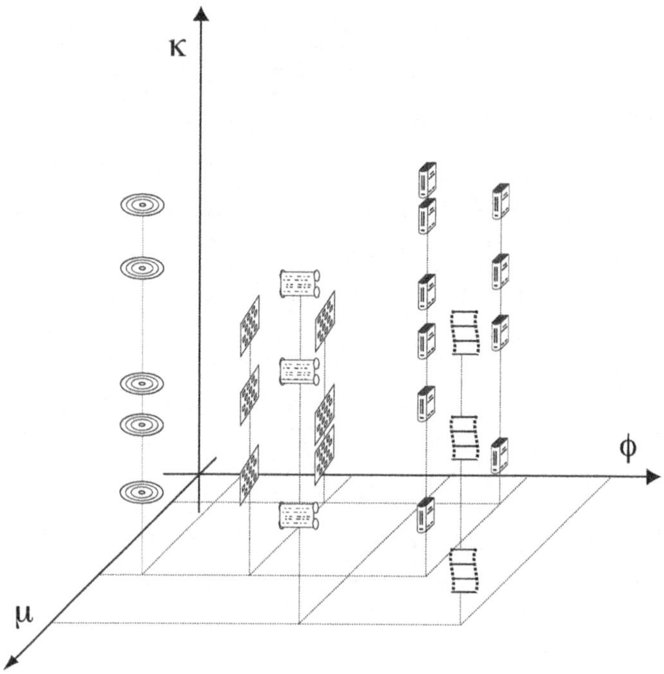

Fig. 20.7. Folding a set of colimit denotators (as shown in Figure 20.8) yields different kinds books or CDs, etc. in rows of shelves.

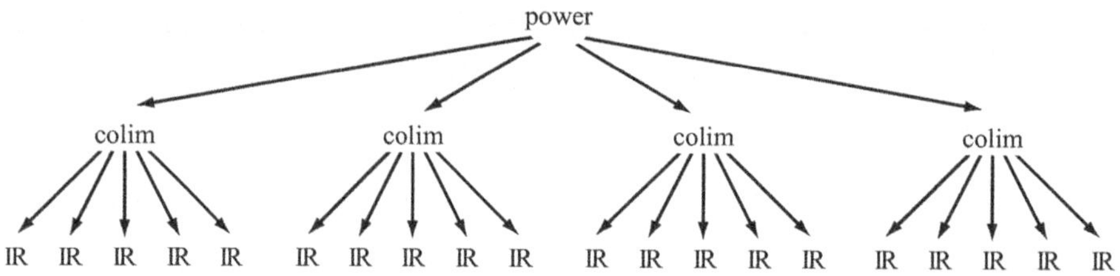

Fig. 20.8. A set of colimit denotators.

Firstly we choose which cofactors we bundle together; this information goes to the first axis. On the second we chose to map the index of the cofactor in question and the actual data is folded in the third axis. By this way we can get different kinds books or CDs, etc. in rows of shelves like in Figure 20.7.

20.3.3 Folding Powersets

The most natural way to deal with a powerset denotator is to view it as the collection or 'container' of its elements and proceed by folding one element after the other recursively with respect to their type. (Note that all elements of the powerset are of the same type.) We call this *powerset disclosure*. The preceding folding of limit and colimit denotator collections ('powerset of limit', 'powerset of colimit') were special cases of this general situation. However, the 'powerset of powerset' case is somewhat special: Here we represent the elements in a $2 \times n$ array, we use their position in their canonically ordered arrangement as first and a their *barycenter*-value as second coordinate. This arrangement is then folded with the $1 \times n$ folding matrix.

Logically there is also an *undisclosed* way of folding a powerset denotator. In this case the barycenters of the n elements are folded with the standard $1 \times n$ folding matrix. This *undisclosed* folding is the default case if a powerset denotator is itself a (co-)factor or element of another denotator.

20.3.4 Folding Circular Denotators

In Sections 6.5 and 6.7, the problems of circular denotators where already discussed. When actually working with those denotators, one has to take care not to run into infinite loops. Therefore we decided to run a three-level strategy: Down to level γ the usual folding takes place, between γ and τ all the real values from the foldings are generalized with the $index(x)$ function[5], and at level τ the folding process stops completely.

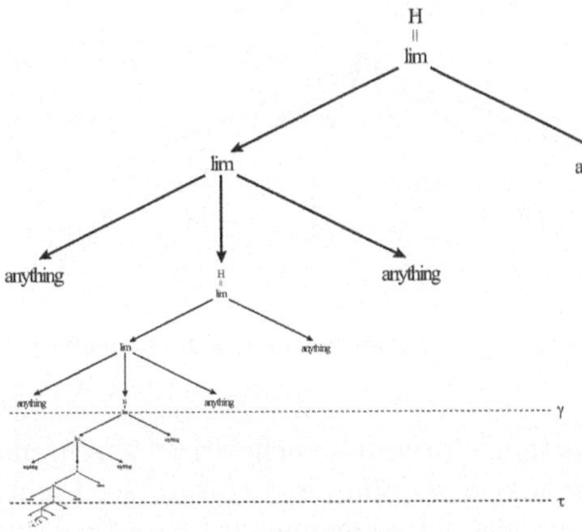

Fig. 20.9. Down to level γ the usual folding takes place, between γ and τ all the real values from the foldings are generalized with the $index(x)$ function, and at level τ the folding process stops completely.

[5] See Appendix D.5.1.

20.4 Compound Parametrized Objects

Even if we are now in the position to be able to transform any denotator into a 3-dimensional vector we often prefer not to fold all those n dimension down to 3 but to keep some more and map them on any graphical, acoustical, haptic, etc. property of our display object. Therefore we need lots of different objects—let's just call them *multimedia objects*—with many different parameters like color, transparency, pitch, stiffness, etc.

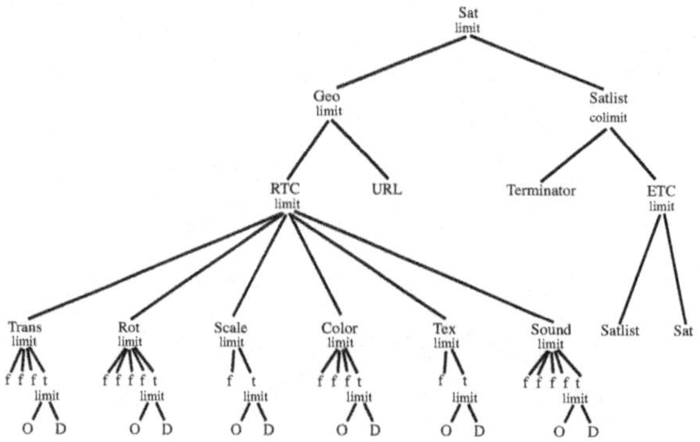

Fig. 20.10. Description of a compound multimedia object as a *circular denotator.*

As it is not recommended to actually program each object again and again, we decided to describe a compound multimedia object as a *circular denotator space* (form) (see Figure 20.10). The *Sat* form (*Satellite*) consists of a list of (sub-)satellites and the actual geometric, acoustical and time-critical information in the *RTC* denotator space, which actually is a limit space according to the above information types.

We can for example construct a Pinocchio puppet (see Figure 20.11) by defining a denotator for the body^0, one for the neck^1, the head^2, the hat^3, the arm^4, the foot^5, and the leg^6. Reusing these denotators we can define the whole Pinocchio. The f-denotator coordinates in the *RTC*-denotators provide the parameter functions to steer the concerning properties respectively. By this construction every part of the Pinocchio can be moved, rotated, scaled, etc. as shown in Figure 20.11. The *Sat* denotator space is recursively stable, which means that the parameters still work if the Pinocchio itself becomes a subpart of a more complex multimedia object.

Fig. 20.11. Construction of a Pinocchio puppet by defining a denotator for the body(^{0}), one for the neck(^{1}), the head(^{2}), the hat(^{3}), the arm(^{4}), the foot(^{5}), and the leg(^{6}). Reusing these denotators we can define the whole Pinocchio.

20.5 Examples

The following example shows a denotator like the one in Figure 20.5; the corresponding limit is 19-dimensional and was folded to seven dimensions to fit the Pinocchio object, which provides four object parameters additionally to the x-y-z-position.

Fig. 20.12. In this figure some objects are shown, with relevant coordinates mapped to the leg length and the body color.

Fig. 20.13. In this figure some 400 objects are visible with different quite random parameter mapping.

Fig. 20.14. This figure shows the same data but with a more sensible parameter mapping. One immediately sees more of the structure; colinear and complanar subsets arise.

Part V

Topologies for Rhythm and Motives

21

Metrics and Rhythmics

... wir operieren mit einem zweiten Grundbegriffe,
dessen Feststellung wir uns nun zuzuwenden haben,
demjenigen des to 1.2verschiedenen Gewichtes der Zeiten,
der ***metrischen Qualität.***
Hugo Riemann [904, p.8]

Summary. Metrics and rhythmics are an excellent elementary test-case for global structures in music. We shall critically review two commonly known approaches: the Riemann and the Jackendoff–Lerdahl theories. We then develop the concepts of global time structures and their topologies, including associated weight functions.

– Σ –

In this chapter, we refer to the introductory discussions of meter and rhythm in section 7.2.2 (local case) and section 13.4.3 (global case). We shall above all concentrate on the metrical aspect since it yields a simple and formally transparent situation from which the rhythmical complexification can easily be unfolded.

21.1 Review of Riemann and Jackendoff-Lerdahl Theories

Summary. We review these theories and trace back the concepts of metrical weight and metrical hierarchies.

– Σ –

The reader should however be aware that we cannot include a philologically valid analysis of these works here, our scope is a modest one: to present the tie between these historically important outlets and our own approach *as it really happened.* The interest of such a tie is to start reflecting about the incredibly fuzzy state of conceptualization in music which Riemann also had to suffer ([904, p.VIII]):

> *Meine musikalische "Dynamik und Agogik" (1884) arbeitet sich noch mühselig durch den Wust falscher Definitionen, die, wie ein undurchdringliches Gewirr der Wurzeln von Sumpfpflanzen unter dem anscheinend klaren mit Blumen geschmück-ten Wasserspiegel, sich jedem Vordringen in den Weg stellen.*

21.1.1 Riemann's Weights

Summary. Hugo Riemann's approach in [904] is discussed, stressing his conceptual specifications of meter and rhythm, in particular the weight idea.

– Σ –

G. Mazzola, *The Topos of Music I: Theory*, Computational Music Science,
https://doi.org/10.1007/978-3-319-64364-9_21

Riemann's concept of rhythm deals with a regular division of time ([904, p.2]):

> *Freilich ist aber doch wohl kein Zweifel, daß in der* erkennbaren Gleichmäßigkeit *der Zeitteilung wirklich ein besonderes lustgebendes Moment liegt und zwar eben gerade dasjenige, welches man Rhythmus nennt...*

The temporal period defining a rhythm is not an abstract one, it is related to the concrete music events of the given composition ([904, p.8]):

> *Welches in jedem Einzelfalle das Grundmaß [des Rhythmus, G.M.] ist, ergibt sich wie bereits betont, nicht aus dem abstrakten absoluten Mittelmaß, sondern aus dem der konkreten Melodie selbst.* Die Zählzeiten *(Schlagzeiten, rhythmischen Grundzeiten)* gewinnen unter allen Umständen erst reale Existenz durch ihre Inhalte.

For Riemann, rhythm is a result of grouping musical events to equitemporally distributed units, and to relate the grouped events to the units as relative instances (*Relativität der rhythmischen Qualität* [904, p.7]). This is why we have chosen the definition of a rhythm as described in section 13.4.3. This is also stressed in the explication of rhythmical unity ([904, p.9]):

> *... als Inhalt einer den Pulsschlag des Rhythmus bildenden Einheit ist prinzipiell eine Mehrheit von wahrnehmbaren Erscheinungen* (also für die Musik: Tongebungen) *anzunehmen.*

However, a detailed study of Riemann's theory is not our scope and should be taken up in the framework of mathematical music theory as it has been taken up in motif theory with Rudolph Reti's approach by Chantal Buteau [163].

In contrast to rhythmical grouping, Riemann views metrical quality as a differentiation among onsets (of notes) which comes from their roles and relations. This quality is stated as a weight ([904, p.8]):

> *... wir operieren mit einem zweiten Grundbegriffe, dessen Feststellung wir uns nun zuzuwenden haben, demjenigen des* verschiedenen Gewichtes der Zeiten, *der* **metrischen Qualität**.

The weight is meant to be a metaphor for the degree of importance:

> *... die Unterscheidung der einander folgenden Zeiteinheiten in wichtige und minder wichtige oder wie man zu sagen pflegt:* schwere *und* leichte ...

21.1.2 Jackendoff-Lerdahl: Intrinsic Versus Extrinsic Time Structures

Summary. The time domain is a prototype of a highly ambiguous dimension where intrinsic structures, such as notes, are mixed with extrinsic structures, such as bar-lines. The former are part of the musical material, the latter are part of the abstract time framework. We discuss the difference by the example of Ray Jackendoff's and Fred Lerdahl's approach in [496].

– Σ –

In this section we shall compare the metrical analysis of the Jackendoff-Lerdahl theory (GTTM, Generative Theory of Tonal Music) with what is proposed by the present mathematical music theory and its implementations in the RUBATO® platform's MetroRUBETTE® for metrical analysis. For a more generic discussion of the GTTM method, see [712].

In order to perform a metrical/rhythmical analysis or interpretation of a musical composition, the basis of such an investigation must be questioned. This one is linked to the "neutral" work data on one hand, and to the analytical approach on the other. Every investigation of the neutral data presupposes a choice of relevant aspects. This choice occurs very early in the propaedeutic evaluation.

For example, in the first Viennese school, bar-lines, period limits, notes, and pauses are recognized. In a metrical/rhythmical analysis, the question of the bar-lines' relevance would be set forth: Will this type of events strongly influence our analysis? As is shown in the MetroRUBETTE® concept, the answer is not necessarily "Yes" or "No", a fuzzy decision between these two extremal values is possible and reasonable.

In this situation, a normative analysis would imply that we decide to include bar-lines as external formal limits of metrical/rhythmical analysis. This bed of Procrustes is what GTTM prescribes in its metrical structure analysis: Bar-lines are always instances of the metrical structure, this is implicit in the fourth rule[1]. of metrical structure[2]. Here are the four MWFRs:

- *MWFR 1. Every attack point must be associated with a beat at the smallest metrical level present at that point in the piece.*
- *MWFR 2. Every beat at a given level must also be a beat at all smaller levels present at that point in the piece.*
- *MWFR 3. At each metrical level, strong beats are spaced either two or three beats apart. (This rule is recognized as idiom-specific in GTTM.)*
- *MWFR 4. The tactus and immediately larger metrical levels must consist of beats equally spaced through the piece. At subtactus levels, weak beats must be equally spaced between strong beats.*

Worse than that: even non-existent onsets are inserted in determined levels in order to enforce periodicity where the text does not show such. This is a consequence of (for example) the axiom MWFR3 which asks that successive metrical levels have duration (period) ratios 1:2 or 1:3. All other ratios are forbidden! The analysis of eight bars from Mozart's *Jupiter* Symphony are a typical example, see Figure 21.1.

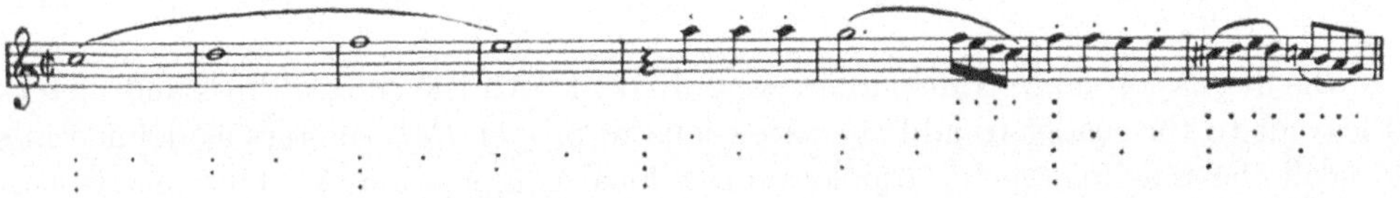

Fig. 21.1. The metrical hierarchy in the finale's eight bar theme of Mozart's *Jupiter* Symphony, according to the GTTM, [496, p.73]. Reprinted with permission of The MIT Press.

It shows that the whole notes in the first four bars have to be enriched by half-note points in order to cope with the succeeding onset structure. It also shows that the bar-lines are points even when there is no such note-onset. Such a procedure is not only rigid, it is also dangerous since the complexity of meter/rhythm is destroyed by secondary instances such as bar-lines, and by a spurious subdivision dogma (1:2, 1:3). And it forces a norm which only cements bad mass taste instead of representing possible compositional intensions.

But even in tonal music, also in modern tonal music, rhythm and metrics are much less tied to bar-lines than the formal score notation suggests. For instance, in contemporary jazz[3], we recognize extremely complex poly- and microrhythms which escape bar-lines and would be eliminated by the said normative standardization as being non-metrical or non-rhythmical music.

Moreover, the GTTM casting of metrical structure makes evident a remarkable drawback of non-computerized musicology[4]: The effective usage of the following metrical analysis method (21.2) would have been fictitious without computers because of its combinatorial complexity. We in fact argue that sometimes, a normative approach is only propagated because operational control of analysis would break down without such a "terrible simplification", so the simple and wrong is preferred to reality which would possibly force one to wait until the power of tools has reached an adequate level. During the design period of the workstation RUBATO® it was therefore mandatory to ask for the following condition:

[1] There are two types of GTTM rules: Well-formedness rules (WFR), and preference rules (PR). The rule system has four parts: G: grouping structure, M: metrical structure, TSR: time-span reduction, and PR: prolongational reduction. Accordingly, the two rule types are symbolized by the part prefix and the type postfix. For example, metrical well-formedness rules are MWFR. The complete index of GTTM rules is given in [496, p.345ff.]

[2] See also [496, p.71]: *Yet metrical intuitions about music clearly include at least one specially designated metrical level, which we are calling the* tactus.)

[3] Listen, e.g., to Steve Coleman's CD *The Sonic Language of Myth*, BMG(RVA Victor 1998).

[4] It is well known that the authors of GTTM do not preconize computer-aided music analysis.

Axiom 3 *We need concrete tools which are adequate to the analyzed work, and which yield an immanent analysis, not merely abstract or brutally normative principles.*

Despite our critical review of the GTTM method, the idea to cover the onsets of a given composition by "levels" has been a fruitful input for our global metrical perspective. It gave us the motivation to introduce global meters from the concept of a local meter, which were introduced in [682, pp.32-34], and from the insight that simultaneous presence of several local meters in a piece of music is the rule and not the exception (see the example [682, Bild 11, p.34]). The crucial difference between the global meter theory and the GTTM metrical structure theory is—besides our non-normative attitude—the construction of level functions from global data (in fact the nerve structure) and not their a priori definition, see the level function defined in section 13.4.3.

21.2 Topologies of Global Meters and Associated Weights

Summary. The hierarchic structure of global meters gives rise to topologies and associated weights, i.e., numerical functions which contain information about the local connection of onsets within their topological neighborhoods. Metrical topologies are the theoretical background for the MetroRUBETTE® module for metric and rhythmic analysis in the RUBATO® software discussed in section 41.1.

– Σ –

To make the discussion more transparent, we only deal with the zero address and meters in the *Onset* domain and leave it to the reader to add the extensions to proper *Para*-meters as defined in section 13.4.3. Let us first recall the maximal meter topology on a local composition X. This one is associated to the interpretation X^{MaxMet} whose charts are all finite intersections of the maximal local meters in X. We know that every closed irreducible set is the closure of any of its maximal points with respect to the dominance relation. So if x, y have the same closure $\overline{\{x\}} = \overline{\{y\}}$, then they are mutually dominating, or equivalently, their simplexes coincide, $Sp(x) = Sp(y)$, or equivalently, their minimal open neighborhoods $U(x), U(y)$ coincide (lemma 20 in section 13.4.3). Recall from appendix F.2.1 the left adjoint construction of the continuous map $i : X \to X^s$ of a topological space X into its associated sober space X^s which sends x to the point $\overline{\{x\}}$. This map is a quasi-homeomorphism (see lemma 96, appendix F.2.1), so the sober topology is coarser than the quotient topology. But i is also surjective, so any open set of the quotient topology lives in the sober topology, i.e., the sober topology and the quotient topology coincide. Let $Sp(X) = \{Sp(x) \in n(X) | \ x \in X\}$. From this information we deduce this result:

Lemma 33 *The sober quasi-homeomorphism $i : X \to X^s$ for the maximal meter topology is open, surjective, and (therefore) the topology of X^s is the quotient topology. We have a bijection*

$$is_X : X^s \xrightarrow{\sim} Sp(X) : i(x) \mapsto Sp(x)$$

of sets.

We may rephrase the sober topology definition to generate a topology on the nerve $n(X^{MaxMet})$ which extends the topology on $Sp(X)$ which is induced by the bijection is_X. By definition, open sets in X^s are given by

$$V^s = \{\overline{\{x\}} | \ x \in X, \text{ there is } y \in V, Sp(y) \subset Sp(x)\}$$

for any open sets V in X. If we use the transition bijection, we have the corresponding open sets

$$is_X(V^s) = \{Sp(x) | \ x \in X, \text{ there is } y \in V, Sp(y) \subset Sp(x)\}$$

and this can be rephrased to extend to the entire nerve by

$$V^{sp} = \{\sigma | \ \sigma \in n(X^{MaxMet}), \text{ there is } y \in V, Sp(y) \subset \sigma\}.$$

We have $V^{sp} \cap Sp(X) = is_X(V^s)$. Let $n(X^{MaxMet})^s$ be the *maximal meter nerve topology* which is generated by the sets V^{sp}. Observe that we have $\bigcup_i V_i^{sp} = (\bigcup_i V_i)^{sp}$ whereas the intersection $V_1^{sp} \cap V_2^{sp}$ need not be a generating open set. However, the restriction of the maximal meter nerve topology to the subspace $Sp(X)$ is the induced topology. So we have this:

Proposition 24 *With the maximal meter and the maximal meter nerve topologies as defined above, the map $Sp : X \to n(X^{MaxMet})^s$ becomes a continuous map whose image (as a subspace with the relative topology) identifies to the associated sober space X^s.*

This justifies the following definition:

Definition 64 *An $\mathbb{R}$-valued* sober weight *on the zero-addressed local composition $X \subset Onset$ is a function $W : X \to \mathbb{R}$ which factorizes through the nerve map Sp. A factor $nW : n(X^{MaxMet}) \to \mathbb{R}$ is called a* nerve weight *for W. A nerve weight nW is said to be* induced *iff its values are determined by the values on the vertexes, i.e., by the restriction $n_0W = nW|_{n_0(X^{MaxMet})}$.*

Example 41 If n_0W is any $\mathbb{R}$-valued function, we may take the sum $nW(\sigma) = \sum_{M \in \sigma} n_0W(M)$ of vertex values. A typical vertex function will be discussed for the MetroRUBETTE® (section 41.1): Fix a real exponent $prof$, weight's "profile", and two natural numbers $min \leqslant max$. The function associates with the maximal local meter $M \in n_0(X^{MaxMet})$ the number

$$n_0W(M) = \begin{cases} l(M)^{prof} \text{ if } min \leqslant l(M) \leqslant max, \\ 0 \text{ else.} \end{cases} \tag{21.1}$$

So musically, a sober weight is a valuation of onset events in X by their simplexes, i.e., their position within the covering by maximal local meters which are particular points in the topological space $(X^{MaxMet})^s$. No systematic treatment of the real vector space of sober weights or, more generally, the cohomology of such weights, has been undertaken to date.

Exercise 42 *In this context, one may consider the covering $MaxMet$ of X and then restrict everything to the integers as a ground ring. Given a prime number l, the functors of $\mathbb{Z}_{l^r}$-forms yields the l-adic cohomology $H_l^\star(X^{MaxMet})$ introduced in equation (19.11). We argue that this information should be important for the classification of global meters, and also, for defining refined sober weights. Elaborate the form functor on the maximal meters and its cohomology.*

The method of induced weights can be generalized to interpretations in the onset domain. Suppose that $X \subset Onset$ is interpreted to yield the global composition X^I. This may happen if we generate this interpretation from a knowledge about the origin of X. For example, it could happen that $X = pr_{Onset}(Y^J)$, the projection of a given global composition into the onset domain, and that we have the charts $Y_j \subset Y$ of a covering J. Then we obtain an interpretation X^I of X by the projections $pr_{Onset}(Y_j)$ of these charts. We shall learn from section 41.1 that this is a very common situation in music research, for instance when collecting different kinds of rhythmically significant objects, such as notes, bar-lines, pauses, etc.

We may then consider the maximal meter topology on each chart X_i of X^I. Suppose we are given a sober weight $W_i : X_i \to \mathbb{R}$ for each chart. Then we may construct an induced weight from the nerve $n(X^I)$. For $x \in X$, let $Sp_I(x)$ be the simplex of x with respect to the covering I. Then we define $W(x) = \sum_{X_i \in Sp_I(x)} W_i(x)$. This enables us to create mixed weights for points in an interpretation and thusly build a refined weighting of onsets with respect to different predicative specifications. In particular, we shall see in 41.1 that the Jackendoff-Lerdahl approach can be simulated (and this in a refined way) by reintroducing the bar-lines with a specific weight to add up to the total weight of an onset.

21.3 Macro-events in the Time Domain

Summary. This section reviews macro-events in the light of topologies of global meters and rhythms and the associated weights.

– Σ –

The previous discussion of temporal denotators was centered around topological relations, such as dominance, of onset events. It is evident that topological relations yield a tool for grouping onset events according to their metrical role. To this end, we review the definitions of local rhythms from section 13.4.3. In that context, a (A-addressed) rhythmic germ is a local composition $G \subset A@Rhythm(Para)$.

We shall deal with germ denotators G which have the shape

$$G : A \rightsquigarrow Para(g_1, \ldots g_m)$$

where

$$Para \underset{Fin(Rhythm(Para)) \rightarrowtail \Omega^F}{\longrightarrow} \mathbf{Power}(Rhythm(Para))$$

with $F = Fun(Rhythm(Para))$. So the form $Para$ is circular and reappears as a parameter space associated with the onset coordinator. We know from section 6.5 that $Para$ exists. Such a denotator is a $Para$-rhythmic germ as well as a macro denotator as defined in formula (6.116). The point is that the germ parameter space coincides with the germ space. Intuitively speaking, such a germ is a local composition of onsets which are parametrized by germs of onsets, etc., recursively, until—in a 'tame' case—the parametrizing germs reduce to the empty set.

This being true, we want to associate such a macro germ with each maximal meter topology on a (zero-addressed) local composition $X \subset Onset$. To this end, consider the Hasse diagram[5] $Dom(X)$ for the partial ordering of the dominance relation on X under the maximal order topology. This defines a macro germ as follows. We take the level function[6] $lev(x), x \in X$, and define by level recursion: $Dom(x) = (x, \emptyset)$ if $lev(x) = 0$, and

$$Dom(x) = (x, \{(Dom(y)_1 - x, Dom(y)_2) |\ y \in pre(x)\}) \tag{21.2}$$

in the other cases with the two components $Dom(y)_1, Dom(y)_2$ of the macro $Dom(y)$. Then we put $Dom(X) = \{Dom(x)|\ x \in Max(X)\}$.

Example 42 Let us make an example, see figure 21.2

Denoting simplexes by $\{I_1, I_5, I_6\} = \Delta(1,5,6), \{I_1, I_5\} = \Delta(1,5)$, etc., we have these Sp function values:

$$\begin{aligned}
Sp(x1) &= \Delta(1,2), Sp(x2) = \Delta(1,3),\\
Sp(x3) &= Sp(x4) = Sp(x6) = \Delta(1), Sp(x5) = \Delta(1,5,6),\\
Sp(x7) &= Sp(x9) = \Delta(3), Sp(x8) = \Delta(3,4),\\
Sp(x10) &= Sp(x11) = \Delta(5), Sp(x12) = Sp(x13) = \Delta(6),\\
Sp(x14) &= Sp(x15) = \Delta(2), Sp(x16) = Sp(x17) = \Delta(4).
\end{aligned}$$

So the points $x1, x2, x5, x8$ are generic, and we have

$$\begin{aligned}
Dom(x1) &= (x1, \{(x3 - x1, \emptyset), (x4 - x1, \emptyset), (x6 - x1, \emptyset), (x14 - x1, \emptyset),\\
&\quad (x15 - x1, \emptyset)\}),\\
Dom(x2) &= (x2, \{(x3 - x2, \emptyset), (x4 - x2, \emptyset), (x6 - x2, \emptyset), (x7 - x2, \emptyset),\\
&\quad (x9 - x2, \emptyset)\}),\\
Dom(x5) &= (x5, \{(x3 - x5, \emptyset), (x4 - x5, \emptyset), (x6 - x5, \emptyset), (x10 - x5, \emptyset),\\
&\quad (x11 - x5, \emptyset), (x12 - x5, \emptyset), (x13 - x5, \emptyset)\}),\\
Dom(x8) &= (x8, \{(x7 - x8, \emptyset), (x9 - x8, \emptyset), (x16 - x8, \emptyset), (x17 - x8, \emptyset)\}).
\end{aligned}$$

[5] See appendix C.2, definition 135.

[6] See appendix C.2.

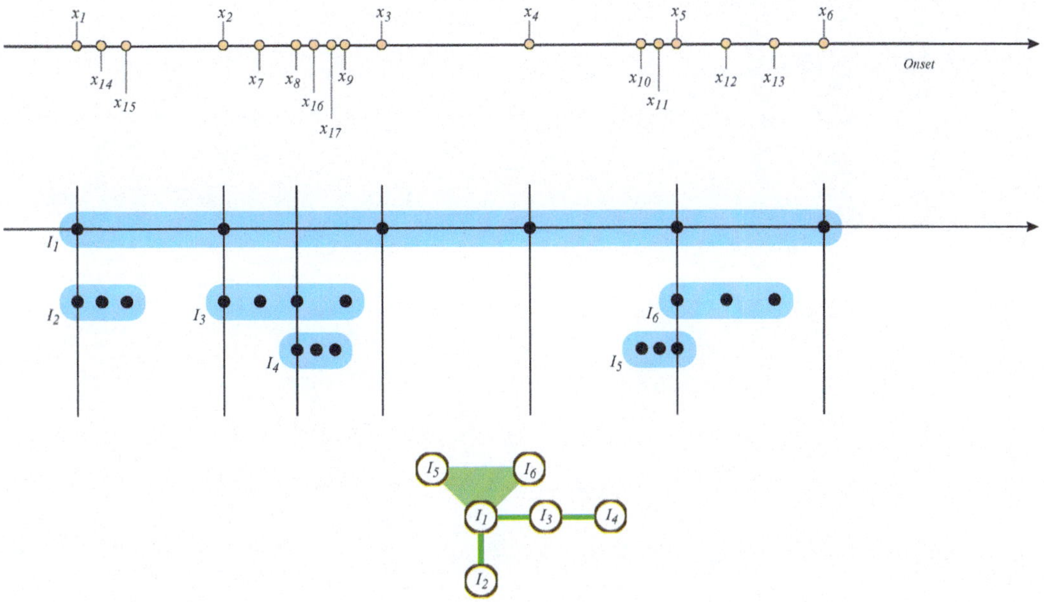

Fig. 21.2. An irregular metrical structure X and its nerve. The associated macro germ is described in the text.

We see that the same point, e.g., $x4$ appears in three different macro ramifications since it is a specialization of three different generic points. But flattening this macro-event eliminates this multiplicity. We then have

$$Dom(X) = \{Dom(x1), Dom(x2), Dom(x5), Dom(x8)\}$$

which gives a complete picture of the dominance hierarchy; make a figure of this dominance hierarchy.

If we are given a macro germ, it is possible to disregard its ramification structure which is deeper than a given limit, this operation of "greeking"[7] can be defined by replacing the macro satellite sets by the empty set from a given depth on.

Exercise 43 *Give a rigorous definition of greeking macro germs.*

[7] An operation known from text applications, meaning that a structure is blurred, recalling the English idiom "that sounds Greek to me".

22

Motif Gestalts

In the organic sphere one cell is different from all the others.
By a magic interplay between these identical yet different cells,
the higher forms of life come into existence.
In an astoundingly analogous way one musical motif,
one theme releases another as
an expression of its own innermost idea,
yet the latter is a being
entirely different from the first.
Rudolph Reti [892, p.359]

Summary. This chapter is not only a good test-case for the mathematization of elementary music concepts, it is above all a refined study of turning fuzzy concept sketches of the humanities into precise and consistent frameworks—without the expected side-effect of "terrible mathematical simplification". In the present case of motives, our topic is—within the general task of grasping motivic phenomena—the construction of Rudolph Reti's immanent motif analysis.

– Σ –

Understanding the motivic, melodic, and thematic organism of a musical composition is one of the most delicate tasks of musicology. It has three principal components which characterize its problematic status: semantic depth, formal complexity, and ill-defined aspects.

Semantic depth is the phenomenon that motives, melodies, and themes are not only formal constructs, i.e., denotator structures with little or no paratextual and easy textual predication. A composition's motif is a germ of a structural hierarchy, unfolding into its most diversified ramifications, variations, and fragmentations. This hierarchy is an expression of meaning, of multiply layered semantic depth which may transcend pure textuality and point at more philosophical, ethical or at least esthetical programs.

Formal complexity appears when one tries to grasp von Ehrenfels' easy-going definition of a gestalt. One recognizes that its characteristics: super-summativity, and transformational invariance, are very complex requirements if rendered in a precise concept framework. We have already discussed some of these issues in Chapter 12. This problem makes clear that the construction of semantic layers from motivic, melodic, or thematic surface is a task which involves some structural complexity.

Ill-defined aspects of motivic, melodic, and thematic analysis show up if one asks for abstraction principles when defining gestalt qualities. It is commonly accepted in motif theory (if that name were really adequate) that gestalt is connected to simplification and class building from the given material. It virtually never happens that the given note information is taken literally to build gestalts. But it is completely unclear which aspects of the given material should be retained, and in what form, or even in what variety of particular forms.

In what follows, we have attempted to give a number of solutions or to make proposals for such, and to discuss the relations between our approaches and the fuzzy but sensitive ideas from traditional musicology.

G. Mazzola, *The Topos of Music I: Theory*, Computational Music Science,
https://doi.org/10.1007/978-3-319-64364-9_22

The application of the following topological methods in motif theory to programming software for motif analysis is important and will be dealt with in Chapter 41.2.

22.1 Motivic Interpretation

Summary. The first step in the present motive analysis framework is the insight that motives cannot be processed without preliminary abstraction. We present musical motivations and case studies.

– Σ –

To begin with, recall the definition of a motif from Section 7.2.3: If A is an address, a (A-addressed) motif M is an objective local composition $M \subset A@F$ whose ambient space $F = Onset \oplus Pitch \times Para$, for any Form $Para$ of additional "parameters", and such that for any pair of different elements in M, their projections to the onset space are different. We shall write M_{OP} for the projection of M onto its OP component, i.e., the local composition $M_{OP} \subset A@(Onset \oplus Pitch)$, and analogously $M_O \subset A@Onset$. We therefore have bijective projection morphisms

$$M \xrightarrow{pr_{OP}} M_{OP} \xrightarrow{pr_O} M_O$$

for motives.

Before entering into the technical discussion, let us have a look at the auxiliary space $Para$. As with meters, we can also use this space to realize macro constructs in the motivic context. More precisely, in this case the motif M is a denotator of Form $Para \underset{Fin(F) \rightarrowtail \Omega^{Fun(F)}}{\longrightarrow} \mathbf{Power}(F)$. The flattening and greeking operations discussed in Chapter 21 can be applied mutatis mutandis. So the parameter space may be an external space regarding duration, sound colors, etc., or a macro refinement construct which englobes hierarchies of onset plus pitch events. We denote by $A@MOT_F$ the set[1] of A-addressed motives of space F, and we write MOT_F for the set zero-addressed motives, and $A@MOT$ (MOT) for the set of A-addressed (zero-addressed) motives in the space $F = Onset \oplus Pitch$ (the parameter space is trivial). For a positive natural number n, we denote by $A@MOT_{F,n}, MOT_{F,n}, MOT_n$, respectively the corresponding sets of motives with cardinality n, we therefore have $A@MOT_F = \coprod_n A@MOT_{F,n}$, etc.

Our first abstraction is a strong one: We shall forget about all parameters from the $Para$ component and concentrate on the OP-projection M_{OP}, i.e., on $A@pr_{OP} : A@MOT_F \to A@MOT$. Although our analyses will deal with the real motives, their properties will be 'filtered' through pr_{OP}. This is not mandatory, any extension of the forthcoming theory to the other parameters is feasible, for instance inclusion of sound colors, loudness or duration in order to describe similarity of motives. So our exposition is just a prototypical one concentrating on the core parameters. We shall describe a compensatory action to this preliminary reduction below.

Definition 65 *A (finite)* motivic interpretation *of a local composition $X \subset A@F$ is an interpretation by a (finite) subset $\mu \subset A@MOT_F$. The motives of μ are called the* covering motives *of the motivic interpretation X^μ, and μ is called the* motif space. *We say that a motivic interpretation (or its motif space)* satisfies the submotif existence axiom (SEA) *iff every submotif of a covering motif is also covering whenever its cardinality equals at least the minimal cardinality of covering motives. The nerve of a motivic interpretation is called the* motivic nerve.

This extends the definition of a melody given in Definition 42, Section 13.4.4. Motivic interpretations occur for local compositions X in score spaces F with events having onset and pitch coordinates. Typically, the covering motives are selected by upper and lower limits min, max of admitted motif cardinality and by the maximal admitted difference *span* of onsets within a motif. We shall deal with this kind of covering satisfying the SEA in the discussion of the MeloRUBETTE® in Section 41.2. A generalization of this selection method

[1] Pay attention to the fact that this is not a functor since distinction of points by their onsets is not functorial!

is given by a preliminary interpretation X^I of X, for example by the charts of the voices in a polyphonic composition. We may then consider all motives within one of these charts, and subjected to the above limits (here the lower limit should be set to 1 in order to guarantee a covering of the given composition). Of course, there is already a strong semantics in the definition of a motivic interpretation. This one may be generated essentially by mathematical rules, by prima vista predicates (such as slurs, beams), by shifter predicates, or by combinations thereof, see Section 18.3.3 for this differentiation where we refer the reader to anchor corresponding semantic questions.

Evidently, the intersections $\bigcap \sigma$ of the simplexes σ of a motivic nerve are also motives, and we obtain a base for a topology on X. In contrast to the maximal meter topology, this one is not interesting since often, the intersection $\bigcap \sigma$ is a singleton, and the topology becomes discrete. We call the simplex $Sp_\mu(x)$ of a point $x \in X$ the *motivic simplex of* x (with respect to μ). As with the metrical theory, the simplex of a point should give information about its role in the zoo of selected motives. The following study of topologies on motif spaces aims at providing us with that information.

This possibility to select motives by use of charts and their predicative background is a certain compensation for the preliminary reduction to onset and pitch. On the other hand, the analytical, cognitive or compositional perspective still does not work with the full information produced by the onset and pitch data. For example, in counterpoint one is interested in the categorization of interval sequences: the cantus firmus and discantus may move in different directions: in parallel, oblique, or contrary motion. And this enforces a very coarse classification of motivic gestalts. In American contour theory [766], pitch is only regarded as a number in an abstract discrete pitch space whereas onset is abstracted to a pure ordering index.

Therefore, a general abstraction concept framework is needed to grasp the variety of perspectives which contribute to the motivic gestalt concept. This is what we want to expose in the following section. It will help us in the construction of topological relations on a motif space as described above.

Although the motif concept works for any address, all the abstractions which are known from musicological/cognitive contexts germinated on zero-addressed motives. In this setting, they all make use of the linear ordering on a motif M which is induced by the bijection with its onset projection. Now, this projection can be given a linear ordering if the address A is such that $A@Onset$ is linearly ordered. For example, if we have $\mathbb{Z}$-addressed motives, we know from Section 7.2.3 that the onset coordinates of their points D are represented by arrows $ON_D \Rightarrow OFF_D$. This data bears a canonical linear lexicographic ordering: first order the onset times, and then the offset times. Moreover, since the pitch coordinates are represented by arrows $PIFF_D \Rightarrow POFF_D$, the same construction works for pitch. This elementary situation can be used to generalize zero-addressed abstractions to $\mathbb{Z}$-addressed generalizations, but there is no intuition or method to do so for general addresses.

The following abstraction examples will therefore be of rather selective nature, restricting to zero and integer addresses, but the theory will nonetheless be carried out on a general address level, starting from the general axioms which we can learn from the given examples.

22.2 Shape Types

Summary. Several abstraction methods are in usage in musicology. We establish the formal concept of *shape types* and present a list of common constructs.

– Σ –

We get off by the formal definition of a shape type and then give representative examples thereof:

Definition 66 *Given a parameter space F and an address A, a* shape type t *is a family $(\Gamma_{t,n})_{n\in\mathbb{N}_+}$ of non-empty sets, together with a set map*

$$t : A@MOT_F \to \Gamma_t \tag{22.1}$$

with codomain $\Gamma_t = \bigcup_n \Gamma_{t,n}$ such that $t(A@MOT_{F,n}) \subset \Gamma_{t,n}$, for each index n. Equivalently, we may give the n-components

$$t_n : A@MOT_{F,n} \to \Gamma_{t,n} \tag{22.2}$$

with the direct sum $t = \coprod t_n$. *The spaces* $\Gamma_{t,n}$ *are called the* spaces of (abstract n-motives of type) t *and their elements are called* abstract n-motives of type t; *the space* Γ_t *is called the* space of t-abstract motives. *For an abstract motif* $m \in \Gamma_t$ *we write* $abcard_t(m)$ *for the minimal index* n *such that* $m \in \Gamma_{t,n}$, *this number is called the* abstract cardinality of m. *If* $M \in A@MOT_F$, *we write* $abcard_t(M) = abcard_t(t(M))$ *and call this number the* t-abstract cardinality of M.

We shall use the following notation: $\Gamma_t|_k$ denotes the subset of abstract motives of abstract cardinality k, and we set $A@MOT_F|_k = t^{-1}(\Gamma_t|_k)$, whence $\Gamma_t = \coprod_k \Gamma_t|_k$, and $A@MOT_F = \coprod_k A@MOT_F|_k$.

Assumption 1 *In the sequel we shall always make the assumption that the abstract cardinality is compatible with inclusions, i.e.,* $M \subset N$ *implies* $abcard_t(M) \leqslant abcard_t(N)$.

All examples which we shall present satisfy this property.

In principle it is possible and reasonable to view the abstract motives as denotators, but we will refrain from this formalism since the denotator theory is not (yet?) primordial here and would only burden the symbolism.

22.2.1 Examples of Shape Types

Summary. We present the rigid, diastematic, elastic, and toroidal shape types.

– Σ –

All the following types will undergo the announced generic abstraction from parameters outside onset and pitch.

22.2.1.1 Rigid Types

This abstraction is a generic one with respect to the others, in fact, it is a one-to-one restatement of the OP-projection M_{OP} of a motif M in terms of the ordering which is induced by the onset projection if this is given at the selected address A. For $A = \mathbb{Z}^r$, we have the well-known lexicographic ordering of $\mathbb{Z}^r@Onset$ denotators $o = (o_0, o_1, \ldots o_r), o_i \in \mathbb{R}$. In this case, the onset ordering is transported to the points of M_{OP}, i.e., the rigid shape type $t = Rg$ is defined by $Rg(M) = (m_0, m_1, \ldots m_{c-1})$ where the sequence $(m_0, m_1, \ldots m_{c-1}) \in \Gamma_{Rg,c} = \mathbb{R}^{2(r+1)\times c}$ is the sequence of all elements of $M_{OP} = \{m_0, m_1, \ldots m_{c-1}\}$ with $c = card(M)$ in their onset ordering. In this setup, we have $abcard_{Rg}(M) = card(M)$, $\Gamma_{Rg} = \coprod_n \Gamma_{Rg,n}$, $\Gamma_{Rg}|_n = \Gamma_{Rg,n}$, and $\mathbb{Z}^r@MOT_F|_n = \mathbb{Z}^r@MOT_{F,n}$. Concretely, $Rg(M) = (m_0, m_1, \ldots m_{c-1})$ is denoted by a $2(r+1) \times c$-matrix

$$Rg(M) = \begin{pmatrix} o_{0,0} & \cdots & o_{0,i} & \cdots & o_{0,c-1} \\ \cdots & & & & \\ o_{r,0} & \cdots & o_{r,i} & \cdots & o_{r,c-1} \\ p_{0,0} & \cdots & p_{0,i} & \cdots & p_{0,c-1} \\ \cdots & & & & \\ p_{r,0} & \cdots & p_{r,i} & \cdots & p_{r,c-1} \end{pmatrix} \text{ with } m_i = \begin{pmatrix} o_{0,i} \\ \cdots \\ o_{r,i} \\ p_{0,i} \\ \cdots \\ p_{r,i} \end{pmatrix} \tag{22.3}$$

and where the upper half $Rg_o(M)$ refers to the onsets whereas the lower half $Rg_p(M)$ refers to the pitch coordinates.

A second shape type ΔRg, the *rigid difference shape type* is directly derived from Rg: We take $\Gamma_{\Delta Rg,c} = \mathbb{R}^{2(r+1)\times(c-1)c/2}$ and define

$$\begin{aligned} \Delta Rg(M) &= (m_j - m_i)_{0 \leqslant i < j \leqslant c-1} \text{ if } 1 < c, \\ \Delta Rg(M) &= 0 \text{ if } c = 1. \end{aligned} \tag{22.4}$$

We shall usually represent the i, j-indexed entities in the Delta rigid difference vector as coefficients $\Delta Rg(M)_{i,j}$ of an upper triangular $c \times c$-matrix (starting at the upper codiagonal).

As a specialization of the rigid difference shape type, we may consider a selected kth codiagonal with index difference $j - i = k$. This *rigid k-difference shape type* will be denoted by $\Delta_k Rg$, we have $\Gamma_{\Delta_k Rg,c} = \mathbb{R}^{2(r+1)\times(c-k)}$, and

$$\Delta_k Rg(M) = (m_{i+k} - m_i)_{0 \leqslant i < c-k} \text{ if } 1 < c,$$
$$\Delta_k Rg(M) = 0 \text{ if } c = 1. \tag{22.5}$$

22.2.1.2 Diastematic Types

The diastematic type $t = Dia$ comes from classical musicology where one is often only interested in the direction of a melodic movement. The *diastematic shape type* is a 'child' of the rigid difference shape type. It resides on the index function[2] index(x) when applied to the coefficients of the rigid difference shapes. We take $\Gamma_{Dia,c} = \mathbb{Z}^{(r+1)\times(c-1)c/2}$ and define

$$Dia(M)_{i,j} = (index(\Delta Rg(M)_{i,j,s}))_{r+1<s\leqslant 2(r+1)} \tag{22.6}$$

with the s-index being the row number of the matrix. And we may evidently carry over the codiagonal procedure to the diastematic situation in defining $\Gamma_{Dia_k,c} = \mathbb{Z}^{(r+1)\times(c-k)}$, and

$$Dia_k(M) = (Dia(M)_{i,i+k})_{0 \leqslant i < c-k} \text{ if } 1 < c,$$
$$Dia_k(M) = 0 \text{ if } c = 1. \tag{22.7}$$

This is the *diastematic k-difference shape type*, the special case of $k = 1$ is also called *diastematic index shape type*. For the special case $r = 0$, the diastematic shape type includes the COM matrix used in AST contour concepts, see [160, 765].

As a function of this data, we may shape contrapuntal step types as follows. Take the address $A = \mathbb{Z}$, a case which we have already illustrated in Section 7.2.3 as being a common contrapuntal situation. We have the values $Dia_{i,j} = (u, v)$ which means that the cantus firmus movement from interval i to interval j has index u whereas the discant movement has index v. Therefore, the product $u.v$ is positive iff we have parallel motion, it is zero iff we have oblique motion, and it is negative iff we have contrary motion. More generally, we may define the *contrapuntal motion shape type* $t = Contra$ as a 'child' of the diastematic shape type as follows: We take $\Gamma_{Contra,c} = \mathbb{Z}^{r\times(c-1)c/2}$ and put

$$Contra(M)_{i,j} = (Dia(M)_{i,j,s}.Dia(M)_{i,j,r+2})_{r+2<s\leqslant 2(r+1)},$$
$$Contra(M) = 0 \text{ if } c = 1. \tag{22.8}$$

In the case $r = 1$ we get the above contrapuntal example. As above, we may also project to the k-th codiagonal and consider the corresponding *contrapuntal k-motion shape type k-Contra*, the details are left to the reader. The special case of the 1-*Contra* shape type really gives us an abstraction which codifies exactly the successive contrapuntal motion type picture.

Exercise 44 *Try to define a* reduced diastematic shape type *by use of the above lexicographic order (end of Section 22.1) on the pitch coordinate vectors and the diastematic values* $-1, 0, +1$ *if pitch decreases, is unchanged, or increases. Describe the shape spaces* $\Gamma_{RedIndia,c}$ *for the codiagonal projection of this reduced diastematic type.*

---

[2] See Appendix D.5.1.

22.2.1.3 Elastic Type

From the cognitive point of view, it might be reasonable to concentrate on geometric configurations up to dilatations and translations. This type $t = Elast$ is codified as a 'child' of the rigid difference shape type modulo dilatations, i.e., the diagonal action of the multiplicative group $\mathbb{R}_+^\times$ on the coefficients of the matrix $\Delta Rg(M)$. The orbits of the matrices can be parametrized by special homogeneous coordinates. We observe that each matrix $\Delta Rg(M)$ has in its onset part (upper half) its positive first coefficient $o_{a(i,j),i,j}$, for each matrix position $(i,j) = (0,1), \ldots (c-2, c-1)$ in (22.4). Take the unique matrix dilatation such that $\sum_j o_{a(i,j),i,j} = 1$, then we have a unique representation of that $\mathbb{R}_+^\times$-orbit. So we first fix a function $a : \{(0,1), \ldots (c-2, c-1)\} \to \{0, 1, \ldots r\}$ and denote $\Sigma a = \sum_{i,j} a(i,j)$. We then define the smallest non-vanishing (positive) coefficients $o_{a(i,j),j}, (i,j) = (0,1), \ldots (c-2, c-1)$ of the (i,j) position, together with the unity sum condition. The other coefficients are all free. So the abstract motif space looks as follows. For a fixed function a, we have the vector of initial values $(o_{a(i,j),i,j})_j \in |\Delta_{c(c-1)/2-1}|^o$, the interior of the affine $c(c-1)/2-1$-simplex. The other free values are distributed below the initial values in the rigid difference matrix (22.4). Each position (i,j) has $2(r+1) - (a(i,j)+1) = 2r - a(i,j) + 1$ free values, which add up to $\sum_{i,j} 2r - a(i,j) + 1 = (2r+1)(c-1) - \Sigma a$ and are therefore in $\mathbb{R}^{(2r+1)(c-1)-\Sigma a}$. This means that we have the shape spaces $\Gamma_{Elast,c} = |\Delta_{c(c-1)/2-1}|^o \times \coprod_a \mathbb{R}^{(2r+1)(c-1)-\Sigma a}$.

22.2.1.4 Toroidal Type

For the toroidal shape type $t = Toroid_\gamma^{m,l}$, we restrict to the zero address. We are given two positive natural numbers m, l, and an *affine grid basis* $\gamma = (o, x, y) \in (Onset \oplus Pitch)^3$, consisting of an origin o, and a linear grid basis x, y which define the grid[3] $G(\gamma) = o + \mathbb{Z}.x + \mathbb{Z}.y \subset Onset \oplus Pitch$. With respect to γ, every point $p \in Onset \oplus Pitch$ has a unique representation $p = \gamma(p) + \xi.x + \eta.y$ with $0 \leqslant \xi, \eta < 1$ and $\gamma(p) \in \gamma$. We then associate with p the grid point $p_\gamma = \gamma(p) + \xi.round(x) + \eta.round(y)$ (see Definition 148 in Appendix D.5.1 for the rounding function). Let $p_\gamma = o + a.x + b.y$ be the representation of p_γ in the grid. Then we have the element $p_{\gamma,m,l} = (a \bmod m, b \bmod l) \in OnPiMod_{m,l}$ whose coordinates are in the discrete torus $\mathbb{Z}_m \oplus \mathbb{Z}_l$ as discussed in formula (6.43) of Section 6.4.1. If a motif $M \in MOT_{F,n}$, this defines a local composition $Toroid_\gamma^{m,l}(M) = \{p_{\gamma,m,l} | \ p \in M_{OP}\}$. It is contained in the set $\Gamma_{Toroid,n} = ObLoc_0^{\leqslant n, OnPiMod_{m,l}}$ of local, zero-addressed compositions in $OnPiMod_{m,l}$ with cardinality at most n, and we have a shape map

$$Toroid_\gamma^{m,l} : MOT_{F,n} \to \Gamma_{Toroid,n} \tag{22.9}$$

on the given cardinality level n.

Exercise 45 *Observe that in this case, the shape spaces are not disjoint, but the inclusion assumption 1, Section 22.2.1, is verified.*

By use of the rigid type, we can also define a *toroidal sequence shape type* $TorSeq$ as follows. Again, we fix an affine grid γ and a pair m, l of positive integers. Take the (zero-addressed) rigid abstract motif $Rg(M) = (m_0, \ldots m_{c-1})$, then the sequence $TorSeq(M) = (m_{0,\gamma,m,l}, \ldots m_{c-1,\gamma,m,l}) \in \Gamma_{TorSeq,c} = 0@(OnPiMod_{m,l})^c \xrightarrow{\sim} (\mathbb{Z}_m \oplus \mathbb{Z}_l)^c$.

22.3 Metrical Similarity

Summary. This section deals with the metrical similarity of abstract motives. This concept is the base of motivic topologies used to understand Reti's thinking.

– Σ –

[3] Recall that the underlying module of $Onset \oplus Pitch$ is $\mathbb{R}^2$.

The abstraction process which yields the shape space Γ_t is only the first identification step. In fact, after this shape abstraction, we need to deal with similarity of motives. This relation will be defined on abstract motives and then retracted to the original motives in the following sense:

Definition 67 *Given a shape type $t : A@MOT_F \to \Gamma_t$, a* distance function *is a sequence $d = (d_n)_{1 \leqslant n}$ of pseudo-metrics d_n which are defined on $\Gamma_t|_n$. A t-*distance function *is the sequence $d_t = (d_{t,n} = d_n|A@MOT_F|_n)_{1 \leqslant n}$ of pseudo-metrics which are induced on the spaces $A@MOT_F|_n$ by the distance function d, i.e., if $M, N \in A@MOT_F|_n$, then $d_{t,n}(M,N) = d_n(t(M), t(N))$. If no confusion is possible, we omit the index n and just write $d_t(M,N)$.*

22.3.1 Examples of Distance Functions

Let us now discuss distance functions for shape types which have been introduced in Section 22.2.1.

22.3.1.1 Distances for Rigid Types

The shape spaces for rigid types (subsubsection 22.2.1.1) are real matrix spaces $\Gamma_{Rg,n} = \mathbb{R}^{2(r+1) \times c}$ and $\Gamma_{\Delta Rg,n} = \mathbb{R}^{2(r+1) \times (c-1)c/2}$. On such a space, we have the usual Euclidean metric which defines the Euclidean distance function Ed and the corresponding Euclidean Rg- or ΔRg- distance function $Ed_{Rg}, Ed_{\Delta Rg}$.

We may also consider the abstract cardinality and define the *relative Euclidean distance function*: Let $abcard(x) = abcard(y) = n$, for two abstract motives, then we set $REd(x,y) = Ed(x,y)/n$ for rigid type Rg, and $REd(x,y) = Ed(x,y)/(n-1)$ for shape type ΔRg. Such a relative distance takes into account that on the *cognitive level*, we would like to say that the distance is a kind of mean value of all the point distances between the first and first, second and second, etc., last and last points of two motives. This reason also holds for the other relative distances to be defined in the sequel.

22.3.1.2 Distances for Diastematic Types

For the diastematic types (subsubsection 22.2.1.2), the abstract motives also lie in matrix spaces, this time with integer coordinates, and we may also take the Euclidean distance function induced from the surrounding real matrix spaces.

A relative Euclidean distance function on Γ_{Dia} which relates to the number of components in the triangular matrix is given by $REd(x,y) = 2Ed(x,y)/(r+1)c(c-1)$. For the codiagonal shape types Dia_k, we set $REd(x,y) = Ed(x,y)/(r+1)(c-k)$. For the contrapuntal shape type $Contra$, we set $REd(x,y) = 2Ed(x,y)/rc(c-1)$.

22.3.1.3 Distances for Elastic Type

As in the previous situations, we can apply the Euclidean distance function to the elastic shape type (Section 22.2.1.3) since its abstract motives are also embedded in a real matrix space. In this case, it may be more adequate to take the metric which is deduced[4] from the 1-norm on n-space since in this case, the component of initial values in the affine simplex has just length 1. Since it is well-known[5] that the topology on n-space is independent of the defining norm, this will have no influence on the following topological considerations.

A relative Euclidean or 1-norm distance can be defined via a division factor $2(r+1)(c-1)$ which takes into account the number of motif points and the dimension of the address-related columns.

[4] See Appendix I.1.1.
[5] See Appendix I.1.2.

22.3.1.4 Distances for Toroidal Types

On the discrete torus $\mathbb{Z}_m \oplus \mathbb{Z}_l$, one defines a metric $steps(x, y)$ by the minimal number of grid steps on all possible walks from x to y which is a kind of geodesic distance on this discrete torus. On the shape space $\Gamma_{TorSeq,c} \xrightarrow{\sim} (\mathbb{Z}_m \oplus \mathbb{Z}_l)^c$, we may take the norm, Euclidean, or uniform distance construction. Given two sequences $x = (x_i), y = (y_i) \in \Gamma_{TorSeq,c}$, we accordingly set

$$d_{1,c}(x, y) = \sum_i steps(x_i, y_i), \tag{22.10}$$

$$d_{2,c}(x, y) = \sqrt{\sum_i steps(x_i, y_i)^2}, \tag{22.11}$$

$$d_{\infty,c}(x, y) = max\{steps(x_i, y_i) |\ i = 0, \ldots c - 1\} \tag{22.12}$$

and get a corresponding distance on toroidal sequences. This construction is also used in counterpoint theory, see Chapter 30.

For the *Toroid* shape type, one can define a distance via the lexicographic ordering among torus points, thereby ordering a local composition lexicographically and then applying the above distance functions for the lexicographic sequences of abstract motives ([163]).

Exercise 46 *Fill in the details of the Toroid distance function and verify that it is a pseudo-metric.*

22.4 Paradigmatic Groups

Summary. Together with abstraction mechanisms, groups of symmetries act on spaces of abstract motives and—in several cases—on the proper motif spaces. These groups are termed *paradigmatic* groups since they relate to Jean-Jacques Nattiez' paradigmatic theme, as discussed in 11.7.1. An important case are those actions which are *equivariant* with respect to the shape maps. The formal definitions of *gestalts* are associated with specific abstractions and group actions. This demonstrates that the concept of gestalt is the result of a multilayered concept hierarchy involving several non-automatic parameter choices.

– Σ –

As a matter of fact, comparison of motives is not restricted to abstraction and similarity, we know from the general discussion of paradigmatic concepts in Chapter 10 that the topological similarity which has been induced (though not executed, this is the scope of the topological theory of motives in Section 22.6) must be complemented by the transformational similarity. The latter deals with group actions on shape types in the following sense:

Definition 68 *A* paradigmatic group P for a shape type $t : A@MOT_F \to \Gamma_t$ *is a left action* $\pi : P \times \Gamma_t \to \Gamma_t$ *on the shape space which leaves the components* $\Gamma_{t,n}$ *invariant.*

An equivariant paradigmatic group P for a shape type $t : A@MOT_F \to \Gamma_t$ *is a pair of group actions*

$$\gamma : P \times A@MOT_F \to A@MOT_F, \pi : P \times \Gamma_t \to \Gamma_t$$

which is equivariant[6] *with respect to the shape map* t *and such that* P *is a paradigmatic group for* t.

We also write

$$\begin{array}{ccc} A@MOT_F & \xrightarrow{\ t\ } & \Gamma_t \\ \underset{P:\gamma}{\circlearrowleft} & & \underset{P:\pi}{\circlearrowleft} \end{array}$$

or more concisely (if the actions must not be specified)

$$A@MOT_F \xrightarrow[\underset{P}{\circlearrowleft}]{\ t\ } \Gamma_t$$

for this situation.

[6] See Appendix C.3.1, Example 100.

Lemma 34 *If P is a paradigmatic group for type t, then the abstract cardinality is an invariant for abstract motives, i.e., the group action leaves invariant the components $\Gamma_t|_n$.*

The proof is an easy exercise. We are now ready to define the gestalt concept.

Definition 69 *The orbits of a paradigmatic group P action on Γ_t are called the* abstract gestalts of shape type t. *The inverse images $t^{-1}G$ of* abstract gestalts G of t *are called the* t-gestalts. *If $M \in A@MOT_F$, then its t-gestalt $t^{-1}(P.t(M))$ is denoted by $Ges_t^P(M)$ or $Ges_t(M)$ if P is clear.*

Sorite 7 *Let P be a paradigmatic group for shape type t. Then every t-gestalt $Ges_t(M)$ is contained in $\Gamma_t|_{abcard_t(M)}$. If the paradigmatic group is equivariant, then the orbit of a motif M is contained in the gestalt: $P.M \subset Ges_t(M)$, i.e., a gestalt is a disjoint union of orbits. We therefore have the commutative diagram*

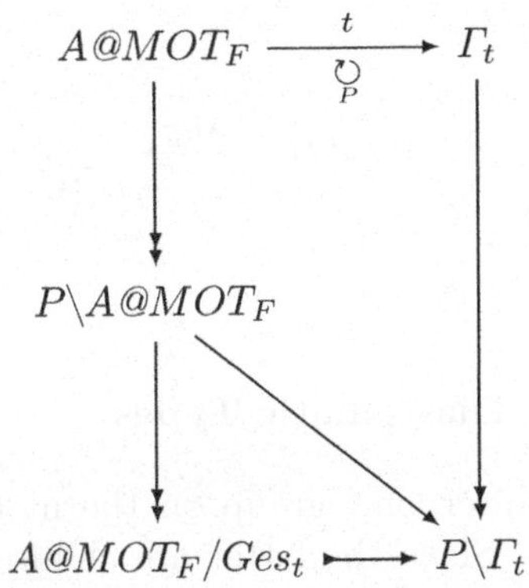

This follows immediately from the definitions and Lemma 34.

Definition 70 *The* cardinality of a gestalt $Ges_t(M)$ *is defined as $abcard_t(N)$ for any $N \in ges_t(M)$.*

22.4.1 Examples of Paradigmatic Groups

Here are some group actions which are of practical use. Preliminary remark: For any group $P \subset \overrightarrow{GL}(\mathbb{R}^2)$ of affine transformations on $Onset \oplus Pitch$, one has the canonical induced action γ on the onset and pitch coordinates of motives in $A@MOT_{F,n}$. Many paradigmatic actions are defined as equivariant actions which issue from this canonical action on the proper motives, i.e., one first transports the motivic action to a motivic shape action and then tries to generalize the latter to any abstract motif.

The most prominent group is the *affine counterpoint group* $CP = T_{2,\mathbb{R}} \rtimes LCP$, the group which is generated by the normal subgroup $T_{2,\mathbb{R}} = e^{\mathbb{R}^2}$ of translations and the linear counterpoint group $LCP = \langle U, K\rangle \xrightarrow{\sim} K_4$ which is a Klein group[7].

22.4.1.1 Paradigmatic Groups for Rigid Types

For the rigid shape type $Rg : \mathbb{Z}^r@MOT_F \to \Gamma_{Rg}$, we have the above canonical action of $p \in CP$ on a motif M: $p(M) = \{p(m)|\ m \in M\}$. Evidently, the retrograde motion reverses the time order of elements in M. Therefore we can define a CP-action π on $\Gamma_{Rg,c}$ by the rule

$$p(x_0, \ldots x_{c-1}) = \begin{cases} (p(x_0), p(x_1), \ldots p(x_{c-1})) & \text{if } p \text{ does not have the retrograde component,} \\ (p(x_{c-1}), p(x_{c-2}), \ldots p(x_0)) & \text{(reversed order) else.} \end{cases} \quad (22.13)$$

[7] The four-element group of plane symmetries generated by inversion U (reflection at zero pitch) and retrograde K (reflection at zero onset), see also Section 8.1.1.

Clearly, this defines an equivariant action

$$\begin{array}{ccc} \mathbb{Z}^r@MOT_F & \xrightarrow{Rg} & \Gamma_{Rg} \\ \circlearrowright_{CP:\gamma} & & \circlearrowright_{CP:\pi} \end{array}$$

for rigid shape type.

For the rigid difference shape type, the affine action of CP reduces to the linear action, i.e., we have the quotient action $\Delta\pi$ via an element $q \in LCP \xrightarrow{\sim} CP/T_{2,\mathbb{R}}$ on $\Gamma_{\Delta Rg,c}$ by the following rule. If we write $d = (d_o, d_p)$ for the onset-pitch decomposition, and $U(d) = (d_o, -d_p)$ we have

$$q((d_{i,j})_{i,j}) = \begin{cases} (U(d_{i,j}))_{i,j} & \text{if } q = U, \\ (d_{c-1-j,c-1-i})_{i,j} & \text{if } q = K. \end{cases} \tag{22.14}$$

This gives the equivariant action

$$\begin{array}{ccc} \mathbb{Z}^r@MOT_F & \xrightarrow{\Delta Rg} & \Gamma_{\Delta Rg} \\ \circlearrowright_{CP:\gamma} & & \circlearrowright_{CP:\Delta\pi} \end{array}$$

for rigid difference shape type.

22.4.1.2 Paradigmatic Groups for Diastematic Types

The canonical action of the affine counterpoint group on the motives also carries over to an equivariant action on the diastematic shape types. Since the latter are children of the rigid difference shape type, we have to see if we can have an equivariant action

$$\begin{array}{ccc} \Gamma_{\Delta Rg} & \xrightarrow{index} & \Gamma_{Dia} \\ \circlearrowright_{LCP:\Delta\pi} & & \circlearrowright_{LCP:Dia\pi} \end{array}$$

between rigid difference and diastematic shape types (in this direction!). Again, we may define the action $Dia\pi$ on the generators and set

$$q((ind_{i,j})_{i,j}) = \begin{cases} ((-ind_{i,j}))_{i,j} & \text{if } q = U, \\ ((-ind_{c-1-j,c-1-i}))_{i,j} & \text{if } q = K \end{cases} \tag{22.15}$$

for a diastematic index family $(ind_{i,j})_{i,j}$. Evidently, this action makes the *index* map equivariant. So as above, the quotient $CP \to LCP$ defines the equivariant action

$$\begin{array}{ccc} \mathbb{Z}^r@MOT_F & \xrightarrow{Dia} & \Gamma_{Dia} \\ \circlearrowright_{CP:\gamma} & & \circlearrowright_{CP:Dia\pi} \end{array}$$

of the affine counterpoint group in the diastematic case. This clearly carries over to the different codiagonal projections since the codiagonals are invariant under the retrograde action.

Exercise 47 *Define an equivariant action of the affine counterpoint group on the contrapuntal motion shape type by use of the diastematic action.*

22.4.1.3 Paradigmatic Groups for Elastic Type

The paradigmatic counterpoint group action is a bit more delicate for the elastic type since we have to take care of the initial value functions a and the change of these function under the counterpoint group. As the elastic type is a child of the rigid difference type, we again want to define an equivariant action of the linear counterpoint group LCP for the elastic map

$$\underset{LCP:\Delta\pi}{\overset{\circlearrowleft}{\Gamma_{\Delta Rg}}} \xrightarrow{elast} \underset{LCP:Elast\pi}{\overset{\circlearrowleft}{\Gamma_{Elast}}}$$

To begin with, the inversion U leaves the initial value function a invariant, whereas the retrograde K maps a to $K.a$ with $K.a(i,j) = a(c-1-j, c-1-i)$. Since the multiplicative action of $\mathbb{R}_+^\times$ commutes with any linear action, and since the condition $\sum_j a(j) = 1$ is invariant under the linear counterpoint group, we can take the rigid difference action on the homogenous coordinates in the elastic shape spaces Γ_{elast}, together with the above initial value function transformation. The K-action permutes the summands of the disjoint union $\coprod_a \mathbb{R}^{(2r+1)(c-1)-\Sigma a}$ according to the rigid difference action and the operation $K.a$ on the disjoint union "index" a.

Exercise 48 *Fill in the details of the previous elastic action.*

22.4.1.4 Paradigmatic Groups for Toroidal Types

This case is a bit different. There is no analogous equivariant action of the affine counterpoint group for toroidal types since the rounding function in the affine frame quantization is not compatible with the inversion and retrograde operations (check this!). We however have an action of the affine group (or any subgroup) $G_{m,l} = \overrightarrow{GL}(\mathbb{Z}_m \oplus \mathbb{Z}_l)$ of the torus $\mathbb{Z}_m \oplus \mathbb{Z}_l$ on the abstract motives. In fact, if $Q \in \Gamma_{Toroid}$, the pointwise action $\pi_{m,l}$ of $g \in G_{m,l}$, $g.Q = \{g.q|\ q \in Q\}$ gives the action on abstract toroidal motives. So here the abstract gestalts are the isomorphism classes of local compositions.

For the toroidal sequence shape type, we have the usual coordinatewise action $\pi_{m,l}$ of $g \in G$ on a sequence $x. = (x_i)_i$ via $g.x. = (g.x_i)_i$.

22.5 Pseudo-metrics on Orbits

Summary. Metrical similarity between abstract motives is extended to gestalts which essentially refer to orbits of abstract motives under paradigmatic groups.

$- \Sigma -$

It is not sufficient to have metrical relations among the abstract and real motives, we also want to tell what it means that two gestalts are metrically neighboring. To this end, we need a preliminary knowledge about actions of groups of isometries. We know from Appendix I.1.1, Lemma 109, that if a pseudo-metric d and an action of a group G by isometries[8] are given on a space V, then the orbit space $G\backslash V$ is naturally provided with a pseudo-metric d^*.

So we only have to look for isometric actions in our examples, and we obtain a pseudo-metric on the abstract gestalt spaces. Let us now check where this is the case.

1. **Rigid Types.**
For the rigid shape type, the distance between two matrices is the Euclidean norm of their difference. On the other hand, the inversion and retrograde are isometries on the onset-pitch space, therefore, the counterpoint group CP acts by isometries.
The same is valid for the rigid difference shape type which has only its triangular matrix entries permuted and is a priori invariant under translations; the same holds for the codiagonal projections and relative distances.

[8] See Appendix I.1.1, Definition 187.

2. **Diastematic Types.**
   The counterpoint group acts via its linear projection $LCP = \langle U, K \rangle$ on the diastematic shape type. Thereby, only signs are altered and indexes of coefficients are permuted (equation (22.15)). This leaves the Euclidean distance invariant. Same result for the codiagonal projections and relative distances.
3. **Elastic Type.**
   Since the action of LCP on the elastic type is the action of this group on the rigid difference shape type with the corresponding homogeneous coordinates, LCP also acts by isometries on the elastic type.
4. **Toroidal Types.**
   Here, we have to observe the step number between two points on the torus $\mathbb{Z}_m \oplus \mathbb{Z}_l$ when an affine automorphism is applied. In general, this not an invariant. For instance, we have the fourth multiplication on $\mathbb{Z}_4 \oplus \mathbb{Z}_{12}$ (diagonal matrix with 5 in its pitch direction) which sends a fifth (minimal number of 5 steps) to the minor second (one step). However, if the numbers m, l are relatively prime and the Euler functions[9] are $\phi(m), \phi(l) \leqslant 2$, then the minimal step number is conserved. This function value is the case for a very small number of cases: $m, l = 2, 3, 4$, and the combinations $(m, l) = (2, 3), (3, 2), (3, 4), (4, 3)$. In this shape type, we should therefore concentrate on selected subgroups of the general affine group in order to preserve distances. One such subgroup is the LCP which acts in the canonical way on the factors.

After these case studies we can define what is an orbit pseudo-metric for shape types. We say that a paradigmatic group P for a shape type t with distance d *consists of isometries* if P acts as group of isometries on each d_n-pseudo-metric shape space $\Gamma_t|_n$. If this is the case, we have *a pseudo-metric* $d^*_{P,n}$ *on each orbit space* $P\backslash\Gamma_t|_n$. In a more sloppy language we shall also say that we have a pseudo-metric d^*_P on the abstract gestalt space $P\backslash\Gamma_t$ when we think of the sequence of pseudo-metrics $(d^*_{P,n})_n$. We shall also use this convention for all pseudo-metrics induced on the following spaces: Reconsider the diagram just before Section 22.4.1. By retraction of the abstract gestalt pseudo-metric d^*_P to all the codomains, we obtain a system of pseudo-metrics which we denote as indicated in the corresponding diagram which now automatically turns into a diagram of (sequences of) pseudo-metrics

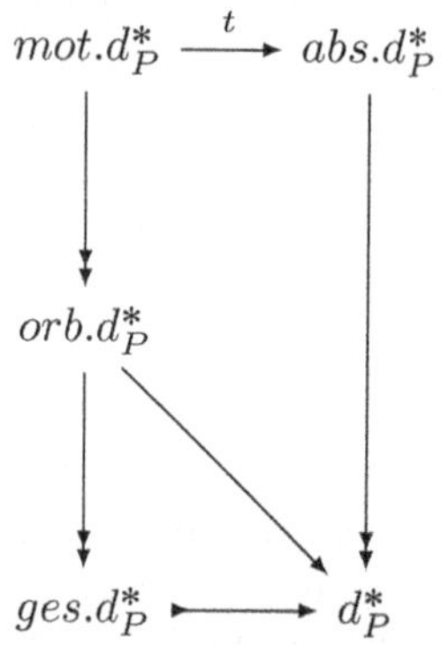

where we have omitted the underlying (real or abstract) motif sets.

22.6 Topologies on Gestalts

Summary. The overall comparison of gestalts (of different cardinalities) is conceived. It is built upon the *inheritance property* which essentially captures the psychological fact that recognition of motives often amounts to recognition of their submotives.

– Σ –

[9] See Appendix C.3.4.1, Proposition 84.

We are now ready to introduce topologies on motif and gestalt spaces. Except for the toroidal type, this approach combines the elements of gestalt and metrical similarity while comparing motives. In the toroidal situation, we shall however introduce a more algebro-geometric idea to topology which deals with specialization instead of metrics.

22.6.1 The Inheritance Property

Summary. This section is devoted to the formal discussion of the inheritance property and to the study of different shape types with respect to this property.

– Σ –

Though metrical properties suggest that we may have a common topology on $A@MOT_F$ defined by use of pseudo-metrics, it turns out that there is an important restriction to this program, a restriction which is of an evident cognitive semantic. We postpone this rationale and first present the formal aspect.

Definition 71 *Given a shape type $t : A@MOT_F \to \Gamma_t$, together with a distance function*[10] *d on Γ_t, we say that this data* has the inheritance property *iff for each motif $M \in A@MOT_F|_n$, any submotif $M^* \subset M$, and any real number $\epsilon > 0$, there is a real number $\delta > 0$ such that if $N \in A@MOT_F|_n$ is such that $d_t(M, N) < \delta$, then there is a submotif $N^* \subset N$ such that $abcard(M^*) = abcard(N^*)$ and $d_t(M^*, N^*) < \epsilon$.*

Intuitively speaking, *neighboring motives have neighboring submotives.* An equivalent definition states that a particular function on the motives is continuous: Suppose that for two motives, M^*, N, $abcard(M^*) \leqslant abcard(N) = n$. We define

$$Sub_{M^*}(N) = \{N^* \subset N |\ abcard(N^*) = abcard(M^*)\},$$

the set of submotives of N with fixed abstract t-cardinality $abcard(M^*)$. Consider the function

$$d_{t,M^*,n} : A@MOT_F|_n \to \mathbb{R} \tag{22.16}$$

defined by $d_{t,M^*,n}(N) = Min_{N^* \in Sub_{M^*}(N)}(d_t(M^*, N^*))$. Then the inheritance property reads

Definition 72 *Given a shape type $t : A@MOT_F \to \Gamma_t$, together with a distance function d, we say that this data* has the inheritance property *iff for any motif M^* and any $abcard(M^*) \leqslant n$, the function $d_{t,M^*,n}$ is continuous in every supermotif M of M^* (i.e., $M^* \subset M$ of abstract cardinality n with respect to the pseudo-metric topology of d_n).*

In fact, suppose that the second version of the inheritance property holds. Then $d_{t,M^*,n}$ is continuous in any supermotif M of M^* of abstract cardinality n. But then, $d_{t,M^*,n}(M) = 0$, and for a given positive ϵ there is δ such that $d_t(M, N) < \delta$ implies $|d_{t,M^*,n}(N) - d_{t,M^*,n}(M)| = d_{t,M^*,n}(N) < \epsilon$, and this is the claim of the first version of the inheritance property. The converse is evident.

This property is not automatic. An illustrative 'pathology' is the diastematic k-difference shape type t_{Dia_k}. Let us take the codiagonal difference $k = 1$, i.e., the diastematic index shape type, see Figure 22.1.

Here, we work in MOT_n, and we look at the motif $M \in MOT_5$ with $Dia_1(M) = (1, -1, 1, -1)$, and the submotif $M^* \subset M$ with $abcard(M) = 3$ and $Dia_1(M^*) = (1, 1)$. Take the motif $N \in MOT_5$ with $Dia_1(N) = Dia_1(M)$. The distance is $d_t(M, N) = 0$, but N has no submotif N^* with $Dia_1(N^*) = Dia_1(M^*)$, and so the inheritance property cannot be fulfilled. For rigid and elastic types we have affirmative results:

Proposition 25 *Rigid shape types $Rg, \Delta Rg$, and corresponding Euclidean distance functions $Ed_{Rg}, Ed_{\Delta Rg}$ have the inheritance property.*

[10] Recall from Definition 67 that this induces a t-distance function d_t on $A@MOT_F$.

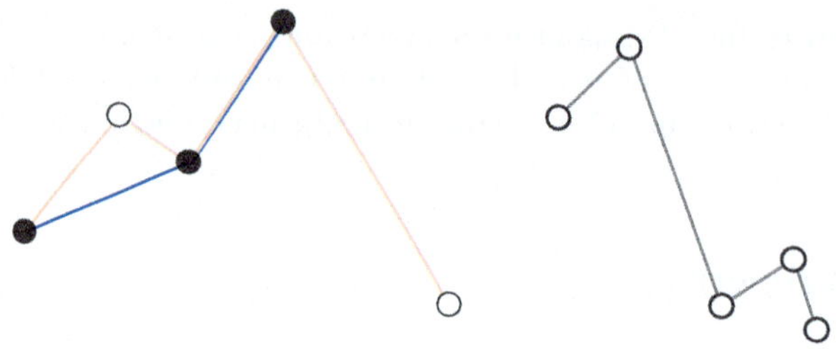

Fig. 22.1. Two motives M, N which are at diastematic distance zero but N has no submotif at this distance which corresponds to the submotif M^* drawn with black points.

Proof. This is evident for the Rg-type since the passage to a nearby submotif of N is reflected by a projection onto a submatrix of the rigid image $Rg(N)$ which is defined by selection of those columns which are defined by the submotif M^*. And this projection is continuous. On the ΔRg-type, we have to observe that the matrix columns of the ΔRg-abstractions of submotives are given by summing up the Rg-difference vectors of their supermotives. But this is a continuous operation, and we are done. QED.

Observe that this property and the following are independent of the chosen metric on the rigid or elastic shape spaces, since the continuity is only a topological property, and we know that the topology on the real n-space is independent of the defining norm.

Proposition 26 *Elastic shape type and Euclidean distance has the inheritance property.*

Proof. The elastic shape type is derived from the rigid difference type by normalization of the initial value function a to $\Sigma a = 1$. This means that we just divide the rigid difference matrix by Σa, and this is a continuous function on the interior $|\Delta_{c(c-1)/2-1}|^o$ of the affine $c(c-1)/2-1$-simplex which defines the shape space, see 22.2.1.3. QED.

Proposition 27 *Diastematic shape type Dia and Euclidean distance has the inheritance property.*

Exercise 49 *Give a proof of Proposition 27.*

Proposition 28 *If we have an equivariant paradigmatic group action*

$$A@MOT_F \xrightarrow[\circlearrowright_P]{t} \Gamma_t$$

which consists of isometries with respect to the distance function d, and if the shape type t and d have the inheritance property, then the gestalt distance $mot.d^*_P$ and shape type t also has the inheritance property.

Proof. Clearly, we have $(mot.d^*_P)_{t,M*,n}(N) \leqslant d_{t,M*,n}(N)$, and then, continuity of $d_{t,M*,n}$ in supermotives of M^*, where the function value is zero, implies continuity of $(mot.d^*_P)_{t,M*,n}$ in supermotives of M^*. QED.

22.6.2 Cognitive Aspects of Inheritance

Summary. This section makes the inheritance property plausible from the psychological and cognitive point of view.

– Σ –

Before we proceed to the construction of motivic topologies, we should briefly inspect the cognitive relevance of the basic inheritance property. Recognition of similarity of motives is a complex task of gestalt recognition. Little if anything is known about recognition of auditory gestalts, in particular, it is not known which shape type could be a cognitively relevant one, and whether there is a kind of combination of different shape types introduced above which could represent good cognitive information.

A particular problem of auditory gestalt recognition is the transitory nature of auditory gestalts: in the common situation of a piece being presented to a listener, motives pass by very fast and comparison of such transient objects must be an economic process, typically taking a time interval around 0.5 *sec*. Now, similarity of longer motives, consisting of eight tones, say, is not likely to be perceived as an irreducible entity. Fast processing must rely on elementary gestalt aspects of such motives. Mathematically speaking, it is plausible that gestalt perception breaks down to the perception of an interpretation of such motives, i.e., to the perception of motivic interpretations, i.e., melodies in the technical sense of global motivic compositions.

If we assume that we are given melodies M^I which are covered by motives M_i of a limited small cardinality c, then similarity of motives M can only be mediated via similarity of their charts M_i. So the present hypothesis is that similarity perception of large motives M breaks down into two subtasks: similarity perception of their charts M_i, and perception of the nerve $n(M^I)$. In this framework, the chart similarity check is related to the inheritance property. In fact, the latter means that similarity of charts is a necessary condition for similarity of the big motives. So if our cognitive performance is not strong enough to check similarity among big motives, it can at least make the chart check and thereby learn whether a necessary condition for similarity of big motives holds.

Now, this necessary condition takes place iff the inheritance property holds: This is the cognitive interpretation of this topological property.

22.6.3 Epsilon Topologies

Summary. Topologies on motif and gestalt spaces are introduced in case we are provided with a pseudo-metric and the inheritance property is fulfilled.

– Σ –

To begin with, we define the 'open discs' which will yield a base for a motivic topology:

Definition 73 *If we have an equivariant paradigmatic group action*

$$A@MOT_F \xrightarrow[\circlearrowleft_P]{t} \Gamma_t$$

which consists of isometries for the distance function d, if $M \in A@MOT_F$, and if ϵ is a positive real number, the ϵ-neighborhood of M *is the set*

$$D_\epsilon(M) = \{N \in A@MOT_F|\ (mot.d^*_P)_{t,M}(N) < \epsilon\} \tag{22.17}$$

which implicitly means that the candidates N have abstract cardinality at least equal to $abcard(M)$.

Proposition 29 *If we have an equivariant paradigmatic group action*

$$A@MOT_F \xrightarrow[\circlearrowleft_P]{t} \Gamma_t$$

which consists of isometries for the distance function d and such that the inheritance property is fulfilled, then the system of the ϵ-neighborhoods $D_\epsilon(M)$, all ϵ, M, is a base of a topology $\mathfrak{T}_{t,P,d}$ on $A@MOT_F$. This topology is called the epsilon topology *(for the data t, P, d).*

Proof. Let $D_{\epsilon_1}(M_1), D_{\epsilon_2}(M_2)$ be two ϵ-neighborhoods and take a motif $O \in D_{\epsilon_1}(M_1) \cap D_{\epsilon_2}(M_2)$. We must look for an $\epsilon_3 > 0$ such that $D_{\epsilon_3}(O) \subset D_{\epsilon_1}(M_1)$. A similar argument yields an $\epsilon_4 > 0$ which does the job for M_2, and the smaller of the two epsilons solves the problem.

By construction, we have $(mot.d^*_P)_{t,M_1}(O) = q < \epsilon_1$. Take a submotif $O_1 \subset O$ with $q = mot.d^*_P(M_1, O_1)$. By the inheritance property, we find a positive ϵ_3 such that $mot.d^*_P(O, Q) < \epsilon_3$ implies $(mot.d^*_P)_{t,O_1}(Q) < \epsilon_1 - q$. Then clearly, $D_{\epsilon_3}(O) \subset D_{\epsilon_1}(M_1)$. QED.

The next concepts relate gestalts and their representatives to motif inclusion:

Definition 74 *If P is a paradigmatic group for shape type t, we say that* gestalts behave well for t *iff for every submotif pair $M^* \subset M$, and motif $M_1 \in Ges_t(M)$, there is a submotif $M_1^* \subset M_1$ with $M_1^* \in Ges_t(M^*)$.*

Definition 75 *Let G^*, G be two gestalts for shape type t. Then we say that G^** is a small gestalt in G *and write $G^* \sqsubset G$ iff there are motives $M^* \in G^*, M \in G$ such that $M^* \subset M$.*

The proofs of the following two propositions are left as an exercise for the reader:

Proposition 30 [163, Prop.3] *If P is a paradigmatic group for shape type t which behaves well, then the small gestalt relation $\sqsubset$ is reflexive, transitive and antisymmetric*[11].

Proposition 31 [163, Prop.8] *Suppose that we are given an equivariant paradigmatic group action*

$$A@MOT_F \xrightarrow[\circlearrowright_P]{t} \Gamma_t$$

which consists of isometries for the distance function d, such that the inheritance property is fulfilled, and such that the pseudo-metric $ges.d_P^*$ on $A@MOT_F/Ges_t$ is a metric. Then gestalts behave well for t.

Exercise 50 *Show that the action of the affine counterpoint group CP, or more generally the equivariant action of any group P with finite orbits in the shape space, defines a metric $ges.d_P^*$ on $A@MOT_F/Ges_t$. In particular, the described actions of CP on rigid difference, diastematic, and elastic types have this property.*

Corollary 7 [163, Prop.9] *Given the epsilon topology $\mathfrak{T}_{t,P,d}$ on $A@MOT_F$, consider the quotient topology $\mathfrak{T}_{t,P,d}/Ges_t$ on the gestalt space $A@MOT_F/Ges_t$ for the gestalt mapping $Ges_t : A@MOT_F \twoheadrightarrow A@MOT_F/Ges_t$. Suppose that $ges.d_P^*$ is a metric. Then Ges_t is open and the system*

$$D_\epsilon(H) = \{G \in A@MOT_F/Ges_t |\ \exists G^* \sqsubset G, ges.d_P^*(G^*, H) < \epsilon\}, H \in A@MOT_F/Ges_t, 0 < \epsilon,$$

forms a base for $\mathfrak{T}_{t,P,d}/Ges_t$. More precisely, this follows from the formulas

$$Ges_t(D_\epsilon(M)) = D_\epsilon(Ges_t(M)), \tag{22.18}$$
$$Ges_t^{-1}(Ges_t(D_\epsilon(M))) = D_\epsilon(M). \tag{22.19}$$

The topology $\mathfrak{T}_{t,P,d}/Ges_t$ is called the *epsilon gestalt topology*. The following is immediate from the previous constructions:

Corollary 8 *For the epsilon topology $\mathfrak{T}_{t,P,d}$ on $A@MOT_F$ and the epsilon gestalt topology $\mathfrak{T}_{t,P,d}/Ges_t$ on $A@MOT_F/Ges_t$, and for a positive integer n, the relative topology*[12] *$\mathfrak{T}_{t,P,d}|_n$ on $A@MOT_F|_n$ is the pseudo-metric topology of $mot.d_P^*|_n$, whereas the relative topology on $\mathfrak{T}_{t,P,d}/Ges_t|_n$ on $A@MOT_F|_n/Ges_t$ is the pseudo-metric topology of $ges.d_P^*|_n$.*

Exercise 51 *Use assumption 1, Section 22.2.1, to show that $N \in D_\epsilon(M)$ implies $abcard(M) \leqslant abcard(N)$. It follows that $\bigcup_{n \leqslant i} A@MOT_F|_i$ is open for any positive index n.*

Exercise 52 *Show that $N \in D_\epsilon(M)$ implies $Ges_t(N) \subset D_\epsilon(M)$.*

Exercise 53 *Show that in the epsilon topology, if we have a submotif $M^* \subset M$ and $0 < \epsilon$, then there is $0 < \delta$ such that $D_\delta(M) \subset D_\epsilon(M^*)$.*

---

[11] See also Appendix C.2.

[12] See Appendix H.1.2.

22.7 First Properties of the Epsilon Topologies

Summary. For the total motif space $A@MOT_F$ as well as for the motif space μ associated with a motivic interpretation X^μ of a finite local composition X, some elementary properties of motivic topologies are discussed.

$-\Sigma-$

Suppose that we are given the motivic topology $\mathfrak{T}_{t,P,d}$ as introduced above. Whenever a motivic interpretation X^μ of a finite local composition X is given according to Definition 65, Section 22.1, the motif space $\mu \subset A@MOT_F$ will be given the relative topology $\mathfrak{T}_{t,P,d}|\mu$ which we also call the *epsilon topology.* Moreover, the image μ/Ges_t of μ under the gestalt map Ges_t is also given the relative topology $\mathfrak{T}_{t,P,d}/Ges_t|\mu/Ges_t$. To ease notation, we shall henceforth write $\mathfrak{T} = \mathfrak{T}_{t,P,d}, \mathfrak{T}_\mu = \mathfrak{T}_{t,P,d}|\mu$, and $\mathfrak{T}/Ges = \mathfrak{T}_{t,P,d}/Ges_t, \mathfrak{T}_{\mu/Ges} = \mathfrak{T}_{t,P,d}/Ges_t|\mu/Ges_t$ if the background assumptions are unambiguous.

Assumption 2 *In this section we always assume that we are given a type t, a pseudo-metric d on the space of t, an equivariant paradigmatic group P for t, which acts by isometries. We further assume that the inheritance property is verified and such that the gestalt distance $ges.d^*_P$ is in fact a metric.*

Proposition 32 [163, Prop.11] *The epsilon topologies $\mathfrak{T}/Ges$ and $\mathfrak{T}_{\mu/Ges}$ are T_0*[13].

Corollary 9 *Irreducible subsets in the epsilon topologies $\mathfrak{T}/Ges$ or $\mathfrak{T}_{\mu/Ges}$ have at most one generic*[14] *point.*

Lemma 35 *If the group T_{Onset} of all onset translations is contained in the paradigmatic group P, the topological spaces $(A@MOT_F, \mathfrak{T})$ and $(A@MOT_F/Ges, \mathfrak{T}/Ges)$ are irreducible.*

However, the motif space of $\mathfrak{T}$ is not sober in this case since it is irreducible, but no motif can be the generic point of the space since such a motif is a proper specialization of all its proper supermotives.

Lemma 36 *Given an epsilon topology $\mathfrak{T}$, if $M, N \in A@MOT_F$, the following statements are equivalent:*

(i) *There is $\epsilon > 0$ with $N \notin D_\epsilon(M)$.*
(ii) $Ges_t(N) \notin D_\epsilon(Ges_t(M))$.
(iii) $Ges_t(M) \not\sqsubset Ges_t(N)$.

Proof. If $abcard(N) < abcard(M)$, the above exercise 51 tells us that both, $N \notin D_{\epsilon(M)}$, and $Ges_t(M) \sqsubset Ges_t(N)$ are the case for any $\epsilon > 0$. Suppose $abcard(N) \leqslant abcard(M)$. If $Ges_t(M) \not\sqsubset Ges_t(N)$, then $\epsilon = (ges.d^*_P)_N(M) > 0$ because the gestalt distance is a metric, and therefore $N \notin D_\epsilon(M)$. Conversely, if there is a $\epsilon > 0$ with $N \notin D_\epsilon(M)$, then by definition of disk neighborhoods, $\epsilon = (ges.d^*_P)_N(M) > 0$, so no submotif of M can be in the gestalt of N. From corollary 7 we know that $N \notin D_\epsilon(M)$ iff $Ges_t(N) \notin D_\epsilon(Ges_t(M))$, and we are done.

Sorite 8 *Given an epsilon topology $\mathfrak{T}$, if $M, N \in A@MOT_F$, the following three statements are equivalent:*

(i) *M dominates N.*
(ii) *$Ges_t(M)$ dominates $Ges_t(N)$.*
(iii) $Ges_t(N) \sqsubset Ges_t(M)$.

Moreover these statements are equivalent:

(i) $\{M\}^- = \{N\}^-$.
(ii) $\{Ges_t(M)\}^- = \{Ges_t(N)\}^-$.
(iii) $Ges_t(M) = Ges_t(N)$.

[13] See Appendix H.1.4, Definition 181.
[14] See Appendix F.2.1.

The following two statements are equivalent:

(i) $\{M\}^- = Ges_t(M)$.
(ii) $Ges_t(M) \in A@MOT_F/Ges_t$ *is closed.*

Moreover, if M is closed, then $abcard(M) = 1$, *and conversely, if the spaces* $\Gamma_{t,n}$ *are mutually disjoint. We have*

$$\{M\}^- = \{N \in A@MOT_F|\ Ges_t(N) \sqsubset Ges_t(M)\}$$

and

$$\{Ges_t(M)\}^- = Ges_t(\{M\}^-) = \{Ges_t(N) \in A@MOT_F|\ Ges_t(N) \sqsubset Ges_t(M)\}.$$

In particular:

$$Ges_t(M) = \{N|\ M \text{ dominates } N\} \cap \{N|\ N \text{ dominates } M\}. \tag{22.20}$$

The only point to observe here is that a motif M with $abcard(M) > 1$ has its singleton subsets as submotives of abstract cardinality 1, and therefore their gestalts are different from $Ges(M)$ and M cannot be closed. The converse is evident.

The last property (22.20) is a purely topological characterization of a motif's gestalt. This means that talking about "gestalts" in our technical sense is absorbed by a purely topological fact, i.e.:

Fact 11 *The entire "gestalt" concept has been absorbed by a topological approach.*

Proposition 33 *If* $A@MOT_F|_1/Ges$ *is finite, then* $(A@MOT_F, \mathfrak{T})$ *is quasi-compact*[15].

Proof. Let C be an open cover of $A@MOT_F$, then

$$A@MOT_F|_1 = \coprod_{i=1,\ldots k} Ges(M_i) = \bigcup_{i=1,\ldots k} \{M_i\}^-$$

is a finite union of its gestalts and of the closures of its member representatives M_i. So there is a finite subset $C' = \{U_1, \ldots U_l\}$ of C that covers all representatives M_i. Since $MOT_{F,1} \subset A@MOT_F|_1$, each singleton motif is in one of C''s open sets U_j. Therefore any motif M is in the union of the open sets in C', and C' is a finite subcovering of C, QED.

22.7.0.1 Relative Topologies

In this general framework we consider the relative topologies $\mathfrak{T}_\mu, \mathfrak{T}_\mu/Ges$ when μ is a motif space for an interpretation X^μ of a finite local composition X. For the relative topology on μ, we have these notations:

- $D^\mu_\epsilon(M) = D_\epsilon(M) \cap \mu$.
- $Ges^\mu_t(M) = Ges_t(M) \cap \mu$.

We have seen in corollary 9 that irreducible closed sets in the epsilon topologies $\mathfrak{T}_\mu, \mathfrak{T}_\mu/Ges$ have at most one generic point. Now, since μ is a finite set, an irreducible closed subset $W \subset \mu$ has a maximal index $n(W)$ such that $W \cap (\bigcup_{n(W)\leqslant i} A@MOT_F|_i) \neq \emptyset$. But by the metric hypothesis on gestalts, this intersection must consist of one single gestalt since this intersection is irreducible and the metric implies T_2 on the abstract cardinality level $n(W)$. So this gestalt is the generic gestalt of W, and we have

Proposition 34 *On a finite space* X, *the gestalt topology* $\mathfrak{T}_\mu/Ges$ *is sober.*

[15] See Appendix H.1.4, Definition 183.

The above procedure also yields [163, prop.21] an algorithm for constructing the irreducible components of μ/Ges. First, start with the gestalts $Ges(M_i), i = 1, \ldots r$ of maximal abstract cardinality in μ. These points define irreducible components. In $\mu/Ges - \bigcup_{i=1,\ldots r}\{Ges(M_i)\}^-$, take the maximal abstract cardinality gestalts $Ges(N_1), \ldots Ges(N_s)$ which define irreducible components $\{Ges(N_j)\}^-$ of μ/Ges, and so on, until the remainder is empty. Now, if any irreducible component C is given, it must be covered by at least two of such irreducible components both of which have proper intersection with C. And it easily follows that C must be in fact one of them.

Proposition 35 *On a finite space X, the gestalt space μ/Ges (with the gestalt topology $\mathfrak{T}_\mu/Ges$) is homeomorphic to the associated sober space*[16] μ^s.

Proof. Let $W \subset \mu$ be irreducible and closed. We know that it is a union of relative gestalts $Ges_t^\mu(M)$. Since $\mathfrak{T}_\mu/Ges$ is the quotient topology, the image $Ges(W)$ is irreducible and closed. Since μ/Ges is sober, $Ges(W)$ has a unique generic point $Ges(N)$. Then by Sorite 8, $W = \{N\}^-$. Therefore, the canonical continuous map $q : \mu \to \mu^s$ is surjective. By Sorite 8, it has the same fibers as the gestalt map. So the canonical[17] continuous map $p_\mu : \mu^s \to \mu/Ges$ is a bijection. Let $V \subset \mu^s$ be an open set. By definition of the sober topology, it is associated to an open set $U \subset \mu, V = U^s$, and we have $q^{-1}(V) = U$. So the image $p_\mu(V) = Ges(U)$ is open, and we are done. QED.

Exercise 54 *If instead, we take the total motif space $A@MOT_F$ with the epsilon topology $\mathfrak{T}$, we have a canonical continuous surjection $p : A@MOT_F^s \to A@MOT_F/Ges_t$ which induces a homeomorphism on the image $q(A@MOT_F)$ of the canonical map $q : A@MOT_F \to A@MOT_F^s$.*

Proposition 36 *Let X^μ satisfy the SEA property, n_{min} being the minimal abstract cardinality of members of μ, and $\mu_{min} = \{M \in \mu |\ abcard(M) = n_{min}\}$. Then the closed points are contained in μ_{min}. If the shape spaces $\Gamma_{t,n}$ are mutually disjoint, then the set of closed points is exactly μ_{min}.*

Proof: Exercise.

22.7.1 Toroidal Topologies

Summary. The special abstraction type of pitch and onset class formation leads to a similarity concept of gestalts which cannot be dealt with by use of (pseudo-)metrics. It is related to the topological specialization type discussed in Section 12.2.2.

– Σ –

In this case we shall consider a completely different kind of topologies which stems from algebraic geometry rather than from Euclidean geometry. Omitted proofs in this section can be found in [163, ch.4]. To fix the ideas, we take the toroidal shape type with shape spaces

$$\Gamma_{Toroid,n} = ObLoc_0^{\leqslant n, OnPiMod_{m,l}}$$

of local, zero-addressed compositions in $OnPiMod_{m,l}$ with cardinality at most n and shape map $Toroid = Toroid_\gamma^{m,l}$ as introduced in formula (22.9). Here, the abstract cardinality is the usual cardinality of a local composition. Also we take the full affine group $P = \overrightarrow{GL}(\mathbb{Z}_m \oplus \mathbb{Z}_l)$ as a paradigmatic group introduced in 22.4.1.4.

[16] See Appendix F.2.1.
[17] See Appendix F.2.1, Proposition 105.

22.7.1.1 Dominance Topology

The analogy to the pseudo-metric topologies on the shape spaces $\Gamma_{Toroid|n}$ is played by the dominance topology introduced in formula (12.4) of Section 12.2.2. This time, we essentially work in $\mathbb{Z}_m \oplus \mathbb{Z}_l$ and consider abstract motives $x, y \subset \mathbb{Z}_m \oplus \mathbb{Z}_l$ of abstract cardinality $abcard(x) = abcard(y) = n$. We say that y *is an abstract specialization of* x (in symbols: $x \twoheadrightarrow y$) iff there is a bijective morphism of local compositions $s : x \to y$; so until now, no topology is involved, and the terminology seems a bit ambiguous, but the reason for this confusion will be given soon. Given two motives $M, N \in A@MOT_F|_n$, we also say that N *is a specialization of* M (in symbols: $M \twoheadrightarrow N$) iff $Toroid(N)$ is an abstract specialization of $Toroid(M)$.

The following lemma makes clear what it means that two abstract motives are mutually abstract specializations.

Lemma 37 *For the toroidal type Toroid, if x, y are two abstract motives, the following statements are equivalent:*

(i) The abstract motives x, y are mutually abstract specializations $x \twoheadrightarrow y, y \twoheadrightarrow x$.
(ii) The local compositions x, y are isomorphic.
(iii) We have $Ges_{Toroid}(x) = Ges_{Toroid}(y)$.

Proof. The equivalence of (ii) and (iii) follows from Sorite 6, Section 8.3.5. Clearly (ii) implies (i); for the converse, suppose we have two bijective morphisms $f : x \to y, f : y \to x$. Then their compositions $g \circ f, f \circ g$ are bijective and there is a positive power n such that $(g \circ f)^n = Id_x, (f \circ g)^n = Id_y$, whence $(g \circ f)^{n-1} \circ g$ is the right and left inverse of g, and x, y are isomorphic. QED.

This implies that the following settings are well defined.

Definition 76 *If M, N and x, y are two motives and abstract motives, respectively, then we say that* the gestalt $Ges(N)$ is a specialization of gestalt $Ges(M)$ *(in symbols $Ges(M) \twoheadrightarrow Ges(N)$) and that* the abstract gestalt $P.y$ is an abstract specialization of the abstract gestalt $P.x$ *(in symbols $P.x \twoheadrightarrow P.y$), respectively, iff $M \twoheadrightarrow N$ and $x \twoheadrightarrow y$, respectively.*

Corollary 10 *Two abstract gestalts (two gestalts) are mutually abstract specializations (mutually specializations) iff they are equal.*

For a fixed abstract cardinality k, we have the following Kuratowski closure operator[18] on $A@MOT_F|_k/Ges$. If $X \subset A@MOT_F|_k/Ges$, then we set

$$\bar{X} = \{y \in A@MOT_F|_k/Ges|\ \exists x \in X \text{ such that } x \twoheadrightarrow y\}.$$

Clearly, in this dominance topology, the sets $[x) = \{y|\ y \twoheadrightarrow x\}$ are the smallest open neighborhoods of x and define a base for the dominance topology on the gestalt space. On the inverse image $A@MOT_F|_k$ of motives having abstract cardinality k, we take the inverse image of the dominance topology, i.e., the base $[M) = \{N|\ Ges(N) \twoheadrightarrow Ges(M)\} = \{N|\ N \twoheadrightarrow M\}$, the latter equality being the case because by Lemma 37, dominance is invariant modulo gestalts. Clearly, the gestalt map $A@MOT_F|_k \to A@MOT_F|_k/Ges$ is open and the dominance topology is the quotient topology.

22.7.1.2 Specialization Inheritance and Specialization Topology

In the next step of our construction, we look at inheritance properties for the dominance topology.

Definition 77 *If for any triple $M, M^*, N \in A@MOT_F$ with $M^* \subset M$ and $M \twoheadrightarrow N$, there is $N^* \subset N$ such that $M^* \twoheadrightarrow N^*$, we say that* specialization is inherited.

If for any triple $M, N^*, N \in A@MOT_F$ with $N^* \subset N$ and $M \twoheadrightarrow N$, there is $M^* \subset M$ such that $M^* \twoheadrightarrow N^*$, we say that specialization is co-inherited.

[18] See Appendix H.1. Kuratowski closure operators are used to define topologies.

Proposition 37 [163, prop.27,cor.5] *Every toroidal shape type $Toroid_\gamma^{m,l}$ is inherited and co-inherited.*

This property guarantees that we may deduce another topology in the spirit of the Epsilon topologies, but now starting from dominance topology:

Proposition 38 [163, def.19,prop.28,cor.6] *The system*

$$El(M) = \{N \in A@MOT_F |\ \exists N^* \subset N \text{ such that } Ges(N^*) \in [Ges(M))\}$$

of elementary neighborhoods *of motives $A@MOT_F$ is a base for a topology on $A@MOT_F$, which is called the* specialization topology $\mathfrak{T}^{m,l,sp}_{Toroid,\gamma} = \mathfrak{T}^{sp}_{Toroid}$. *Elementary neighborhoods are invariant under gestalts.*

The quotient topology $\mathfrak{T}^{sp}_{Toroid}/Ges$ on the gestalt space $A@MOT_F/Ges_{Toroid}$ is also called specialization topology. *Under these topologies, the gestalt map is an open continuous map.*

The specialization topology $\mathfrak{T}^{sp}_{Toroid}/Ges$ has a base which consists of all elementary neighborhoods (in fact images of the elementary neighborhoods of the motif space) of shape

$$El(G) = \{H \in A@MOT_F/Ges |\ \exists H^* \sqsubset H \text{ such that } H^* \in [G)\}$$

for $G \in A@MOT_F/Ges$.

In the specialization topology, we have analogous results as in the epsilon topology, regarding the closure of a point:

Proposition 39 [163, prop.31,prop.39,prop.41,prop.42] *Ff $G \in A@MOT_F/Ges$ for the specialization topology $\mathfrak{T}^{sp}_{Toroid}/Ges$, then*

(i) $El(G) = \{H \in A@MOT_F/Ges |\ H \text{ dominates } G\}$.
(ii) $\{G\} = \bar{G} \cap El(G)$.

For the specialization topology $\mathfrak{T}^{sp}_{Toroid}$, if $M \in A@MOT_F$ is a motif, then

(i) $El(M) = \{N \in A@MOT_F |\ N \text{ dominates } M\}$.
(ii) $Ges(M) = \bar{M} \cap El(M)$.

The specialization topology $\mathfrak{T}^{sp}_{Toroid}/Ges$ is T_0, irreducible, and quasi-compact.

Many of the properties for the specialization topologies $\mathfrak{T}^{sp}_{Toroid,\mu}$ on a motif space μ of a finite local composition X, and $\mathfrak{T}^{sp}_{Toroid,\mu/Ges}$ on μ/Ges are analogous to those which we have discussed for the epsilon topologies. We refer to Buteau's work [163, ch.4] for the details. In particular, the irreducible components for $\mathfrak{T}^{sp}_{Toroid,\mu}$ can be calculated by a recursive procedure which is similar to the one discussed in Section 22.7.0.1.

Exercise 55 *Show that the specialization topology $\mathfrak{T}^{sp}_{Toroid,\mu}$ still homeomorphic to the associated sober space of the specialization topology on gestalts.*

Remark. In [163, ch.4], it is shown that specialization relations, together with different subgestalt relations can be used to define a *gestalt specialization category* $\mathfrak{Ge}(A@MOT_F/Ges_{Toroid})$.

Principle 21 *As an overall result from the construction of motivic topologies, we can now consider* generic points *and related dominance structures from the theory of sober spaces. This is the basis for a topological realization of Reti's ideas in the sense that the* semantic charge *connoted with the very concept of a motif turns out to be a precise instance of the genericity properties of the topological motif space which is associated with a given score. In other words: Cognitive and structural motivic semantics is related to and reduced from the topological dominance configurations in the motivic topologies.*

We come back in Section 22.9 to the more detailed technique of motif weights to make motivic semantics explicit on a quantitative level.

22.8 Rudolph Reti's Motivic Analysis Revisited

Summary. After the precise construction of immanent motivic "organisms" associated with determined scores, we conclude this chapter with a critical review of Rudolph Reti's classical texts in the light of the presented theory. For more details, see [163, ch.5.2].

– Σ –

22.8.1 Review of Concepts

Complete works about motivic analysis are practically non-existent, therefore Reti's book [892] is still a good reference to the methodology of traditional musicological motif theory. Despite his accurate overview on the evolution of a motif through a composition, Reti does not use rigorous terminology, i.e., he uses the same word for different concepts or different words for the same concept. It seems very difficult to grasp its contents and substance. This vague terminology produces hazy deductions, not to say contradictions. However, behind his words, Reti's conception of a first sketch for a thematic theory of music is considerable.

Reti is not even concerned with reliable terminology:

> [892, p.12] *In general, the author does not believe in the possibility or even desirability of enforcing strict musical definitions.*

But it is undeniable that before any attempt to interpret Reti's motivic analysis within our mathematical model, we must take a closer look at his terminology whose nuances are essential for bringing consistency into Reti's motivic score analysis.

Let us get off ground with Reti's definition of a motif:

> [892, p.2-3] ... *any musical element, be it a melodic phrase or fragment or even only a rhythmical or dynamical feature which, by being constantly repeated and varied throughout a work or a section, assumes a role in the compositional design somewhat similar to that of a motif in the fine arts.*

Since Reti asserts that he handles the words "motif" and "theme" without categorical distinction, we are aware that in some of the following quotations or examples, the word *theme* can be, for our purposes, replaced by *motif*. All the same, Reti points out a difference between the two concepts by defining a theme as

> [892, p.3] ... *a fuller (compared to a motif) group or "period" which acquires a "motivic" function in a composition's course.*

According to Reti, both a "motif" and a "theme" have the same function and therefore are reduced to almost the same concept. A first distinction between these two notions is length. Then, Reti adds that a theme is composed of few motives. Moreover, the difference of the length has consequences on how and where motives and themes can occur in the composition. A motif being short, it can be easily repeated throughout the composition. It can also be modified and inserted in parts having different moods. These frequent occurrences create a certain unity within the piece. A theme has a similar function but cannot, as easily as the motif, be modified or/and inserted in the composition.

We now state a crucial idea in Reti's book. Reti asserts that abstractions have to be made in order to compare themes and motives:

> [892, p.243] *Similar* liberties *were taken in our analysis on several other occasions...by* rearranging *the design..., by exchanging the octaves in which some phrases were notated in the score.*

In a more general way, when talking about classical symphonies, Reti claims that

> [892, p.13-14] *He (the classical composer) strives toward homogeneity in the inner essence... toward variety in the outer appearance. Therefore he changes the surface but maintains the substance of his shapes... Tempo, rhythm, melodic detail, in fact the whole character and mood are altered and adjusted to the form in which the composer conceived them fitting to the new movement.*

Reti clearly affirms that we have to make abstraction from some qualities of a motif in order to make the connection with its transformations in other parts of the piece. As a matter of fact, Reti always makes abstraction when comparing sequences of notes. For example, it can be the simple abolition of note loudness or more complex procedures such as the exchange of octave positions inside a given sequence of notes.

We should remark that Reti never uses the word "gestalt"[19]. Instead, he is concerned with comparisons of "shapes". The word "shape" is not chosen at random and is met throughout the book:

> [892, p.205] ... *a theme is that shape around which...*

The importance of the shape concept is that it englobes the idea of abstraction. Nevertheless, Reti does not have a well-defined usage of this concept. In similar situations, such as in Examples [892, 328,p.210] and [892, 334,p.213], both words "motif" and "shape" are used with the same meaning. The reason for this lack of rigor is simply that when comparing motives, natural abstractions like the abolition of loudness are somehow unconsciously operated. For Reti, starting a comparison with shapes seems trivial.

According to Reti, the shape of a motif is incessantly heard, literally or modified, through the piece. The shape is imitated, varied or transformed:

> [892, p.240]
> 1. imitation, *that is literal repetition of shapes, either directly or by inversion, reversion, and so forth;*
> 2. varying, *that is, changing of shapes in a slight, well traceable manner;*
> 3. transformation, *that is, creating essentially new shapes, though preserving the original substance.*

In other words, a motif has a "smooth shape", i.e., its shape can be imitated or smoothly modified and remains the same, or as Reti asserts, it remains "identical"[20]. However the accepted transformations of shapes can reach a limit:

> [892, p.355] *...his (Beethoven's) most impressive thematic constructions lie on the border line between being matchless master strokes of transformation or utterances wherein the thematic bond has almost dissolved in that very transformation process by which they were created.*

Reti states that a shape can not be endlessly transformed and still remain identical with its initial shape: identity has a 'limit of transitivity'. This is by itself a contradiction since any accepted concept of "identity" is transitive. However, there are subtle identification problems which, in case the analyst is not precise enough, i.e., does more identification than the situation requires, cause contradictions while the essential points drop out. Unfortunately, Reti did not realize the implications that bears his use of the concept of "identity".

Reti tries to avoid this contradiction by constant reference to the first appearance of a motif in the score. Reti usually affirms that a shape is a transformation, imitation or variation of a motif by, most of the time, comparing the shape of the first motif appearance with the considered transformation. In a sense, this "identity" is an order relation, the first appearance A being "greater" than all the later appearances, $B, C, \ldots : A > B, A > C, \ldots$, but transitivity is not relevant here, because, besides $A > B, A > C \ldots$, no relation $B > \ldots, C > \ldots$ etc is considered. So the transitivity question $A > B \;\&\; B > C \stackrel{?}{\Rightarrow} A > C$ is of no interest. But again, we are confronted with a non-sense definition of the identity: it is not even symmetric! In any case the common concept of identity is violated. In the next section, we shall clarify how to use "identity of motives" in a consistent way.

Although Reti was imprecise he was not irreversibly wrong. The abuse of the term "identity" is somewhat clarified in the concept of a motif, unfortunately hidden by a fluffy terminology. For Reti, a motif conceives

[19] Nonetheless, the gestalt concept is omnipresent. We shall see that Reti's implicit definition of a gestalt is first the choice of an abstraction perspective, and second all the immalleable transformations (the imitations) of the motif's shape in the chosen perspective.

[20] The words "identity" and "identical" are used over and over in the book and their meaning is not the one that one is accustomed to from logic! For example, see [892, pp.13,21,38,102,167,243,...].

> [163, p.108] *the first appearance of a certain group of notes with shape abstraction such that, throughout the whole composition, there are "many" note groups having respective shapes (i.e., shape abstractions) which are imitation, variation or transformation of the first shape.*

To distinguish the different concept levels in our discussion, we denote by *Reti-motif* the above "definition" of a motif.

22.8.2 Reconstruction

We now proceed in reconstructing Reti's concept framework from the point of view of the present motif theory.

22.8.2.1 Choice of Parameters

We first fix the parameter space $F = Onset \oplus Pitch \oplus Loudness \oplus Duration$ since for the context of Reti's book, no other relevant tone parameters are of use. To be clear, this framework does not include pauses or bar-lines as relevant tone objects, only 'sounding' notes are considered. Here are the detailed settings (see also Appendix A for these standard values):

- The first note has *Onset* value 0.
- Duration values are taken by the prescription that 1.0 in the *Onset* coordinate corresponds to the literal mathematical value of 4/4 duration.
- For the *Pitch* values, we select the usual gauge with middle $C = 0$, and the chromatic pitch set being parametrized by the integers, i.e., $C\sharp = D\flat = 1$, $D = 2$, etc.
- *Loudness* is gauged by the usual setting where $mf \sim 0$, $f \sim 1$, $ff \sim 2$, etc., and $p \sim -1$, $pp \sim -2$, etc.
- By *toroidal type* we intend $Toroid_{\gamma}^{12\times 12}$ with a default grid[21] $o = (0,0), x = (1,0), y = (0,1)$.

Furthermore, as might be expected, we always deal with the zero-address in Reti's context.

22.8.2.2 Shapes, Imitations and Transformations

Let $X \subset 0@F$ be the local composition consisting of all notes within the given composition, and with respect to the above gauge of the four parameters. Then let μ^* be the set of all motives within X.

Reti's implicit first step in his motivic analysis of a composition is the choice of a determined set $\mu \subset \mu^*$ of (mathematical) motives, i.e., "sequences of notes" which he wants to compare with each other in order to find imitations, variations and transformations of the shape of a certain sequence of notes which he will then call a motif (a Reti-motif in our terminology). Call our mathematical motives $M \in MOT_F$ *Math-motives* in order to distinguish them from Reti's more involved Reti-motives. In fact, it is remarkable that most of the Math-motives in μ are no way qualified for becoming contributors to Reti-motives. They are just a kind of motivic 'raw material'. Furthermore, Reti naturally allows himself to compare parts (subsequences) of the chosen Math-motives (sequences). He normally does not regress to the single notes, thus he gives himself a lower limit of motif cardinality. This all means that in the mathematical model the SEA property must hold with respect to the lower limit for μ. Most of the time Reti applies the limit of the intervals, i.e., $n_{min} = 2$ in SEA. Let us now reconstruct Reti's motif definition

> *...the first appearance of a certain group of notes with shape abstraction such that, throughout the whole composition, there are "many" note groups having respective shapes (i.e., shape abstractions) which are imitation, variation or transformation of the first shape.*

[21] This grid is by no means mandatory. According to the concrete situations, we shall define better adapted quantization data.

within the mathematical model:
"... with shape abstraction..." As seen in the previous part of this section, the choice of an abstraction of the sequences of notes is the starting point for comparison. To 'simulate' Reti's abstraction procedure by our model, we select a shape type $t : MOT_F \to \Gamma_t$, and we model the shape abstraction of a Math-motif $M \in \mu$ by its associated abstract motif $t(M) \in \Gamma_t$.
"... which are imitation..." We must emphasize that declaring that a shape is an "imitation of another" suggests the choice of a collection of admissible imitation transformations, a choice which we associate with a choice of a paradigmatic group P in the mathematical model. For example, most of the time strict repetitions (translations in time) are admissible, and it is clear that the choice depends on the selected shape abstractions. The implicitly conceived gestalt of a sequence of notes in Reti's book perfectly corresponds to a gestalt of a Math-motif in a motif space.

According to Reti's analysis, three examples [892, example 35, p.33] (first scene of Schumann's Kinderszenen, shape I), [892, example 20b, p.21] (second version of Beethoven's Ode an die Freude shape), and [892, example 1, p.11] (descending third in measure 19 of the first movement of Beethoven's Ninth Symphony) of imitations seems to coincide with the counterpoint group.

The choice of abstraction (the choice of the types) greatly influences our view on a composition, and this is a strong point of this analysis since it takes into consideration different qualities and characters of the composition.

We now introduce the distance $ges.d^*_P$ (defined in Section 22.5) of Math-motives into our model of Reti's motivic analysis. In a sense the distance $ges.d^*_P$ between motives, i.e., two sequences of notes such that their shapes have the same number of "elements" (same abstract cardinality in the selected shape type), gives a measure for their diversity. For Reti, two sequences of notes with shape having the same number of elements are identical (imitated), closely alike (varied or transformed) or different (not related). In the mathematical model, this kind of relation is more subtle. The shapes of two sequences of notes "being identical" means having distance 0, "being closely alike" requires to know "how close" they are from each other and being different requires to know "how far" they are from each other. There is no longer only three rough categories but instead, we are given subtly nuanced relations between sequences of notes.

Remark 5 The $ges.d^*_P$ distance is a priori defined on MOT_F and not on μ. This means that the calculation of $ges.d^*_P$ for two sequences of notes *within* a given composition X is determined by first looking at their shapes (following a given abstraction), and then, given a paradigmatic group P, by comparing not only all their imitations within the composition but also all the a priori imaginable imitations. According to rigid type, we easily see that a similar distance defined *only* on μ would yield absurd results. Moreover, these 'imaginary' shapes are naturally, but probably also unconsciously, used by Reti.

"...variation or transformation..." Variations and transformations of a sequence of notes are more delicate to translate into the mathematical model. Since the difference between variation and transformation is simply a question regarding the level of change between shapes, we merely consider transformations, i.e., variations are viewed as special cases of transformations. There seem to be no disadvantageous consequences to this identification.

We recall that, according to the analyst Reti, there is a limit to a transformation of a shape such that it still relates to the original shape, i.e., such that it remains "closely alike". In mathematical terminology, there is a maximal admissible distance ϵ_{max} between the Math-motives. In the motivic topology, this limit corresponds to a fixed positive real number ϵ_{max}. The latter is a border line such that if two shapes with the same abstract cardinality are less than ϵ_{max} distant, then they are considered as being "closely alike". However, transformations do not necessarily maintain the number of elements. In the mathematical model, transformations of shapes can be translated to three different situations corresponding to the change of number of elements of the shapes:

1. If two shapes have the same number of elements and are closely alike, then, in the motivic space, their Math-motives M and N are in the other's ϵ_{max}-neighborhood, respectively, which is $M \in D_{\epsilon_{max}}(N)$ and $N \in D_{\epsilon_{max}}(M)$. Example: [892, examples 85,86, p.62] (Palestrina's Missa Marcelli, comparison of beginnings of the two first sections in Kyrie Eleison and the Christe Eleison).

2. If in the terminology of Reti, a shape A is "closely alike" inside the shape B, then, in the mathematical model, for a given shape type, this means that a Math-motif M with shape $t(M) = B$ is in the ϵ_{max}-neighborhood of the Math-motif N with $t(N) = A$: $M \in D^{\mu}_{\epsilon_{max}}(N)$. Intuitively speaking, this means that the shape of N is closely alike to the shape of a shape of a submotif of M. Example: [892, example 118, p.80] (first movement of Brahms' Second Symphony).
3. If a part A^* of a shape A is closely alike inside shape B (and conversely in Reti's terminology), then, in the motivic space, there is a submotif M^* of M with shape A^* and A, respectively, and such that N with shape B is in the ϵ-neighborhood of M^*: $N \in D^{\mu}_{\epsilon_{max}}(M^*)$. Example: [892, example 218, p.143] (Brahms' First Rhapsody).

If the relation between a shape and its transformation is musically too hidden inside the transformation, in the sense that it is hardly heard or even not at all heard, then this is, what Reti calls, an "indirect affinity". This phenomena is similarly translated into motif space terminology as a transformation with possibly larger ϵ_{max}.

22.8.2.3 Reti's Identity Relation Revisited

In the previous section we mentioned the inconsistent use by Reti of the concept of "identity" in the sense of imitations, variations and transformations. However in the mathematical model, identity of gestalt can be compared with Reti's use of the concept of Identity. In fact, we know from Sorite 8, that, under certain conditions, $Ges_t(M) = Ges_t(N)$ for two Math-motives M and N iff $M \in D^{\mu}_{\epsilon}(N)$ and $N \in D^{\mu}_{\epsilon}(M)$, for all $\epsilon > 0$.

This gives us the possibility to review the gestalt identity concept under a purely topological perspective. We can see now that Reti's violation of the strict concept of identity is related to violation of one or several parts of the above neighborhood condition for identity of gestalt. It is undeniable that the identity of motives should be limited to imitations, i.e., the identification of motives with their gestalts yields consistent statements. The identification with transformations creates a system where every Math-motif can be identified to any other Math-motif! This is certainly not what Reti wanted to state unrestrictedly. A reconstruction of Reti's theory (and other fuzzy attempts, such as [870]) should be undertaken on the basis of topological concepts as modeled above.

22.9 Motivic Weights

Summary. Motivic weight functions are a means for understanding motivic topologies by use of real 'coordinate functions'. We describe the technique and its meaning for semantic purposes, in particular, regarding abstraction from motif parameters.

– Σ –

In the analysis of metrical structure of Section 21.2, Definition 64, we have introduced sober weights. This can be carried over to motivic structure without any further modification; recall (see Section 21.2) that for the motivic interpretation X^{μ}, we have the nerve map $Sp : X \to n(X^{\mu})$. Also recall that an $\mathbb{R}$-valued sober (motif) weight on the A-addressed local composition $X \subset A@F$ is a function $W : X \to \mathbb{R}$ which factorizes through the nerve map Sp. A factor $nW : n(X^{\mu}) \to \mathbb{R}$ is called a nerve weight for W. A nerve weight nW is said to be induced iff its values are determined by the values on the vertexes, i.e., by the restriction $n_0W = nW|_{n_0(X^{\mu})}$.

In the implementation of motif theory on the RUBATO®'s MeloRUBETTE®, such induced sober motif weights have been used for performance oriented motivic analysis. We shall discuss this approach in Chapter 41.2. In the present discussion, we want to concentrate on the semantics of such weights. To calculate the weight $W(x)$ means that we give the tone event x a value which stems from the weight values of motives $n_0W(M), x \in M$, including some additional configuration functions which take care of the simplex structure $Sp(x)$. Basically, this means that we validate x in its motivic environment: "What is the motivic position of x

within the given motif space μ of X?" So we give each tone event $x \in X$ a 'motivic coordinate'. Isn't this too coarse? The point is that for performance, complex coordinates are not useful since the instrumentalist really needs numeric information of how strongly that key should be stroked. And that may be ok for performance, but it is not the yoga of motivic analysis. There is a more faithful approach to motivic coordinates if we step over to the nerve weight nW, which we suppose to be induced, to make the ideas more transparent.

Restating the precedent question, a nerve weight fixes the motivic coordinate $nW(M)$ of any motif $M \in \mu$. Why are we—beyond applications to performance theory—interested in such coordinates? The point is that the motivic epsilon or toroidal topologies are not immediately understandable in terms of classical Euclidean coordinates, in fact, these topologies have more of an algebro-geometric flavor, including generic points, specializations, and related 'far-out' structures. It has been shown in the discussion of Reti's theory that these tools are adequate. But the prize was a strong deviation from numerical, quantitative representation. The immediate semantical potential of motivic topologies is not a numeric one. On this basis, the introduction of nerve weights should provide us with a reconstruction of quantitative aspects of topological perspectives, possibly with a bunch of numerical invariants which could in the ideal case approach a complete set of invariants.

One approach to this goal has been realized by the MeloRUBETTE®. Let us just have a look at that methodology. Fix a shape type t, an equivariant paradigmatic group P of isometries, the standard situation of our motif theory. Given a positive limit number ϵ, we look at the $\mathfrak{T}$ neighborhood configuration on μ as seen from a fixed motif M. What does this object 'know' from its topological position? We could be interested in knowing how many motives are in the disk neighborhood $D^{\mu}_{\epsilon}(M)$. This would yield a kind of 'presence' $pr_{\mu,\epsilon}(M)$ of M within larger motives, up to some tolerance ϵ of shape similarity. It could also be interested in knowing how much motives M participate in their 'presence', thereby yielding a kind of 'content' $ct_{\mu,\epsilon}(M)$ of M. This data can be used to give the motif a measure for its relevance or weight in the present motif space μ, a weight function

$$nW_{\epsilon}(M) = \omega(pr_{\mu,\epsilon}(M), ct_{\mu,\epsilon}(M)), \tag{22.21}$$

for example with $\omega(x, y) = x.y$. This is only one approach to grasp topological information via numerical values, but we just want to give a concrete idea of how to proceed in general. Given these weights $nW_{\epsilon} : \mu \to \mathbb{R}, \epsilon > 0$, we obtain a map

$$nW : \mu \to \mathbb{R}^{\mathbb{R}_+} \tag{22.22}$$

on the vertexes of the nerve $n(X^{\mu})$, being extended to the entire nerve $nW : n(X^{\mu}) \to \mathbb{R}^{\mathbb{R}_+}$ via a given induction function, i.e., $nW(\sigma) = f(nW(M), M \in \sigma)$. Since the motif space is finite, we may restrict to a finite sequence $\epsilon. = \epsilon_1 < \epsilon_2 \ldots < \epsilon_k$ of limit numbers to grasp all changes within these weight functions. This means that we consider the weight $nW_{\epsilon.} : \mu \to \mathbb{R}^k$, and its induced extension $nW_{\epsilon.} : n(X^{\mu}) \to \mathbb{R}^k$. It is a major task in motif theory to construct weight functions following the scheme of $nW_{\epsilon.}$ in order to investigate the abstract motivic topologies via their images in real n-space.

Remark 6 The AST-ideas on similarity, as developed by Morris, Lewin, Rahn, and others may be looked up in Morris' work *[765, p.103ff.]*.

Part VI

Harmony

23

Critical Preliminaries

Aber atonal wird man irgend ein Verhältnis von Tönen sowenig nennen können, als man ein Verhältnis von Farben aspektral oder akomplementär bezeichnen dürfte. Diesen Gegensatz gibt es nicht.
Arnold Schönberg [948, p.240]

Summary. The present part on harmony is a traditionally dominant and extended portion of music theory. Therefore, it is adequate to review some of the important approaches to harmony. This chapter is however far from a complete synthesis of harmony and its history. We have selected three representative approaches which are systematically elaborate and theoretically founded: H. Riemann, P. Hindemith, and H. Schenker. The following overview concentrates on the divergence between claim and realization, and it does, once again, lay bare the enormous difficulty to set up a precise discourse about music *without*—hélas—the power of mathematical language. Also this critique is not thought to be a preliminary to something which in the subsequent chapters of this part will be perfectly solved by mathematical music theory. The discourse simply tries to persuade music theorists that a) the commonly cultivated status quo of the subject is scientifically unacceptable, and b) that mathematically sharpened concepts, constructs, and models can show ways to more in-depth and precise understanding of harmony—without banning it to history and "atonal" negation. Generic harmony is a universal perspective of music, and it is unscientific as well as near-sighted if not anti-musical to abandon harmonic paradigms *instead of* embedding them into a diachronically and synchronically open, unified, and universal concept framework. To be clear, the *main question* is not to defend or instantiate *any* ideology of harmony—this is the unhappy business of Pythagorean fundamentalists—but to investigate its possible semiotic functions in musical works and their communicative explication, to develop an adequate language, and to propose consistent and sound models of harmonic processes.

– Σ –

The theory of harmony as a theory of chords and chord progressions comprises a wide range of different approaches, methods and questions. Thereby the historic context is only one factor among others to explain these differences. Harmonic research is situated in the intersection of various fields of interests, which correspond to different approaches and questions. For instance Paul Hindemith's interest in *Unterweisung im Tonsatz* [467] is that of a composer—so his focus is to explain, in how far and why diverse combinations of tones as chords differ in their esthetic effects. Arnold Schönberg's *Harmonielehre* [948] aims at students of composition as well. In comparison to this perspective Schenker and his adepts analyze the harmonic structure of pieces of music in order to explain the impression of "inner coherence" in this piece which includes a description of different hierarchic levels. This approach results in a completely distinct definition of what is called a "chord" compared to, for instance, Hindemith's theory. In the following preliminaries to the mathematical theory of harmony, we don't want to give a complete summary of all these different investigations, but to sketch some of the different ways of classifying chords and thereby shed light at least on some typical questions concerning chords and chord progressions.

G. Mazzola, *The Topos of Music I: Theory*, Computational Music Science,
https://doi.org/10.1007/978-3-319-64364-9_23

23.1 Hugo Riemann

Summary. This section briefly reviews Hugo Riemann's concept of triad-based tonality as it was developed in his writings, such as *Musikalische Logik* [900] and *Musikalische Syntaxis* [901], from Rameau's early adumbration. Riemann's construction of tonality from a chord progression is described. It is based on the concept of *thesis* (German "These"); the tonal coherence of a composition is constituted by a succession of theses. To conclude, we discuss Riemann's unfinished program to assign a harmonic function to strictly *every* possible chord, given a specific tonality.

– Σ –

Rameau was one of the first theorists to deal with the classification of the variety of possible combinations of tones as chords in order to find systematic laws of combination. He classified chords on the basis of thirds, which means to understand the combinations of thirds which build chords. Thus a fourth-sixth-chord can be traced back to a triad by an inversion, i.e., by adding an octave to the lowest tone. The triad is the starting point not only of the horizontal, but also of the vertical organization of the tones in a piece of music. The system of tonal functions is adumbrated, but not fully exposed.

This exposition can be found in Riemann's *Musikalische Logik* [900]. It involves a classification of chords into the three categories *tonic, dominant, subdominant* and the discussion of the cadence ("Kadenz") as a specific progression of chords when subsumed under these categories. His system includes a classification of chords on the basis of the minor and the major triad. This method classifies every chord by tracing it back to a triad (or a combination of triads) it was derived from (e.g., by altering the chord's single tones). Consequently, it involves the problem that the preimage from which a chord is derived is often far from unequivocal. The classification of all chords into the three categories tonic, dominant, and subdominant, is ambiguous as well.

Another important question Riemann deals with in *Musikalische Syntaxis* [901] is that of tonality. How does the phenomenon of tonality arise within a specific chord progression such that one chord is classified as a tonic? Riemann discusses examples of such chord progressions, but in some cases there is more than one possible triadic solution for the suspected tonality—a fact which Riemann does not reflect. So the question, how to find criteria for deciding upon possible tonalities in non-trivial ambiguous situations, remains open.

23.2 Paul Hindemith

Summary. This section briefly reviews Paul Hindemith's harmony as developed in *Unterweisung im Tonsatz* [467]. In order to explain the tension/relaxation within a chord progression, he constructs a hierarchy of intervals which is driven by their consonance character. The chords are then classified according to the hierarchical position of their constituting intervals.

– Σ –

Hindemith's classification of chords starts with a discussion of the intervals with respect to their stability and consonant or dissonant character. He critizes the approach to alter all chords into triadically composed chords. Chords are classified according to the intervals they consist of in such a way that stable or consonant intervals generate stable chords. Among all chords, the triads are the most stable ones. Chord progressions are analyzed in terms of different stabilities of the consecutive chords. For instance, if a given chord consists of "more consonant" intervals and the next one of "less consonant" intervals, an effect of tension is produced. By this definition, Hindemith creates curves of tension within long chord successions. The concept of tension is an instrument to explain possible effects or impressions when listening to such a chord successions.

23.3 Heinrich Schenker and Friedrich Salzer

Summary. This section reviews Friedrich Salzer's *Structural Hearing* [930] as an updated version of tonal theory following Heinrich Schenker's ideas, as exposed in [939] and [940]. This theory considers a hierarchy of structural layers from the background ("Ursatz des Hintergrundes"), traversing the transformational middleground ("Mittelgrund der Verwandlungsschichten"), and developing into concrete music on the foreground ("Vordergrund"). Structural chords on a higher hierarchical level are complemented by (relative) prolongational chords on a lower level, expressing harmonic "motion" between structural instances. We discuss the question of ambiguities and indeterminacies in the case-specific definition of such a hierarchy.

– Σ –

Salzer (as representative of Schenkerian analyses) critizes those harmonic approaches, which structure a given piece of music into a progression of chords and assign functions to them by analyzing the specific chord gestalts. Instead, he proposes the classification of chords into two groups—structural and prolongational chords—in order to explain phenomena like 'musical motion' or 'coherence'. Structural chords support the harmony, while harmonic meaning is not assigned to prolongational chords—they mediate between structural chords and support the 'motion' from one harmonic function to the next one. This approach starts with a grouping of the piece into periods and themes, followed by the specification of the structural chords within these periods. There are however no general arguments or criteria for the process which decides about the chords belonging to structural or prolongational categories. Also, the problem of which tones form a chord remains unsolved as well.

Remark 7 We should mention Sigfrid Karg-Elert's harmony *[521]* which is based on a strictly polar symmetry between major and (aeolian) minor modes, as well as Arthur von Oettingen's symmetry-based modulation theory *[806]*. A thorough discussion and understanding of their works is still unsettled.

24

Harmonic Topology

Tyger! Tyger! burning bright
In the forests of the night,
What immortal hand or eye
Could frame thy fearful symmetry?
William Blake (1757–1827)

Summary. This chapter introduces a systematic correspondence between chords and symmetries. It lays the morphological fundament for a semantic theory of harmonic functions which will be exposed in the following chapter. Essentially, the idea of such a correspondence is to carry over the structural discourse on harmony to richer objects by an intermediate switch to "richer" addresses.

– Σ –

In Sections 11.3.6 and 11.3.7, we have presented a preliminary study of symmetries of chords, and of the classification of self-addressed chords in simple ambient spaces of modules with finite length. In this chapter, the idea of switching between chords and symmetries will be developed a bit more systematically and then applied to the construction of harmonic semantics as suggested by Noll [802].

The basic idea is that to understand the full morphological potential of chords, it is not sufficient to look at their points, rather one should include the information given by their inner symmetries. Now, there are two ways to deal with this idea. The first is an ontological switch. It resembles the concept from quantum mechanics, stating that the measurable physical reality is a trace—in fact a bunch of eigenvalues—of more abstract hermitian (linear) operators: the observables. The higher reality of these operators is the semantic source of microscopic physics whose seemingly contradictory phenomenology of wave and particle are just projections of the consistent higher reality of observables [318]. In this vein, chords would be replaced by endomorphisms of the underlying form space.

The drawback of this approach is that it enforces a new reality which is not that of addressed points but resides in abstract morphisms between form spaces. But music should not have its reality split into what has meaning and what is audible. It should avoid doing so as much as physics should. We cannot speak for the physical ontology, but for music, there is an elegant solution which has been prepared via the very approach to musical objects which is based on addressed denotator substance. We contend that the richness of ontological reference which is offered on each address is essential and sufficient to condense the morphic flavor of symmetries on the level of address ontology (see also our discussion in 8.3.4, in particular principle 3)—provided that "generic" addresses can be found for a given form space.

There is a further advantage of the 'address condensation' of morphic structures: With this tool it becomes possible to consider global compositions on the basis of local compositions on a generic address, a construction which Noll calls "Morphemfeld" in [800], and which he uses to (re)build harmonic semantics in the spirit of Hugo Riemann.

We are aware that the following approach is not evaluated at all its implications, in fact, the harmonic topology is but a first application of a more general topological interpretation of denotator morphology. We

G. Mazzola, *The Topos of Music I: Theory*, Computational Music Science,
https://doi.org/10.1007/978-3-319-64364-9_24

nevertheless feel obliged to sketch this general perspective to make the stream of ideas evident and their coherence plausible.

24.1 Chord Perspectives

Summary. This section deals with chords in Euler and pitch class spaces, together with a comparative study while applying the enharmonic projection of self-addressed Euler spaces onto pitch class spaces.

– Σ –

We start our discourse from concrete situations in harmony which arise from chord descriptions in different ambient spaces for pitch and pitch classes. Noll's term "perspective" for pitch denotators at different addresses seems adequate: it emphasizes the ontological perspective offered on each address, as well as the change of perspective induced by a change of address. These examples also will prepare the reader for the semantic discussion of chord morphology in the next chapter.

24.1.1 Euler Perspectives

Summary. This section discusses zero-addressed and self-addressed chords in the context of pitch classes in just tuning.

– Σ –

Recall from Section 7.2.1 that the coordinators of *EulerModule* and *EulerPlane* are $\mathbb{Q}^3$ and $\mathbb{Q}^2$, with the canonical decompositions $\mathbb{Q}^3 = \mathbb{Q}\cdot o \oplus \mathbb{Q}\cdot q \oplus \mathbb{Q}\cdot t$ and $\mathbb{Q}\cdot q \oplus \mathbb{Q}\cdot t$, respectively. Within these spaces, one identifies the subspaces $EulerZModule = EulerModule|_{\mathbb{Z}}$ and $EulerZPlane = EulerPlane|_{\mathbb{Z}}$ which are defined by integer coordinates, i.e., the lattices generated by the canonical bases o, p, q and p, q, respectively. We shall mainly start from these lattices when talking about harmony in just tuning.

For example, the C-chromatic scale of Vogel, from which one octave is shown in the lower half of Figure 7.6, is a zero-addressed chord χ in the *EulerZModule*. In order to have a better representation of this χ, we transform it under the linear space automorphism

$$S = \begin{pmatrix} 1 & 2 & 2 \\ -2 & -3 & -4 \\ 0 & 0 & 1 \end{pmatrix} \tag{24.1}$$

and get the more transparent configuration $\chi^\star = S.\chi$ as shown in Figure 24.1. To understand this chord better, observe that it contains (the heads and tails of) four $\mathbb{Z}$-addressed points $g \Rightarrow e_\flat$, $g \Rightarrow b$, $g \Rightarrow f_\sharp$, $g \Rightarrow a_\flat$ with common tail g. Then $g \Rightarrow e_\flat + g \Rightarrow b$ is a zero length arrow, in fact a major third 'coupling', and $g \Rightarrow f_\sharp + g \Rightarrow a_\flat$ is a zero length arrow, a minor second coupling. A similar situation holds for exactly one other set of four $\mathbb{Z}$-addressed points, i.e., $f \Rightarrow a$, $f \Rightarrow d_\flat$, $f \Rightarrow c$, $f \Rightarrow b_\flat$ with common tail f. Again, we have these relations: $f \Rightarrow a + f \Rightarrow d_\flat$ is a zero length arrow, a minor third coupling, and $f \Rightarrow b_\flat + f \Rightarrow c$ is a zero length arrow, a fourth coupling. These two four-point configurations are each contained in one of two parallel planes. Besides these two four-element $\mathbb{Z}$-addressed chords, there is one $\mathbb{Z}$-addressed point $g \Rightarrow d$ in the first plane, and one $\mathbb{Z}$-addressed point $f \Rightarrow e$ in the second plane.

Now, if $u/1 : \chi^\star \xrightarrow{\sim} \chi^\star$ is an automorphism, it must either exchange the two $\mathbb{Z}$-addressed chords or leave them invariant, and it must also exchange the two additional points or leave them fixed. It is easily seen that an automorphism which leaves the planes fixed must be the identity since it must leave the above couplings fixed. But there is an automorphism which exchanges the two planes, and by the previous remark, this will be the unique non-trivial automorphism. The automorphism has this effect:

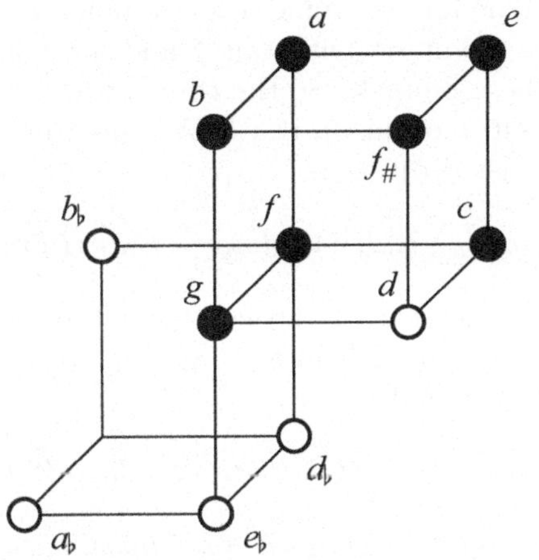

Fig. 24.1. The transformed chromatic scale makes its inner symmetries evident.

$$u/1 : b \rightsquigarrow d_\flat$$
$$u/1 : d \rightsquigarrow e$$
$$u/1 : e_\flat \rightsquigarrow a$$
$$u/1 : f_\sharp \rightsquigarrow c$$
$$u/1 : a_\flat \rightsquigarrow b_\flat$$
$$u/1 : g \rightsquigarrow f$$

and on the original chromatic scale χ (!) is given by the affine *EulerZModule* automorphism $U = e^s.U_0$ with translation $s = (-5, 2, 1)$ and matrix

$$U_0 = \begin{pmatrix} 11 & 18 & 24 \\ -4 & -7 & -8 \\ -2 & -3 & -5 \end{pmatrix} \tag{24.2}$$

which is automatically an involution since it exchanges the planes of the two chords which are centered around g and f. Also observe that the $\mathbb{Z}$-addressed point $g \Rightarrow f$ is reversed to $f \Rightarrow g$ under $u/1$. The interesting thing about this involution is that it yields an involution $e^6.5$ of $OnMod_{12}$ if we transport the tone names as usual, i.e., $c \sim 0, d_\flat \sim 1, \ldots b \sim 11$.

Moreover, if we consider the translation χ to $\Xi = e^{-h}\chi$ by the vector $h = (-3, 1, 1)$, the automorphism of this scale chord is the conjugate $V = e^{-h}.U.e^h$. But the elements of Ξ are the just interval quantities of the chromatic chord χ when viewed from h. The above dichotomy which was given by the two parallel planes now corresponds to the two following six-by-six element dichotomy of Ξ:

$$\Xi_K = \{(0,0,0), (0,1,-1), (4,0,-2), (-2,1,0), (6,-1,-2), (2,0,-1)\},$$
$$\Xi_D = \{(7,-2,-2), (1,-1,0), (3,-2,0), (3,-1,-1), (7,-3,-1), (5,-2,-1)\},$$

and its interpretation in terms of differences of numbers in $\mathbb{Z}_{12}$ as above, e.g., $b - b \sim 11 - 11 = 0, d - b \sim 2 - 11 = 3, e_\flat - b \sim 3 - 11 = 4$, etc. yields the consonance-dissonance dichotomy $K = \{0, 3, 4, 7, 8, 9\}, D = \{2, 5, 10, 1, 6, 11\}$ of classical counterpoint. The literally transformed automorphism is the involution $e^2.5$ of $OnMod_{12}$, the (unique) *autocomplementarity function* of the counterpoint model from which the rules of elementary counterpoint will be deduced in part VII. We come back in Section 24.1.3 to this phenomenon.

Instead of looking at a chord of $\mathbb{Z}$-addressed points with tail in g, one can also take a $\mathbb{Z}^3$-addressed 'dominant seventh' point D^7 with tail in g, and three arrowheads in b, d, f. The automorphism U is completely determined by this self-addressed point since it defines an affine base of the ambient space. So U is uniquely determined by $U(D^7)$. The same is true for the 'tonic seventh' point T^7 with tail g and arrowheads in $c, e, b_\flat$; same for the 'subdominant seventh' point S^7 with tail f and arrowheads $a, c, e_\flat$ (observe that the major seventh would not work in any of these points!). So the three previous self-addressed seventh points X^7 are all sufficient to describe the entire automorphism, i.e., $U(X^7)$ determines U.

Consider the canonical pitch class projection

$$mod_o : EulerModule \to EulerPlane$$

defined in 7.11. This induces the projection of the corresponding chord spaces $mod_o : EulerChord \to$ *o-ClassChord*. Since mod_o is a projection with integer matrix, we have a corresponding integer situation $mod_o : EulerZModule \to EulerZPlane$ and

$$mod_o : EulerZChords \to o\text{-}ClassZChord$$

with $EulerZChord = EulerChord|_{\mathbb{Z}}, o\text{-}ClassZChord = o\text{-}ClassChord|_{\mathbb{Z}}$, the restrictions to $\mathbb{Z}$-valued points of chords.

Two A-addressed chords C_1, C_2 in *EulerModule* are called *an inversion of each other* iff $mod_o(C_1) = mod_0(C_2)$. Inversion equivalence classes of chords, i.e., pitch class chords, are central objects in harmony. The present just pitch class space is spanned by the fifth q and by the third class t. Vogel's zero-addressed scale *C-chromatic* which is the projection $mod_o(\chi)$ of the chromatic chord χ discussed before, gives rise to an infinity of zero-addressed scales

$$X\text{-}chromatic = e^X.C\text{-}chromatic$$

in *EulerZPlane* for any zero-addressed class X in *EulerZPlane*.

Proposition 40 *Consider the lattice* $L = \mathbb{Z}.(1,0,0) + \mathbb{Z}.(0,-4,1) + \mathbb{Z}.(0,0,3)$. *Then the* L*-translates of* χ *define a tiling of EulerZModule.*

Proof. The $mod_o(L)$ translates of *C-chromatic* define a tiling of *EulerZPlane*, and the octave translates fill up the space in octave direction. Observe, that the vector $(0,-4,1)$ corresponds to the fifth comma defined in 6.33, whereas the third generator of the lattice is $Kq + 19.o + 3.Kt$, so *the lattice is generated by the octave and the two commata.* We shall discuss this lattice in a moment.

The zero-addressed major scale *C-major* and natural minor scale *C-minor* give rise to corresponding shifted scales

$$X\text{-}major = e^X.C\text{-}major, X\text{-}minor = e^X.C\text{-}minor.$$

In the same way, the just zero-addressed harmonic minor scale *X-harm* stems from the harmonic minor scale *C-harm* which differs from *C-minor* by a non-flattened b, whereas the zero-addressed melodic minor scale *X-mel* stems from the melodic minor scale *C-mel*, and this one differs from *C-minor* by non-flattened a, b.

For the time being, we shall use the same tone names for pitch classes as for their original pitches in *EulerZModule*. As with the *EulerZModule*, we have prominent $\mathbb{Z}^2$-addressed, i.e., self-addressed points in *EulerZPlane* which determine any endomorphism $V : EulerZPlane \to EulerZPlane$ because their tail and arrowheads are an affine base. Here are such candidates: the 'major triad' point C^5 with tail $C^5(0) = c$ and heads $C^5(e_1) = g(= q)$, $C^5(e_2) = e(= t)$, the 'minor triad' point c^5, same as C^5, except $c^5(e_2) = e_\flat(= q - t)$.

24.1.1.1 Just Mutation

Summary. On the *EulerZModule* a Lie algebra structure can be introduced in order to give a "processual" interpretation of the enharmonic projection (see 24.1.3).

– Σ –

For the following Lie algebra theory, see Appendix E.4.4. The *EulerZModule* is charged with a Lie algebra structure in the following way. We represent the module of *EulerZModule* in the basis $o, q, r = Kt - q$ ($Kt = t - 4q - 4o$ being the syntonic comma, see 6.33) and induce the Lie ($\mathbb{Z}$-)algebra structure from the module isomorphism

$$sl_2 : \mathbb{Z}.o \oplus \mathbb{Z}.q \oplus \mathbb{Z}.r \xrightarrow{\sim} \mathfrak{sl}(2, \mathbb{Z})$$

defined[1] by

$$sl_2(o) = \begin{pmatrix} 1 & 0 \\ 0 & -1 \end{pmatrix}, sl_2(q) = \begin{pmatrix} 0 & 1 \\ 0 & 0 \end{pmatrix}, sl_2(r) = \begin{pmatrix} 0 & 0 \\ 1 & 0 \end{pmatrix},$$

The Lie bracket $[x, y]$, i.e., Noll's *mutation* ([802]), on this basis reads as follows: $[o, q] = 2q, [o, r] = -2r, [q, r] = o$, the other values are mandatory by skew symmetry.

With this Lie bracket, the adjoint endomorphisms $ad(q), ad(r)$ are nilpotent, by definition, and we have the Lie algebra automorphisms

$$Q = exp(ad(q)) = Id + ad(q) + ad(q)^2/2, \tag{24.3}$$
$$R = exp(ad(r)) = Id + ad(r) + ad(r)^2/2. \tag{24.4}$$

It is well known that on the level of $\mathfrak{sl}(2, \mathbb{Z})$, these automorphisms Q, R identify to the conjugation by

$$exp(sl_2(q)) = \begin{pmatrix} 1 & 1 \\ 0 & 1 \end{pmatrix}, exp(sl_2(r)) = \begin{pmatrix} 1 & 0 \\ 1 & 1 \end{pmatrix},$$

respectively, a generator set of $SL(2, \mathbb{Z})$. Consider the group

$$Mut = \langle Q, R \rangle \subset Aut(EulerZModule)$$

generated by these operators and the center $Z(SL(2, \mathbb{Z})) = \{\pm Id\}$. Then we have this result:

Theorem 27 [802, III.4.3, Theorem 1] *We have a group diagram*

$$\begin{array}{ccccccccc}
 & & 1 & & 1 & & 1 & & \\
 & & \downarrow & & \downarrow & & \downarrow & & \\
1 & \rightarrow & 1 & \rightarrow & (SL(2,\mathbb{Z}), SL(2,\mathbb{Z})) & \rightarrow & (Mut, Mut) & \rightarrow & 1 \\
 & & \downarrow & & \downarrow & & \downarrow & & \\
1 & \rightarrow & Z(SL(2,\mathbb{Z})) & \rightarrow & SL(2,\mathbb{Z}) & \rightarrow & Mut & \rightarrow & 1 \\
 & & \downarrow & & \downarrow & & \downarrow & & \\
1 & \rightarrow & \mathbb{Z}_2 & \rightarrow & \mathbb{Z}_{12} & \rightarrow & \mathbb{Z}_6 & \rightarrow & 1 \\
 & & \downarrow & & \downarrow & & \downarrow & & \\
 & & 1 & & 1 & & 1 & &
\end{array}$$

[1] See Appendix E.4.4, Example 110, for the definition of $\mathfrak{sl}(2, \mathbb{Z})$.

with exact columns and rows and the canonical homomorphisms that are related to the conjugacy operation of the special linear group on $\mathfrak{sl}(2,\mathbb{Z})$.

In fact, in [802, III.4.3], it is proven that $SL(2,\mathbb{Z})/(SL(2,\mathbb{Z}), SL(2,\mathbb{Z})) \xrightarrow{\sim} \mathbb{Z}_{12}$, but the center generator $-Id$ is not in the commutator, and therefore projects to the element 6 of order 2 in $\mathbb{Z}_{12}$, so we are done.

Furthermore, we may define a set map

$$Ex : EulerZModule \to SL(2,\mathbb{Z})$$

by $Ex(a.o + b.q + c.r) = exp(sl_2(q)^b.exp(sl_2(r)^c$ (o goes to the identity). By the above theorem, this map induces a surjective group homomorphism

$$ENH : EulerZModule \to \mathbb{Z}_{12}.$$

Consider the subform *CommaZModule* of *EulerZModule* which is spanned by the Pythagorean and Syntonic commata Kq, Kt, and by the octave o. Then we have

Theorem 28 [802, III.4.3, Theorem 2]
The kernel $Ker(ENH)$ *equals the subform* $CommaZModule$.

We shall discuss this result in Section 24.1.3 in the context of enharmonic identification methods. However, the point of the preceding construction is that it essentially introduces a group of space automorphisms on *EulerZModule* which reside on an "exponentiation lifting" from zero-addressed points o, q, r to self-addressed points, i.e., Id, Q, R, related to a linear Lie algebra interpretation of the underlying module. So it not only lifts zero-addressed points to constant self-addressed points via address change projection, but gives them a non-trivial lifting. At present, no musicological justification of this Lie algebra construction is known—except that it does give a new interpretation to enharmonic identification. It is in particular not known whether there is a more general and/or not arbitrary, viz., canonical construction behind this idea. It is also not clear why the special linear group should be invoked as a lifting of the mutation operator group *Mut*, and a musicological interpretation of the mutation operators Q, R is not known.

24.1.2 12-Tempered Perspectives

Summary. This section discusses constant and self-addressed chords in the context of pitch classes in 12-tempered tuning.

– Σ –

This situation stems from the 12-tempered part of the *EulerModule*, i.e., the subform 12-*Temp* of *EulerModule* of the one-dimensional $\mathbb{Z}$-submodule $\mathbb{Z}.\frac{1}{12}.o \subset \mathbb{Q}.o \oplus \mathbb{Q}.q \oplus \mathbb{Q}.t$, together with its projection

$$mod : 12\text{-}Temp \to 12\text{-}PitchClass \xrightarrow{\sim} PiMod_{12}$$

of octave identification. So we have to keep in mind that this construction is completely separated from the just tuning spaces. We are going to discuss the relation between these two approaches in Section 24.1.3 below.

Beyond the general Yoneda principle, and to our knowledge of mathematical music theory, the historically[2] first passage from zero-addressed tones in $PiMod_{12}$ to self-addressed tones, i.e., tones in $\mathbb{Z}_{12}@PiMod_{12}$, was suggested by the question in [682, section 4.2.3.2.2] concerning the deduction of important chords of music theory from basic principles of the underlying pitch spaces. In the foundations of harmony, they try to deduce the structure of important chords, above all the major and minor triad, from the acoustic phenomenon of partials in Fourier analysis, for example with Rameau [876], Schönberg [948], or Hindemith [467].

[2] See Noll's motivation in [802, I.1.4].

A slightly different deduction is offered by Oettingen [806] or Vogel [1089] by use of the *EulerZPlane* where fundamental, third, and fifth define a special generating zero-addressed local composition.

"Deduction" means that it is attempted to justify the eminent role of those chords from an ontological point of view, as an a priori condition that precedes their role as structural fundamentals of harmony. In any case, this foundation is based on just tuning and contradicts the practice of well-tempered tuning. Mostly, the argument in favor of this contradiction is that we are living a compromise, a neat refusal of the tempered tuning being the exception, such as with Vogel who even ran a political initiative to banish tempered instruments from public schools.

Our concern is however not a dogmatic one, we rather would offer an approach to fundamental harmony which also works for tempered tuning and—at the same time—evokes the rationales in just tuning ideology. In his approach, Schönberg views the major triad as an "imitation of its fundamental tone by partials". Starting from c, he denotes the sequence of partials $c', g', c'', e'', g'', b''_\flat, c''', \ldots$ where the first three octave pitch classes are c, g, e (in this order!), i.e., an inversion equivalent of the major triad. There is no immanent reason to stop at this stage, except that one step further, the catastrophe happens: the "natural seventh" (German: "Naturseptime"), with tonality $b_\flat$, where after 2,3,5 the new prime 7 appears in the frequency ratio, and we have to extend the *EulerModule*. So the triad is not really given as a closed unit, but as a first segment of an infinite chain.

For the 12-tempered tuning, the 'partial' argument does not work, and we have to look for other mechanisms. Whereas the partial frequencies are positive multiples of the fundamental frequency, the 12-tempered tuning offers endomorphisms on 12-*Temp* and on $PiMod_{12}$. Let us look at the pitch classes in $PiMod_{12}$. If $k \in \mathbb{Z}_{12}$ is a zero-addressed pitch class, and if $f = e^x.n$ is an endomorphism, its effect on k is $f(k) = e^{k.n}x$, the k-fold translation of n from x. So the action of f on all pitch classes $k = 0, 1, 2, 3, \ldots 11$ is just the sequence

$$x, x + n, x + 2.n, \ldots x + 11.n$$

which in music theory is called the *n-circle from x* (the number n corresponding to multiplication in $\mathbb{Z}_{12}$).

Exercise 56 *An instructive example is the "circle of fourths" and the "circle of fifths" (just run through the circle of fourths in the opposite direction). Major scales are usually arranged in this order:*

$$\text{ascending: } C, F, B_\flat, E_\flat, A_\flat, D_\flat, \text{ descending: } C, G, D, A, E, B, F_\sharp$$

where $F = e^5.C$, etc., and $G = e^{-5}.C$, etc. It is well-known that the alteration signs increase one-by-one on the fourth circle, whereas they increase one-by-one on the fifth circle when we transgress the scales in this order. Why? Since this question only regards the white and black keys, we may really work in tempered terminology.

The musician's common argument is that we may go from C to F by alteration $b \mapsto b_\flat$, and than proceed analogously. But the black keys do not shift along the transition from C to F. The exact reason is that we may multiply everything by 5 and then work in the chain of fourths. Thereby, C switches to $5.C = \{b, e, a, d, g, c, f\}$, an uninterrupted sequence of fourths. And the shifted scale $F = e^5.C$ switches to $5.F = 5.e^5.C = e^1.(5.C)$, the shift of the switched scale $5.C$ by one unit. This shifting operation is repeated in every fourth step of scales, and therefore, each new scale adds one black key, i.e., one flattening of a white key. Check the details of this argument. Try to convince musicians that the argument does not work for other white-black-key distribution such as those which would define "white = melodic C minor, black = remaining keys".

The 12-tempered simulation of the partial argument is this: We start from the fundamental tone $c = 0 \in \mathbb{Z}_{12}$ and take an endomorphism $S = e^7.3$ of $PiMod_{12}$. We consider the monoid $\langle S \rangle$ generated by S and let it act on the fundamental. We have this orbit:

$$\langle S \rangle(c) = \{c, S(c) = g, S^2(c) = e, S^3(c) = g, \ldots\} = I_C$$

which is precisely the major C triad. Observe that the order of images is the same as with partials: c, g, e, except that it is completed on this triad, no further tones must be cut in order to obtain the triad.

Exercise 57 *Show that the minor triad I_{c_m} is a similar orbit $\langle T\rangle(g)$. Show that there are exactly two endomorphisms W such that $\langle W\rangle(x) = I_C$ for a selected tone x.*

This suggests the question of whether this method produces other prominent chords and whether there are chords which look essentially different from those defined by this orbit method.

The second question is easily answered: Yes, take the dominant seventh chord $V_C^7 = \{g, b, d, f\}$. This one cannot be defined as an orbit of a cyclic monoid $\langle S\rangle$ which acts on a single tone. One needs at least two generators for such an orbit, e.g., $R = e^8.3, T = e^2.0$. We then have

$$\langle R\rangle(g) = \{g, R(g) = f, R^2(g) = b, R^3(g) = f, \ldots\},$$
$$\langle T\rangle(g) = \{g, T(g) = const = d \ldots\},$$
$$R(d) = d, \text{ and therefore}$$
$$\langle R, T\rangle(g) = V_C^7.$$

Possibly, this bigeneric monoid simulates what Oettingen called "Bissonanz", however here this one is not given by two unrelated triads (as with Oettingen) but by two symmetries acting on one fundamental tone.

Before we set forth a systematic discussion of this generative approach to chord theory, let us answer the first of the above questions. We call a zero-addressed chord of shape $Ch = \langle S\rangle(x)$ a *circle chord* (generated by S and x). Any chord which is isomorphic to a circle chord is also a circle chord. In fact, if $h : PiMod_{12} \xrightarrow{\sim} PiMod_{12}$ is an automorphism inducing an isomorphism $h : \langle S\rangle(x) \xrightarrow{\sim} X$, then we have $X = \langle h.S.h^{-1}\rangle(h(x))$. So we are left with the problem to describe all circle chords for representatives of conjugacy classes of symmetries of $PiMod_{12}$. The table of these sixteen chord classes in Figure 24.2 shows that—except one class number 33 (see the classification of chords in Appendix N.1)—all classes are *common chords*.

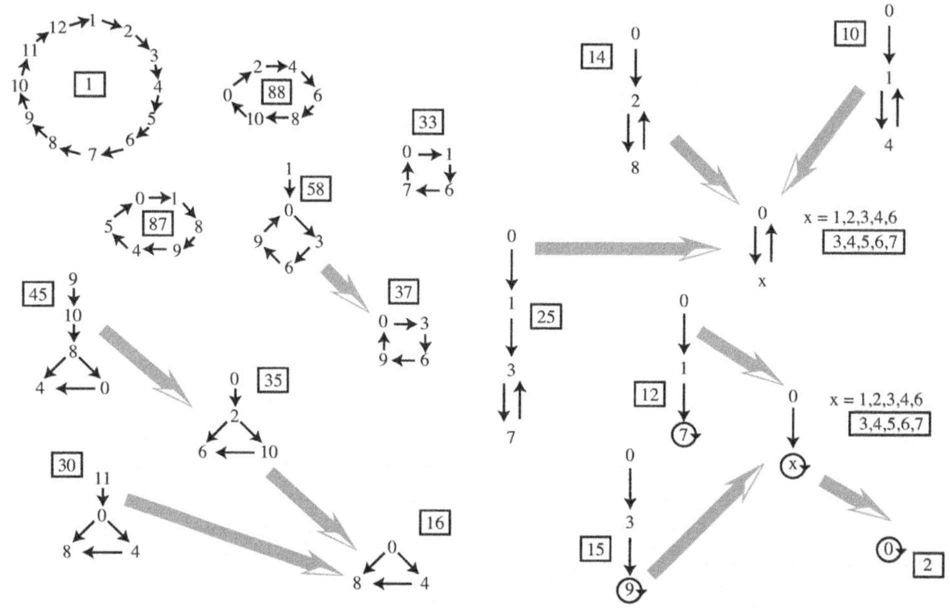

Fig. 24.2. The 16 circle chords in $PiMod_{12}$, together with their generating symmetry action (arrows). The numbers refer to the classification of chords. A large arrow from one chord to another means that the target chord is isomorphic to a subchord of the start chord, and that the subchord is generated by a suborbit of the same symmetry action as for the superchord.

Fact 12 *Circle chords really yield a generative fundament for basic chords in harmony, and one may refrain from evoking just temperament and tuning compromises when laying this basis to harmony.*

24.1.3 Enharmonic Projection

Summary. There is an *enharmonic* projection from Euler pitch classes to 12-tempered pitch classes which relates constructions in the two spaces. These relations are discussed.

– Σ –

Just tuning and well-tempered tuning are fundamentally different choices of pitch spaces within the *EulerSpace*. The first is a free $\mathbb{Z}$-submodule of rank one generated by $\frac{1}{12}o$, whereas the second is free of rank three, i.e., the *EulerZModule* and intersects the well-tempered module only in the octave-submodule $\mathbb{Z}.o$. The usual practical argument for switching between the two tunings is that the pitch differences between the corresponding Euler points are relatively small.

The theoretical counterpart of this switch resides on the so-called enharmonic identification under the comma module *CommaZModule*. Classically, two points in the *EulerZModule* are called *enharmonic (with each other)* iff their difference lies in the submodule $\mathbb{Z}.Kt \oplus \mathbb{Z}.Kq \oplus \mathbb{Z}.o$ of the *CommaZModule*. We shall more generally say that they are enharmonic iff their difference is in the *EulerZModule*, i.e., they are identified by the enharmonic projection

$$enh : EulerZModule \rightarrow EulerZModule/CommaZModule \xrightarrow{\sim} \mathbb{Z}_{12}$$

(see Section 24.1.1.1, Theorem 28) which besides the fifth and third commata annihilates the octave. Enharmonic identification is a construction which resides on the cognitive rationale of commata, i.e., their role for auditory identification, but it has nothing to do with tempered tuning. *The enharmonic projection is not a tuning, but an abstraction.* So making music theory on this abstraction is one business, whereas realizing this abstraction on the 12-tempered tuning is another.

In his processual interpretation of enharmonic identification via Theorem 28, Noll charges the enharmonic projection by a cognitive dimension in the following sense. Look at the commutative diagram issued from Section 24.1.1.1:

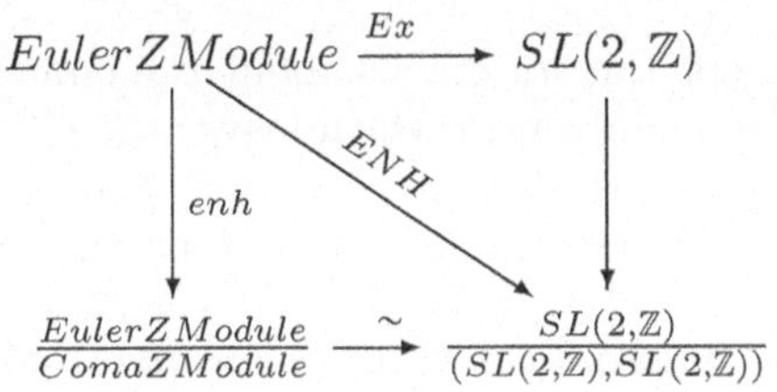

where the upper horizontal map Ex is the exponential map, and the lower isomorphism identifies both quotient groups to $\mathbb{Z}_{12}$. The lifting map Ex interprets the enharmonic homomorphism enh as follows: A point in the *EulerZModule* is lifted to a product of two-by-two matrices, starting from the lifting of the fifth to the upper triangular generator, whereas the third is lifted to the lower triangular generator. The general Euler point is lifted to a product of such matrices. This cognitively means that the cognitive identification of an Euler point is given by a sequence of steps of either fifth or third generator type. The non-commutativity of the concatenation of fifth and third generators is resolved by identification of commutators $x.y.x^{-1}.y^{-1}$ with the identity. And this precisely yields the modular group $\mathbb{Z}_{12}$. If instead, we stay in the modular group $SL(2,\mathbb{Z})$, the non-commutativity of the fifth and third has been interpreted by Noll [802] as a rationale of the enharmonic identification in the same sense that non-commutativity of operators gives rise to the uncertainty relation in quantum mechanics: enharmonicity resembles uncertainty. As we already stated, the nature of the modular lifting—up to now—has neither formal nor musicological or cognitive justifications; not every mathematical structure yielding $\mathbb{Z}_{12}$ in some way is automatically musicologically justified.

However, the enharmonic projection helps understand the above remark concerning the automorphism U of the chromatic scale χ (see the linear part U_0 of U in formula 24.2) when it is transferred to pitch classes via "literal transformation".

Definition 78 *An endomorphism Φ of the EulerZModule is called* enharmonic *iff it commutes with the enharmonic projection, i.e., iff there is a (necessarily unique) endomorphism ϕ of $PitchMod_{12}$ such that the diagram*

$$\begin{array}{ccc} EulerZModule & \xrightarrow{\Phi} & EulerZModule \\ {\scriptstyle enh}\downarrow & & \downarrow{\scriptstyle enh} \\ PitchMod_{12} & \xrightarrow{\phi} & PitchMod_{12} \end{array} \tag{24.5}$$

commutes.

Here is a characterization of enharmonic endomorphisms:

Proposition 41 *(i) Every translation $\phi = e^s$ is enharmonic.*

(ii) An endomorphism Φ of the EulerZModule is enharmonic iff its linear part Φ_0 is, i.e., iff $\Phi_0(CommaZModule) \subset CommaZModule$.

(iii) The enharmonic endomorphisms define a multiplicative monoid

$$End_{enh}(EulerZModule) \subset End(EulerZModule).$$

The invertible elements among the enharmonic endomorphisms define a subgroup, the enharmonic group *$\overrightarrow{GL}_{enh}(EulerZModule)$.*

(iv) Rewriting the Euler points in the basis o, Kt, q, and Φ_0 as a matrix

$$\begin{pmatrix} a_{11} & a_{12} & a_{13} \\ a_{21} & a_{22} & a_{23} \\ a_{31} & a_{32} & a_{33} \end{pmatrix}$$

in this basis, Φ is enharmonic iff $a_{21} \equiv a_{23} \equiv 0 \mod 12$.

Proof: Exercise for the reader, except the statement about the enharmonic group $\overrightarrow{GL}_{enh}(EulerZModule)$: Observe that an invertible linear endomorphism which leaves the *CommaZModule* invariant automatically induces an automorphism of *CommaZModule* since its index in *EulerZModule* is finite. This statement also follows directly from the calculation of the inverse of a matrix and the criterion (iv) of the proposition.

Corollary 11 ([802, III.4.1]) *The autocomplementarity involution U of the just chromatic chord χ and its conjugate V, the autocomplementarity involution of Ξ, are both enharmonic.*

In fact, the first follows from the matrix criterion above, and the second follows from the fact that translations, and therefore conjugation with translations, are enharmonic.

Corollary 12 ([802, III.4.1]) *The chromatic chord χ and its translate Ξ define two sections $OnPiMod_{12} \to EulerZModule$ to the enharmonic projection. The enharmonic autocomplementarity involutions U and V of χ and Ξ, respectively, induce the autocomplementarity involutions $e^6.5$ and $e^2.5$ on χ and Ξ, respectively.*

This follows from Corollary 11 and from Proposition 40.—So the literal translation is in fact induced by the enharmonic projection, and we easily recognize that the tiling lattice in Proposition 40 is the *CommaZModule*. So the autocomplementarity function U permutes the tiling lattice and therefore also the tiling as such. Conversely, modulo *CommaZModule*-translations, any affine permutation of the chromatic tiling identifies to the autocomplementarity involution.

Theorem 29 *The group of affine automorphisms of the χ-chromatic tiling of the EulerZModule is the semi-direct product $e^{CommaZModule} \rtimes \langle U \rangle$, of the autocomplementarity involution U with the CommaZModule-translations. The same holds with, mutatis mutandis, χ being replaced by Ξ, and U by the Ξ-involution V.*

Fact 13 *Musicologically, this means that the autocomplementarity of consonant and dissonant intervals (see Chapter 30 for this subject) is induced by a unique inner symmetry on the level of just tuning and its (discrete) interpretation by Vogel's chromatic chord.*

The moral of this discussion is that in both cases, 12-tempered and just tuning, certain endomorphisms of the underlying pitch space are crucial structures for understanding zero-addressed harmonic, and—as will be shown in the counterpoint theory—contrapuntal phenomena. Intuitively, the above example of the automorphism groups of chromatic interpretations (by Vogel's χ or by its 'intervallic shift' Ξ) suggests that this order two group is composed of two agents: the identity for the consonant, and the autocomplementarity involution for the dissonant 'character'. In the same spirit, the preceding generators of zero-addressed circle chords in 12-tempered pitch class space suggested that these agents should be viewed as if they were tones of a more general nature. In the sequel we shall set up a more systematic context for such agents and their relations.

24.2 Chord Topologies

Summary. Chord topologies are built upon chord sets which are invariant under collections of symmetries. These sets define a topological basis.

– Σ –

Although this section deals with and stems from investigations of harmonic structures, it turns out that some fundamental buildings can be carried over to completely general local compositions. We prepend this setup before specializing to proper harmony.

24.2.1 Extension and Intension

Summary. The general technical setup for intension-extension topologies (to be presented in the next section) in terms of local compositions is presented.

– Σ –

Let F be any form space (which we identify with its functor if no confusion is likely), A an address, $a \subset @A \times F$ an A-addressed local composition in F, and $f : F \to F$ an endomorphism. We shall say in this context that f *is an endomorphism of* a if $f|a/1 : a \to a$ is a morphism, i.e., if $Im(a) \subset a$. Denote by $SemiEnd(F), MonEnd(F)$, respectively, the set of subsemigroups, submonoids of $End(F)$ (with the identity Id_F as neutral element), respectively.

Definition 79 *Let* $M \subset End(F)$ *a set of endomorphisms of* F *and* A *an address. The set*

$$Ext_A(M) = \{a \in A@\Omega^F |\ f \text{ is an endomorphism of } a \text{ for all } f \in M\} \tag{24.6}$$

is called a basic extension of F at address A. *The sets*

$$Int(M) = \{m \in SemiEnd(F) |\ M \subset m\}, \tag{24.7}$$
$$Int^e(M) = Int(M) \cap MonEnd(F), \tag{24.8}$$

are called basic (semigroup, monoid, respectively) intensions of F.

For a set $E \subset A@\Omega^F$*, the monoid*

$$Int(E) = \{f |\ f \in End(F), f \text{ is an endomorphism of every } a \in E\} \tag{24.9}$$

is called the intension[3] of E.

[3] Terminology introduced by Noll [800] in the case of self-addressed chords of pitch classes and to be justified below in Chapter 25.

Sorite 9 *With the above notation, we have the following facts:*

(i) *For* $M \subset N \subset End(F)$, *we have* $Ext_A(N) \subset Ext_A(M)$, $Int(N) \subset Int(M)$, *and* $Int^e(N) \subset Int^e(M)$.

(ii) *For* $M \subset End(F)$, $Ext_A(M) = Ext_A(\langle M\rangle) = Ext_A(\langle M\rangle_e)$[4], $Int(M) = Int(\langle M\rangle)$, *and* $Int^e(M) = Int^e(\langle M\rangle) = Int^e(\langle M\rangle_e)$.

(iii) *For* $M, N \subset End(F)$,

$$Ext_A(M) \cap Ext_A(N) = Ext_A(M \cup N) = Ext_A(\langle M, N\rangle) = Ext_A(\langle M, N\rangle_e),$$

and

$$Int(M) \cap Int(N) = Int(M \cup N) = Int(\langle M, N\rangle),$$
$$Int^e(M) \cap Int^e(N) = Int^e(M \cup N) = Int^e(\langle M, N\rangle) = Int^e(\langle M, N\rangle_e).$$

(iv) *For a set* $E \subset A@\Omega^F$, *we have* $E \subset Ext_A(Int(E))$, *and for a set* $M \subset End(F)$, *we have* $M \subset Int(Ext_A(M))$.

(v) *For an A-addressed local composition* $a \subset @A \times F$, *we set* $Int(a) = Int(\{a\})$ *and call it* the intension of a. *Then we have*

$$Ext_A(Int(a)) = \{b|\ f/1 : a \to a \text{ implies } f/1 : b \to b\}.$$

(vi) *For a set* $M \subset End(F)$ *and an address change* $\alpha : A \to B$, *we have*

$$Ext_B(M).\alpha \subset Ext_A(M).$$

If α *is a retraction, i.e., there is a right inverse* $\beta : B \to A$, $\alpha.\beta = Id_B$, *then, for any local composition* $a \subset @B \times F$, $Int(a.\alpha) = Int(a)$, *whence*

$$\alpha^{-1}(Ext_A(M)) = Ext_B(M).$$

(vii) *For a set* $M \subset End(F)$, *and* $a, b \in Ext_A(M)$, *we have* $a \cap b, a \cup b \in Ext_A(M)$.

(viii) *If* $a \subset A@F$ *is an objective local composition, the set of endomorphisms* $f/1 : a \to a$ *qua objective local composition coincides with the intension of its functorial counterpart* $\hat{a}$.

Proof. All statements except (vi), (vii) and (viii) are straightforward. Statement (viii) follows from the fact (see Proposition 3 in Chapter 8) that objective local compositions build a full subcategory of the functorial local compositions, and that underlying form endomorphisms of the functorial versions carry over to the objective versions. To prove statement (vi), let $a \in Ext_B(M)$. Then, by definition, we have a cartesian diagram

$$\begin{array}{ccc} a.\alpha & \longrightarrow & a \\ \downarrow & & \downarrow \\ @A \times F & \xrightarrow{@\alpha \times Id_F} & @B \times F \end{array} \tag{24.10}$$

and therefore, for any $m \in M$, we have a unique vertical arrow ! making the diagram

[4] $\langle M\rangle$ is the semigroup generated by M, whereas $\langle M\rangle_e$ is the monoid generated by M, see Appendix C.2.3.

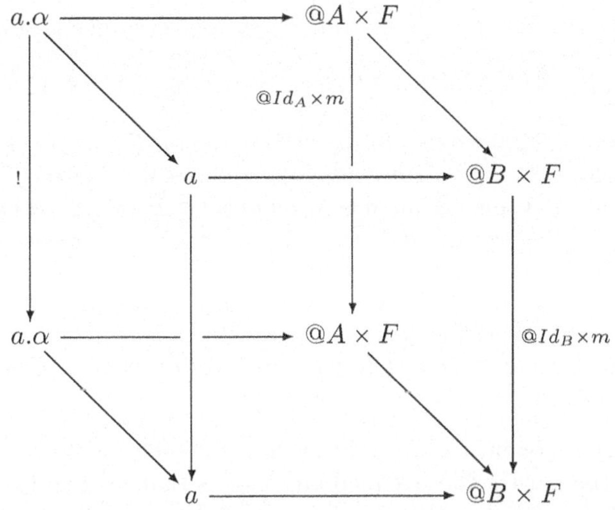

commute. This shows that $a.\alpha \in Ext_A(M)$. To prove the second part of (vi), consider $a.\alpha \in Ext_A(M)$. Since under the hypothesis on α, the transformation $@\alpha \times Id_F$ is epi, and since the upper horizontal square is cartesian, the morphism $a.\alpha \to a$ is also epi. Therefore, there is a vertical arrow $a \to a$ making the entire cube commute, and we conclude $a \in Ext_B(M)$ and therefore, $Int(a.\alpha) = Int(a)$. By the first part of (vi), we conclude that $\alpha^{-1}(Ext_A(M)) = Ext_B(M)$. QED.

24.2.2 Extension and Intension Topologies

Summary. Intension and extension topologies on local compositions are presented.

– Σ –

With the above technicalities, we may now introduce different topological structures.

Definition 80 *The collection of basic semigroup, monoid, respectively, intensions of F define bases for two topologies on $SemiEnd(F), MonEnd(F)$, respectively, the* (semigroup, monoid, respectively) intension topologies

$$SemInTop(F), InTop(F)$$

of F, *respectively.*

Lemma 38 *The projection*

$$?_e : SemiEnd(F) \to MonEnd(F)$$

(see Appendix C.2.3, exercise 84) is continuous and open onto the (basic) open subspace $MonEnd(F) = Int(Id_F)$ of $SemiEnd(F)$.

Proof. Clearly, $MonEnd(F) = Int(Id_F) \subset SemiEnd(F)$. To see that $?_e$ is continuous, take a basic monoid intension $Int^e(M)$. Its inverse image under $?_e$ is $\{s|\ M \subset s_e\}$. Suppose first that $Id_F \notin M$. Then $M \subset s_e = s \cup \{Id_F\}$ implies $M \subset s$, and conversely. So $?_e^{-1}(Int^e(M)) = Int(M)$. If $Id_F \in M$, then $s \in ?_e^{-1}(Int^e(M))$ iff $M \backslash \{Id_F\} \subset s_e \backslash \{Id_F\}$. This implies that $?_e^{-1}(Int^e(M)) = Int(M \backslash \{Id_F\})$, so the map is continuous. It is open since we have $?_e(Int(M)) = Int^e(M)$, and we are done.

Sorite 10 *The intension topologies have these properties:*

(*i*) *With the preceding notation, if* $s \in SemiEnd(F), m \in MonEnd(F)$, *respectively, then*

$$Int(s) = \{t \in SemiEnd(F), s \subset t\}, Int^e(m) = \{n \in MonEnd(F), m \subset n\}$$

are the smallest open neighborhoods of s, m *in* $SemInTop(F), InTop(F)$, *respectively.*

(*ii*) *A subset* $X \subset SemiEnd(F), X \subset MonEnd(F)$, *respectively is closed iff for a subsemigroup* t, *submonoid* n, *respectively,* $t \subset s, s \in X$ *or* $n \subset m, m \in X$ *implies* $t \in X, n \in X$, *respectively.*

(*iii*) *The closure of a point* $s \in SemiEnd(F), m \in MonEnd(F)$, *respectively, is* $\bar{s} = \{t \in SemiEnd(F) | \; t \subset s\}, \bar{m} = \{n \in MonEnd(F) | \; n \subset m\}$, *respectively.*

(*iv*) *A closed subset* $X \subset SemiEnd(F), X \subset MonEnd(F)$, *respectively, is irreducible iff for* $m, n \in X$, $\langle m, n\rangle \in X, \langle m, n\rangle_e \in X$, *respectively.*

(*v*) *If every monoid in* $MonEnd(F)$ *is finitely generated (for example, if End(F) is finite), then* $SemInTop(F)$ *and* $InTop(F)$ *are sober.*

Proof. We prove the monoid case, the other is completely analogous. The first three statements are straightforward. As to the fourth, the irreducibility means that for any two $m, n \in X$, their smallest neighborhoods $Int(m), Int(n)$ intersect in X, i.e., there is $r \in X, n, m \subset r$, therefore the monoid $\langle m, n\rangle_e$ is in X since X is closed. As to the last statement, if X is closed and irreducible, there is at most one generic point. In fact, $\bar{x} = X = \bar{y}$ means $x \subset y \subset x$. To find the generic point, consider the monoid $U = \langle m, m \in X\rangle_e$. Since U is finitely generated, there are finitely many $m_1, \ldots m_k$ such that $U = \langle m_1, \ldots m_k\rangle_e$. But by the one but last statement, this is a member of X, and we have the generic point. QED.

Definition 81 *The collection of basic extensions of* F *at address* A *defines a base for a topology on the set* $A@\Omega^F$ *of* A*-addressed local compositions in* F, *the* extension topology $ExTop_A(F)$ of F at address A.

Corollary 13 *For a local composition* $a \in ExTop_A(F)$, *the set*

$$\begin{aligned} Ext_A(Int(a)) &= \{b | \; f/1 : a \to a \text{ implies } f/1 : b \to b\} \\ &= \{b | \; Int(a) \subset Int(b)\} \end{aligned}$$

is the smallest neighborhood of a, *whereas the topological closure of* a *is*

$$\begin{aligned} \bar{a} &= \{b | \; f/1 : b \to b \text{ implies } f/1 : a \to a\} \\ &= \{b | \; Int(b) \subset Int(a)\}. \end{aligned}$$

This follows from Sorite 9, statement (v).

Corollary 14 *The three topologies* $ExTop_A(F), SemInTop(F), InTop(F)$ *are irreducible with generic points* $A@F^\frown, End(F), End(F)$, *respectively (non-unique). The topology* $ExTop_A(F)$ *is the inverse image topology of* $InTop(F)$ *under the map*

$$Int_A : ExTop_A(F) \to InTop(F) : a \mapsto Int(a) \tag{24.11}$$

which therefore is automatically continuous. The fibers $Int_A^{-1} Int_A(a)$ *consist of the* A*-addressed local compositions* b *in* F *with* $\bar{b} = \bar{a}$.

Proof. The irreducibility claims are clear. For a set $M \subset End(F)$, we have $a \in Ext_A(M)$ iff $M \subset Int(a)$ iff $Int(a) \in Int^e(M)$, in other words, the inverse image $Int_A^{-1} Int^e(M)$ equals $Ext_A(M)$.

Corollary 15 *For an address change* $\alpha : B \to A$ *which is a retraction, the natural map* $.\alpha : ExTop_A(F) \to ExTop_B(F)$ *is continuous,*

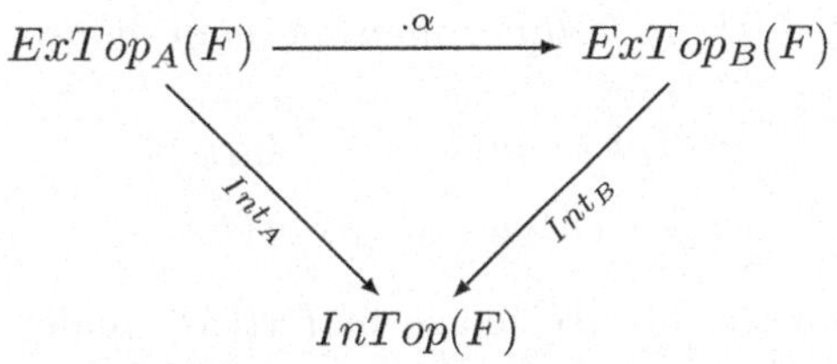

(24.12)

commutes.

This follows from Sorite 9, statement (vi), in particular, we have $Int(a.\alpha) = Int(a)$ in this case.

Proposition 42 *The induced topology on the image*

$$Int_A(ExTop_A(F)) \subset InTop(F)$$

is homeomorphic to the induced topology of the canonical image

$$i(ExTop_A(F)) \subset ExTop_A(F)^s$$

in the sober topology of $ExTop_A(F)$. If all submonoids in $MonEnd(F)$ are finitely generated, the map Int_A factorizes through $ExTop_A(F)^s$.

Proof. We already know that the fibers of Int_A and of i coincide. A base for the topology of $Int_A(ExTop_A(F))$ is the collection of sets $I(M) = \{Int(a)|\ M \subset Int(a)\}$, $M \subset End(F)$. A base for the induced topology on $i(ExTop_A(F))$ in the sober topology is given by $E(M) = \{\bar{a}|\ Ext_A(M) \cap \bar{a} \neq \emptyset\}$. But this set, by corollaries 13, 14, equals the set $\{\bar{a}|\ \exists b, Int(b) \subset Int(a), \text{ and } M \subset Int(b)\}$, and this equals $I(M)$. If all submonoids are finitely generated, we know from Sorite 10, statement (v), that the topology $InTop(F)$ is sober, and the statement follows from the universal property of the associated sober space. QED.

Remark 8 The fact from Proposition 42 which relates the closure of an A-addressed local composition a in the extension topology, or, equivalently: the associated point in the sober topology, with its intension $Int(a)$ justifies the terminology of Noll in [802] where $Int(a)$ is called "Hüllakkord", in English "chord closure" or "closure chord". The difference to our point of view is, of course, that the intension of a chord is a set of endomorphisms, and not a chord. But this will be clarified in the following section.

24.2.3 Faithful Addresses

Summary. This section discusses the relation between endomorphisms and their "trace" on local compositions which live at addresses that capture enough information to produce a faithful image of endomorphisms. We deduce topological properties of such addresses.

– Σ –

Definition 82 *Let A be an address, F a form space, and $z \in A@F$. The point z is called* faithful, full, fully faithful, *respectively, if the map $.z : End(F) \to A@F : f \mapsto f.z$ is injective, surjective, bijective, respectively. We also say that the* address A is faithful, etc., *if there is a point z at that address which has the corresponding property.*

The classical example of a fully faithful point is the case of a representable space $F = @M$ for a module M, and $z = Id_M$.

Definition 83 *Let A be an address, F a form space, and $z \in A@F$. Then we have a map*

$$ex_z : SemInTop(F) \to ExTop_A : m \mapsto m.z^\frown$$

which associates with a semigroup m the functorial local composition of the m-orbit of z; we set $monex_z = ex_z|InTop(F)$ for the restriction of ex_z to the monoids.

Proposition 43 *If the point $z \in A@F$ is faithful, then the map $monex_z$ is a continuous section of Int_A, i.e.,*

$$Int_A \cdot monex_z = Id_{InTop},$$

and the induced topology on $Im(monex_z)$ identifies to $InTop(F)$.

Proof. If $M \subset End(F)$, then the inverse image of $Ext(M)$ under $monex_z$ is the set of monoids m such that $M \subset End(ex_z(m))$. But the latter is m. In fact, by Sorite 9, statement (viii), the endomorphisms of $ex_z(m)$ coincide with those of the objective trace $m.z$. But if f is such an endomorphism, we have $f.Id_F.z = \mu.z, \mu \in m$. Since z is faithful, $f = \mu$. Since trivially, $m \subset End(ex_z(m))$, the claim follows, and therefore, $M \subset End(ex_z(m))$ means $M \subset m$, i.e., the inverse image of $Ext(M)$ is $Int(M)$. Moreover, by the above, we also have $Int_A \cdot ex_z(m) = End(ex_z(m)) = m$ QED.

This means that the topological space $InTop(F)$ of monoids of F can be viewed as a retraction of the topological space $ExTop_A$ of local compositions at a faithful address A (as a function of the chosen faithful point). In particular, if B is any address, and if z is a faithful point at address A, we have the composed continuous map

$$intex_{z,B} = ex_z \cdot Int_B : ExTop_B \to ExTop_A \tag{24.13}$$

which associates with any B-addressed local composition the "objective trace" of its endomorphism monoid at the faithful address A. This means that the closure of a local composition a can be reinterpreted in terms of the objective local composition $intex_{z,B}(a)$, the *z-objective (or, less precisely, A-objective) closure of a.* This is, what in the elementary case of $F = @\mathbb{Z}_{12}, B = 0, z = Id_{\mathbb{Z}_{12}}$ was studied by Noll in [802]. Here, we have a fully faithful point and therefore a complete identification of endomorphisms and points. We should, however keep in mind the ontological difference of endomorphisms and points. But mathematically, this very special case of pitch class theory is indifferent to ontological or more general mathematical perspectives.

So far, if $z \in A@F$ is faithful and $\alpha : A \to B$ is a retraction, we have a diagram with these continuous maps

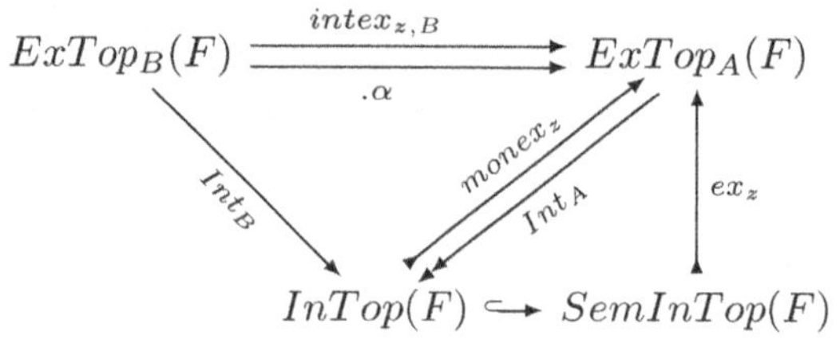

where a local composition a in $ExTop_B$ is mapped to its natural image $a.\alpha$ and to its objective closure $intex_{z,B}(a)$. In this general setup, the two images have nothing in common. We shall now discuss a context where these images are intimately related.

To this end, we suppose that first, z is fully faithful. This means that $.z$ is bijective, call $/z$ its inverse. We further suppose that the image $B@F^\wedge.\alpha = B@F.\alpha^\wedge$ of the maximal B-addressed objective local composition in F corresponds to a set of mutually right-absorbing endomorphisms of F under $/z$, i.e., for all $u, v \in B@F$,

$$u.\alpha/z \cdot v.\alpha/z = u.\alpha/z.$$

Again, the classical example with $F = @M, A = M, B = 0, z = Id_M$, and $\alpha : M \to 0$ the zero-homomorphism, yields the embedding of zero-addressed local compositions as self-addressed local compositions with points $m \in M$ corresponding to the constant points $e^m.0$. These are evidently mutually right-absorbing.

A point z that is fully faithful and induces a set of mutually right-absorbing endomorphisms of F on B-addressed points of F is called *B-absorbing*[5]. In the following calculations, we shall use the Yoneda notation $x : X \to F$ for the point $x \in X@F$ corresponding to $x : @X \to F$.

[5] More precisely: "α-absorbing".

Lemma 39 *Let z be a B-absorbing point and $\alpha : A \to B$ a retraction. If $g \in B@F$, then $g_\alpha := g.\alpha/z$ is constant on B-valued points of F, i.e., $g_\alpha.h = g$ for all $h : B \to F$.*

Proof. Since α is an epimorphism, it suffices to show $g_\alpha.h.\alpha = g.\alpha$. But $g_\alpha.h.\alpha = g_\alpha.h_\alpha.z = g_\alpha.z = g.\alpha$, QED.

Corollary 16 *Let z be a B-absorbing point, $\alpha : A \to B$ a retraction, $b = b^\diamond \subset @B \times F$ a non-empty objective local composition at B, and $x : B \to F$. Then x_α is an endomorphism of b (or, equivalently $\check{b}$), iff $x \in \check{b}$.*

Corollary 17 *Let z be a B-absorbing point, $\alpha : A \to B$ a retraction, and $m \in SemInTop(F)$ a semigroup. Denote by $|m|_\alpha$ the inverse image $(\alpha^{-1}(ex_z(m)^\vee))^\wedge$ in $B@F$, and call it* the B-constant part *of m (to be precise, we would have to specify the absorbing point and the retraction). Then we have*

$$m \subset Int(|m|_\alpha).$$

Proof. Let $g \in m, x \in |m|_\alpha$, and denote $y = g.x \in B@F$. Taking over the notation of Lemma 39, $y.\alpha = g.x.\alpha = g.x_\alpha.z = g_x.z, g_x \in F@F$. By hypothesis, $x_\alpha \in m$, and so is g_x, i.e., $y.\alpha = g_x.z$, whence $y \in |m|_\alpha$. QED.

Definition 84 *For an address change $\alpha : A \to B$, form space F, and a local composition $a \subset @A \times F$, we denote*

$$a|\alpha = a \cap B@F^\wedge.\alpha$$

and call it the α-restriction *of a.*

Observe that even for objective local compositions, their restriction need not be objective!

Proposition 44 *Let z be a B-absorbing point, $\alpha : A \to B$ a retraction, and $b = b^\diamond \subset @B \times F$ a non-empty objective local composition at B. Then we have*

$$(intex_{z,B}(b)|\alpha)^\vee = \check{b}.\alpha. \qquad (24.14)$$

In particular, for two objective, non-empty B-addressed local compositions b_1, b_2, we have

$$intex_{z,B}(b_1)|\alpha = intex_{z,B}(b_2)|\alpha$$

iff $b_1 = b_2$.

Proof. We have

$$\begin{aligned}(intex_{z,B}(b)|\alpha)^\vee &= (intex_{z,B}(b))^\vee \cap B@F.\alpha \\ &= End(\check{b}).z \cap B@F.\alpha\end{aligned}$$

and the claim follows from Corollary 16, QED.

Therefore, if z is an absorbing point and $\alpha : A \to B$ a retraction, any objective B-addressed local composition b in F fulfills $b.\alpha \subset intex_{z,B}(b)$. This turns the terminology of an "objective closure" into an operator that adds new elements to the given original local composition. The closure gives us back the original local composition by restriction to the "B-constant" points, i.e., the natural image of $B@F$ under the address change α.

Compared to the original approach [802], this presentation is independent of the specific properties of the basic form space F. If restricted to the objective local compositions, we have this fact:

Fact 14 *Let $ObExTop_B(F)$ the subspace of $ExTop_B$ consisting of objective local compositions. Under the conditions of Proposition 44, we have a continuous injection*

$$obintex_{z,B} : ObExTop_B(F) \to ObExTop_A(F), \qquad (24.15)$$

and we get the original objects back via α-restriction.

24.2.4 The Saturation Sheaf

Summary. Sheaves of monoids and monoid algebras are associated with each topological space $ExTop_A(F)$.

$- \Sigma -$

For an address A and space form F, if $M \subset End(F)$, the basic open set $Ext_A(M)$ identifies to $Ext_A(\widehat{M})$ if we define

$$\widehat{M} = \bigcap_{a \in Ext_A(M)} Int_A(a), \tag{24.16}$$

the *A-saturation of M*. In fact, $a \in Ext_A(M)$ iff $M \subset Int_A(a)$, and therefore, $M \subset \widehat{M}$, whence $Ext_A(\widehat{M}) \subset Ext_A(M)$, the converse is clear. The saturation of M is the largest monoid defining the given basic open set. This gives rise to a sheaf of monoids as follows.

Proposition 45 *Select an address B and space form F. On an open set $U \subset ExTop_B(F)$ the presheaf Sat_B^F defined by*

$$Sat_B^F(U) = \begin{cases} \bigcap_{u \in U} Int(u) \text{ if } U \neq \emptyset, \\ 1 \text{ if } U = \emptyset, \end{cases}$$

together with the transition inclusion maps $i_V^U : Sat_B^F(U) \hookrightarrow Sat_B^F(V)$ for open sets $V \subset U$, is a sheaf of monoids, the saturation sheaf of F at address B.

If $\alpha : A \to B$ is a retraction, the continuous map $.\alpha : ExTop_B(F) \to ExTop_A(F)$ canonically extends to a map of sheaves $.\alpha : Sat_B^F \to Sat_A^F$.

Proof. This presheaf is a sheaf because the restrictions are inclusions (everything within the big monoid $End(F)$), and because any two open sets intersect by the irreducibility of $ExTop_B(F)$. Moreover, in the case of a retraction α, we know from Sorite 9, statement (vi), that for any local composition $a \in ExTop_B(F)$, we have $Int(a) = Int(a.\alpha)$. To define the sheaf morphism, we can restrict to basic open sets $Ext_A(M)$. We have to construct a morphism of monoids $Sat_A^F(Ext_A(M)) \to Sat_B^F(\alpha^{-1}Ext_A(M))$. By Sorite 9, statement (vi), we also know that $\alpha^{-1}Ext_A(M) = Ext_B(M)$. So we need $Sat_A^F(Ext_A(M)) \to Sat_B^F(Ext_B(M))$. But we have

$$Sat_A^F(Ext_A(M)) = \bigcap_{a \in Ext_A(M)} Int(a) \subset \bigcap_{b \in Ext_B(M)} Int(b.\alpha),$$

and

$$\bigcap_{b \in Ext_B(M)} Int(b.\alpha) = \bigcap_{b \in Ext_B(M)} Int(b) = Sat_B^F(Ext_B(M)),$$

so the inclusion map of monoids $Sat_A^F(Ext_A(M)) \hookrightarrow Sat_B^F(Ext_B(M))$ does the job. QED.

Remark 9 Given any commutative ring R, the above constructions canonically extend to sheaves ${}_RSat_A^F$ and morphisms of R-algebras via the monoid algebras ${}_RSat_A^F(U) = R\langle Sat_A^F(U)\rangle$. We leave the details to the reader.

The saturation sheaf will be used in the discussion in Chapter 25 of the tonal function concept as it was developed in [802, 801].

Exercise 58 *The stalk of the saturation sheaf Sat_B^F at a point $a \in ExTop_B(F)$ is the monoid $Int(a)$.*

25

Harmonic Semantics

"Logik" ist in der Funktionstheorie
ein fundamentaler, aber dunkler Begriff.
Carl Dahlhaus [231, p.95]

Summary. This chapter is about "understanding" aggregates of pitches in their combination on the syntagmatic/paradigmatic axes. This requires constructions of targets of such an understanding, i.e., harmonic *semantics.* The present theory succeeds in a (re)construction of function-theoretic semantics which is based on the paradigms associated with classical tonal functions. We discuss different approaches, among others the morphological theory of Noll and the approach of Mazzola based on chains of triads.

– Σ –

We have already seen in the previous discussion of metrical, rhythmical, and motivic issues that the question concerning the position of a compositional item leads to topological considerations. This will also hold for the harmonic perspective. The topology comes in by the fact that the syntagmatic network of a composition is filtered by a paradigmatic assimilation of units such that this simplified structure can be read as a logical one. It is not the scope of this book to establish dogmatic theories of how this logic should look. We rather want to make clear what general processes are involved when logical constructions are elaborated from the crude pitch configurations.

There are two general differentiations of such logical methodologies: First, the harmonic functions must be related to the basic address where harmonic facts take place, i.e., the address of the local and/or global compositions which we investigate. This will in fact turn out to generate fundamentally different harmonic ontologies, such as it has been made clear by Noll's dissertation [802]. Second, harmonic functions germinate from prototypical harmonic units, such as major or minor triads, chains of thirds and the like, where one starts on a firm harmonic (pre)judice and then wants to extend the commonly accepted to less common chords. The point here is the extension approach from common germs to general situations. But it seems that the topological flavor of all the known harmonies is the common denominator, and that, beyond this, the basic constructors of harmony have very little to do with the special pitch domains. We have therefore anticipated the chapter on harmonic topologies, in fact, as already stressed in that introduction, a chapter which is only accidentally directed towards harmony.

We are not yet in the position to claim that all these topological approaches converge in the common topos-theoretic logic which of course specializes to the logic of open sets of topologies. But it is nevertheless important to open the path to a purely topological interpretation of musical paradigmatics, be it via combinatorial topology or via more or less exotic (but nevertheless canonically defined) sober topologies.

The topos logic of predicates as described in Section 18.3.2 does in fact suggest the following reinterpretation of harmonic topologies. In the truth form defined in formula (18.7)

$$TRUTH(I) \underset{Id}{\longrightarrow} \mathbf{Power}(Val(I)), \tag{25.1}$$

G. Mazzola, *The Topos of Music I: Theory*, Computational Music Science,
https://doi.org/10.1007/978-3-319-64364-9_25

we took a special simple form $Val(I)$ associated with a truth module I. Instead, we could have taken any form F and considered the F-truth form

$$TRUTH(F) \underset{Id}{\longrightarrow} \mathbf{Power}(F), \tag{25.2}$$

whose values stem from F instead of $Val(I)$. This means that the set $\mathbb{T}^A_F$ of truth denotators at address A are the A-addressed local compositions $a \subset @A \times F$. We have the usual Heyting algebra structure of subobjects in $\mathbb{T}^A_F$, together with its logic, as discussed in 18.3.2. But now, we have more: the extension topology $ExTop_A(F)$ with its natural Heyting logic of open sets (see Appendix C.2.2 and [639, p.51]). What we should keep in mind in the following discourse is this:

Principle 22 *The extension topologies are a natural topos-theoretic framework for predicate logic.*

One the one hand, this principle preconizes a seamless transition from logic to geometry as already indicated in principle 12 in Section 18.3.3.3, but on the other, it does also confirm the fuzzy character of harmonic logic in so far as the extension topology is quite contrary to a common geometric space. Rather could it be associated with the irreducible ring spectra in algebraic geometry.

25.1 Harmonic Signs—Overview

Summary. This section gives an overview on levels of harmonic semantics.

– Σ –

Generally speaking, the semantics of harmonic signs has always been built via ontological "superspaces", such as the metaphysical level of the Pythagorean tetractys: Greek musicology was based upon transcendence, and the core activity of Pythagorean scholars was the introspection of this mystic unit via instrumental practice. It seems that—at least in European tradition—semantics of musical harmony has been attached to a transcendence of one or another type.

In modern times, especially with Hugo Riemann's never accomplished program after Rameau's incipit, harmonic semantics has been focused on the attribution of tonic, dominant, and subdominant meaning to chords. This semantics is a very mysterious one since it is quite formalistic on one hand, but carries with it a certain amount of romantic connotation, in the spirit of Moritz Hauptmann's dialectic overloading of octave, fifth, and third [1090], say.

We shall refrain from such transcendental semantics and restrict our analysis to the structural framework which can be recognized as a machinery for attributing to given chord configurations their harmonic functionalities.

The basic setup is always this (concrete examples follow below):

1. A pitch space form F is chosen.
2. At a specific address A, a set $StandardChords$ of local compositions in F is selected, playing the role of *standard chords* where the hoped-for semantics will be based, and where one knows from experience/history/practice what to expect.
3. A set of interpretations S^I of local "scale" compositions S at A by determined subsets of $I \subset StandardChords$ is defined which will define the admitted *tonalities* in the specific setup.

On this background, the harmonic semantics of any chord, i.e., any local composition c at selected (envisaged by the specific approach) addresses B in F, will be constructed by one of these two strategies:

Degree Theory This approach relates c to the given standard chords and/or tonalities by coverings, i.e., c is included in a specific selection of standard chords and/or tonalities. This means that a chord is viewed as either a part of a standard structure, or else a part of a covering by standard structures, e.g., the seventh chord $G^7 = V^7_C$ is covered by two triadic degrees V_C, VII_C. *The moral of this approach is that the semantical process is a completion of a partial sign to fit in a standard object.*

Function Theory This approach relates c to the given standard chords and/or tonalities by topological proximity to a specific selection of standard chords and/or tonalities. Such a procedure presupposes a topological environment, in fact, we shall see that the extension topologies are the formal realization of this task. *The moral of this approach is that the semantical process uses similarity to standard objects without necessarily being part of them as in degree theory, i.e., commonality instead of completion.*

Evidently, both approaches imply degrees of fuzziness: Completion, because this can be achieved in different ways. For example, an interval $\{c, g\}$ may be completed to a major third degree $C = \{c, e, g\}$ or else a minor third degree $Cm = \{c, e_\flat, g\}$, whereas this interval may represent the tonic function in C or the subdominant function in G, for example. But this is the very nature of the harmonic approach which is a genuinely paradigmatic one. The problem of harmonic semantics is not its polyvalence, but the formally clear theory of such polyvalence, because only the precise analysis of polyvalence can qualify or disqualify a harmonic theory.

We should stress that the degree-theoretic and function-theoretic approaches do not imply that harmonic functions can only be generated by the second approach. The degree-theoretic approach may also be used to attribute functional values, but these values will then be calculated from the complete standard chords instead of functions which are calculated from similarity to standard objects.

Remark 10 We have to note another source of fuzziness which is often applied to both strategies: Instead of the given chord, one admits its deformation by a shift c' of c which is induced by an alteration (see Section 7.5). As we have already discussed this deformation process, we shall restrict ourselves here to non-alterated harmonic semantics. The decisive point here is that one should keep in mind the alteration Theorem 1 in Section 7.5.1 which states that—in its negative interpretation—any two chords may be shifts of each other if the alteration interval is large enough. In contrast to this not very elegant deformation rhetorics stands the empirical use of alterations in existing musical styles which sometimes is quite subtle. In jazz harmony, for example, the fifth of a dominant seventh chord can be altered without problems (Domiant7/-5 or Dominant7/+5), but the seventh not (an alteration of the minor seventh invariably results in a loss of the dominant function). In a better Approach to alterations, one should not develop harmony on the given form, but on a "tangent form $F[\epsilon]$" (corresponding to the tangent module $M[\epsilon]$ in the representable case). Presently, no further theory is known in this direction.

25.2 Degree Theory

Summary. We give three short accounts of computer programs for degree theory.

– Σ –

25.2.1 Chains of Thirds

Summary. Software for analysis by chains of thirds: Mazzola's *presto*® chord analyzer.

– Σ –

The *presto*® software[1] analyzes zero-addressed chords in $PiMod_{12}$. The chords in *presto*® are abstracted from any local compositions in the four-dimensional space $OPLD_{\mathbb{Z}} = Onset \oplus Pitch \oplus Loudness \oplus Duration|_{\mathbb{Z}}$ via projection to $PiMod_{12}$. The program also retains the local composition's lowest pitch, see Figure 25.1. The user may then select a reference pitch class; by default the class of the retained lowest pitch is chosen. This means that in fact, the chord is not taken as a local composition, but as a global one with two charts: the given chord, and an additional chart, the singleton of the reference pitch, the latter being a representative of the so-called "fundamental" pitch in classical harmony. This is an interpretative activity and cannot be justified without a deeper analysis of the context of the given chord. So a priori, the reference pitch is a

[1] See Chapter 50 for details.

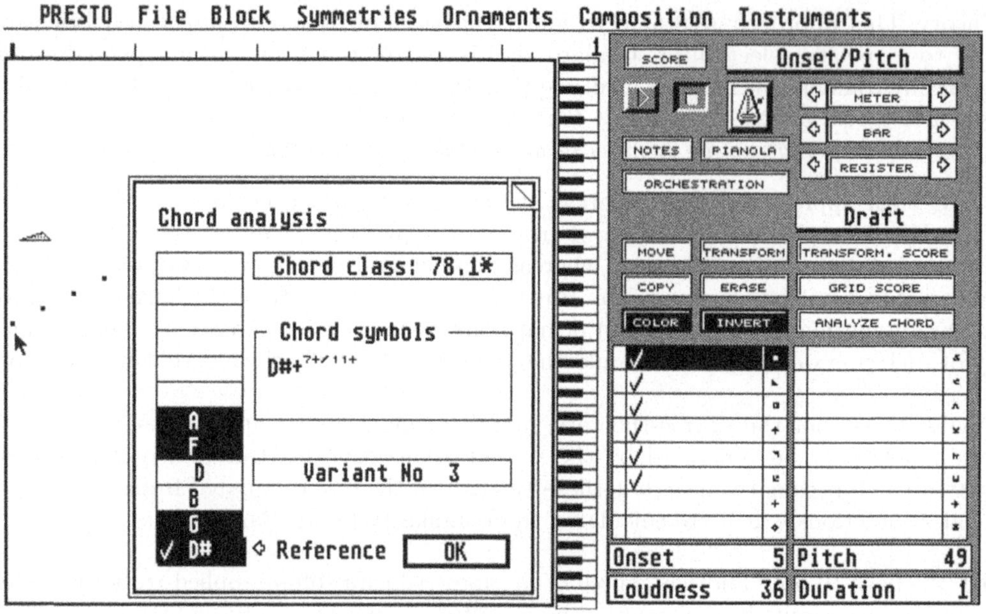

Fig. 25.1. The sound-event set of ascending small filled squares to the left is analyzed by the software's chordmaster routine (middle small window entitled "Chord analysis"). This one shows the chord's pitch classes in third chain sequence from below. The black notes are those of the chord, the white ones in between are the added pitches from the embedding third chain. The isomorphism class of this chain (78.1*) is indicated, as well as the standard denotation of the chain in jazz chord symbols. The variant number (No 3 here) is the number of the chain out of the total number of possible chains. The reference pitch is the lowest pitch, it can be changed by the user to any other pitch class (even outside the chord).

variable which has to be integrated within a context-sensitive calculus. The projected local composition S is then confronted with all chains of thirds (see Section 13.4.2) which start from the reference pitch and have a minimal cardinality such that they contain S. For example, the chord $\{0, 1, 2\}$ has 23 minimal third chains when calculated from the reference 1. The list of all possible chains of thirds is shown in Appendix N.2. Among these third chains the chains with absolute minimal cardinality can be found, it is the set $3Chain(S)$ defined in 13.4.2. This set is independent of the reference assumptions.

Although it does not determine the isomorphism class of S in general, it gives a lot of information for the harmonic semantics of the chord. The absolutely minimal third chains of the chord are not yet function-theoretic constructs, but they can and will be used to build function-theoretic values, see Section 25.3.3.

The complete list of third chains in Appendix N.2 also contains the symbols from the American jazz notation. It is well known that there are many and not really optimal variants of this notation. The *presto*® implementation of our list has been charged with one variant of the jazz lead-sheet notation that can be justified by the total number of third chains (including 12-chains, not as with traditional (and not the new!) jazz practice, where chords only have a restricted number of notes). Basically the procedure is this. We have chosen one of the four 12-tone chains to build the reference interval step succession. It is reasonable to choose the one which is lexicographically first (most packed to the left) with respect to the "white keys", i.e., to the diatonic scale from C. This maximal chain of thirds has its initial subchains which all are named with the usual numbers. So the first chain from C is $C = \{0, 4, 7\}$, the second is $C7 = \{0, 4, 7, 10\}$, etc., until the last and total chain

$$C23 = \{0, 4, 7, 10, 2, 5, 8, 11, 3, 6, 9, 1\} \tag{25.3}$$
$$= \{C, E, G, B_\flat, D, F, A_\flat, B, E_\flat, G_\flat, A, D_\flat\}. \tag{25.4}$$

All degree names will be related to this reference chain, e.g., if we have a "13" in a symbol, this relates to $A_\flat$ with respect to the C chain, and to the corresponding transposition in the general case. For the triadic chains, we use the traditional four types of major C, minor $Cm = \{0, 3, 7\}$, diminished $C0 = \{0, 3, 6\}$ and augmented $C+ = \{0, 4, 8\}$ chains. The larger chains are built upon these basic cases, and the added symbols refer to the above degree codification. For example, the symbol $Cm7 + /11$ means that we add the major seventh 11 and go until the 11-chain, including the none ("9" adding pitch class 2) and the undecime ("11" adding pitch class 5). So the deviations from the normal chain are indicated by $m, 0, +$ for the triadic initial piece, and then by the alterations $x+, x$- for the other deviations from the norm. In some cases, the single alterations are not sufficient, for example, the chain nr. 46: $\{0, 4, 8, 11, 3, 7\}$ would have the notation C+7+/9+/11++. Alternatively, we can write $C+7+/9+/13+/\widehat{11}$ to indicate that 11 has been omitted and instead, 13+ was added. This notation (well known to mathematicians) can also be used to describe precisely the pitch classes of the given chord: we add a hat for every omitted ingredient, and we write this information after the standard symbol. For example, the fifth $\{C, G\}$ would read as $C/\widehat{3}$ or $Cm/\widehat{3}$-.

The set of third chains $3Chain(S)$ of a non-empty chord S evidently is a simplex in the nerve of the third chain interpretation of $PiMod_{12}$. We shall see in Section 25.3.2 that harmonic functions starting from degree theory can be viewed as fuzzy truth functions on this nerve, i.e., functions defined on each vertex and then extended to the entire chain in the affine way to the geometric realization of the nerve. This type of functions was already encountered in the metric and melodic analysis under the label of "weight functions". So the topological flavor of the degree theory reduces to the combinatorial nerve of the standard chords (the third chains in this case). The question of the function values of standard chords with respect to specific tonalities will be discussed in Section 25.3.2.

25.2.2 American Jazz Theory

Summary. John Amuedo's MAX programs CHORD-CLASSIFIER, SCALE-FINDER, and SCALE-MONITOR for analysis of AST-based jazz harmonies.

– Σ –

In his music theory thesis [29], Amuedo has programmed a bunch of LISP routines for AST and jazz-theoretic degree classification of chords. We have already mentioned the AST part of this work in Section 11.5.2.3 and Section 16.3. This software-part is used in the three harmonic analysis routines. The 76 standard ("fundamental" in Amuedo's wording) chords as well as the 18 standard scales of this approach are retrieved from relevant American jazz literature[2]. Like *presto*®'s chord analyzer, the present chord-analysis works on pitch classes (*pcs*) in $PiMod_{12}$, however, the class of the lowest pitch in the original integer pitch space is retained to specify the chord and scale names.

The chord names or symbols are derived from seven groups of "foundation chords". In the table below is their list (see the list in Section 11.5.2.3 for the definition of the different "norm" symbols). This list is complemented by Amuedo's "Shortest List" of 18 "essential jazz scales" shown below.

[2] Above all, it is [152], for further literature, see [29, Bibliography: Jazz harmony]

Lead-sheet symbol	DNR	ZNF	Representative
Major Type (Ma)			
Ma	317.0	0,4,7	C,E,G
Ma6	437.3	0,3,5,8	C,E,G,A
Ma7	416.1	0,1,5,8	C,E,G,B
Dominant Type (Do)			
Do	439.3	0,3,6,8	C,E,G,Bb
Dob5	430.3	0,2,6,8	C,E,Gb,Bb
Do5#	425.2	0,2,4,8	C,E,Ab,Bb
Augmented Type (Au)			
Au	318.0	0,4,8	C,E,Ab
Au7	434.2	0,3,4,8	C,E,Ab,B
Minor Type (mi)			
mi	314.0	0,3,7	C,Eb,G
mi6	428.2	0,2,5,8	C,Eb,G,A
mi7	437.2	0,3,5,8	C,Eb,G, Bb
miM7	413.1	0,1,4,8	C,Eb,G, B
Diminished Type (di)			
di	313.0	0,3,6	C,Eb,Gb
di6	440.0	0,3,6,9	C,Eb,Gb,A
di7	428.1	0,2,5,8	C,Eb,Gb,Bb
diM7	412.1	0,1,4,7	C,Eb,Gb,B
Major Flat Five Type (Mb5)			
Mb5	316.0	0,4,6	C,E,Gb
Mb5M7	415.1	0,1,5,7	C,E,Gb,B
Sustained Four Type (Su)			
Su2	308.0	0,2,5	C,D,F
Su5	310.2	0,2,7	C,F,G
Su7	310.1	0,2,7	C,F,Bb
Added to the common foundation chords by Amuedo			
Sustained Root Type (R)			
R25	310.0	0,2,7	C,D,G
R27	307.1	0,2,4	C,D,Bb
Stacked Perfect-Fourth/Perfect-Fifth (P4, P5)			
P4*3	427.1	0,2,5,7	C,F,Bb,Eb
P5*3	427.2	0,2,5,7	C,G,D,A

Name	DNF	Representative
Alt_Dominant+5	834	C,Db,Eb,E,F#,G,Ab,Bb
Augmented	646	C,Eb,E,G,Ab,B
Bebop_Major	832	C,D,E,F,G,G#,A,B
Bebop_Seventh	817	C,D,E,F,G,A,Bb,B
Blues_(6-note)	656	C,Eb,F,Gb,G,Bb
Blues_(7-note)	708	C,Eb,E,F,Gb,G,Bb
Diminished	835	C,D,Eb,F,Gb,Ab,A,B
Harmonic_Minor	738	C,D,Eb,F,G,Ab,B
Hung/Gypsy/Byz	729	C,D,Eb,Gb,G,Ab,B
Locrian+n9	815	C,Db,D,Eb,F,Gb,Ab,Bb
Major	743	C,D,E,F,G,A,B
Major_b6	742	C,D,E,F,G,Ab,B
Melodic_Minor	739	C,D,Eb,F,G,B
Penta_Dom	549	C,D,E,G,Bb
Penta_Lydian	550	C,E,F#,Ab,B
Penta_Major	5501	C,D,E,G,A
Penta_Minor	555	C,D,Eb,G,A
Whole_Tone	674	C,D,E,Gb,Ab,Bb

The CHORD-CLASSIFIER routine takes the input chord with integer coefficients (also, like *presto*®, not necessarily as a 'vertical' chord), retains the *pcs* and the lowest note's *pc*. The output is a list of seven entries. Apart the general AST-classification routine SET-SLAVE, this information is deduced from the standard chord list and from the essential scale list. The last four entries are just different labelings in terms of AST norms, such as DNR, DNF, DPF, INF, we refrain from this accounting overhead. The first entry: "RootPos." exhibits the lowest pitch class, together with a standard chord from the above list, possibly extended to higher third-chain notes. If this task fails, MAX' LISP-routine inserts the chord's DNR on this position. The second entry "Inversion:" exhibits other possible chords from the standard list which could describe the given chord by use of another fundamental note when applying inversions. The third entry "PC-Aliases:" lists all possible chord names irrespectively of the given DNF.—The SCALE-FINDER routine finds scales which are "compatible" with a given chord in the sense of *pc*-subset relations. This routine outputs all scale types, together with possible transpositions such that they contain the chord's *pcs*.—The third routine SCALE-MONITOR is more pedagogical and just for exercises in selected scales.

On this basis, a harmonic analysis is reduced to a set of degree symbols (lead-sheet notation) for chords (together with fundamental notes) which are grouped according to typical initial triad, as used in *presto*® and other degree notation. The enrichment with respect to *presto*®'s notation consists in a more function-theoretic flavor of the names, in the case of "dominant" chord types, for example. This information is—however—also present in *presto*®'s fundamental note selection option, such that the distinction between $Am7$ and $C6$ is managed by the distinction between the fundamental A and C in the two cases, respectively, and also with the consequence that the minimal third chains in these cases are not the same.

In both cases, we end up with a determined list of standard chords (including standard scales in the shape of "large chords") which describe the given chord and reduce its study to the harmonic semantics of these empirically or systematically retrieved standard objects.

25.2.3 Hans Straub: General Degrees in General Scales

Summary. Hans Straub has initiated a general harmony, including modulation theory, as a generalization of Mazzola's original approach [669]. Here we just give a summary of Hans Straub's approach to scales and degrees.

– Σ –

In the author's original approach to harmony in [669], the diatonic scale X was covered by triadic degree chords to yield the triadic interpretation $X^{(3)}$ discussed in Section 13.4.2. The corresponding cadence and modulation theory was subsequently generalized by Daniel Muzzulini [779], and this one still more generalized by Straub [1018].

In Straub's general approach, a scale is not given as a starting structure, but as a result, in fact the union, of a set S of "degree" zero-addressed chords $s \subset 0@PiMod_{12}$. No further requirement upon similarity or structure of the chosen chords is imposed. For example, the "Hora Cero" interpretation consists of the set of seven three-element chords

$$S_{\text{Hora Cero}} = \{C, Dm, Em, F, G, Am, B_\flat\}$$

with a nerve that looks like a belt which has been pinched to a point (the G vertex) in a position. Another interpretation is called "Dorico Flamenco" because of its Flamenco origin, its seven chord atlas is

$$S_{\text{Dorico Flamenco}} = \{Dm, C, B0, Am, G, F, E\}$$

both generating scales of type 17.1 in the chord classification Appendix N.1.

The spirit in Straub's approach is not so much the corresponding generality of cadence and modulation theory (which will be taken up in the following chapters), but the insight that much of existing music traditions, in particular popular music or American jazz, as we have seen in Section 25.2.2, is defined by a grown chord vocabulary that is not subjected to a scale, on the contrary, it generates 'scales' that issue from the chord vocabulary, however not scales that will play a melodic role, only harmonically driven background material.

This approach is however a purely degree-theoretic one, even in a very radical vein: No approach to theoretical background structures to allowed chords is imminent. You are just allowed to play a number of chords per "tonality" and to use cadences and modulations according to determined rules, but no further function theory is addressed.

The question of harmonic semantics is indeed not explicit here. Although the jazz idiom has a strong affinity to European function semantics, it is not true that this and other ethnological/cultural contexts are or can be cast in this kind of abstraction. Some ideosyncratic usages of chord vocabularies happen to look like absolute pitch perception: No comparison between different pitches leads to identification, each pitch is an individual. In harmony, this would mean that some usages of chords relates to a kind of "absolute harmonic position", every used chord is irreducible, in fact a *semantic atom* of harmony.

25.3 Function Theory

Summary. We discuss the topological similarity approach to harmonic functions, as it has been prepared in the previous Chapter 24.

– Σ –

In this section, we consider two addresses, A, B, a retraction $\alpha : A \to B$, a form F, and a B-absorbing point $z : A \to F$.

In traditional harmony, for example in Schönberg's treatise [948], the harmonic band between two chords is the set of common notes. We have seen in Section 13.4.2 that the Möbius strip shaped nerve of the triadic interpretation of a diatonic scale represents the information about harmonic bands among certain sets of triadic degrees. In that naive setup, the harmonic relation was given by common notes. Following the empirical insight 11.3.6, the deeper semantics of a (zero-addressed) chord may be recognized from its symmetries [802], or, to put it in the wording of harmonic topology, from the self-addressed objective closure (24.2.3). Putting this together with our empirical insight, one may consider the harmonic bands between the objective closures of chords, instead of the original, usually zero-addressed objects.

This motivates Noll's definition of allomorphic chords for a given semigroup of symmetries ([802]). Here is our rephrasing of that definition (Noll does not refer to topological structures) in the general terms of harmonic topologies.

Definition 85 *Let m be a monoid in $InTop(F)$. Then the B-*allomorphic extension *of m is the open set*

$$AllExt_B(m) = \bigcup_{n<m, n\neq 1} Ext_B(n)$$

*of all points in $ExTop_B(F)$ which contain some non-trivial submonoid n of m, i.e., some non-closed monoid n which is topologically dominated by m. The elements of $AllExt_B(m)$ are called the B-*allomorphs *of m.*

Evidently, $Ext_B(m) \subset AllExt_B(m)$. In order to make the common endomorphisms precise, we have locally closed[3] subsets

$$AllExt_B(n, m) = Ext_B(n) - \bigcup_{n<l<m, n\neq l} Ext_B(l)$$

for all non-trivial dominance relations $n < l < m$ for monoids l strictly larger than n. This gives a partition

$$AllExt_B(m) = \coprod_{n<m, n\neq 1} AllExt_B(n, m) \tag{25.5}$$

of an allomorphic extension into locally closed subsets. Such a subset $AllExt_B(n, m)$ is called the *strict B-extension of n in m*. Except the open part $AllExt_B(m, m) = Ext_B(m)$, all strict extensions have empty interiors. Observe that in general, the intersection of two allomorphic extensions is *not* an allomorphic extension, so these open sets do not define a basis of a topology.

Example 43 We take the classical situation $B = 0, A = \mathbb{Z}_{12}, F = @A$, however with the usual identification of elements of F as multiples of fifths. Then Noll's "second dominant morpheme in C major" is $D^\star = \langle \alpha^\star = e^8.3, \beta = e^9.8\rangle_e$, a bigeneric monoid.

Allomorphic extensions are a means for defining harmonic functions via extensions, this is the idea in [802, I.3.1, Definition 4]. Given specific monoids $M_{m,n}$ which "signify" certain harmonic functions (see below for details), the representation of these functions extends to the (zero-addressed) chords in $AllExt_0(M_{m,n})$. This approach reveals the thoroughly intuitionistic logic of topological spaces [639, I.7/8]. In fact, by the irreducibility of $ExTop_B(F)$, extensions—and a fortiori allomorphic extensions—are never disjoint. So there is little chance to have a unique functional representation for a given chord. Even on the level of strict extensions, we may have non-empty intersections $AllExt_B(n, m) \cap AllExt_B(n', m')$ for different pairs (n, m), (n', m'). We have this lemma on empty strict extensions:

Lemma 40 *With the above notation, if the defining monoid m is saturated, i.e., equal to its saturation $\hat{m}$, and its non-trivial submonoid n is not, then $AllExt_B(n, m) = \emptyset$.*

Proof. From the construction of the saturation sheaf it is clear that a monoid inclusion $n \subset m$ implies $\hat{n} \subset \hat{m}$. Therefore, if $m = \hat{m}$ and n is not saturated, $x \in Ext_B(n)$ implies $x \in Ext_B(\hat{n})$ and therefore $AllExt_B(n, m) = \emptyset$.

[3] Intersections of open and closed sets, see Appendix H.1.4, Definition 182.

Remark 11 In general, it is difficult to give these functional specifications a quantitative weight, in terms of percentages, say. The only evident measure of the role of a chord X in $AllExt_B(m)$ is the position of its strict extension—the unique non-trivial submonoid $n \subset m$ such that $X \in AllExt_B(n, m)$—in the lattice $\bar{m}$ (the closure of m) of submonoids. If the entire descent from monoid to submonoids is finite, one may for example take the relative depth of descent from top to define a numerical weight of a submonoid.

We still do not know very much about the relations between the saturation sheaf and the topological properties of the extension topology. But here are some exercises for straightforward connections:

Exercise 59 *With the above notation, prove these statements:*

1. *If $U \subset ExTop_B(F)$ is open, then the basic extension of the monoid of sections $Sat^F_B(U)$ satisfies $U \subset Ext_B(Sat^F_B(U))$.*
2. *If for two monoids $n < m$, then $AllExt_B(n) \subset AllExt_B(m)$.*
3. *We have $Sat^F_B(AllExt_B(m)) = \bigcap_{n<m,\ saturated\ and\ non\text{-}trivial} n$. If m verifies the descending chain condition*[4] *then*

$$Sat^F_B(AllExt_B(m)) = \bigcap_{minimal\ among\ saturated\ and\ non\text{-}trivial\ n<m} n.$$

25.3.1 Canonical Morphemes for European Harmony

Summary. We discuss the construction principle for bigeneric morphemes, the crucial objects for Noll's reconstruction of traditional European harmony.

– Σ –

We have so far prepared the semantic field via extension and allomorphic extension concepts. So we are ready to look for monoids which may recover the harmonic semantics known from classical function theory.

As above in Example 43, we are now working in the situation $B = 0, A = \mathbb{Z}_{12}, F = @A$, the case studied in [802] with the elements of F being interpreted as multiples of fifths. Following the suggestion from Section 24.1.2 concerning the structure of certain common chords, in particular circle chords, the construction of special monoids (or their objective trace under an absorbing point $z = Id_{\mathbb{Z}_{12}}$ in the present case) has been set up following a bigeneric approach. In fact, the circle chords are the zero-constant traces of self-addressed chords $monex_z(\langle f, e^c.0\rangle_e)$ associated with a bigeneric monoid, whereas the example V^7_C is the zero-objective trace of $monex_z(\langle e^8.3, e^2.0, e^7.0, \rangle_e)$ which is associated with a trigeneric monoid.

The bigeneric approach then takes two generators $\alpha_d = e^d.3, \beta_a = e^a.8$ (d for "diminished", a for "augmented", will be justified below) and considers the "morpheme" monoids[5]

$$M_{d,a} = \langle \alpha_d, \beta_a \rangle_e \tag{25.6}$$

generated by α_d, β_a for any coefficients $a, d \in \mathbb{Z}_{12}$. The experimental evidence of the crucial role of these monoids is given by the following proposition. It uses the affine Lie bracket $[xy] = x.y - y.x$ which is defined on any monoid of endomorphisms $X@X$ of a module X. Further, we need to encode the translation classes of some zero-addressed chords in $\mathbb{Z}_{12}$. We need the class numbers 2,5,6,7,10, and 14, see Appendix N.1 for the class numbers. Class 2 contains two translation classes, 2.0 (=2 in the class table), and 2.1. Classes 5,6,7 are translation classes. Class 10 splits into four translation classes, two for each subclass 10, 10.1 of the table. We denote them again by 10.1, 10.2, 10.3, 10.4 according to the lexicographic order. Class 14 splits into two translation classes 14.1, 14.2.

Proposition 46 [802, III.3.3, Lemma 2] *Given two coefficients $a, d \in \mathbb{Z}_{12}$, the translation class of the zero-constant part $|M_{d,a}|_0$ of the morpheme monoid $M_{d,a}$ is a function of the Lie bracket $[\alpha_d\beta_a] = e^{5d+2a}$, and we have these translation classes:*

[4] Every descending chain of submonoids is stationary, for example for a finite endomorphism monoid $End(F)$.
[5] In [802], only the generated semigroups are considered, but the results are the same.

Translation Class of constant part of $M_{d,a}$	*Lie invariant* $5d + 2a$
2.0	*0*
5	*3 or 9*
6	*4 or 8*
7	*6*
10.1	*1*
10.2	*11*
10.3	*5*
10.4	*7*
14.1	*10*
14.2	*2*

The point of this proposition is that these classes are very common. In fact, the Lie invariant $5d + 2a$ yields these chord types for the constant parts of the above morphemes:

1. $5d + 2a = 1$ gives a major chord such as $\{c, e, g\}$,
2. $5d + 2a = -1$ gives a minor chord such as $\{c, e_\flat, g\}$,
3. $5d + 2a = -2$ gives a (major) seventh chord without fifth, such as $\{g, b, f\}$,
4. $5d + 2a = 2$ gives a (minor) sixth chord without tonic, such as $\{e_\flat, g, a\}$.

This means that we get very classical chords for the Lie invariants $\pm 1, \pm 2$. These Lie invariants are taken to define the functional semantics of tonic, dominant, and subdominant values in [802].

Before we step over to the tonal function calculus in 25.3.2, we should understand the general principle which governs the construction of bigeneric morphemes. In [802], the situation was uniquely $F = @\mathbb{Z}_{12}$, and no systematic account was given for the "bi" in the bigeneric approach. It was also not clear which should be the relation between the two generators. This is what we want to complete now. To this end, recall from Appendix C.2.3, Definition 139, the idempotent component $\sqrt{x}$ of an idempotent element $x \in End(F)$ of the endomorphism monoid of a form F. Recall also from Appendix C.2.3, Proposition 78, that for a representable form $F = @Q$, the idempotent components are all conjugate under translations to idempotent components of linear idempotents $x_{U,V}$ which correspond to direct decompositions $Q = U \oplus V$.

In $\mathbb{Z}_{12}$, we have the four direct decompositions

$$\begin{aligned}
\mathbb{Z}_{12} &= \mathbb{Z}_{12} \oplus (0), \\
\mathbb{Z}_{12} &= (0) \oplus \mathbb{Z}_{12}, \\
\mathbb{Z}_{12} = 3.\mathbb{Z}_{12} \oplus 4.\mathbb{Z}_{12} &\xrightarrow{\sim} \mathbb{Z}_4 \oplus \mathbb{Z}_3, \\
\mathbb{Z}_{12} &= 4.\mathbb{Z}_{12} \oplus 3.\mathbb{Z}_{12}
\end{aligned}$$

which are pairwise coupled by the exchange of factors, corresponding to the exchange $x \mapsto 1 - x$ of idempotents. The second group of idempotents is the Sylow decomposition of $\mathbb{Z}_{12}$, the only non-trivial direct decomposition. The above selection of generators α_d, β_a corresponds exactly to the Sylow decomposition; we take $\alpha_d \in \sqrt{e^x.pr_{\mathbb{Z}_4}}, \beta_a \in \sqrt{e^y.pr_{\mathbb{Z}_3}}$. More precisely,

$$\alpha_d = e^d.\pi_{\mathbb{Z}_4}, \pi_{\mathbb{Z}_4} = -pr_{\mathbb{Z}_4} = (3),$$

and

$$\beta_a = e^a.\pi_{\mathbb{Z}_3}, \pi_{\mathbb{Z}_3} = -pr_{\mathbb{Z}_3} = (8),$$

twice the negative projections as linear factors, and any shift coefficients a, d, which in fact can be arbitrary because of the choice of the linear part!

So the recipe is this: Take any non-trivial direct decomposition of the underlying module $Q = U \oplus V$ and the associated idempotents p_U, p_V, take two elements $\alpha \in \sqrt{e^x.p_U}, \beta \in \sqrt{e^y.p_V}$ in idempotent components which are conjugate to the components $\sqrt{p_U}, \sqrt{p_V}$, and build the bigeneric monoid $\langle \alpha, \beta \rangle_e$. So the *bi*generic approach corresponds to the direct decomposition into two factors, the relation among these generators is the complementary role of the direct factors in the idempotent relation $x \mapsto 1 - x$.

Let us apply this generalization to the just tuning space $EulerZModule = \mathbb{Z}.q \oplus \mathbb{Z}.t$. This rank-two $\mathbb{Z}$-module has infinitely many direct decompositions into two rank-one factors, for example the above one, defining the projections $p_1, p_2 = 1 - p_1$ onto the factors $\mathbb{Z}.q, \mathbb{Z}.t$, respectively, or the decomposition $\mathbb{Z}.q \oplus \mathbb{Z}.Kt$ via syntonic comma Kt. Let us look at two canonical generators for the idempotents p_1, p_2. Set $\pi_1 = -p_1, \pi_2 = -p_2$, and then

$$\alpha_{a,\xi} = e^{\binom{a}{\xi}}.\pi_1, \beta_{\eta,b} = e^{\binom{\eta}{b}}.\pi_2.$$

This entails

$$\alpha^2_{a,\xi} = e^{\binom{0}{\xi}}.p_1, \beta^2_{\eta,b} = e^{\binom{\eta}{0}}.p_2,$$

and $\alpha^3_{a,\xi} = \alpha_{a,\xi}, \beta^3_{\eta,b} = \beta_{\eta,b}$. Then we have this zero-constant chord:

$$|\langle \alpha_{a,\xi}, \beta_{\eta,b} \rangle_e|_0 = \{\begin{pmatrix} a-\eta \\ \xi \end{pmatrix}, \begin{pmatrix} \eta \\ b-\xi \end{pmatrix}, \begin{pmatrix} \eta \\ \xi \end{pmatrix}\},$$

which yields the major triad $I_C = \{\binom{0}{0}, \binom{1}{0}, \binom{0}{1}\}$ for $\eta = \xi = 0, a = b = 1$. So the direct decomposition approach not only yields standard chords in 12-tempered space, but also in just tuning. No further developments have been investigated to date in the just tuning space.

So if ever the form is representable, $F = @Q$ for a module Q, if $Q = U \oplus V$ is a direct decomposition with corresponding projections p_U, p_V, and if $x, y \in Q$, we have the generators of a bigeneric monoid $\alpha^U_x = e^x. - p_U, \beta^V_y = e^y. - p_V$ which live in the idempotent components $\sqrt{e^{x - p_U(x)}.p_U}, \sqrt{e^{y - p_V(y)}.p_V}$, respectively. If $s = e^t.s_0 \in \overrightarrow{GL}(Q)$ is any invertible endomorphism of Q, then the conjugate generators are

$$s.\alpha^U_x.s^{-1} = \alpha^{s(U)}_{s_0(x)},$$
$$s.\beta^V_y.s^{-1} = \beta^{s(V)}_{s_0(y)},$$

and we have the conjugate monoid

$$s.\langle \alpha^U_x, \beta^V_y \rangle_e.s^{-1} = \langle \alpha^{s(U)}_{s_0(x)}, \beta^{s(V)}_{s_0(y)} \rangle_e$$

whose zero-constant part is

$$s(|\langle \alpha^U_x, \beta^V_y \rangle_e|_0).$$

This means:

Proposition 47 *The transformations via $s \in \overrightarrow{GL}(Q)$ of all zero-constant chords in $@Q$ derived from bigeneric morphemes are the zero-constant chords derived from s-conjugate generators. In particular, all translates of zero-constant chords derived from bigeneric morphemes are again of this type.*

In particular, the full translation classes of common zero-addressed chords in $@\mathbb{Z}_{12}$ derived from bigeneric morphemes are derived in this way, this is the situation encountered in [802].

Generalizations of this approach are obvious: One may step over to direct decompositions into three, four, etc., factors and corresponding multigeneric monoids, one may also take into account more general bigeneric chords with not so common zero-constant parts, and, last, but not least, one may reconsider the whole theory for more general addresses B, not just the zero address, and for more general forms (representable or not).

Remark 12 At present, we do not know how far the canonical transitive $GL(2, \mathbb{Z})$-action on the set of direct, non-trivial decompositions $\mathbb{Z}^2 = U \oplus V$ could be connected to the idea of the abelian group $SL(2, \mathbb{Z})/(SL(2, \mathbb{Z}), SL(2, \mathbb{Z}))$ being identified with the set $\mathbb{Z}_{12}$ of pitch classes (see Section 24.1.1.1).

25.3.2 Riemann Matrices

Summary. The Riemann program suggests a special matrix associated with every possible chord. The coefficients of this *Riemann matrix* are accounts of the role of the chord within the coordinates of tonalities, modes, and function values. We give the formal definition of the Riemann matrix and its construction principles.

– Σ –

We are now ready for considerations of tonal functions built upon the preceding mechanisms from degree and function theory[6]. The essence is this:

Principle 23 *Any function theory should aim at defining tonal functions for any existing chord*[7]*, i.e., relations between instances of the concept of a tonality and "semantic pointers", the harmonic function values, which provide the syntactic structure with means for understanding harmonic coherence.*

This statement is not meant as a cryptic sentence to cheat humanities, rather do we delimit our subject from levels of understanding which are not the scope of exact science, viz. textual predication in the sense of Section 18.2. In this vein, functional values are just names that are attributed to chords by means of precise evaluation modalities. So the only intrinsic meaning we can offer is extensionality, the determination of fibers lying over specific functional value symbols. The names are not really relevant, except that they may evoke some paratextual mechanism. Our approach yields textual predicates which give every chord a truth value in some truth module, and such that this evaluation is built upon the structure theory thus far developed.

The collection of tonal functions is defined by three domains: A name set TON of "tonalities", a name set VAL of "tonal function values", and a form space $TRUTH(I)$ for of "truth denotators" in a specific truth module (or, more generally, a truth form) I and at a specific address B for a form F.

Formally speaking (see also formula (18.7)), a tonal function $TF_{f,t}(f,t) \in \mathrm{VAL} \times \mathrm{TON}$, is a parametrized set of objective textual predicates which are encoded by maps

$$TF_{f,t} : ExTop_B(F) \to \mathbb{T}_I^B$$

on the supporting domain $ExTop_B(F)$. For such a collection of predicates, the *Riemann matrix* is the matrix $TF = (TF_{f,t})$, $(f,t) \in \mathrm{VAL} \times \mathrm{TON}$ of predicates. Given such a matrix, for any chord $a \in ExTop_B(F)$, its Riemann matrix is by definition the matrix[8] of all values, i.e.,

$$TF(a) = (TF_{f,t}(a))_{(f,t)\in \mathrm{VAL}\times\mathrm{TON}} \in \mathbb{M}_{\mathrm{VAL},\mathrm{TON}}(\mathbb{T}_I^B), \tag{25.7}$$

and therefore, we have the Riemann matrix map

$$TF : ExTop_B(F) \to \mathbb{M}_{\mathrm{VAL},\mathrm{TON}}(\mathbb{T}_I^B).$$

Here, a tonality symbol $t \in \mathrm{TON}$ is codified by its value vector $TF_t(a) = (TF_{f,t}(a))_{f\in\mathrm{VAL}})$ on all chords $a \in ExTop_B(F)$. This is a fuzzy evaluation of tonal roles of chords with respect to the function symbols in VAL.

25.3.3 Chains of Thirds

Summary. This section gives an elementary construction of a Riemann matrix, based on chains of thirds and minimal embeddings of chords in such chains.

[6] Recall that we are still given two addresses A, B, a form F, and an absorbing retraction $\alpha : A \to B$.

[7] See also Dahlhaus' paper [231] on Riemann theory.

[8] The set of all matrices with rows in a set m, columns in a set n, and coefficients in a space X, is denoted by $\mathbb{M}_{m,n}(X)$.

– Σ –

This setup works in the zero-address and form F with $Fun(F) = @\mathbb{Z}_{12}$. The tonality set is the set of "tonics" with just one tonic per pitch class

$$\text{TON} = \{C, D_\flat, D, E_\flat, E, F, G_\flat, G, A_\flat, A, B_\flat, B\}.$$

The value set is the classical six-element set for tonic, dominant, subdominant, each in major and minor mode:

$$\text{VAL} = \{T, D, S, t, d, s\},$$

with the minor problem of symbol confusion (D is doubly coded, but this is the tradition). The truth module is the already discussed fuzzy group $I = S_1 = \mathbb{R}/\mathbb{Z}$ (see Example 34), and the values will be just the denotators associated with the half open interval subsets $\phi\hat{} = [0, \phi[$ of S_1.

From Section 25.2.1 we learned that this approach deals with the third chain simplex $3Chain(c)$ of a chord c. Given a tonality symbol f in our context, and a third chain or singleton a, we shall define the function values $TF_{f,t}(a)$ and then extend this evaluation to any chord c by the formula

$$TF_{f,t}(c) = at(\frac{1}{card(3Chain(c))}\Sigma_{a\in 3Chain(c)} at^{-1}(TF_{f,t}(a))) \qquad (25.8)$$

which includes the case of third chains and singletons as zero simplexes of the third chain interpretation. Here, the auxiliary function

$$at : [0, \infty[\to [0, 1[$$

is supposed to be a strictly increasing bijection[9], for example $at(x) = \frac{2}{\pi}\arctan(x)$. The reader may generalize this formula by more sophisticated contributions of the simplex vertexes, we just show the principle here. The principle is that a chord's tonal function is some kind of average of the tonal function taken over all minimal third chains, i.e., the "standard" chords covering the—possibly exotic—given chord.

We are left with the definition of the tonal function for singletons and non-trivial third chains. Again, this is a prototypical formalism, to be adopted by the reader for specific needs. Suppose that a third chain a is given by the sequence $a_1, a_2, \ldots a_k$, then we suppose given the non-negative values

$$TF_{f,t}(a_i)$$

and then set

$$TF_{f,t}(a) = at(3CH(at^{-1}(TF_{f,t}(a_1)), at^{-1}(TF_{f,t}(a_2)), \ldots at^{-1}(TF_{f,t}(a_k))))$$

with a non-negative function $3CH$ on the singleton arguments, e.g., the sum of their second powers. One may in particular implement singleton values which yield good truth values for major, minor and similar common triads.

The formulas for the singletons x should be such that the values for basic tonality symbol $f = C$ yield the values for general tonality symbols, according to the formula

$$TF_{f,t}(x) = TF_{C,t}(e^{-t}(x)) \qquad (25.9)$$

if the symbol f corresponds to the tonic t. So this method is based on an elementary truth evaluation on the singletons *relative to one reference tonality* and then uses different constructors via third chain evaluation and then minimal covering families of third chains. This method has been implemented in RUBATO®'s HarmoRUBETTE®, see Section 41.3 for details.

The above approach seems to be numerically oriented. It is however still a logical approach in the fuzzy logic module S_1. Nonetheless, such a specific truth value assignment is not mandatory, one may think of much more abstract truth modules J and then, after having defined a method that works on J, transform such abstract values to more 'practical' values via appropriate truth module morphisms $J \to I$.

[9] This normalization is used to cast the raw values in the unit interval of the circle group.

25.3.4 Tonal Functions from Absorbing Addresses

Summary. This section discusses the tonal functions of chords associated with absorbing addresses, generalizing the idea of self-addressed morphemes in [802].

– Σ –

This approach does not take values of a numerical nature, but gets off on a purely topological level of truth levels. The philosophy is that the evaluation is not so much information about truth or falsity, but about the esthetic position within a 'neighborhood system of harmonic centers'. Harmonic evaluation in this spirit follows a 'logic' where the location is more relevant than the decision between dichotomic alternatives. It is in fact an open question whether Riemann's harmonic logic ever was meant to be a restatement of classical binary logic, and whether its role was ever limited to yield decisions between correct and false musical syntax.

We start on the truth value form functor $I = ExTop(F)$ for a pitch form F, for example

$$F = PiMod_{12}, EulerZModule),$$

defined on the subcategory of addresses with retractions as morphisms. This generates the truth form $TRUTH(I)$ and, by definition, its functor values $B@TRUTH(I) = B@\Omega^{ExTop(F)}$ at address B. So a subset $U \subset ExTop_B(F)$, i.e., an element $U \in B@2^{ExTop(F)}$, gives rise to a truth value $U^\wedge \in B@TRUTH(I)$, and we have integrated topologically deduced subsets of the B-addressed extension topology of form F.

On this basis, the evaluation $TF_{f,t}(c)$ of a chord c in $ExTop_B(F)$ follows this scheme: The tonality symbol t is associated with an interpretation $S_t^{M_t}$ of an A-addressed scale $S_t \subset ExTop_A(F)$. The elements of M_t are images $U_m = monex_z(m)$ of monoids[10] $m \in InTop(F)$. The monoids are deduced from the interpretation $S_{t_0}^{M_{t_0}}$ for a basic tonality t_0 by conjugation with the translation[11] that corresponds to the tonic shift from t_0 to t, or another transformation for more involved collections of tonal symbols.

This method's function symbols relate to the basic interpretation $S_{t_0}^{M_{t_0}}$ and its charts. If m is such a chart's monoid, some specific specialization monoids, i.e., non-trivial submonoids $n_{m,i}, i = 1, \ldots q(m)$ are chosen to define the function symbols $f_{m,i}$. For a general tonality symbol t, the function symbols are associated with the correspondingly transformed monoids and submonoids. This being true, the function value $TF_{f,t_0}(c)$ for a symbol f corresponding to the specialization pair $n_{m,i} \prec m$ is the locally closed allomorphic extension set $AllExt_B(n_{m,i}, m)$ iff c is an element of this set, and $\emptyset$ else.

In what follows, we want to give some examples of tonal functions from absorbing addresses. They are all focused on the classical context of pitch form $F = PiMod_{12}$ with fifth identifier, addresses $A = \mathbb{Z}_{12}, B = 0$, and absorbing address retraction $z : A \to 0$.

Example 44 Bigeneric Major Tonality *[802]*. We take up the bigeneric morpheme monoids $M_{d,a}$ from equation (25.6) and their associated self-addressed chords $M_{d,a}^z = monex_z(M_{d,a})$. We are given the usual twelve tonality symbols

$$\text{TON} = \{C, D_\flat, D, E_\flat, F, G_\flat, G, A_\flat, A, B_\flat, B\}$$

and calculate the pitch class p corresponding to tonality symbol X in multiples of fifths, e.g., G corresponds to $p = 1$, we write $G = e^1C$, to be short. The $X = e^pC$-major tonality is defined by the twelve morpheme charts with the respective functional symbols as follows: We choose $d = 1$ and set

Function Symbol	T	D	S	t	d	s
Consonant Mode	$M^z_{1,p+4}$	$M^z_{11,p+9}$	$M^z_{3,p+11}$	$M^z_{9,p+1}$	$M^z_{7,p+6}$	$M^z_{11,p+8}$
Dissonant Mode	$M^z_{10,p+4}$	$M^z_{8,p+9}$	$M^z_{0,p+11}$	$M^z_{0,p+1}$	$M^z_{10,p+6}$	$M^z_{2,p+8}$

[10] The "harmonic morphemes" according to the terminology of Noll [800].

[11] See the calculations of morpheme transformations for bigeneric morphemes in Section 25.3.1.

The corresponding predicates are related to the extensions (the unique open allomorph) $Exto(M^{z}_{d,a})$ and take symbols T, D, S, t, d, s for the "consonant mode", and $T^\star, D^\star, S^\star, t^\star, d^\star, s^\star$ for the "dissonant mode"[12], i.e.,

$$\mathrm{VAL} = \{T, D, S, t, d, s, T^\star, D^\star, S^\star, t^\star, d^\star, s^\star\},$$

and the value

$$TF_{e^pC,v}(c) = \begin{cases} Exto(M_{a(p,v),d(p,v)}) \text{ if } c \in Exto(M_{a(p,v),d(p,v)}), \\ \emptyset \text{ else,} \end{cases}$$

with $v \in \mathrm{VAL}$, and

$$\begin{aligned} (a(p,T), d(p,T)) &= (1, p+4), \\ (a(p,D), d(p,D)) &= (11, p+9), \\ &\dots \\ (a(p,s^\star), d(p,s^\star)) &= (2, p+8), \end{aligned}$$

according to the above table. Recall that the zero-constant parts of these morpheme charts are classical triads, as discussed in Proposition 46 of Section 25.3.1 and the remarks following that proposition. We also see that the terminology of a "consonant" or "dissonant" mode is now plausible by the triads in the dissonant mode being related to seventh and tritonus intervals.

This example can be enriched by adding selected allomorphs to the above open allomorphs. A priori, we would preconize the full allomorphic extension of each of the above morphemes, and then build a harmonic function theory upon the complete 'allomorphic spectrum'.

Exercise 60 *Calculate and draw the nerve of the twelve-chart interpretation by the semigroup morphemes*[13] $ex_z(\langle\alpha_d, \beta_a\rangle)$ *and the sub-nerve of its zero-constant trace by the bigeneric morphemes in Example 44. The nerve of the semigroup submorphemes tells more about the interSection configurations because the identity of a monoid 'collapses' all information about other common points.*

Example 45 Third Degree Tonality. We have already defined the triadic interpretation $X^{(3)}$ of a diatonic scale X in Section 13.4.2. This was a zero-addressed situation which we now 'blow up' in the self-addressed context[14]. Again, we have the twelve pitch class related tonalities

$$\mathrm{TON} = \{C, D_\flat, D, E_\flat, E, F, G_\flat, G, A_\flat, A, B_\flat, B\}.$$

We first stick to the basic $X = C$ and start on seven monoid morphemes, $M_I = Int(I_C), M_{II} = Int(II_C), \dots M_{VI} = Int(VI_C), M_{VII} = Int(VII_C)$. We have $Int(I_C) = M_{1,4}$, i.e., the generators α, β generate the entire monoid $Int(I_C)$, which is the semigroup generated by α, β, united with the identity[15].

[12] Terminology from [802].

[13] These can also be defined as being the intersection of the monoid morphemes with the semi-group of all non-invertible elements of $End(F)$.

[14] Recall that we work in pitch space $PiMod_{12}$ with the fifth identifier.

[15] We know from Appendix C.2.3 that

$$\begin{aligned} \mathbb{Z}_{12}@\mathbb{Z}_{12} &= \sqrt{Idempot(\mathbb{Z}_{12}@\mathbb{Z}_{12})} \\ &= \sqrt{1} \sqcup \\ &\quad \sqrt{4} \sqcup \sqrt{e^3 4} \sqcup \sqrt{e^6 4} \sqcup \sqrt{e^9 4} \sqcup \\ &\quad \sqrt{3} \sqcup \sqrt{e^4 3} \sqcup \sqrt{e^8 3} \sqcup \\ &\quad \sqrt{0} \sqcup \sqrt{e^1 0} \sqcup \sqrt{e^2 0} \sqcup \sqrt{e^3 0} \dots \sqcup \sqrt{e^{11} 0} \end{aligned}$$

is the idempotent decomposition of $\mathbb{Z}_{12}@\mathbb{Z}_{12}$ into invertible endomorphisms, those above linear factor 4, those above linear factor 3, and those above linear factor zero. Since these linear factors commute, the components, which are conjugate with each other above a fixed linear idempotent (Appendix C.2.3, Proposition 78), are all semigroups.

The analog is true for the other monoids $Int(II_C), \ldots Int(VI_C)$ since degrees II to VI are isomorphic to degree I. However, $Int(VII_C)$ is more complicated. The symmetry group of VII_C is a Klein group and the sub-semigroup of non-invertible elements is not bigeneric in the sense discussed above. Here it is:

$$\begin{aligned} Int(VII_C) - Sym(VII_C) = \{&e^1 10, e^2 0, e^5 0,\\ &e^5 6, e^1 16,\\ &e^8 3, e^8 9,\\ &e^1 2, e^7 2, e^3 10, e^9 10, e^1 8, e^7 8, e^{10} 8,\\ &e^3 4, e^6 4, e^9 4\} \end{aligned}$$

We denote the monoid generated by these non-invertible elements of $Int(VII_C)$ by $Int_0(VII_C)$. The charts of our self-addressed triadic interpretation are the six charts $M_I = intex_z(I_C), M_{II} = intex_z(II_C), \ldots M_{VI} = intex_z(VI_C)$, and the degree seven chart $M_{VII} = monex_z Int_0(VII_C)$. The function values are symbolized according to all possible allomorphs of the seven chart monoids, which yield the truth values as in the preceding example. We leave the details to the reader.

The intersection configuration is remarkable. The nerve of the objective traces of the ex_z-charts from the sub-semigroups of non-invertible elements[16] in the monoids

$$Int(I_C), \ldots Int(VI_C), Int_0(VII_C)$$

is a priori a simplicial complex containing the nerve of the zero-constant part. In this case, no new intersection configuration intervenes: *As in the zero-constant case, the nerve of the self-addressed extension of the triadic interpretation of a diatonic scale is again a Möbius strip.* Figure 25.2 shows the intersection configuration in detail.

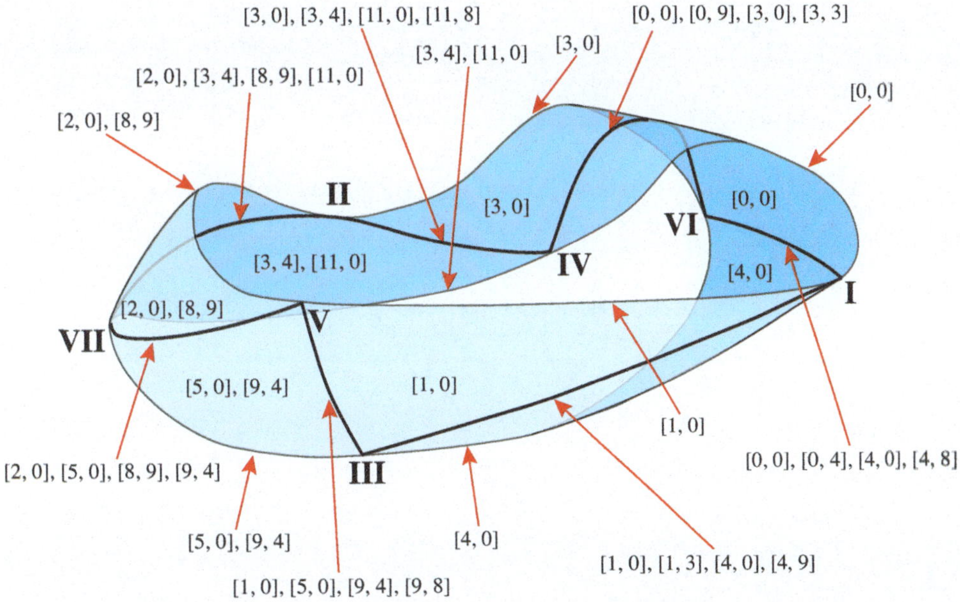

Fig. 25.2. The non-invertible elements of the self-addressed charts built upon the endomorphisms of classical triads define a nerve which is identical to the nerve from the constant parts of the self-addressed triads! The figure shows the self-addressed chart intersections for C tonality, everything calculated in the fifth-identifier form $PiMod_{12}$.

[16] These are all elements except the identity.

Remark 13 The passage from zero-addressed to self-addressed degrees maintains the nerve, but if we take one more step from the self-addressed to the functorial setup, the Möbius strip is enriched by new simplexes. More concretely, we take the self-addressed degrees $X \subset \mathbb{Z}_{12}@\mathbb{Z}_{12}$ and consider the associated functorial local compositions $\hat{X} \subset (\mathbb{Z}_{12}@\mathbb{Z}_{12})\hat{}$ and the global composition Z^I defined by this covering I. In this setup, the nerve of the self-addressed configuration is of course part of the nerve of the functorial configuration Z^I, but we encounter new simplexes. Specifically:

Exercise 61 *Show that the functorial degrees $I\hat{}$ and $II\hat{}$ have a non-empty intersection and thereby extend the Möbius strip of the self-addressed triadic covering. Hint: Find a morphism $f : \mathbb{Z}_{12} \to \mathbb{Z}_{12}$ which yields an f-slice $f@I\hat{} \cap f@II\hat{} \neq \emptyset$.*

26

Cadence

Was ist C-Dur?
Wahrlich, Herr Mazzola, eine gute Frage!
Rudolf Wille [1130]

Summary. Cadences are shorthand representations of tonalities. (We do not discuss the other meanings—e.g., the solo cadence in the sense of a concert climax—of this typically homonymic term.) There is a variety of approaches to realize such a representation. We give an explicit definition of the concept of a cadence with respect to varied addresses and ambient spaces. In particular, we present the very classical cadences, those related to self-addressed function theory, and more exotic self-addressed cadences which relate to symmetries rather than to tones or sets of such objects.

– Σ –

The concept of a cadence as a final cadence or a cadential formula to consolidate a tonality goes back to Italian theorists of the 16th century. In this chapter, we shall discuss this meaning of the word and not that of the concert solo cadence.

In this sense, a cadence is a good example for the construction of sign structures in music. The signifier expression is fixed in a cadential formula, such as *I-IV-I-V-I*, which is supposed to obtain its signification in a specific context, and thereby points to its signified content: the "uniquely determined tonality" or the "composition's conclusion", respectively. We shall only deal with the former meaning because this one is also the structural substrate of the latter: the conclusion is expected because a specific, uniquely determined tonality is recognized from a cadential formula.

If this tonality (whatever the technical meaning of this word may be) is the significate of the given cadence formula, a mathematical function of type "cadence formula=function(tonality)" is given. The characteristic of this function is that it is *injective*. Two tonalities are associated with different cadence formulas: I_C-IV_C-I_C-V_C-I_C differs from I_D-IV_D-I_D-V_D-I_D.

Implicitly, the existence of a cadence function depends upon two conditions which usually are not noticed by theorists because they seem straightforward:

- The first deals with the very concept of a tonality. Making this concept precise is a condition for a unique determination of its instances whatsoever. The recognition of a tonality from its cadence presumes that you are able to distinguish it from musical objects that are not tonalities or are so, but share a different flavor which you would eliminate in the present context. For example, the recognition of the European C-major tonality from the classical Rameau cadence I_C-IV_C-I_C-V_C-I_C presupposes that we would not search in a set of Turkish or other non-European tonalities. So we may suppose being given a set X of global compositions which define the extension of the concept of a tonality.
- The second condition relates to the range of the possible cadence formulas. In fact, searching for a formula presupposes that one will find it in a delimited domain of musical cadence expressions. In other words, one should know what is a cadence formula in the present context and what is not. Call P the set of

G. Mazzola, *The Topos of Music I: Theory*, Computational Music Science,
https://doi.org/10.1007/978-3-319-64364-9_26

admitted cadence formulas, or "cadence parameters" as the elements of P parametrize the domain of formulas. We shall see that for all known situations, P can be chosen to be a set of global compositions.

26.1 Making the Concept Precise

Summary. This section describes the concept of a *cadence*.

– Σ –

Definition 86 *Given two sets X, P of tonalities and cadence parameters, respectively, as introduced above, and a map $\kappa : X \to P$, we say that κ is* cadential *in a tonality $x \in X$ if the fiber $\kappa^{-1}\kappa(x)$ is the singleton $\{x\}$, i.e., κ is injective in x. The map κ is called* cadential *or* a cadence, *iff it is cadential in every tonality $x \in X$.*

We should however be aware that the general nature of a cadence map cannot be restrained to mathematical or even functorial properties. A cadence is a relation which can issue from any motivation, be it historical, systematic or arbitrary. Essentially, the above definition is the one used in [669, 670, 682]. In the present theory language, we would prefer restating that definition in terms of textual predicates. Disregarding the fact that global compositions are not denotators, we would have to extend the cadence map to a 'global predicate' *Cad* outside X by the value $\kappa(y) = y/Cad = \emptyset$ for $y \notin X$. Inside X, we then would have the values x/Cad in $Glob^A_F$, the set of A-addressed global compositions with charts living in $\mathbb{T}^A_F$ for a specific *cadence* truth form F. Whenever we use this wording for cadences, keep in mind that no extension of the category *Den* of denotators to global objects has been defined so far, except for powerset denotators, i.e., local compositions. Nonetheless, we shall use this sloppy language because the environments are evident.

Before discussing the examples, we should emphasize the dramatic ambiguity of these concepts between truth and beauty. The truth values for a cadence are global compositions, i.e., objects of musical esthetics rather than poor truth denotators. A cadence gives you a parametrization of musical objects—tonalities in this case—in an auxiliary space of musical objects in order to obtain a shorthand representation of these objects. Rather than evaluating these objects by truth or falsity, the cadence positions them in an esthetic parameter space.

Observe also that we did not give a specific delimitation of the tonality concept, it is just a word which will be filled with meaning in the following examples. Nonetheless, we should recall the discussion of function theory in chapter 25.3. There, "tonality" was introduced as a symbol in an abstract set TON. Its usage was the construction of a bunch of textual predicates $TF_{f,t}$ which give each chord a tonal function (in this case: an allomorphic set in the topological extension space). These predicates may be used to define cadences, but the effort is considerable, we leave this as an exercise[1]. The link between the function predicates and the cadence predicates is that the function predicates are evaluated by use of global compositions (such as the triadic interpretation of a scale). These global compositions correspond to the tonalities in the set X for cadences. Notice that the scope of tonal functions is not the identification of tonalities, but the attribution of a tonal value for any given chord, and this is much more than mere identification.

26.2 Classical Cadences Relating to 12-Tempered Intonation

Summary. This first example reviews the classical cadences of European tonal music in relation to pitch classes in $PiMod_{12}$.

– Σ –

[1] Speaking of a cadence in the context of tonal function predicates, this cadence would be formalized by the function vector TF_t discussed in sec Section 25.3.2.

The first group of cadences relates to zero-addressed (commutative) compositions, more precisely, to compositions stemming from the pitch class ambient spaces $\mathbb{Z}_{12}$ for 12-tempered tuning and $\mathbb{Z}^2$ for just temperament. These cases are also prominent since we have extensive modulation theorems for them.

These approaches are also prototypical in so far as the used atlases consist of quite special charts (typically triadic coverings) which could be generalized to richer and more realistic coverings (and partly have been by Straub's investigations). But it is just this simplistic perspective which shows that even under elementary conditions, interesting and experimentally significant structures may emerge. The experimental investigations will be explicated in Chapter 80 dedicated to modulation phenomena in classical scores from European literature.

26.2.1 Cadences in Triadic Interpretations of Diatonic Scales

Summary. This section describes the cadences related to triadic degrees.

– Σ –

We start with the classical example. Let $X = Dia^{(3)}$ be the twelve-element set of zero-addressed triadic interpretations of all major scales in $\mathbb{Z}_{12}$. Take the cadence parameter set $P = ObGlob_0^{\mathbb{Z}_{12}}$ of global, zero-addressed compositions with charts in $\mathbb{Z}_{12}$. For a given sequence $J = 1 \leqslant j_1 < j_2 < \ldots j_k \leqslant 7$ of indices for triadic charts, we define the map $\kappa_J(S^{(3)}) = \{J_{j_1}, J_{j_2}, \ldots J_{j_k}\}$ with image the sub-interpretation of the covering defined by the charts from the subsequence J. For example, if $J = 1, 4$, we get the value $\kappa_J(E^{(3)}) = \{I_E, IV_E\}$, a two-chart-interpretation of the six-element pitch class set $\{e, g_\sharp, b, a, c_\sharp\}$. The extension of such a function to other triadic interpretations, for example of melodic minor scales, would evidently destroy cadentiality. On the other hand, ignorance of the precise formula (first and fourth triadic degrees) would make it impossible to retrace the argument. For example, if we had also at our hands the first and fifth triadic degrees, it could not be decided whether the interpretation is the image $\{I_E, IV_E\}$ or $\{I_A, V_A\}$.

Although the function κ_J is a cadence for the given tonality domain $Dia^{(3)}$, the usual selection method for 'good' sequences asks for more. The idea is that you only know that the associated interpretation is defined by some J, but one is not sure about which one has been applied. So the requirement is that the tonality should be uniquely determined whatever J may be the defining selection parameter. This leads to the concept of a cadential set of triadic degrees.

Definition 87 *Given a set X of interpretations of objective, local A-addressed compositions of form space F, a set $C \subset I$ of charts of an element $S^I \in X$ is called* cadential in S^I *iff S^I is the only interpretation T^J in X such that $C \subset J$. The cadential set C is called* minimal cadential in S^I *iff it has no proper subset which is also cadential in S^I.*

A family $(C_x)_{x \in X}$ of (minimal) cadential sets C_x for each interpretation $x \in X$ is called a (minimal) cadential family *for X.*

The classical example $X = Dia^{(3)}$ yields five index sequences for minimal cadential families:

$$J_1 = (II, III), J_2 = (II, V), J_3 = (III, IV), J_4 = (IV, V), J_5 = (VII).$$

This means that the family J_1, for example, defines its set $\{II_{S^{(3)}}, III_{S^{(3)}}\}$ for the triadic interpretation $S^{(3)}$. The fourth one is the reminiscence to Rameau's cadence *I-IV-I-V-I*, except that the first degree triad is superfluous since no tonic is stressed in this triadic interpretation.

Remark 14 Observe that in the present definition of cadential sets, no syntactical position is defined, so the sequence in the Rameau cadence *I-IV-I-V-I* is not reflected, only the set $\{I, IV, I, V, I\}$ is of interest in this context. A further specification of syntactical relations is necessary for the concrete applications, but this is another subject which we do not open up here.

In [779], Daniel Muzzulini calculated all minimal cadential sets for the following tonality sets X (by computer). We take the translation orbits $X = e^{\mathbb{Z}_{12}}.S^{(3)}$ of triadic interpretations of seven-element scales $S \subset \mathbb{Z}_{12}$ by the seven degrees consisting of those three-element subsets such that every second pitch class in the clock order is selected. The number of cadential sets varies between 5 and 21, the first number being taken by the triadic major scale interpretations $Dia^{(3)}$ while the maximum of 21 cadence sets is taken by the orbit of harmonic minor scale interpretations $Har^{(3)} = e^{\mathbb{Z}_{12}}.c_h^{(3)}$, in fact, here every pair (X, Y) of two different degree indexes defines a minimal cadence set, and we have $\binom{7}{2} = 21$ sets.

A more modern definition of cadential sets runs as follows. We consider a set X of A-addressed interpretations

$$x = S^I \to S \to @A \times F$$

in form space F. A cadential interpretation for S^I is an interpretation

$$C^J \to C \to @A \times F$$

which factorizes through S^I, i.e., there is a morphism

$$\overset{\iota}{f}/1 : C^J \to S^I$$

such that the triangle

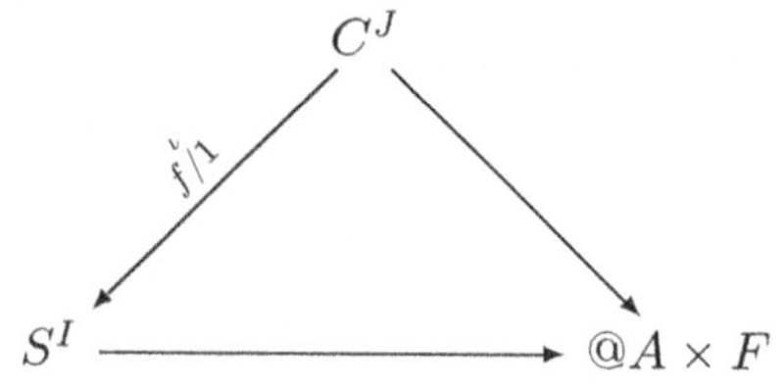

(26.1)

commutes. This implies that each chart of C^J is associated with a chart of S^I, and that this is done in such a way that the points of the charts are mapped into each other in accordance with the underlying embeddings of the local support compositions S, C. And we ask that there is no factorization through an interpretation $y \neq x$ in the set X. Morally (which in this book reduces to topos theory), such a cadential interpretation is a point in the full sieve $@x$ which characterizes x within the collection X. So cadential sets relate to selecting 'separating' points c_x in full sieves $@x, x \in X$. A general theory of such constructs has not been developed to date.

26.2.2 Cadences in More General Interpretations

Summary. This section describes the cadences related to generalized degrees in general scales, as investigated by Hans Straub.

– Σ –

Starting from the generalization of minimal cadence sets for the seven-element scales studied in [779], Hans Straub [1018] has undertaken a vast generalization of cadence set types on seven-element scales. A (minimal) cadential set is the same as above, but it need not cover the scale, it is a more abstract representation not of the single scale notes, but of the degree representation of the scale. For example, one may consider as degrees all the triadic chords of minor, major, diminished, or augmented types, which are contained in a given scale. Complete computer-aided (here, computers become definitively mandatory) calculations have been carried out for minimal cadence sets consisting of a number of three-element degrees of any type, no classical triadic chords are preferred.

If the degrees are all the triadic chords of minor, major, diminished, or augmented types in a scale, the nerve dimensions may vary from zero to seven, and not every scale is covered by this degree system. But this degree system has the advantage that all degree chords are classical, which is not the case in Muzzulini's generalization. However, in Straub's general concept framework, one finds degree systems consisting of seven triadic chords which have no cadence sets at all, for example for the scales of type 60 and 62 (in our classification table N.1).

It turns out that 21 is the absolute maximum of possible minimal cadence sets for systems of seven three-element chords in seven-element scales, and that it is achieved by the majority of scales.

In the very general setup, Straub recognizes that a particular type of degree systems is interesting. According to the degree-theoretic approach which includes coverings by standard structures (as sketched in Section 25.1), Straub therefore considers *irreducible* degree systems DEG in the sense that no element of DEG can be written as a union of other elements of DEG. Evidently, systems consisting of elements of equal cardinality, such as triadic systems, are irreducible. Also, a system is always built by a unique irreducible subsystem. A classification of all irreducible degree systems has not been accomplished to date. It is however important to embed the classical European degree concepts in this variational environment to understand how special or singular the classical choice is. According to Straub's investigations, the distinguished position of the classical diatonic, harmonic, and melodic scales does not emerge on the general level of degree systems and the associated cadence systems, but only with respect to characteristic properties in modulation theory (see Chapter 80 for this discussion). This latter aspect has however not been investigated in depth.

Exercise 62 *Show that a degree system of a finite scale is always built by a unique irreducible subsystem.*

26.3 Cadences in Self-addressed Tonalities of Morphology

Summary. Relating to interpretation of self-addressed tonalities, minimal cadential sets have been calculated by Thomas Noll, we present and discuss the results.

– Σ –

In [800], the basic form is the classical 12-tempered pitch class space $F = PiMod_{12}$, with fifth identifier. In the vein of our Example 45 in Chapter 25, major and minor tonalities are constructed as interpretations of $\mathbb{Z}_{12}$-addressed local compositions. Elements $e^a.b$ are denoted by $[a, b]$. We restrict to the description of C-major and C-minor, the other 22 tonalities being defined via transposition.

The basic local composition (the 'scale') has 23 elements, eight of them being constant, i.e., stemming from zero-addressed tones:

$$\begin{aligned} C\text{-major} = \{ & [0,0],[1,0],[2,0],[3,0],[4,0],[5,0],[8,0],[11,0],\\ & [1,3],[2,3],[3,3],[8,3],[11,3],\\ & [0,4],[3,4],[9,4],\\ & [4,8],[8,8],[9,8],[11,8],\\ & [0,9],[4,9],[8,9]\}, \end{aligned}$$

whereas the interpretation is given by 10 charts which are related to the function symbols T, D, S, D^*, s, s^* in major and t, d, s, s^*, D, D^* in minor. Each of these main function symbols is associated with a "Gegenklang" and a "Parallelklang" symbol g, p, respectively. We have these major tonality charts[2]:

$$\begin{aligned} & Tg=[1,3],[4,9],[1,0],[4,0], && Tp=[4,8],[0,4],[0,0],[4,0]\\ & Sg=[3,3],[0,9],[0,0],[3,0], && Sp=[11,8],[3,4],[11,0],[3,0]\\ & Dg=[11,3],[8,9],[2,0],[5,0], && Dg^*=[8,3],[8,9],[11,0],[5,0]\\ & Dp=[9,8],[9,4],[1,0],[5,0] && \\ & sG=[11,3],[8,9],[11,0],[8,0], && sG^*=[2,3],[8,9],[2,0],[8,0]\\ & sP=[8,8],[0,4],[0,0],[8,0] && \end{aligned}$$

[2] The symbol construction in [800] is not thoroughly logical.

which are motivated by the allomorphs to the bigeneric morpheme construction from Section 25.3.1 (see also [800]).

Exercise 63 *Draw the nerve of this interpretation, and the nerve of the constant subcharts. Compare them to the nerve (harmonic strip) of the triadic interpretation from Example 45 in Chapter 25.*

With these settings, the set X of tonalities is defined by the 24-element set of 12 major tonalities C-major, $D_\flat$-major,... B-major, and 12 minor tonalities C-minor, $D_\flat$-minor,... B-minor. The cadence maps are given by 56 minimal cadential sets. Half this number defines the minimal cadential sets and are applied to major tonalities only (!), the other half operates on minor tonalities. This means that the 28 candidates for major tonalities characterize any major tonality within X, whereas the other half does the job for the minor tonalities. The following list from [800] shows the candidates for major tonalities. With each chart Z, the number in brackets indicates the number of zero-addressed chords with up to four tones, which is contained in the extension $Ext_0(Z)$.

Observe that in general, replacing the chart by its zero-addressed extension is not cadential. The point is that in a concrete score, only zero-addressed chords will appear, at least without any additional constructions. So the question of how to represent a self-addressed chord on the zero-addressed level arises. To look for zero-addressed extensions means that one switches to the topological framework and seeks zero-addressed chords in the extension of a chart Z. A refined method would consist in the consideration of allomorphs as discussed in the function-theoretic logic above. This method should however be worked out in more detail. In fact, it is possible that a zero-addressed chord is contained in several extensions, and, worse than that, in extensions stemming from different cadential sets. So the overall information is that a zero-addressed chord is contained in some extensions for some cadential sets. The problem is this: How can one decide which tonality is the best candidate for the placement of the given chord?—Such a question becomes important if one needs recognition algorithms in a systematic tonality analysis of given scores, on a software platform such as RUBATO®, for example.

Still more generally set, a zero-addressed chord need not be viewed as a local object, but it may be interpreted by subchords which live in some extension sets, much like the degree concepts in non-irreducible chords of Straub's above approach. No systematic research has been carried out to date.

Minimal cadence sets in major self-addressed tonalities.

Dg^* (26)	Tg (12)	
Dg^* (26)	Tp (32)	
Sg (12)	Dg^* (26)	
Sg (12)	Dg (12)	
Sg (12)	Dp (32)	
Sg (12)	sG^* (26)	
sG^* (26)	Tg (12)	
sG^* (26)	Tp (32)	
Sp (32)	Dg^* (26)	
Sp (32)	Dp (32)	
sP (32)	Dg (12)	
sP (32)	Sg^* (26)	
Sg (12)	sG (12)	Tg (12)
Sg (12)	sG (12)	Tp (32)
Sg (12)	sP (32)	Tg (12)
Sg (12)	sP (32)	Tp (32)
sG (12)	Dg (12)	Tg (12)

sG (12)	Dg (12)	Tp (32)
sG (12)	Dp (32)	Tg (12)
sG (12)	Dp (32)	Tp (32)
sP (32)	Dg (12)	Tg (12)
sP (32)	Dg (12)	Tp (32)
sP (32)	Dp (32)	Tg (12)
sP (32)	Dp (32)	Tp (32)
Sp (32)	sG (12)	Tg (12)
Sp (32)	sG (12)	Tp (32)
Sp (32)	sP (32)	Tg (12)
Sp (32)	sP (32)	Tp (32)

The corresponding minor list is deduced from the preceding one by these exchanges: $Tg \to tG, Tp \to tP, Sg \to dG, Sp \to dP, Dg \leftrightarrow sG, D^*g \leftrightarrow s^*G, Dp \leftrightarrow sP$.

26.4 Self-addressed Cadences by Symmetries and Morphisms

Summary. More *exotic* cadences can be introduced by use of symmetries instead of chords. This example will be applied in the analysis of L. van Beethoven's "Hammerklavier"-Sonata in Section 28.2.

– Σ –

The preceding examples have dealt with different addresses, but they had in common that the addresses of the tonality sets and of the cadence parameter sets were the same in each case. One can also try to associate tonalities and cadences on different addresses. One example of such a procedure, which we shall use in the discussion of the modulation architecture in Beethoven's op. 106 (Section 28.2), will be discussed hereafter.

We start on the space $ExTop(PiMod_{12})$ of zero-addressed local compositions on pitch classes. To each such local composition Ch, the continuous injection in formula (24.13)

$$intex_{z,B} : ExTop_0(PiMod_{12}) \to ExTop_{\mathbb{Z}_{12}}(PiMod_{12})$$

associates the 'self-addressed chord' $intex_{z,B}(Ch)$ of all endomorphisms. The idea is that one may associate to this fractal chord a self-addressed chord defined by a specific sub-monoid of endomorphisms. Since $intex_{z,B}$ is injective, it is a cadence map on any set of 'tonalities' $X \subset ExTop_0(PiMod_{12})$. So if X is any set of zero-addressed local compositions of pitch classes, playing the role of scales, we have a first example of a cadence via $intex_{z,B}|_X$. The example we are aiming at is the selection of a specific group of automorphisms in $intex_{z,B}(Ch)$. For example, we may consider a group G of automorphisms of $PiMod_{\mathbb{Z}12}$ and then consider the intersection (in the naive sense, not as functors) $intex_{z,B,G}(Ch) = monex_z(G) \cap intex_{z,B}(Ch)$. For example, take the group $G = TI_0$ of inversions and transpositions on $\mathbb{Z}_{12}$ and then take the intersection $intex_{z,B,TI_0}(Ch)$, i.e., the group of 'counterpoint pitch symmetries' on $PiMod_{\mathbb{Z}12}$ which induce automorphisms of Ch. Or else take the full automorphism group $Aut(PiMod_{\mathbb{Z}12}) \xrightarrow{\sim} \overrightarrow{GL}_1(\mathbb{Z}_{12})$.

Clearly, such a map is not injective on all of $ExTop_0(PiMod_{12})$ since the automorphisms of a zero-addressed chord are the same as those of its complement. If we do however restrict the cadence domain to the domain X_6, those objective zero-addressed local compositions with up to six elements, then the map

$$intex_{z,B,\overrightarrow{GL}_1(\mathbb{Z}_{12})} : X_6 \to ExTop_{\mathbb{Z}_{12}}(PiMod_{12})$$

has a nice property of cadentiality:

Lemma 41 *The map $intex_{z,B,\overrightarrow{GL}_1(\mathbb{Z}_{12})}$ on the domain X_6 is cadential exactly in the three diminished seventh chords (of class number 37, pitch classes in semitone steps):*

$$\mathfrak{M}_0 = \{1,4,7,10\},$$
$$\mathfrak{M}_1 = e^1.\mathfrak{M}_0 = \{2,5,8,11\},$$
$$\mathfrak{M}_2 = e^2.\mathfrak{M}_0 = \{0,3,6,9\}.$$

A proof results from the contemplation of the symmetry groups of chord classes in the class list of Appendix N.1.

This means that the diminished seventh chords are the only ones that can be recovered from their symmetries! As with the self-addressed tonalities above, these maps have the drawback that their codomain values are somewhat above naive reality of score objects. But it is possible to give a more down-to-earth restatement of this map in terms of orbits!

To this end, recall that the set $Dia^{(3)}$ of triadic interpretations of diatonic scales is the orbit of the counterpoint pitch group TI_0. Any subgroup $G \subset TI_0$ generates the partition $orb(G)$ of G-orbits on $Dia^{(3)}$, an element of the set $Part(Dia^{(3)})$ of partitions of $Dia^{(3)}$. Let this set be the new parameter set $P = Part(Dia^{(3)})$ of our map, and take again $X = X_6$ as domain, i.e., $\kappa_{orb} : X_6 \to Part(Dia^{(3)})$. We define

$$\kappa_{orb}(Ch) = orb(intex_{z,B,TI_0}(Ch))$$

with the obvious restatement of $\mathbb{Z}_{12}$-addressed elements as automorphisms. Then:

Lemma 42 *The map κ_{orb} is cadential exactly in the three diminished seventh chords*

$$\mathfrak{M}_0, \mathfrak{M}_1, \mathfrak{M}_2.$$

We leave the proof as an exercise for the reader, but observe that there are exactly two orbits in the three cadential chords: We have

$$\kappa_{orb}(\mathfrak{M}_0) =$$
$$\{W = \{B_\flat^{(3)}, D_\flat^{(3)}, E^{(3)}, G^{(3)}, A^{(3)}, C^{(3)}, E_\flat^{(3)}, G_\flat^{(3)}\},\ W* = \{B^{(3)}, D^{(3)}, F^{(3)}, A_\flat^{(3)}\}\},$$

the others are transpositions thereof. These two orbits, the large W ("world") and the small $W*$ ("antiworld") will play a crucial role in the analysis of the modulation architecture of Beethoven's op.106.

26.5 Cadences for Just Intonation

Summary. For use in just modulation theory as it has been worked out by Hildegard Radl [871], we present minimal cadence sets in third-fifth and Pythagorean tuning.

– Σ –

We recall the context from just and "justest" tuning introduced in Section 13.4.2.2. We are working in the *EulerPlane* space spanned by fifth and third axes, as well as in the Pythagorean subspace *PythagorasLine* spanned by the fifth axis alone, and related to the *EulerPlane* by the projection modulo third comma. In all the following cases, we shall only discuss the triadic interpretations of the C scale since the other scale interpretations are deduced from this one by transpositions, and the corresponding cadential sets are the transposed sets, so the names of degrees in cadential sets just vary by the scale symbol. We shall therefore omit this specification in order to keep everything readable.

26.5.1 Tonalities in Third-Fifth Intonation

Summary. The minimal cadential sets of degrees are described for major, (natural) minor, harmonic minor, and melodic minor tonalities.

– Σ –

The just triadic *major* scale interpretation $C^{(3)}$ has the seven degrees[3]

$$I, II, III, IV, V, VI, VII$$

described in Figure 13.10. The minimal cadential sets are these:

$$J_1 = (II), J_2 = (III, VI), J_3 = (III, IV), J_4 = (IV, V), J_5 = (VII), J_6 = (V, VI). \tag{26.2}$$

The just triadic *natural minor* tonality $c^{(3)}$ is the isomorphic image of $C^{(3)}$ under the 180 degree rotation around the middle between tones c and g, see also Figure 7.5 which shows the scale and the seven degrees. The minimal cadential sets follow from the transformation of the above major cadence sets 26.2 and are these[4]:

$$J_1 = (VII), J_2 = (III, VI), J_3 = (V, VI), J_4 = (IV, V), J_5 = (II), J_6 = (III, IV). \tag{26.3}$$

The just triadic *harmonic minor* scale interpretation $c_h^{(3)}$ works as follows: The harmonic minor scale and its degrees are shown in Figure 27.4. The scale c_h is selected out from the chromatic scale shown in Figure 7.5, the degrees are uniquely determined by the tone names given in the chromatic scale.

The minimal cadential sets are these:

$$J_1 = (III), J_2 = (II), J_3 = (VII), J_4 = (I, IV), \tag{26.4}$$
$$J_5 = (I, V), J_6 = (I, VI), J_7 = (IV, V), J_8 = (IV, VI), J_9 = (V, VI). \tag{26.5}$$

The just triadic *melodic minor* scale interpretation $c_m^{(3)}$ works as follows: The melodic minor scale and its degrees are shown in Figure 27.5. The scale c_m is selected out from the chromatic scale shown in Figure 7.5, the degrees are uniquely determined by the tone names given in the chromatic scale.

The minimal cadential sets are these:

$$J_1 = (I), J_2 = (II), J_3 = (III), J_4 = (IV, V), J_5 = (VI), J_6 = (VII). \tag{26.6}$$

26.5.2 Tonalities in Pythagorean Intonation

Summary. The minimal cadential sets of degrees are described for Pythagorean major, (just) minor, harmonic minor, and melodic minor tonalities.

– Σ –

The Pythagorean scale interpretations are defined by the comma projection *EulerPlane* $\to$ *PythagorasLine*, applied to the synonymous just tonalities discussed above.

Exercise 64 *Draw all the above major and minor scales in the Pythagorean tuning.*

Here are the minimal cadential sets:
Major:

$$J_1 = (VII), J_2 = (II, III), J_3 = (II, V), J_4 = (III, IV), J_5 = (IV, V).$$

Natural Minor, here, the degrees are coupled by a reflection on *PythagoreanLine*[5] and accordingly yield these minimal cadential sets:

$$J_1 = (II), J_2 = (VII, VI), J_3 = (VII, IV), J_4 = (VI, V).$$

[3] We do not consider the alternative degrees II^* and VII^* from Figure 13.10.

[4] The rotation transforms major degrees into minor degrees as follows: $I \mapsto I, II \mapsto VII, III \mapsto VI, IV \mapsto V, V \mapsto IV, VI \mapsto III, VII \mapsto II$.

[5] The symmetry $e^1.(-1)$ on the line transforms major degrees into minor degrees as follows: $I \mapsto I, II \mapsto VII, III \mapsto VI, IV \mapsto V, V \mapsto IV, VI \mapsto III, VII \mapsto II$.

Harmonic minor:

$$J_1 = (III) \text{and all pairs of any two different degrees} \neq III.$$

Melodic minor:

$$J_1 = (III), J_2 = (I, II), J_3 = (I, V), J_4 = (I, VII), \tag{26.7}$$
$$J_5 = (II, IV), J_6 = (II, VI), J_7 = (IV, V), J_8 = (IV, VII), \tag{26.8}$$
$$J_9 = (V, VI), J_9 = (VI, VII). \tag{26.9}$$

Check for updates

27

Modulation

Denn das Wesentliche an einer Modulation
ist nicht das Ziel, sondern der Weg.
Arnold Schönberg [948]

Summary. This chapter deals with the central issue of modulation between two given tonalities. It involves explicit models of tonalities, of cadences and—even more crucial—of the transition process from one tonality to its successor. The present model involves the analogy to elementary particle physics: Modulation is viewed as a 'force interaction' between two 'tonality particles' which is mediated by a 'modulation quantum'. The model allows for a complete *calculation of fundamental degrees of modulation* in congruence with Arnold Schönbergs harmony [948]. The model is realized for diatonic tonalities in 12-tempered and just tuning. It has been extended to all 7-element scales in 12-tempered tuning and to a number of scales in just tuning. The 12-tempered extension reveals a privileged position of the diatonic scale with regard to this modulation theory. We conclude the chapter with a discussion of the basic role of modulation models and their application to optimize harmonic paths in the sense of Section 27.2.

– Σ –

In the historical development of mathematical music theory, the modulation models were the initial major topic. The first results were traced in [668, 669]. The original problem was to understand the role of the diminished seventh chord in the modulation strategy of Beethoven's "Hammerklavier" sonata op.106. It turned out that the inner symmetries of that chord seemed to govern the possible modulatory transitions and their specific stamp. This entailed the idea that modulation could be described by use of 'modulation symmetries' between start and target tonality, and that the pivotal chords (those which enforce the switch between the involved tonalities) could be extracted from a set of 'modulatory tones' which admits such modulation symmetries as *inner* symmetries. The first (not so satisfactory) attempts in fact focused on the diminished seventh chords, but then, the definition of such modulatory sets turned out to work perfectly when coupled to cadential sets for the target tonalities instead of the rather artificially chosen diminished seventh chord.

Interestingly enough, no essential extensions of those early methods have been proposed to date. Which does not mean that the results have not been greatly extended to many general scales and to just tuning. But the general setup of the involved concepts is still the old one.

This is why we shall start this subject with a general view on the modulation methodology thus far developed in the following Section 27.1.3. This will also reveal a fascinating perspective towards a "grand unification" of harmony and counterpoint as it has been suggested by Noll's definition of Riemann consonances and dissonances [802].

It is important to review the crucial role of the very concept of modulation in music. The medieval concept of "modulatio", as it is encountered in Augustinus' famous early definition *musica est ars bene modulandi*, has the quite general meaning of rational organization of tones and rhythms or, in the later medieval tradition, and more generally, musical composition. The modern meaning of a tonal modulation is

G. Mazzola, *The Topos of Music I: Theory*, Computational Music Science,
https://doi.org/10.1007/978-3-319-64364-9_27

a specialization of the original concept and was only introduced in the 18th century (see article "modulatio" in [907]).

The following approach essentially relates to tonal modulation, but the mathematical formalism and related canonical generalizations suggest a broader conceptualization. This means that the idea of a tonal modulation is in fact a special case of a potential theory of transitions between instances of a particular musical structure. Transition between tonalities can be paralleled by transitions between motives, rhythms, global objects, and by transitions between more sophisticated representations of tonalities (such as the self-addressed setup), motives, rhythms, and global objects. So the theory would have to describe transition mechanisms and their expressions in terms of compositional instances of practical use.

In this spirit, and it is the mathematical universality which preconizes the development, modulation should be rephrased in the understanding of medieval tradition, i.e., as a rational construction of music in all its structural streams. Unfortunately, this latter perspective is by no means a contemporary vein of musicology or even computer music science. But it turns out that any future system of music composition technology should deal with the general problem of transitional structures in all musical layers, such as harmony (in its most general setup), motivic, and rhythmical layout—including their interactions on local and global levels.

27.1 Modeling Modulation by Particle Interaction

Summary. This section introduces the motivation for developing modulation models. These models are designed to convey a harmonic transformation process. We discuss the 'particle interaction' model which allows us to simulate A. Schönberg's fundamental modulation degrees *wherever* he described direct modulations. The model is thoroughly formalized for 12-tempered and just tuning. The corresponding modulation theorems are proven and commented.

– Σ –

To begin with, this typical case of a model in musicology should be understood with care: It is *not* a dogmatic prescription of what a modulation should be, and a fortiori not a declaration of what kind of tonalities, cadences and modulation mechanisms should be englobed. Everyone can modulate as she likes. Modulation is even and—hélas quite often in popular music—processed without any reflection: In the commercial shortcuts, after a song unit (following the four-period blues scheme, for example) a song is suddenly transposed by a number of semitone steps and restarts without any preparation or transitory passage on a new shifted tonality. In our approach, we evidently are more interested in the description of possible relations between adjacent tonalities in musical compositions, and in the question of how such relations should be expressed explicitly on the material level of harmonic syntax.

27.1.1 Models and the Anthropic Principle

Summary. We sketch the role of mathematical models as a means to investigate the anthropic principle of music history, i.e., the historical selection of distinguished solutions of theoretical and practical approaches to musical rule systems within a variety of fictitious alternatives.

– Σ –

The following modulation model is a first, and a very simple construction which however yields very good results and has the typical property of mathematical models in music: To enable a quasi-automatic generalization to situations where the classical music theory for which the model was constructed has no answer. In the case of modulation which originally was modeled for major scales, the generalization extends to arbitrary scales.

The property of extensibility of a mathematical model relocates the existing music theory (which it models) in a field of potential, fictitious theories. This puts the historically grown facticity into a relation with the potential 'worlds of music'. The purely historic justification of existing modulation rules, for example, does not give us reasons for this choice, and this makes the purely historical approach a poor knowledge basis: We know that something is the case, but not why, and why other possibilities are not. In contrast, the mathematical approach gives us a field of potential theories wherein the actual one can be asked for its possible special properties with respect to non-existing variants. This differentia specifica is a remarkable advantage of mathematical methodology against the historical approach of musicology which cannot, in its poor intellectual performance, embed the facts in a variety of fictions and thereby understand the selection of what is against what is not.

This evokes Leibnitz idea that the existing world is the best of all possible worlds: Is the existing music theory the best possible choice? Or is it at least a distinguished one? In cosmology, this idea has been restated under the title of the "anthropic principle" [83]. It says that the physical laws are the best possible for the existence of humans, more precisely (and less radically), it is the theorem stating that a slight variation of the fundamental constants, such as the gravitational constant, or the electric charge of electrons and protons, would make any higher molecular complexity as is necessary for carbon-based biochemistry impossible.

27.1.2 Classical Motivation and Heuristics

Summary. We describe Schönberg's tripartite modulation scheme and the task of formalization.

– Σ –

In his classical text on harmony [948], Schönberg has described tonal modulation from one tonality to another as a three-part process of neutralization, turning point to the new tonality, and cadential confirmation, see Figure 27.1. The central statement is the list of modulation degrees in the middle turning point

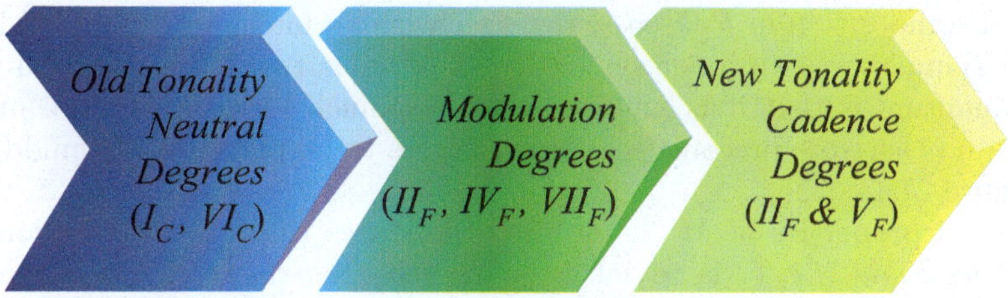

Fig. 27.1. The three parts of Schönbergs modulation scheme in the case of a modulation from C-major to F-major.

part. A mathematical model must start with a precise restatement of the musicological terms and facts in a rigid mathematical terminology. So we have to answer the questions

1. What is the set of admitted tonalities?
2. What is a degree?
3. What is a cadence?
4. What is the underlying modulation mechanism?
5. How can the modulation degrees be deduced from a solution of the preceding questions?

To begin with, we want to work in the context of well-tempered tuning, and we shall only look for pitch classes, i.e., for for zero-addressed points in $PiMod_{12}$. So our pitch class set identifies to the cyclic group $\mathbb{Z}_{12}$. We consider twelve possible diatonic scales, $C = \{0, 2, 4, 5, 7, 9, 11\}$ and its eleven translates $F, B_\flat, E_\flat, ..., E, D, G$, in the cycle of fourth.

To define tonalities, we then consider the set

$$Dia^{(3)} = C^{(3)}, F^{(3)}, B_\flat^{(3)}, E_\flat^{(3)}, ..., E^{(3)}, D^{(3)}, G^{(3)}$$

of triadic interpretations. The degrees in tonality $T^{(3)}$ are identified with the triadic charts $X_T, X = I, II, \ldots VII$, everything being visualized by means of the nerve, i.e., the harmonic strip of a triadic interpretation.

This means that a seventh chord, e.g., $\{d, f, a, c\}$ in $C^{(3)}$, is not simply II_C, but must be 'interpreted' as a union $II_C \cup IV_C$ of degrees in C. And the chord $\{d, f_\sharp, a\}$ is not a second degree in C with "sharpened third"; degrees are strictly triadic degrees within the respective tonalities!—We should recall the Yoneda philosophy from Section 9.3 in this context. Accordingly, an object X is identified by the functor of its 'perspectives', i.e., by the system of all morphisms $f : Y \to X$ in the given category. In the category *Glob* of global compositions, where we are presently working, our perspective is the canonical morphism $X^{(3)} \to X$ onto the basic scale X. Another perspective, such as a modal interpretation $X_f \to X$ (see Section 13.4.2), or a combined perspective of triads and modal tone $X_f^{(3)} \to X$, would yield a variant of the present approach.

Remark 15 For the traditionally educated in harmony, this self-control is hard to understand. It seems that one can no longer understand everything. But understanding everything amounts to understanding nothing. We had already seen in Section 13.4.2.1 that Riemann's attempt to define tonal functions on all chords had to fail. And the technique of alterations is questionable in harmony since it does not lead to contradictions like function theory, but produces meaningless statements. We know from Mason's theorem in Section 7.5.1 that two sharps and three flats suffice to interpret any chord in a seven-note scale as an alteration of a chord in any other seven-note scale! Mason's result tells us that with a sufficiently high number of alteration signs, nothing is in fact said if one interprets a chord in this way as an alteration of a chord in a determined scale.

With these definitions, and with the concept of a cadence being related to minimal sets of degrees (see definition 87 in Section 26.2.1) defining uniquely the surrounding tonality, the mechanism of modulation from tonality $S^{(3)}$ to tonality $T^{(3)}$ is defined by a symmetry which transforms $S^{(3)}$ into $T^{(3)}$. As in particle physics, we will materialize such a transformation force by a modulation quantum M. By definition, this is a zero-addressed local composition in $PiMod_{12}$, together with an interpretation $M^{(3)}$ by triadic chords such that the modulation symmetry is an inner symmetry of M, plus two more technical properties. The point of this modulation quantum is that it is 'musical matter'—much like the physical modulation quanta are materialized forces—and therefore preconizes a set of degrees to be played in the middle part of Schönberg's process shown in Figure 27.1.

As to the technical properties of $M^{(3)}$, the first one states that the intersection $M \cap T$ is rigid, i.e., the group of symmetries $Sym(M \cap T)$ is trivial. It guarantees that the modulation symmetry $S^{(3)} \to T^{(3)}$ is uniquely determined. The second property requires that the modulation quantum be a minimal set of pitch classes which satisfies the other properties. This is an economical condition designed to exclude superfluous tones. The precise definitions will be given below in Section 27.1.4.

The central statement of this modulation model is that such modulation quanta exist for all pairs of tonalities, and that they yield exactly the modulation degrees described by Schönberg [948]. Figure 27.2 shows such a quantum $M^{(3)}$ for the modulation from C to $E_\flat$. We recognize the two harmonic strips (light) as well as the connecting quantum (dark) in its nerve representation. The modulation degrees are exactly those vertexes of $M^{(3)}$ which lie in the harmonic band (four degrees in the example of Figure 27.2) of the target tonality. The statements of this model and its background mechanism, viz. the inherent symmetry transformation, will also be tested on the repertoire of classical music, such as Beethovens op. 106, for example (see Section 28.2). For the latter, it is possible to obtain a deeper understanding of its well-known intricate and complex modulation plan, switching back and forth between a "world" around the main tonality $B_\flat$-major, and the "anti-world" around tonality B-minor (theses of Erwin Ratz and Jürgen Uhde, see Section 28.2.2).

The above modulation model illustrates perfectly the "anthropic principle" subject which emerges from mathematical modeling. The model will be extended to any seven-tone scale and corresponding harmonic bands, see Section 27.1.5, and one can thereby locate the case of the major scale in a world of potential variants. The main result in this variational context is that the major scale occupies a singular position in that its modulation arsenal turns out to be the minimal possible variety among all scales which admit modulations between any two tonalities. So the historical dominance of the major scale receives a supplementary

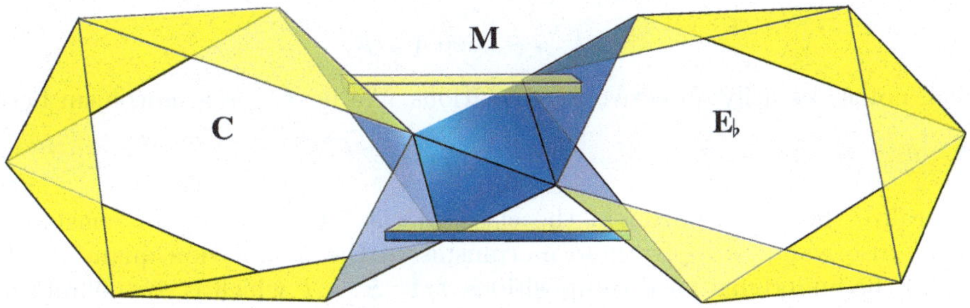

Fig. 27.2. Visualization of a modulation quantum $M^{(3)}$ for the transition from C to $E_\flat$. The nerve representation of $M^{(3)}$ shows the connecting middle structure in dark color, in particular it contains two five-dimensional simplexes (symbolized by horizontal bars) above and below a three-dimensional simplex (the tetrahedron).

justification in a perfectly systematic framework, and one which the purely historical approach can a priori not provide.

27.1.3 The General Background

Summary. The general situation is sketched in order to establish a connection to the topos-theoretic perspective and grand unification options.

– Σ –

In the preceding classical model of modulation, the space form was $PiMod_{12}$, and we were concerned with a collection $X = Dia^{(3)}$ of triadic interpretations $S^{(3)}$ of scales S, i.e., zero-addressed local compositions in $PiMod_{12}$. Moreover, these interpretations were all isomorphic with each other under symmetries which are given on the space $PiMod_{12}$; more precisely, we considered a full orbit $X = G.C^{(3)}$ under the action of a group G of symmetries on the space form. Further, the cadence was chosen to be the one with minimal cadence sets (according to definition 87). Finally, the modulation quantum $M^{(3)}$ was a triadic interpretation of a zero-addressed local composition in $PiMod_{12}$. We have not yet proven *that* such quanta exist, and *when* they do so in the specific setups. We shall discuss this topic in detail later in Sections 27.1.4, 27.1.6. For the time being, we are only interested in an adequate generalization of this conceptual setup in terms of the theory of global compositions. We should again stress that this generalization has not yet been completed by corresponding theorems as they exist in some classical cases.

The evident drawback of the classical model is the restriction to isomorphic interpretations. In general, transitions between instances of a type of musical structures cannot be restricted to isomorphic instances, proper deformations and specializations should be possible. In particular, tonal modulation should also work between non-isomorphic tonalities, modulations between non-isomorphic motives should be the common case, and modulations between completely different rhythms are just what any interesting composition should englobe after a century of jazz. We shall come back to rhythmical modulations between isomorphic rhythms in Section 28.3.2, this example uses the present theory in the onset dimension.

It is remarkable that modulation between non-isomorphic tonalities has not yet been treated in a systematic way in traditional music theory. The reason may be that in traditional music cultures, the selection of tonalities is a quite stable preference which identifies the traditional music culture and therefore would contradict its own message while being allowed to change its isomorphism class.

Let us now set up a more general approach to modulation. We first have to identify the basic objects. The obvious generalization for tonalities and modulation quanta is that they are interpretations of local compositions at an address A in a form space F. We first have to consider a set X of interpretations. Such an interpretation S^I maps epi onto its local support S which means that we have a composed arrow (not a mono, in general!)

$$S^I \rightarrow S \rightarrowtail @A \times F \tag{27.1}$$

of global and local, not necessarily objective, compositions. We next give a morphism between such global compositions

$$S^I \rightarrow T^J \tag{27.2}$$

which should generalize the isomorphism in the classical case. But we cannot consider 'abstract' global morphisms since we also want to apply these morphisms to the modulation quantum. Therefore, we are given a morphism of the underlying local compositions $f/1 : S \rightarrow T$ which is, by definition, induced from a "symmetry" $h : F \rightarrow F$ of the form space. Notice that it is not necessary to suppose h being an isomorphism. So we are looking for global morphisms $\overset{\iota}{f/1} : S^I \rightarrow T^J$ which extend the local morphism $f/1$ to the chart configuration, i.e., the diagram

$$\begin{array}{ccccc} S^I & \longrightarrow & S & \longrightarrow & @A \times F \\ {\scriptstyle \overset{\iota}{f/1}}\downarrow & & {\scriptstyle f/1}\downarrow & & \downarrow{\scriptstyle 1\times h} \\ T^J & \longrightarrow & T & \longrightarrow & @A \times F \end{array} \tag{27.3}$$

with the evident horizontal arrows commutes. Next, we suppose that we are given an interpreted cadence composition, also A-addressed in F, for each tonality S^I:

$$C^T_{S^I} \rightarrow C_{S^I} \rightarrowtail @A \times F \tag{27.4}$$

which lives in S^I, i.e., we have a factorization of interpretations of local compositions in F:

$$\begin{array}{ccccc} C^K_{S^I} & \longrightarrow & C_{S^I} & \longrightarrow & @A \times F \\ {\scriptstyle \overset{\tau}{1/1}}\downarrow & & 1/1\downarrow & & \downarrow{\scriptstyle 1} \\ S^I & \longrightarrow & S & \longrightarrow & @A \times F \end{array} \tag{27.5}$$

and this means that not only is the support C_{S^I} a subcomposition of S, but the degrees of the cadence are also associated with degrees of this tonality via the map τ. In the above classical case, this means that the triadic degrees of the cadence set must be triadic degrees in the englobing tonality. Cadentiality in this situation means that whenever we have two such factorizations of $C^K_{S^I}$ through S^I, T^J in the tonality collection X, then we must have $S^I = T^J$.

The last of our objects is the modulation quantum. This is again an A-addressed interpretation

$$M^L \rightarrow M \rightarrowtail @A \times F \tag{27.6}$$

and we ask that the modulation morphism be an endomorphism of this quantum, more precisely, the diagram

$$\begin{array}{ccccc} M^L & \longrightarrow & M & \longrightarrow & @A \times F \\ {\scriptstyle \overset{\lambda}{m/1}}\downarrow & & m/1\downarrow & & \downarrow{\scriptstyle 1\times h} \\ M^L & \longrightarrow & M & \longrightarrow & @A \times F \end{array} \tag{27.7}$$

commutes under a suitable morphism $\overset{\lambda}{m/1}$, a condition which reduces to the condition that we find a chart map λ which is compatible with the restriction endomorphism m of the ambient 'symmetry' transformation $1 \times h$. So in the setup of the harmonic topologies, we just ask that $M \in Ext_A(h)$, and that this local condition be compatible with the charts of M's interpretation.

The classical condition that the target tonality T^J as well as the modulation quantum should contain the cadence set now means that we have a commutative diagram

$$\begin{array}{ccc} C^K_{T^J} & \longrightarrow & M^L \\ \downarrow & & \downarrow \\ T^J & \longrightarrow & T^J \cap M^L \end{array} \tag{27.8}$$

where the intersection[1] $T^J \cap M^L$ is the fiber product $T^J \times_{@A \times F} M^L$. The rigidity condition now states that the only invertible symmetry $k : F \to F$ which induces an endomorphism of the intersection $T^J \cap M^L$ is the identity. This condition may be enriched by the additional conditions, for example, that k must also induce an isomorphism which transforms interpretations within X into each other.

In the classical definition of the modulation quantum $M^{(3)}$, there is a last requirement, i.e., that the intersection $T \cap M$ be covered by the degrees of $T^{(3)}$ which are contained in M. In the general setup this reads as follows: For each subset J^* of the given atlas J of T^J, we have a sub-composition $T^{J^*} \rightarrowtail T^J$ and therefore a corresponding fiber product diagram

$$\begin{array}{ccc} T^{J^*} \cap M^L & \longrightarrow & M^L \\ \downarrow & & \downarrow \\ T^{J^*} & \longrightarrow & @A \times F \end{array} \tag{27.9}$$

Take the unique maximal sub-composition $T^{J_{max}}$ such that the projection $T^{J^*} \cap M^L \to T^{J^*}$ has a section $\sigma : T^{J_{max}} \to T^{J^*} \cap M^L$. (The empty sub-composition evidently has this property, so existence is evident.) Intuitively, this means that we take all charts in T which also live in M. Then we ask that the canonical composed map

$$T^{J_{max}} \xrightarrow{\sigma} T^{J^*} \cap M^L \to T \cap M$$

is an *epimorphism*.

The philosophy of this approach is that a modulation looks like a perspective from the old tonality into the new one, and this is quite the same as the approach of Noll to Riemann's concept of consonance and dissonance (see Section 30.2.1) which views this qualification as an expression of the morphisms between two chords, i.e., as a perspective view of relative consonance/dissonance of one chord with respect to another. The point in modulation theory is that we do not consider the perspectives (morphisms) as an ultimate structure, but want them to be an expression of a materialization (the quantum) in order to grasp the material trace of 'modulation forces'.

27.1.4 The Well-Tempered Case

Summary. The 12-tempered modulation theorem is proven for diatonic scales. We visualize the nerves of the modulation quanta and compare to Schönberg's lists in [948].

– Σ –

This technical section follows the general modulation theorem as exposed in Muzzulini's paper [779]. This modulation theorem generalizes the earlier approaches in [669, 670, 682].

Muzzulini considers a seven-element scale S in $PiMod_{12}$ and its interpretation $S^{(3)}$ by seven three-element degrees $X_S \subset S$ defined by taking all subsets of form $X_S = \{s_i, s_{i+2}, s_{i+4}\}$ with indexes modulo 7 and referring to any indexation $S = \{s_0, s_1, \ldots s_6\}$ of the S-elements along the circle line. The naming of these degrees is not relevant for the interpretation, however, one usually enumerates the degrees as with ordinary diatonic scales, i.e.,

$$I_S = \{s_1, s_3, s_5\}, II_S = \{s_2, s_4, s_6\}, \ldots VII_S = \{s_6, s_1, s_3\},$$

although the naming is a function of the indexation of S. By construction any such triadic interpretation has a Möbius strip as its geometric nerve. From the classification table Appendix N.1, we see that there are

[1] Observe that even when the factors are objective, their intersection will not be so in general!

66 translation classes, 38 translation-inversion classes, and 25 affine classes of seven-element scales. The first two groups also conserve interpretations and therefore are also orbits of triadic interpretations. All minimal cadential sets have been calculated by computer for the 66 translation classes (this can be done on the basis of the 38 translation-inversion classes). The number of these minimal cadential sets are indicated in the third column of table Appendix P.1.1. We already know the five minimal cadential sets for major scales (No. 38.1 in the classification Appendix N.1) from the cadence discussion in Section 26.2.1. The maximal number of cadential sets appears for harmonic minor scales (and only here): Here, each pair of different triadic degrees is a minimal cadential set, so we have 21 such sets. Besides the major diatonic scale, there are two other scale orbits with this minimal cadence set number: No.52 and No. 62. The melodic minor scale has 15 minimal cadence sets.

Call a scale S is *rigid* iff its symmetry group $Sym(S)$ is trivial. If an inner symmetry of S induces an automorphism of the triadic interpretation $S^{(3)}$, we say that it is an inner symmetry of $S^{(3)}$. If the group of inner symmetries of $S^{(3)}$ is trivial, $S^{(3)}$ is also called rigid. The following lemma from [779] is clear:

Lemma 43 *No translation of a seven-element scale is an inner symmetry, and any inversion in $Sym(S)$ is an inner symmetry of $S^{(3)}$. The only possible non-trivial inner symmetries of $S^{(3)}$ are inversions. In particular, $S^{(3)}$ is rigid iff $Sym(S)$ contains no inversions.*

Given a pair $S^{(3)}, T^{(3)}$ of different triadic interpretations in the same translation orbit $T_{12}.X^{(3)}$, a *modulator* is a symmetry g which induces an isomorphism $g : S^{(3)} \xrightarrow{\sim} T^{(3)}$. A modulator must be of type $g = e^t.h$, where h is an inner symmetry of $S^{(3)}$. By the preceding lemma, h must be an inversion or the identity. *Hence a modulator must be a translation or an inversion.*

Definition 88 *Given a pair $S^{(3)}, T^{(3)}$ of different triadic interpretations in the same translation orbit $T_{12}.X^{(3)}$, a* modulation *from $S^{(3)}$ to $T^{(3)}$ is a pair (g, μ) consisting of a modulator for $S^{(3)}, T^{(3)}$, together with a minimal cadential set μ for the target tonality $T^{(3)}$.*

Given a modulation (g, μ) from $S^{(3)}$ to $T^{(3)}$, the axiomatic properties of a modulation quantum $Q \subset PiMod_{12}$ are these:

Property 1

$$(\pounds_\mu)\begin{cases} (1) & \textit{The modulator } g \textit{ is an element of } Sym(Q). \\ (2)_\mu & \textit{All triads in } \mu \textit{ are subsets of } Q. \\ (3) & \textit{The intersection } TI_{12} \cap Sym(T \cap Q) \textit{ is trivial, and} \\ & T \cap Q \textit{ is covered by triads of } T^{(3)}. \\ (4) & \textit{The quantum } Q \textit{ is a minimal set with properties } (1) \textit{ and } (2)_\mu. \end{cases}$$

The main problem of this construction is the existence of a modulation quantum. A modulation is said to be *quantized* iff it has at least one modulation quantum. The modulation theorem below will give an exhaustive answer to this problem. If a modulation (g, μ) has a quantum Q, denote by $(Q \cap T)^{(3)}$ the interpretation of the trace of Q in T by the triads of T which cover $T \cap Q$. The interpretation of Q by triads of $S^{(3)}$ and $T^{(3)}$ is denoted by $Q^{(3)}$. The degrees of $(Q \cap T)^{(3)}$ are called the *pivots* of this quantized modulation. Observe that, by property $(\pounds_\mu)$, (3), the modulator is uniquely determined by the quantum.

In the following modulation theorem [779], we distinguish the 38 TI_{12}-orbits of seven-element scales.

Theorem 30 *In the following, let $0 \neq p \in \mathbb{Z}_{12}$, and $T^{(3)} = e^p.S^{(3)}$.*

- *Scales with rigid triadic interpretation: For each scale S of the 28 TI_{12}-orbits with rigid interpretations, and for arbitrary p, there exists at least one quantized modulation from $S^{(3)}$ to $T^{(3)}$.—The maximum of 226 quantized modulations occurs for the orbit of scale type No.54.1 (the harmonic minor scale), while the minimum of 53 quantized modulations occurs for the orbit of scale type No.41.1.*

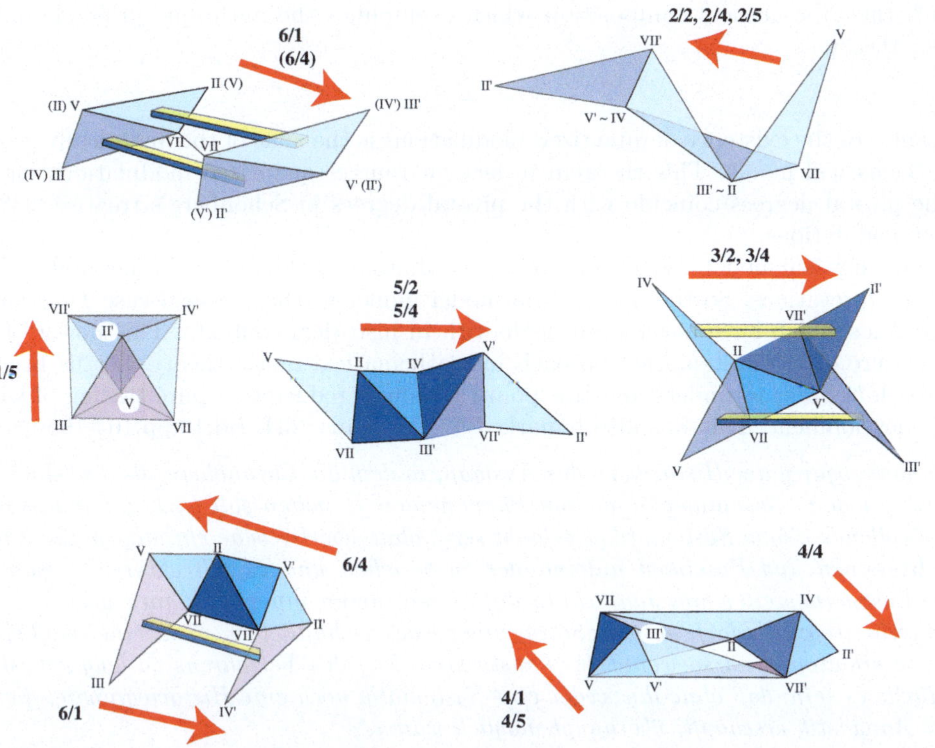

Fig. 27.3. Visualization of all nerves of interpreted modulation quanta $Q^{(3)}$ for modulations on $Dia^{(3)}$. Certain nerves are higher than three-dimensional. There, we have symbolized the higher-dimensional simplexes by bars which connect all vertexes of the related lower-dimensional simplexes. For example, the bars of the quantum for 3/2, 3/4 symbolize a five-dimensional simplex connecting the vertexes of two triangular surfaces, each.

- *Scales with non-rigid triadic interpretations: For orbit of scale type No.52 and No.55, there exist quantized modulations except for $p = 1$ and $p = 11$. For the orbit of scale type No.38 and No.62, there exist quantized modulations except for $p = 5$ and $p = 7$. The six remaining orbits have at least one quantized modulation for each p.—The maximum of 114 quantized modulations occurs at the orbit of scale type No.47.1 (melodic minor scale). Among the scales with quantized modulations for each p, the minimum of 26 occurs at the orbit of scale type No.38.1 (diatonic major scale).*

Idea of proof. The proof [779] used a computer program to calculate quanta and pivots as follows: For a given pair $S^{(3)}, T^{(3)} = e^p.S^{(3)}$, choose a corresponding modulation (g, μ). Then there is exactly one candidate Q for a quantum which fulfills properties (1), $(2)_\mu$, and (4) of $\pounds_\mu$. In fact, Q is the orbit of the tones of μ under the group of symmetries generated by the modulator. This candidate is rejected if property (3) does not hold, and we may proceed with a new modulation.

Figure 27.3 shows the nerves of all quanta for modulations in the orbit $Dia^{(3)}$ of diatonic major scales.

For lists of modulation data, such as quanta, cadence sets, pivot sets, for different scale classes, please consult the tables P.1.1, P.1.2, P.1.3, and P.1.4 in Appendix P.1.

27.1.5 Reconstructing the Diatonic Scale from Modulation

Summary. We discuss and comment the modulation theorem 30 for 12-tempered diatonic 7-tone scales. The overall behavior of the classical diatonic and melodic or harmonic minor scales for the present modulation model turn out to be in a distinguished position among all 7-tone scales. They can thus be reconstructed from their role in modulation processes; this gives modulation theory a new perspective within historical development of scales for musical composition which exemplifies the anthropic principle of mathematical models in music theory.

– Σ –

The guarantee of the existence of quantized modulations is the alias of the historically grown rule canon within the mathematical model. This theorem in fact guarantees quantized modulations for all couples in $Dia^{(3)}$, and the pivotal degrees coincide with the pivotal degrees in Schönberg's treatise [948] wherever he considers *direct* modulations.

But the mathematical model also applies to the non-diatonic scales of seven tones and to any translation pair of triadic interpretations. So the modulation model immerses the classical case $Dia^{(3)}$ in a variety of modulation scenarios which have never been dealt with in historical contexts. This power of variation is a characteristic feature of the systematic approach in mathematical music theory. Let us briefly review the historical approach in order to understand this point. It cannot reduce to a pure tracing of what is the case in a diachronic development. This was also remarked by Dahlhaus [234, Bd.10,pp.104-105]:

> *Ist demnach sogar unter Historikern das Ausmaß, in dem die Chronologie als "substantiell" — als "wesentlich" für den "Zusammenhang" von Phänonemen — gelten soll (...), zumindest umstritten, so sollte es vollends einem Systematiker erlaubt sein, ohne übertriebene chronologische Zaghaftigkeit Ideen, Institutionen und Praktiken aufeinander zu beziehen und zu "Strukturen" zusammenzufassen, deren Enstehungszeiten um Jahrzehnte differieren, deren innerer Konnex aber keiner strikten äußeren Gleichzeitigkeit bedarf, um den Status einer wissenschaftlichen, nicht einer bloß spekulativen Hypothese zu erhalten. Und so wenig der konstruktive Zug des Verfahrens zu leugnen ist, so unbestreitbar dürfte es sein, daß ohne ihn weder eine Systematik noch eine Historiographie, die sich nicht in schierer Annalistik erschöpft, überhaupt möglich wären.*
> *(...)*
> *Erst die systematische Konstruktion öffnet den Blick dafür, welche Tatsachen einer Geschichte angehören, die zu erzählen lohnend erscheint.*

In the terminology of Saussurean semiology, Dahlhaus recognizes that the diachronic dimension of music must also be understood as an expression and extension of the system of music. And that this extension only becomes worth being considered qua instance of an overall system.

Clearly, the surface character of the diachronic axis with respect to the systematic background is paralleled by the surface character of the synchronic axis. In fact, this axis refers to the ethnological dimension of music, i.e., its cross-cultural variability. It would be as annoying as pure historiography to repertorize ethnological diversification if the fundamental question of systematic connections over the cultural localities could not steer ethnological musicology. In other words, the spatio-temporal extension and diversification of music could not be understood without its characteristic nature qua system. A semiotic system cannot be reduced to a bunch of records over time and space.—This is the basic axiom of Saussure's original linguistic approach—even if it turns out to be hard to recognize the system's laws.

So why is the modulation theorem in its present form (theorem 30) necessary to the comprehension of the music system? Wouldn't it suffice to have the original theorem for $Dia^{(3)}$? In the synchronic axis it would in fact: It gives the different transposition distances a common basis from which pivotal degrees can be deduced. The quantum model of modulation is the system, and the different transposition distances are synchronic places in a certain time slice of music history.

But that does not explain why the European corpus of classical compositions could grow to such a prominent and voluminous status, a status which is strongly based upon some standard scales, the diatonic major, the melodic, and the harmonic minor scales, and an associated harmony whose top performance crystallizes in tonal modulation.

We do not claim that the modulation theorem explains all reasons for the prominent status of these scales, but we claim that it gives the prominent scales a prominent position in modulation processes. From this, we deduce that the historical success of these scales could relate to their modulatory excellence.

The position of the harmonic minor scale is such that its number of 226 quantized modulations is maximal among all (seven-tone) scales with rigid triadic interpretation. Among the scales with non-rigid triadic interpretation and quantized modulations for each translation, the maximal number of 114 quantized modulations occurs for the melodic minor scale. So these minor scales have a maximal freedom of modulatory actions. Among the scales with non-rigid triadic interpretation and quantized modulations for each translation, the diatonic major scale is charged with the minimum of 26 quantized modulations. The scale No. 41.1 with a minimum of 53 quantized modulations has no interpretation in the historical scale repertory, to our knowledge.

To understand these extremal positions, observe that the modulation theorem does not prefer any of these scales in its concept framework or hypotheses. So it is a priori not expectable that the three scales of diatonic major, melodic, and harmonic minor could play any role in their variety of quantized modulations. The scale No. 41.1 is also in a distinguished position in this context, but we do not have any interpretation of this fact.

However, the historic "choice" of scales does coincide with three of four scales that are exhibited by this generic modulation model. So the two final questions remain:

- How can we recognize the modulation model in the historic traces?
- If one accepts the model's presence in the systematic background of the historic development, to what degree has the historic development been guided by system parameters which are related to the given modulation model?

The first point will be attacked in the experimental chapter 28 on applications of the theorem to classical compositions of the European tradition. The second question is hard to tackle since the question relates to the innermost reflections or mere intuitions which may have guided composers and—to a lesser degree—theorists.

Independently of its role as an explanatory mechanism for historical phenomena, the present model can very well be used to produce modulations in new contexts. For instance, one may try to write compositions in exotic scales and to use modulation rules from this model. Or one may also apply this model to other parameter spaces. In fact, the modulation model does not preconize the pitch domain in its structural premises! One may also apply the modulation rules to time, loudness or other musical parameters in order to cope with Augustinus' philosophy that music is the art of good 'modulation'. This latter approach has been tested in a composition *Synthesis* [681] for piano, percussion, and base. In the first movement of that composition, modulation was applied to the time domain, i.e., to rhythmical modulation instead of tonal modulation. We shall also discuss this experiment in Section 28.3.2.

The table in Appendix P.1.5 shows examples of direct schematic and short modulations (as usual in the modulation texts) for every fourth translation $p = 1, 2, 3, 4, 5, 6, 7, 8, 9, 0, 11$, starting from C-major. The degrees, which are separated by commata, are meant to be played in time order, one after the other.

27.1.6 The Case of Just Tuning

Summary. This section is devoted to the proof of the modulation theorem for just tuning, presented in [673, 682] for major scales, and elaborated by Hildegard Radl [871] to harmonic, melodic minor, and other scales. The theorem sheds some light on the particular chromatic scale in just tuning, as considered by Martin Vogel [1089].

– Σ –

27.1.6.1 Just Scales and their Triadic Interpretations

We shall first of all deal with just scales in the *EulerPlane* as introduced in Section 7.2.1.4. The scales S which we shall encounter in the sequel will all be generating, i.e., $\mathbb{Z}S = \mathbb{Z}^2$, the ambient module whose first axis encodes fifths while the second encodes thirds. Therefore, it will be equivalent to speak about automorphisms and about inner symmetries. For any zero-addressed local composition C in the *EulerPlane*, we say that it is *rigid* iff $Aut(C) = Id$.

We are going to discuss these scale types (refer to Section 7.2.1.4 for just scales and Section 13.4.2.2 for triadic interpretations):

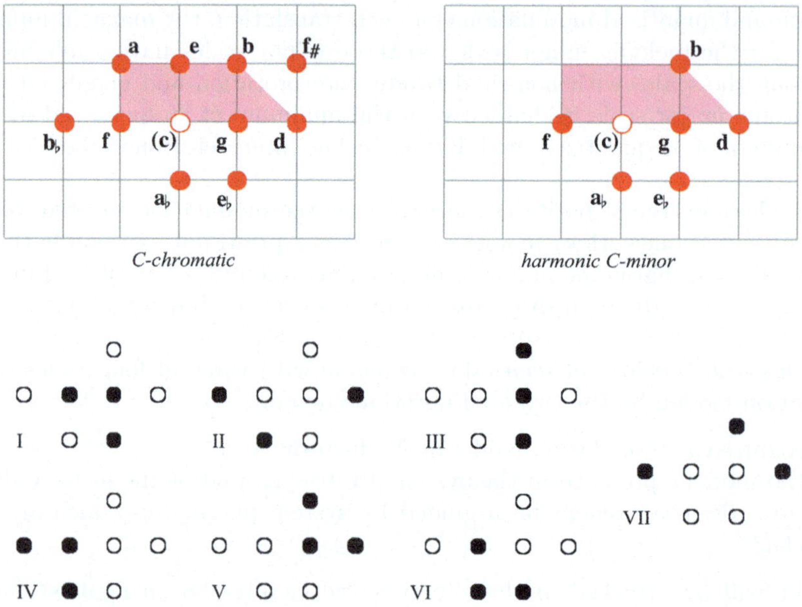

Fig. 27.4. The harmonic minor scale and its triadic interpretation.

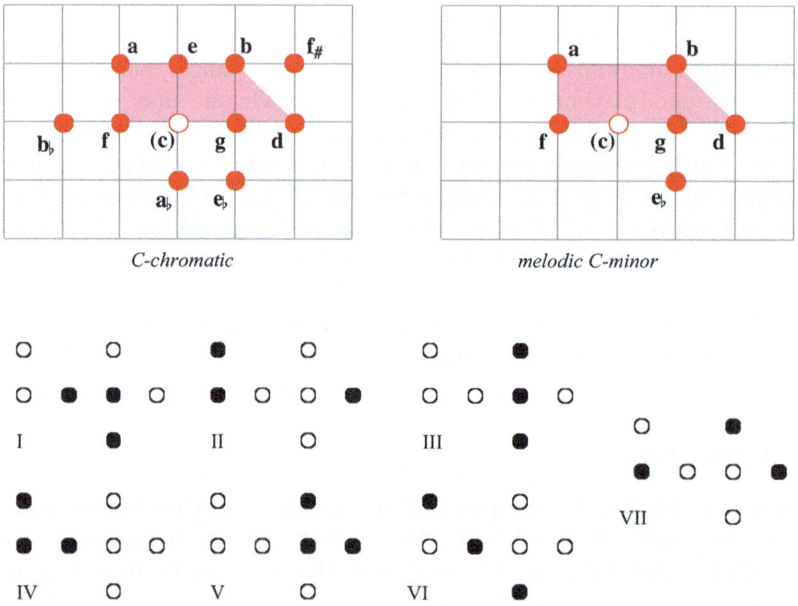

Fig. 27.5. The melodic minor scale and its triadic interpretation.

- The set Maj of major scales, i.e., the translates $X = e^t C$ of just C-major C, together with the set $Maj^{(3)}$ of triadic interpretations by the seven degrees $I, II, \ldots VII$, excluding the variants II^*, VII^*.
- The set $NatMin$ of natural minor scales, i.e., the translates $X = e^t c_{nat}$ of natural C-minor c_{nat}, together with the set $NatMin^{(3)}$ of triadic interpretations.
- The set $HarMin$ of harmonic minor scales, i.e., the translates $X = e^t c_{har}$ of harmonic C-minor c_{har}, see Figure 27.4, together with the set $HarMin^{(3)}$ of triadic interpretations.
- The set $MelMin$ of melodic minor scales, i.e., the translates $X = e^t c_{mel}$ of melodic C-minor c_{mel}, see Figure 27.5, together with the set $MelMin^{(3)}$ of triadic interpretations.

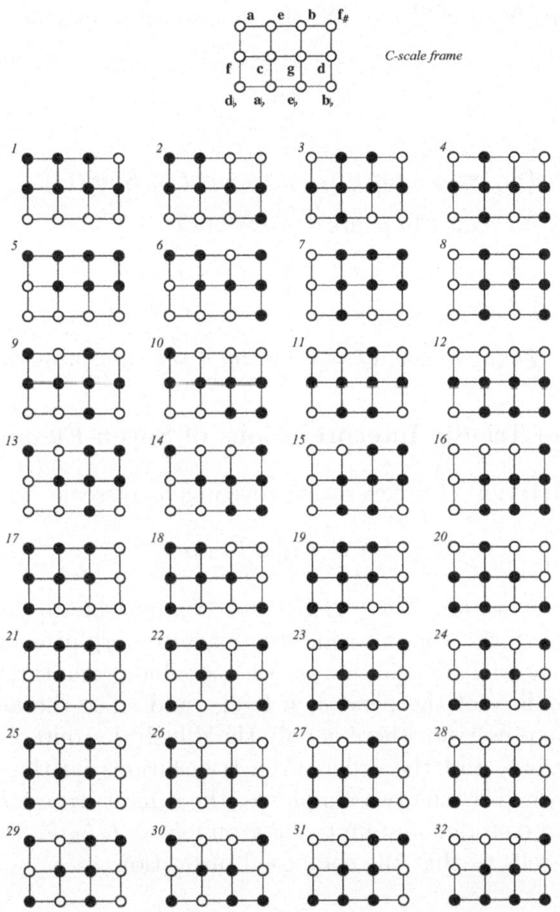

Fig. 27.6. The 32 altered scales are derived from the C-major frame on top.

- The 32 translation orbits $AltMaj(i)$ of seven-element scales $C(i)$ which are defined as follows, see Figure 27.6: Take the fifth pairing c, g. Take the a priori *C-scale frame* shown on top of Figure 27.6. Then, choose any additional five tones which are either the major tones f, d, a, e, b or one of their alterations, whence we obtain $2^5 = 32$ variants. Of course, these known special cases are included: $C = C(1)$, $c_{nat} = C(5)$, $c_{mel} = C(9)$, $c_{har} = C(11)$. And the corresponding triadic interpretation sets $AltMaj(i)^{(3)}$.

On the one hand, we shall discuss modulations within the orbits

$$Maj^{(3)}, NatMin^{(3)}, HarMin^{(3)}, MelMin^{(3)}, AltMaj(i)^{(3)},$$

on the other, we shall also switch between major and natural minor scales, this latter case meaning that we take the larger orbit under translations and 180°-rotations.

27.1.6.2 Modulations and Quanta

We now have to adapt the crucial definitions from 12-tempered modulation to just modulation.

Definition 89 *Fix a subgroup* $G \subset \overrightarrow{GL}(\mathbb{Z}^2)$. *Given a pair* $S^{(3)}, T^{(3)}$ *of different triadic interpretations of one of the above scale types which live in the same orbit* $G.X^{(3)}$, *a* modulation from $S^{(3)}$ to $T^{(3)}$ *is a pair* (g, μ) *consisting of a modulator* $g \in G$ *for* $S^{(3)}, T^{(3)}$, *together with a minimal cadential set* μ *for the target tonality* $T^{(3)}$.

Given a modulation (g, μ) from $S^{(3)}$ to $T^{(3)}$, the axiomatic properties of a modulation quantum $Q \subset EulerPlane$ are these:

Property 2

$$(\pounds^{just}_{\mu}) \begin{cases} (1) & \text{The modulator } g \text{ is an element of } Sym(Q). \\ (2)_{\mu} & \text{All triads in } \mu \text{ are subsets of } Q. \\ (3) & \text{The intersection } T \cap Q \text{ is rigid, and} \\ & T \cap Q \text{ is covered by triads of } T^{(3)}. \\ (4) & \text{The quantum } Q \text{ is a minimal set with properties (1) and } (2)_{\mu}. \end{cases}$$

27.1.6.3 Automorphisms of Triadic Interpretations of Seven-Element Scales

Let $S^{(3)}$ be the triadic interpretation of a generating seven-element scale

$$S \subset EulerPlane,$$

for example the above types in Section 27.1.6.1. Let $\sigma \in Sym(S)$ be an automorphism which induces an automorphism $\sigma \in Sym(S^{(3)})$. Since this also induces an automorphism of the nerve $n(S^{(3)})$ which is a Möbius strip, we have combinatorial information on the possible permutations induced by σ. Clearly, σ is of finite order. Since the boundary of the nerve is a circle, and since the adjacency relations of the seven vertexes on this circle must be preserved, there is only the dihedral group of the regular heptangle, i.e., the identity, the order-two reflections, and the order-seven translations on the vertexes, and hence, since the elements of the scale are recovered by the two-simplexes, the same is true for the scale.

On the other hand, if we consider any finite subgroup $H \subset GL_2(\mathbb{Z})$, then Minkowski's theorem (see Appendix C.3.5, theorem 51) tells us that the composed projection

$$H \rightarrowtail GL_2(\mathbb{Z}) \to GL_2(\mathbb{Z}_3)$$

is injective. But we have $card(GL_2(\mathbb{Z}_p)) = (p-1)^2 p(p+1)$ for any prime p (see Appendix C.3.4) which is 48 for $p = 3$. Hence, if we apply this fact to the linear part of our automorphism, order 7 is forbidden for our automorphisms, and we are left with orders 1 and 2. This means that if $\sigma = e^a.A$, then we have $A^2 = Id, A.a + a = 0$ for $A \in GL_2(\mathbb{Z}), a \in \mathbb{Z}^2$.

Suppose we are given a modulator $\phi : S^{(3)} \to T^{(3)}$ between two just tonalities, then it has the shape $\phi = \tau.\sigma$ with a translation $\tau = e^b$ and an automorphism $\sigma = e^a.R$ of $S^{(3)}$ as above. Since the modulation quantum must be a full orbit of the cadence μ's tones under the group $\langle\phi\rangle$ generated by ϕ, it is important to know the shape of an orbit $\langle\phi\rangle(x)$ of a point $x \in \mathbb{Z}^2$ of the $EulerPlane$ space. Here is the formula ([871, p.12]):

$$\langle\phi\rangle(x) = e^{\mathbb{Z}.(R+Id)b}x \cup e^{\mathbb{Z}.(R+Id)b}.\phi(x). \tag{27.10}$$

So we have one 'branch' of this orbit which is just the translation orbit of x under translation $e^{(R+Id)b}$, and the other 'branch' which is the translation orbit of $\phi(x)$ under the same translation. If $\sigma = Id$, both branches are shifted against each other by b, and they are generated by a translation orbit under e^{2b}.

Exercise 65 *Draw an image of the orbit of C-major under* $b = (9, 2)$ *and* σ *being the unique non-trivial automorphism A of C-major, as described in formula 8.6.*

27.1.6.4 Finiteness of Modulation Domains

Definition 90 *The* modulation domain *of a tonality* $S^{(3)}$ *is the set of shift vectors* $b \in \mathbb{Z}^2$ *such that there is a quantized modulation from* $S^{(3)}$ *to* $T^{(3)} = e^b S^{(3)}$.

In a good number of cases, the modulation domain of a tonality is finite. We investigate this situation.

In order to control the possible intersections between the target tonality and the orbit $\langle\phi\rangle(X)$ of a set X of tones in the target tonality which stands for the potential modulation quantum, we should control the cases where the different transforms of a scale are disjoint from each other.

Proposition 48 *With the above notation, if C is the major scale in EulerPlane, then the transforms*

$$e^{z.(R+Id)b}C, e^{w.(R+Id)b}C, e^{z.(R+Id)b}.\phi(C), e^{w.(R+Id)b}.\phi(C)z, w \in \mathbb{Z},$$

are mutually disjoint except for a finite number of shift vectors b.

Proof. The translation vector $b = (b_1, b_2)$ induces a translation vector $(R + Id)b = b_2.(-1, 2)$. So clearly, the two branches $e^{\mathbb{Z}.(R+Id)b}C$ and $e^{\mathbb{Z}.(R+Id)b}.\phi(C)$ are disjoint if $|b_1|$ is sufficiently large. On the other hand, the vertical shift component by b_2 of the scale C to the second branch $e^{\mathbb{Z}.(R+Id)b}.\phi(C)$ under ϕ cannot hit the first branch for $1 < |b_2|$ since the vertical component of $b_2.(-1, 2)$ is twice the vertical shift component b_2, and thus, the branches cannot intersect in the vertical progression. QED.

Corollary 18 *For modulations among just major tonalities, the modulation domain is finite.*

Proof. Consider the minimal cadence sets $J_1, \ldots J_6$ as defined in formula (26.2). It is easily seen that the union of every cadence set, except $J_1 = \{II\}, J_5 = \{VII\}$, has no rigid superset in the scale. So only $J_1 = \{II\}, J_5 = \{VII\}$ can be candidates for cadence sets in the quantum construction ($\pounds^{just}_{\mu}$). But these two degrees are not rigid. Now, the quantum, if it exists, must be the orbit of either of these degrees under the group $\langle\phi\rangle$. But this is just one of these two degrees in every transformed instance of the scale. So, if the transformed instances are all disjoint from each other, the trace of the quantum in the target scale is either II or VII, so it is not rigid. Now, by the above proposition, this is the case for all but a finite number of shift values b, and we are done. QED.

27.1.7 Quantized Modulations and Modulation Domains for Selected Scales

Summary. We present the quantized modulations for modulations between major scales, between natural minor scales, between major and natural minor scales, between harmonic minor scales, and between melodic minor scales. We finally discuss Radl's results concerning the 32 altered scales introduced in 27.1.6.1 and their modulation domains.

– Σ –

In the following sections, we present the modulation domains and the pivotal degrees for a number of standard scales. The proofs are omitted since they are a tedious and lengthy, but not very illuminating work of case distinction. The details are found in Radl's work [871] which in its algorithmic part should be incorporated in a computer-aided algorithm.

Theorem 31 *This variety of modulation results, i.e., the existence of modulation quanta is not written in the closed form of a theorem. Nevertheless, it should be read as a highly segmented existence theorem of just modulation theory. We shall refer to this list of results when referring to the modulation theorem for just tuning.*

The main ideas are these:

- to use the finiteness statement of corollary 18 and to digress on the different cases in this finite set of possibilities;
- to use the fact that a modulation by a shift b can be used to calculate the modulation in the negative direction $-b$.

Exercise 66 *Calculate the $-b$-case from the results for the b-case.*

We should also mention that Radl has equally calculated some modulation results for other situations in the Euler context, such as scales in the Pythagorean tuning, and in the three-dimensional EulerSpace from third, fifth and seventh components. We withdraw from a discussion of these ramifications.

27.1.7.1 Modulation Between Major Tonalities

Suppose that we have to modulate from $S^{(3)}$ to $T^{(3)} = e^b S^{(3)}$. We have two modulators in this case:

$$\Phi_1 = e^b, \tag{27.11}$$
$$\Phi_2 = A.e^b. \tag{27.12}$$

The modulations are listed in detail in table Appendix P.2.1. Here, we just present the modulation domain in Figure 27.7. We see that this domain covers all the tones of the major scale (except the tonic itself which

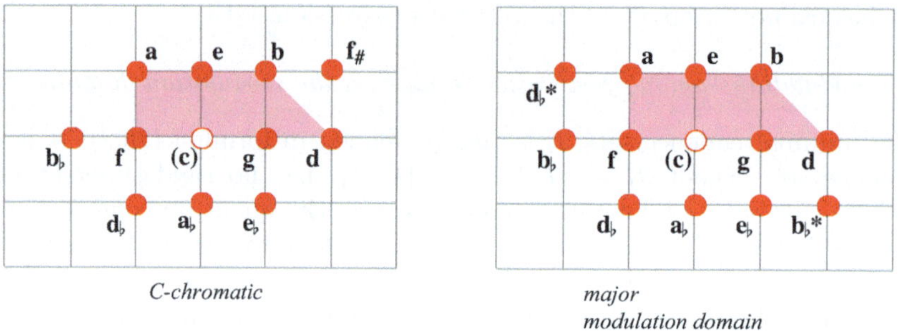

Fig. 27.7. The modulation domain from just C-major.

is excluded by definition) as well as six external tones. The modulation domain is very similar to Vogel's chromatic. From this compositional point of view, Vogel's chromatic is optimally chosen as a modulatory domain. Other chromatics, such as Michel's [737] or Roederer's [914] are less adapted to this modulation domain.

27.1.7.2 Modulation Between Natural Minor Tonalities

We have the two modulators

$$\Phi_1 = e^b,$$
$$\Phi_2 = e^b.A.$$

The modulations are listed in detail in table Appendix P.2.2. We just present the modulation domain in Figure 27.8.

As may be expected from the symmetry relation between major and natural minor type, this domain is the same as that of major tonalities, however only absolutely. Relative to the scale, the domain is a rotated one.

27.1.7.3 Modulation From Natural Minor to Major Tonalities

We have the two modulators

$$\Phi_1 = e^b.A,$$
$$\Phi_2 = e^b.B, B = \begin{pmatrix} 1 & 1 \\ 0 & -1 \end{pmatrix}.$$

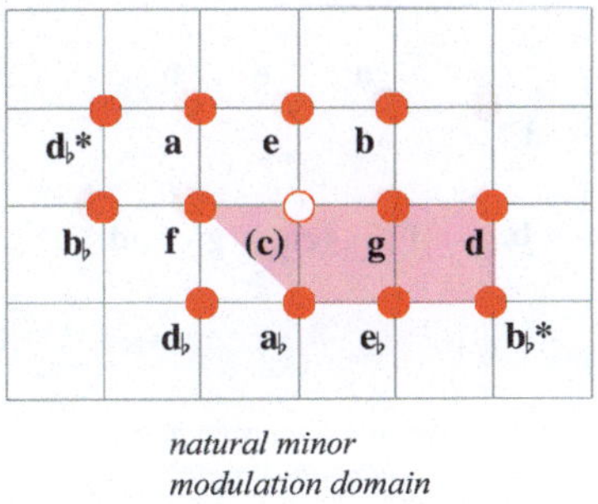

Fig. 27.8. The natural minor modulation domain from natural C-minor.

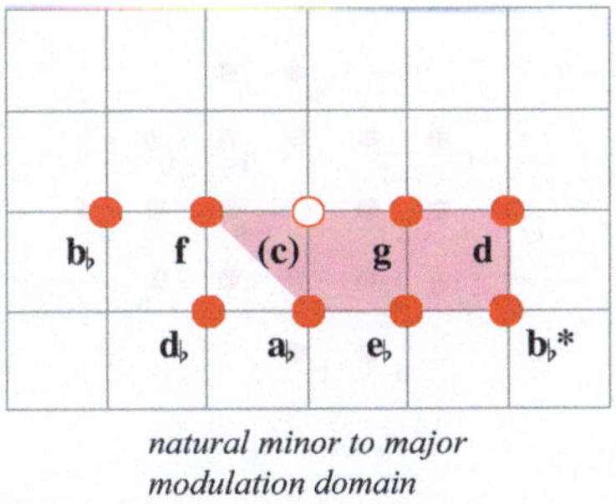

Fig. 27.9. The modulation domain from natural C-minor to major tonalities.

The modulations are listed in detail in table Appendix P.2.3. We just present the modulation domain in Figure 27.9.

This is a remarkable result since in the well-tempered situation, there is no distinction of the scales for major and natural minor, it is just the tonic which differs. So here, we have a modulation between different translation classes.

27.1.7.4 Modulation Steps From Major to Natural Minor Scales

We have the two modulators (same as above).

$$\Phi_1 = e^b.A,$$

$$\Phi_2 = e^b.B, B = \begin{pmatrix} 1 & 1 \\ 0 & -1 \end{pmatrix}.$$

The modulations are listed in detail in table Appendix P.2.4. We just present the modulation domain in Figure 27.10.

This is the symmetric result to the above one.

27.1.7.5 Modulation Steps Between Harmonic Minor Scales

We have the unique translation modulator

$$\Phi = e^b.$$

The modulations are listed in detail in table Appendix P.2.5. We just present the modulation domain in Figure 27.11.

This modulation domain is extremely large compared to the others. Perhaps this is a reason why harmonic minor is so prominent in classical European harmony, and also, to some degree, in composition.

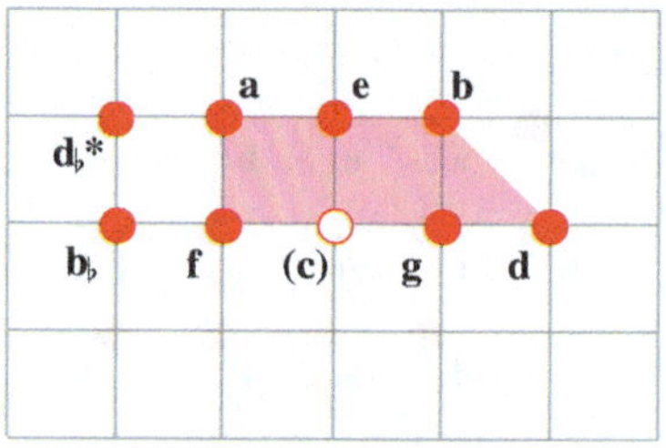

major to natural minor
modulation domain

Fig. 27.10. The modulation domain from C-major to natural minor tonalities.

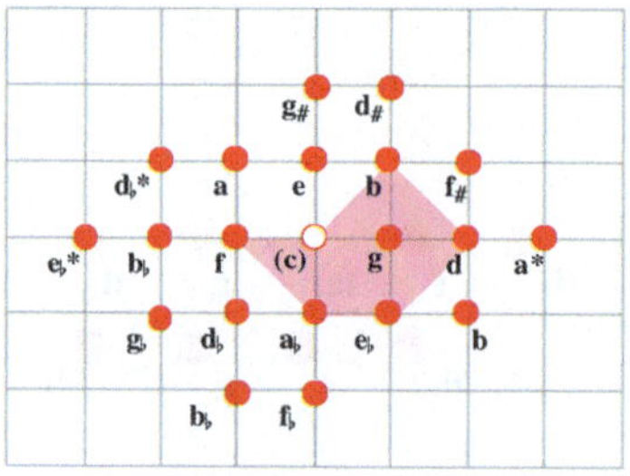

harmonic minor
modulation domain

Fig. 27.11. The modulation domain from C harmonic minor tonalities.

27.1.7.6 Modulation Steps Between Melodic Minor Scales

We have the two modulators

$$\Phi_1 = e^b,$$
$$\Phi_2 = e^b.A.$$

The modulations are listed in detail in table Appendix P.2.6. We just present the modulation domain in Figure 27.12.

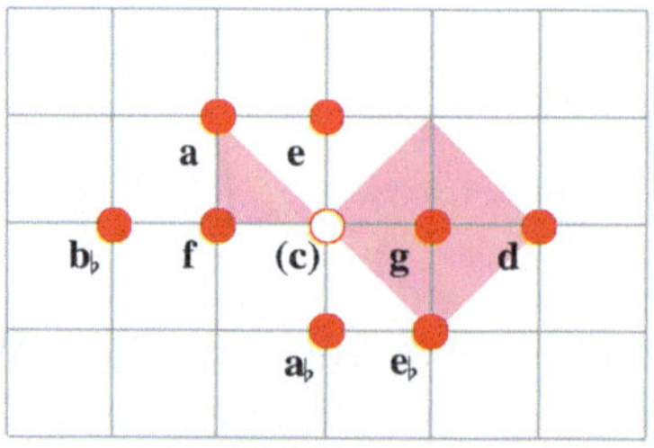

melodic minor
modulation domain

Fig. 27.12. The modulation domain from C melodic minor tonalities.

27.1.7.7 General Modulation Behavior for 32 Altered Scales

This investigation in Radl's work was undertaken to get a global view of possible scale behavior in modulation when deriving scales from the major scale by determined alterations as described above. From the table P.2.7, we can distinguish these cases.

- *has no modulations* if its modulation domain is empty (always excluding the start tonality!),
- *has infinite modulations* if its modulation domain is infinite,
- *has modulations* if its modulation domain is not empty (always excluding the start tonality!), *has limited modulations* if the transitive closure (all tonics which can be reached by successive modulations from relative modulation domains) of its modulation domain is not total space.

The cases (15, 19) with no modulation are pathological. The cases with infinite modulations (21, 22, 24, 29, 30, 32) are too indeterminate since we can modulate in a set of tonalities which cannot be distinguished from their perceivable pitch structure, see also the discussion of valence theory in Appendix B.2. The cases (8, 25) of limited modulations means that modulations cannot be concatenated to yield any target tonality. The remaining cases are the reasonable ones. There, we have reduction to modulations with just four target tonalities of the cases 2, 7, 13, 14, 23, same for the cases 10, 16, 17, 18, 20, 26, 27, 31. So these cases have very small domains against the remaining cases which all reduce to one of the special tonality types of major, natural minor, harmonic minor or melodic minor which have relatively large modulation domains (eight and more targets against four in the previous cases 2,7,...). The only case which remains is number 6, and this is a variant of the melodic scale. So we may summarize that the known scales have a much larger number of target domains than the other scales in the frame of the reasonable scales. *This is also a confirmation of Muzzulini's results on the distinguished position of major, melodic minor, and harmonic minor in the well-tempered case.*

27.2 Harmonic Tension

Summary. Harmonic tension regards the valuation of syntagmatic chord complexes. This concept adds to the isolated semantics of chords a semantic charge of complex syntagmatic arrangements. We discuss a distance-based concept of harmonic tension and related mathematical structures, such as the Riemann tensor algebra. The chapter concludes with a technical discussion of optimal harmonic paths, including algorithms for optimal path search.

– Σ –

The harmonic evaluation of isolated chords does not give the full information about a preferable choice of harmonic evaluation in the syntagmatic context. Classical harmony has not dealt with this problem except in the rudimentary concept of cadences. We shall here set up a formalism to grasp harmonic context structures from given Riemann matrix data. This is an important subject since it is very difficult to define a global syntagmatic equilibrium on the harmonic level. The Riemann tensor algebra will give a first approach to the formal treatment of the problem of local and global harmonic tension. This approach has been used in the HarmoRUBETTE®and will be discussed on a more practical level in Section 41.3.

27.2.1 The Riemann Algebra

Summary. We introduce the Riemann algebra as a quiver algebra that is built on the Riemann matrix.

– Σ –

We take up the formalism of Riemann matrices introduced in Section 25.3.2, including the basic data of that context, i.e., a name set TON of "tonalities", a name set VAL of "tonal function values", and a form space $TRUTH(I)$ for of "truth denotators" in a specific truth module (or, more generally, a truth form) I and at a specific address B for a form F. To make the ideas concrete, and also related to the HarmoRUBETTE®module of RUBATO®(see Section 41.3 for this tool) we shall work in the truth module $I = \mathbb{R}$. The form F can be any form which englobes the parameters for harmony (this is however not a formal condition in the following discussion). The address B will be any real vector space.

We first need a mathematical construction: the Riemann algebra.

Definition 91 *Given value and tonality sets* VAL, TON, *a space form* F *with address* B *(a real vector space), the* Riemann algebra *(over the real numbers) is the quiver algebra*[2] $Rie(\mathrm{VAL}, \mathrm{TON}, F, B)$ *over the reals for the complete quiver*[3] *(the* Riemann quiver*) over the set* $\mathrm{VAL} \times \mathrm{TON} \times ExTop_B(F)$.

The definition is motivated by the usual situation in harmony: When one analyzes a piece of music, there is a sequence of chords which are given certain function values with respect to certain tonalities. Such sequences are formally restated as being paths in the Riemann quiver. The Riemann algebra is needed in order to make calculations on different harmonic paths in the Riemann quiver, and in order to evaluate the best ones among a given selection of such paths. This is needed in order to decide which path should be taken when we have a certain way of giving weights to special sub-paths according to specific approaches to harmony.

The basic data in such a calculation are real-valued weights which we assign to certain paths. For example, one may exhibit a classical cadence path of shape $I_X \to IV_X \to V_X \to I_X$ and give it a high positive weight in order to express that such sequences are good paths. Or one may like to express that a passage $V_X \to I_X$ is a good relaxation movement, where as the other direction $I_X \to V_X$ is not and receives negative values, say. So these weight assignments regard more or less long paths in the Riemann quiver and should help in determining the quality, i.e., weight of a given path through a piece of music.

Before delving into details, observe that this approach is still a relative one in the following sense. Suppose we are given a sequence of chords $a_0, a_1, a_2, \ldots a_n$, and suppose that we have two harmonic interpretations of them in terms of two Riemann paths

$$(v_0, f_0, a_0) \to (v_1, f_1, a_1) \to \ldots (v_n, f_n, a_n),$$
$$(v_0^*, f_0^*, a_0) \to (v_1^*, f_1^*, a_1) \to \ldots (v_n^*, f_n^*, a_n)$$

which we want to compare: Which one is harmonically preferable? One may therefore calculate "weights" for these paths and decide upon these weights. This is what we shall achieve. But there is also a question concerning absolute weights: What is the harmonic weight of the last chord a_n, compared to the first one a_0? Is it legitimate to say that this is just the weight of a selected Riemann path? It is certainly not because the local values of the weights could add up to some path weight which always gives the last chord a higher weight than the first one.

This question is in fact the question about the global development of harmonic tension. Up to date, no musicological theory has ever dealt with this problem: Is it possible to calculate global harmonic tension from local knowledge? It is certainly not true that every piece of music should have a global slope of zero, i.e., the beginning tension being equal to the ending tension. For example, in a Scriabin sonata, such as op. 72 *Vers la flamme*, the final tension is expected to be much higher than the beginning one. So how should one calculate global 'absolute' tension? We shall come back to this issue in the discussion of the HarmoRUBETTE® software module in Section 41.3.

27.2.2 Weights on the Riemann Algebra

Summary. Weights on the Riemann algebra are predicates which measure the quality of a harmonic development path.

– Σ –

So let us now work on the weight functions on harmonic paths. Suppose that we are given a harmonic, real-valued *default weight function* w on all paths $(v_0, f_0, a_0) \to (v_1, f_1, a_1) \to \ldots (v_i, f_i, a_i)$ of lengths $i = i_1, i_2, \ldots i_k$. Then we may choose

- a set of coefficients $\mu_i, i = i_1, i_2, \ldots i_k$, and,
- for each such i, a system λ^i of coefficients in the sub-path operators $\sqsubset_i^{\lambda^i}$,

[2] See Appendix D.1.1.1 for this concept.

[3] See Appendix C.2.2, definition 95.

and define this linear form on the Riemann algebra:

$$\Omega(p) = \sum_{i=i_1,i_2,\ldots i_k} \mu_i.\omega(\sqsubset_i^{\lambda^i}(p)) \tag{27.13}$$

where the linear form ω extends the original w on the spaces spanned by the paths in the images under the endomorphisms $\sqsubset_i^{\lambda^i}$. The idea of this definition is that we suppose that certain paths are already weighted (in fact by ω). We then extend the weights to all paths by use of the sub-path operators. These operators take the given weights into account, each with its weight coefficient from the systems $\sqsubset_i^{\lambda^i}$. These systems give a sub-path its specific weight according to its position within the argument p. This is reasonable since we will in general give a sub-path a different harmonic meaning in a path according to whether it stays in the beginning, the middle, or the end. After this, the different contributions of sub-paths of specified lengths $i = i_1, i_2, \ldots i_k$ are weighted against each other according to the role of the sub-paths of these lengths in the present harmonic analysis. For example, if the harmonic concept gives more importance to short sub-paths of length 2, against lengths above 5, say, then the coefficient μ_2 should be significantly larger than the coefficients above μ_5. It may even happen that certain lengths play a completely different role than others, as we shall see in a moment.

So let us make a concrete example concerning the default weight function. The address is $B = 0$, and we are working in the function-theoretic context discussed in Section 25.3.3. This means

$$\begin{aligned} Fun(F) &= @\mathbb{Z}_{12}, \\ \mathrm{TON} &= \{C, D_\flat, D, E_\flat, E, F, G_\flat, G, A_\flat, A, B_\flat, B\}, \\ \mathrm{VAL} &= \{T, D, S, t, d, s\} \end{aligned}$$

and the circle group truth module $I = S_1$ for the function theory, assigning a half open interval $TF_{f,t}(a) = \phi(a)\hat{} = [0, \phi(a)[$ of S_1 to each chord $a \in ExTopo_0(F)$, function value $f \in \mathrm{VAL}$, and tonality $t \in \mathrm{TON}$.

Based on this information, we want to define the linear form ω on all paths of the Riemann quiver. By definition, the present form will vanish on all paths of length larger than 1. So this tension theory is very local, just limited to contributions from isolated chords and from transitions between two successive chords.

Let us first deal with the vertexes, i.e., the isolated chords. We have to define the values $\omega(f, t, a)$ for function values $f \in \mathrm{VAL}$, tonalities $t \in \mathrm{TON}$, and chords $a \in ExTopo_0(F)$. Suppose that $TF_{f,t}(a) = \phi(a)\hat{}$ with $\phi(a) \neq 0$. Then we set $\omega(f, t, a) = \ln(\phi(a))$. If $\phi(a) = 0$, the corresponding predicate is $\phi(a)\hat{} = \emptyset$. This means that in a given harmonic path, we encounter a chord which has "False" as its harmonic truth value. Such a path should fail to be a competitor in the selection of a best harmonic path. More generally, we would like to eliminate paths containing chords with a value $\phi(a) \leqslant \phi_{min}, 0 \leqslant \phi_{min}$. So we replace $\omega(f, t, a) = \ln(\phi(a))$ by $\omega(f, t, a) = -\infty$ if $\phi(a) \leqslant \phi_{min}$. Of course, this infinity value is incompatible with the definition of a common linear form. But if we read the linear form codomain $\mathbb{R}$ as the set of intervals $]-\infty, r[, r \in \mathbb{R}$, together with the transported structure of $\mathbb{R}$-module, we can extend the set by the empty interval $\emptyset$ and add the rules $s.\emptyset = \emptyset, s \in \mathbb{R}$ and $\emptyset +]-\infty, r[=]-\infty, r[+ \emptyset = \emptyset$. This structure is still a semigroup under addition and has a bilinear scalar multiplication. The empty interval is however an absorbing element: Every linear expression involving the empty set is absorbed by this one and reduces to the empty set. So the values of our "linear form" live in the truth space $\mathbb{T}^0_{\mathbb{R}}$. We have in fact defined the first piece of a predicate on tuples of chords, given by the predicates $TF_{f,t}$ and the limiting value ϕ_{min}.

The second piece of our predicate uses the linear form on paths of length 1. This contribution deals with the tension between two positions in the Riemann matrix. We are given a path $(f_1, t_1, a_1) \to (f_2, t_2, a_2)$. We would like to measure the harmonic tension of this transition as such, i.e., independently of the involved chords. The value $\omega((f_1, t_1, a_1) \to (f_2, t_2, a_2))$ will therefore be just a function of the arrow $(f_1, t_1) \to (f_2, t_2)$ in the complete quiver over the Riemann matrix index set $\mathrm{VAL} \times \mathrm{TON}$, call this quiver the *Riemann index quiver*. Two points of this quiver share two components: the function values and the tonality symbols. In our case, the function values are also split into a set $\mathrm{VAL}_{type} = \{T, D, S\}$ of function value types and a set $\mathrm{VAL}_{mode} = \{min, maj\}$ of major or minor mode. So we shall restate the index quiver as the complete quiver over the triple product $\mathrm{VAL}_{type} \times \mathrm{VAL}_{mode} \times \mathrm{TON}$.

Given a path $(f_1^{type}, f_1^{mode}, t_1) \to (f_2^{type}, f_2^{mode}, t_2)$, we have to take into account the component steps in each coordinate. For each of these three coordinates, we set up a real-valued matrix to express the respective tension component. For the function type, the 3×3-matrix is

$$T_{\mathrm{VAL}_{type}} = (t_{XY}^{type})_{X,Y \in \mathrm{VAL}_{type}}.$$

This means that in a 1-path where the type changes from X to Y enforces a tension of quantity t_{XY}^{type}. The corresponding meaning for mode changes is quantized by the 2×2-matrix

$$T_{\mathrm{VAL}_{mode}} = (t_{mn}^{mode})_{m,n \in \mathrm{VAL}_{mode}}.$$

Finally, the change of tonality is codified by the 12×12-matrix

$$T_{\mathrm{TON}} = (t_{st}^{\mathrm{TON}})_{s,t \in \mathrm{TON}}.$$

Observe that neither of these matrices will be symmetric or even distance like, i.e., having zero diagonal etc. For instance, the change from major to minor is supposed to be lower valued than the converse change in classical European music.

With these component contributions, we may finally define the transition default values as follows:

$$\begin{aligned}&\omega((type_1, mode_1, tonality_1) \to (type_2, mode_2, tonality_2))\\&= (t_{type_1 type_2}^{type})^2 + (t_{mode_1 mode_2}^{mode})^2 + (t_{tonality_1 tonality_2}^{\mathrm{TON}})^2.\end{aligned}$$

Of course this Euclidean approach can be replaced by any equivalent distance definition, this one is only one typical solution. So we never have minus infinity here, the only "False" contributions stem from the values on isolated chords as explained above.

To put everything together, the zero-length and one-length components are combined to yield the formula (27.13), i.e.,

$$\Omega(p) = \mu_0.\omega(\sqsubset_0^{\lambda^0}(p)) + \mu_1.\omega(\sqsubset_1^{\lambda^1}(p)) \tag{27.14}$$

where the μ-coefficients give a relative weight to the path-length contributions. We shall see in Section 41.3 that the HarmoRUBETTE® implements exactly this type of formula with the special values $\lambda^0_{..} = \lambda^1_{..} = 1$, and zero else, except that the nominal values in that implementation are the exponential values e^v of our values v.

This construction yields an objective tension predicate

$$Tension(\lambda^{\cdot}, \mu_{.}, \omega, \phi_{min})$$

on n-tuples $(a_0, a_1, \ldots a_n)$ of zero-addressed chords of the given form in the truth space $\mathbb{T}^0_{\mathbb{R}}$. The predicate yields the empty set as soon as one chord in the sequence has truth value "False". This evaluation can be used to calculate best paths in the harmonic development.

27.2.3 Harmonic Tensions from Classical Harmony?

Summary. We discuss a particular "harmonic perspective" of harmonic tension proposed by Anja Fleischer in [324].

– Σ –

The above construction of a tension predicate has the advantage that it works for any chord as proposed by Hugo Riemann. But it is not the usual approach in traditional harmony. In this latter context, it happens that most chords are not given any specific weight. We present one such approach as it was explicated in a diploma thesis by Anja Fleischer in [324]. In that approach, she took the classical binary logic and gave selected paths values of either "True" or "False" in the circle group truth module. Moreover, the value domains are chosen more in the flavor of Riemannian Harmony as it is taught in music schools.

The function list is this:

$$\mathrm{VAL} = \{T, t, Tv, tv, D, d, Dv, dv, S, s, Sv, sv\}$$

which are major tonical, dominant, and subdominant values, together with their minor and substitute (German: "Vertreter") variants.

The tonality list is this:

$$\begin{aligned}\mathrm{TON} = \\ &\{DurC, DurCis, DurD, DurDis, DurE, DurF, DurG, DurGis,\\ &DurA, DurB, DurH, Mollc, Mollcis, Molld, Molldis,\\ &Molle, Mollf, Mollg, Mollgis, Molla, Mollb, Mollh\}.\end{aligned}$$

The 24 symbols relate to German major (=Dur), minor (=Moll), and the twelve tonality names. Here is the list of core chords plus values in the sense that these values are those where a chord takes the "True" value in the harmony predicate. The chord data in the first coordinate relate to the same form $F = PitchMod_{12}$ as before.

$$\begin{aligned}True \sim \\ &\{(0,2,3,tv),(0,2,4,t),(0,2,5,s),(0,2,6,sv),(0,2,7,t),(0,2,7,tv)\\ &(0,2,9,s),(0,2,9,sv),(0,3,7,Tv),(0,4,5,s),(0,4,7,T),(0,4,9,t),\\ &(0,4,9,d),(0,4,10,t),(0,4,11,t),(0,5,7,s),(0,5,7,sv),(0,5,8,Sv)\\ &(0,5,9,S),(0,7,9,t),(0,7,10,tv),(0,7,11,t),(0,7,11,tv),(0,9,11,t),\\ &(0,9,11,d),(1,4,9,tv),(1,4,9,sv),(1,7,9,tv),(1,9,11,tv),(1,9,11,dv),\\ &(2,4,5,s),(2,4,6,sv),(2,4,7,t),(2,4,7,d),(2,4,8,tv),(2,4,8,dv),\\ &(2,4,9,s),(2,4,9,sv),(2,4,11,t),(2,4,11,tv),(2,4,11,d),(2,4,11,dv),\\ &(2,5,7,d),(2,5,7,dv),(2,5,9,s),(2,5,11,D),(4,5,7,t),(4,5,7,d),\\ &(4,5,8,tv),(4,5,8,dv),(4,5,8,sv),(4,5,9,s),(4,5,11,t),(4,5,11,tv),\\ &(4,5,11,d),(4,5,11,dv),(4,7,9,t),(4,7,9,tv),(4,7,9,d),(4,7,9,dv),\\ &(4,9,11,t),(4,9,11,tv),(4,9,11,d),(4,9,11,dv),(7,9,10,dv),(7,9,11,d)\}.\end{aligned}$$

The other chords of same interval structure are given the corresponding values relative to the related translation operations. This is a kind of historical "kernel" of harmonic knowledge. The question is whether one should just give "False" values to all chords that are not covered by this list. I would not, but it is difficult to establish a transparent rule set to evaluate the other chords (and these are most of the total chord system). The choice of falsity is the worst case scenario, but the present list does not give a good alternative, one just did not find general rules to be applied in the negative cases. We have included this example to make evident the dramatic gap between what is known and what is fuzzy and problematic in historical contexts.

27.2.4 Optimizing Harmonic Paths

Summary. This section deduces algorithms for valuations of entire paths under modulation constraints. The latter subject reveals the exorbitant complexity of harmonic analysis and throws new light on the question of complexity and depth in the humanities.

– Σ –

In the preceding optimization procedure, changes of tonalities (for example in the distance matrix T_{TON}) could be accounted for by numerical weights such that a tonality change is much more expensive than a conservation of the current tonality. Nonetheless, there is no finer tuning of the neighborhood of a harmonic development when a tonality change is recognized. The following algorithm is adequate for such a purpose. Let us reconsider the above case described in formula (27.14). Each chord of our chord sequence $(a_0, a_1, \ldots a_n)$ has its Riemann matrix evaluation, and the replacement by the "False" value for values below ϕ_{min}. Suppose that every chord of the sequence has a Riemann matrix with at least one non-false coefficient. Call the sequence *coherent* iff there is at least one tonality symbol such that each Riemann matrix in the sequence has at least one non-false coefficient at this tonality. A coherent sequence allows a harmonic path without "false" chord values living in one and the same tonality. Suppose further that one is interested in getting a minimum of tonality changes. This requirement is met as follows. Start with the maximal coherent subsequence $(a_0, a_1, \ldots a_{m_1})$ starting from the first chord (no gaps allowed). Then consider all maximal coherent subsequences (no gaps allowed) starting from $a_1, \ldots a_{m_1}, a_{m_1+1}$ (if possible). From these sequences, take the sequence with maximal last index: $a_{i_2}, \ldots a_{m_1}, a_{m_1+1}, \ldots a_{m_2}$, then repeat this procedure starting from index i_2 and so on. After a finite iteration, we end up with the covering of the whole sequence by coherent subsequences $(a_0, a_1, \ldots a_{m_1})$, $(a_{i_2}, \ldots a_{m_2})$, ... $(a_{i_k}, \ldots a_n)$. In each sequence, we may select a best path in a common tonality. This results in a spline-like configuration of paths $p_1, p_2, \ldots p_k$ with possibly overlapping parts. These parts can be taken as modulatory regions, more precisely: regions of neutralization which are followed by regions of pivotal chords and cadences. A deeper analysis of such splining regions with regard to the modulation theorems (Sections 27.1.4 ff. and 27.1.6 ff.) is pending.

28

Applications

Aus der Tatsache, daß sich die von den Theoretikern geschaffenen Strukturen und Formen meist als künstlich und zuweilen als abwegig herausgestellt haben, folgt nicht, daß es überhaupt keine allgemeine Struktur gibt, die nicht eines Tages von einer besseren Analyse der Musik, die alle ihre Erscheinungsformen in Zeit und Raum berücksichtigt, freigelegt werden könnte.
Claude Lévi-Strauss [597]

Summary. This chapter deals with illustrations of the modulation model in chapter 80. It treats short and longer examples from Bach to modern jazz. This exercise should make clear the methodological background of these musicological experiments and illustrate the theory exposed in chapter 4.

– Σ –

Before working out the concrete examples, we should emphasize that this approach is not meant to be a confirmation of any classical music theory. It is rather a discussion of a selection of empirical material in the light of the previous modulation theorems. The delicate point here is that the modulation theorems do not preconize any historical specification, they couldn't care less. These theorems give a list of pivotal degrees and not the answer to the question of historical adequacy. For such questions, one would have to look for other criteria.

A principal branch of such criteria comes from the communicative dimension of musical topography. When a work is analyzed, it is first taken in its neutral setup as a determined denotator of a more or less complex structure. As such, it has several pointers to poietic determinants, for example by the composer's name or/and the year of publication. When considering the poietic coordinate, one could ask for the adequacy of the modulation model in question with respect to the historical position of the composer and the composer's knowledge base. Questions as this one arise: "Did the composer (Beethoven, for instance) use the type of reflections as described in the modulation model to construe the modulations in the given score (op.106. for example)?" Up to unknown and highly unlikely secret historical legacies, the answer is clearly "No!". But this is definitely not the discourse of our enterprise. We are not discussion conscious compositional strategies. Rather we are confronting the result of compositional efforts: the score (denotator) with the present modulation models. We are, so to speak, in the physicist's position to investigate nature's laws without ever knowing if and what and how a divine being could have built nature.

The default answer of a musicologist would be that we are not physicists in so far as we know quite a bit about Beethoven's approach to modulation or at least, we know that he was *not* in a state of considering those mathematical tools used in the mathematical model of modulation. And therefore—this the argumentation of the musicologist—any such modeling remains disconnected from the historical reality.

G. Mazzola, *The Topos of Music I: Theory*, Computational Music Science,
https://doi.org/10.1007/978-3-319-64364-9_28

But it is rightly this latter concept of historical reality which we have to question in view of Dahlhaus' request for a systematic (re)construction of the dia- and synchronic system of music cited in Section 27.1.5. Reality is not just a set of phenomena, but the totality of agents which shape the historical dynamics of the system. It is precisely the invisible and counter-intuitive strata of a system which usually are responsible for its unfolding. The absence of conscious traces of Beethoven's modulatory constructs does not explain these modulations and it does a fortiori not explain the global architecture of these modulations in the development of a large piece, such as Beethoven's op. 106. Moreover, the incredible organization and process of such masterworks cannot be explained by biographic factors, these things are of a much finer nature than the gross facts of a composer's life. And even if the composer had made statements about his poietic activity, this would just be one perspective upon the work's neutral (objective) data. There is no guarantee that this perspective must be the best possible. It is very likely that the intuitive and ingenious forces cover a strong majority of the work's characteristic and outstanding instances.

So what can we retain from an application of a mathematical model of modulation if not the historical adequacy? We can retain the explanatory power, i.e., the adequacy of model and instance, theory and experiment. How much of the given example can the model explain, and in which detail? This is what we shall explain in the following examples, and not, whether or not the composer would have thought in the same lines.

A last and personal remark about composers: From all composers that I have known, the great majority—also those who work with explicit constructive methods—was never in a state of explaining their own works to a satisfactory degree. Good works have always been far from algorithmic runs—even with composers of a knowledgeable mathematical flair or training, such as Tom Johnson or Jan Beran or myself. This is not a dogmatic statement, of course. It is the insight that good music is an enormously complex affair for which the present knowledge is not yet in a state of giving exhaustive construction algorithms.

28.1 First Examples

Summary. This section discusses short example from Johann Sebastian Bach, Wolfgang Amadeus Mozart, and Claude Debussy.

– Σ –

In all subsequent examples, we shall refer to the modulation theorem in the well-tempered case. This seems legitimate in the historical context, starting from Bach. But it is not a priori clear which theorem should be applied in general.

28.1.1 Johann Sebastian Bach: Choral from "Himmelfahrtsoratorium"

Summary. The example exposes the first eight bars of the composition and shows a remarkable congruence of the model with a piece of "nature" which is far from being poietically conceived according to the model or its expression in music theories.

– Σ –

In this example, we consider bars 1 through 8 (see Figure 28.1). For our analysis we strictly take the perspective of triadic degree interpretations, i.e., the global objects contained in $Dia^{(3)}$. Chords are always zero-addressed and are deduced from onsets in the sense that whenever there is a new onset, one considers the chord consisting of all pitch classes of events which start here or did not end at this onset or before.

According to our triadic degree perspective, such a chord will not remain uninterpreted, it will be covered by degrees of a triadic degree interpretation in the sense already explicated in Section 27.1.2. For chords which are not triadic degrees, we choose an interpretation by a minimal number of triadic degrees. For example, the interval chord $\{d, a\}$ could be embedded in the major triad $\{d, f_\sharp, a\}$ or in the minor triad $\{d, f, a\}$. Since this embedding is not unique, we have listed our choices in Figure 28.2. We shall view this

Fig. 28.1. Johann Sebastian Bach: Choral from the *Himmelfahrtsoratorium* [56, p.41] (with kind permission of the Bärenreiter Publishers).

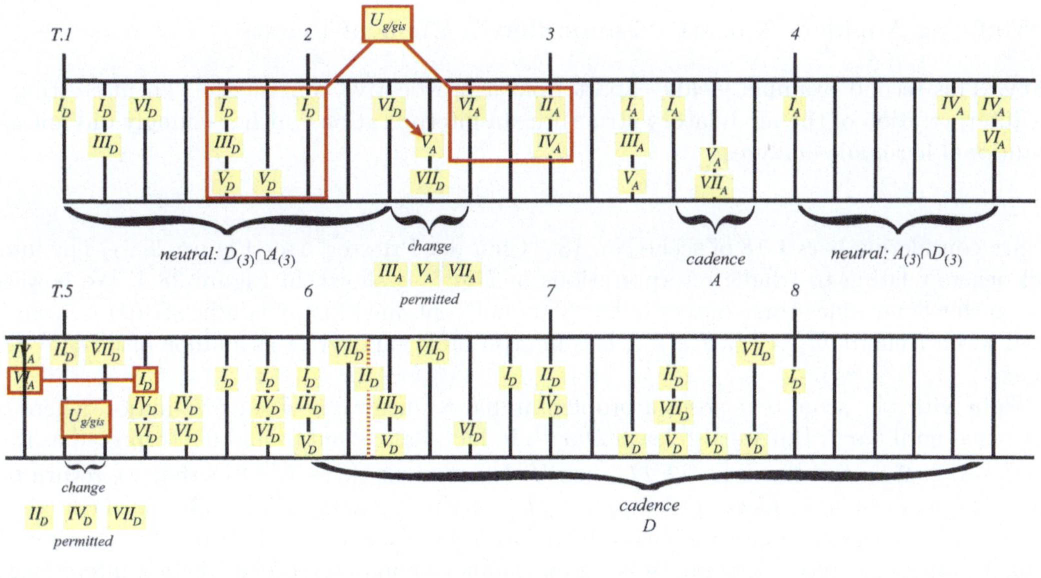

Fig. 28.2. Two successive modulations with Bach from D-major to A-major, and back. In both cases, the pivotal degrees correspond to those predicted by the (well-tempered) modulation theorem. Moreover, within the symmetric degree neighborhood of the turning point of the first modulation, the modulator, i.e., the inversion $U_{g/g\natural}$, becomes evident.

process as a sequence of two modulations, the first from $D^{(3)}$ to $A^{(3)}$, the second back from $A^{(3)}$ to $D^{(3)}$. The scope is to compare the harmonic development with instances of the well-tempered[1] modulation theorem. The modulator is the inversion $U_{g/g\natural}$ (recall that we denote $U_{g/g\natural}$ for the inversion U_3, a reflection at the point between g and $g\natural$). The pivot degrees for the modulation $A^{(3)} \rightsquigarrow D^{(3)}$ are III_A, V_A, VII_A whereas the modulation $D^{(3)} \rightsquigarrow A^{(3)}$ admits the pivotal degrees II_D, IV_D, VII_D (see Appendix P.1). The modulator

[1] We often use the sloppy wording "well-tempered" instead of "12-tempered", but are conscious that there are subtle differences in the terms.

symmetry $U_{g/g_\sharp}$ gives rise to the following correspondences:

$$\begin{array}{ccc} I_D \sim VI_A & VII_D \sim VII_A & VI_D \sim I_A \\ II_D \sim V_A & & V_D \sim II_A \\ III_D \sim IV_A & & IV_D \sim III_A. \end{array}$$

The first bar and the first two onsets of bar 2 are *neutral* since all appearing *neutral* degrees I, III, V, VI of $D^{(3)}$ are also degrees of $A^{(3)}$. The third onset in bar 2 presents the pivots V_A, VII_A according to table Appendix P.1.2 and may therefore be viewed as the turning point of this modulation to $A^{(3)}$. The following onsets until the fermata at the beginning of bar 4 are a cadence of the target tonality $A^{(3)}$ with minimal cadence set $\{VII_A\}$, or, distributed among several onsets, with $\{II_A, V_A\}$. Beyond this it is interesting to observe the neighborhood of the turning point: it makes the modulator evident. The last four onsets before the turning point present $I \cup III \cup V, V, I$, and (less weighted as onset of the fifth to the delayed V_A) VI_D. After the turning point and before the cadence towards the end of bar 3, appear two onsets of $VI, IV \cup II$ of $A^{(3)}$. These three degrees of $A^{(3)}$ correspond (under the modulator) to the degrees I, III, V of $D^{(3)}$ before the turning point.

The modulation in the reversed direction brings the neutral degrees in bar 4, then in the first two onsets of bar 5 the turning point with the admitted pivots II_D and VII_D, and finally, in bar 6, follows the cadence according to the same pattern as observed in the first modulation.

28.1.2 Wolfgang Amadeus Mozart: "Zauberflöte", Choir of Priests

Summary. This second example yields—apart from a provocative congruence—an interesting text-based *semantic* interpretation of the modulatory structure, an interpretation which seemingly did not appear from more traditional harmonic analyses.

– Σ –

We are considering bars 1-18 of scene No. 18: "Chor der Priester" (see Figure 28.3). The interpretation via chord onsets relating to triadic interpretations in $Dia^{(3)}$ is shown in Figure 28.4. We however have to comment on this figure since three onsets in bar 8 are not contained in any triadic structure from $Dia^{(3)}$: the diminished seventh chord $\mathfrak{M}_2 = \{f_\sharp, a, c, e_\flat\}$. It is interpretable via degrees as a union of $VII_G, VII_{B_\flat}, VII_{D_\flat}$, and VII_E.

To begin with, the score text poses a problem in bar 8 since we now leave D-major, where the degrees were positioned until bar 7. But we also leave $Dia^{(3)}$. In fact, bars 8 and 9 can be interpreted as these degrees in $g_h^{(3)}$ (see also 26.2.1): five times $II \cup VII$, then VII, and eight times I. After that we return to D-major, and we have no more such conflicts until degree $VII_{E_\flat}$ at the beginning of bar 14, which can also be viewed as $VII_{E_\flat}$ or II_C in harmonic minor tonalities. If we interpret the strict neighborhood of the diminished triad as lying in A-major, it appears as being isolated and cannot be understood easily. It is interesting that these two places are in a special position when related to the text semantics. The diminished seventh chord, when exiting the orbit $Dia^{(3)}$, is the moment of "düst'rer Nacht", the diminished triad $VII_{E_\flat}$ corresponds to the text which tells about the dark night, death, the polar concept against "life".

Let us try to set up a connection between the text-semantic extra position of $\mathfrak{M}_2$ and of $VII_{E_\flat}$ within the modulatory configuration. To this end, we write $III_{E_\flat}$ in bar 9, and in bar 14, we write II_E after $VII_{E_\flat}$. Bar 8, i.e., $\mathfrak{M}_2$, is not interpreted by degrees, but as a representation of modulators. Here is our modulation scheme (the pivots are bold-faced):

Fig. 28.3. Wolfgang Amadeus Mozart: *Zauberflöte*, Pianoexcerpt (Soldan), second Scene No. 18, *Chor der Priester* [769, p.118] (with kind permission of the C.F. Peters Publishers, Frankfurt/Main).

Modulator:	$U_{e/f}$	$U_{e/f}$	$U_{g/g\natural}$	$U_{g\natural}$	$U_{b/c}$	$U_{g\natural/a}$
Tonality:	$D \rightsquigarrow$	$E_\flat \rightsquigarrow$	$D \rightsquigarrow$	$A \rightsquigarrow$	$E_\flat \rightsquigarrow$	$E \rightsquigarrow A$
Circle of fourths:	5.	7.	11.	6.	5.	1.
Pivots:	*II*	*II*		*II*	**II**	*II*
	III		*III*	*III*	*III*	
		IV		(*IV*)		*II*
	V	**V**	**V**	*V*	*V*	**IV**
	VII	**VII**	*VII*	**VII**	*VII*	*VII*
Appearance:	*b*.9	*b*.10	*b*.13	*b*.14	*b*.14	*b*.16

In Figure 28.5, part a), the modulation paths are drawn on the circle of fourths. One recognizes the two segments in $E_\flat$ as a pole to A an its symmetrically positioned neighbors D, E. The two semantic poles "night" (first $E_\flat$) and "life" (second $E_\flat$) are placed on distinguished tonality positions in front of the contextual tonalities D, A, E. Also, the modulators appear in a neat symmetric way: Part b) of Figure 28.5 shows the axes of five different inversions and the diminished seventh chord $\mathfrak{M}_2$. These five modulators are

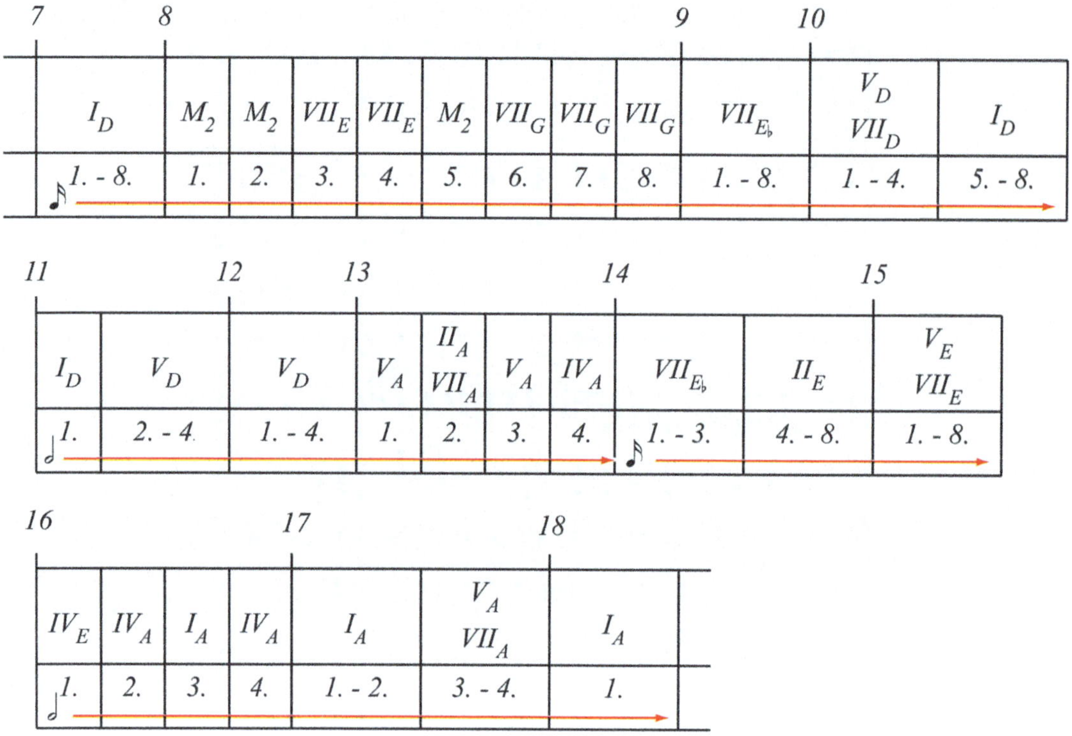

Fig. 28.4. The configuration of triadic degrees as it is shown here from the perspective of modulation theory. In bar 8, the diminished seventh chord plays a special role since it is not interpretable within $Dia^{(3)}$.

divided into three groups: The pair of orthogonal axes (minor thirds!) associated with $U_{e/f}$ and $U_{g/g_\sharp}$ is contained in the symmetry group of $\mathfrak{M}_2$. It controls the first three modulations, i.e., the 'edges' $D - E_\flat$ and $D - A$ in Figure 28.5, part a). Hence, $\mathfrak{M}_2$ appears as a 'support' of two modulators. These $\mathfrak{M}_2$-symmetries are in fact visible in bar 8. On the change $a \mapsto f_\sharp = U_{g/g_\sharp}(a)$, from the first to the second eighth onset, on the change $c \mapsto a = U_{e/f}(c)$ from the second to the third, on the change

$$\{a, e_\flat\} \mapsto \{f_\sharp, c\} = U_{g/g_\sharp}(\{a, e_\flat\}) = U_{e/f}(\{a, e_\flat\}) = e^3(\{a, e_\flat\})$$

from the fifth to the sixth, and on the change $c \mapsto a = U_{e/f}(c)$ from the sixth to the seventh eighth onset.

Symmetrically to this, the two 'edges' $E - E_\flat$ and $E - A$ in part b) of Figure 28.5 are mediated by the equally orthogonal axes associated with $U_{b/c}$ and $U_{g_\sharp/a}$. And the middle modulation $A \rightsquigarrow E_\flat$ can be viewed as 'symmetry edge' on part b) of Figure 28.5; its modulator $U_{g_\sharp}$ is equally inserted in a symmetric way between the double pair of orthogonal axes. This beautiful symmetry, which is semantically accompanied by the linguistic text, has the advantage that we can recognize an inner and semantic connectivity among the four tonalities $D, A, E, E_\flat$ instead of five tonalities $D, A, E, E_\flat$, and g_h, without any semantical connectivity.

28.1.3 Claude Debussy: "Préludes", Livre 1, No.4

Summary. This analysis again produces a possibly unexpected congruence with the model and succeeds in giving a very explicit interpretation to the title's "parfum".

– Σ –

It is particularly important to recall from the introduction to this chapter that we do not make any statements about the poietic rationales of these compositions in our examples. And that our analysis is

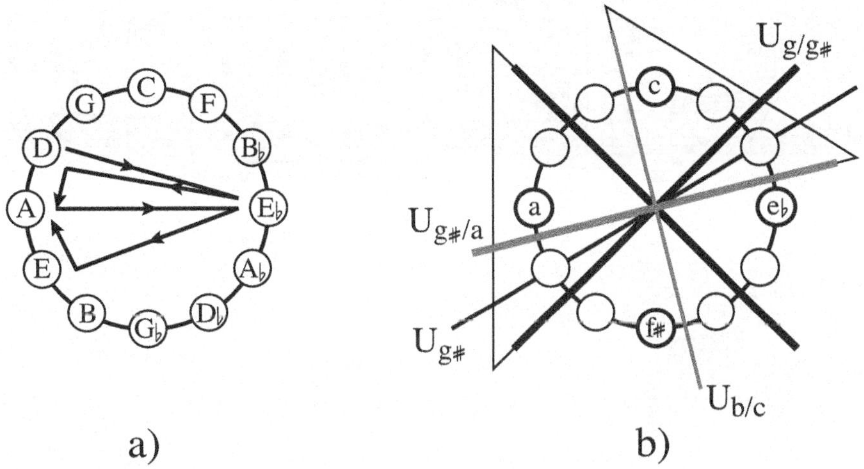

Fig. 28.5. Part a) of this figure shows the modulation paths within the circle of fourths. Here, $E_\flat$ is the pole to A, and to its neighboring tonalities D, E which are symmetrically positioned with respect to the polar axis $A, E_\flat$. The two semantic poles "night" (first $E_\flat$) and "life" (second $E_\flat$) are placed on distinguished tonality positions in front of the contextual tonalities D, A, E. Part b). The five modulators of the present modulations are partitioned into three groups: Two pairs of orthogonal axes, $U_{e/f}$ and $U_{g/g_\natural}$, as well as $U_{b/c}$ and $U_{g_\natural/a}$, and the axis $U_{g_\natural}$ which lies symmetrically with respect to these two pairs.

nonetheless a scientifically legitimate and valid approach, much as human physical models of nature. We would strongly suggest to set up comparative discourses between traditional music analysis, jazz analysis, and the mathematical approach. For obscure reasons, this type of discourse has been applied very rarely[2] . We would however prefer to lead such a discourse on the level of neutral analysis, and not under the petitio principii perspective that knowing the composer's innermost agitations (who will ever know that?!) preconizes a distinguished analysis. We would rather like to see the objective advantages in the understanding of what is written down in the score.

We want to study a piece in Debussy's "Préludes", Livre 1. It is No. 4 "Les sons et les parfums tournent dans l'air du soir", a title which is taken from Baudelaire's "Harmonie du soir", third verse. We want to deal with the modulation from $A_\flat$-major to A-major (this is what the signature suggests), bars 33-37, see Figure 28.6. The analysis starts on the last fourth (indication: *serrez...*) of bar 33 and terminates on the arpeggio at the beginning of bar 37.

The degree onsets are shown in Figure 28.7. In contrast to the preceding example, there is no problem here to immerse chords within triadic interpretations. Rather do we encounter two problems: the mixed variety of tonalities and the ambiguity of most onsets with respect to tonal specification. It is particularly remarkable that a cadence of A-major does not occur. Moreover, the rhythmically simplistic sequence of mixed chords, which alters between I_C, V_C, and V_A, I_A (bars 35 to 2/3 of bar 36, indication: *la basse un peu appuyée*) and are close to shocking in their strangeness to the context.

If we suppose that there is a modulation from $A_\flat$ to A, what could be the role of C? One can testify the presence of C by the cadence IV, V, in fact the mentioned I_C-V_C sequence, by the arpeggio of the chord $I_C \cup VI_C$ of the left hand, and by the melody line of the right hand. The role of C cannot be that of an intermediate tonality since C incessantly interplays with the target tonality. Also it is impossible that modulation targets to the intersection of A and C since the end clearly points away from C. The only

[2] An exceptional case is found in *Musiktheorie*, Vol. 4/2000.

Fig. 28.6. Claude Debussy: **Les sons et les parfums...** "Préludes", Livre 1, Nr. IV, [248] (with kind permission of the Henle Publishers).

possibility which remains is that C does not play the role of a modulatory station, but of a modulator, if we stick to the present model.

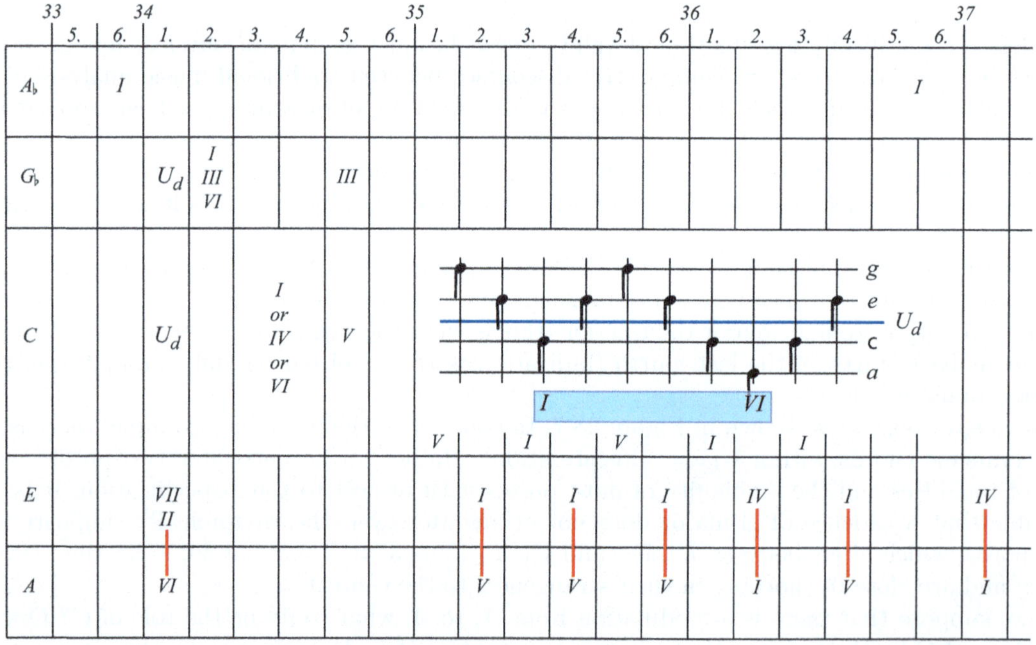

Fig. 28.7. In the modulation from $A_\flat$ to E (and A), the modulator, i.e., the inner symmetry of C and $G_\flat$ is evidenced by chordical and motivic representation of this double scale material.

The triadic interpretation $C^{(3)}$ has exactly one non-trivial inner symmetry, namely the inversion U_d. There is a hint for this symmetry in the line of the left-hand voice in bars 35-36: Here, we have the mentioned arpeggio of $I_C \cup VI_C$, a chord which is symmetrical with respect to U_d, a fact which is felt as a to-and-fro movement within the chordically completed arpeggio (indication to play the bass!). Moreover, we observe three tones $d_\flat, e_\flat, f$ at the second onset of bar 34, which lie in $IV \cup V$ of $G_\flat^{(3)}$, and the last onset of the same bar can also be specified as III of $G_\flat^{(3)}$. However, relating to symmetries, $G_\flat$ and C are equivalent. Also, regarding inner symmetries, $G_\flat$ and C are equivalent; i.e., the inversion U_d is 'represented' by both, C and $G_\flat$.

But if U_d is the modulator, then it operates via $A_\flat^{(3)} \rightsquigarrow E^{(3)} = U_d(A_\flat^{(3)})$, and not to target $A^{(3)}$. This should happen with the pivots II, III, V, VII or with II, IV, V, VII (second cadence option). This is exactly what happens in the beginning of bar 34: $II_E \cup VII_E$ appear as unambiguous pivots towards $E^{(3)}$. The other degrees which do not pertain to C or to $G_\flat$ are drawn in Figure 28.7. They are all in E as well as in A, except the last onset $I_{A_\flat}$ in bar 36, which—together with the onsets in $G_\flat$—concludes the tonality $A_\flat$ in an atmospheric flavor the tonality $A_\flat$ which is in symmetrical position to E.

This not only helps to understand the modulatory function of the tritonally related tonalities C and $G_\flat$, it also makes evident the ambiguity of the modulation—and thereby the charm of this passage: The piece modulated to E, but the cadence of E did not take place, one remains within the intersection $E \cap A$, and the first degree of A in bar 37 is a fact that can only be interpreted in an ambiguous way; it could also be IV_E. So tonality E has been introduced in a defined way, but without cadencing, rather by neutralizing it with respect to A.

In this view, the title receives a structural reading. The tones are really turned within the modulation (via U_d), the "parfume" of $A_\flat$ remains mixed with the "parfumes" of A and E until the end, it keeps returning like a memory.

28.2 Modulation in Beethoven's Sonata op.106, 1st Movement

Summary. This longer example illustrates complex connections of modulatory processes within the Allegro movement of the "Hammerklavier" Sonata. We give an interpretation and precision of the famous theses by Erwin Ratz [879] and Jürgen Uhde [1066] with respect to the overall architecture of the modulation processes.

– Σ –

28.2.1 Introduction

Summary. This introduction sketches the key position of the "Hammerklavier" Sonata in the history of European piano music.

– Σ –

In contrast to the preceding examples, this one will not deal with short modulatory passages but with *the overall organization of tonalities and architecture of modulatory passages.* The methods will be the same as in the preceding examples, but we shall only sketch the details—except for some punctual zooms. A detailed discussion, including motivic aspects ("motivisch-thematische Arbeit") and their connection to harmony can be found in [670]; some analysis of the motivic aspects have also been discussed in Example 22 of Section 14.2.

We want to investigate the Allegro movement in Beethoven's "Hammerklavier" sonata op.106 [96], a very famous and equally difficult late work which deaf Beethoven had composed 1817–1819. In his description of op.106 [517], Joachim Kaiser starts with the lapidary phrase: "Grösste Sonate der Musikgeschichte." The sonata's public premiere was executed in 1836 in Paris by Franz Liszt. But the "Hammerklavier" sonata is not only a technical and mnemotechnical challenge for the interpreter, it is also a mysterium for the sonata theory—developed by Adolf Bernhard Marx [659] in 1850s—which established the sonata form as an accepted norm.

This sonata form ("Sonatenhauptsatzform") is an architectural and processual scheme of a large musical form which is framed by a syntactic subdivision into exposition, development, recapitulation, and coda (see Figure 28.8).

The comparison with the classical sonata form will only be a starting point of our analysis of the present *individual* composition. It is well known that the sonata scheme is not adequate for more than a statistical estimation of really existing sonata movements. However as a starting point, we can use the scheme in order to evaluate knowledge and limits of traditional musicology with the complexity of tonal and modulatory structures of the present sonata.

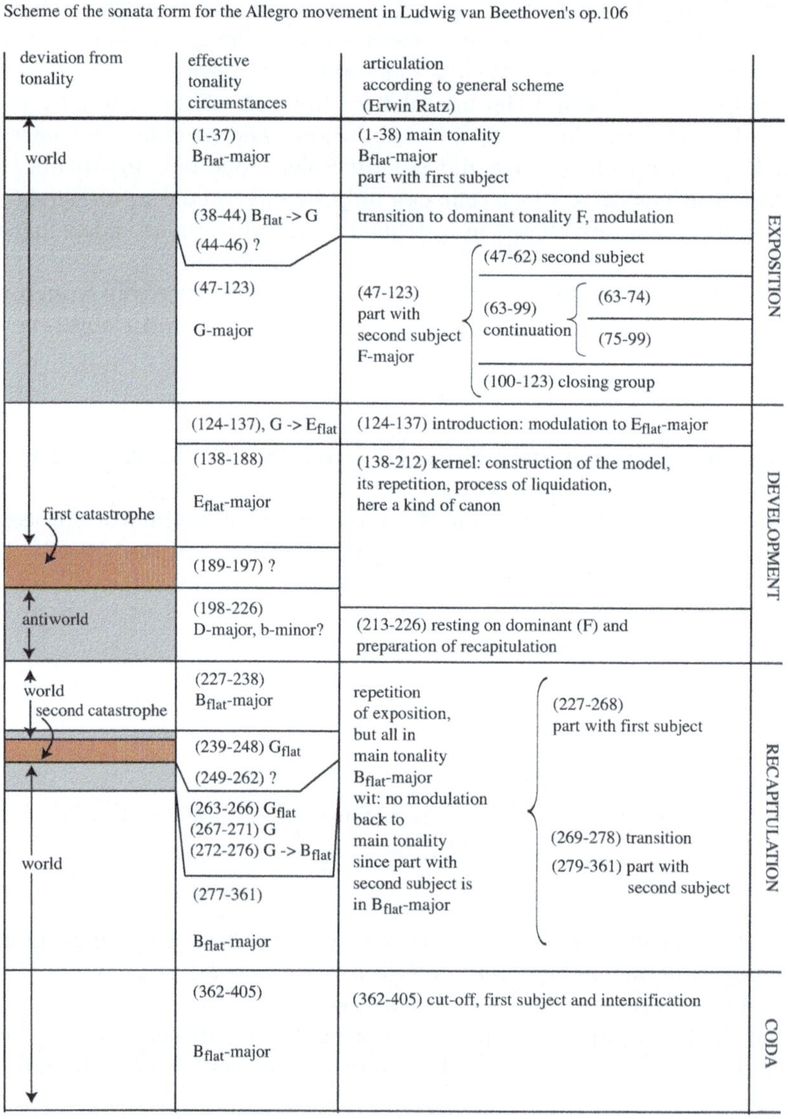

Fig. 28.8. The sonata scheme of the first movement of the "Hammerklavier" sonata op.106, compared to the normed sonata scheme.

Our analysis is not thought of as a contradiction but as a confirmation and elaboration of existing analyses. In particular, we take over the fundamental *world-antiworld thesis* of Erwin Ratz [879] and the *"catastrophe theory"* of Jürgen Uhde [1066] which relates to the specific modulatory role of the diminished seventh chords in the sonata. With regard to the formal explicitness of our model, the take-over will however include precision, completion, and unification of aforesaid world-antiworld thesis and catastrophe theory.

28.2.2 The Fundamental Theses of Erwin Ratz and Jürgen Uhde

Summary. This section presents the thesis of Ratz and its weltanschauung, as well as the thesis of Uhde. A restatement of these theses in group-theoretic terms is given. As its consequence, a particular modulation architecture is predicted. This will be verified in Section 28.2.3.

– Σ –

In [879], Ratz expressed the idea that the "empire of tonalities" in op.106 is polarized into a "world" around the pole $B_\flat$-major, the sonata's main tonality, and an "antiworld" around the counterpole b-minor. In [1066], Uhde supplemented this thesis in so far as two so-called "catastrophes" occur in the Allegro movement when the "world" is left in order to enter the "antiworld". Catastrophes are dramatic modulatory processes which differ significantly from normal modulations (see Figure 28.9). We discuss the following thesis:

A

B

Fig. 28.9. Ludwig van Beethoven: Two modulations in the first movement of op.106 [96] (with kind permission of the C.F. Peters Publishers, Frankfurt, Frankfurt/Main). The first (A), from G to $E_\flat$, is a common one, whereas the second (B), from $E_\flat$ to b-minor, is a catastrophe modulation in the sense of Uhde.

Thesis 4 *Let $\mathfrak{M}_0 = \{c_\sharp, e, g, b_\flat\} \subset PiMod_{12}$ the local composition of the diminished seventh chord. Then the modulation structure of op.106 is determined by the modulators in the sense of Theorem 30 which are contained in $Aut(\mathfrak{M}_0)$.*

What is the relation of this thesis with Ratz' world/antiworld? To begin with, the group $Aut(\mathfrak{M}_0)$ solves the following purely group-theoretic problem: We look for a *maximal* subgroup M of the group of inversions and translations TI_{12} on $PiMod_{12}$ which acts as modulator group on $Dia^{(3)}$, but under the restriction that no group element transforms $B_\flat^{(3)}$ to $D^{(3)}$. Here, the interpretation $D^{(3)}$ stands for the minor tonality b of ionian mode. Under the action of such a subgroup, the set $Dia^{(3)}$ is divided in at least two orbits (one for $B_\flat$, and one for b). A priori, it is not clear whether there are several such subgroups, and how many orbits such maximal subgroups will produce. But it turns out that the subgroup is uniquely determined, viz., $M = Sym(\mathfrak{M}_0) \cap TI_{12}$, these are the inversions and translations which leave $\mathfrak{M}_0$ invariant. We have the surjection $Sym(\mathfrak{M}_0) \to Aut(\mathfrak{M}_0)$, and it is easily seen that its restriction

$$M \to Aut(\mathfrak{M}_0)$$

is an isomorphism. So we may identify these groups. Moreover, there are exactly two orbits under this automorphism group, namely the world

$$W = Aut(\mathfrak{M}_0).B_\flat = \{B_\flat, D_\flat, E, G, A, C, E_\flat, G_\flat\},$$

and the antiworld

$$W^* = Aut(\mathfrak{M}_0).D = \{D, F, A_\flat, B\}.$$

We are now in state to make more precise Ratz' hypothesis: The tonalities of W are those of Ratz' "world", whereas the complementary set W^* defines Ratz' "antiworld". This casts our possibilities in two ways, we are given a precise dichotomy W/W^* and the admitted modulators, i.e., the elements of the automorphism group $Aut(\mathfrak{M}_0)$.

The modulatory situation here is that of a *restriction of modulators* to the group $Aut(\mathfrak{M}_0)$. What is the meaning of this condition for the modulation model? As long as the modulator which is described by the modulation theorem is contained in $Aut(\mathfrak{M}_0)$, there is no problem to apply the theorem and to look for pivots. But a modulation to the third circle of fourths, like $C \rightsquigarrow E_\flat$ or $A \rightsquigarrow C$ only admits translations in our restricted framework. In this case we must refrain from rigidity condition (3) (the triviality of the intersection $TI_{12} \cap Sym(T \cap Q)$) in property 1 of Section 27.1.4. In fact, this condition which guaranteed the uniqueness of the modulator is superfluous now: there is only this translation symmetry, no other candidate! But then, there is a corresponding theorem where the modulators are the translations $e^{\pm 3}$.

Therefore, one should modulate in the sense of restricted modulators and pivots within W and within W^*, whereas modulation between two tonalities living in different worlds should yield a catastrophe in the sense of Uhde, whaich means in particular that in such a situation, diminished seventh chords—which are the 'creators' of the catastrophe—should become visible at the surface of the score.

28.2.3 Overview of the Modulation Structure

Summary. We present the modulatory architecture and the modulations which split into "ordinary" cases and "catastrophes" in the sense of Uhde.

– Σ –

We now have to test these postulates in the total plan of modulations of the Allegro movement. The modulation plan looks like this:

$$\overset{W}{B_\flat} \underset{e^{-3}}{\rightsquigarrow} \overset{W}{G} \underset{U_g}{\rightsquigarrow} \overset{W}{E_\flat} \underset{!}{\rightsquigarrow} \overset{W^*}{D/b} \underset{U_{d/d_\sharp}}{\rightsquigarrow} \overset{W^*}{B} \underset{!}{\rightsquigarrow} \overset{W}{B_\flat} \underset{U_{b_\flat}}{\rightsquigarrow} \overset{W}{G_\flat} \underset{U_{a_\flat/a}}{\rightsquigarrow} \overset{W}{G} \underset{e^3}{\rightsquigarrow} \overset{W}{B_\flat}.$$

In order to view these modulations on the circle of fourths and relative to the world/antiworld dichotomy, see Figure 28.10. For each modulation, the modulators from $Aut(\mathfrak{M}_0)$ are indicated. Here are the single modulations in detail:

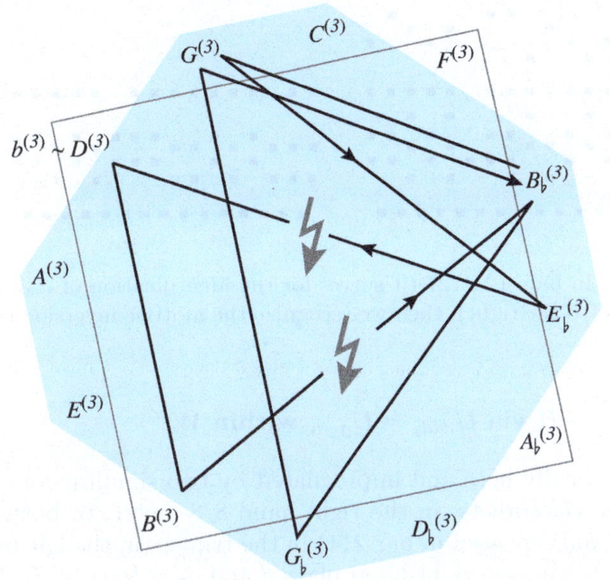

Fig. 28.10. The graph shows the modulation plan in the Allegro movement of Beethoven's op.106 in the tonality system arranged on the circle of fourths. The start switches from $B_\flat$ to G. The inverse modulation occurs at the end, and both follow the same procedure, see Sections 28.2.4 and 28.2.10.1. Except these initial and terminal movements, the modulation plan is perfectly symmetric around the symmetry axis between $B_\flat^{(3)}/E_\flat^{(3)}$ and $A^{(3)}/E^{(3)}$.

28.2.4 Modulation $B_\flat \rightsquigarrow G$ via e^{-3} in W

The first modulation $B_\flat \rightsquigarrow G$ in the transition (bars 39-46) to the second subject could in principle be performed by use of a "pedal modulation" [948]. We do however not encounter this modulation, but 'merely' a sequence of VII_G-degrees whose top notes are shifted by minor thirds from each other, i.e., exactly the situation of the pivot VII and the third translation, as predicted by the modulation under restricted modulators.

28.2.5 Modulation $G \rightsquigarrow E_\flat$ via U_g in W

This modulation is bipartite (first part: bars 124-127, second part: bars 128-129). Before we encounter the pivots $VII - V - VII$ of $E_\flat$ according to modulation table Appendix P.1 in part two, we hear tone g as an octave interval: pedal and stationary voice. Here the inversion at g is made evident (Figure 28.11). On the one hand, this section is a cadence of c_{mel} whose inner symmetry is U_g. One recognizes a bipartite contrapuntal motive structure of which the second part (bars 126-127) is the inversion of the first part (bars 124-125) in pitch $c_\sharp$. But modulo octave (in the pitch class space), this inversion is the same as U_g. Since $c_\sharp$ is not contained in the present scales, it is delegated to the octave in g. This means that the modulator U_g is motivically evidenced in this first modulatory section preceding the pivots and the cadence.

28.2.6 Modulation $E_\flat \rightsquigarrow D/b$ from W to W^*

This modulation is a catastrophe in the sense of Uhde since it leads to the antiworld W^*. As we may recognize already from the score (B) in Figure 28.9, bars 189-197 are of a dramatic shape. Any elaborate motivic, rhythmic or harmonic effort is postponed in favor of a pertinent presentation of diminished seventh chords. An approach to modulation $VI_{E_\flat}, I_D$ (bars 189-192) fails, the resolution of all alteration signs indicates the exit from tonal space. We hear the "generator" of the catastrophe, the diminished seventh chord as such.

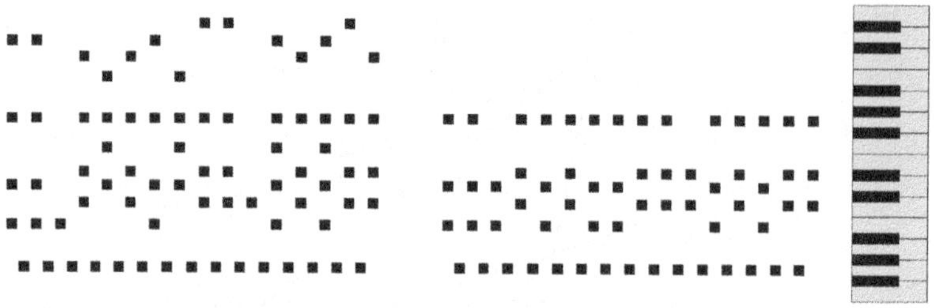

Fig. 28.11. If we omit tone f in bars 124-127 (it serves for the identification of c_{mel}) and transpose all pitches into one octave between the two gs (to the right), then we recognize the motivic inversion symmetry between bars 124-125 and bars 126-127.

28.2.7 Modulation $D/b \rightsquigarrow B$ via $U_{d/d_\sharp} = U_{g_\sharp/a}$ within W^*

The process is resolved chordically here and impregnated by two simultaneous local meters (Figure 28.12): in the left hand a 6/4 meter via triplets, in the right hand 8/8 meter. In both expressivo indications (bars 205, 209) appears (bar 207, and repeated in bar 211) in the triplets in the left hand the motif $f_\sharp - b - f_\sharp$ and $b - f_\sharp - b = U_{d/d_\sharp}(f_\sharp - b - f_\sharp)$. And the right hand plays d and $d_\sharp = U_{d/d_\sharp}(d)$ in bars 205-209 as well as e and $c_\sharp = U_{d/d_\sharp}(e)$. In bar 209 appears at the beginning degree I_B and after that in the right hand a chromatic sequence which comprises the seven pitches $B^* = \{c_\sharp, d, d_\sharp, e, f, f_\sharp, g\}$. Now, $B = U_{d/d_\sharp}(D) = e^5.11(D)$. But by a concatenation with the fifth circle symmetry we produce $e^5.5 = U_{d/d_\sharp}.e^0.7$ out of $U_{d/d_\sharp}$. This result lies in $Sym(\mathfrak{M}_0)$ (!), and we exactly get

$$B^* = e^5.5(B),$$

so that we really are situated in B—up to an inner symmetry of $\mathfrak{M}_0$.

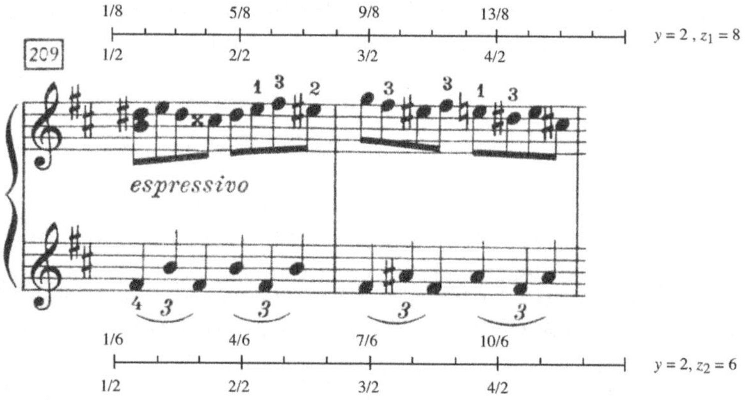

Fig. 28.12. Example of two simultaneous local meters, corresponding to the left and right hand in bars 209-210 of the first movement of Beethoven's op.106 [96].

28.2.8 Modulation $B \rightsquigarrow B_\flat$ from W^* to W

The next modulation leads us back from the antiworld W^* to the world W (bars 214-226). Corresponding to the short stay in W^* the return is an easy business. There follows a sequence of arpeggiated intervals which is strongly based upon $\mathfrak{M}_0$ and ends on $I_{B_\flat}$:

twice	$e, c_\sharp$ within $\mathfrak{M}_0$
six times	g, e within $\mathfrak{M}_0$
four times	$g_\sharp, e$
four times	$b_\flat, e$ within $\mathfrak{M}_0$
four times	$b_\flat, f$ $I_{B_\flat}$

28.2.9 Modulation $B_\flat \rightsquigarrow G_\flat$ via $U_{b_\flat}$ within W

For this modulation within the world W, we can apply the normal modulator. The modulation is a fast one (bars 238-239), although the involved tonalities $B_\flat$ and $G_\flat$ are separated by four fourths. At the end of bar 238, the neutral degree $I_{B_\flat}$ is followed by its inversion $VI_{G_\flat} = U_{b_\flat}(I_{B_\flat})$, so the modulator is again put into evidence. Immediately after that, at the beginning of bar 239, follows $\{b_\flat, f, a_\flat\}$ which lies in $III \cup V$ of $G_\flat$ and corresponds to the pivots.

28.2.10 Modulation $G_\flat \rightsquigarrow G$ via $U_{a_\flat/a}$ within W

This second Uhde catastrophe (bars 249-262) is highly dramatic. At the beginning (bars 249-250), we remain within the large orbit W: I of $G_\flat$. As with the first catastrophe, we then encounter a pronounced series of diminished seventh chords. In bars 259-262, it terminates in the intersection $B \cap G_\flat$: we do not know whether the change or orbits has been successful or not. The decision is only taken in bar 263 with pitch f which is not in B, but in $G_\flat$. So we actually did not really leave the world, at least not unambiguously. This 'delusion' is particularly refined since we hear the "fanfare" with III_G which could also be viewed as I of b-minor. But this third degree corresponds to a pivot of our model, the inversion of the bass in the moment of the forte onset at the end of bar 266: $b = U_{a_\flat/a}(f_\sharp)$ shows the same symmetry as for the above modulation $D/b \rightsquigarrow B$ within the antiworld after the first catastrophe. So we do not move from B to D as it is suggested by the change of signature in bar 267 and from the pitch material before the appearance of f in bar 264. This superficial impression is an "allusion" to the antiworld situation, unveiled however as being an illusion and is resolved in a consistent way. We have moved from $G_\flat$ to G.

28.2.10.1 Modulation $G \rightsquigarrow B_\flat$ via e^3 within W

This last modulation proceeds completely regularly according to the scheme that we already know from the inverse modulation in the transition to the second subject as discussed above.

28.3 Rhythmical Modulation in "Synthesis"

Summary. The modulation model does not restrict to time from its logical structure. This fact is exploited in a composition of rhythmical modulation for percussion ensemble and piano.

– Σ –

The harmonic modulation model is based upon the pitch class construction modulo octave, i.e., the space $PiMod_{12}$ based on the module $\mathbb{Z}_{12}$ derived from the semi-tone pitch space based on the integers, modulo the submodule generated by the octave quantity 12. We have viewed the tonalities as being scales, together with their triadic interpretations by the seven well-known degrees. But there is nothing substantial to the choice of these spaces. The mathematical framework is completely insensitive to the forms and denotators which implement the underlying parameter and class spaces.

So why not apply the modulation theory to a concept framework which is mathematically the same as for harmony, but which is semantically different, more precisely: the new approach presents a space and its denotators in the context of onset time and its rhythms.

We shall develop this switch from pitch to time in the next section and apply this theory to a rhythmical modulation in movement No. 1 of Guerino Mazzola's jazz concert "Synthesis" [681] for piano, large percussion ensemble, and e-bass.

Before delving into technicalities, we should however observe a fundamental difference between two time qualities involved in musical composition. We have known time as a form related to onset and duration. These dimensions are part of what defines events in scores. They are comparable to other event parameters such as pitch, loudness etc. A local or global composition is a configuration of events which represent the composition's substance. This material data is without any logical or strategic specification. It does not include the composer's poietic construction plans or the analyst's structural evaluation.

But modulation is a structure that is not only defined on the level of the effective neutral, pivotal, and cadence degrees, it is rather built on the tripartite strategy Neutralization—Turning Point—Cadence as defined by Schönberg. This is a logical construction on the syntactical level: a sequence of three functional units in the syntagmatic string of musical development. Like abstract logical schemes, such as modus barbara ("a implies b" and "b implies c" implies "a imples c"), this is not a priori a syntagmatic string in the material musical time. It regards logical time rather than material time. In harmonic modulation, we have three logical stages in the pitch domain, and their unfolding on the material time line is only a representation of an abstract process, not the process as such.

Substantially, in harmonic modulation we have an excellent example for Augustinus' definition of music as an art of instantiation of good rational strategies. As already mentioned in the previous discussion of Augustinus' definition, modulation could also regard rhythmical strategies, for example. It is very important to distinguish logical time from material time in this case since here, the syntagmatic string which embodies the logical time is superposed to the material time of rhythmical structures.

28.3.1 Rhythmic Modes

Summary. This section describes the transfer of harmonic modality to rhythmic categories, and from this derives the modulations in the rhythm domain.

– Σ –

The transfer of harmonic dimensions to rhythmic ones must deal with the semantic specifications of these dimensions. In fact, pitch is a space category which carries a strong connotation of sound quality, i.e., of instrumental realization. No abstract pitch quality has ever been used in harmony as soon as realistic music parameters are to be described. We could of course say that the abstraction from instrumental realities is only a question of habituation, and that abstract time could also be thought of in an abstract comprehension which already worked for harmony.

However, it is easy to imagine a more or less good orientation in the pitch space whereas it is difficult to imagine an orientation in an abstract time space, since the distinction between later and earlier time onsets is only relative and risks failure if no supplementary time-dependent structures can be found. Moreover, the superposition of logical and material time layers asks for a clear marking of material time events.

This is why we would like to add supplementary, logically redundant markers of time such that one can use such redundancies to shape time in a rhythmical modulation theory. The idea is as follows: Suppose the onset time is parametrized by integers, $Onset_{\mathbb{Z}} \underset{Id}{\longrightarrow} \mathbf{Simple}(\mathbb{Z})$. Suppose further (in analogy with the octave period) that we have the translation e^{12} in time on $Onset_{\mathbb{Z}}$ in the sense of a time period. This could be realized by a 12/8 meter, for example, where the eighth duration stands for the unit time in $Onset_{\mathbb{Z}}$. As a marker parameter, we select the form $Percussion \underset{Id}{\longrightarrow} \mathbf{Simple}(\mathbb{Z})$ whose zero-addressed denotators are percussion instruments as represented in a list with integer indexes. This could for example be a list of percussive sounds in an MIDI-environment where the numbers stand for the program changes (see below for more concrete setups). Our events are situated in the combined space

$$PerOns \underset{Id}{\longrightarrow} \mathbf{Simple}(\mathbb{Z} \oplus \mathbb{Z}) \stackrel{\sim}{\rightarrow} PO \underset{Id}{\longrightarrow} \mathbf{Limit}(Onset_{\mathbb{Z}}, Percussion)$$

and will be called *percussion events* in this context. But we want more, i.e., macro percussion events, in order to construct profiled markers. So we consider these macro percussion event spaces as they were introduced in Section 13.4.3.1:

$$\begin{aligned} &Knot_{PerOns} \xrightarrow[Id]{} \mathbf{Limit}(PerOns, Makro_{PerOns}) \\ &\text{with } Makro_{PerOns} \xrightarrow[f:F \xrightarrow{\sim} 2^{FK} \rightarrowtail \Omega^{FK}]{} \mathbf{Power}(Knot_{PerOns}) \\ &\text{and } F = Fun(Makro_{PerOns}), FK = Fun(Knot_{PerOns}). \end{aligned}$$

A macro rhythmic germ R^* is a finite local composition in the ambient space $Knot_{PerOns}$. And an infinite macro local Para-rhythm is paraphrased by the 12-periodic local composition $e^{[-\infty,\infty]12}R^*$. In the sequel, we shall only consider germs R^* which are in bijection to their time projections R^*_{Ons}, and such that R^*_{Ons} are contained in the period interval $[0, 12[$. This means that we may equivalently consider the rhythm classes modulo the time period 12. These objects are the residual classes of the macro rhythmic germs R^* which we now identify with the germs because of our additional assumptions. So we are working in the macro class space:

$$\begin{aligned} &Knot_{PerOns_{12}} \xrightarrow[Id]{} \mathbf{Limit}(PerOns_{12}, Makro_{PerOns_{12}}) \\ &\text{with } Makro_{PerOns_{12}} \xrightarrow[f:F \xrightarrow{\sim} 2^{FK} \rightarrowtail \Omega^{FK}]{} \mathbf{Power}(Knot_{PerOns_{12}}) \\ &\text{and } F = Fun(Makro_{PerOns_{12}}), FK = Fun(Knot_{PerOns_{12}}), \end{aligned}$$

deduced from

$$PerOns_{12} \xrightarrow[Id]{} \mathbf{Simple}(\mathbb{Z}_{12} \oplus \mathbb{Z})$$

in an analogue construction to scales in pitch class spaces. Intuitively, these macro rhythmic germs in $Makro_{PerOns_{12}}$ are just time scales whose points are loaded with a rhythmic satellite object each.

As to modulation theory with such objects, we can apply the modulation model for 12-tempered tuning, but we have to take care of the satellites! There may be many rhythmic scales with the same time projection! This means that the translation and time-inversion symmetries will have to carry over the (unaltered) satellites. This is as if the harmonic modulation would be carried out with "colored" pitch classes, whereas the inversion of pitch classes would preserve colors.

28.3.2 Composition for Percussion Ensemble

Summary. This final section discusses the concrete realization of the rhythmical modulation model, essentially by use of a large ensemble of percussive instruments.

– Σ –

The composition in question here is a rhythmical modulation in movement No. 1 of Mazzola's jazz concert "Synthesis" [681]. The modulation takes place after the exposition in the sonata scheme of this movement (3:18-5:48 on piece # 1 of [681]). This developmental start is written in 12/8 measure, in fact the measure which we need for the rhythmical modulation in a 12-periodic scheme as exposed above. The modulation is built upon the rhythmic scale (the time projection of the macro germ) corresponding to the complement $\widehat{62}$ of No. 62 the chord classification list in Appendix N.1. The macro germ G^* above $\widehat{62}$ has the following shape. For each onset $x \in \widehat{62}$, we have a corresponding denotator $((x, p_x), Sat_x)$ of form $Knot_{PerOns_{12}}$. The p_x-coordinate is an integer value for a percussion sound, whereas the satellite Sat_x is a zero-addressed denotator of form $Makro_{PerOns_{12}}$. We choose for each satellite a zero-addressed three-element motif of form $PerOns_{12}$,

$$Sat_x = \{(0,0), (do^1_x, dp^1_x), (do^2_x, dp^2_x)\},$$

with first element the origin $(0, 0) \in \mathbb{Z}_{12} \oplus \mathbb{Z}$ (see also Figure 28.13). The choice of these three-element

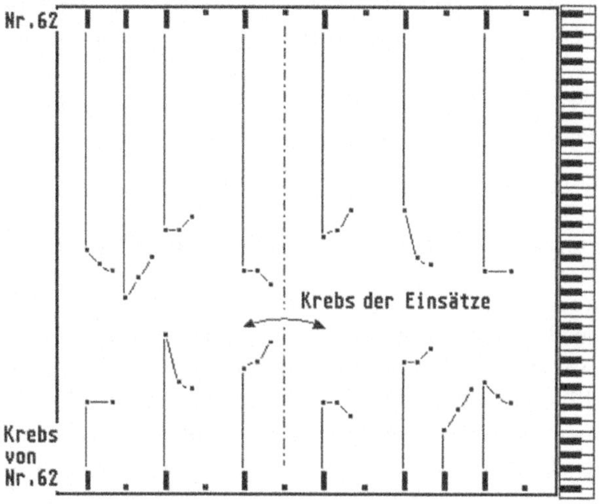

Fig. 28.13. Above: The rhythmic macro germ G^* of the rhythmical modulation in the first movement of the "Synthesis" concert [681]. Below: The retrograde of G^* with the satellites being carried over to the transformed onset-percussion events.

satellites is due to the general construction principle of the "Synthesis" concert, i.e., the use of the 26 isomorphism classes of three-element motives for all melodic and rhythmic structures. This principle was already encountered in Section 11.6.3 where we discussed the third movement of the concert.

The modulation goes from the rhythmic macro germ G^* to its symmetric image $H^* = R(G^*)$ under the retrograde motion R which fixes the seventh tone (first tone of the seventh degree). The lower half of Figure 28.13 shows the retrograde germ with the transported satellites. This is the germ for the "rhythmic target scale". According to the modulation theory (well-tempered case), there is a determined set of pivots related to a selected cadence in the target scale, and we can get off with the explicit construction.

In the "Synthesis" concert, we first set up six bars in order to define the start rhythm scale. In each bar 1-5, a new tone of the start rhythm scale is added, and two tones are added in bar 6, so we have the complete scale. Observe that this addition of tones means that each added tone can be repeated in the successive bars and thereby enriches the previous rhythms.

From bar 7 to bar 12, the tones of G^* are successively removed so that we have a neutralization process here. The modulator is made evident by a rhythmic motif which is built around the above retrograde symmetry R, including its repetition during the whole second modulation segment presenting the pivots, from bar 13 to bar 23.

Already in bar 21, the new tonic of H^*, which is (by definition) the image of the old tonic—with its characteristic rhythmic motif satellite—, is played and remains alive until the end of the modulatory process.

This means that the new first onset is no more the old one, but occurs on onset 9/8 (modulo 12/8) of the old bar onset. This change is realized by a new bar onset at onset 9/8 of bar 24. From this point on, we have 5 bars to confirm the target rhythm and to terminate the modulation.

Both cadences, that of the start and that of the target rhythm, are accompanied by a kind of regular falling drop sequence of two eights duration tones in order to stress the beginning of the respective bar units.

Recall from Section 13.4.3.1 that the macro rhythms can be flattened down to the sets in the effective form $PerOns_{12}$ of rhythmic events. But the construction of the effective material had to be performed on a higher conceptual level of macro events. Probably this kind of conceptual upgrading of modulatory processes could also be used to understand existing and construct new harmonic modulations, or even melodic modulations if a modulation model of motivic structures (in $OnPiMod_{n,m}$, say) is available.

Part VII

Counterpoint

29

Melodic Variation by Arrows

punctus est cuius pars nulla est
Euclid [455, book I]

Summary. The ideas of tangent objects described in Section 7.5 are applied to define contrapuntal intervals as tangential "arrows" from the cantus firmus to the discantus tone. This formalism fits with the idea that the discantus is a kind of *melodic variation around the cantus firmus line.* The ring structure of the set of such arrows is discussed and motivated from the musical perspective.

– Σ –

The mathematical theory of counterpoint is an excellent subject to illustrate the idea of mathematical conceptualization of musical and musicological objects. It shows that mathematical subtleties are no formal overhead and can help to grasp music(ologic)al differences at the highest level. After the introduction of arrows as a formal restatement of the contrapuntal interval concept, we shall see in Section 30.2.1 that this setup provides us with an astonishing relation between harmonic and contrapuntal objects, a relation which was never observed by musicological tradition.

This insight could have deep consequences for the understanding of the hitherto unresolved transition process from polyphony to homophony, from counterpoint to harmony as a basis of musical composition.

29.1 Arrows and Alterations

Summary. Arrows are a conceptual refinement of ordered intervals in the pitch domain. We compare these and related concepts and work out their specific differences.

– Σ –

Most of the common modules of simple space forms describe aspects of elementary music objects which can be related to an abstraction from physical events. However, musical reflection is a mental and symbolic approach, and this is not only the case on the higher level of compound object interaction as it was developed in the theory of global compositions. We already observe such a symbolism on the most elementary understanding of what we mean when talking about a tone.

We are referring to the fact that in systematic musicology, some German specialists (see e.g., [436]) distinguish between "Ton" (tone) and "Tonort" (a tone's place), when discussion alterations. For example, in the 12-tempered system, the Tonort of $f_\sharp$ is the same as the Tonort of $g_\flat$, whereas these two tones are thought to be different since their origin, one f, one g, expresses a musical thinking that cannot be traced anymore on the level of the common Tonort.

This understanding is based on the concept of musical objects as being grown, dynamic entities and not merely static states. In this approach, a tone is not a point but a *variation* of a point, something that has a

G. Mazzola, *The Topos of Music I: Theory*, Computational Music Science,
https://doi.org/10.1007/978-3-319-64364-9_29

direction from past to present, in short: an 'arrow'. This unveils a fundamental stream of musical thinking: that of varied structures.

In this part, we want to discuss a central instance of this type of thinking, namely counterpoint. The objective is to interpret contrapuntal intervals as "arrows" and to deduce the basic rules of counterpoint. We cannot present a general theory of varied structures yet. Instead, we shall give a prototypical sketch of how such a theory could look.

In the following, we suppose that we are working in a form space S_q of simple type, module[1] M and identifier automorphism $q : M \xrightarrow{\sim} M$ (giving rise to the functorial identifier automorphism $@q$). For example: $S_{Id} = EulerModule$ with $M = \mathbb{Q}^3$, or $S_1 = PiMod_{12}$ with $M = \mathbb{Z}_{12}$, or $S_{(7)} = FiPiMod_{12} = PiMod_{12,(7)}$, the fifth identifier pitch class space, etc. In contrapuntal reasoning, one is not primarily interested in single pitches, but in "arrows" starting from a basic pitch x: the *cantus firmus* and ending in a target pitch y, the *discantus*. In order to catch this information, one could come back to the arrow approach exposed in Section 7.2.3, i.e., to the $\mathbb{Z}$-addressed denotators $D : \mathbb{Z} \rightsquigarrow S_q(ON = x, OFF = y)$. For the following reason we refrain from this point of view[2]: The $\mathbb{Z}$-addressed denotators have too little canonical algebraic structure for our purposes. We shall see this in a moment.

The formalization we choose for counterpoint is the dual numbers approach described in Section 7.5. The module M is enriched to the dual numbers module $M[\varepsilon]$, and we also admit the presence of the identifier module automorphism $q : M \xrightarrow{\sim} M$ which induces the canonical automorphism $q[\varepsilon]$ of $M[\varepsilon]$; call the corresponding space $S_q[\varepsilon]$. For example we would have $EulerModule_q[\varepsilon] \underset{@q}{\longrightarrow} \mathbf{Simple}(\mathbb{Q}^3[\varepsilon])$, more generally, with the notations of Appendix G.5.3:

$$S_q[\varepsilon] \underset{Id(S_q)[\varepsilon]=@(q[\varepsilon])}{\longrightarrow} \mathbf{Simple}(coord(S_q)[\varepsilon])$$

for an automorphism q of the coordinator module $coord(S_q)$. In this setting, if T is an address, a *T-addressed contrapuntal interval of space form S_q* is a T-addressed denotator $D : T \rightsquigarrow S_q[\varepsilon](x + \varepsilon.i)$, $x + \varepsilon.i \in T@M = T@coord(S_q)$. If $q = Id_M$, we also omit the q-index, and without any further address specification, we tacitly suppose the zero address. There are four canonical surjections [3]

$$p_{cf}, p_{int}, \alpha_+, \alpha_- : S_q[\varepsilon] \to S_q$$

associated with the first (cantus firmus) and second (interval) projections $M[\varepsilon] \to M$ and with the synonymous alterators of sweeping and hanging orientation α_+, α_- introduced and motivated musicologically in Section 7.5. In Section 29.5 we shall also consider more general addresses (in particular self-addressed contrapuntal intervals), but for the time being, we stick to the zero-addressed contrapuntal intervals.

The enrichment of the dual number modules $M[\varepsilon]$ does not lie in the R-linear structure, it is based on the R-algebra structure of $R[\varepsilon]$ with $\varepsilon^2 = 0$. In other words, the well-known possibility of address killing

$$X@B\tilde{@}S_q = X \boxtimes B@S_q$$

described in Section 11.2 could lead to an identification of $\mathbb{Z}$-addressed arrows in S_q with zero-addressed arrows in $\mathbb{Z}@S_q$, but we need an additional algebraic structure on the latter form space.

29.2 The Contrapuntal Interval Concept

Summary. This section presents a formal definition of contrapuntal intervals or *arrows*, together with the two possible orientations of *sweeping* and *hanging* counterpoint.

[1] Always over a commutative ring R in this context.

[2] This point of view is however the adequate for harmonic considerations as we have already learned in the self-addressed theory.

[3] Pay attention to the fact that these surjections are operating on the form functors and not on the frames.

– Σ –

The point of departure of this theory is the fact that the known physical and mathematical concepts of consonance and dissonance are not adequate to the musical paradigm of an antagony of interval categories since they only conceptualize gradual changes in sonance. This unsatisfactory situation was also recognized by Hugo Riemann [906] in his critique of Euler and Helmholtz. Our problem is to turn this antagonistic paradigm into a mathematical concept framework in order to deduce the announced counterpoint model.

Our requirements would not be satisfied if we thought of an interval as being a set of two pitch events. This would in particular not do justice to the concept of voices. The cantus firmus could not be distinguished from the discantus, and the crossing of voice could not be conceived. It is adequate to contrapuntal reasoning to distinguish a basic cantus firmus tone from the dependent discantus tone, i.e., to consider arrows as defined above, and to indicate the orientation.

In the sense of this musical requirement, we shall now describe *an oriented contrapuntal interval* formally as a pair

$$(\alpha_{\pm}, D : 0 \rightsquigarrow S_q[\varepsilon](x + \varepsilon.i))$$

of a sweeping or hanging orientation plus a contrapuntal interval D as defined above. When the rest is clear, we shall also reduce the description to the simple pair $(\alpha_{\pm}, x + \varepsilon.i)$ or even $x + \varepsilon.i$, if the orientation is clear. Whereas here, x is the cantus firmus of the contrapuntal interval, the quantity $x \pm i$ is the discantus of the interval.

Within a given orientation, the two-part style "note against note" is now interpreted as a sequence

$$x_s + \varepsilon.i_s, s = 1, 2, 3, \ldots$$

of arrows. If a change of orientation happens, the indexes have to be split into regions of constant orientation. The historical development of the counterpoint of many parts is now reflected in the historically variable characteristic types of geometric variations of the Gregorian chant of the cantus firmus' melodic line by arrows. Other tone attributes, such as duration or loudness, are neglected in this elementary motivic exposition. Except some minor remarks concerning just pitch spaces, we shall focus our discussion to the 12-tempered pitch class space $PiMod_{12,q}$ and its contrapuntal space $PiMod_{12,q}[\varepsilon]$ of pitch class arrows, q being an automorphism , e.g., one of the four circle automorphisms $q = (1), (5), (7), (11)$, of $\mathbb{Z}_{12}$. This is quite congruent with the contrapuntal reasoning. For example, the octave extension of a perfect consonance is again perfect, idem for imperfect consonances and for dissonances [924]. So after this reduction we are left with 144 contrapuntal intervals for each orientation, the arrows in $PiMod_{12,q}[\varepsilon]$.

29.3 The Algebra of Intervals

Summary. The set of arrows is canonically provided with the structure of an algebra. The formal definition and first mathematical properties are exposed.

– Σ –

The module $\mathbb{Z}_{12}[\varepsilon]$ of $PiMod_{12,q}[\varepsilon]$ is not only a $\mathbb{Z}_{12}$-module but also a $\mathbb{Z}_{12}$-algebra, i.e., a module with a bilinear, associative and unitary multiplication defined by the nilpotent element ε with $\varepsilon^2 = 0$ (see also example 106 in Appendix D.1.1). The product of contrapuntal intervals $a + \varepsilon.b, c + \varepsilon.d$ is defined by

$$(a + \varepsilon.b).(c + \varepsilon.d) = ac + \varepsilon.bc + \varepsilon.ad + \varepsilon^2.bd = ac + \varepsilon.(bc + ad)$$

which yields a new contrapuntal interval. An invertible element of $\mathbb{Z}_{12}[\varepsilon]$ is an element $a + \varepsilon.b$ with $a = 1, 5, 7, 11$, and the inverse element is $(a + \varepsilon.b)^{-1} = a - \varepsilon.b$. Corresponding to the symmetries in $\mathbb{Z}_{12}$, the symmetries in $\mathbb{Z}_{12}[\varepsilon]$ are the affine endomorphisms (over the ring $\mathbb{Z}_{12}[\varepsilon]$)

$$e^{a+\varepsilon.b}(u + \varepsilon.v)$$

which is invertible iff $u = 1, 5, 7, 11$.

29.3.1 The Third Torus

Summary. The third torus is the geometric representation of $\mathbb{Z}_{12}$ associated with its Sylow decomposition[4] . We discuss the mathematical structure and its metrical properties.

– Σ –

The module isomorphism $T : \mathbb{Z}_{12} \xrightarrow{\sim} \mathbb{Z}_3 \oplus \mathbb{Z}_4 : z \mapsto (z(mod3), -z(mod4))$ with its inverse $T^{-1}(z_3, z_4) = 4.z_3 + 3.z_4$ identify the cyclic module $\mathbb{Z}_{12}$ with a discrete torus. Under this isomorphism, we have an isomorphism of forms $T : PiMod_{12} \xrightarrow{\sim} PiThirds_{3,4}$ with the third class form $PiThirds_{3,4} \underset{Id}{\longrightarrow} \mathbf{Simple}(\mathbb{Z}_3 \oplus \mathbb{Z}_4)$. Hereby, the first component of a T-transformed pitch class is the number of its major thirds while the second component is the number of minor thirds that add to the given pitch when counted from zero. For example, $T(7) = (1, 1)$, i.e., "fifth = major third plus minor third", see Figure 29.1.

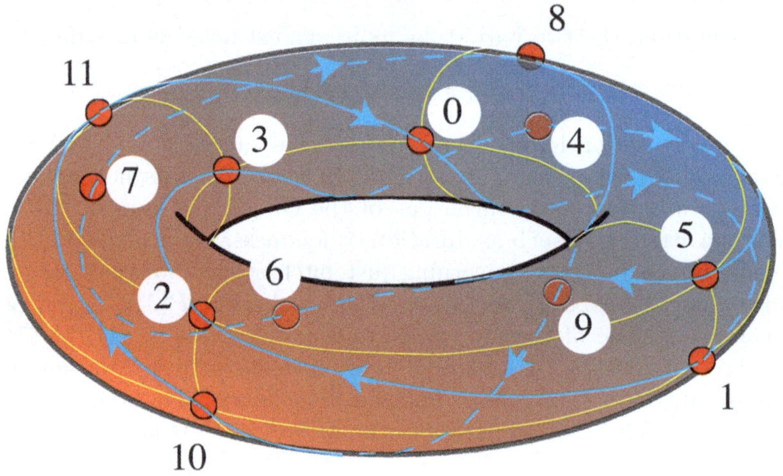

Fig. 29.1. On the torus of thirds, the third relations among pitch classes and intervals are represented in a geometric way. The sequence of semi-tone steps appears as an entwined closed spiral.

To describe 'pure intervals' in pitch classes, starting from the form

$$IntMod_{12} \underset{Id}{\longrightarrow} \mathbf{Syn}(PiMod_{12})$$

for interval quantities in pitch classes, we also use the third torus structure as a synonym for interval quantities in terms of thirds, i.e.,

$$IntThirds_{3,4} \underset{Id}{\longrightarrow} \mathbf{Syn}(PiThirds_{3,4}).$$

[4] See Theorem 49 in Appendix C.3.

As in differential geometry, one has the space of contrapuntal intervals $IntMod_{12,q}[\varepsilon]$ or the corresponding third tangent torus

$$IntThirds_{3,4,q}[\varepsilon]$$

in terms of a cantus firmus torus where twelve "tangent" tori are attached: The contrapuntal intervals are viewed as tangents to their cantus firmus points: A tangent torus $I_x = x + \varepsilon.\mathbb{Z}_{12}$ is attached at each of its points x, see Figure 29.2.

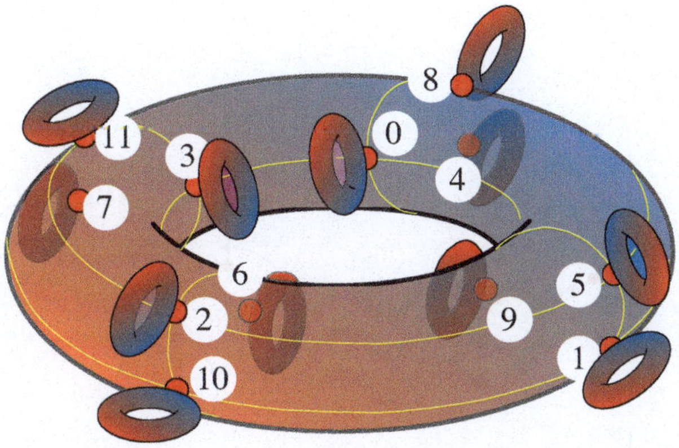

Fig. 29.2. The contrapuntal intervals viewed as tangents to their cantus firmus points.

Let us review the affine automorphism group $\overrightarrow{GL}(\mathbb{Z}_{12})$ in terms of the Sylow representation of $\mathbb{Z}_{12}$ as a discrete torus. The group is evidently generated by these four symmetries:

$$c_1 = e^0.11 \text{ multiplication by } -1; \text{ inversion}$$
$$c_2 = e^0.5 \text{ multiplication by 5; make fourth circle}$$
$$c_3 = e^3.1 \text{ addition of constant 3; minor third chain}$$
$$c_4 = e^4.1 \text{ addition of constant 4; major third chain}$$

This system is not only musically meaningful and therefore turns the group $\overrightarrow{GL}(\mathbb{Z}_{12})$ into a musically meaningful group by the concatenation principle 2, it also shows that all automorphisms preserve the third distance in the following sense: Let $x, y \in \mathbb{Z}_3 \oplus \mathbb{Z}_4$. Define $d(x,y)$ as the minimal number of ascending or descending minor or major third steps on the third torus to move from x to y; for example: $d(0,9) = 1$ since $0 - 3 = 9$, $d(0,1) = 2$ since $0 + 4 - 3 = 1$, etc. The distance is just the minimal number of edges on the discrete third torus that connects the two points x and y. This distance, in fact a metric, is left invariant under each of the four generators c_1 to c_4, see Figure 29.3 for the evidence.—Summarizing:

Proposition 49 *The group $\overrightarrow{GL}(\mathbb{Z}_{12})$ leaves invariant the third distance, i.e., it is a group of isometries of the interval torus $IntThirds_{3,4}$.*

29.4 Musical Interpretation of the Interval Ring

Summary. This section deals with musical interpretations of the operations in the interval ring. We take this occasion to deepen the topic of mathematical "overhead" structures and their role in building musical concepts.

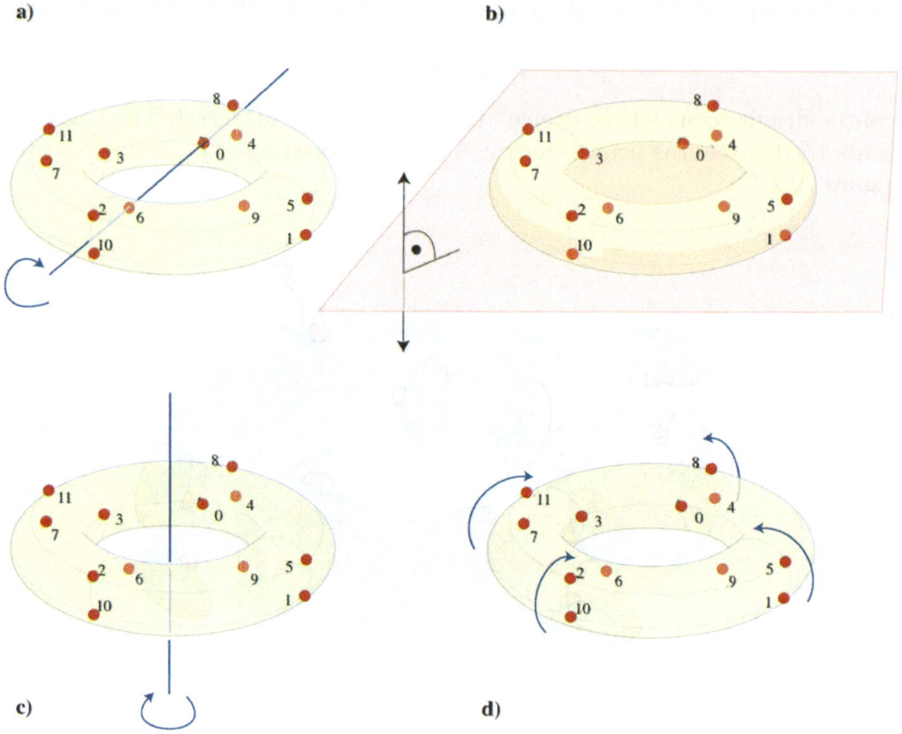

Fig. 29.3. The elementary symmetries on the third torus. The generator c_1 is a 180°-rotation around the axis through 0 and 6; c_2 is a reflection at the torus' equatorial plane; c_3 is a 90°-rotation around the polar axis; c_4 is a tilting movement in by 120°. Each of these generators preserves the third distance on the torus.

– Σ –

In this section we only deal with invertible symmetries.

We want to understand what could be the musical meaning and interpretation of multiplication in $PiMod_{12}[\varepsilon]$, or, mathematically equivalent, in $IntMod_{12,q}[\varepsilon]$. This is indeed a crucial situation. A priori, music theory never considered such a multiplication of intervals, and a conservative attitude could very well prohibit such an extension as an inadmissible mathematical overhead. But this is a classical situation with mathematical objects in the sciences, be it in physics, chemistry or musicology: Mathematical structures have more properties than their application needs. So there is always a degree of mathematical overhead. The question is rather whether ingredients of that overhead can be enriched by meaning within the applying science. More precisely, we are not preconizing the power of a conceptual oracle with mathematical properties. We simply associate mathematical properties with musical properties (in our case) and try to profit from the semantical added value in favor of a richer mathematization of musical phenomena.

As already exposed in Section 8.3, we shall again use the concatenation principle 2 from Section 8.3 here. This means that understanding a symmetry in $e^t.g \in \overrightarrow{GL}(\mathbb{Z}_{12}[\varepsilon])$ means understanding its factors in a determined factorization. As in Section 8.3, we shall present a system of 'elementary', musically understandable symmetries which generates all symmetries.

To begin with, we shall treat the translations $e^t = e^t.1$ and the multiplications g separately. For a translation quantity $t = a + \varepsilon.b$, we have $e^t = e^{a+\varepsilon.b} = e^a.e^{\varepsilon.b}$. Furthermore, $g = u + \varepsilon.v = u.(1 + \varepsilon.uv) = u.(a + \varepsilon.1)^{uv}$.

This yields the following system:

1. $s_1 = e^a$,

2. $s_2 = e^{\varepsilon.b}$,
3. $s_3 = u, u = 1, 5, 7, 11$,
4. $s_4 = 1 + \varepsilon.1$.

Here are the musical interpretations of the system's symmetries:

1. $s_1 = e^a$. Symmetry s_1 causes a transposition if the interval by a since we have $s_1(x+\varepsilon.y) = (a+x+)+\varepsilon.y$.
2. $s_2 = e^{\varepsilon.b}$. Symmetry s_2 causes a transposition of the discant voice by b while the cantus firmus remains unaltered, i.e., $s_2(x + \varepsilon.y) = a + \varepsilon.(y + b)$. this constant enlargement of the distance between cantus firmus and discantus corresponds to a traditional technique of double counterpoint.
   An example of an interesting symmetry, a concatenation of types 1 and 2, is given for $t = a \pm \varepsilon.1$. For $t = a + \varepsilon.1$, minor intervals are transformed into corresponding major intervals under the transposition by a. Conversely, $t = a + \varepsilon.1$ transforms major intervals into corresponding minor ones under the transposition by a. This means that also not 'strict' parallels of thirds, such as

$$2 + \varepsilon.3, 7 + \varepsilon.4 = e^{5+\varepsilon.1}(2 + \varepsilon.3), 11 + \varepsilon.3 = e^{4-\varepsilon.1}(7 + \varepsilon.4)$$

   are viewed as translation symmetries.
3. $s_3 = u, u = 1, 5, 7, 11$. Symmetries of this type multiply the cantus firmus, and the distance between the voices, and hence also the discantus, by u. Since we cannot reduce this discussion to the involved pair of tones, and hence not to the basic pitch domain $\mathbb{Z}_{12}$, we must have a closer look at this situation.

3.1 $u = 11$. The cantus firmus is reflected at the origin $c = 0$ whereas the voice distance $11.b$ means the octave complement to b. For example, the sweeping minor third $d, f(= 2 + \varepsilon.3)$ is transformed into the sweeping major sixth $b_\flat, g(= 10+\varepsilon.9)$. Observe that, by definition, symmetries do not alter the underlying orientation.

3.2 $u = 5$. Musically, the distance b of the interval tones can be of equal interest as a multiple of minor seconds $(= 1)$ or of fourths $(= 5)$. The symmetry $u = 5$ connects intervals of equal values in these two perspectives. In a multiplication by 5, the *minor second distance* b of an interval $a + \varepsilon.b$ is transformed into the *fourth distance* $5.b$.

3.3 $u = 7$. For this value, we have an analogous argument as for the preceding case, with fifths instead of fourths. One could also invoke the concatenation principle and use the factorization $7 = 5.11$ to reduce this case to the cases 3.1 and 3.2.

4 $s_4 = 1 + \varepsilon.1$. With this symmetry, the cantus firmus remains fixed whereas the voice distance increases by the distance of the cantus firmus from the origin $c = 0$. Thereby, the cantus firmus acquires a new function, i.e., to appear itself as a 'discantus' with respect to the origin. Thus its distance to the origin is added to the given discantus as a *fixed reference quantity.* If we repeat the application of this symmetry to the resulting interval, the same reference quantity is again added to the discantus, etc. Repeated application of s_4 therefore generates a circle on the discantus of the original interval, a circle whose step width is defined by the cantus firmus' distance to the origin. Musically speaking, the origin could be imagined as being a tonic. The appearance of the origin as a reference pitch is only justified by our choice of the generator s_4 in the given system. If we took $e^{-\varepsilon.b}.s_4$ instead, then b would play the role of the reference pitch.

Example 46

$$\begin{aligned} 3 + \varepsilon.4 \mapsto\ 3 + \varepsilon.7 &= (1 + \varepsilon.1).(3 + \varepsilon.4) \mapsto \\ 3 + \varepsilon.10 &= (1 + \varepsilon.1).(3 + \varepsilon.7) \mapsto \\ 3 + \varepsilon.1 &= (1 + \varepsilon.1).(3 + \varepsilon.10) \mapsto \\ 3 + \varepsilon.4 &= (1 + \varepsilon.1).(3 + \varepsilon.1) \end{aligned}$$

Example 47

$$6 + \varepsilon.3 \mapsto 6 + \varepsilon.9 = (1 + \varepsilon.1).(6 + \varepsilon.3) \mapsto$$
$$6 + \varepsilon.3 = (1 + \varepsilon.1).(6 + \varepsilon.9)$$

By the way, this symmetry type is the only one within our generator system which connects the cantus firmus and the discantus components in an irreducible manner. This symmetry type crystallizes an essential difference to the common reasoning in the pitch class space $PichMod_{12}$.

29.5 Self-addressed Arrows

Summary. Since arrows formally behave much like tones, self-addressed arrows can be introduced as a natural generalization of self-addressed tones and (ordinary) arrows via *address change* according to 8.3.4. This extension is described—together with a canonical projection which will play a major role in the theory of consonances and dissonances in Section 30.2.1.

– Σ –

In this section, we want to concentrate on the algebraic relations between self-addressed tones and contrapuntal arrows. Fix an arbitrary commutative ring R and consider the canonical R-linear injection $i : R \to R[\varepsilon]$. We also have the functorial R-linear injection $[\varepsilon] : R@_R R \to R[\varepsilon]@_{R[\varepsilon]}R[\varepsilon] : f \mapsto f[\varepsilon]$ defined in Section 7.5. Furthermore, we have two R-linear address change injections $c : R \xrightarrow{\sim} 0_R@_R R \to R@_R R, c[\varepsilon] : R[\varepsilon] \xrightarrow{\sim} 0_R@_R R[\varepsilon] \to R[\varepsilon]@_{R[\varepsilon]}R[\varepsilon]$. This entails the following commutative diagram

$$\begin{array}{ccc} R & \xrightarrow{i} & R[\varepsilon] \\ {\scriptstyle c}\downarrow & & \downarrow{\scriptstyle c[\varepsilon]} \\ R@_R R & \xrightarrow{[\varepsilon]} & R[\varepsilon]@_{R[\varepsilon]}R[\varepsilon] \end{array} \tag{29.1}$$

of R-linear injections. Whereas the left lower corner parametrizes the self-addressed tones in a simple form of module R, and the right upper corner parametrizes the contrapuntal intervals in the module R, the right lower corner parametrizes the self-addressed contrapuntal intervals, as opposed to the zero-addressed tones in the module R, or the prime counterpoint intervals, respectively.

The four-dimensional R-module $R[\varepsilon]@_{R[\varepsilon]}R[\varepsilon]$ can also be viewed as a left $R[\varepsilon]$-module by the ordinary composition $a.x = e^0.a.x$ with linear endomorphisms $e^0.a, a \in R[\varepsilon]$. Then we have a direct decomposition in one-dimensional $R[\varepsilon]$-modules:

$$R[\varepsilon]@_{R[\varepsilon]}R[\varepsilon] = R[\varepsilon].e^1 0 \oplus R[\varepsilon].e^{-\varepsilon}1 \tag{29.2}$$

i.e., two two-dimensional R-modules. Moreover, this decomposition is also a kernel-image decomposition with respect to the idempotent[5] right multiplication endomorphism $?.e^{\varepsilon}0$, of $R[\varepsilon]$-modules, i.e.,

$$Ker(?.e^{\varepsilon}0) = R[\varepsilon].e^{-\varepsilon}1, Im(?.e^{\varepsilon}0) = R[\varepsilon].e^1 0.$$

Now, the image is exactly the image of $R[\varepsilon]$ under $c[\varepsilon]$, whereas $?.e^{\varepsilon}0$ maps the image of $R@_R R$ under $[\varepsilon]$ isomorphically (as R-module) onto the image $Im(?.e^{\varepsilon}0)$. in other words:

Theorem 32 *With respect to the embeddings of diagram (29.1), the projection $?.e^{\varepsilon}0$ associates bijectively the self-addressed tones with the contrapuntal intervals and leaves the original tones in R invariant. This means that self-addressed tones and contrapuntal intervals are put under a canonical algebraic correspondence in the large space of self-addressed arrows*

[5] See also Appendix C.2.3, Example 98.

29.6 Change of Orientation

Summary. Orientation within a contrapuntal sequence may change, and enforce a special treatment of contrapuntal steps from sweeping orientation to hanging orientation or vice versa. We show how such a change can be reduced to an orientation preserving situation by means of the regular embedding of the algebra of dual numbers in the linear endomorphism ring.

– Σ –

The left-regular embedding[6]

$$(?) : R[\varepsilon] \rightarrowtail \mathbb{M}_{2,2}(R)$$

identifies the two-dimensional dual number algebra with a subspace of the ring of two-by-two matrices over R. In this embedding, the dual number algebra is generated by the dual number multiplication

$$(\varepsilon) = \begin{pmatrix} 0 & 0 \\ 1 & 0 \end{pmatrix}$$

with the relation $(\varepsilon)^2 = 0$. The sweeping and hanging orientations are interpreted as a projection

$$\alpha_+ = \begin{pmatrix} 1 & 1 \\ 0 & 0 \end{pmatrix}, \alpha_- = \begin{pmatrix} 1 & -1 \\ 0 & 0 \end{pmatrix}$$

which are related by the multiplicative relation

$$\alpha_- = -\alpha_+(1+\varepsilon)^{-2}. \tag{29.3}$$

The matrix algebra $\mathbb{M}_{2,2}(R)$ is generated by two indeterminates $(\varepsilon), \alpha_+$ with relations $(\varepsilon)^2 = 0, \alpha_+^2 = \alpha_+$, and $\alpha_+.(\varepsilon) + (\varepsilon).\alpha_+ = (1+\varepsilon)$, and it is spanned by the linear basis $1, (\varepsilon), \alpha_+, \alpha_+.(\varepsilon)$.

Relation (29.3) can be used to reinterpret hanging counterpoint in terms of sweeping counterpoint as follows: Suppose that we have a sequence $x_1 + \varepsilon.i_1, x_2 + \varepsilon.i_2$ with the first interval in sweeping, but the second one in hanging orientation. This means that the evaluation via orientation projections produces the two discantus instances $\alpha_+(x_1 + \varepsilon.i_1), \alpha_-(x_2 + \varepsilon.i_2)$. In order to change the interpretation of orientations, use (29.3) and rephrase the second discantus as

$$\alpha_-(x_2 + \varepsilon.i_2) = -\alpha_+(1+\varepsilon)^{-2}(x_2 + \varepsilon.i_2) = \alpha_+(-(1+\varepsilon)^{-2}(x_2 + \varepsilon.i_2))$$

so that we are dealing with sweeping orientation related to the new interval

$$-(1+\varepsilon)^{-2}(x_2 + \varepsilon.i_2).$$

This technique will be used to deduce contrapuntal steps while orientation changes. Observe that the mediator factor $(1+\varepsilon)$ in formula (29.3) enhances the musical meaning of the fourth generator symmetry s_4 in Section 29.4: *The multiplication by generator s_4 helps to reinterpret hanging orientation in terms of sweeping orientation.*

---

[6] See Appendix D.1.

30

Interval Dichotomies as an Expression of Contrast

enim vero sicut vitium mala virtus
a nullo umquam morali philosopho dictum fuit,
ita nec musicus umquam litteratus
discordiantiam malam concordantiam nuncupavit.
Johannes Tinctoris [1053, p.90]

Summary. For contrapuntal composition and theory, consonant and dissonant intervals are a dichotomic concept. We present the mathematical restatement and fundamental properties of the basic concept of an *interval dichotomy*. For the classification of dichotomies, a strong condition on unique *symmetries of polarity* between the two halves of dichotomies is added. It reveals a distinguished role of the *consonance/dissonance* dichotomy of classical counterpoint and of the *major* dichotomy (associated with the major scale). We discuss evidence of the consonance/dissonance dichotomy from theoretical and empirical points of view and open the discourse to an intercultural perspective guided by the classification of dichotomies, as investigated by Jens Hichert [466].

– Σ –

Remark 16 In the following counterpoint chapters, we shall tacitly work in the fifth pitch and interval spaces, i.e., we take the automorphism $q = (7)$ in the identifiers. We therefore shall—for example—speak of the fifth when addressing the unit 1. Whenever we deviate from this convention, the reader should be warned.

We shall also stick to the forms built upon $\mathbb{Z}_{12}$ and tacitly carry over the distance structures of the isomorphic Sylow torus representation $\mathbb{Z}_3 \oplus \mathbb{Z}_4$ in order to keep notation simpler.

30.1 Dichotomies and Polarity

Summary. The technical definition of *interval dichotomies* in $IntMod_{12,q}$ and in the counterpoint arrow form $IntMod_{12,q}[\epsilon]$ is given[1]. We discuss canonical *polarities*, in fact automorphisms of such dichotomies, as well as their topological behavior which is measured by *diameter* and *span* of a dichotomy. This is used to classify the system of dichotomies. Among the 26 classes, six classes admit unique polarities. These classes of *strong* dichotomies show a topologically distinguished position of the *consonance/dissonance* and the *major* dichotomy, the latter being defined by the six tonic-rooted (proper) intervals of major tonality. Among the strong dichotomies, the consonance-dissonance dichotomy itself (not only its class) is distinguished by a number of characteristics.

[1] This means the interval form with the identifier defined by the automorphism q of $\mathbb{Z}_{12}$, i.e., $IntMod_{12,q} \underset{Id}{\longrightarrow} \mathbf{Syn}(PiMod_{12,q})$, etc.

G. Mazzola, *The Topos of Music I: Theory*, Computational Music Science,
https://doi.org/10.1007/978-3-319-64364-9_30

$-\Sigma-$

Let A be an addressed $\mathbb{Z}$-module. Then an objective, A-addressed local composition X in ambient space $IntMod_{12,q}$ is called a (A-addressed) *marked interval dichotomy* iff it is equipollent[2] to its complement $C(X)$ in the total local composition $A@IntMod_{12,q}$. Observe that the latter is not necessarily finite. Hence we have the *complement action* $i \mapsto C^i(X)$ of $\mathbb{Z}_2$ on the set $DiM(A, IntMod_{12,q})$ of A-addressed marked dichotomies. An *interval dichotomy* is a $\mathbb{Z}_2$-orbit of marked interval dichotomies, the orbit set is $Di(A, IntMod_{12,q})$.

A second left action is defined by the automorphism group $\overrightarrow{GL}(\mathbb{Z}_{12})$ of the ambient space $IntMod_{12,q}$. If $g \in \overrightarrow{GL}(\mathbb{Z}_{12}), X \in DiM(A, IntMod_{12,q})$, then we set $g.X = \{g.x|x \in X\}$, and the set $CiM(A, IntMod_{12,q})$ denotes the orbit space of this action, its elements are called the (A-addressed) *marked dichotomy classes.* Clearly the two actions commute, i.e., we have a left action of $\mathbb{Z}_2 \times \overrightarrow{GL}(\mathbb{Z}_{12})$ on $DiM(A, IntMod_{12,q})$, an induced action of $\overrightarrow{GL}(\mathbb{Z}_{12})$ on $Di(A, IntMod_{12,q})$, and one of $\mathbb{Z}_2$ on the space of marked dichotomy classes $CiM(A, IntMod_{12,q})$. The total orbit space

$$Ci(A, IntMod_{12,q}) = \mathbb{Z}_2 \times \overrightarrow{GL}(\mathbb{Z}_{12})\backslash DiM(A, IntMod_{12,q})$$

is the space of (A-addressed) *dichotomy classes.*

The same constructions (mutatis mutandis) can be made on the space of counterpoint arrows $IntMod_{12,q}[\epsilon]$. Let A be an addressed $\mathbb{Z}$-module. Then an objective, A-addressed local composition X in ambient space $IntMod_{12,q}[\epsilon]$ is called a (A-addressed) *marked counterpoint dichotomy* iff it is equipollent to its complement $C(X)$ in the total local composition $A@IntMod_{12,q}[\epsilon]$. On the set $DcM(A, IntMod_{12,q}[\epsilon])$ of A-addressed marked counterpoint dichotomies, we have the *complement action* $i \mapsto C^i(X)$ of $\mathbb{Z}_2$. A *counterpoint dichotomy* is a $\mathbb{Z}_2$-orbit of marked counterpoint dichotomies, the orbit set is $Dc(A, IntMod_{12,q}[\epsilon])$.

A second left action is defined by the automorphism group $\overrightarrow{GL}(\mathbb{Z}_{12}[\epsilon])$ of ambient space $IntMod_{12,q}[\epsilon]$. If $g \in \overrightarrow{GL}(\mathbb{Z}_{12}[\epsilon]), X \in DcM(A, IntMod_{12,q}[\epsilon])$, then we set $g.X = \{g.x|x \in X\}$, and $CcM(A, IntMod_{12,q}[\epsilon])$ denotes the orbit space of this action, its elements are called the (A-addressed) *marked counterpoint dichotomy classes.* Clearly the two actions commute, i.e., we have a left action of $\mathbb{Z}_2 \times \overrightarrow{GL}(\mathbb{Z}_{12}[\epsilon])$ on $DcM(A, IntMod_{12,q}[\epsilon])$, an induced action of $\overrightarrow{GL}(\mathbb{Z}_{12}[\epsilon])$ on $Dc(A, IntMod_{12,q}[\epsilon])$, and one of $\mathbb{Z}_2$ on the space of marked counterpoint dichotomy classes $CcM(A, IntMod_{12,q}[\epsilon])$. The total orbit space

$$Cc(A, IntMod_{12,q}[\epsilon]) = \mathbb{Z}_2 \times \overrightarrow{GL}(\mathbb{Z}_{12}[\epsilon])\backslash DcM(A, IntMod_{12,q}[\epsilon])$$

is the space of (A-addressed) *counterpoint dichotomy classes.*

Definition 92 *A marked interval dichotomy X is called* autocomplementary *if it is isomorphic to its complement $C(X)$, i.e., iff its dichotomy class coincides with its marked dichotomy class. The marked dichotomy X is called* rigid *if its symmetry group is trivial. It is called* strong *if it is autocomplementary and rigid.*

If a marked dichotomy X is autocomplementary, so is its complement. If a marked dichotomy is rigid, so is its complement. Hence, if X is strong, so is its complement. So *autocomplementarity, rigidity, and strength are invariants of the dichotomy classes.* From the classification of zero-addressed objective local compositions in $PiMod_{12}$ in Appendix N.1, we know that there are 34 classes of zero-addressed marked interval dichotomies (26 classes numbers 63 to 88, complements not counted twice, count twice the 8 numbers without *). There are 26 interval classes, 8 autocomplementary classes, and 6 strong classes.

We often denote a marked dichotomy by $(X/C(X))$ and its class by $[C/C(X)]$, whereas a dichotomy is denoted by $(X|C(X))$ and its class by $[C|C(X)]$. Since the group of symmetries acts on the set of dichotomies, one can say that a symmetry stems not only from an isomorphism of the underlying marked dichotomies but is also associated with an isomorphism of the associated dichotomy. More precisely, if we are given a symmetry $f : (X/Y) \to (U/V)$ between two marked dichotomies, we get an isomorphism of the interpretations that are associated with the partitions $X \sqcup Y$ and $U \sqcup V$. Conversely, an isomorphism between such two interpretations is induced by two isomorphisms on two pairs of charts. But each such isomorphism gives automatically rise

[2] ...has same cardinality as...

to an isomorphism of the other chart pair—the only thing which we lose is the order of charts, i.e., we are left with a transformation among the (non-marked) dichotomies. In this sense the unique non-trivial inner symmetry p of a strong dichotomy $(X|Y)$, i.e., $p(X/Y) = (p(X)/p(Y)) = (Y/X)$ is also called its *polarity.*

Example 48 Here are the six strong (marked and unmarked) dichotomies, the numbers referring to the classification table of local compositions in Appendix N.1. The polarity of dichotomy number n and identifier number q is denoted by $p_{q,n}$, or p_n for $q = 7$, if the context is clear. If an index is omitted, we tacitly suppose the fifth representation with $q = 7$.

1. The dichotomy Nr. 64
$$\Delta_{7,64} = (I_7/J_7) = (\{1,2,3,4,5,11\}|\{0,6,7,8,9,10\})$$
with polarity $p_{7,64} = 7.p_{1,64}.7 = e^{11}.11$ corresponding to the dichotomy
$$\Delta_{1,64} = (I_1/J_1) = (\{2,4,5,7,9,11\}|\{0,1,3,6,8,10\}),$$
in the semitone representation ($q = 1$), with polarity $p_{1,64} = e^5.11$. The dichotomy $\Delta_{7,64}$ arises when considering all proper (non-vanishing) intervals in a major scale when counted from the tonic.
2. The dichotomy Nr. 68
$$\Delta_{7,68} = (\{0,2,7,8,9,11\}|\{1,3,4,5,6,10\}),$$
with polarity $p_{7,68} = 7.p_{1,68}.7 = e^6.5$ corresponding to the dichotomy
$$\Delta_{1,68} = (\{0,1,2,3,5,8\}|\{4,6,7,9,10,11\})$$
in the semitone representation with $q = 1$, with polarity $p_{1,68} = e^6.5$.
3. The dichotomy Nr. 71
$$\Delta_{7,71} = (\{0,2,7,8,9,11\}|\{1,3,4,5,6,10\}),$$
with polarity $p_{7,71} = 7.p_{1,71}.7 = e^5.11$ corresponding to the dichotomy
$$\Delta_{1,71} = (\{0,1,2,3,6,7\}|\{4,5,8,9,10,11\})$$
in the semitone representation with $q = 1$, with polarity $p_{1,71} = e^{11}.11$.
4. The dichotomy Nr. 75
$$\Delta_{7,75} = (\{0,2,4,7,8,11\}|\{1,3,5,6,9,10\}),$$
with polarity $p_{7,75} = 7.p_{1,75}.7 = e^5.11$ corresponding to the dichotomy
$$\Delta_{1,75} = (\{0,1,2,4,5,8\}|\{3,6,7,9,10,11\})$$
in the semitone representation with $q = 1$, with polarity $p_{1,75} = e^{11}.11$.
5. The dichotomy Nr. 78
$$\Delta_{7,78} = (\{0,2,4,6,7,10\}|\{1,3,5,8,9,11\}),$$
with polarity $p_{7,78} = 7.p_{1,78}.7 = e^3.11$ corresponding to the dichotomy
$$\Delta_{1,78} = (\{0,1,2,4,6,10\}|\{3,5,7,8,9,11\})$$
in the semitone representation with $q = 1$, with polarity $p_{1,78} = e^9.11$.
6. This is the classical dichotomy Nr. 82
$$\Delta_{82} = \Delta_{7,82} = (K_7/D_7) = (\{0,1,3,4,8,9\}|\{2,5,6,7,10,11\}).$$
of contrapuntal consonances (left) and dissonances (right) in the fifth system. Its polarity is 'the' *autocomplementarity function* $p_{7,82} = 7.e^2.5.7 = e^2.5$ deduced from the known[3] autocomplementarity function $p_{1,82} = e^2.5$ of the consonance-dissonance dichotomy
$$\Delta_{1,82} = (K_1/D_1) = (\{0,3,4,7,8,9\}|\{1,2,5,6,10,11\})$$
in the semitone representation, as discussed in Section 24.1.1.

[3] See [678], for example.

These strong dichotomies can also be represented as partitions of the interval torus in the Sylow representation $IntThirds_{3,4,q} \underset{(q)}{\longrightarrow} \mathbf{Syn}(IntThirds_{3,4})$ with the q-identifier (usually the fifth). Recall that we have the third distance $d(x,y)$ on this torus, and that it is the same as the distance with identifier $q = 1$.

Definition 93 *Let (X/Y) be a strong marked dichotomy in $IntThirds_{3,4,q}$. Then its* diameter *is defined by*

$$\delta(X/Y) = \frac{1}{2} \sum_{u,v \in X} d(u,v).$$

By use of the polarity of a strong dichotomy one sees that $\delta(X/Y) = \delta(Y/X)$ so that we may define this number as the diameter $\delta(Y|X)$ of the strong dichotomy $(X|Y)$. *Since evidently all dichotomies of a dichotomy class have the same diameters (recall that symmetries of the torus are isometries for the distance), we may define the diameter $\delta[X|Y]$ of a dichotomy class $[X|Y]$ by the diameter of any of its representatives.*

The diameter measures the average distance between points of one half of the dichotomy.

Definition 94 *Let (X/Y) be a strong marked dichotomy in $IntThirds_{3,4,q}$ with polarity p. Then its* span *is defined by*

$$\sigma(X/Y) = \sum_{u \in X} d(u, p(u)).$$

Since any polarity is an involution in our context, one sees that $\sigma(X/Y) = \sigma(Y/X)$ so that we may define this number as the span $\sigma(X|Y)$ of the strong dichotomy $(X|Y)$. *Since evidently all dichotomies of a dichotomy class have the same span (recall that symmetries of the torus are isometries for the distance), we may define the span $\sigma[X|Y]$ of a dichotomy class $[X|Y]$ by the span of any of its representatives.*

Diameter and span of our six strong classes are visualized in Figure 30.1. Intuitively, the minimality of $\delta(K/D)$ means that the subsets K and D are separated in an optimal way on the torus (Figure 30.1). The maximality of $\sigma(I/J)$ means that I and J are optimally mixed on the torus. If we stay on a point on I, and we want to go to another point of I on a shortest path, then we often have to traverse a point of J, a phenomenon which never happens for (K/D): Between any two consonant intervals there is always a shortest path which does not leave the consonant half. By the given polarities, all these statements about K and I are also valid for D and J. The possible connections within K and within I (and the complements, respectively) are shown in Figure 30.2.

In contrast to the graph of I, the graph of K has no inner symmetry, i.e., every consonant interval is uniquely determined by its position on the graph. This means that the consonance-dissonance dichotomy has a privileged position among all strong dichotomies. But this does not yet exhibit the precise representative (K/D) within the class $[K|D]$. The selection of the marked dichotomy (K/D) is in fact realized by an algebraic condition: to require that the first half X of an element (X/Y) of the class $[K|D]$ be a multiplicative monoid. This condition was discovered by Noll [802]. This in fact exhibits the marked half of consonances against dissonance, these two parts are not equivalent from this point of view. Until now it is however not clear what are the structural consequences of the predicates which uniquely exhibit the consonances.

30.2 The Consonance and Dissonance Dichotomy

Summary. The consonance-dissonance dichotomy is in a canonical bijection with the Riemann consonances. We also give empirical evidence of this dichotomy from brain research. The section concludes with a cognitive interpretation of the specific role of music for the individual psyche.

– Σ –

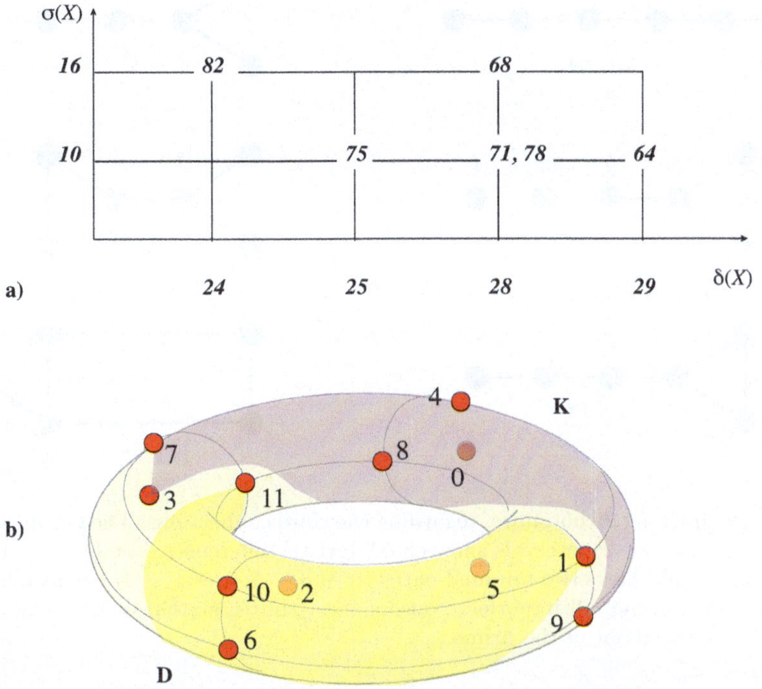

Fig. 30.1. (a) Span and diameter of the six strong dichotomies. The polar position of the consonance-dissonance dichotomy (Nr. 82) against the major dichotomy (Nr. 64) is visible. (b) The geometric meaning of the minimal diameter and the maximal span in (K/D) is evidenced by an optimal separation of the two halves of the dichotomy.

Recall from Section 24.1.1 that the consonance-dissonance dichotomy also appears in the context of just intonation and relates to the 12-tempered case by means of the enharmonic projection. We now want to investigate further remarkable properties of the (K/D) dichotomy. To begin with, we define a (marked) dichotomy $(X[\varepsilon]/Y[\varepsilon])$ of contrapuntal intervals for every (marked) dichotomy (X/Y) on the space form S by the rule $(X[\varepsilon]/Y[\varepsilon]) = (coord(S) + \varepsilon.X/coord(S) + \varepsilon.Y$, where $coord(S)$ is the coordinator module of S.

30.2.1 Fux and Riemann Consonances Are Isomorphic

Summary. We present the Riemann dichotomy and its 1-1 correspondence with the Fux dichotomy based on the diagram discussed in 29.5.

– Σ –

Recall the commutative diagram (29.1) of $\mathbb{Z}_{12}$-linear injections for the special ring $R = \mathbb{Z}_{12}$.

$$\begin{array}{ccc} \mathbb{Z}_{12} & \xrightarrow{\ i\ } & \mathbb{Z}_{12}[\varepsilon] \\ {\scriptstyle c}\downarrow & & \downarrow{\scriptstyle c[\varepsilon]} \\ \mathbb{Z}_{12}@_{\mathbb{Z}_{12}}\mathbb{Z}_{12} & \xrightarrow{[\varepsilon]} & \mathbb{Z}_{12}[\varepsilon]@_{\mathbb{Z}_{12}[\varepsilon]}\mathbb{Z}_{12}[\varepsilon] \end{array} \tag{30.1}$$

We may vew this configuration as a realization of determined forms as follows:

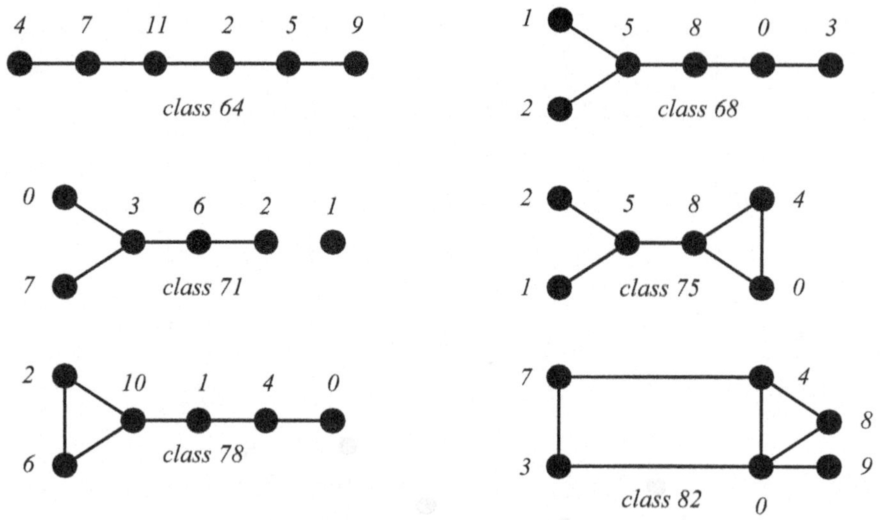

Fig. 30.2. The graphs of all strong dichotomies, regarding the third connections. The graphs among the I (major, class 64) and the K (consonance, class 82) intervals show characteristic differences. While in I, all intervals are positioned on a line, the consonance graph shows two minimal paths ($0 \to 3 \to 7; 0 \to 4 \to 7$) between $0, 7$, for example. Observe that when starting from the prime, all imperfect consonances (thirds, sixths) can be reached directly, whereas the perfect fifth is the only non-neighbor of the prime.

$$\begin{array}{ccc} 0@PitchMod_{12,q} & \xrightarrow{i} & 0@IntMod_{12,q}[\epsilon] \\ {\scriptstyle c}\downarrow & & \downarrow{\scriptstyle c[\varepsilon]} \\ \mathbb{Z}_{12}@PitchMod_{12,q} & \xrightarrow{[\varepsilon]} & \mathbb{Z}_{12}[\varepsilon]@IntMod_{12,q}[\epsilon] \end{array} \tag{30.2}$$

Observe that the space in the right bottom corresponds to the larger space $\mathbb{Z}_{12}[\varepsilon]@\mathbb{Z}_{12}[\varepsilon]$. Also, the lower horizontal arrow is not a natural one in terms of ambient morphisms f/α for local compositions.

To begin with, let us look at the *Riemann dichotomy* which was first defined in [802]. For a justification of this naming that relates to Riemann's concept of relative consonances and dissonances, see [802]. The Riemann dichotomy is the monoid $Trans(D,T)$ generated by the "transporter" set $Tr(D,T)$ consisting by definition of all not necessarily invertible symmetries $f : D = \{1,2,5\} \to T = \{0,1,4\}$ from the dominant triad D to the tonic triad T (in the fifth system). The monoid $Trans(D,T)$ has 72 elements. So it naturally gives rise to a dichotomy on $\mathbb{Z}_{12}@PitchMod_{12,q}$ and also defines a point in the intension topology $InTop(PitchMod_{12,q})$.

Exercise 67 *Show that* $card(Trans(D,T)) = 72$.

On the other hand, consider the consonant contrapuntal intervals $K[\varepsilon] = \mathbb{Z}_{12} + \varepsilon.K$ in $0@IntMod_{12,q}[\epsilon]$. The intension $Int(K[\varepsilon])$ is canonically identified to a local composition in $\mathbb{Z}_{12}[\varepsilon]@IntMod_{12,q}[\epsilon]$ since the address $\mathbb{Z}_{12}[\varepsilon]$ is faithful (take the identity). The intersection W of $Int(K[\varepsilon])$ with the subspace $\mathbb{Z}_{12}@PitchMod_{12,q}$ consists of those endomorphisms $e^x y$ of $PitchMod_{12,q}$ that induce endomorphisms of $K[\varepsilon]$. Using the isomorphism

$$\nu = ?.e^{\varepsilon}0|_{\mathbb{Z}_{12}@_{\mathbb{Z}_{12}}\mathbb{Z}_{12}} : \mathbb{Z}_{12}@_{\mathbb{Z}_{12}}\mathbb{Z}_{12} \xrightarrow{\sim} \mathbb{Z}_{12}[\varepsilon]$$

from Section 29.5, W identifies to the monoid $R \in InTop(PitchMod_{12,q})$ of left stabilizers of $\nu^{-1}(K[\varepsilon])$. In [802] it is shown that $Trans(D,T) = W$. More precisely, with the above identification, we have this theorem:

Proposition 50 *Let* $e^x.y$ *be an endomorphism of* $PitchMod_{12,q}$. *Then it is in the left stabilizer*[4] *of* $\nu^{-1}(K[\varepsilon])$ *iff it is a product* $e^x.y = \prod_i e^{x_i}.y_i$ *of endomorphisms*[5] *which transport the dominant triad* $D = \{1, 2, 5\}$ *into the tonic triad* $T = \{0, 1, 4\}$. *In particular, every element of* $\nu^{-1}(K[\varepsilon])$ *can be written as such a product of transporter endomorphisms.*

The last statement follows from the fact that the consonance interval numbers K are a multiplicative monoid, and therefore, each element $e^x.k, k \in K$ is a stabilizer of $\nu^{-1}(K[\varepsilon])$. This implies the following:

Corollary 19 *Let* $f = 0 + \varepsilon.1$ *the fifth interval at the C tonic 0. For any consonant interval* $c = x + \varepsilon.k$, *there is a sequence* $t_i = e^{x_i}.y_i, i = 1, \ldots m$ *of transporter endomorphisms* $t_i : D \to T$ *such that* $c = t_m.t_{m-1}.\ldots.t_1(f)$.

Evidently, there is a deep relation between Riemann theory (as it appears in Noll's perspective) and the Fux dichotomy of consonance-dissonance. Presently, we do not know more about the harmonic/contrapuntal implications of the above results. The straightforward hope is that the transition from polyphonic counterpoint to homophonic harmonic relates to the above mathematical facts. But neither the systematic nor the historical consequences of these facts are at reach. It is however true that musicology has never understood the theoretical relations between counterpoint and harmony. There must be a fundamental relation because the development of homophony out of the polyphonic tradition cannot be a rupture without any inner coherence. Even if such a rupture were a historic fact, it would be a primordial question of systematic musicology to explain the structural, system-immanent rationales for such a rupture. The present results give first hints for the explanation of this lacuna.

30.2.2 Induced Polarities

Summary. We describe the autocomplementary functions on the contrapuntal dichotomies $(X[\varepsilon]/Y[\varepsilon])$ which are deduced from strong dichotomies (X/Y).

– Σ –

Suppose that we are given a strong dichotomy (X/Y) which bears the polarity $e^u.v$. Then the contrapuntal dichotomy is no longer rigid, but still autocomplementary. The precise situation is described as follows.

Proposition 51 *Let* $\Delta = (X/Y)$ *be a strong dichotomy with polarity* $p_\Delta = e^u.v$. *Choose a cantus firmus point* x. *Then there is exactly one symmetry* p_Δ^x *on the counterpoint interval space* $IntMod_{12,q}[\varepsilon]$ *which is a polarity of* $(X[\varepsilon]/Y[\varepsilon])$ *and fixes the "tangent space"* $I_x = x + \varepsilon.\mathbb{Z}_{12}$ *at* x. *Call this the* polarity at x. *We have*

$$p_\Delta^x = e^{x(1-v)+\varepsilon.u}.v, \text{ and } p_\Delta^{x+y} = e^x.p_\Delta^y.e^{-x},$$

and under this polarity, a tangent space I_y *is mapped onto the tangent space* $I_{x+v(y-x)}$.

Exercise 68 *Give a proof of Proposition 51.*

30.2.3 Empirical Evidence for the Polarity Function

Summary. A review of neurophysiological verifications of the presence of the Fux polarity in human depth EEG is exposed.

– Σ –

[4] I.e., for all endomorphisms $e^t.k$ with consonant component k, $e^x.y.e^t.k = e^s.k'$ has also a consonant component k'.
[5] Including the "empty" product, i.e., the identity.

Although this book is not a report on physiological or psychological correlates to music structures, it is important to give an overview of a pronounced evidence of electrophysiological correlates of the consonance-dissonance dichotomy. This is by no means a justification of even a proof of the adequacy of the mathematical investigations, but it must be considered as a fundamental relativization of traditional consonance-dissonance theories. These ideas never produce dichotomies but yield degrees of consonance or dissonance, a quality which is completely irrelevant to the musical counterpoint dichotomy.

The aim of a project at the Neurology Department of the Zürich University Hospital which the epileptologist Heinz-Gregor Wieser, the author, and their collaborators conducted during the years 1984–1988 was to test mathematical principles of classical counterpoint by means of depth EEG responses to musical stimuli. In particular, it was planned to test the relative results of the EEG to different musical inputs (consonances vs. dissonances) and not the relation of responses to musical stimuli versus non-musical stimuli. The latter problem has been investigated with much success by Hellmuth Petsche and his collaborators [843]. It is important to stress that from our results, we do not draw any kind of conclusions concerning a possible genetic nature of musical understanding or a possible universal validity of classical European interval categories. Our investigations show that in some defined regions of the brain of some European humans, certain significant reactions take place—nothing more and nothing less. There is no reason to generalize whatsoever, but there are enough reasons to try to repeat these investigations in other research sites with a comparable infrastructure. This is all the more desirable since the qualitative results of the investigations by Wieser and the author (namely the prominent role of limbic structures for the judgment of musical pleasantness) have been confirmed by others; see [127], for example.

For a more complete report of our results, we refer to [678, 679, 695, 1122, 1123, 1124]. In this short review, we shall restrict ourselves to the two subtests concerning (a) isolated successive intervals and (b) the polarity between simultaneous consonances and dissonances. Our results confirm our hypothesis on (1) a significant differentiation of EEG responses to consonant vs. dissonant intervals in limbic and auditory brain areas and (2) a pronounced sensitivity of these areas in EEG responses to the fundamental polarity between consonances and dissonances. In particular, the quantitative measurement of these responses by use of the "spectral participation vector" has confirmed our belief that this vector may carry some of the semantic charge of EEG signals.

30.2.3.1 The EEG Test

The test concerned different contexts of consonances and dissonances as well as the test of the polarity $e^2.5$. We used EEG from the scalp (Hess system), stereotactic depth EEG following [1121], and multipolar foramen ovale recordings [1125]. The tests were applied to the rare cases of patients suffering from medically intractable complex partial epilepsy seizure of suspected mediobasal temporal lobe origin and underwent presurgical evaluation with a view towards surgical epilepsy therapy. None of the total 13 patients considered the voluntary 30 minute music test through monophonic earphones as being disagreeable.

There are several reasons why, despite the particular state of epileptics, the tests remain comparable to tests with normal humans which cannot be conducted for evident reasons. First, our tests were performed during interictal periods. Second, localization of the focus gives a good estimation of its possible influence. Third, epileptiform potentials are easily distinguished from others by the expert.

For each patient, we recorded 700 time windows for fast Fourier transform (FFT) spectral analysis with 256 samples per second, each window for different EEG channels, different power windows ($\delta = 0 - 4Hz, \theta = 4 - 8Hz, \alpha = 8 - 14Hz, \beta = 14 - 40Hz$) and repetitions, totally 11'000 raw spectral data per patient. Unfortunately, the project could not be completed for extrascientific reasons and hence, only two patients have been thoroughly evaluated. These patients were C. J.-L., a 35-year-old academic, and V.S., a 31-year-old artisan. Both are male Europeans ad prefer standard classical, light, and folklore music. Figure 30.3 shows the positions of bipolar depth EEG recordings which we are going to discuss. Notice that C. j.-L.'s recordings RCA, RH, LCA are homologous to V.S.'s recordings 4, 10, 14. Recording RCA lies within the right hippocampus, recording RH is positioned within Heschl's gyrus (auditory cortex), and recording LCA lies within the left hippocampus.

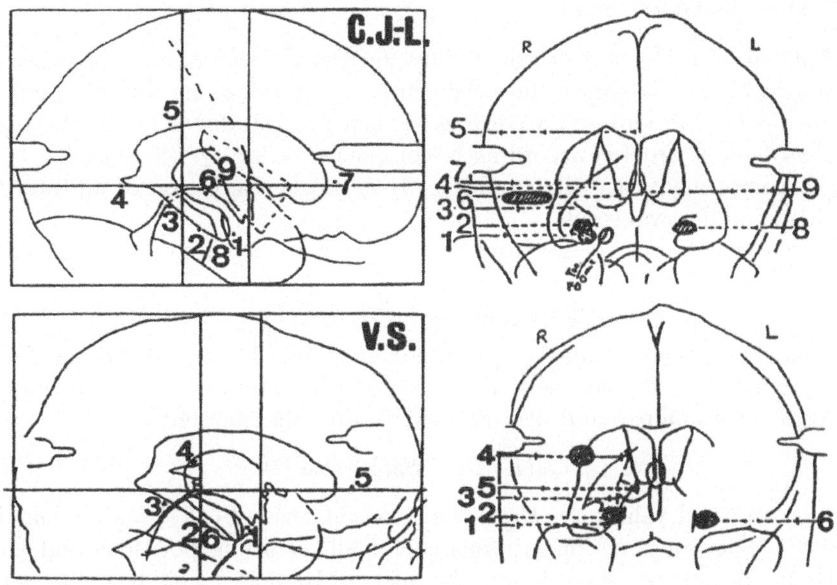

Fig. 30.3. Implantation scheme for the two patients C. J.-L. and V.S. following X-rays showing the topographic position of depth electrodes. The three homologous regions are indicated by dashed ovals. For C. J.-L., RCA is electrode 2/1-3, RH is 6/5-6, and LCA is 8/1-3. For V.s., 4 is electrode 2/1-2, 10 is 4/5-6, and 14 is 6/1-2. (Numbers after the slashes indicate precise positions on the electrodes, where 1 = deepest position and 10 = position near surface.

Nevertheless, we have been able to observe a great deal of visual evidence for EEG response to music stimuli in neo- and archicortical regions of all patients; for details, see [678, 1122].

30.2.3.2 Analysis by Spectral Participation Vectors

We used four sounds for these tests: piano, sine wave, cello (without vibrato), and "test", a clear, organ-like sound, all synthesized from a Yamaha TX7 synthesizer and CX5M voicing program in order to avoid possible emotional artifacts associated with natural sounds from sociocultural premises. The music program was written on a precursor of the commercial composition software *presto*® [680].

The spectral analysis was executed on a CDC Cyber computer. We made use of the *spectral participation vector*

$$S(E) = (P(E), P(E)/P_\delta(E), P(E)/P_\theta(E), P(E)/P_\alpha(E), P(E)/P_\beta(E))$$

of an event E and its associated *participation value*

$$v(E) = P(E)/P_\theta(E) + P(E)/P_\alpha(E) + P(E)/P_\beta(E)$$

which is a length measure. Here, $P(E)$ is the total spectral power [863], $P_\delta(E)$ is the δ-power of event E, $P_\theta(E)$ is the θ-power, $P_\alpha(E)$ is the α-power, and $P_\beta(E)$ is the β-power, see also [678].

From the results obtained thus far we conclude that this representation is well suited to give an adequate picture of possible semantic charge of EEG signals. It is a measure motivated by, among others, the well-known vigilance-related α-participation $S_\alpha = P(E)/P_\alpha(E)$ observed by Berger. We also use the delta participation $S_\delta = P(E)/P_\delta(E)$, the θ-participation $S_\theta = P(E)/P_\theta(E)$, and the β-participation $S_\beta = P(E)/P_\beta(E)$, the latter has been recognized as being strongly related to higher cognitive brain activity by Giannitrapani (see [365]) and Petsche et al. (who also focus on the γ-band $30 - 50Hz$) [844].

30.2.3.3 Isolated Successive Intervals

We first focus our attention on a subtest concerning musically isolated successive intervals, i.e., the two tones of an interval are played one after the other without interruption. All the intervals were played in three orders: (1) all consonances, ordered according to their size; all dissonances, also ordered according to their size. (2) All consonances, ordered according to complementarity size (if possible); then all dissonances, ordered according to complementarity (if possible). (3) A mixed succession of all intervals according to a particular dodecaphonic all-interval series.

Having fixed a frequency band, θ, say, and an interval with first tone event E_1 and second event E_2, we consider the quotient

$$Q_\theta(E_1, E_2) = S_\theta(E_1)/S_\theta(E_2)$$

of the theta participations of the first and second tones. If $Q_\theta(E_1, E_2) > 1$ or $Q_\theta(E_1, E_2) < 1$, respectively, then theta participation lowers or increases, respectively, from the first to the second tone. In order to compare these ratios for consonances and dissonances, we take the quotient

$$Q_\theta(K/D) = S_\theta(K)/S_\theta(D)$$

of the mean value $S_\theta(K)$ of all values $Q_\theta(E_1, E_2)$ for consonances (E_1, E_2) and the analogous mean value $S_\theta(D)$ for dissonances. This construction is repeated for all recording positions and all frequency bands, including θ, α, β; band δ is omitted since it may be affected by noise. This test was performed four times with patient C. J.-L. and six times with V.S. A one-sided Wilcoxon test shows significantly higher quotients for consonances compared to dissonances, i.e., $Q_?(K/D)$ is significantly larger than 1 for many recording positions and frequency bands, see Figure 30.4, and observe the similarity of distribution of these quotients for our patients with respect to homologous recordings and frequency bands. This means that for consonances, participation lowers more when the second tone appears than for dissonances.

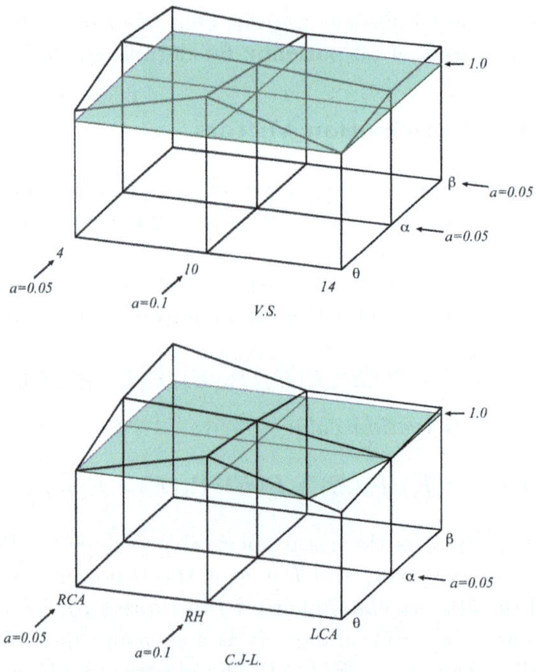

Fig. 30.4. Graphical representation of the quotients $Q_\theta(K/D)$, $Q_\alpha(K/D)$, and $Q_\beta(K/D)$ for all locations and patients C. J.-L. and V.S. These numbers show that for consonances, participation lowers more when the second tone appears, compared to dissonances. The 1-level is indicated in the graphics, observe the places where this level is exceeded.

30.2.3.4 Polarity

To test the response to the (K/D)-polarity, we confronted each consonant interval X with all the dissonant intervals Y, and we looked for particular responses in cases where Y was the interval which should correspond to X according to the polarity formula $Y = 2X + 5$. Here, we looked at simultaneous intervals. For each given consonance X, we played a sequence of six confrontations, i.e., immediate successions $(X, Y_1), (X, Y_2), \dots (X, Y_6)$ of X with each of the six dissonances $Y_1, Y_2, \dots Y_6$, see also Figure 30.5. The

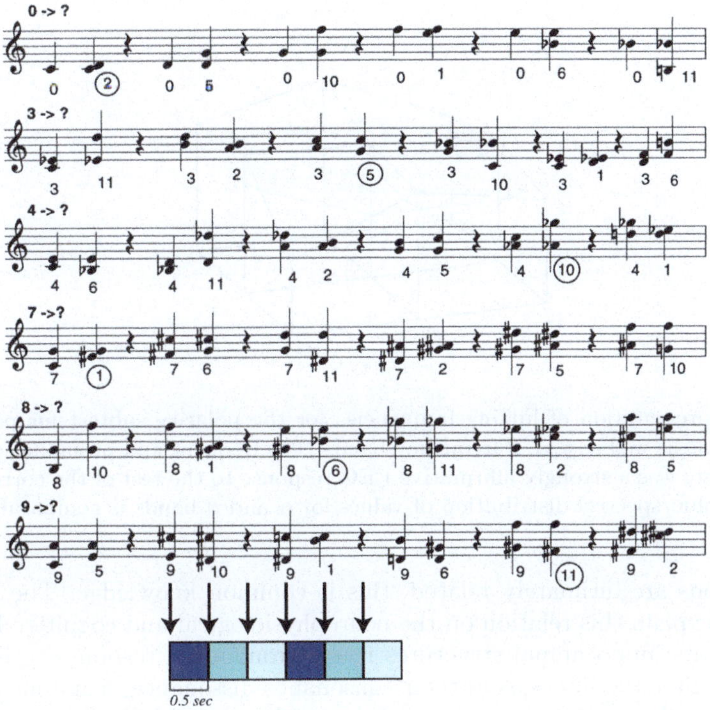

Fig. 30.5. The score of the confrontation test of each consonance with all dissonances; duration of a note in the test is 0.68 seconds.

duration of each interval was 0.68 seconds. We looked for the least participation values among the dissonant intervals $Y_1, Y_2, \dots Y_6$, when confronted with a fixed consonance X. We then compared the effective hits to the a priori chance to hit the correct dissonance. This method was applied to every recording position and to the three above frequency bands. Figure 30.6 shows the numbers compared to 100%, the measure for a priori chance. Due to the small number of samples for this subtest, we did not apply any statistical test here. However, as Figure 30.6 shows, the results are remarkable and similar for both patients, and we conclude that this pilot investigation strongly supports the presence of the (K/D)-polarity as a foundation of contrapuntal processes.

30.2.4 Music and the Hippocampal Gate Function

Summary. The neurophysiological results, in particular their localizations in the emotional brain and the auditory cortex, are interpreted from the cognitive perspective. The gate function of the hippocampal formation suggests a key function of music in opening subconscious—preferredly emotional—memory contents.

– Σ –

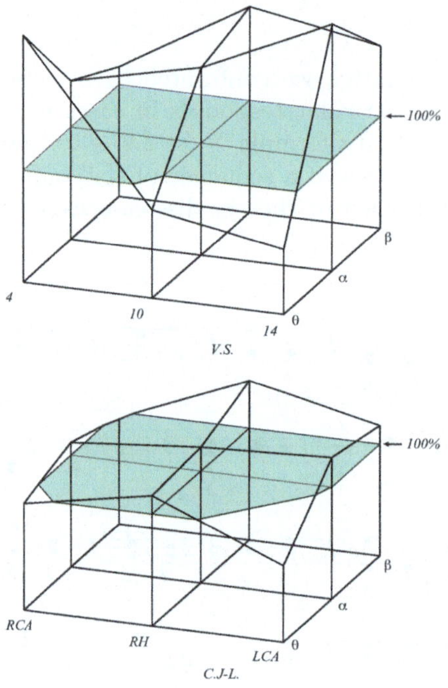

Fig. 30.6. Graphical representation of hitting frequencies for the polarity subtest for patients C. J.-L. and V.S., three homologous locations as well as θ, α, β frequency bands. The frequent and pronounced values above 100% show that for both patients there was a strongly affirmative EEG response to the test of the correct values for the polarity. In addition, the topographic/spectral distribution of values for α and β bands is comparable for these patients.

Music and emotions are intimately related, this is common knowledge. The above results suggest a mechanism that could explain this relation on the neurophysiological and cognitive level. We have seen that the emotional brain in its hippocampal structures has a pronounced response to elementary structures of harmony and counterpoints: the intervals in their consonance-dissonance dichotomy, and this is so independently of any sound color physics. Now, the classical thesis of Papez and MacLean [640] states that the limbic system, a prominent part of the archicortex, is responsible for emotional human behavior, this is why it is also called the emotional brain. So the hippocampal sensitivity to consonances vs. dissonances could relate to the emotional function of music, i.e., of musical intervals in our case. The question is, how musical signs which are by no means emotions by themselves (although Schönberg and other prominent music experts constantly evoke the notes' emotional and erotic life) can evoke and signify emotions in humans, and why this is done in such a way that the same music may evoke a great variety of such reactions and significations. Evidently these outputs are the result of a determined sample of music *plus* an individual human ingredient.

The point is that the hippocampal formation has been recognized as a key structure for memory [1000]. The neuroscientist Jonathan Winson has proposed a more specific theory of the hippocampal memory function [1134], in that he argues that the hippocampus performs a gate function to the subconscious (he even evokes Freud's "Unbewusstes"), i.e., to memory contents of emotional character. This means that the hippocampus is a structure that plays the role of a gateway to hidden memory contents. It is well known that humans do not have a free or controlled access to their memory contents, in particular not on the level of long-term and emotional memory, concerning early childhood, for example.

This suggests that special mechanisms must be activated in order to open the hippocampal gate to unveil locked memory contents. It is straightforward from our neurophysiological findings and the gate function of the hippocampus that its musical stimulation could yield such a "key" to open the gate to hidden memory contents. If this were the case, two specifica of the relation of music and emotion would be explained at once: (1) The emotional contents are not generated by music, they are merely retrieved and evoked from a memory database, whence the individual emotional response to one and the same music would receive a logical explanation. (2) The musical stimulation of the hippocampus is very probably not independent of the human individual who undergoes this process, in other words: If the music is a key, each individual is likely to have his/her individual key to the "subconscious". This would explain why there are so many different musical tastes—beyond musical education and culture. This would also explain why it is often a specific tune or musical mood that is the personal preference: If this tune played a role in the encoding of a specific emotional memory content, the same key-tune could play a role in the decoding process. Summarizing, we have this thesis:

Thesis 5 *Consonant and dissonant intervals and associated harmonic or contrapuntal structures evoke a hippocampus state/process which activates a gateway to mainly subconscious memory contents. In other words, Winson's gate hypothesis of the hippocampal formation must also be stated in the sense of the* existence of a musicogenic key to the gate.

This thesis does not mean that music produces emotions, it only retrieves and reactivates them from a memory database. So it acts on the brain like a drug and produces psychic effects. In this metaphor, the 'chemical formula' of the music drug corresponds to the involved musical structure.

31

Modeling Counterpoint by Local Symmetries

Der Rangunterschied zwischen den perfekten und den imperfekten Konsonanzen ermöglichte die Formulierung genereller Konsonanzfolgeregeln für eine indeterminata positio.

Klaus-Jürgen Sachs [924, p. 114]

Summary. This chapter presents the counterpoint model in form of a counterpoint theorem which guarantees the existence and exhibits an arsenal of admitted contrapuntal steps that come in extremely close to the rules of classical counterpoint. The theorem is based on the concept of a contrapuntal symmetry and follows the paradigm of local symmetries as a rationale for forces in physics. Because of its generic concept framework, the theorem, which in this general form[1] was proved by Jens Hichert [466], is also valid for non-European scales. We discuss these extensions.

– Σ –

31.1 Deformations of the Strong Dichotomies by Contrapuntal Symmetries on $IntMod_{12,q}[\varepsilon]$

Summary. In the core theory of counterpoint [924], the concept of contrapuntal "tension" between successive, perfect and imperfect consonant intervals plays a crucial role. This idea is made precise in the framework of *contrapuntal* symmetries which deform the strong dichotomy. The *separation property* of contrapuntal symmetries is proven.

– Σ –

The following counterpoint theorem (Section 31.3) is concerned with the elementary and core situation of classical counterpoint: "note-against-note". It is a theory which describes a system of rules for "allowed" sequences $(x_s + \varepsilon.i_s, \alpha_s)_s$ of oriented intervals, a system which essentially boils down to a set of rules for allowed successor pairings $(x_s + \varepsilon.i_s, \alpha_s), (x_{s+1} + \varepsilon.i_{s+1}, \alpha_{s+1})$ within such contrapuntal sequences. We shall also restrict our investigation to sequences of constant sweeping orientation $\alpha_s = \alpha_+ =$ const., changes in the orientation are handled as described above in 29.6.

The first elementary rule of counterpoint "note-against-note" says that we are not allowed to take other intervals than the consonances $\xi \in K[\varepsilon]$. This seems evident, but it imposes a strong obstruction against another, more hidden directive: the idea of creating a tension between each interval and its successor. More precisely, the meaning of "contra" is not only that of a vertical opposition between cantus firmus and

[1] The original theorem is presented in [678, 682] and deals with the classical consonances and dissonances, when applied to European diatonic scales.

G. Mazzola, *The Topos of Music I: Theory*, Computational Music Science,
https://doi.org/10.1007/978-3-319-64364-9_31

discantus. As Sachs has remarked in [924], the preposition "contra" equally means a horizontal opposition between successive intervals in the given sequence. This requirement is not very explicit, but it is reflected in the distinction between perfect consonances (prime, fifth, octave) and the others, the imperfect sixths and thirds, and the idea of changing between perfect and imperfect consonances in order to create tension. This conceptual distinction seems to evoke a dissonant ingredient in the consonant character, although it does not really abolish the consonance, it is a kind of coloring effect.

So, the idea of contrapuntal tension is in some sense a contradictory requirement against the primordial rule of forbidden dissonances: We should like to behave as if there were dissonances within consonances and to create a tensed movement from consonances to dissonances and vice versa. In order to solve this requirement in our mathematical remake of the contrapuntal rules, we introduce this technique: Given a symmetry g of $IntMod_{12,q}[\varepsilon]$, we may apply g to the consonance-dissonance dichotomy $(K[\varepsilon]/D[\varepsilon])$ and "deform" it to the dichotomy $g(K[\varepsilon]/D[\varepsilon])$. In general, the deformed dichotomy will have its parts in such a position that some real consonances ξ are also *g-deformed consonances*, i.e., $\xi \in g(K[\varepsilon])$, and some are *g-deformed dissonances*, i.e., $\xi \in g(D[\varepsilon])$. This implies that we may restate the directive of creating contrapuntal tension in the sense that for a given pair of successive consonances ξ, η, the first one is a g-deformed dissonance while the other is a g-deformed consonance (or vice versa) for a determined symmetry g, in which case we say that *the (unordered) pair ξ, η is g-polarized.* We shall see below in Section 31.2 that the symmetries which we shall exhibit for this role are indeed local symmetries which in physics are responsible for creation of forces, i.e., deformational tension.

Of course it is not evident that there is always a symmetry g that polarizes two consonances ξ, η in the above sense. Let us first discuss this topic. It shows that the strong dichotomies are exactly what is needed to guarantee this polarization property.

Proposition 52 *Let $(X[\varepsilon]/Y[\varepsilon])$ be a strong dichotomy and let ξ, η be two different intervals. Then there is a symmetry g such that the pair ξ, η is g-polarized.*

If ξ, η lie in different halves of $(X[\varepsilon]/Y[\varepsilon])$, then $g = Id$ does the job. So we may suppose that both, ξ and η lie one half, say in X, the other case is settled by a transformation of the pair of intervals to a pair within X via the polarity of $(X[\varepsilon]/Y[\varepsilon])$. Let $\xi = i + \varepsilon.j, \eta = k + \varepsilon.t$ with $j, t \in X$. The symmetry is $g = e^{l+\varepsilon.m}(n + \varepsilon.o), n^2 = 1$. Then we have these applications:

$$g.\xi = l + ni + \varepsilon(m + jn + oi), g.\eta = l + nk + \varepsilon(m + tn + ok).$$

We want the coefficients of ε to stay in X and Y, respectively. If we try with $o = 0$, this means that $e^m.n(j)$ and $e^m.n(t)$ stay in different halves. Suppose that both are always in the same half. Try further $n = 1$. Then we have $j + m, t + m \in X$ or $j + m, t + m \in Y$ for all $m \in \mathbb{Z}_{12}$. Take any $a \in X$ and set $m = a - j$. Then $j + m = a$, and $t + m = (t - j) + j + m = (t - j) + a$. So adding the difference $t - j$ to any $a \in X$ is again an element of X. But then the symmetry e^{t-j} is a non-trivial automorphism of X, contradicting the rigidity of X, and we are done. QED.

We shall give another proof of this fact below. Let us now formulate the properties of symmetries which we want to use for the deformations for a counterpoint rule set:

Definition 95 *Let $\Delta[\varepsilon] = (X[\varepsilon]/Y[\varepsilon])$ be a strong dichotomy. Let $\xi = x + \varepsilon.i \in X[\varepsilon]$. A symmetry g is* contrapuntal for *ξ iff*

(i) $\xi \notin g(X[\varepsilon])$,
(ii) p^x_Δ *is a polarity of* $g.\Delta[\varepsilon]$,
(iii) The cardinality of $g(X[\varepsilon]) \cap X[\varepsilon]$ *is maximal among those g which have properties* (i) *and* (ii).

The reason for these requirements is this. We have seen that every pair of intervals can be polarized by a specific symmetry. But we are interested in a rule set which guarantees more than the mere possibility of separation. We want to have those symmetries which admit a maximal number of polarized couples starting from a fixed interval. The second condition is introduced in order to relate the polarizing symmetry g to the given polarity p^x_Δ of the dichotomy at the cantus firmus x. It can be shown [675] that for $x = 0$ this property is also equivalent to the commutativity condition $g.p^0_\Delta = p^0_\Delta.g$, so it is a generalized commutativity condition.

Definition 96 *If a strong dichotomy $\Delta[\varepsilon] = (X[\varepsilon]/Y[\varepsilon])$ and an interval $\xi \in X[\varepsilon]$ are given, an interval η is called an* admitted successor of ξ *if it is contained in an intersection $g(X[\varepsilon]) \cap X[\varepsilon]$ for a contrapuntal symmetry g for ξ.*

31.2 Contrapuntal Symmetries Are Local

Summary. A closer look at contrapuntal symmetries shows their local character. In complete analogy with modern physics, local symmetries produce the looked-for forces of contrapuntal tension. Thus, the melodic variation of the cantus firmus in counterpoint is perfectly interpreted as a deformation *caused* by the "forces" from local symmetries.

– Σ –

Before we deal with the counterpoint theorem, we should explain the *local* character of contrapuntal symmetries. We shall consider the example of classical consonances and dissonances, i.e., the dichotomy $\Delta[\varepsilon] = (K[\varepsilon]/D[\varepsilon])$. We compare two interpretations of the local composition of all zero-addressed intervals $I = 0@IntMod_{12,q}[\varepsilon]$. The first I^{C_1} is induced by Δ an has the atlas

$$I_x, K_x = x + \varepsilon.K, D_x = x + \varepsilon.D, x \in \mathbb{Z}_{12}.$$

The second I^{C_g} is the analogous construction built on the deformed dichotomy $g.\Delta[\varepsilon]$, i.e., its atlas is

$$I_x, gK_x = I_x \cap g.K[\varepsilon], gD_x = I_x \cap g.D[\varepsilon], x \in \mathbb{Z}_{12}.$$

Although the autocomplementary function p^x_Δ for the cantus firmus x is also a polarity of the g-deformed dichotomy $g.\Delta[\varepsilon]$ according to definition 95, its action on the two interpretations is qualitatively different (see also Figure 31.1).

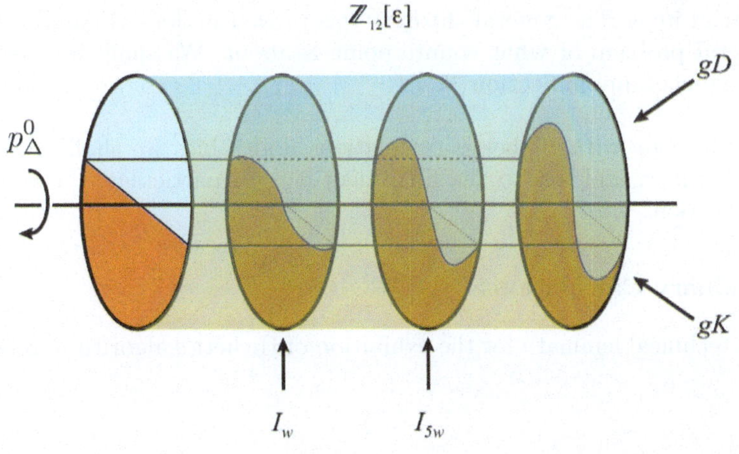

Fig. 31.1. The autocomplementary symmetry on the deformed consonance-dissonance dichotomy is a local symmetry, whereas it is a global one on the original dichotomy. This fact resembles physical forces being induced by local symmetries.

Consider the example $g = e^{\varepsilon.8}(5 + \varepsilon.4)$ for the consonance $\xi = \varepsilon.9$, i.e., at cantus firmus $x = 0$. We have

$$g.K[\varepsilon] = (1 - \varepsilon.4)\mathbb{Z}_{12} + \varepsilon.e^8.5K.$$

For a cantus-firmus point w, this means

$$gK_w = w + \varepsilon.(5K + 8 - 4w).$$

But under p^0_Δ, w is transported to $5w$, and we have

$$gK_{5w} = 5w + \varepsilon.(5K + 8 + 4w).$$

So we recognize the following: On the first interpretation, p^0_Δ acts on K_w via *translation* on the cantus firmus: $K_{5w} = e^{4w}(K_w)$, followed by the autocomplementarity symmetry on I_{5w}. But on the second interpretation, p^0_Δ does not operate on gK_w by a translation plus autocomplementarity symmetry on I_{5w}, in fact, gK_{5w} is different from $e^{4w}(gK_w)$! This can be understood in the sense that p^0_Δ acts on the second interpretation via g-deformation in the spirit of physics, i.e., as a *local symmetry*, instead of a global symmetry as it is the case for the first interpretation. The latter can be viewed as a 'spiral turn': 'rotation' (=autocomplementarity symmetry on each chart I_w) plus translation ($K_{5w} = e^{4w}(K_w)$) 'along the rotation axis'. The former however does not shift the part gK_w of I_w to $e^{4w}(gK_w)$, but deforms it to

$$gK_{5w} = 5w + \varepsilon.(5K + 8 + 4w) = e^{\varepsilon.8w}(e^{4w}(gK_w))$$

by the factor $\varepsilon.8w$. This phenomenon is analogous to physics in the sense that forces in physics are induced by local symmetries [335]. In this understanding, local symmetries on the interval space seem to be responsible for the tension which controls the progression from interval to interval in our model.

31.3 The Counterpoint Theorem

Summary. This section presents the counterpoint theorem and its corollaries.

– Σ –

In this section, we prove the general counterpoint theorem. "General" means that we deal with all strong dichotomies and construct the lists of admitted interval successors for these cases, including the classical case as a special item. The general theorem was proved in Jens Hichert's thesis [466] and sheds a new light on the general problem of what counterpoint is about. We shall discuss the special case of the consonance-dissonance dichotomy in Section 31.4.1.

Remark 17 In order to cope with Hichert's calculations and tables we shall use the interval space relative to the identity identifier, and not to the fifth identifier. In particular, the consonance quantities are $\{0, 3, 4, 7, 8, 9\}$ in this section.

31.3.1 Some Preliminary Calculations

Summary. We prove technical lemmata for the exhibition of Hichert's algorithm to be introduced in Section 31.3.3.

– Σ –

We fix a marked dichotomy $\Delta = (X/Y)$. Let us first come back to the rigidity property of $X[\varepsilon]$:

Lemma 44 *The symmetry group of* $X[\varepsilon]$ *is* $Sym(X[\varepsilon]) = e^{\mathbb{Z}_{12}}$.

Proof. Clearly $e^{\mathbb{Z}_{12}} \subset Sym(X[\varepsilon])$. Therefore,

$$e^{z+\varepsilon.t(u+\varepsilon.v}X[\varepsilon] = X[\varepsilon] \text{ iff } e^{\varepsilon.t(u+\varepsilon.v)}X[\varepsilon] = X[\varepsilon].$$

This implies $t + vz + uk \in X$ for all $z \in \mathbb{Z}_{12}, k \in X$. Whence $t = 0, u = 1$ since X is rigid. Therefore, for $z = 1$, we imply $v + k \in X$, all $k \in X$, i.e., $v = 0$. QED.

Proposition 53 *Let $H = e^{\varepsilon.\mathbb{Z}_{12}}.GL(\mathbb{Z}_{12}[\varepsilon])$. Then $X[\varepsilon]$ is H-rigid, i.e., the orbit application*

$$H \to ObLoc_0^{\mathbb{Z}_{12}[\varepsilon]} : g \mapsto g.X[\varepsilon]$$

$\varepsilon.t$

Exercise 69 *Show that each consonant part $X_x = x + \varepsilon.X$ of the tangent space I_x is rigid.*

Lemma 45 *With the notation of corollary 53, if $g = \varepsilon.t(u + \varepsilon.v) \in H$, and if $z \in \mathbb{Z}_{12}$, we set $g^{(z)} = g.e^{\varepsilon.vz}$ Then we have*

(i) $(g^{(z_1)})^{(z_2)} = g^{z_1+z_2}$,
(ii) $e^z.g.X[\varepsilon] = g^{(z)}.X[\varepsilon]$, *and* $e^z.g.Y[\varepsilon] = g^{(z)}.Y[\varepsilon]$.

Proof. The first formula is clear. The second follows from $e^z g = e^{z+\varepsilon.t}(u + \varepsilon.v) = e^{\varepsilon.t}(u + \varepsilon.v).e^{z(u-\varepsilon.v)} = g^{(z)}.e^{zu}$. QED.

Corollary 20 *For $g \in \overrightarrow{GL}(\mathbb{Z}_{12}[\varepsilon])$, there is a symmetry $h \in H$ such that $g.X[\varepsilon] = h.X[\varepsilon]$.*

In fact, there is a $u \in H$ such that $g = e^z.u$, so by Lemma 45, $g.X[\varepsilon] = u^{(z)}.X[\varepsilon]$, and we have the solution $h = u^{(z)}$.

Lemma 46 *Let $\xi = x + \varepsilon.k, g \in \overrightarrow{GL}(\mathbb{Z}_{12}[\varepsilon])$, and $z \in \mathbb{Z}_{12}$. Then, if $\xi \notin g.X[\varepsilon]$ and $p^x_\Delta : g.X[\varepsilon] \xrightarrow{\sim} g.Y[\varepsilon]$, we also have $e^z\xi \notin e^z.g.X[\varepsilon]$, $p^{z+x}_\Delta : e^z.g.X[\varepsilon] \xrightarrow{\sim} e^z.g.Y[\varepsilon]$, and $e^z.g.X[\varepsilon] \cap X[\varepsilon] = e^z.(g.X[\varepsilon] \cap X[\varepsilon])$, in particular $card(g.X[\varepsilon] \cap X[\varepsilon]) = card(e^z.g.X[\varepsilon] \cap X[\varepsilon])$.*

Proof. It is clear that $e^z\xi \notin e^z.g.X[\varepsilon]$, while by Lemma 51,

$$p^{z+x}_\Delta.e^z.g.X[\varepsilon] = e^z.p^x_\Delta.e^{-z}.e^z.g.X[\varepsilon] = e^z.p^x_\Delta.g.X[\varepsilon] = e^z.g.Y[\varepsilon],$$

and finally, $e^z.g.X[\varepsilon] \cap X[\varepsilon] = e^z.g.X[\varepsilon] \cap e^z.X[\varepsilon] = e^z.(g.X[\varepsilon] \cap X[\varepsilon])$. QED.

Proposition 54 *The contrapuntal symmetries can be calculated if one knows the contrapuntal symmetries $g \in H$ at cantus firmus $x = 0$. More precisely, if $\xi = x + \varepsilon.k \in X[\varepsilon]$ and if g is any symmetry, such that properties* (i) *through* (iii) *of g in definition 95 are true, then they are also true with unchanged set $g.X[\varepsilon] = h.X[\varepsilon]$ for a symmetry $h \in H$. Furthermore, to check this property for h, we may verify the properties* (i) *through* (iii) *for the interval $\varepsilon.k$, the symmetry $h^{(-x)} \in H$, and the polarity $p^{(0)}_\Delta$. Finally, the intersection $h.X[\varepsilon] \cap X[\varepsilon]$ coincides with the translate $e^x(h^{(-x)}.X[\varepsilon] \cap X[\varepsilon])$, which means that we may just look for relative cantus firmus steps when building the rules of admitted steps.*

Proof. The replacement of g by h follows from corollary 20. By Lemma 45 we have $e^{-x}.h.X[\varepsilon] = h^{(-x)}.X[\varepsilon]$, and by Lemma 46, with $z = -x$, we can verify the contrapuntality of h on $e^{-x}\xi = \varepsilon.k$, and on $p^{(-x+x)}_\Delta = p^{(0)}_\Delta$. The last statement follows from the translation formula for intersections in Lemma 45. QED.

31.3.2 Two Lemmata on Cardinalities of Intersections

Lemma 47 *Let* (X/Y) *be a strong dichotomy in* $0@IntMod_{12,1}$*, and* $0 \leqslant i \leqslant 6$ *and integer. Setting* $G_i = \{g \in \overrightarrow{GL}(\mathbb{Z}_{12}) |\ card(g.X \cap X) = i\}$*, we have* $card(G_i) = card(G_{6-i})$*.*

Proof. Let p be the polarity of (X/Y), the right multiplication with p induces a permutation of order 2 $?p : \overrightarrow{GL}(\mathbb{Z}_{12}) \xrightarrow{\sim} \overrightarrow{GL}(\mathbb{Z}_{12})$. For $g \in G_i$, consider the dichotomy $g(X/Y)$ and its intersection with X. This gives $X = g.X \cap X \cup g.p.X \cap X$, whence $card(g.p.X \cap X) = 6 - card(g.X \cap X)$. Therefore, $g.p \in G_{6-i}$, and by the order two permutation $?p$ and the finiteness of all the involved sets, we have $p.G_i = G_{6-i}$. QED.

The next lemma basically guarantees the existence of admitted contrapuntal successor intervals, as we shall see in the next section.

Lemma 48 *Let* K *be a zero-addressed objective local composition in a finite cyclic group* $M \xrightarrow{\sim} \mathbb{Z}_n$ *and let* $U \in GL(M)$*. Then*

$$\sum_{m \in M} card(e^m.U(K) \cap K) = k^2.$$

Proof. Let $U(K) = \{u_1, \ldots u_k\}$ the image set with its $k = card(K)$ elements. Let r be a generator of M. Then

$$\begin{aligned}
&\sum_{m \in M} card(e^m.U(K) \cap K) \\
&= \sum_{t=0}^{n-1} card(e^{tr}.U(K) \cap K) \\
&= \sum_{t=0}^{n-1} \sum_{s=1}^{k} card(\{e^{tr}u_s\} \cap K) \\
&= \sum_{t=0}^{n-1} \sum_{s=1}^{k} \chi_K(e^{tr}u_s) \\
&= \sum_{s=1}^{k} (\sum_{t=0}^{n-1} \chi_K(e^{tr}u_s))
\end{aligned}$$

with the characteristic function χ_K for an element being in K or not. Since r is a generator of M, the expression $\sum_{t=0}^{n-1} \chi_K(e^{tr}u_s)$ adds up to k, and we have the result. QED.

31.3.3 An Algorithm for Exhibiting the Contrapuntal Symmetries

Summary. This section discusses Hichert's algorithm for the calculation of all contrapuntal symmetries and admitted contrapuntal steps by use of a specific software.

– Σ –

This section restates the three conditions (i) through (iii) of the contrapuntal symmetries in order to provide an algorithm for the calculation of all admitted contrapuntal interval successors. Given the strong dichotomy $\Delta = (X/Y)$, and according to Proposition 54, we may start from an interval $\xi = \varepsilon.k, k \in X$ and restrict to symmetries $g = e^{\varepsilon.t}(u + \varepsilon.uv) \in H$ (uv instead of v without restriction). Let us first reformulate conditions (i) and (ii). We have

$$
\begin{aligned}
g.X[\varepsilon] &= \bigcup_{x\in\mathbb{Z}_{12}} g.(x+\varepsilon.X)\\
&= \bigcup_{x\in\mathbb{Z}_{12}} ux+\varepsilon.(uvx+t)+\varepsilon.uX\\
&= \bigcup_{y\in\mathbb{Z}_{12}} y+\varepsilon.(vy+t)+\varepsilon.uX\\
&= \bigcup_{y\in\mathbb{Z}_{12}} y+\varepsilon.(e^{vy+t}u.X).
\end{aligned}
$$

Setting $h(y) = e^{vy+t}u$, we have

$$g.X[\varepsilon] = \bigcup_{y\in\mathbb{Z}_{12}} y+\varepsilon.h(y).X.$$

Therefore, we have

$$g.X[\varepsilon] \cap X[\varepsilon] = \bigcup_{y\in\mathbb{Z}_{12}} y+\varepsilon.(h(y).X\cap X).$$

This means that condition (i) is equivalent to $k \notin h(0).X$ which means $k \in h(0).p.X$, and this is equivalent to

$$\exists s \in X \text{ such that } k = h(0).p(s) = u.p(s)+t.$$

This is equivalent to the statement that there is an $s \in X$ such that

$$g.X[\varepsilon] = \bigcup_{y\in\mathbb{Z}_{12}} y+\varepsilon.(e^{vy+k-u.p(s)}u.X) \tag{31.1}$$

so that we have

$$card(g.X[\varepsilon]\cap X[\varepsilon]) = \sum_{y\in\mathbb{Z}_{12}} card(e^{vy+k-u.p(s)}u.X\cap X). \tag{31.2}$$

Further, condition (ii) means $p^0_\Delta.g = g.p^0_\Delta$, see the remarks after definition 95. If we have $p = e^r.w$, this means that

$$wt + r = ur + t. \tag{31.3}$$

Therefore, conditions (i) and (ii) are equivalent to equations 31.1 and 31.3, whereas the maximal number is calculated upon formula 31.2.

In order to calculate the number 31.2, we have to distinguish three cases concerning the values of v:

1. v is invertible. Then we have
$$card(g.X[\varepsilon]\cap X[\varepsilon]) = \sum_{y\in\mathbb{Z}_{12}} card(e^{y}.u.X\cap X). \tag{31.4}$$
2. $v = 0$. This gives
$$card(g.X[\varepsilon]\cap X[\varepsilon]) = 12card(e^{k-u.p(s)}.u.X\cap X). \tag{31.5}$$
3. $v = \pm\rho, \rho \in \{2,3,4,6\}$. This gives
$$card(g.X[\varepsilon]\cap X[\varepsilon]) = v\sum_{j=1}^{12/v} card(e^{(j-1)v+k-u.p(s)}.u.X\cap X). \tag{31.6}$$

If we recall Lemma 15, the first case with formula 31.4 implies:

Fact 15 *There are always at least 36 successors for a fixed given interval $\xi \in X[\varepsilon]$.*

We are now ready to state the algorithm which was implemented on Turbo-Pascal by Hichert [466]. The algorithm starts from the fixed value $k \in X$ and first calculates all the possible coefficients of g. This means that we have to go through the loop which transgresses all $g = e^{\varepsilon.t}(u + \varepsilon.uv)$ via $u \in \{1, 5, 7, 11\}, s \in X, v \in \mathbb{Z}_{12}, t = k - up(s)$, and then for each such value set calculate $card(g.X[\varepsilon] \cap X[\varepsilon])$ according to the three cases 31.4, 31.5, 31.6. For each case, we update the set C_k of intermediate candidates for contrapuntal symmetries, i.e., we add a new g to the existing set if its intersection number is maximal among the already given candidates in C_k, and we remove all previous g where intersection cardinalities are smaller than the actual maximum.

Exercise 70 *Write a C or Java program which implements the above algorithm. If you do not speak C or Java, write a Mathematica or Maple program. If you do not speak these languages either, interrupt reading this book, learn one of these languages, and proceed.*

The complete lists of all contrapuntal symmetries, together with their intersection configurations, can be found in Appendix Q.1. The admitted successors will be listed below. This information yields the following

Theorem 33 *Let $\Delta = (X/Y)$ be a strong dichotomy, and let $\xi \in X[\varepsilon]$. The number of admitted successors of ξ is always at least equal to 36, in particular, there exists always a contrapuntal symmetry g for ξ. An admitted successor of ξ also always exists if one prescribes the cantus firmus of the successor interval.—For each of the six strong dichotomy classes, a list of forbidden successor intervals is exposed below, after this theorem.*

We shall say that *an interval $\xi \in X$ is a "cul-de-sac" under determined conditions* if there is no admitted successor under these conditions.

The following table is meant as follows: Each table section is related to a fixed representative (X/Y) of the indicated class. The first column indicates the interval quantity k of the "consonant" start interval $\xi = x + \varepsilon.k \in X[\varepsilon]$. The obstructions to successor interval $\eta = y + \varepsilon.l \in X[\varepsilon]$ are visible in columns 2–13 where the difference $d = y - x = 0, 1, 2, \ldots 11$ leads the column in the first row. For each couple k, d, we see the forbidden interval quantities l of the target interval. For example, in class 64, the steps $x + \varepsilon.4 \mapsto x + 6 + \varepsilon.l$ is forbidden for $l = 2, 4, 7, 9$ and admitted for all other l.

k	0	1	2	3	4	5	6	7	8	9	10	11
Forbidden Successors for Dichotomy Class Nr. 64												
2	2	-	2	-	2	-	2	-	2	-	2	-
4	2,4,7,9	-	2,4,7,9	-	2,4,7,9	-	2,4,7,9	-	2,4,7,9	-	2,4,7,9	-
5	5,11	5,11	5,11	5,11	5,11	5,11	5,11	5,11	5,11	5,11	5,11	5,11
7	2,4,7,9	-	2,4,7,9	-	2,4,7,9	-	2,4,7,9	-	2,4,7,9	-	2,4,7,9	-
9	9	-	9	-	9	-	9	-	9	-	9	-
11	11	11	11	11	11	11	11	11	11	11	11	11
Forbidden Successors for Dichotomy Class Nr. 68												
0	0	0	0	0	0	0	0	0	0	0	0	0
1	1	-	1	-	1	-	1	-	1	-	1	-
2	2,8	2,8	2,8	2,8	2,8	2,8	2,8	2,8	2,8	2,8	2,8	2,8
3	1,3,5	-	1,3,5	-	1,3,5	-	1,3,5	-	1,3,5	-	1,3,5	-
5	5	-	5	-	5	-	5	-	5	-	5	-
8	8	8	8	8	8	8	8	8	8	8	8	8
Forbidden Successors for Dichotomy Class Nr. 71												
0	0,3	3	0,3	3	0,3	3	0,3	3	0,3	3	0,3	3
1	1,2,7	-	1,2,7	-	1,2,7	-	1,2,7	-	1,2,7	-	1,2,7	-
2	2,3	-	2,3	-	2,3	-	2,3	-	2,3	-	2,3	-
3	2,3	-	2,3	-	2,3	-	2,3	-	2,3	-	2,3	-
6	3,6	3	3,6	3	3,6	3	3,6	3	3,6	3	3,6	3
7	1,2,7	-	1,2,7	-	1,2,7	-	1,2,7	-	1,2,7	-	1,2,7	-
Forbidden Successors for Dichotomy Class Nr. 75												
0	0	-	-	-	-	-	0	-	-	-	-	-
1	0,1,4,5	-	0,1,4,5	-	0,1,4,5	-	0,1,4,5	-	0,1,4,5	-	0,1,4,5	-
2	2,5	-	-	2,5	-	-	2,5	-	-	2,5	-	-
4	0,1,4,5	-	0,1,4,5	-	0,1,4,5	-	0,1,4,5	-	0,1,4,5	-	0,1,4,5	-
5	2,5	-	-	2,5	-	-	2,5	-	-	2,5	-	-
8	2,8	2	2	2,8	2	2	2,8	2	2	2,8	2	2
Forbidden Successors for Dichotomy Class Nr. 78												
0	0,2,6	-	0,2,6	-	0,2,6	-	0,2,6	-	0,2,6	-	0,2,6	-
1	1,2	-	1,2	-	1,2	-	1,2	-	1,2	-	1,2	-
2	1,2	-	1,2	-	1,2	-	1,2	-	1,2	-	1,2	-
4	1,4	-	-	1,4	-	-	1,4	-	-	1,4	-	-
6	0,2,6	-	0,2,6	-	0,2,6	-	0,2,6	-	0,2,6	-	0,2,6	-
10	1,10	-	1	1,4	1	-	1,10	-	1	1,4	1	-
Forbidden Successors for Dichotomy Class Nr. 82												
0	0	-	-	-	-	-	0	-	-	-	-	-
3	3,9	-	-	3,9	-	-	3,9	-	-	3,9	-	-
4	0,4,8	-	0,4,8	-	0,4,8	-	0,4,8	-	0,4,8	-	0,4,8	-
7	7	7	7	7	7	7	7	7	7	7	7	7
8	8	-	-	-	-	-	8	-	-	-	-	-
9	3,9	-	-	3,9	-	-	3,9	-	-	3,9	-	-

31.3.4 Transfer of the Counterpoint Rules to General Representatives of Strong Dichotomies

Summary. The above counterpoint theorem was made explicit for one selected representative of a strong dichotomy. Following [466], we give rules to transfer these results and tables to arbitrary representatives of strong dichotomies.

– Σ –

Let $\Delta = (X/Y)$ be a strong dichotomy with polarity p_Δ. Take an interval $\xi = \varepsilon.k \in X[\varepsilon]$, and let $\mathfrak{R} \subset H$ be the set of contrapuntal symmetries of Δ and ξ. Take any symmetry $g = e^t u \in \overrightarrow{GL}(\mathbb{Z}_{12})$, set $g_\varepsilon = e^{\varepsilon.t}u$, and consider the transformed dichotomy $(L/M) = g.\Delta$. We have $g_\varepsilon.X[\varepsilon] = L[\varepsilon], g_\varepsilon.Y[\varepsilon] = M[\varepsilon]$.

Proposition 55 [466, Satz 3.2] *With the above notation, the conjugate set $\mathfrak{R}_g = g_\varepsilon.\mathfrak{R}.g_\varepsilon^{-1}$ is exactly the set of contrapuntal symmetries of $g.\Delta$ and $\xi_g = g_\varepsilon\xi$. The number of admitted successors of ξ and of ξ_g coincide.*

Attention: In general, the fact that η is a successor of ξ does not imply that $g_\varepsilon\eta$ is a successor of $g_\varepsilon\xi$!

This result can be used to transform the tables of forbidden successors as given in the table after the counterpoint theorem 33 for other representatives of the dichotomy classes. To this end, suppose that in such a table (for dichotomy representative $\delta = (X/Y)$), the couple $\xi = \varepsilon.k, \eta = b + \varepsilon.j$ is forbidden. This means that in our table, on the location of row with interval quantity k and column with interval quantity b, the coefficient j appears as a forbidden quantity. According to Proposition 55, in the transformed table, the row $g.k$ and the column $g.b$ must show a forbidden quantity $g.j$. So we have a recipe for transforming a table for Δ under a symmetry $g = e^t u$:

1. Permute the 12 columns of the given table by a multiplication of the column head numbers $0, 1, 2, 3, \ldots 11$ by u (mod 12) and rearrange the new numbers by increasing values.
2. Replace the leading column interval numbers k by $g.k$ and rearrange the corresponding 6 rows by increasing values of the leading numbers.
3. Replace each forbidden item j by the item $g.j$.

31.4 The Classical Case: Consonances and Dissonances

Summary. This section deals with the classical case of the counterpoint theorem for the consonance-dissonance dichotomy of Palestrina-Fux Theory.

– Σ –

We give a specialized counterpoint theorem for the consonance-dissonance dichotomy, the sweeping orientation, and relating to ecclesiastical modes as defined in Section 13.4.2.

Theorem 34 *Let $\Delta = (K/D)$ be the consonance-dissonance dichotomy, and let $\xi \in K[\varepsilon]$. The number of admitted successors of ξ is always at least equal to 36, in particular, there exists always a contrapuntal symmetry g for ξ. An admitted successor of ξ also always exists if one prescribes the cantus firmus of the successor interval.—If one restricts the admitted pitch classes to an ecclesiastical mode (see Section 13.4.2), then an admitted successor of ξ always exists, even if the cantus firmus is prescribed. So, under these conditions, there is no cul-de-sac.*

Parallels of fifth ($x+\varepsilon.7 \mapsto y+\varepsilon.7$) are generally forbidden. For all other parallels, no general obstruction exists.

The admitted relative progressions are listed in the table after Theorem 33, Class Nr. 82, whereas the progressions for the C-major scale are listed in Appendix Q.2.

We shall discuss below the relation between strong dichotomies and scales, in particular the appearance of culs-de-sac.

31.4.1 Discussion of the Counterpoint Theorem in the Light of Reduced Strict Style

Summary. The concluding section gives an overview of the (strong) commonalities with and (weak) diversities from the classical rules of the Palestrina—Fux system and its reduction to pitch classes.

– Σ –

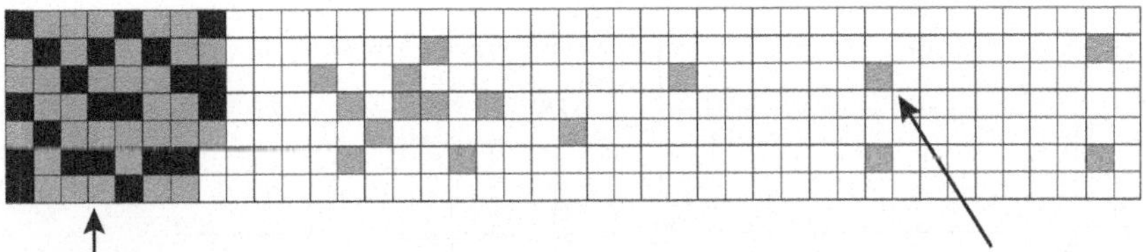

Fig. 31.2. Out of the 287 possible progressions with a mode (modulo translations of the cantus firmus), the reduced strict style exhibits 54 inadmissible cases. According to the mathematical model of counterpoint, 37 progressions are not admitted. Out of these, 21 cases are inadmissible in Fux' sense. intuitively, the commonalities of the two approaches can be described by use of a probabilistic argument: If somebody tries to hit at least 21 of the 54 inadmissible cases of the reduced strict style without knowing anything about counterpoint by 37 trials, the chance is less than 2.10^{-8}.

In this discussion, we refer to the codification of the strict style by Fux [349] (see also [1054]) since its explicit and rigorous rule system is particularly useful for a qualitative and quantitative comparison of the mathematical model with the classical counterpoint rules. In order to establish a basis of comparison for the mathematical model, we first have to transfer the Fux rules to the interval space $IntMod_{12,1}$. This yields a model of Fux' rules 'modulo octave', a rule system which we call the *reduced strict style.*

The upshot of a detailed investigation [684][2] states this:

Fact 16 *In the reduced strict style, only the rule of forbidden fifth parallels and the tritone rules have an unrestricted validity.*

Within an ecclesiastical mode, there are 287 a priori possible progressions [684]. According to the consonance-dissonance counterpoint Theorem 34, 37 of them are forbidden. Among them, 21 coincide with the 54 Fux-inadmissible progressions. The remaining 16 forbidden progressions of the mathematical model deal with progressions which are bad or allowed by Fux. Out of the ten allowed ones, four concern tritone movements of the cantus firmus. Three of the remaining obstructions (in the mathematical model) concern progressions from the major third into the prime, from which one may lead to an "ottava battuta". The remaining three obstructions deal with progressions which leave unaltered the pitch material, see also Figure 31.2.

We should stress that the mathematical model does not formalize a switch between perfect and imperfect consonances. The concrete shape of a polarization which induces a determined progression is redefined for each individual progression. Further, the model can be applied to any scale. This unveils an interesting fact concerning the dominant role of the major scale: In the analysis for the three seven-element scale classes which consist exclusively of minor and major second steps (classes $\widehat{38.1}$: major, $\widehat{47.1}$: melodic minor, and $\widehat{62}$: whole-tone scale, extended by one pitch class) the major scale is by far optimal for the degree of freedom in the choice of admitted successors. There is no cul-de-sac. Only for two progressions with prescribed start

[2] The paper [684] was accepted for publication in the journal *Musiktheorie*, but never published for marketing reasons (!), an anecdote about German musicology which shares a particular flavor.

interval and cantus firmus progression is the successor uniquely determined. The melodic minor scale has less successor freedom, but there are no culs-de-sac. In 16 cases there is only one possible solution. In the extended whole-tone scale, there are 18 culs-de-sac. The freedom of choice is minimal. We shall discuss this item in more generality in Section 31.4.2.

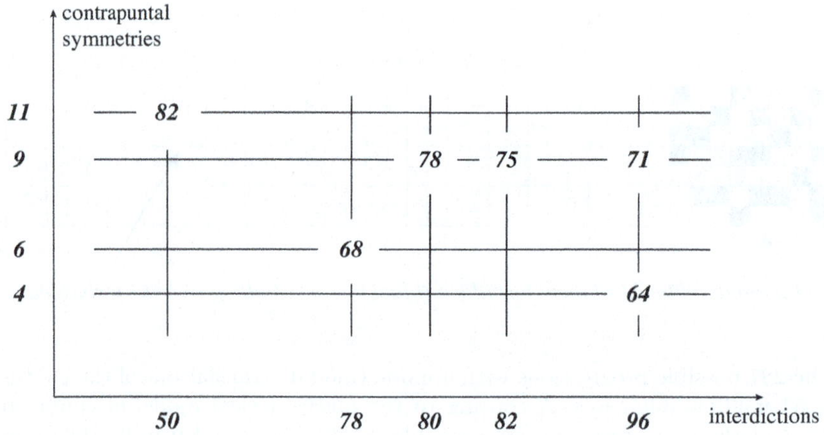

Fig. 31.3. From the proof of the counterpoint theorem, one deduces the numbers of contrapuntal symmetries and interdictions. These again position the major and consonance-dissonance dichotomies in a polar relation.

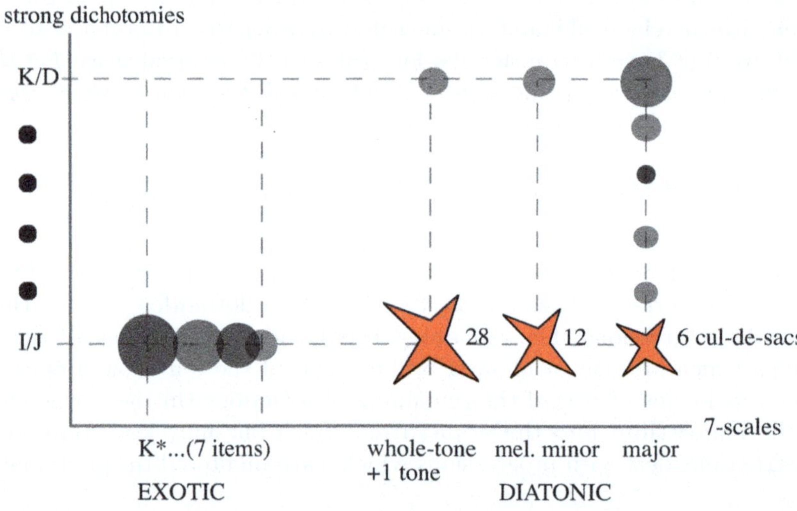

Fig. 31.4. Polarity between seven-tone scales and strong dichotomies shows a polarization between European and exotic (in particular: akin to Indian) scales. The size of darkened disks shows the number of progressions while the stars show culs-de-sac.

31.4.2 The Major Dichotomy—A Cultural Antipode?

Summary. Among the six strong dichotomy classes, we look for others than the Fux dichotomy and discuss their relative positions. Among the possible alternatives we especially focus on the major dichotomy, a topological antipode to the Fux dichotomy. The possibility to associate the major dichotomy with classical Indian scales is discussed.

– Σ –

In his thesis [466], Hichert has observed a number of interesting topological properties of the six strong dichotomies. Here is a representative record of these observations:

The major and consonance-dissonance dichotomies are not only polar with respect to diameter (definition 93 in Section 30.1) and span (definition 94 in Section 30.1), see Figure 30.1. They are also polar with regard to their number of contrapuntal symmetries and interdictions, see Figure 31.3.

In other words:

Fact 17 *The consonance-dissonance dichotomy has a maximum of contrapuntal symmetries and a minimum of interdictions, as opposed to the polar major dichotomy.*

We have already observed in Section 31.4.1 that, among the 'diatonic' scales (only minor and major second steps), the major scale has a maximum of freedom of choice for the consonance-dissonance dichotomy. Conversely, the major scale has only culs-de-sac for the major dichotomy.

So we are also interested in the scales with seven tones with respect to the major dichotomy! It is a further fact that among these scales with *no cul-de-sac for the major dichotomy* no European scale appears. There is one such scale, namely $K^* = \{0, 3, 4, 7, 8, 9, 11\}$ of class $\widehat{60}$, which is very interesting. To begin with, it is the consonance set K plus an added 'leading' note 11. So if the major scale is good for consonances and dissonances, the major dichotomy is good for a scale which is intimately related to consonances!

This scale really shows a character which relates to Indian raga music. The basic melakarta framework for ragas is known [234, Bd.8, p.265ff] to be built by the 72 melas. The small number of seven of these mela scales have been used until the present days. One of them, "mayamalavagaula", $Nr.15 = \{c, d_\flat, e, f, g, a_\flat, h\}$, which is class $\widehat{61}$ in our chord classification, can be represented by $\{0, 3, 4, 7, 8, 9, 1\}$. But this is very similar to the above scale K^*, we only have to switch the 11 to the 1, as opposed to 11 in 0. So we are led to the question of how far there is a polarity not only in mathematical relations between consonances-dissonance and the major scale as opposed to the major dichotomy and a 'consonance-dissonance' scale K^* which is akin to mela Nr.15 in raga music, but also a global polarity in musical cultures between European and Indian tradition. See also Figure 31.4 for this polarization.

This concluding discussion is an excellent example of the anthropic principle on the level of scale and interval interplay. Seemingly, the historical selection tends to optimize certain abstract a priori properties of topological and transformational character. The possibility to compare such different musical cultures as European and Indian traditions opens a wide field of comparative musicology (relative ethnology) which is based upon systematic results instead of historiographic and ethnographic contingencies.

31.4.3 Software for Counterpoint and Theoretical Extentions

The contrapuntal composition method for all of the six strong dichotomy classes has been implemented by Julien Junod in the RUBATO® Composer software, please see Section 69.2.1 for more information.

The present theory for 12-periodic pitch classes has been extended to general pitch classes in $\mathbb{Z}_{2n}$ by Octavio Alberto Agustín-Aquino, see [16] for a detailed exposition of this fascinating extension.

Part XXIV

References and Index

References

1. Abel V and Reiss P: MUTABOR II - Software Manual. Mutabor Soft, Darmstadt 1991
2. Abraham R and Marsden J: Foundations of Mechanics. Benjamin, New York et al. 1967
3. Abramowitz M and I A Stegun: Handbook of mathematical functions. With formulas, graphs and mathematical tables. Dover, 1965
4. Ackermann Ph: Computer und Musik. Springer, Wien and New York 1991
5. Aczel P: Non-well-founded Sets. No. 14 in CSLI Lecture Notes. Center for the Study of Language and Information, Stanford 1988
6. Ackermann Ph: Developing Object-Oriented Multimedia Software. dpunkt, Heidelberg 1996
7. Adorno Th W: Fragment über Musik und Sprache. Stuttgart, Jahresring 1956
8. Adorno Th W: Zu einer Theorie der musikalischen Reproduktion. Suhrkamp, Frankfurt/M. 2001
9. Adorno Th W: Der getreue Korrepetitor (1963). Gesammelte Schriften, Bd. 15, Suhrkamp, Frankfurt am Main 1976
10. Agawu V K: Playing with Signs. Princeton University Press, Princeton 1991
11. Agmon E: A Mathematical model of the diatonic system. JMT, 33, 1-25, 1989
12. Agmon E: Coherent Tone-Systems: a study in the theory of diatonicism. JMT, 40(1), 39-59, 1996
13. Agon A: OpenMusic: Un langage visuel pour la composition musicale assistée par ordinateur. PhD Dissertation, Université Paris VI, Paris 1998
14. Agustín-Aquino O A: Counterpoint in $2k$-tone equal temperament. Journal of Mathematics and Music, Vol 3, nr. 3, pp. 153-164, 2009
15. Agustín-Aquino O: Extensiones Microtonales de Contrapunto. PhD Thesis, Facultad de Ciencias, UNAM, Mexico City, 2012
16. Agustin-Aquino, O A, J Junod, G Mazzola: Computational Counterpoint Worlds. Springer Series Computational Music Science, Heidelberg 2014
17. Akmajian A et al.: Linguistics. MIT Press, Cambridge MA 1995
18. Alain: Alain Citation on Almada's painting in the Gulbenkian Foundation Center. Lisbon 1968
19. Albert R S and M A Runco: A History of Research on Creativity. In: R J Sternberg (Ed.): Handbook of Creativity. Cambridge University Press, New York 1999
20. Aleinikov A G: On Creative Pedagogy. Higher Education Bulletin, 12, 29-34, 1989
21. D'Alembert J Le Rond: Einleitung zur Enzyklopädie (1751). (German Translation) Fischer, Frankfurt/Main 1989
22. Alforhs L V: Complex Analysis. McGraw-Hill, New York et al. 1966
23. Alpaydin R and G Mazzola: A Harmonic Analysis Network. Musicae Scientiae, Vol. 19(2) 191–213, 2015
24. Alunni Ch: Diagrammes & catégories comme prolégomènes à la question: Qu'est-ce que s'orienter diagrammatiquement dans la pensée? In: Batt, N (éd): Penser par le diagramme. Presses Universitaires de Vincennes no.22-2004, Vincennes 2004
25. Alunni Ch: Relativités et puissances spectrales chez Gaston Bachelard. Revue de synthèse, 1999, 120-1, p. 73-110, 1999
26. Alunni Ch: Diagrammes et catégories comme prolégomènes à la question : qu'est-ce que s'orienter diagrammatiquement dans la pensée ? In: Théorie Littérature Enseignement no 22 - Penser par le diagramme de Gilles Deleuze à Gilles Châtelet, p. 83-93, 2004

G. Mazzola, *The Topos of Music I: Theory*, Computational Music Science,
https://doi.org/10.1007/978-3-319-64364-9

27. Alunni Ch: Le Lemme de Yoneda : enjeux pour une conjecture philosophique ? (variations sous la forme prolemmatique, mais en prose), in Andreatta M, F Nicolas and C Alunni (Eds.): A la lumière des mathématiques et à l'ombre de la philosophie. Coll. Musique/Sciences, ENS, Delatour, Ircam, Centre Pompidou, p. 195-211, 2012
28. American Mathematical Society (AMS): Mathematical Subject Classification 2010, 47 pages `http://www.ams.org/msc/pdfs/classifications2010.pdf`
29. Amuedo J: Computational Description of Extended Tonality. Master's Thesis, U Southern California, Los Angeles 1995
30. Andreatta M: Group-theoretical Methods Applied to Music. Independent Study Dissertation, University of Sussex, Sussex 1997
31. Andreatta M: Méthodes algébriques en musique et musicologie du XXème siècle : aspects théoriques, analytiques et compositionnels. PhD thesis, Ircam/EHESS 2003
32. Andreatta M, A Ehresmann, R Guitart, G Mazzola: Towards a Categorical Theory of Creativity for Music, Discourse, and Cognition, in J. Yust, J. Wild, and J.A. Burgoyne (Eds.): MCM 2013, LNAI 7937, pp. 19-37, 2013
33. Andreatta M: La théorie mathématique de la musique de Guerino Mazzola et les canons rythmiques. Mémoire DEA, EHESS Paris IV & IRCAM, Paris 1999
34. Andreatta M and G Mazzola: From a Categorical Point of View: K-nets as Limit Denotators. Perspectives of New Music, volume 44, number 2 (Summer 2006)
35. `http://www.staps.uhp-nancy.fr/bernard`
36. `http://www.staps.uhp-nancy.fr/bernard/revues_corps.htm`
37. Ansermet E: Die Grundlagen der Musik im menschlichen Bewusstsein. 4th Ed. Piper, München and Zürich 1986
38. Apfel E: Diskant und Kontrapunkt in der Musiktheorie des 12. bis 15. Jahrhunderts. Heinrichshofen, Wilhelmshafen 1982
39. Apollonius: Les coniques d'Apollonius de Perge, trad. Paul Ver Eecke. Blanchard, Paris, 1959
40. Aristoteles: Topik (Organon V, 345 b.C.). Rolfes E (German transl.), Meiner, Hamburg 1992
41. Artin M: Algebra. Birkhäuser, Basel et al. 1993
42. Artin M, Grothendieck A, Verdier J L: Théorie des Topos et Cohomologie Etale des Schémas (Tomes 1,2,3). Springer Lecture Notes in Mathematics 269, 270, 305, Springer, New York et al. 1972-1973
43. Askenfelt A et al.: Musical Performance. A Synthesis-by-Rule Approach. Computer Music J. 7/1, 1983
44. Assayag G: Du calcul secret au calcul visuel. In: Delalande F and Vinet H (Eds.): Interface homme-machine et création musicale. Hermes, Paris 1999
45. Assayag G, G Bloch, M Chemillier: OMax-Ofon. Sound and Music Computing (SMC), Marseille 2006
46. Assayag G and S Dubnov: Using Factor Oracles for Machine Improvisation. Soft Computing 8, pp. 1432-7643, September 2004
47. Assmann J: Schöpfungsmythen und Kreativitätskonzepte im Alten Ägypten. In: R M Holm-Hadulla (Ed.): Kreativität. Springer, Heidelberg 2000
48. Auroux S: La sémiotique des encyclopédistes. Payot, Paris 1979
49. Babbitt M: Some Aspects of Twelve-Tone Composition. In: Hays W (Ed.): The Score and IMA Magazine, 12, 53-61, 1955 (reprinted in "Twentieth Century Views of Music History", 364-371, Scribner, New York 1972)
50. Babbitt M: Twelve-Tone Invariants as Compositional Determinants. Musical Quarterly, 46, 245-259, 1960
51. Babbitt M: Set Structure as a Compositional Determinant. JMT, 5(2), 72-94, 1961
52. Babbitt M: Twelve-Tone Rhythmic Structure and the Electronic Medium. PNM, 1(1), 49-79, 1962
53. Babbitt M: The Structure and Function of Music Theory. College Music Symposium, Vol.5, 1965 (reprinted in Boretz and Cone, 1972, 10-21)
54. Babbitt M: Words about Music. Dembski S and Straus J N (Eds.), University of Wisconsin Press, Madison 1987
55. Bach J S: Krebskanon. In: *Musikalisches Opfer*, 70. (BWV 1079), Neue Gesamtausg. sämtl. Werke Ser. VIII, Bd.1, Bärenreiter, Kassel 1978
56. Bach J S: Choral No.6, In: *Himmelfahrtsoratorium* (BWV 11), Neue Gesamtausg. sämtl. Werke Ser. II, Bd.8, Bärenreiter, Kassel 1978
57. Bachelard G: La Valeur inductive de la Relativité. 257 p., Vrin, Paris 1929
58. Bacon F: De dignitate et augmentis scientiarum. 1st Ed. 1605, improved 1623
59. Bacon F: Novum Organum, trad. Malherbe et Pousseur. coll. épiméthée, PUF, Paris, 1986
60. Badiou A: Court traité d'ontologie transitoire. Seuil, Paris, 1998

61. Badiou A, F Tarby: La philosophie et l'événement. Entretiens. Suivis d'une courte introduction à la philosophie d'Alain Badiou. Germina, 2010
62. Badiou A: La République de Platon. Pluriel, 2012
63. Badiou A: Le Séminaire – Parménide. L'être 1 - Figure ontologique. 1985-1986. Fayard, 2014
64. Badiou A: Le Séminaire – Heidegger. L'être 3 - Figure du retrait. 1986-1987. Fayard, 2015
65. Badiou A: Éloge des mathématiques. Flammarion, 2015
66. Baer J and J C Kaufman: Creativity Research in English-Speaking Countries. J C Kaufman and R J Sternberg (Eds.): The International Handbook of Creativity. Cambridge University Press, Cambridge, pp. 10-38, 2006
67. Bagchee S N: Understanding Raga Music. BPI, New Delhi 1996
68. Bahle J: Der musikalische Schaffensprozess. Psychologie der schöpferischen Erlebnis- und Antriebsformen. S Hirzel, Leipzig 1936
69. Baker J, Beach D, Bernard J: Music Theory in Concept and Practice. Eastman Studies in Music, University of Rochester Press, 1997
70. Bakhtin M: Toward a Philosophy of the Act. Translated by Vadim Liapunov. University of Texas Press, Austin 1993
71. Balzano G: The group-theoretic description of 12-fold and microtonal pitch systems. CMJ, 4, 66-84, 1980
72. Bandemer H and Gottwald H: Fuzzy Sets, Fuzzy Logic, Fuzzy Methods. Wiley, New York et al. 1995
73. Banter H: Akkord-Lexikon. Schott, Mainz 1982
74. Barbin E: Histoire et enseignement des mathématiques : pourquoi ? comment ?. Bulletin de l'Association Mathématique du Québec, XXXVII-1, 20-25
75. Barbin E: La révolution mathématique du XVIIème siècle. Ellipses, 2006
76. Barbin E et R Guitart: La pulsation entre les conceptions optiques, algébriques, articulées et projectives, des ovales cartésiennes, in: L'Océan indien au carrefour des mathématiques arabes, chinoises, européennes et indiennes. IUFM de La Réunion, 1998, pp. 359-394
77. Barbin E et R Guitart: Algèbre des fonctions elliptiques et géométrie des ovales cartésiennes. Revue d'histoire des mathématiques, SMF, n7, 2001, pp. 161-205
78. Barbin E: Gaston Tarry et la doctrine des combinaisons. Labyrinthes et carrés magiques, conférence au colloque *Les travaux combinatoires entre 1870 et 1914 et leur actualité*, Guéret, 30 Sept. - 2 Oct. 2015
79. Barbin E et J-P Cléro (Eds.): Les mathématiques et l'expérience : ce qu'en ont dit les philosophes et les mathématiciens. Hermann, Paris 2015
80. Barbin E: La révolution mathématique du XVIIème siècle. Ellipses, 2006
81. Bareil H: Analyse de "La pulsation mathématique", *A.P.M.E.P.* 427, March-April 2000
82. Barlow K: Über die Rationalisierung einer harmonisch irrationalen Tonhöhenmenge. Preprint, Köln 1985.
83. Barrow J and F Tipler: The Anthropic Cosmological Principle: Oxford University Press, New York 1986
84. Bars I: Extra Dimensions in Space and Time (Multiversal Journeys). Springer, Heidelberg 2009
85. Bartenieff I: The Roots of Laban Theory: Aesthetics and Beyond. In: Bartenieff et al.: Four Adaptations of Effort Theory. Dance Notation Bureau, New York 1970
86. Barthes R: Eléments de sémiologie. Communications 4/1964
87. Barwise J and L Moss: Hypersets. Mathematical Intelligencer, 13(4):3141, 1991
88. Basten A: Personal e-mail communciaction to G.M. November 22, 1996
89. Bastiani A (= A Ehresmann): Théorie des ensembles. CDU, 328 p., Paris, 1970
90. Bätschmann O: Einführung in die kunstgeschichtliche Hermeneutik. Wissenschaftliche Buchgemeinschaft Darmstadt, Darmstadt 1986
91. Batt N (éd): Penser par le diagramme. Presses Universitaires de Vincennes no.22-2004, Vincennes 2004
92. Bauer M: Die Lieder Franz Schuberts. Breitkopf und Härtel, Leipzig 1915
93. Bazelow A and F Brickel: A Partition Problem Posed by Milton Babbitt. PNM, 14(2), 15(1), 280-293, 1976
94. Becker M: Nineteenth Century Foundations of Creativity Research. Creativity Research Journal, Vol. 8, pp. 219-229, 1995
95. Bednarik R G: A Figurine from the African Acheulian. Current Anthropology, Vol. 44:3, pp. 405-413, 2003
96. Beethoven L van: *Grosse Sonate für das Hammerklavier* op.106 (1817-1818). Ed. Peters, Leipzig 1975
97. Békésy G von: Experiments in Hearing. McGraw-Hill, New York 1960
98. Bell J L: From absolute to local mathematics. *Synthese* 69, 409-426, 1986
99. Bellal A: Il faut sortir des mathématiques (humour) (see also [437]), `http/::environ-energie.org/category/humourpastiches/`
100. Benveniste É: La nature des pronoms. In: Benveniste É: Problèmes de linguistique générale, I. Gallimard, Paris 1966
101. Beran J: *Cirri.* Centaur Records, 1991

102. Beran J: Statistics in musicology. Chapman & Hall (CRC Press), New York 2003
103. Beran J and G Mazzola: *Immaculate Concept*. SToA music 1002.92, Zürich 1992
104. Beran J and G Mazzola G: Rational composition of performance. In Auhagen W, R Kopiez (Eds.): Regulation of creative processes in music, Staatliches Institut für Musikforschung (Berlin), pp. 37-67, Lang Verlag, Frankfurt/New York, 1997
105. Beran J and G Mazzola: Analyzing Musical Structure and Performance—a Statistical Approach. Statistical Science. Vol. 14, No. 1, 47-79, 1999
106. Beran J and G Mazzola: Visualizing the Relationship Between Two Time Series by Hierarchical Smoothing. Journal of Computational and Graphical Statistics, Vol. 8, No. 2, 213-238, 1999
107. Beran J and G Mazzola: Timing Microstructure in Schumann's "Träumerei" as an Expression of Harmony, Rhythm, and Motivic Structure in Music Performance. Computers and Mathematics with Applications, Vol. 39, Issue 5/6, 99-130, 2000
108. Beran J and G Mazzola: Musical composition and performance - statistical decomposition and interpretation. Student, Vol. 4, No.1, 13-42, 2001
109. Beran J: Maximum likelihood estimation of the differencing parameter for invertible short and long-memory ARIMA models. J.R. Statist. Soc. B, 57, No.4, 659-672, 1995
110. Beran J, R Goswitz, G Mazzola, P Mazzola: On the relationship between tempo and quantitative metric, melodic and harmonic information in Chopin's Prélude op. 28, No. 4: a statistical analysis of 30 performances. Journal of Mathematics and Music, Vol 8., No. 3, p. 225-248, 2014
111. Beranek L L: Acoustics, 1954 -1993, The Acoustical Society of America, ISBN: 0-88318-494-X
112. Berg A: Die musikalische Impotenz der Neuen Ästhetik Hans Pfitzners. Musikblätter des Anbruchs. Jg 02, S. 400, Wien 1920
113. Berger M: Geometry I, II. Springer, Berlin et al. 1987
114. Bernard J: Chord, Collection, and Set in Twentieth-Century Theory. In: Baker J et al. [69], 11-52
115. Béziau J-Y: La Théorie des Ensembles et la Théorie des Catégories : Présentation de deux Sœurs Ennemies du Point de Vue de leurs Relations avec les Fondements des mathématiques. Boletin de la Asociación Matemática Venezolana, Vol. IX, No 1, pp. 45-52, 2002
116. Bilkey D and K M Douglas: Amusia Is Associated with Deficits in Spatial Processing. Nature Neuroscience 10, pp. 915 - 921, 2007
117. Biok H-R: Zur Intonationsbeurteilung kontextbezogener sukzessiver Intervalle. Bosse, Regensburg 1975
118. Biot J-B: Essai de géométrie analytique, 8 ième éd. (1ère éd. 1802), Bachelier, Paris 1834
119. Birkhoff G: Universal Algebra. In: Rota G-C and J.S. Oliveira J S (Eds.): Collected papers of G. Birkhoff, pp. 111-115, Birkhäuser, Basel
120. Blind Boy Fuller: `ttp://www.allaboutbluesmusic.com/blind-boy-fuller/` 2015
121. Block I: Oscar Peterson Lays Down the Story of a Life Lived in the Swing of Things. The Gazette (Montreal), August 31, 2002
122. Blumenthal L M: Distance Geometry. Chelsea 1953
123. Boden M A: Conceptual Spaces. In [753], pp. 235-244
124. Boden M A: The Creative Mind: Myths and Mechanisms. Routledge, London 2004
125. Boden M A: Creativity & Art. Oxford University Press, Oxford 2010
126. Blumenthal L M: Distance Geometry. Chelsea 1953
127. Blood A, J Zatorre, P Bermudez, A C Evans: Emotional response to pleasant and unpleasant music correlates with activity in paralimbic brain regions. Nature Neuroscience, Vol. 2, No. 4, 382-387, April 1999
128. Boi L: Le problème mathématique de l'espace. Une quête de l'intelligible. Springer, 1995
129. Boirel R: Comment résoudre aisément les problèmes de mathématiques. Initiation au dynamisme opératoire mathématique. éd. Culture humaine, Sermaise, 1964
130. Boissière A: Geste, interprétation, invention selon Pierre Boulez. Revue DEMéter, Dec. 2002, Univ. Lille-3
131. Borceux F: Handbook of Categorical Algebra: Volume 1. Cambridge University Press, Cambridge 1994
132. Borceux, F: Handbook of Categorical Algebra: Volume 2. Cambridge University Press, Cambridge 1995
133. Borgo D: Sync or Swarm. Continuum International Publishing Group 2006
134. Boretz B and E T Cone: Perspectives on Contemporary Music Theory. W.W. Norton and Company, New York 1972
135. Bose R C and S S Shrikhande: On the construction of sets of mutually orthogonal Latin squares and the falsity of a conjecture of Euler. Trans. Amer. Math. Soc. 95, 191-209, 1960
136. Bose R C, S S Shrikhande and E T Parker: Further results on the construction of mutually orthogonal Latin squares and the falsity of Euler's conjecture. Canadian J. of Math. 12, 189-203, 1960

137. Bott J, J Crowley, J LaViola Jr: Exploring 3D Gestural Interfaces for Music Creation in Video Games. CFDG 2009 April 2630, 2009, Orlando 2009
138. Boulez P: Musikdenken heute I,II; Darmstädter Beiträge V, VI. Schott, Mainz 1963, 1985
139. Boulez P: Le timbre et l'écriture, le timbre et le langage. In: Bourgeois Chr (Ed.): Le timbre, métaphore pour la composition. IRCAM, Collection Musique/Passé/Présent, Paris 1991
140. Boulez P: Jalons (dix ans d'enseignement au Collège de France). Bourgeois, Paris 1989
141. Boulez P: L'Ecriture du geste. Bourgeois, Paris 2002
142. Boulez P: Structures, premier livre. UE, London 1953
143. Boulez P: Structures, deuxièxime livre. UE, London 1967
144. Bourgeois Chr (Ed.): Le timbre, métaphore pour la composition. IRCAM, Collection Musique/Passé/Présent, Paris 1991
145. Bourbaki N: Eléments de Mathématique, Algèbre, Ch.1-9. Hermann, Paris 1970-1973
146. Bourbaki N: Eléments de Mathématique, Algèbre Commutative, Ch.1-7. Hermann, Paris 1961-65
147. Bourbaki N: Eléments de Mathématique, Topologie Générale, Ch.1-4. Hermann, Paris 1971
148. Bourbaki N: Foundations of mathematics for the working mathematician. Journal of Symbolic Logic, 14, pp. 1-8, 1949
149. Boyer C B: Newton as an Originator of Polar Coordinates. American Mathematical Monthly 56: 73-78, 1949
150. Braem P B and T Braem. A Pilot Study of the Expressive Gestures Used by Classical Orchestra Conductors. In: Emmorey K and H Lane (Eds.): The signs of Language Revisited. Lawrence Erlbaum Associates, 2000
151. Brändle L: Die "Wesentlichen Manieren" (Ornamente in der Musik). Oesterreichischer Bundesverlag, Wien 1987
152. Brandt C and C Roemer: Standardized Chord Symbol Notation: Roerick Music, Sherman Oaks, CA 1976
153. Brecht B et al.: Conductor Follower. In: ICMA (Ed.): Proceedings of the ICMC 95, San Francisco 1995
154. Brockwell P J and R A Davis: Time Series: Theory and Methods. Springer, New York 1987
155. Brofeldt H: Piano Music for the Left Hand Alone. `http://www.left-hand-brofeldt.dk` Non-dated
156. Bruhn H: Harmonielehre als Grammatik der Musik. Psychologie Verlags Union, München et al. 1988
157. Bruter C-P: Comprendre les mathématiques. Les 10 notions fondamentales. Ed. Odile Jacob, Paris, 1996
158. Bulwer J: Chironomia. Thomas Harper, London 1644
159. Buteau Ch and G Mazzola: From Contour Similarity to Motivic Topologies. Musicae Scientiae 2, pp. 125-149, 2000
160. Buteau Ch and G Mazzola: From Contour Similarity to Motivic Topologies. Musicae Scientiae, Vol. IV, No. 2, 125-149, 2000.
161. Busch H R: Leonhard Eulers Beitrag zur Musiktheorie. Bosse, Regensburg 1970
162. Buser P and M Imbert: Audition. Hermann, Paris 1987
163. Buteau Ch: Motivic Topologies and their Signification in Musical Motivic Analysis. Master's Thesis, U Laval/Québec 1998
164. Buteau Ch: Reciprocity Between Presence and Content Functions on a Gestalt Composition Space. Tatra Mt. Math. Publ. 23, 17-45, 2001
165. Butsch C and H Fischer: Seashore-Test für musikalische Begabung. Huber, Bern/Stuttgart 1966
166. Byers W: How Mathematicians Think. Using Ambiguity, Contradiction and Paradox to Create Mathematics. Princeton U. Press, 2007
167. Cabral B and L Leedom: Imaging vector fields using line integral convolution. Computer Graphics 27 (SIGGRAPH'93 Proceedings), 263-272, 1993
168. Cadoz C and M M Wanderley: In: Gesture - Music. Trends in Gestural Control of Music, M M Wanderley and M Battier eds., Ircam - Centre Pompidou, 2000
169. Cadoz C: Le geste canal de communication homme-machine. La communication 'instrumentale'. Sciences Informatiques, numéro spécial: Interface homme-machine. 13(1): 31-61, 1994
170. Cage J: ASLSP. `http://www.john-cage.halberstadt.de`
171. Calvet O et al.: Modal synthesis: compilation of mechanical sub-structure and acoustical sub-systems. In: Arnold S, Hair G, ICMA (Eds.): Proceedings of the 1990 International Computer Music Conference. San Francisco 1990
172. Campos D G: Peirce on the Role of Poietic Creation in Mathematical reasoning, trans. Charles S. Peirce Society: A Quaterly J. in American Philosophy, Vol. 43, No 3, pp. 470-489, Summer 2007
173. Franco Ferrara's conducting technique, Personal conversation with maestro C. Caruso at State Music Conservatory of Palermo, 2014
174. Carnot L: Mémoire sur la Relation qui existe entre les distances respectives de cinq points quelconques dans l'espace, suivi d'un Essai sur la théorie des transversales. 1806

175. Castellengo M and N H Bernardoni: Interplay between harmonics and formants in singing: when vowels become music. Proceedings of ISMA, International Conference on Noise and Vibration Engineering 2014
176. Camuri A et al.: Toward a cognitive model for the representation and reasoning on music and multimedia knowledge. In: Haus G and Pighi I (Eds.): X Colloquio di Informatica Musicale. AIMI, LIM-Dsi, Milano 1993
177. Carey N and D Clampitt: Aspects of well-formed scales. MTS, 11, 187-206, 1989
178. Cardelli L and P Wegner: On Understanding types, data Abstraction and polymorphism. Computing Surveys, Vol. 17,4, 1985
179. Castagna G: Foundations of Object-oriented programming. ETAPS, Lisbon 1998
180. Castine P: Set Theory Objects. Lang, Frankfurt/Main et al. 1994
181. Cavaillès J: Méthode axiomatique et formalisme, p. 178, 1938. In: Oeuvres complètes de Philosophie des sciences. Hermann, 1994
182. Cayley A: On a Theorem in the Geometry of Position. Camb. Math. Jour., t. II, pp. 267-271, 1841
183. Celibidache Video: `http://www.youtube.com/watch?v=_BAsS5SmiNU`
184. Celibidache at Wikipedia: `http://en.wikipedia.org/wiki/Sergiu_Celibidache`
185. CERN. Internet Information via http://www.cern.ch 1995
186. Certeau M de: L'étranger ou l'union dans la différence. 1ère éd. 1969, Le Seuil, 2005
187. Châtelet G: Arithmétique et algèbre modernes, II, PUF, Paris 1956
188. Chastel A: Le geste dans l'art. Liana Levi, Paris 2001
189. Châtelet G: Les enjeux du mobile. Seuil, Paris 1993
190. Châtelet G: Figuring Space. Kluwer 2000 (translation of the French original *Les enjeux du mobile*, [189])
191. Chatterjee S and B Price: Regression Analysis by Example. Wiley, 2nd ed., New York 1995
192. Chauviré C: Peirce et la signification. PUF, Paris 1995
193. Chauviré C: L'œil mathématique. Essai sur la philosophie mathématique de Peirce. éd. Kimé, 2008
194. Cherlin M: Memory and Rhetorical Trope in Schoenberg's String Trio. Journal of the American Musicological Society, 51/3, pp. 559-602, fall 1998
195. Cherry D & E Blackwell: Mu (Parts I and II). Actuel 1969
196. Choi I: Cognitive Engineering of Gestural Primitives for Multi-Modal Interaction in a Virtual Environment. In: Proceedings of the 1998 IEEE International Conference on Systems, Man and Cybernetics (SMC'98), 1998
197. Chomsky N and M Halle: The Sound Pattern of English. Harper and Row, New York 1968
198. Chowning J: The Synthesis of Complex Audio Spectra by Means of Frequency Modulation. Journal of the Audio Engineering Society 21 (7), 1985
199. Chowning J: The synthesis of complex audio spectra by means of frequency modulation. Journal of the Audio Engineering Society, 21, 1973.
200. Clarke E: Imitating and Evaluating Real and Transformed Musical Performances. Music Perception 10/3, 317-341, 1993
201. Clayton M and G Fatone, L Leante, and M Rahaim: Imagery, Melody, and Gesture in Cross-Cultural Perspective. In: King E: New Perspectives on Music and Gesture (2016), Routledge 2011
202. Clough J and Myerson G: Variety and multiplicity in diatonic systems. JMT, 29, 249-270, 1985
203. Clough J and J Douthett: Maximally even sets. JMT, 35, 93-173, 1991
204. Clough J: Diatonic Interval Cycles and Hierarchical Structure. PNM, 32(1), 228-253, 1994
205. Clynes M: Sentics. The Touch of Emotions. Anchor Doubleday, New York 1977
206. Clynes M: Secrets of life in music. In: Analytica, Studies in the description and analysis of music in honour of Ingmar Bengtsson, Royal Swedish Academy of Music, No 47 pp. 3-15, 1985
207. Colbourn C J and J H Dinitz: Handbook of Combinatorial Designs. Chapman & Hall/CRC, 2007
208. Cont A, S Dubnov and G Assayag: On the Information Geometry of Audio Streams with Applications to Automatic Structure Discovery. IEEE Transactions on Audio, Speech and Language Processing 2009
209. Cohn R: Introduction to Neo-Riemannian Theory: A Survey and a Historical Perspective. JMT 42(2), 167-180, 1998
210. Conen H: Formel-Komposition. Schott, Mainz et al. 1991
211. Cook P: `http://www.cs.princeton.edu/ prc/SingingSynth.html`
212. Cook N: Playing God: Creativity, Analysis, and Aesthetic Inclusion. In: I Deliège and G A Wiggins (Eds.): Musical Creativity: Multidisciplinary Research in Theory and Practice. Psychology Press, Hove/New York, pp. 9-24, 2006
213. Coolidge J L: The origin of polar coordinates. American Mathematical Monthly, 59 : 78-85, 1952
214. Cooley J W and J W Tuckey: An Algorithm for the Machine Calculation of Complex Series. Math. Comput. 19, 297301 1965
215. Cope D: Virtual Music: Computer Synthesis of Musical Style. MIT Press, Cambridge, MA 2001

216. Cope D: Computer Models of Musical Creativity. MIT Press, Cambridge, MA 2005
217. Corbin A, J-J Courtine, G Viagarello (Eds.): Histoire du corps, T3, Les mutations du regard. Le XX siècle. Seuil, Paris 2006
218. Cox A: Hearing, Feeling, Grasping Gestures. In: Gritten A and E King (Eds.): Music and Gesture. Aldershot: Ashgate, 2006
219. Cooper K N D et al.: Handwritten Music-Manuscript Recognition. In: ICMA (Ed.): Proceedings of the ICMC 96, San Francisco 1996
220. Couasnon B et al.: Using a Grammar For a Reliable Full Score Recognition System. In: ICMA (Ed.): Proceedings of the ICMC 95, San Francisco 1995
221. Creutzfeldt O D: Cortex Cerebri. Springer, Berlin et al. 1983
222. Cros S C H: Théorie de l'homme intellectuel et moral, vol.2. Bachelier, 1842
223. Csíkszentmihályi M: Beyond Boredom and Anxiety. Jossey-Bass, San Francisco 1975
224. Csíkszentmihályi M: Society, Culture, and Person: A Systems View of Creativity. In: R J Sternberg (Ed.): The Nature of Creativity: Contemporary Psychological Perspectives. Cambridge University Press, New York, pp. 325-339, 1988
225. Csíkszentmihályi M: The Psychology of Optimal Experience. Harper and Row, New York 1990
226. Csíkszentmihályi M: Creativity: Flow and the Psychology of Discovery and Invention. HarperCollins, New York 1996
227. Cusick S: Feminist Theory, Music Theory, and the Mind/Body Problem. Perspectives of New Music 32/1 (Winter 1994): 8-27
228. Czerny C: Pianoforte Schule. 1840
229. Dacey J S and K Lennon: Understanding Creativity: The Interplay of Biological, Psychological, and Social Factors. Josey-Bass, San Francisco, 1998
230. Dahlhaus C: Zur Theorie des klassischen Kontrapunkts. Kirchenmusikalisches Jb 45, 1961
231. Dahlhaus C: Über den Begriff der tonalen Funktion. In: Vogel M (Ed.): Beiträge zur Musiktheorie des 19. Jahrhunderts. Bosse, Regensburg 1966
232. Dahlhaus C and H H Eggebrecht: Was ist Musik? Heinrichshofen, Wilhelmshaven et al. 1985
233. Dahlhaus C: Untersuchung über die Entstehung der harmonischen Tonalität. Bärenreiter, Kassel et al. 1967
234. Dahlhaus C et al.: Neues Handbuch der Musikwissenschaft, Bd. 1-13: Athenaion and Laaber, Laaber 1980-1993
235. Dahlhaus C and G Mayer: Musiksoziologische Reflexionen. In: Dahlhaus, C and de la Motte-Haber H (Eds.): Neues Handbuch der Musikwissenschaft, Bd. 10: Systematische Musikwissenschaft. Laaber, Laaber 1982
236. Dahlhaus C: Ludwig van Beethoven und seine Zeit. Laaber, Laaber 1987
237. D'Alembert J Le Rond: Einleitung zur 'Enzyklopädie'. Fischer, Frankfurt/Main 1989
238. D'Alembert, J Le Rond: Einleitung zur Enzyklopädie (1751). (German Translation) Fischer, Frankfurt/Main 1989
239. Dallos P: The active cochlea. J. Neurosci. Dec;12(12):4575-85, 1992
240. Dannenberg R B: An on-line algorithm for real-time accompaniment. In: ICMA (Ed.): Proceedings of the ICMC 84, San Francisco 1984
241. Dannenberg R B et al.: Automatic Ensemble Performance. In: ICMA (Ed.): Proceedings of the ICMC 94, San Francisco 1994
242. Alighieri, D: La Divina Commedia (1321). CreateSpace Independent Publishing Platform 2014
243. Danuser H et al. (Eds.): Neues Handbuch der Musikwissenschaft, Bd. 11: Interpretation. Laaber, Laaber 1992
244. Darwin Ch: The Origin of Species. London 1859
245. Daval R and G-T Guilbaud: Le raisonnement mathématique. PUF, 1945
246. de Bruijn N G: Pólya's Theory of Counting. In: Beckenbach E F (Ed.): Applied Combinatorial Mathematics, Ch.5. Wiley, New York 1964
247. de Bruijn N G: On the number of partition patterns of a set. Nederl. Akad. Wetensch. Proc. Ser. A 82 = Indag. Math. 41, 1979
248. Debussy C: *Préludes*, Livre I (1907-1910). Henle, München 1986
249. Dechelle F et al.: The Ircam Real-Time Platform and Applications. In: ICMA (Ed.): Proceedings of the ICMC 95, San Francisco 1995
250. Delalande F: La gestique de Gould. In: Guertin G (Ed.): Glenn Gould — pluriel. Corteau, Verdun 1988
251. Degazio B: A Computer-Based Editor For Lerdahl and Jackendoff's Rhythmic Structures. In: ICMA (Ed.): Proceedings of the ICMC 96, San Francisco 1996
252. de la Motte D: Harmonielehre. Bärenreiter/dtv, Kassel 1976
253. de la Motte-Haber H and H Emons: Filmmusik. Eine systematische Beschreibung. München, Hanser 1980
254. de la Motte-Haber H: Handbuch der Musikpsychologie. Laaber-Verlag, 2.Ed., Laaber 1996

255. de la Motte-Haber H: Rationalität und Affekt. In: Götze H and Wille R (Eds.): Musik und Mathematik. Springer, Berlin et al. 1985
256. de la Motte-Haber H: Musikalische Hermeneutik und empirische Forschung. In: Dahlhaus, C and de la Motte-Haber H (Eds.): Neues Handbuch der Musikwissenschaft, Bd. 10: Systematische Musikwissenschaft. Laaber, Laaber 1982
257. de la Motte-Haber H: Die Umwandlung der Interpretationsparameter in Struktureigenschaften. In: "Das Paradox musikalischer Interpretation", Symposion zum 80. Geburtstag von K. von Fischer, Univ. Zürich 1993
258. Deleuze G: Francis Bacon. La logique de la sensation. Éditions de la Différence, Paris 1981
259. Demazure M and P Gabriel: Groupes Algébriques. Masson & Cie./North-Holland, Paris/Amsterdam 1970
260. Dennett, D C: Quining qualia. In Marcel A and Bisiach E (Eds.): Consciousness in Contemporary Science. Oxford University Press, 1988
261. Desain P and Honing H: The Quantization of Musical Time: A Connectionist Approach. Computer Music Journal 13 (3), 56-66, 1989
262. Descartes R: Musicae Compendium. Herausgegeben und ins Deutsche übertragen als "Leitfaden der Musik" von J. Brockt, Wiss. Buchgesellschaft, Darmstadt 1978
263. Descartes R: Principia philosophiae (Principles of Philosophy). Translation with explanatory notes by Valentine Rodger and Reese P. Miller (Reprint ed.). Dordrecht: Reidel
264. Descartes R: Œuvres. Vrin, Paris, 1989. See also in English: The Geometry of René Descartes, transl. by E Smith and M L Latham, Dover Publ., 1954
265. Detienne M: Les Maîtres de Vérité dans la Grèce archaïque. 1ère éd. 1967, Le livre de poche/Références, no 611, Paris, 2012
266. De Tranquelléon: Sur l'intersection de deux cônes. Nouvelles Annales de mathématiques, 2ème série, tome III, pp. 539-540, 1864
267. Deyoung L: Pitch Order and Duration Order in Boulez' Structure 1a. PNM 16/2: 27-34, 1978
268. Dieudonné J: Foundations of Modern Analysis. Academic Press, New York et. al. 1960
269. Dieudonné J: La conception des mathématiques chez Valéry. In: Robinson-Valéry J: Fonctions de l'esprit. Treize savants redécouvrent Paul Valéry. Hermann, p. 183-191, Paris 1983
270. Dieudonné J: Mathematics – Music of Reason. Springer, 1992
271. Discogs: Horace Parlan. `http://www.discogs.com/artist/253065-Horace-Parlan` Not-dated
272. Döhl F: Webern - Weberns Beitrag zur Stilwende der Neuen Musik. Katzbichler, München et. al. 1976
273. Donald M: Mimesis and the Executive Suite: missing links in language evolution. In: J R Hurford, M Studdert-Kennedy, Ch Knight (Eds.): Approaches to the evolution of language, Cambridge U Press 1998
274. Dominguez M, D Clampitt and Th Noll: Well-formed Scales, Maximally Even Sets, and Christoffel Words. In: Proceedings of the MCM 2007, Springer, Heidelberg et al. 2009
275. Douady R: Jeux de cadres et dialectique outil-objet dans l'enseignement des mathématiques. Une réalisation dans le cursus primaire. Thèse, Paris 7, 1984
276. Douthett J and R Steinbach: Parsimonious Graphs: A Study in Parsimony, Contextual Transformations, and Modes of Limited Transposition. Journal of Music Theory 42.2, pp. 241-263, 1998
277. Doyle P, J Conway, J Gilman, B Thurston: Geometry and the Imagination in Minneapolis, version 0.91 dated 12 April 1994
278. Drake R M: Drake Musical Aptitude Tests. Science Research Associates, Chicago 1954
279. Dreiding A et al.: Classification of Mobile Molecules by Category Theory. In: Symmetries and Properties of Non-Rigid Molecules. Studies in Physical and Theoretical Chemistry, 23, 1983
280. Drossos C A: Sets, Categories and Structuralism. In: Sica, G. (Ed.): What is Category Theory?, Polimetrica, Monza 2005
281. Duffy M and S Tsuda: Shield Your Eyes Sonic Anarchy. The Sonic Alchemists, Minneapolis 2010
282. Dufourt H: Les difficultés d'une prise de conscience théorique. In: Le compositeur et l'ordinateur. 6-12, Ircam, Centre Georges Pompidou, Paris 1981
283. Duvignau K, B Gaume, J-L Nespoulous: Proximité sémantique et stratégies palliatives chez le jeune enfant et l'aphasique. Revue Parole, numéro spécial: 'Handicap langagier et recherches cognitives: Apports mutuels', J-L Nespoulous and J Virbel (Eds.), UMH Belgique, vol. 31-32: 219-255, 2004
284. Eberle G: "Absolute Harmonie" und "Ultrachromatik". In: Kolleritsch O (Ed.): Alexander Skrjabin. Universal Edition, Graz 1980
285. Ebisch S J H, M G Perrucci, A Ferretti, C Del Gratta, G Luca Romani, V Gallese: The Sense of Touch: Embodied Simulation in a Visuotactile Mirroring Mechanism for Observed Animate or Inanimate Touch. Journal of Cognitive Neuroscience 20:1-13, 2008
286. Eco U: Kunst und Schönheit im Mittelalter. Hanser, Wien 1991

287. Eco U: The Search for the Perfect Language. Wiley-Blackwell 1997
288. Eco U: Die Suche nach der vollkommenen Sprache. Beck, München 1994
289. Eco U: Le Signe. biblio essais n4159, éd. Labor 1988
290. Edelsbrunner H and J L Harer: Computational Topology. AMS 2010
291. Eggebrecht H H: Interpretation. In: "Das Paradox musikalischer Interpretation", Symposion zum 80. Geburtstag von K. von Fischer, Univ. Zürich 1993
292. Eggebrecht H H: Musik im Abendland. Piper, München and Zürich 1996
293. Ehrenfels Chr von: Über Gestaltqualitäten. Vierteljahresschrift für wissenschaftliche Philosophie XIV, 1890
294. Ehresmann C: Œuvres complètes et commentées. Part. IV-2 : esquisses et structures monoïdales fermées. ed. par Andrée Ch. Ehresmann, Amiens, 1983
295. Ehresmann A C, J-P Vanbremeersch: Memory Evolutive Systems. Hierarchy, Emergence, Cognition. Elsevier Amsterdam, 2007
296. Ehresmann A and J Gomez-Ramirez: Conciliating Neuroscience and Phenomenology via Category Theory. Progress in Biophysics and Molecular Biology, Vol. 19, issue 2, Elsevier, 2015
297. Eitz C: Das mathematisch reine Tonsystem. Leipzig 1891
298. Eimert H: Grundlagen der musikalischen Reihentechnik. Universal Edition, Wien 1964
299. Eisenbud D: Commutative Algebra with a View Toward Algebraic Geometry. Springer New York et al. 1996
300. Eisenbud D, and J Harris: The Geometry of Schemes. Springer, New York 2000
301. Engström B: Stereocilia of sensory cells in normal and hearing impaired ears. Scand. Audiol. Suppl. 19, 1-34, 1983
302. Engquist B and W Schmid (Eds.): Mathematics Unlimited — 2001 and Beyond. Springer, 2001
303. Ericson K: Heinz Holliger. Spurensuche eines Grenzgängers. Peter Lang, Bern 2004
304. Erlewine S: Movin' & Groovin' - Horace Parlan.
`http://www.allmusic.com/album/movin-groovin-mw0000381525` 2015
305. Essl K: Strukturgeneratoren. Beiträge zur elektronischen Musik 5, IEM, Graz 1996
306. Euclid: Euclid's Elements. Translated with introduction and commentary by Sir Th L Heath, vol. 1, 2 and 3, Dover, 1956
307. Euler L: Tentamen novae theoriae musicae (1739). In: Opera Omnia, Ser.III, Vol.1 (Ed. Bernoulli, E et al.). Teubner, Stuttgart 1926
308. Euler L: Conjecture sur la raison de quelques dissonances générales reçues dans la musique (1764). In: Opera Omnia, Ser.III, Vol.1 (Ed. Bernoulli, E et al.). Teubner, Stuttgart 1926
309. Euler L: De harmoniae veris principiis per speculum musicum representatis (1773). In: Opera Omnia, Ser.III, Vol.1 (Ed. Bernoulli, E et al.). Teubner, Stuttgart 1926
310. Euler L: Recherches sur une nouvelle espèce de quarrés magiques. Opera Omnia, Ser. I, Vol. 7: 291-392, 1782
311. Fauré G: Requiem in D minor, Op. 48. Edited by P. Legge 2000
312. Feldman J et al.: Force Dynamics of Tempo Change in Music. Music Perception, *10*, 1992
313. Felver Chr: Cecil Taylor - All The Notes. A Chris Felver Movie, 2003
314. Ferretti R and G Mazzola: Algebraic Varieties of Musical Performances. Tatra Mt. Math. Publ. 23, 59-69, 2001
315. Feulner J et al.: MELONET: Neural Networks that Learn Harmony-Based Melodic Variations. In: ICMA (Ed.): Proceedings of the ICMC 94, San Francisco 1994
316. Fichte J G: Doctrine de la science. Nova methodo (1799), Livre de poche 4621, LGF, 2000
317. Fichtner R: Die verborgene Geometrie in Raffaels "Schule von Athen". Oldenburg 1984
318. Fink E: Grundlagen der Quantenmechanik. Akademische Verlagsgesellschaft, Leipzig 1968
319. Finke R A, Th B Ward, S M Smith: Creative Cognition: Theory, Research, and Applications. MIT Press, Cambridge, MA 1992
320. Finscher L: Studien zur Geschichte des Streichquartetts. Bärenreiter, Kassel 1974
321. Finsler P: Über die Grundlegung der Mengenlehre. Erster Teil. Die Mengen und ihre Axiome. Math. Z. 25, 683-713, 1926
322. Finsler P: Aufsätze zur Mengenlehre. Unger G (Ed.), Wiss. Buchgesellschaft, Darmstadt 1975
323. Fisher G and J Lochhead: Analyzing from the Body. In: Theory and Practice, Journal of the Music Theory Society of New York State, Vol. 27
324. Fleischer A: Eine Analyse theoretischer Konzepte der Harmonielehre mit Hilfe des Computers. Magisterarbeit, MWS, HU Berlin 1996
325. Fleischer A, G Mazzola, Th Noll: Zur Konzeption der Software RUBATO für musikalische Analyse und Performance. Musiktheorie, Heft 4, 314-325, 2000
326. Florida R: The Rise of the Creative Class. Basic Books, 2003

327. Floris E: La rupture cartésienne et la naissance d'une philosophie de la culture dans les œuvres juvéniles de J.-B. Vico. Thèse de 3ème cycle, dir. P. Ricœur, Université Paris X-Nanterre, 22 juin 1974
328. Flusser V: Gesten — Versuch einer Phänomenologie. Fischer, Frankfurt am Main 1994
329. Forster M: Technik modaler Komposition bei Olivier Messiaen. Hänssler, Neuhausen-Stuttgart 1976
330. Forte A: Structure of Atonal Music: Practical Aspects of a Computer-Oriented Research Project. In: Musicology and the Computer. Musicology 1966-2000. A Practical Program. Three Symposia. American Musicological Society, NY 1970
331. Forte A: Structure of Atonal Music. Yale University Press, New Haven 1973
332. Forte A: La Set-complex theory: Elevons les enjeux! Analyse musicale, 4e trimestre, 80-86, 1989
333. Foster W C: Singing Redefined. Recital Publications 1998
334. Frank H: RUBATO® Broadcast. ORF2: Modern Times, Jan. 10, 1997
335. Freedman D Z and P van Nieuwenhuizen: Supergravitation und die Einheit der Naturgesetze. In: Dosch H G (Ed.): Teilchen, Felder und Symmetrien. Spektrum der Wissenschaft, Heidelberg 1984
336. Friberg A: Generative Rules for Music Performance: A Formal Description of a Rule System. Computer Music Journal, Vol. 15, No. 2, 1991
337. Friberg A et al.: Performance Rules for Computer-Controlled Contemporary Keyboard Music. Computer Music Journal, Vol. 15, No. 2, 1991
338. Friberg A et al.: Recent Musical Performance Research at KTH. In: Sundberg J, (Ed.): Generative Grammars for Music Performance. KTH, Stockholm 1994
339. Friberg A: A Quantitative Rule System for Musical Performance. KTH PhD-Thesis, Stockholm 1995
340. Fripertinger H: Enumeration in Musical Theory. Beiträge zur elektronischen Musik 1, Hochschule für Musik und Darstellende Kunst, Graz 1991
341. Fripertinger H: Die Abzähltheorie von Pólya. Diplomarbeit, Univ. Graz 1991
342. Fripertinger H: Endliche Gruppenaktionen in Funktionenmengen—Das Lemma von Burnside—Repräsentantenkonstruktionen—Anwendungen in der Musiktheorie. Doctoral Thesis, Univ. Graz 1993
343. Fripertinger H: Untersuchungen über die Anzahl verschiedener Intervalle, Akkorde, Tonreihen und anderer musikalischer Objekte in n-Ton Musik. Magisterarbeit, Hochschule für Musik und Darstellende Kunst, Graz 1993
344. Fripertinger H: Anwendungen der Kombinatorik unter Gruppenaktionen zur Bestimmung der Anzahl "wesentlich" verschiedener Intervalle, Chorde, Tonreihen usw. Referat an der Univ. Innsbruck, Math. Institut d. Karl-Franzens-Univ., Graz 1996
345. Fripertinger H: Enumeration of Mosaics. Discrete Mathematics, 199, 49-60, 1999
346. Fripertinger H: Enumeration of Non-isomorphic Canons. Tatra Mt. Math. Publ. 23, 47-57, 2001
347. Frova A: Fisica nella musica. Zanichelli, Bologna 1999
348. Fussi F and S J Ghiotti: Il canto fra terminologia e didattica, 2015 `http://www.voceartistica.it/it-IT/index-/?Item=ghiotti`
349. Fux J J: Gradus ad Parnassum (1725). Dt. und kommentiert von L. Mitzler, Leipzig 1742
350. Gabriel P: Personal Communication. Zürich 1979
351. Gabriel P: Des catégorie abéliennes. Bull. Soc. Math. France 90, 1962
352. Gabriel P: Representations of Finite-Dimensional Algebras. In: Kostrikin, A.I.; Shafarevich, I.R. (Eds.): Encyclopaedia of Mathematical Sciences, Vol. 73, Springer Berlin 1992
353. Gabrielsson A: Music Performance. In: Deutsch D (Ed.): The Psychology of Music (2nd ed.). Academic Press, New York
354. Gabrielsson A: Expressive Intention and Performance. In: Steinberg R (Ed.): Music and the Mind Machine. Springer, Berlin et al. 1995
355. Gaither C C and A E Cavazos-Gaither: Mathematically Speaking. A Dictionary of Quotations. IOP Publishing Ltd, 1998
356. Galenson D W: Old Masters and Young Geniuses: The Two Life Cycles of Creativity. Princeton University Press, Princeton, NJ 2006
357. Gamer C: Some combinatorial resources of equal-tempered systems. JMT, 11, 32-59, 1967
358. Gamma E, R Helm, R Johnson, J Vlissides: Design Patterns, Elements of Reusable Object-Oriented Software. Addison-Wesley, Reading, MA et al., 1994
359. Gardner H: Kreative Intelligenz: Was wir mit Mozart, Freud, Woolf und Gandhi gemeinsam haben (Translated by Andreas Simon). Campus Verlag, Frankfurt/New York 1999
360. Gasquet S: L'illusion mathématique. Le malentendu des maths scolaires. Syros, 1997
361. Geisser H, G Mazzola, S Onuma: Imaginary Time. Video, Vimeo 2015

362. Gentilucci M & and M C Corballis: From manual gesture to speech: A gradual transition. Neuroscience & Biobehavioral Reviews Vol. 30, Issue 7, pp 949-960 2006
363. Gerwin Th: IDEAMA, Zentrum fur Kunst und Medientechnologie, Karlsruhe 1996
364. Geweke J: A comparison of tests of independence of two covariance-stationary time series. J. Am. Statist. Assoc., 76, 363-373, 1981
365. Giannitrapani D: The Electrophysiology of Intellectual Functions. Karger, Basel 1985
366. Gilson E: Introduction aux arts du beau. Vrin, Paris 1963
367. Ginzburg C: Leçon de Montaigne. Sud-Ouest, 17 octobre 2015
368. Ginzburg C: L'estrangement. In à distance, Neuf essais sur le point de vue en histoire. Traduction française par Pierre Antoine Fabre, éditions Gallimard NRF, Paris 2001
369. Giocanti S: L'art sceptique de l'estrangement dans les *Essais* de Montaigne, in *L'estrangement, Retour sur un thème de Carlo Ginzburg* [573, p. 19-35]
370. Godement R: Topologie algébrique et théorie des faisceaux. Hermann, Paris 1964
371. Godøy R I and M Leman (Eds.): Musical Gestures—Sound, Movement, and Meaning. Routledge, New York and London 2010
372. Goeller S: Object Oriented Rendering of Complex Abstract Data. PhD. thesis, Universität Zürich, 2004
373. Goethe J W von: Brief an Zelter. 9. Nov. 1829
374. Goffman E: The Presentation of Self in Everyday Life. University of Edinburgh, Edinburgh 1956
375. Goffman E: Behavior in Public Places. The Free Press, New York 1963
376. Goldblatt R: Topoi. North-Holland, Amsterdam et al. 1984
377. Goldin-Meadows S: Hearing Gesture: How Our Hands Help Us Think. Harvard University Press 2003
378. Goldstein J L: An Optimum Processor Theory for the Central Formation of the Pitch of Complex Tones. J. Acoust. Soc. Am. 54, 1496 1973
379. Gomes Teixeira F: Traité des courbes spéciales remarquables planes et gauches. 3 vol., (Coïmbre, Impr. de l'Université, 1908-1915), éd. Jacques Gabay, 1995. Traduction de *Tratado de las curvas especiales notables*, Ac. Sc. Madrid, 1899
380. Interview with Emilia Gómez: "PHENICX is a new way of experiencing a classical music concert", `http://www.upf.edu/enoticies/en/entrevistes/1232.html#.VbfU7s63QUU`
381. Gorenstein D: Classifying the finite simple groups. Bull. A.M.S. 14, 1-98, 1986
382. Gottschewski H: Tempohierarchien. Musiktheorie, Heft 2, 1993
383. Götze H and R Wille R (Eds.): Musik und Mathematik. Springer, Berlin et al. 1985
384. Gould G: The Glenn Gould Reader. Alfred A. Knopf, New York 1984
385. Goupillaud P, A Grossmann, J Morlet: Cycle-octave and related transforms in seismic signal analysis. Geoexploration, 23, 85-102, 1984-1985
386. Grabusow N: Vielfalt akustischer Grundlagen der Tonarten und Zusammenklänge - Theorie der Polybasiertheit. Musiksektion des Staatsverlags, Moskau 1929
387. Graeser W: Bachs "Kunst der Fuge". In: Bach-Jahrbuch 1924
388. Graeser W: Der Körpersinn. Beck, München 1927
389. Gramain A: Géométrie élémentaire. Hermann, 1997
390. Greimas A J: Les actants, les acteurs et les figures. In: Chabrol C (Ed.): Sémiotique narrative et textuelle. Larousse, Paris 1974
391. Greub W: Linear Algebra. Springer, Berlin et al. 1967
392. Grisey G: Tempus ex machina: a Composer's Reflections on Musical Time. Contemporary Music Review 2, no. 1:238-275, 1987
393. Grondin J: L'herméneutique. Coll. Que sais-je ? no 3758, PUF, Paris 2006
394. Gross D: A Set of Computer Programs to Aid in Music Analysis. Ph. Diss. Indiana Univ. 1975
395. Grothendieck A and J Dieudonné: Eléments de Géométrie Algébrique I. Springer, Berlin et al. 1971
396. Grothendieck A and J Dieudonné: Eléments de Géométrie Algébrique I-IV. Publ. Math IHES no. 4, 8, 11, 17, 20, 24, 28, 32, Bures-sur-Yvette 1960-1967
397. Grothendieck A: Correspondence with G. Mazzola. April 1, 1990
398. Grothendieck A: Récoltes et Semailles. Université Univ. Sci. et Tech. Languedoc et CNRS, Montpellier, 1st edn. 1985
399. Guevara R C L et al.: A Modal Distribution Approach to Piano Analysis and Synthesis. In: ICMA (Ed.): Proceedings of the ICMC 96, San Francisco 1996
400. Guillaume P: La psychologie de la forme. Journal de Psychologie, XXII, p. 768-800, 1925
401. Guillaume P: La psychologie de la forme. Flammarion, 1979

402. Guitart R: Sur l'ébauche des structures. Proc. 3d Congress of Bulgarian math., summaries part II, p. 354, Varna, Sept. 6-15, 1972
403. Guitart R: Sur le foncteur diagrammes, CTGD XIV-2, pp. 181-182, 1973
404. Guitart R: Remarques sur les machines et les structures, CTGD XV-2, pp. 113-144, 1974
405. Guitart R: Remarques sur les machines et les structures. CTGD XV-2, pp. 113-144, 1974
406. Guitart R and L Van den Bril: Décompositions et lax-complétions. CTGD XVIII-2, pp. 333-407, 1977
407. Guitart R: Des machines aux bimodules. polycopié, Univ. Paris 7, 24 p., 1978
408. Guitart R: Relations et carrés exacts. Ann. Sc. Math. Qué., vol. IV-2, p. 103-125, juillet 1980
409. Guitart R and C Lair: Calcul syntaxique des modèles et calcul des formules internes, *Diagrammes*, vol. 4, déc. 1980, 106 p.
410. Guitart R: On the geometry of computations, I and II. Cahiers Top. Géo. Diff. Cat. XXVII,4, 107-136, 1986; XXIX,4, p. 297-326, 1988
411. Guitart R: Mathématiques. 1ère année. Polycopié U. Paris 7; 27 chap., 432 p. 1992-1993
412. Guitart R: La pulsation mathématique. L'Harmattan, Paris, 1999
413. Guitart R: évidence et étrangeté. PUF, Paris, 2000
414. Guitart R: Sur les places du sujet et de l'objet dans la pulsation mathématique. Questions éducatives, revue du Centre de Recherche en éducation de l'Université Jean Monnet de Saint-Étienne, no. 22-23, pp. 49-81, décembre 2002
415. Guitart R: Toute théorie est algébrique et topologique. Cahiers Top. Géo. Diff. Cat. vol. XLIX-2 (2008), p. 83-128. 2008 (first partial version: "Toute théorie est algébrique", in Journée Mathématique en l'Honneur d'Albert Burroni, vendredi 20 septembre 2002, Université Paris 7, Institut de Mathématiques de Jussieu, Prépublication 368, p. 79-102, avril 2004
416. Guitart R and Ch Ehresmann: Au carrefour des structures locales et algébriques. Colloque International Ch Ehresmann : 100 ans, Amiens 7-9 octobre 2005; Cahiers Top. Géo. Diff. Cat., vol. XLVI-3, p. 172-175, 2005
417. Guitart R: L'évidemment des objets et le dehors comme substance. Colloque Le lemme de Yoneda, "Enjeux mathématiques et philosophie", ENS, Paris 18 juin 2007
418. Guitart R: An anabelian definition of abelian homology. Cahiers Top Géo Diff Cat, XXXXVIII, 4, pp. 261-269, 2007
419. Guitart R: Mathematical Pulsation at the root of invention. Anais do Congresso Htem 4, Rio de Janeiro, de 5 a 9 de maio de 2008
420. Guitart R: Sur la représentation de l'herméneutique comme commerce gratuit des interprétations, conférence au *CERCI*, Université de Nantes, 7 mars 2009
421. Guitart R: Cohomological Emergence of Sense in Discourses (As Living Systems Following Ehresmann and Vanbremeersch). Axiomathes, 19(3), 245-270, 2009
422. Guitart R: Les coordonnées curvilignes de Gabriel Lamé, représentations des situations physiques et nouveaux objets mathématiques, 12 p. In: Actes du Colloque International Gabriel Lamé. Les pérégrinations d'un ingénieur au XIXème siècle, 15-16-17 janvier 2009, Nantes. SABIX, no 44, p. 119-129, octobre 2009
423. Guitart R: Sense and Signs in the Mathematical Invention of Coordination (in the Perspective of Arrows, Relations and Sketches). Talk at the colloquium *Semiotic Approaches to Mathematics, the History of Mathematics and Mathematics Education (SemMHistEd)*, 3rd Meeting Aristotle University of Thessaloniki, July 16-17, 2009
424. Guitart R: Modélisation qualitative catégoricienne : modèles, signes et formes. In: Andreatta M, F Nicolas and Ch Alunni (Eds.): à la lumière des mathématiques et à l'ombre de la philosophie, Coll. Musique/Sciences, ENS, Delatour, Ircam, Centre Pompidou, pp. 133-147, Paris 2012
425. Guitart R: Bachelard et la pulsation mathématique. Revue de synthèse, tome 136, 6ème série, n. 1-2, p. 33-74, 2015
426. Guitart R: Note sur deux problèmes pour une épistémologie transitive des mathématiques, *Revue de synthèse*, tome 136, 6ème série, n. 1-2, p. 237-279, 2015
427. Guitart R: Nietzsche face à l'expérience mathématique, in: Barbin E and J-P Cléro éds, *Les mathématiques et l'expérience : ce qu'en ont dit les philosophes et les mathématiciens*, Hermann, p. 247-278, 2015
428. Guitart R: History in mathematics according to André Weil. ESU7, Copenhagen, July 2014
429. Guitart R: L'évidement et la nécessité de la contingence. Conférence au Colloque *Les fins de cure : la poursuite du désêtre*, 7/8, AECF Lille, novembre 2015
430. Gurlitt et al. (Eds.): Riemann Musiklexikon/Sachteil. Schott, Mainz 1967
431. Gusdorf G: La révolution galiléenne, tomes I et II. Payot, Paris, 1969
432. Hába A: Neue Harmonielehre. (Reprint from 1927) Universal Edition, Wien 1978
433. Habermann R: Elementary Applied PDEs, Prentice Hall 1983

434. Hadamard J: An Essay on the Psychology of Invention in the Mathematical Field. Princeton University Press, 1945, Dover, 1954
435. Halsey D and E Hewitt: Eine gruppentheoretische Methode in der Musiktheorie. Jahresber. d. Dt. Math.-Vereinigung 80, 1978
436. Handschin J: Der Toncharakter. Zürich 1948
437. Hansen P: Pourquoi il faut sortir des mathématiques. Mercredi 1er avril 2015 (see also [99]). `http://www.energie-crise.fr/spip.php?article176`
438. Hanslick E: Vom Musikalisch-Schönen. Breitkopf und Härtel (1854), Wiesbaden 1980
439. Hardy G H: A Mathematician's Apology. Cambridge U. Press, 1967
440. Hashimoto K: D-Brane. Springer, Heidelberg et al. 2012
441. Hardt M: Zur Zahlenpoetik Dantes. In: Baum R. and Hirdt W (Eds.): Dante Alighieri 1985. Stauffenburg, Tübingen 1985
442. Harkness R: Songs of Seoul. University of California Press 2014
443. Harris C and A R Brinkman: A unified set of software tools for computer-assisted set-theoretic and serial analysis of contemporary music. Proc. ICMC 1986, ICA, San Francisco 1986
444. Hartshorne R: Algebraic Geometry. Springer, New York et al. 1977
445. Hashimoto S and H Sawada: Musical Performance Control Using Gesture: Towards Kansei Technology for Art. In: Kopiez R and Auhagen W (Eds.): Controlling Creative Processes in Music. Peter Lang, Frankfurt am Main et al. 1998
446. Hatten R: Interpreting Musical Gestures, Topics, and Tropes. Indiana University Press 2004
447. Hawking S: A Brief History of Time: From the Big Bang to Black Holes. Bantam Books, New York 1988
448. Haugh L D: Checking for independence of two covariance stationary time series: a univariate residual cross correlation approach. J. Am. Statist. Assoc., 71, 378-385, 1976
449. Haus G et al.: Stazione di Lavoro Musicale Intelligente. In: Haus G and Pighi I (Eds.): X Colloquio di Informatica Musicale. AIMI, LIM-Dsi, Milano 1993
450. Hayes J R: Cognitive Processes in Creativity. In: J A Glover, R R Roning, C R Reynolds (Eds.): Handbook of Creativity. Plenum, New York 1989
451. Hebb D O: Essay on mind. Hillsdale, New Jersey, Lawrence Erlbaum Associates, 1980
452. Hegel G W F: Wissenschaft der Logik I (1812). Felix Meiner, Hamburg 1963
453. Hegel G W F: Wissenschaft der Logik, Erster Band, Erstes Buch. Schrag, Nürnberg 1812 Online `http://www.deutschestextarchiv.de/book/show/hegel_logik0101_1812` Modern German online version: `http://www.zeno.org/Philosophie/M/Hegel,+Georg+Wilhelm+Friedrich/Wissenschaft+der+Logik`
454. Hegel G W F: Hegel's Science of Logic. Online `http://www.marxists.org/reference/archive/hegel/works/hl/hlconten.htm`
455. Heiberg J L and H Menge (Eds.): Euclidis opera omnia. 8 vol. & supplement, in Greek. Teubner, Leipzig 1883-1916.
456. Heijink H, P Desain, H Honing, L Windsor: Make me a match: An evaluation of different approaches to score-performance matching. Computer Music Journal, 24(1), 43-56, 2000
457. Helmholtz H von: Die Lehre von den Tonempfindungen als physiologische Grundlage der Musik (1863). Nachdr. Darmstadt 1968
458. Henck H: Karlheinz Stockhausens Klavierstück IX. Verlag für systematische Musikwissenschaft, Bonn-Bad Godesberg 1978
459. Hentoff N: Liner notes to Coltrane's last album "Expression". Impulse AS-9120, New York 1967
460. Hecquet S and A Prokhoris: Fabriques de la danse. PUF, Paris 2007
461. Herbort H J: Keine Ausweispflicht für cis. Die Zeit Nr. 43, 21. Oktober 1988
462. Hermann J: Consideratio curvarum in punctum positione datum projectarum, et de affectionibus earum inde pendentibus. Commentarii Academiae Petropolitanae, IV, 37-46, 1729
463. Hesse H: Das Glasperlenspiel (1943). Fretz und Wasmuth, Zürich 1943
464. Hesse H: Materialien zu H. Hesses "Das Glasperlenspiel", Vol. I. : Vom Wesen und von der Herkunft des Glasperlenspiels (1934). Suhrkamp, Frankfurt am Main 1981
465. Heussenstamm G: Norton Manual of Music Notation. Norton & Comp., New York 1987
466. Hichert J: Verallgemeinerung des Kontrapunkttheorems für die Hierarchie aller starken Dichotomien in temperierter Stimmung. Diplomarbeit, TU Ilmenau 1993
467. Hindemith P: Unterweisung im Tonsatz. Schott, Mainz 1940
468. Hilbert D and S Cohn-Vossen: Geometry and the Imagination. AMS Chelsea Pub. (transl. by Paul Nemenyi from *Anschauliche Geometrie*, 1932), 1952

469. Hiller L and P Ruiz: Synthesizing sounds by solving the wave equation for vibration objects. J. of the Audio Engineering Soc. 19: 463-470, 542-551, 1971
470. Hilton P J and S Wylie: Homology Theory. Cambridge University Press, London et al. 1967
471. Hjelmslev L: La Stratification du Langage. Minuit, Paris 1954
472. Hjelmslev L: Prolégomènes à une théorie du langage. Minuit, Paris 1968-71
473. Hjelmslev L: Nouveaux essays. PUF, Paris 1985
474. Hofstädter D: Gödel, Escher, Bach. New York: Basic Books, New York 1979
475. Hofstadter D R: Staring Emmy Straight in the Eye—and Doing my Best not to Flinch. In: D Cope (Ed.): Virtual Music: Computer Synthesis of Musical Style. MIT Press, Cambridge, MA 2001
476. Hodgson J: Mastering Movement: the life and work of Rudolf Laban. Routledge, New York 2001
477. Holden S P: Right-Handed and Right On. The New York Times, No. 3, June 2, 1999
478. Holm-Hadulla R M: Kreativität. Konzept und Lebensstil. Vandenhoeck & Ruprecht, Göttingen 2005
479. Hong Y: Testing for independence between two covariance stationary time series. Biometrika, 83, No.3, 615-626, 1996
480. Honing H: Expresso, a strong and small editor for expression. In: ICMA (Ed.): Proceedings of the ICMC 92, San Francisco 1992
481. Hooft G 't: Symmetrien in der Physik der Elementarteilchen. In: Dosch H G (Ed.): Teilchen, Felder und Symmetrien. Spektrum der Wissenschaft, Heidelberg 1984
482. Hörnl D et al.: Learning Musical Structure and Style by Recognition, Prediction and Evolution. In: ICMA (Ed.): Proceedings of the ICMC 96, San Francisco 1996
483. Horry Y: A Graphical User Interface for MIDI Signal Generation and Sound Synthesis. In: ICMA (Ed.): Proceedings of the ICMC 94, San Francisco 1994
484. Hosken D: An Introduction to Music Technology. Routledge New York 2011
485. Howe H: Some combinatorial properties of Pitch-Structures. PNM, 4(1), 45-61, 1965
486. Howe M J A: Prodigies and Creativity. In: R J Sternberg (Ed.): Handbook of Creativity. Cambridge University Press, New York 1999
487. Hu S-T: Mathematical Theory of Switching Circuits and Automata. University of California Press, Berkeley and Los Angeles 1968
488. Hudak P et al.: Haskore Music Notation - An Algebra of Music. J. Functional Programming Vol. 6 (3) 1996
489. Hudspeth A J and D P Corey: Sensitivity, Polarity, and Conductance Change in the Response of Vertebrate Hair Cells to Controlled Mechanical Stimuli. Proc. Nat. Acad. Sci. Am. 74(6), 2407-2411, 1977
490. Hull P et al.: The Music Business and Recording Industry. Third Edition. Routledge, New York and London 2011
491. Humphreys J E: Introduction to Lie Algebras and Representation Theory. Springer, New York et al. 1972
492. Hung R et al.: The Analysis and Resynthesis of Sustained Musical Signals in the Time Domain. In: ICMA (Ed.): Proceedings of the ICMC 96, San Francisco 1996
493. Hunziker E and G Mazzola: Ansichten eines Hirns. Birkhäuser, Basel 1990
494. Husmann H: Einführung in die Musikwissenschaft. Heinrichshofen, Wilhelmshaven 1975
495. Ilchmann A: Kritik der Übergänge zu den ersten Kategorien in Hegels Wissenschaft der Logik. Hegel-Studien, Bd. 27, Bouvier, Bonn 1992
496. Jackendoff R and F Lerdahl: A Generative Theory of Tonal Music. MIT Press, Cambridge MA, 1983
497. Jackson J D: Classical Electrodynamics. Wiley, New York 1998
498. Jairazbhoy N A: What Happened to Indian Music Theory? Indo-Occidentalism? Ethnomusicology 52 (3): 34977, 2008
499. Jakobson R and M Halle: Fundamentals of Language. Mouton, Le Hague 1957
500. Jakobson R: Linguistics and Poetics. In: Seboek, TA (Ed.): Style in Language. Wiley, New York 1960
501. Jakobson R: Language in relation to other communication systems. In: Linguaggi nella società e nella tecnica. Edizioni die Communità, Milano 1960
502. Jakobson R: Six leçons sur le son et le sens. Ed. de Minuit, Paris 1976
503. Jakobson R: Hölderlin, Klee, Brecht. Suhrkamp, Frankfurt/Main 1976
504. Jakobson R and K Pomorska: Poesie und Grammatik. Dialoge. Suhrkamp, Frankfurt/Main 1982
505. Jakobson R: Semiotik - Ausgewählte Texte 1919-1982. Holenstein E (Ed.), Suhrkamp, Frankfurt/Main 1988
506. Jauss H R: Rückschau auf die Rezeptionstheorie—Ad usum Musicae Scientiae. In: Danuser H and Krummacher F (Hsg.): Rezeptionsästhetik und Rezeptionsgeschichte in der Musikwissenschaft. Laaber, Laaber 1991
507. Jeans J H: Science and Music. 1938, reprinted by Dover, 1968
508. Jeppesen K: Kontrapunkt. Breitkopf und Härtel, Wiesbaden 1952
509. John F: Partial Differential Equations. Springer, Heidelberg et al. 1978

510. Johnson T: See his web page at http://www.tom.johnson.org
511. Johnson-Laird P N: Freedom and Constraint in Creativity. In: R J Sternberg (Ed.): The Nature of Creativity: Contemporary Psychological Perspectives. Cambridge University Press, New York, pp. 202-219, 1988
512. Junod J: Counterpoint Worlds and Morphisms. Dissertation, University of Zurich, Zurich 2010
513. Julia G: Sur l'itération des fonctions rationnelles. J. de Math. Pure et Appl. 8, 1918
514. Jurek T: The Complete Blue Note Horace Parlan Sessions. AllMusic 2015. `http://www.allmusic.com/album/the-complete-blue-note-horace-parlan-sessions-mw0001007267`
515. Kagel M: Translation - Rotation. Die Reihe Bd.7, Universal Edition, Wien 1960
516. Kahle W: Taschenatlas der Anatomie, Bd.3: Nervensystem und Sinnesorgane. Thieme/dtv, Stuttgart 1979
517. Kaiser J: Beethovens 32 Klaviersonaten und ihre Interpreten. Fischer, Frankfurt/Main 1979
518. Kan D: Adjoint functors. Transactions of the American Mathematical Society, 87: 294-329, 1958
519. Kant I: Kritik der reinen Vernunft. Meiner, Hamburg, 2nd edn. 1957
520. Kantor J-M (Ed.): Ou en sont les mathématiques?. Vuibert/SMF, 2002
521. Karg-Elert S: Polaristische Klang- und Tonalitätslehre (1931). Out of print, cf.: Schenk P.: Karg-Elerts polaristische Harmonielehre. In: Vogel, M (Ed.) Beiträge zur Musiktheorie des 19. Jahrhunderts. Bosse, Regensburg 1966
522. Kasner E and J Newman: Mathematics and the imagination. Simon and Schuster, New York, 1940
523. Katayose H et al.: Demonstration of Gesture Sensors for the Shakuhachi. In: ICMA (Ed.): Proceedings of the ICMC 94, San Francisco 1994
524. Katsman R: Anthropoetic Gesture: A Key to Milorad Pavić's Poetics (Landscape Painted With Tea). Academic Electronic Journal in Slavic Studies No. 16, Univ. Toronto, 2006
525. Katsman R: Gestures Accompanying Torah Learning/recital among Yemenite Jews. Gesture 7 (1): 119, 2007
526. Kayser H: Akróasis. Die Lehre von der Harmonik der Welt. Schwabe, Basel 1946
527. Kelley J L: General Topology. Van Nostrand, Princeton et al. 1955
528. Kempe A B: A memoir on the theory of mathematical form. Philosophical Transactions of the Royal Society of London, 177, 1-70, 1886
529. Kendon A: How Gestures Can Become like Words. Crosscultural Perspectives in Nonverbal Communication. 1988
530. Kendon A: Gesture: Visible Action as Utterance. Cambridge University Press, Cambridge 2004
531. Keyser C J: The Meaning of Mathematics. Scripta Mathematica, 1933, p. 15-28
532. Kiczales G, J Rivieres, D G Bobrow: The Art of the Metaobject Protocol. The MIT Press, Boston 1991
533. Kinderman W: Artaria 195. University of Illinois Press, Urbana and Chicago 2003
534. King E: New Perspectives on Music and Gesture. Routledge 2016
535. Kirsch E: Wesen und Aufbau der Lehre von den harmonischen Funktionen. Leipzig 1928
536. Klemm M: Symmetrien von Ornamenten und Kristallen. Springer, Berlin et al. 1982
537. Klumpenhouwer H: Deep Structure in K-net Analysis with special reference to Webern's opus 16,4. Unpublished manuscript
538. Klumpenhouwer H: A Generalized Model of Voice-Leading for Atonal Music. Ph.D. Thesis, Harvard University 1991
539. Knapp J: Franz Liszt. Berlin 1909
540. Köhler E: Brief an Guerino Mazzola. Hamburg 1988
541. Kollmann A: An Essay on Musical Harmony. London 1796
542. Komparu K: The Noh Theatre. Weatherhill/Tankosha, New York et al. 1983
543. Koenig Th: Robert Schumanns *Kinderszenen* op.15. In: Metzger H-K und Riehn R (Hrg.): Robert Schumann II. edition text+kritik, München 1982
544. Kononenko N O: Ukrainian minstrels and the blind shall sing. Armonk, N.Y.: M E Sharpe, 1998
545. Kopiez R and J Langner: Entwurf einer neuen Methode der Performanceanalyse auf Grundlage einer Theorie oszillierender Systeme. In: Behne K-E and de la Motte H (Eds.): J.buch der D. Ges. für Musikpsychologie. **12** (1995), Wilhelmshaven 1996
546. Kopiez R: Aspekte der Performanceforschung. In: de la Motte H: Handbuch der Musikpsychologie. Laaber, Laaber 1996
547. Kopiez R: Mensch - Musik - Maschine. Musica, 50 (1), 1996
548. Kopiez R: "The most wanted song/The most unwanted song" – Klangfarbe als wahrnehmungsästhetische Kategorie. Musicology Conference, Halle 1998
549. Kopiez R, J Langner, Ch Stoffel: Realtime analysis of dynamic shaping. Talk at the 6th International Conference on Music Perception and Cognition (ICMPC), Keele, England, August 5-10, 2000

550. Kopiez R, J Langner and B Feiten: Perception ad Representation of Multiple Tempo Hierarchies in Musical Performance ad Composition: Perspectives from a New Theoretical Approach. In: Kopiez R and W Auhagen (Eds.): Controlling Creative Processes in Music. Peter Lang, Frankfurt am Main et al. 1998
551. Körner T W: Fourier Analysis. Cambridge University Press 1988, Cambridge
552. Kostelanetz R: John Cage. Praeger, New York 1968
553. Kouneiher J, D Flament, P Nabonnand, J-J Szczeciniarz (Eds.): Geometrie au XXème siècle, 1930-2000. Histoire et horizons. Hermann, Paris 2005
554. Kouzes R T et al.: Collaboratories: Doing Science on the Internet. Computer, August 1996
555. Kozbelt A, R A Beghetto, M A Runco: Theories of Creativity. In: J C Kaufman and R J Sternberg (Eds.): The Cambridge Handbook of Creativity. Cambridge University Press, New York, pp. 20-47, 2010
556. Krömer R: La théorie des catégories : un outil de musicologie scientifique aux yeux de la critique philosophique. In: Andreatta M, F Nicolas et C Alunni (Eds.): Á la lumière des mathématiques et à l'ombre de la philosophie. Coll. Musique/Sciences, ENS, Delatour, Ircam, Centre Pompidou, p. 213-222, Paris 2012
557. Kronland-Martinet R: The Wavelet Transform for Analysis, Synthesis, and Processing of Speech and Music Sounds. Computer Music Journal, 12 (4), 1988
558. Kreiman J and D Sidtis: Foundations of Voice Studies. Wiley-Blackwell 2013
559. Kronland-Martinet R: The Wavelet Transform for Analysis, Synthesis, and Processing of Speech and Music Sounds. Computer Music Journal, 12 (4), 1988
560. Kronman U and J Sundberg: Is the Musical Ritard an Allusion to Physical Motion? In: Gabrielsson A (Ed.): Action and Perception in Rhythm and Meter. Bibl. of the Royal Swedish Acad. of Sci. 55, Stockholm 1987
561. Kuratowski K and A Mostowski: Set Theory. North Holland, Amsterdam 1968
562. Kühner H: Virtual Table. CG TOPICS 3/97
563. Laban R: Language of Movement: A guidebook to Choreutics, ed. Lisa Ullmann, Plays, Boston 1974
564. Laban R: Die Welt des Tänzers. W. Seifert, Stuttgart 1920
565. Laban R: The Laban Sourcebook. McCaw D (Ed.), Routledge, New York 2011
566. Laban R: The Mastery of Movement, 4th ed. Macdonald & Evans, Plymouth 1980
567. Laban R: The Rhythm of Effort and Recovery I. LAMG Magazine, in Hodgson, Mastering Movement, November 1959
568. Lachenmann H: Mouvement (- vor der Erstarrung) für Ensemble. Breitkopf und Härtel, 1984
569. Lages Lima E: Fundamental Groups and Covering Spaces. Peters, Natick, MA 2003
570. Lakoff G and R Núñez: Where mathematics comes from: How the embodied mind brings mathematics into being. Basic Books, New York 2000
571. Lamé G: Examen des différentes méthodes employées pour résoudre les problèmes de géométrie (1903). Hermann, Paris 1918
572. Lamé G: Leçons sur les coordonnées curvilignes, et leurs diverses applications. Mallet-Bachelier, 1859
573. Landi S (Ed.): L'estrangement, Retour sur un thème de Carlo Ginzburg. Essais, revue interdisciplinaire d'humanités, No Hors série, école doctorale Montaigne-Humanités, Univ. Michel de Montaigne, Bordeaux 3, 2013
574. Landi S (Ed.): Avant-propos. In: L'estrangement, Retour sur un thème de Carlo Ginzburg [573, p. 9-17]
575. Landry, E and J-P Marquis: Categories in Context: Historical, Foundational, and Philosophical. Philosophia Mathematica 13.1: 1-43, 2005
576. Lang S: Introduction to Differentiable Manifolds. Interscience, New York et al. 1962
577. Lang S: Elliptic Functions. Addison-Wesley, Reading, Mass. 1973
578. Lang S: $SL_2(\mathbb{R})$. Addison-Wesley, Reading, Mass. et al. 1975
579. Langley P, H A Simon, G L Bradshaw, J M Zytkow: Scientific Discovery: Computational Explorations of the Creative Process. MIT Press, Cambridge, MA, 1987
580. Langner G: Evidence for Neuronal Periodicity Detection in the Auditory System of Guinea Fowl: Implications for Pitch Analysis in the Time Domain. Exp. Brain Res. 52, 333-355 1983
581. Langner G: Periodicity coding in the auditory system. Hear Res. Jul;60(2):115-42, 1992
582. Langer S: Feeling and Form, Routledge and Kegan Paul, London 1953
583. Lawson R A: The First Century of Blues: One Hundred Years of Hearing and Interpreting the Music and the Musicians. Southern Cultures, pp. 39-61, 2007
584. Langner J, R Kopiez, Ch Stoffel, M Wilz: Realtime analysis of dynamic shaping. In: Woods C et al. (Eds.): Proceedings of the Sixth International Conference on Music Perception and Cognition, Keele, UK, 2000
585. Lawvere F W: An elementary theory of the category of sets. Proc. Natl. Acad. Sci. 52, 1506-1511, 1964
586. Lawvere F W and R Rosebrugh: Sets for mathematics, Cambridge U. Press, 2003

587. Lazier R: Making Schoenberg Dance. In: Sound Moves, Proceedings of the Int. Conf. on Music and Dance, Roehampton University 2005
588. Leach J L: Towards a Universal Algorithmic System for Composition of Music and Audio-Visual Works. In: ICMA (Ed.): Proceedings of the ICMC 96, San Francisco 1996
589. Ledderose L: Kreativität und Schrift in China. In: R M Holm-Hadulla (Ed.): Kreativität. Springer, Heidelberg 2000
590. Lee, J M: Introduction to Topological Manifolds. Springer, Heidelberg et al. 2000
591. Legrand M: Revue critique de "La pulsation mathématique". Repères-IREM, 39, avril 2000
592. Le Guin E: Boccherini's Body. University of California Press, Berkeley and Los Angeles 2006
593. Leman M: Schema Theory. Springer, Berlin et al. 1996
594. Lerdahl F: Timbral Hierarchies. Contemporary Music Review, vol.2, no.1, 1987
595. Lerdahl F: Cognitive Constraints on Compositional Systems. In: ed. J Sloboda (Ed.): Generative Processes in Music. Oxford University Press, Oxford 1988. Reprinted in Contemporary Music Review 6, no. 2 (1992):97-121
596. Levy M: Intonation in North Indian Music. Biblia Impex, 1982
597. Lévi-Strauss C: Le cru et le cuit: Mythologies I. Plon, Paris 1964
598. Lewin D: Re: Intervallic Relations between Two Collections of Notes. JMT, 3(2), 298-301, 1959.
599. Lewin D: Klumpenhouwer Networks and Some Isographies that Involve Them. Music Theory Spectrum 12/1, 1990
600. Lewin D: The Intervallic Content of a Collection of Notes, Intervallic Relations between a Collection of Notes and Its Complement: An Application to Schoenberg's Hexachordal Pieces. JMT, 4(1), 98-101, 1960
601. Lewin D: Forte's Interval Vector, My Interval Function, and Regener's Common-Note Function. JMT, 21(2), 194-237, 1977
602. Lewin D: A Formal Theory of Generalized Tonal Functions. JMT 26(1), 32-60, 1982
603. Lewin D: On Formal Intervals between Time-Spans. Music Perception, 1(4), 414-423, 1984
604. Lewin D: Musical Form and Transformation: 4 Analytic Essays. Yale University Press, New Haven and London 1993
605. Lewin D: Generalized Musical Intervals and Transformations (1987). Cambridge University Press 2007
606. Lewis C I: Mind and Word Order. Dover, New York 1956
607. Leyton M: Symmetry, Causality, Mind. MIT Press, Cambridge/MA and London 1992
608. Leyton M: A Generative Theory of Shape. Springer, Berlin et al. 2001
609. Lichtenhahn E: Romantik: Aussen- und Innenseiten der Musik. In: "Das Paradox musikalischer Interpretation", Symposion zum 80. Geburtstag von K. von Fischer, Univ. Zürich 1993
610. Lidov D: Mind and Body in Music. Semiotica 66/1 (1987), pp. 77-78
611. Ligeti G: Pierre Boulez: Entscheidung und Automatik in der Structure Ia. Die Reihe IV, UE, Wien 1958
612. Ligeti G and G Neuweiler: Motorische Intelligenz. Wagenbach, Berlin 2007
613. Lippe E and A H M ter Hofstede: A Category Theory Approach to Conceptual Data Modeling. Informatique Theorique et Applications, Vol 30, 1, 31-79, 1996
614. Liszka J: A general introduction to the Semeiotic of Charles S. Peirce. Bloomington M.A. : Harvard University Press, 1996
615. Liszt F: Mephisto-Walzer I. Ed. Peters, No. 7203, (Edited by E. Von Sauer) 1913-17
616. Locke J: An essay concerning human understanding (first publication 1690). Pennsylvania State University 1999
617. Loomis L H and S Sternberg: Advanced Calculus. Addison-Wesley, Reading, MA et al. 1968
618. Loria G: Perfectionnements, Evolution, Métamorphoses du concept de "coordonnées" Contribution à l'histoire de la géométrie analytique. Osiris, Vol. 8, pp. 218-288, 1948
619. Loria G: A. L. Cauchy in the History of Analytic Geometry. Scripta Mathematica, p. 123-128, 1933
620. Losada C: Isography and Structure in the Music of Boulez. Journal of Mathematics and Music 2008, Vol. 2, no. 3, pp. 135-155, 2008
621. Luck G, P Toiviainen, M Thompson: Perception of Expression in Conductors' Gestures: A Continuous Response Study. Music Perception, 20(1), 47-57, 2010
622. Louis A K et al.: Wavelets. Teubner, Stuttgart 1994
623. Lubet A: (Paralyzed on One) Sideman: Disability Studies Meets Jazz, through the Hands of Horace Parlan. Critical Studies in Improvisation/Études critiques en improvisation 6.2. 2010 `http://www.criticalimprov.com/article/view/1268`
624. Lubet A: Music, Disability, and Society. Temple University Press, Philadelphia 2011
625. Lubet A: Social Confluence and Citizenship: A View from the Intersection of Music and Disability. In: Hirschmann N J, and B Linker (Eds.): Civil Disabilities: Citizenship, Membership, and Belonging. University of Pennsylvania Press, Philadelphia 2015

626. Lubet A: Oscar Peterson's piano prostheses: Strategies of performance and publicity in the post-stroke phase of his career. Jazz Research Journal, 7(2), pp. 151-182, 2015
627. Lubet A: A Case of the Blues (misc). Unpublished, 2014
628. Lucas E: Récréations mathématiques. Gauthier-Villars, Paris, 1882
629. Lüdi W: Fax to G Mazzola. Malans 01/23/1991
630. Ludwig H: Marin Mersenne und seine Musiklehre. Olms, Hildesheim 1971
631. Lussy M: Traité de l'expression musicale. Paris 1874
632. Lutz R: Le poète dans son oeuvre. Seminar Jung, Univ. Zürich, 1980
633. Lyttleton H: The Best of Jazz. Robson Books, London 1998
634. Machado A: Chant XXIX, Proverbios y cantarès, Campos de castilla, (1917), Ed. Cátedra, Madrid, 1989, 2015
635. Mac Lane S: Homology. Springer, Heidelberg et al. 1963
636. Mac Lane S: The development of mathematical ideas by collision: The case of Categories and Topos Theory. In: Categorical Topology and its relation to analysis, algebra and combinatorics, 1989
637. Mac Lane S: Categories for the Working Mathematician. Springer, New York et al. 1971
638. Mac Lane S: Mathematics, Form and Function. Springer, New York 1986
639. Mac Lane S and Moerdijk I: Sheaves in Geometry and Logic. Springer, New York et al. 1994
640. MacLean P D: The triune brain, emotion, and scientific bias. In: Schmitt F O (Ed.): The Neurosciences: Second Study Program, 336-348, Rockefeller Univ. Press, New York 1970
641. Maeder R: Programming Mathematica. Addison-Wesely, Reading, Mass. 1991
642. Manin Y: Interview in the Notices of the American Mathematical Society, November 2009, p. 1268
643. Mannone M and G Mazzola: Hegel's Conceptual Group Action on Creative Dynamics in Music. Gli spazi della musica, University of Torino 2014
644. Mannone M and G Compagno: Characterization of the degree of Musical non-Markovianity. Unpublished
645. Mannone M and S Lo Franco and G Compagno: Comparison of non-Markovianity criteria in a qubit system under random external fields. Phys. Scr. T, 2013
646. Mannone M: Segmentation des séries temporelles pour l'orchestration automatique. IRCAM - UPMC, Paris 2013 `http://www.atiam.ircam.fr/Archives/Stages1213/MANNONE_Maria.pdf`
647. Mannone M: Dalla Musica all'Immagine, dall'Immagine alla Musica - Relazioni matematiche fra composizione musicale e arte figurativa. Compostampa 2011
648. Mannone M and G Mazzola: Hypergestures in Complex Time: Creative Performance Between Symbolic and Physical Reality. Proceedings of the MCM 2015 conference, Springer, Heidelberg 2015
649. Mannone M and G Mazzola: Hypergestures in Complex Time: Creative Performance Between Symbolic and Physical Reality. Talk at MCM conference 2015, Queen Mary U, London 2015
650. Mansouri B: Normes internationales élargies et dépassement du calcul classique de la valeur. Thèse, Doctorat ès Sciences de gestion, Université de Bordeaux, 321 p., 3 juillet 2015
651. Marek C: Lehre des Klavierspiels. Atlantis 1977
652. Marino G, M-H Serra, J-M Raczinski: The UPIC system: Origins and innovations. Perspectives of New Music, 31(1):258–69, 1993
653. Marquis J-P: Category theory and the foundations of mathematics: philosophical excavations, *Synthese*, 103, pp. 229-254, 1995
654. Marrou H-I: De la connaissance historique. 1954, nouv. éd. coll. Points/Histoire, H21, Paris, 1975
655. Martinet A: Eléments de Linguistique Générale. Colin, Paris 1960
656. Martin B: CeDell Davis: The Last Man Standing. Huffington Post, 18 February 2015 `http://www.huffingtonpost.com/barrett-martin/cedell-davis-the-last-man_b_6696710.html`
657. Martino D: The Source Set and Its Aggregate Formations. JMT, 5(2), 224-273, 1961
658. De Martino E: La terra del rimorso. Bellaterra edizioni, 1999
659. Marx A B: Lehre von der musikalischen Komposition. Breitkopf & Härtel, Leipzig 1837-1847
660. Mason R M: Enumeration of Synthetic Musical Scales (...). J. of Music Theory 14, 1970
661. Massey W S: A Basic Course in Algebraic Topology. Springer, New York et al. 1991
662. Mathiesen Th J: Transmitting Text and Graphics in Online Database: The Thesaurus Musicarum Latinarum Model. In: Hewlett W B and Selfridge-Field E (Eds.): Computing in Musicology Vol.9, CCARH, Menlo Park 1993-94
663. Matisoff S: The legend of Semimaru, blind musician of Japan. Columbia University Press, New York 1978
664. Mattheson J: Der vollkommene Kapellmeister. Hamburg 1739
665. Maxwell H J: An Expert System for Harmonizing Analysis of Tonal Music. In: Balaban M et al. (Eds.): Understanding Music with AI: Perspectives on Music Cognition, MIT Press. Cambridge, MA 1992
666. May J P: A Concise Course in Algebraic Topology. The University of Chicago Press, Chicago 2007

667. Mazzola G: *Akroasis—Beethoven's Hammerklavier-Sonate in Drehung (für Cecil Taylor)*. LP, recorded August 30, 1979, WERGO/Schott, Mainz 1980
668. Mazzola G: Musique et théorie des groupes. Conference at Institut de Mathématique, U Genève, January 31, 1980
669. Mazzola G: Die gruppentheoretische Methode in der Musik. Lecture Notes, Notices by H. Gross, SS 1981, Mathematisches Institut der Universität, Zürich 1981
670. Mazzola G: Gruppen und Kategorien in der Musik. Heldermann, Berlin 1985
671. Mazzola G et al.: Rasterbild-Bildraster. Springer, Berlin et al. 1986
672. Mazzola G: Die Rolle des Symmetriedenkens für die Entwicklungsgeschichte der europäischen Musik. In: Symmetrie, Katalogband Vol.1 zur Symmetrieausstellung, Mathildenhöhe, Darmstadt, 1986
673. Mazzola G: Obertöne oder Symmetrie: Was ist Harmonie?. In: Herf F R (Ed.): Mikrotöne, Helbling, Innsbruck 1986
674. Mazzola G: Mathematische Betrachtungen in der Musik I,II. Lecture Notes, Univ. Zürich 1986/87
675. Mazzola G: Der Kontrapunkt und die K/D-Dichotomie. Manuscript, University of Zürich 1987
676. Mazzola G: Die Wahl der Zahl - eine systematische Betrachtung zum Streichquartett. In: Dissonanz 17, 1988
677. Mazzola G and G R Hofmann: Der Music Designer MD-Z71 - Hardware und Software für die Mathematische Musiktheorie. In: Petsche H (Ed.): Musik - Gehirn - Spiel, Beiträge zum 4. Herbert-von-Karajan-Symposion. Birkhäuser, Basel 1989
678. Mazzola G et al.: A Symmetry-Oriented Mathematical Model of Classical Counterpoint and Related Neurophysiological Investigations by Depth-EEG. In: Hargittai I (Ed.): Symmetry II, CAMWA, Pergamon, New York 1989
679. Mazzola G et al.: Hirnelektrische Vorgänge im limbischen System bei konsonanten und dissonanten Klängen. In: Petsche H (Ed.): Musik - Gehirn - Spiel, Beiträge zum 4. Herbert-von-Karajan-Symposion. Birkhäuser, Basel 1989
680. Mazzola G: *presto* Software Manual. SToA music, Zürich 1989-1994
681. Mazzola G: *Synthesis*. SToA 1001.90, Zürich 1990
682. Mazzola G: Geometrie der Töne. Birkhäuser, Basel 1990
683. Mazzola G and Muzzulini D: Tempo- und Stimmungsfelder: Perspektiven künftiger Musikcomputer. In: Hesse H P (Ed.): Mikrotöne III. Edition Helbling, Innsbruck 1990
684. Mazzola G and Muzzulini D: Deduktion des Quintparallelenverbots aus der Konsonanz-Dissonanz-Dichotomie. Accepted for publication in: Musiktheorie, Laaber 1990, but unpublished (!)
685. Mazzola G: Mathematische Musiktheorie: Status quo 1990. Jber. d.Dt. Math.-Verein. 93, 6-29, 1991
686. Mazzola G: Mathematical Music Theory—An Informal Survey. Edizioni Cerfim, Locarno 1993
687. Mazzola G: RUBATO at SMAC KTH, Stockholm 1993
688. Mazzola G and O Zahorka: Tempo Curves Revisited: Hierarchies of Performance Fields. Computer Music Journal **18**, No. 1, 1994
689. Mazzola G and O Zahorka: The RUBATO Performance Workstation on NeXTSTEP. In: ICMA (Ed.): Proceedings of the ICMC 94, San Francisco 1994
690. Mazzola G and O Zahorka: Geometry and Logic of Musical Performance I, II, III. SNSF Research Reports (469pp.), Universität Zürich, Zürich 1993-1995
691. Mazzola G et al.: Analysis and Performance of a Dream. In: Sundberg J (Ed.): Proceedings of the 1995 Symposium on Musical Performance. KTH, Stockholm 1995
692. Mazzola G et al.: The RUBATO Platform. In: Hewlett W B and E Selfridge-Field (Eds.): Computing in Musicology 10. CCARH, Menlo Park 1995
693. Mazzola G and O Zahorka: The PrediBase Data Base System of RUBATO on NEXTSTEP. In: Selfridge-Field E (Ed.): Handbook of Musical Codes. CCARH, Menlo Park 1995
694. Mazzola G: Inverse Performance Theory. In: ICMA (Ed.): Proceedings of the ICMC 95, San Francisco 1995
695. Mazzola G et al.: Neuronal Response in Limbic and Neocortical Structures During Perception of Consonances and Dissonances. In: Steinberg R (Ed.): Music and the Mind Machine. Springer, Berlin et al. 1995
696. Mazzola G and O Zahorka: RUBATO und der Einsatz von Hypermedien in der Musikforschung. Zeitschrift des Deutsches Bibliotheksinstituts, Berlin Jan. 1996
697. Mazzola G and O Zahorka: Topologien gestalteter Motive in Kompositionen. In: Auhagen W et al. (Eds.): Festschrift zum 65. Geburtstag J.P. Fricke. Köln 1995 `http://www.uni-koeln.de/phil-fak/muwi/publ/fs_fricke`
698. Mazzola G: Towards Big Science. Geometry and Logic of Music and its Technology. In: Enders B and Knolle N (Eds.): Symposionsband Klangart '95, Rasch, Osnabrück 1998

699. Mazzola G et al.: The RUBATO Homepage. http://www.rubato.org, Univ. Zűrich, since 1996
700. Mazzola G: Objective C and Category Theory. Seminar Notes, Multimedia Lab, CS Department, U Zurich, Zurich 1996
701. Mazzola G et al.: *Orbit.* Music & Arts CD-1015, Berkeley 1997
702. Mazzola G and J Beran: Rational Composition of Performance. In: Kopiez R and W Auhagen (Eds.): Proceedings of the Conference "Controlling Creative Processes in Music". Lang, Frankfurt and New York 1998
703. Mazzola G: Semiotic Aspects of Musicology: Semiotics of Music. In: Posner R et al. (Eds.): A Handbook on the Sign-Theoretic Foundations of Nature and Culture. Walter de Gruyter, Berlin and New York 1998
704. Mazzola G: Semiotic Aspects of Musicology: Semiotics of Music. - Art. 152, In: Posner et al. (eds): Semiotics (Handbook, Vol.III). W. de Gruyter, Berlin 1999
705. Mazzola G: Music@EncycloSpace. In: Enders B and J Stange-Elbe (Eds.): Musik im virtuellen Raum (Proceedings of the Klangart congress'98). Rasch, Osnabrück 2000
706. Mazzola G: Humanities@EncycloSpace. FER-Studie Nr.XX. Schweizerischer Wissenschaftrat, Bern 1998 http://www.swr.ch
707. Mazzola G: Die schöne Gefangene – Metaphorik und Komplexität in der Musikwissenschaft. In: Kopiez R et al. (Eds.): Musikwissenschaft zwischen Kunst, Aesthetik und Experiment. Königshausen & Neumann, Würzburg 1998
708. Mazzola G: L'Essence du Bleu (sonate pour piano). Acanthus, Rüttenen 2002
709. Mazzola G: Les mathématiques de l'interprétation musicale. Séminaire MaMuX, IRCAM, Paris, May 18, 2002
710. Mazzola G, G Milmeister, J Weissmann: Comprehensive Mathematics for Computer Scientists, Vols. I, II; 700 pp. Springer, Heidelberg et al. 2004
711. Mazzola G: The Topos Geometry of Musical Logic. Proceedings of the Fourth Diderot Symposium of the European Math. Soc., Springer, Heidelberg et al. 2002
712. Mazzola G: Degenerative Theory of Tonal Music. In: Proceedings of the Klangart congress'99. Universität Osnabrück 2002
713. Mazzola G: Classifying Algebraic Schemes for Musical Manifolds. Tatra Mt. Math. Publ. 23, 71-90, 2001
714. Mazzola G et al.: The Topos of Music. Birkhäuser, Basel 2002
715. Mazzola G, G Milmeister, K Morsy, F Thalmann: Functors for music: The Rubato Composer System. In: Digital Art Weeks Proceedings, ETH Zürich, 2006
716. Mazzola G: Global Networks in Computer Science? Invited Talk at the Computer Science Department Colloquium, ETH Zurich, January 2006
717. Mazzola G: Elemente der Musikinformatik. Birkhäuser, Basel et al. 2006
718. Mazzola G: La vérité du beau dans la musique. Delatour/IRCAM, Paris 2007
719. Mazzola G and M Andreatta: From a Categorical Point of View: K-nets as Limit Denotators. Perspectives of New Music, Vol. 44, No. 2, 2006
720. Mazzola G and M Andreatta: Diagrams, Gestures, and Formulas in Music. Journal of Mathematics and Music 2007, Vol. 1, no. 1, 2007
721. Mazzola G and P B Cherlin: Flow, Gesture, and Spaces in Free Jazz—Towards a Theory of Collaboration. Springer Series Computational Music Science, Heidelberg et al. 2009
722. Mazzola G et al.: Topos Theory for a Creative Analysis of Boulez's Structures. In: Naimpally S, G Di Maio (Eds.): Quaderni di Matematica, Theory And Applications of Proximity, Nearness and Uniformity, Vol. 23, 2009
723. Mazzola G: Categorical Gestures, the Diamond Conjecture, Lewin's Question, and the Hammerklavier Sonata. Journal of Mathematics and Music Vol. 3, no. 1, 2009
724. Mazzola G, C Losada, F Thalmann, S Tsuda: Topos Theory for a Creative Analysis of Boulez's Structures. In: Somashekhar Naimpally, Giuseppe Di Maio (Eds.): Quaderni di Matematica, Theory And Applications of Proximity, Nearness and Uniformity, Vol 23, 2009
725. Mazzola G: Musical Performance. Springer, Heidelberg et al. 2011
726. Mazzola G, J Park, F Thalmann: Musical Creativity. Springer, Heidelberg 2011
727. Mazzola G: Singular Homology on Hypergestures. Journal of Mathematics and Music Vol. 6, Nr. 1, 2012
728. Mazzola G: Manifolds and Singular Homology of Compositions, Networks, and Gestures: Classification, Connecting Functors, Examples. Unpublished
729. Mazzola G and F Thalmann: The BigBang Rubette: Gestural Music Comoposition With Rubato Composer. Proceedings of the ICMC 2008, ICMA, Ann Arbor 2008
730. Mazzola G and F Thalmann: Musical Composition and Gestural Diagrams. In Agon C et al. (Eds.): Proceedings of the Third Conference on Mathematics and Computation in Music (MCM), Springer, Heidelberg, 2011

731. Mazzola G, M Mannone, Y Pang: Cool Math for Hot Music. Springer, Heidelberg 2016
732. Mesnage M: La Set-Complex Theory: de quels enjeux s'agit-il? Analyse musicale, 4e trimestre, 87-90, 1989
733. Messiaen O: Technique de mon langage musical. Leduc, Paris 1944
734. Meyer J: Akustik und musikalische Aufführungspraxis. Verlag Das Musikinstrument, Frankfurt/Main 1980
735. Meyer-Eppler W: Grundlagen und Anwendungen der Informationstheorie. Springer, Berlin 1959
736. Meyer-Eppler W: Zur Systematik der elektrischen Klangtransformationen. In: Darmstädter Beiträge III, Schott, Mainz 1960
737. Michels U: dtv-Atlas zur Musik I,II. dtv/Bärenreiter, München/Kassel 1977, 1985
738. Miklaszewski K: A case Study of a Pianist Preparing a Musical Performance. Psychology of Music, 17, 95-109, 1989
739. Milmeister G: The Rubato Composer Music Software: Component-Based Implementation of a Functorial Concept Architecture. Springer Series Computational Music Science, Heidelberg 2009
740. McKay B D and I M Wanless: On the number of Latin squares. arXiv:0909.2101v1 [math.CO] 11 Sep 2009
741. McNeill D: Gesture and Thought. University of Chicago Press, Chicago 2005
742. McPartland M: Marian McPartland's Piano Jazz. National Public Radio, 9 November 1997 `http://www.npr.org/player/v2/mediaPlayer.html?action=1&t=1&islist=false&id=17622315&m=90469170`
743. Mead V H: More than Mere Movement: Dalcroze Eurhythmics. Music Educators Journal, Feb. 1986
744. Mendes R S et al.: Universal patterns in sound amplitudes of songs and musical genres. Phys. Rev. E, Vol. 83, pp. 017101 ff., 2011
745. Menez M: L'expérience mathématique. In: Dorier J-L, S Coutat (Eds.): Enseignement des mathématiques et contrat social : enjeux et défis pour le 21ème siècle — Actes du Colloque EMF20123 (GT4, pp. 632-640)
746. Meredith D, G Wiggins, K Lemstrom: Pattern induction and matching in polyphonic music and other multidimensional datasets. In: Callaos N et al. (Eds.): Proceedings of the 5th World Multiconference on Systemics, Cybernetics and Informatics. Orlando 2001
747. Meredith D, K Lemstrom, G Wiggins: Algorithms for discovering repeated patterns in multidimensional representation of polyphonic music. Journal of New Music Research, Vol 4, 2002
748. Merker B H: Layered Constraints on the Multiple Creativities of Music. In: I Deliège and G A Wiggins (Eds.): Musical Creativity: Multidisciplinary Research in Theory and Practice. Psychology Press, Hove/New York, pp. 25-41, 2006
749. Merleau-Ponty M: The Phenomenology of Perception in The Body: Classic and Contemporary Readings, (ed. Donn Welton), Blackwell, Oxford 1999
750. Merleau-Ponty M: Phenomenology of Perception. Trans. Colin Smith, Routledge & Kegan Paul, London 1978 [1962]
751. Merleau-Ponty M: Phénoménologie de la perception. Gallimard 1945
752. Merleau-Ponty M: The prose of the world, translation by J O'Neill. Northwestern University Press, Evanston 1973 (La prose du monde, Gallimard, 1969)
753. Meusburger P, J Funke, E Wunder (Eds.): Milieus of Creativity: An Interdisciplinary Approach to Spatiality of Creativity. Springer, 2009
754. Miller M: A Jazz Legend Handles his Own Legacy with Care. Globe and Mail (Canada), 23 February: C3. 1999
755. Misch C and Wille R: Stimmungslogiken auf MUTABOR: Eine Programmiersprache. In: Herf F R (Ed.): Mikrotöne II. Edition Helbling, Innsbruck 1988
756. `http://forumnet.ircam.fr/701.html`
757. Mokkapati V: Modal Action: An Indo-European Perpective on Gregorian Chant. Phd Thesis, Wesleyan University 1996
758. Molino J: Fait Musical et Sémiologie de la Musique. Musique en Jeu 17 1975
759. Monti G: Théorie des catégories et méthode conditionnelle appliquées aux comptes en normes internationales élargies. Thèse, Doctorat ès Sciences de gestion, Université de Bordeaux 2017
760. Montiel-Hernandez M: El denotador: Su estructura, construcción y papel en la teoría matemática de la musica. Master's thesis, UNAM, Mexico City, 1999
761. Moog R: MIDI, Musical Instrument Digital Interface. Audio Eng. Soc. 34, Nr.5, 1986
762. Moon T: A Master of Improv is Renewed by Melody. Philadelphia Inquirer, 29 September: H02, 2002
763. Moore C-L: The Harmonic Structure of Movement, Music and Dance According to Rudolf Laban. Edwin Meller Press, Lewiston 2009
764. Moritz R E: On Mathematics. A collection of witty, profound, amusing passages about mathematics and mathematicians. Copyright 1914 Robert Edouard Moritz, 410 p., Dover 1958
765. Morris R D: Composition with Pitch-Classes. Yale University Press, New Haven et al. 1987
766. Morris R D: Compositional Spaces and Other Territories. PNM 33, 328-358, 1995

767. Morris R D: K, Kh, and Beyond. In: Baker J et al. [69], 275-306
768. Morse Ph M and H Feshbach: Methods of Theoretical Physics, Part I. McGraw-Hill, New York et al. 1953
769. Mozart W A: *Die Zauberflöte.* (Klavierauszug) Ed. Peters 1932
770. Müller W: Darstellungstheorie von endlichen Gruppen. Teubner, Stuttgart 1980
771. Müller S: Pianist's Hands: Synthesis of Musical Gestures. PhD Thesis, University of Zurich, Zurich 2004
772. Müller S and G Mazzola: Constraint-based Shaping of Gestural Performance. Proceedings of the ICMC 03, ICMA, Ann Arbor 2003
773. Müller S and G Mazzola: The Extraction of Expressive Shaping in Performance. Computer Music Journal, Vol. 27(1), MIT Press, 2003
774. Mumford D: Lectures on Curves on an Algebraic Surface. Princeton University Press, Princeton 1966
775. Mumford D and K Suominen: Introduction to the theory of moduli. In: Oort F (Ed.): Algebraic Geometry Oslo 1970. Wolters-Noordhoff, Groningen 1972
776. Murenzi R: Wavelets. Combes J M, Grossman A, Tchmitchian P (Eds.), Springer Berlin et al., 1988
777. MusicKit Online-Documentation. Version 4.0 1994
778. Muzzulini D: Konsonanz und Dissonanz in Musiktheorie und Psychoakustik. Lizenziatsarbeit, MWS Univ. Zürich 1990
779. Muzzulini D: Musical Modulation by Symmetries. J. for Music Theory, Vol. 39, No. 2, pp. 311-328, 1995
780. Muzzulini D: Tempo Modifications and Spline Functions. NF-Report 1993, Univ. Zürich 1993
781. Narayan S S, Temchin A N, Recio A, Ruggero M A: Frequency tuning of basilar membrane and auditory nerve fibers in the same cochleae. Science. 282(5395):1882-4, Dec 4 1998
782. Nattiez J-J: Fondements d'une Sémiologie de la Musique. Edition 10/18 Paris 1975
783. Negro G Del: The Passeggiata and Popular Culture in an Italian Town: Folklore and the Performance of Modernity. Montréal; McGill-Queen's University Press 2004
784. Neumaier W: Was ist ein Tonsystem? Lang, Frankfurt/Main et al. 1986
785. Neuman D A: A House of Music: The Hindustani Musician and the Crafting of Traditions. PhD Thesis, Columbia University 2004
786. Neuwirth G: Josquin Desprez, "Erzählen von Zahlen". In: Musik-Konzepte 26/27, edition text+kritik, München 1982
787. Newlove J and J Dalby: Laban for All. Routledge, New York 2004
788. NeXTSTEP Online-Documentation; Version 3.3. NeXt Inc., Redwood City 1995
789. Nicolas F: René Guitart : la pulsation d'une intellectualité mathématique étrangement évidente. Conférence à la *Journée en l'honneur de René Guitart pour ses 65 ans*, 9 novembre 2012, Université Paris 7 Denis Diderot, 8 p, inédit. See a summary in: `http://www-lmpa.univ-littoral.fr/ sic`
790. Nicolas F: Le Monde-Musique III. Le musicien et son intellectualité musicale. Aedam Musicae, 384 p., juin 2015
791. Nicotra E: Introduzione alla tecnica della direzione d'orchestra. Edizioni Curci 2007
792. Nieberle R: IRCAM Signal Processing Workstation. Keyboards, Dec. 1992
793. `http://www.nime.org`
794. Noether E: Hyperkomplexe Grössen und Darstellungstheorie. Math. Zeitschr., Vol. XXX 1929
795. Noland C: Agency and Embodiment: Performing Gestures/Producing Culture. Harvard University Press, Cambridge 2009
796. Noll J: Musik-Programmierung. Addison-Wesley, Bonn 1994
797. Noll Th: Morphologische Grundlagen der abendländischen Harmonik. Doctoral Thesis, TU Berlin 1995
798. Noll Th: Fractal Depth Structure of Tonal Harmony. In: ICMA (Ed.): Proceedings of the ICMC 95, San Francisco 1995
799. Noll Th: `http://www.cs.tu-berlin.de/ noll/ChordDictionary.sea.hqx`, TU Berlin 1996
800. Noll Th: Harmonische Morpheme. Musikometrika 8, 7-32, 1997
801. Noll Th: The Consonance/Dissonance-Dichotomy Considered from a Morphological Point of View. In: Zannos I (Ed.): Music and Signs. ASCO Publ., Bratislava 1999
802. Noll Th: Morphologische Grundlagen der abendländischen Harmonik. Doctoral Thesis, TU Berlin 1995
803. Noll Th: Wie blättert man in einem hypermedialen Gebärdenlexikon? In: Schmauser C, Th Noll (Eds.): Körperbewegungen und ihre Bedeutungen. Berlin Verlag Arno Spitz, Berlin 1998
804. Noll Th et al.: Computer-aided transformational analysis with tone sieves. Contribution au congrés: SMC 2006
805. Nunn D et al.: Acoustic Quanta. In: ICMA (Ed.): Proceedings of the ICMC 96, San Francisco 1996
806. Oettingen A von: Das duale Harmoniesystem. Leipzig 1913
807. Okopenko A: Lexikon-Roman. Residenz Verlag, Salzburg 1970
808. Opcode: MAX. http://www.opcode.com/products/max/, Opcode Systems Inc. 1997

809. Marco Orsini conducting Le Nozze di Figaro 2014 https://www.youtube.com/watch?v=M7QuLXy_61U
810. Osgood C E et al.: The Nature and Measurement of Meaning. Psychological Bulletin 49, 1952
811. Osher S and R Fedkiw: Level Set Methods and Dynamic Implicit Surfaces. Springer, 2003
812. Owens Th: Bebop: The Music and the Players. Oxford University Press, New York 1995
813. Pakkan M and V Akman: Hypersets. Information Sciences: An International Journal, 85(1):4361, 1996
814. Pallasmaa J: The Thinking Hand. Wiley 2009
815. Palmer R: Liner notes, Oscar Peterson: A Night in Vienna. DVD. Verve 02498 6253-4, 2004
816. Pappas T: The Music of Reason: Experience the Beauty of Mathematics Through Quotations. Wide World Publishing/ Tetra, 1995
817. Parker W: Conversations. Rogueart, 2011
818. Parlan H: The Maestro. CD. SteepleChase SCCD 31167, 1995
819. Parlan H: Horace Parlan by Horace Parlan. DVD. Image ID1505R5DVD, 2000
820. Parlan H: In Copenhagen. CD. Storyville 101 8521, 2008
821. Parlan H: Musically Yours. CD SteepleChase SCS 31141, 2010
822. Parmentier M: Connaissances, vérités et idées mathématiques dans l'*Essai sur l'entendement humain* de Locke. In: Barbin E et J-P Cléro (éds): Les mathématiques et l'expérience : ce qu'en ont dit les philosophes et les mathématiciens, p. 95-114, Hermann, Paris 2015
823. Parncutt R: The Perception of Pulse in Musical Rhythm. In: Gabrielsson A (Ed.): Action and Perception in Rhythm and Music. Royal Swedish Adademy of Music, No.55
824. Parncutt R: Recording Piano Fingering in Live Performance. In: Enders B, Knolle N (Eds.): KlangArt-Kongress 1995, Rasch, Osnabrück 1998
825. Parncutt R et al.: Interdependence of Right and Left Hands in Sight-read, Written, and Rehearsed Fingerings. Proc. Euro. Sco. Cog. Sci. Music, Uppsala 1997
826. Parncutt R: Modeling Piano Performance: Physics and Cognition of a Virtual Pianist. In: ICMA (Ed.): Proceedings of the ICMC 97, San Francisco 1997
827. Patras F: La pensée mathématique contemporaine. PUF, Paris, 2001
828. Patras F: La possibilité des nombres. PUF, Paris, octobre 2014
829. Peck R: Klein-Bottle Tonnetze. MTO, Vol. 9, no. 3, August 2003
830. Penrose R: The Road to Reality. Vintage, London 2002
831. Perle G: Serial Composition and Atonality: An Introduction to the Music of Schoenberg, Berg and Webern. 5th ed., revised, University of California Press, Berkeley 1981
832. Peterson O: An Oscar Peterson Christmas. CD, Telarc 515956K, 1995
833. Peterson O: Oscar in Paris. Telarc 2CD-83414, 1997
834. Peterson O: A Tribute to Oscar Peterson: Live at the Town Hall. Telarc 2CD-83401
835. Peterson O: A Summer Night in Munich. CD. Telarc CD-83450, 1999
836. Peterson O: A Night in Vienna. DVD, Verve 02498 6253-4, 2004
837. Peterson O, R Brown, M Jackson: The Very Tall Band: Live at the Blue Note. CD. Telarc CD-83443, 1999
838. Peterson O, R Brown, M Jackson: What's Up? The Very Tall Band. CD. Telarc CD-83663, 2007
839. Peterson O and B Green: Oscar and Benny. CD. Telarc CD-83406, 1998
840. Peterson O, R Hargrove, R Moore: Oscar Peterson Meets Roy Hargrove and Ralph Moore. CD. Telarc, CD-83399, 1996
841. Peterson O, and M LeGrand: Trail of Dreams: A Canadian Suite. CD. Telarc CD-83500, 2000
842. Peterson O, and I Perlman: Side by Side. CD. Telarc, CD-83341, 1994
843. Petsche H et al.: EEG in Music Psychological Studies. In: Steinberg R (Ed.): Music and the Mind Machine. Springer, Berlin et al. 1995
844. Petsche H: Private correspondence. Vienna, March 2001
845. Piaget J: Morphismes et categories. Comparer et transformer, Delachaux et Niestlé, Neuchâtel, 1990
846. Pfeifer R and J Bongard: How the Body Shapes the Way We Think: A New View of Intelligence. MIT Press 2006
847. Peirce C S: Peirce, *Collected Papers*, vol. III, vol. IV The simplest mathematics, 1960
848. Pierce A: Deepening Musical Performance through Movement: The Theory and Practice of Embodied Interpretation (Musical Meaning and Interpretation). Indiana University Press, Bloomington 2010
849. Plato: Phaidron.
850. Plomp R and W Levelt: Tonal Consonance and Critical Bandwidth. J. Acoust. Soc. Am. 38, 548, 1965
851. Plotinus: The Six Enneads, in *Great Books of the Western World*, Vol. 17. Encyclopædia Britannica, Inc., Chicago, 1952
852. Poincaré H: La valeur de la science. Flammarion, Paris 1905

853. Pólya G: Mathematical Discovery. On Understanding, Learning and Teaching Problem Solving (Combined Edition). Wiley, New York, 1962, 1965, 1981
854. Pólya G: Comment poser et résoudre un problème. Dunod, 1965
855. Polyanin A D: Handbook of Linear Partial Differential Equations for Engineers and Scientists. Chapman & Hall/CRC, 2002; see also: `http://eqworld.ipmnet.ru`, section 3.2.
856. Pope S T: Music Notations and the Representation of Musical Structure and Knowledge. Perspectives of New Music, Spring-Summer 1986
857. Popelard M-D and A Wall: Des Faits et gestes. éditions Bréal, Rosny-sous-Bois 2003
858. Popper K R: Conjectures and Refutations. Routledge & Kegan Paul, London 1963
859. Popper K: The open society and its enemies. Vol. 2, Routledge & Kegan, London, 1945
860. Posner R: Strukturalismus in der Gedichtinterpretation. In: Blumensadth H (Ed.): Strukturalismus in der Literaturwissenschaft. Kiepenheuer & Witsch, Köln 1972
861. Prame E: Measurements of the Vibrato Rate of 10 Singers. In: Friberg A et al. (Eds.): Proceedings of the 1993 Stockholm Music Acoustic Conference. KTH, Stockholm 1994
862. Pribram K H: Brain and Perception. L Erlbaum, Hillsdale, NJ 1991
863. Priestley M B: Spectral Analysis of Time Series. Academic Press, London 1981
864. Prinz W: Modes of linkage between perception and action. In W Prinz & A-F Sanders (Eds.): cognition and motor processes (pp. 185–193), Springer, Berlin 1984
865. Promies W: Symmetrie in der Literatur. In: Symmetrie, Katalogband Vol.1 zur Symmetrieausstellung. Mathildenhöhe, Darmstadt, 1986
866. Promies W: Stolbergs Gedicht im Poesiegarten In: Symmetrie, Katalogband Vol.3 zur Symmetrieausstellung. Mathildenhöhe, Darmstadt, 1986
867. Puckette M and Lippe C: Score Following in Practice. In: ICMA (Ed.): Proceedings of the ICMC 92, San Francisco 1992
868. Pulkki V et al.: DSP Approach to Multichannel Audio Mixing. In: ICMA (Ed.): Proceedings of the ICMC 96, San Francisco 1996
869. Puttke M: Learning to Dance Means Learning to Think! In: Bläsing B, Puttke M, Schack Th. (Eds.): The Neurocognition of Dance. Psychology Press, Hove and New York 2010
870. Quinn I: Fuzzy Extensions to the Theory of Contour. Music Theory Spectrum. Vol. 19/2, 1997
871. Radl H: Versuch über die Modulationstheorie Mazzolas in reiner Stimmung. Diploma Thesis, U Augsburg, Augsburg 1998
872. Rahaim M: Musicking Bodies. Wesleyan U P, Middletown, CT 2012
873. Rahn J: Basic Atonal Theory. Longman, New York 1980
874. Rahn J: Review of D. Lewin's "Generalized Musical Intervals and Transformations". JMT, 31, 305-318, 1987
875. Raffman D: Language, Music, and Mind. MIT Press, Cambridge et al. 1993
876. Rameau J-Ph: Traité de l'Harmonie, Réduite à ses Principes Naturels. Paris 1722
877. Ramachandran V S: Mirror Neurons and imitation learning as the driving force behind the "great leap forward" in human evolution, 2015. See: `http://www.edge.org/3rd.culture/ramachandran/ramachandran p6.html`
878. Ramstein C: Analyse, Représentation et Traitement du Geste Instrumental. PhD thesis, Institut National Polytechnique de Grenoble, December 1991
879. Ratz E: Einführung in die musikalische Formenlehre. Universal Edition, Wien 1973
880. Rauchfleisch U: Musik schöpfen, Musik hören. Ein psychologischer Zugang. Vandenhoeck und Ruprecht, Göttingen and Zürich 1996
881. Rav Y: Why do we prove theorems?. Philosophia mathematica, 7 (1), p. 5-41, 1999
882. Read G: Music Notation. Crescendo Publ., Boston 1969
883. Reeves H: Patience dans l'azur. L'évolution cosmique. Seuil, Paris 1981
884. Reichardt J F: Vermischte Musikalien. Riga 1777
885. Reidmeister K: Elementare Begründung der Knotentheorie. Abh. Math. Sem. Univ. Hamburg, 5, 24-32, 1926
886. Repp B: Diversity and Commonality in Music Performance: An Analysis of Timing Microstructure in Schumann's "Träumerei". J. Acoustic Soc. Am. 92, 1992
887. Repp B: e-mail communication of tempo data. Haskins Laboratories, New Haven, June 2, 1995
888. Repp B: Patterns of note onset asynchronies in expressive piano performance. J. Acoustic Soc. Am. **100**, 1996
889. Repp B: Pedal Timing and Tempo in Expressive Piano Performance: A Preliminary Investigation. Psychology of Music **24**, 1996
890. Repp B: The Art of Inaccuracy: Why Pianists' Errors Are Difficult to Hear. Music Perception, *14*, 2 1997
891. Repp B: Expressive Timing in a Debussy Prélude: A Comparison of Student and Expert Pianists. Musicae Scientiae 1, 1997

892. Reti R: The Thematic Process in Music (1951). Greenwood Press, Westport, 2nd ed. 1978
893. Reti R, commented by Kopfermann M: Schumanns Kinderszenen: quasi Thema mit Variationen. In: Musik-Konzepte Sonderband Robert Schumann II, edition text + kritik, München 1982
894. Riguet J: Systèmes de coordonnées relationnels. *CRAS*, t. 236, p. 2369-2371, séance du 22 juin 1953
895. Riguet J: Systèmes de coordonnées relationnels II. Applications à la théorie des groupes de Kaloujnine. *CRAS* t. 238, p. 435-437, séance du 25 janvier 1954
896. Riguet J: Systèmes de coordonnées relationnels III. Fermetures et systèmes symétriques. *CRAS*, t. 238, p. 1763-1765, séance du 3 mai 1954
897. Rhode W S and L Robles: Evidence from Mössbauer Experiments for Nonlinear Vibration in the Cochlea. J. Acoust. Soc. Am. 48, 988, 1970
898. Richter P et al.: How Consistent are Changes in EEG Coherence Patterns Elicited by Music Perception? In: Steinberg R (Ed.): Music and the Mind Machine. Springer, Berlin et al. 1995
899. Riemann B: Über die Hypothesen, welche der Geometrie zugrunde liegen (1854). Gött. Abh. No.13 (published in 1867)
900. Riemann H: Musikalische Logik. Leipzig 1873
901. Riemann H: Musikalische Syntaxis. Leipzig 1877
902. Riemann H: Vereinfachte Harmonielehre oder die Lehre von den tonalen Funktionen der Akkorde. London 1893
903. Riemann H: Über Agogik. In: Präludien und Studien II. Leipzig 1900
904. Riemann H: System der musikalischen Rhythmik und Metrik. Breitkopf und Härtel, Leipzig 1903
905. Riemann H: Handbuch der Harmonielehre. Leipzig 6/1912
906. Riemann H: Grundriss der Musikwissenschaft. Leipzig 1928
907. Riemann H and H H Eggebrecht: Riemann Musiklexikon, Sachteil, 12. Auflage. Schott, Mainz 1967
908. Ries F: Biographische Notizen über L. van Beethoven (1838). New edition by Kalischer A Ch, 1906
909. Risset J-C: Real-time: Composition or performance? Reservations about real-time control in computer music and demonstration of a virtual piano partner. In: Kopiez R and W Auhagen (Eds.): Controlling creative processes in music. Peter Lang, Frankfurt/Main 1998
910. Rizzolati G and M Fabbri-Destro: Mirror neurons: from discovery to autism. Exp. Brain Res. 200:223-237, 2010
911. Roads C: The Computer Music Tutorial. MIT Press, Cambridge, Mass. and London 1998
912. Rodet X et al.: Xspect: a New X/Motif Signal Visualization, Analysis and Editing Program. In: ICMA (Ed.): Proceedings of the ICMC 96, San Francisco 1996
913. Rodet X: Synthesis and Processing of the Singing Voice. Proc.1st IEEE Benelux Workshop on Model based Processing and Coding of Audio (MPCA-2002), Leuven, Belgium, November 15, 2002
914. Roederer J G: Physikalische und psychoakustische Grundlagen der Musik. Springer, Berlin 1977
915. Rogers C R: Toward a theory of creativity, *ETC: A review of General Semantics*, Vol 11, 1954, 249-260
916. Rossi A: Die Geburt der modernen Wissenschaft in Europa. Beck, München 1997
917. Rostand F: Souci d'exactitude et scrupules des mathématiciens. Vrin, Paris 1960
918. Rüetschi U-J: Denotative Geographical Modelling—an attempt at modelling geographical information with the Denotator system. Diploma thesis, University of Zürich, 2001
919. Rufer J: Die Komposition mit zwölf Tönen. Bärenreiter, Kassel 1966
920. Runco M A and I Chand: Cognition and Creativity. Educational Psychology Review, Vol. 7:3, pp. 243-267, 1995
921. Ryan R A: Introduction to Tensor Products of Banach Spaces. Springer, Heidelberg 2002
922. Ruwet N: Langage, Musique, Poésie. Seuil, Paris 1972
923. *S-Plus*. MathSoft Inc., Seattle, Washington 1995
924. Sachs K-J: Der Contrapunctus im 14. und 15. Jahrhundert. AMW, Franz Steiner, Wiesbaden 1974
925. Sacks O W: Musicophilia: tales of music and the brain. (Rev. and expanded 1st ed.) Vintage Books, New York 2008
926. Saiber A and H S Turner (Eds.): Mathematics and the Imagination. Configurations, 17, 195 pp. 2009
927. Sandall R: Miraculously, Magically, the Years Fell Away. Daily Telegraph (London), 1 July: 21 2006
928. Sander E: Notre pensée progresse par analogies. La conscience. La Recherche Hors-série, no 15, pp. 54-57, octobre-novembre 2015
929. Sattler D E (Ed.): Friedrich Hölderlin – Einhundert Gedichte. Luchterhand, Frankfurt/Main 1989
930. Salzer F: Structural Hearing: Tonal Coherence in Music. (German: Strukturelles Hören. Noetzel, Wilhelmshaven 1960) Dover, New York 1962
931. Sarasúa A and E Guaus: Dynamics in Music Conducting: A computational Comparative Study Among Subjects. Proceedings of the International Conference on New Interfaces for Musical Expression, NIME'14, Goldsmiths London, UK, 2014

932. Sauvageot F: Pierre Deligne. *Images des mathématiques*, CNRS, 12 mai 2013 http://images.math.cnrs.fr/Pierre-Deligne.html
933. Saussure F de: Cours de Linguistique Générale. Payot, Paris 1922
934. Saussure F de: Linguistik und Semiotik - Notizen aus dem Nachlass. Fehr J (Ed.), Suhrkamp, Frankfurt/Main 1997
935. Savan D: An introduction to C.S. Peirce's Full System of Semeiotic. Toronto Semiotic Circle, Toronto 1988
936. Sawyer R K: Improvisation and the creative process: Dewey, Collingwood, and the aesthetics of spontaneity. The Journal of Aesthetics and Art Criticism, Vp;. 58, pp. 149-61, 2000
937. Sawyer R K: Group Creativity: Music, Theater, Collaboration. Lawrence Erlbaum Associates, Mahwah 2003
938. Scarborough D L et al.: Connectionist Models for Tonal Analysis. Computer Music Journal Vol. 13, 1989
939. Schenker H: Fünf Urlinien-Tafeln. Universal Edition, Wien 1932
940. Schenker H: Theorien und Phantasien III: Der freie Satz. Universal Edition, Wien 1935
941. Scherchen H: Manuale del direttore d'orchestra. Edizioni Curci 2009
942. Schewe K-D: Type concept in OODB Modelling and its Logical Implications. In Kawaguchi E et al. (Ed.), Information Modelling and Knowledge Bases XI, pp. 256–274. IOS Press, 2000
943. Schmalz R: Out of the mouth of mathematicians. A quotation book for philomaths. Mathematical Association of America, 1993
944. Schmidt J (Ed.): Celibicache. Book and Movie PARS (ISBN 3-9803265-1-9), München 1992
945. Schmidt-Biggemann W: Topica universalis. Meiner,Hamburg 1983
946. Schmitt, J-C: La raison des gestes dans l'occident médiéval. Gallimard, Paris 1990
947. Schneps L and P Lochak (Eds.): Geometric Galois Actions. Cambridge University Press, Cambridge, 1997
948. Schönberg A: Harmonielehre (1911). Universal Edition, Wien 1966
949. Schönberg A: Die Komposition mit zwölf Tönen. In: Style and Idea, New York 1950
950. Schönberg A: Fundamentals of Musical Composition. Edited by Gerald Strang, with an introduction by Leonard Stein. St. Martin's Press, New York 1967
951. Schopenhauer A: De la quadruple racine du principe de raison suffisante. éd. complète (1813-1847), Vrin, 1991
952. Schubert F: *Auf dem Wasser zu singen*, op. 72 (1823). Bärenreiter/Henle München/Kassel 1982
953. Schubert H: Kategorien I, II. Springer, Berlin et al. 1970
954. Schumann R: Kinderszenen, op. 15 (1839). Henle, München 1977
955. Schumann R: Dritter Quartett-Morgen. NZ f. Musik 1838
956. Schweitzer A: Johann Sebastian Bach. (1908) Breitkopf & Härtel, Wiesbaden 1979
957. Scriba C J and P Schreiber: 5000 Years of Geometry: Mathematics in History and Culture. Birkhäuser, 2015
958. Seashore C E: Psychology of Music. McGraw-Hill, New York 1938
959. Shepard R N and L A Cooper: Mental Images and Their Transformations. MIT Press 1982
960. Shepp A: The Geneva Concert. DVD, inakustic LC13921, 2006
961. Shepp A and H Parlan: Goin' Home. CD. SteepleChase (D) SCCD-31079, 1985
962. Shepp A and H Parlan: Trouble in Mind. CD. SteepleChase ? SCCD 31139, 1986
963. Shepp A and H Parlan: En Concert: First Set. CD. 52e Rue Est (F) RECD 015, 1987
964. Shepp A and H Parlan: En Concert: Second Set. CD. 52e Rue Est (F) RECD 016, 1987
965. Shepp A and H Parlan: Reunion. CD. L+R CDLR 45003, 1987
966. Shepp A and H Parlan: Goin' Home. CD. SteepleChase (D) SCCD-31079, 2010
967. Shepp A and H Parlan: Trouble in Mind. CD. SteepleChase ? SCCD 31139, 2010
968. Siebers T: Disability Aesthetics (pp. 2-3). University of Michigan Press, Ann Arbor 2010
969. Serre J-P: Faisceaux Algébriques Cohérents. Annals of Mathematics, 61 (2): 197-278, 1955
970. Serre J-P: Trees. Springer, New York 1980
971. Serres M (Ed.): Eléments d'histoire des sciences. Bordas, Paris 1989
972. Sessions R: Questions About Music. W. W. Norton and Company, 1971
973. Shaffer L H: Musical Performance as Interpretation. Psychology of Music, Vol. 23, 1995
974. http://www.fau.edu/bodymindculture
975. Shusterman R: Somaesthetics: A Disiplinary Proposa. The Journal of Aesthetics and Art Criticism. Vol.57, No.3, pp. 299-313, 1999
976. Simondon G: Du mode d'existence des objets techniques. Aubier, Paris 1989
977. Simonton D K: Thematic fame, melodic originality, and musical zeitgeist: A biographical and transhistorical content analysis. Journal of Personality and Social Psychology, Vol. 38, pp. 972-83, 1980
978. Simonton D K: Thesaurus of scales and melodic patterns. Coleman-Ross Company, New York 1947
979. Siron J: La partition intérieure. Outre Mesure, Paris 1992
980. Slawson W: The Musical Control of Sound Color. Canadian Univ. Music Review, No.3, 1982

981. Slawson W: Sound Color. U. Cal. Press, Berkeley 1985
982. Sloan D: From DARMS to SMDL, and back again. In: Haus G and I Pighi (Eds.): X Colloquio di Informatica Musicale. AIMI, LIM-Dsi, Milano 1993
983. Sloboda J: The Musical Mind: An Introduction to the Cognitive Psychology of Music. Calderon Press, Oxford 1985
984. Sloboda, A S and A C Lehmann: Tracking Performance Correlates of Changes in Perceived Intensity of Emotion During Different Interpretations of a Chopin Piano Prelude. Music Perception, Summer 2001, Vol 19, No. 1, pp. 87-120, 2001
985. Sloterdijk P: Selbstexperiment. Hanser, München & Wien 1996
986. Sluming V et al.: Broca's Area Supports Enhanced Visuospatial Cognition in Orchestral Musicians. The Journal of Neuroscience, Vol. 27(14), pp. 3799-3806, April 4, 2007
987. Smart M A: Mimomania: Music and Gesture in Nineteenth-Century Opera. New Ed edition, University of California Press, Berkeley 2005
988. SMDL committee: Standard Music Description Language Draft, Pittsford, NY 1995
989. Smith III J O: Synthesis of bowed strings. In: Strawn J, Blum T, ICMA: Proceedings of the International Computer Music Conference, San Francisco 1982
990. Smith III J O: Physical Modeling of Musical Instruments, part I. Computer Music Journal, 16 (4), 1992
991. Smith III J O: Physical Modeling Synthesis Update. Computer Music Journal, 20 (2), 1996
992. Sodomka A: KUNSTRADIO at documentaX. http://www.culture.net/orfkunstradio/BIOS/sodomkabio.html, Vienna 1997
993. Spanier E H: Algebraic Topology. McGraw-Hill, New York et al. 1966
994. Späth H: Eindimensionale Spline-Interpolations-Algorithmen. Oldenburg, München et al. 1990
995. Späth H: Zweidimensionale Spline-Interpolations-Algorithmen. Oldenburg, München et al. 1991
996. Spellman A B: Four Lives in the Bebop Business. Pantheon 1966
997. Sperry R W: Neurology and the mind-body problem. American Scientist, 40, 291-312, 1952
998. Spicker V: Das Abstraktmotiv bei Cecil Taylor. MusikTexte, 73/74 March 1998
999. Spivak M: Calculus on Manifolds. Benjamin, New York 1965
1000. Squire R L and Butters N (Eds.): Neuropsychology of Memory. D. Guildford Press, NewYork and London 1984
1001. Stahnke M: Struktur und Ästhetik bei Boulez. Wagner, Hamburg 1979
1002. Stange-Elbe J, G Mazzola: Cooking a Canon with RubatoPerformance Aspects of J.S. Bach's "Kunst der Fuge". In: ICMA (Ed.): Proceedings of the ICMC 98, 179-186, San Francisco 1998.
1003. Stange-Elbe J: Analyse- und Interpretationsperspektiven zu J.S. Bachs "Kunst der Fuge" mit Werkzeugen der objektorientierten Informationstechnologie. Habilitationsschrift, Osnabrück 2000
1004. Stange-Elbe J: Instrumentaltechnische Voraussetzungen für eine computergestützte Interpretation. In: Enders B (Ed.): Proceedings of the Klangart congress '01, Osnabrück
1005. Starr D and R Morris: A General Theory of Combinatoriality and the Aggregate. PNM, 16(1), 364-389, 16(2), 50-84, 1977-78
1006. Starr D: Sets, Invariance and Partitions. JMT, 22(1), 1-42, 1978
1007. Stein S and S Sandor: Algebra and Tiling: Homomorphisms in the Service of Geometry.: Math. Assoc. Amer., Washington, DC 1994
1008. Steinberg R (Ed.): Music and the Mind Machine. Springer, Heidelberg 1995
1009. Steinberg R and A Fairweather (Eds.): Blues-Philosophy for Everyone: Thinking Deep About Feeling Low. John Wiley 2012
1010. Sternberg R J and T I Lubart: The Concept of Creativity: Prospects and Paradigms. In: Sternberg R J (Ed.): Handbook of Creativity. Cambridge University Press, New York 1999
1011. Stinson D R: Combinatorial designs: constructions and analysis. Springer, 2004
1012. Stollberg A, J Weissenfeld, F H Besthorn (Eds.): DirigentenBilder—Musikalische Gesten — verkörperte Musik. Schwabe, Basel 2015
1013. Stravinsky I: The Poetics of Music. Harvard University Press 1947
1014. Steele G L: Common Lisp: The Language, 2nd Edition. Digital Press, 1990
1015. Stone P: Symbolic Composer. http:/www.xs4all.nl/~psto/
1016. Stopper B: Gleichstufig temperierte Skalen unter Berücksichtigung von Oktavstreckungen. In: Hosp I (Hrsg.): Bozener Treffen 1991 — Skalen und Harmonien. Südtiroler Kulturinstitut, Bozen 1992
1017. Straub H: Beiträge zur modultheoretischen Klassifikation musikalischer Motive. Diplomarbeit ETH-Zürich, Zürich 1989
1018. Straub H: Kadenzielle Mengen beliebiger Stufendefinitionen. Unpublished manuscript, Zürich 1999/2000
1019. Straus J N: Voice Leading in Atonal Music. In: Baker J et al. [69], 237-274

1020. Stumpf C: Tonpsychologie. Leipzig 1883-1890
1021. Sundberg J, A Askenfelt, L Frydén: Musical Performance: A Synthesis-by-Rule Approach, Computer Music Journal, 7, 37-43, 1983
1022. Sundberg J and V Verillo: On the Anatomy of the Retard: A Study of Timing in Music. J. Acoust. Soc. Am. 68, 772-779, 1980
1023. Sundberg J: Music Performance Research. An Overview. In: Sundberg J, L Nord, R Carlson (Eds.): Music, Language, Speech, and Brain. London 1991
1024. Sundberg J (Ed.): Generative Grammars for Music Performance. KTH, Stockholm 1994
1025. Sundin N-G: Musical Interpretation in Performance. Mirage, Stockholm 1984
1026. Suzuki T et al.: Musical Instrument Database with Multimedia. In: ICMA (Ed.): Proceedings of the ICMC 96, San Francisco 1996
1027. Suppan W: Zur Verwendung der Begriffe Gestalt, Struktur, Modell und Typus in der Musikethnologie. In: Stockmann D and Steszwski J (Eds.): Analyse und Klassifikation von Volksmelodien. PWM Edition Krakau 1973
1028. Swan, R G: Vector Bundles and Projective Modules, *Transactions of the American Mathematical Society* 105 (2): 264-277
1029. Sylvester D: Interviews with Francis Bacon. Pantheon Books, New York 1975
1030. Szczeciniarz J-J: Réflexion sur Riemann, philosophe mathématicien, in [79], p. 195-221
1031. Symmetrie. Katalogband Vol.1 zur Symmetrieausstellung, Mathildenhöhe, Darmstadt, 1986
1032. Takemitsu T: Confronting Silence. Fallen Leaf Press, Berkeley 1995
1033. Takhtajan L A: Quantum Mechanics for Mathematicians. American Mathematical Society, Providence 2008
1034. Teschl G: Mathematical Methods in Quantum Mechanics. American Mathematical Society, Providence 2009
1035. Tanaka A: Interaction, Experience, and the Future of Music. In: O'Hara K and B Brown (Eds.): Consuming Music Together. Springer, Dordrecht 2006
1036. Tarry G: Le problème des 36 officiers. AFAS, Compt 1900–1901
1037. Tatarkiewicz W: A History of Six Ideas: An Essay in Aesthetics. Polish Scientific Publishers, Warszawa 1980
1038. Taylor Y and H Barker: What's It Like to Be Blind? Faking It. `fakingit.typepad.com` 2007
1039. Taylor C: Burning Poles. VHS, Mystic Fire 1991
1040. Taylor C: Silent Tongues. Freedom 1974
1041. Thalmann F: Musical composition with Grid Diagrams of Transformations. Master's Thesis, University of Bern, Bern 2007
1042. Thalmann F: Gestural Composition with Arbitrary Musical Objects and Dynamic Transformation Networks. PhD thesis, University of Minnesota, 2014
1043. Thalmann F and G Mazzola: The Bigbang Rubette: Gestural music composition with Rubato Composer. In: Proceedings of the International Computer Music Conference, International Computer Music Association, Belfast 2008
1044. Thalmann F and G Mazzola: Affine musical transformations using multi-touch gestures. Ninad, 24:58–69, 2010
1045. Thalmann F and G Mazzola: Gestural shaping and transformation in a universal space of structure and sound. In: Proceedings of the International Computer Music Conference, International Computer Music Association, New York City 2010
1046. Thalmann F and G Mazzola: Poietical music scores: Facts, processes, and gestures. In: Proceedings of the Second International Symposium on Music and Sonic Art, MuSA, Baden-Baden 2011
1047. Thalmann F and G Mazzola: Using the creative process for sound design based on generic sound forms. In: MUME 2013 proceedings, AAAI Press, Boston 2013
1048. Thalmann F and G Mazzola: Visualization and transformation in general musical and music-theoretical spaces. In: Proceedings of the Music Encoding Conference 2013, MEI, Mainz 2013
1049. Tormoen D, F Thalmann, G Mazzola: The composing hand: Musical creation with leap motion and the Bigbang Rubette. In: Proceedings of 14th International Conference on New Interfaces for Musical Expression (NIME), London 2014
1050. Terhardt E: Zur Tonhöhenwahrnehmung von Klängen. II. Ein Funktionsschema. Acustica 26, 187-199, 1972
1051. Thomas d'Aquin: Première question disputée. La vérité (De Veritate). (teaching in the years 1256-1259), Paris, Vrin, 2002
1052. Thom R: Stabilité structurelle et morphogénèse. Benjamin, Reading, MA 1972
1053. Tinctoris J: Opera Theoretica, Vol. 2: Liber de arte contrapuncti. Seay A (Ed.): Amer. Institute of Musicology 1975. gopher:`//IUBVM.UCS.INDIANA.EDU/00/tml/15th/tincon2.text`
1054. Tittel E: Der neue Gradus. Doblinger, Wien-München 1959

1055. Titze I R and D T Talkin: A theoretical study of the effects of various laryngeal configurations on the acoustics of phonation. Acoustical Society of America Journal, Vol. 66, No. 60, 1979
1056. Titze I R: Speed, accuracy, and stability of laryngeal movement in singing. Acoustical Society of America Journal, Vol. 115, p. 2590 ff., 2004
1057. Todd N P M: A Model of Expressive Timing in Tonal Music. Music Percep. **3**, pp. 33-58, 1985
1058. Todd N P M: Towards a Cognitive Theory of Expression: The Performance and Perception of Rubato. Contemporary Music Review **4**, 1989
1059. Todd N P M: A Computational Model of Rubato. Contemporary Music Review **3**, 1989
1060. Todd N P M: The Dynamics of Dynamics: A Model of Musical Expression. J. Acoustic Soc. Am. **91**, pp. 3540-3550, 1992
1061. Todoroff T: A Real-Time Analysis and Resynthesis Instrument for Transformation of Sounds in the Frequency Domain. In: ICMA (Ed.): Proceedings of the ICMC 96, San Francisco 1996
1062. Tomasello M: Why Don't Apes Point? In: Enfield N J & S C Levinson (Eds.): Roots of Human Sociality: Culture, Cognition and Interaction. Berg, pp. 506-524, Oxford & New York
1063. Tomasello M, M Carpenter, U Liszkowski: A New Look at Infant Pointing. Child Development, Vol. 78, No. 3. pp. 705-722, May/June 207
1064. Tymoczko D: The Geometry of Musical Chords. Science, Vol 313, 2006
1065. Tymoczko D: Mazzola's Model of Fuxian Counterpoint. In: Agon, C, G. Assayag, E. Amiot, J. Bresson, J. Mandereau (Eds.): Mathematics and Computation in Music. LNAI 6726, Springer, Heidelberg 2011
1066. Uhde J: Beethovens Klaviermusik III. Reclam, Stuttgart 1974
1067. Uhde J and R Wieland: Forschendes Üben. Bärenreiter 2002
1068. Uhde J and R Wieland: Denken und Spielen. Bärenreiter, Kassel et al. 1988
1069. Ungvary T and R Vertegaal: The Sentograph: Input Devices and the Communication of Bodily Expression. ICMA 1995
1070. Utgoff P E and P B Kirlin: Detecting Motives and Recurring Patterns in Polyphonic Music. Proceedings of the International Computer Music Conference, ICMA 2006
1071. van Dalen D: Logic and Structure. Springer, Berlin et al. 1997
1072. Valéry P: Leçon Inaugurale du Cours de Poétique au Collège de France. Gallimard, Paris 1945
1073. Valéry P: Cahiers I-IV (1894-1914). Celeyrette-Pietri N and J Robinson-Valéry (Eds.), Gallimard, Paris 1987
1074. Valéry P: Philosophie de la danse. Oeuvres, tome I, Variété, "Théorie poétique et esthétique", pp. 1390-1403, Gallimard, 1956
1075. Valéry P: Analecta (1926), in *Tel quel*, tome II, Collection Idées, no 241, Gallimard, 1971
1076. Vallès J: "Penser c'est oublier" *Images des mathématiques*. CNRS, `http://images.math.cnrs.fr/Penser-c-est-oublier.html`, 10 juillet 2015
1077. Valtieri S: La Scuola d'Athene. Mitteilungen des Konsthistorischen Instituts in Florenz 16, 1972
1078. van der Waerden B L: Algebra I, II. Springer, Berlin et al. 1966
1079. van der Waerden B L: Algebra I. (English tranlsation) Springer, New York 2003
1080. van der Waerden B L: Die Pythagoreer. Artemis, Zürich 1979
1081. Varèse E: Erinnerungen und Gedanken. In: Darmstädter Beiträge III. Schott, Mainz 1960
1082. Varèse E: Rückblick auf die Zukunft. edition text+kritik, München 1978
1083. Vercoe B: The Synthetic Performer in the Context of Live Performance. In: ICMA (Ed.): Proceedings of the ICMC 84, San Francisco 1984
1084. Veyne P: Comment on écrit l'histoire, 1959. nouv. ed., Points/Histoire 226, Paris 1996
1085. Vico J-B: La méthode des études de notre temps. Présentation et traduction de A. Pons, (De nostri temporis studiorum ratione, 1708), online, site du Réseau Intelligence de la Complexité, 1981
1086. Vieru A: The Book of Modes. Editura Muzicală, Bucarest 1993
1087. Vinci A C: Die Notenschrift. Bärenreiter, Kassel 1988
1088. Violi P: Beyond the body: towards a full embodied semiosis. In: Dirven R, and R Frank (Eds.): Body, Language and Mind. Mouton de Gruyter, Berlin 2006
1089. Vogel M: Die Lehre von den Tonbeziehungen. Verlag für systematische Musikwissenschaft, Bonn-Bad Godesberg 1975
1090. Vogel M: Arthur v. Oettingen und der harmonische Dualismus. In: Vogel M (Ed.): Beiträge zu Musiktheorie des 19. Jahrhunderts. Bosse, Regensburg 1966
1091. Vogel M: Berechnung emmelischer und ekmelischer Mehrklänge. In: Herf F R (Ed.): Mikrotöne II. Helbling, Innsbruck 1988
1092. Vogt H: Neue Musik seit 1945. Reclam, Stuttgart 1972

1093. Vossen G: Datenbankmodelle, Datenbanksprachen und Datenbank-Management-Systeme. Addison-Wesley, Bonn et al. 1994
1094. Vuza D T: Sur le rythme périodique. Revue Roumaine de Linguistique—Cahiers de Linguistique Théorique et Appliquée 22, no. 1 1985
1095. Vuza D T: Some Mathematical Aspects of David Lewin's Book *Generalized Musical Intervals and Transformations*. Perspectives of New Music, vol. 26, no. 1, 1988
1096. Vuza D T: Supplementary Sets and Regular Complementary Unending Canons (Part One). Perspectives of New Music, vol. 29, no. 2, 1991
1097. Vuza D T: Supplementary Sets and Regular Complementary Unending Canons (Part Two). Perspectives of New Music, vol. 30, no. 1, 1992
1098. Vuza D T: Supplementary Sets and Regular Complementary Unending Canons (Part Three). Perspectives of New Music, vol. 30, no. 2, 1992
1099. Vuza D T: Supplementary Sets and Regular Complementary Unending Canons (Part Four). Perspectives of New Music, vol. 31, no. 1, 1993
1100. Vygotsky L S: Thought and Language (1934). Edited and translated by E Hanfmann and G Vakar, revised and edited by A Kozulin. MIT Press, Cambridge 1986
1101. Wanske H: Musiknotation. Schott, Mainz 1988
1102. Waldvogel M. et al.: *presto* source code. TRIMAX, Zürich 1986-1993
1103. Wallas G: The Art of Thought. Harcourt, New York 1926
1104. Wallis J: Tractatus de Sectionibus Conicis. 1655
1105. Walker J S: Fourier Analysis and Wavelet Analysis. Notices of the AMS, vol.44, No.6, July 1997
1106. Ward J: Gregorian Chant according to the Principles of André Mocquereau of Solesmes. The Catholic Education Press 1923 (Reprint)
1107. Waugh W A O'N: Music, probability, and statistics. In: Encyclopedia of Statistical Sciences, **6**, 134-137, 1985
1108. Webern A: *Variationen op.27* (mit Interpretationsanweisungen). UE 16845, Wien 1980
1109. Wegner P: Interactive Foundations of Object-Oriented Programming. Computer, IEEE Computer Soc., October 1995
1110. Weidman A: Singing the Classical, Voicing the Modern: The Postcolonial Politics of Music in South India. Duke University Press 2006
1111. Weil A: Œuvres scientifiques. Collected papers, Vol. I, Springer, New York, 1979
1112. Weisberg R W: Creativity and Knowledge: A Challenge to Theories. In: R J Sternberg (Ed.): Handbook of Creativity. Cambridge University Press, New York 1999
1113. Weyl H: Symmetrie. Birkhäuser, Basel 1955
1114. Weinberger N M: Musica Research Notes. Music and Science Information Computer Archive, Center for Neurobiology of Learning and Memory, U California Irvine, Vol. III, Issue II, Fall 1996
1115. Whiteside A: Indispensables of Piano Playing. Coleman-Ross, New York 1955
1116. Wicinski A A: Psichologyiceskii analiz processa raboty pianista-ispolnitiela nad muzykalnym proizviedieniem. Izviestia Akademii Piedagogiceskich Nauc Vyp., 25 [Moscow], 171-215, 1950
1117. Widmer G: Learning Expression at Multiple Structural Levels. In: ICMA (Ed.): Proceedings of the ICMC 94, San Francisco 1994
1118. Widmer G: Modeling the Rational Basis of Musical Expression. Computer Music J. 18, 1994
1119. Widmer G and W Goebl: Computational Models of Expressive Music Performance: The State of the Art. Journal of New Music Research 33(3), 203-216, 2004
1120. Wiedemann M and E Zimmermann: Kreativität und Form. Springer Vieweg, Berlin Heidelberg 2012
1121. Wieser H-G: Electroclinical features of the psychomotor seizure. Fischer, Stuttgart and Butterworth, London 1983
1122. Wieser H-G and G Mazzola: Musical consonances and dissonances: are they distinguished independently by the right and left hippocampi? Neuropsychologia 24 (6):805-812, 1986
1123. Wieser H-G and G Mazzola: EEG responses to music in limbic and auditory cortices. In: Engel J Jr, Ojemann G A, Lüders H O, Williamson P D (Eds.): Fundamental mechanisms of human function. Raven, New York 1987
1124. Wieser H-G: Musik und Gehirn. Revue Suisse Méd. 7, 153-162, 1987
1125. Wieser H-G and S Moser: Improved multipolar foramen ovale electrode monitoring. J Epilepsy 1: 13-22, 1986
1126. Wiggins G et al.: A Framework for Evaluation of Music Representation Systems. Computer Music Journal 17:3, 1993
1127. Wiil U K (Ed.): Computer Music Modeling and Retrieval (CMMR 2003). Springer, Heidelberg 2004
1128. Wiil U K (Ed.): Computer Music Modeling and Retrieval (CMMR 2004). Springer, Heidelberg 2005

1129. Wille R: Symmetrie — Versuch einer Begriffsbestimmung. In: Symmetrie, Katalogband Vol.1 zur Symmetrieausstellung, Mathildenhöhe, Darmstadt, 1986
1130. Wille R: Personal Communication. Darmstadt 1982
1131. Wille R: Musiktheorie und Mathematik. In: Götze H and R Wille (Eds.): Musik und Mathematik. Springer, Berlin et al. 1985
1132. Wille R: Bedeutungen von Begriffsverbänden. Preprint Nr. 1058, TH Darmstadt 1987
1133. Wille R: Personal Communication. Darmstadt 1985
1134. Winson I: Brain and Psyche. The Biology of the Unconscious. Anchor Press/Doubleday, Garden City, NY 1985
1135. Wilson F R: The Hand: How Its Use Shapes the Brain, Language, and Human culture. Pantheon Books, New York 1998
1136. Witten E: Magic, Mystery, and Matrix. Notices of the AMS, Vol. 45, Nr. 9, 1998
1137. Wittgenstein L: Tractatus Logico Philosophicus – Logisch-Philosophische Abhandlung, Wilhelm Ostwald (Ed.), Annalen der Naturphilosophie, 14, 1921
1138. Wittgenstein L: Tractatus Logico-Philosophicus (1918). Suhrkamp, Frankfurt/Main 1969
1139. Wolfram S: Software für Mathematik und Naturwissenschaften. In: Chaos und Fraktale. Spektrum der Wissenschaft Verlagsgesellschaft, Heidelberg 1989
1140. Xenakis I: Formalized Music. Indiana Univ. Press, Bloomington 1972
1141. Yaglom A M and I M Yaglom: Wahrscheinlichkeit und Information. Deutscher Verlag der Wissenschaften, Berlin, 1967
1142. Yoneda N: On the homology theory of modules. J. Fac. Sci. Univ. Tokyo, Sct. I,7, 1954
1143. Ystad S et al.: Of Parameters Corresponding to a Propagative Synthesis Model Through the Analysis of Real Sounds. In: ICMA (Ed.): Proceedings of the ICMC 96, San Francisco 1996
1144. Yusuke A et al.: Hyperscore: A Design of a Hypertext Model for Musical Expression and Structure. J. of New Music Research, Vol. 24, 1995
1145. Zahorka O: Versuch einer Charakterisierung des altrömischen Melodiestils. Seminar Lütolf, MWS Univ. Zürich 1993
1146. Zahorka O: From Sign to Sound—Analysis and Performance of Musical Scores on RUBATO. In: Enders B (Ed.): Symposionsband Klangart '95, Schott, Mainz 1997
1147. Zahorka O: PrediBase—Controlling Semantics of Symbolic Structures in Music. In: ICMA (Ed.): Proceedings of the ICMC 95, San Francisco 1995
1148. Zahorka O: RUBATO – Deep Blue in der Musik? Animato 97/3, 9-10, Zürich 1997
1149. Zalamea F: Synthetic Philosophy of Contemporary Mathematics. Urbanomic, Falmouth, UK, & Sequence Press, New York, USA 2011
1150. Zarca B: L'univers des mathématiciens. L'ethos professionnel des plus rigoureux des scientifiques. Presses Univ. Rennes, 2012
1151. Zarlino G: Istitutioni harmoniche. Venezia 1558
1152. Zbikowski Z: Conceptualizing Music: Cognitive Structure, Theory, and Analysis. Oxford University Press 2005
1153. Zbikowski Z: Music, Emotion, Analysis. Music Analysis 29/1-3(2010): 37-60
1154. Zekl G: Topos. Meiner, Hamburg 1990
1155. Zlotin B and A Susman: Creative Pedagogy. Journal of TRIZ. no.4, pp. 9-17 (in Russian) 1991
1156. Zweifel P F: Generalized Diatonic and Pentatonic Scales: A Group-theoretic Approach. PNM, 34(1), 140-161, 1996
1157. Zwicker E and H Fastl: Psychoacoustics. Facts and Models. Springer, Berlin, et al. 1999
1158. Zwicker E and M Zollner: Elektroakustik (2nd edition). Springer, New York et al. 1987
1159. Zwiebach B: A First Course in String Theory. Cambridge University Press, Cambridge et al. 2004

Index

G. Mazzola, *The Topos of Music I: Theory*, Computational Music Science,
https://doi.org/10.1007/978-3-319-64364-9

A

C

D

F

G

J

N

Q

T

U

V

W

X

Y

Z

The manufacturer's authorised representative in the EU is Springer Nature Customer Service Centre GmbH, Europaplatz 3, 69115 Heidelberg, Germany. If you have any concerns regarding our products, please contact ProductSafety@springernature.com

Printed and bound by CPI Group (UK) Ltd, Croydon, CR0 4YY
01/10/2026
02225992-0004